THE ENVIRONMENT AND YOU

Norm Christensen

PEARSON

Boston · Columbus · Indianapolis · New York · San Francisco · Upper Saddle River
Amsterdam · Cape Town · Dubai · London · Madrid · Milan · Munich · Paris · Montréal · Toronto
Delhi · Mexico City · São Paulo · Sydney · Hong Kong · Seoul · Singapore · Taipei · Tokyo

VP/Editor in Chief: Beth Wilbur
Executive Director of Development: Deborah Gale
Executive Editor: Chalon Bridges
Project Editor: Leata Holloway
Development Editors: Melissa Parkin, Julia Osborne
Art Development Editors: Jay McElroy, Elisheva Marcus
Assistant Editor: Rachel Brickner
Marketing Manager: Lauren Rodgers
Editorial Media Producer: Lee Ann Doctor
Managing Editor, Environmental Science:
   Gina M. Cheselka
Project Manager, Production: Ed Thomas
Full Service/Composition: PreMediaGlobal

Full Service Project Manager: Jenna Gray
Senior Art Specialist: Connie Long
Art Studio and Cartographer: Kevin Lear, Spatial Graphics
Design Supervised by: Dorling-Kindersley Books
Photo Manager: Maya Melenchuk
Photo Researcher: Kristen Piljay
Text Permissions Manager: Beth Wollar
Text Permissions Researcher: Tamara Efsen
Operations Specialist: Jeff Sargent
Design Manager: Marilyn Perry
Cover Design: Yvo Riezebos Design
Front Cover and Title Page Photo Credit: Photos
   copyright Shutterstock.

Credits and acknowledgments borrowed from other sources and reproduced, with permission, in this textbook appear on the appropriate page beginning on p. C1.

Library of Congress Cataloging-in-Publication Data
Christensen, Norman L.
   The environment and you / Norm Christensen. -- 1st ed.
      p.   cm.
   Includes bibliographical references and index.
   ISBN 978-0-321-73438-9 (alk. paper)
   1. Environmentalism.   2. Environmental policy.   3. Environmental protection.   I. Title.
   GE195.C575 2012
   304.2'8--dc23

                                           2011041619

1 2 3 4 5 6 7 8 9 10—CRK—15 14 13 12 11

www.pearsonhighered.com   ISBN-10: 0-321-73438-6; ISBN-13: 978-0-321-73438-9

About the Author

Norm Christensen is research professor of Ecology and founding dean of Duke University's Nicholas School of the Environment. He earned his undergraduate and master's degrees in biology at California State University, Fresno, and his doctorate at the University of California, Santa Barbara. A central theme in Norm's career has been ecosystem change from both natural and human causes. His research includes studies of the causes and consequences of fire in grasslands, shrublands, and forests; of the impacts of human land use and land abandonment on ecosystem change and species conservation; and of the influence of global warming patterns of ecosystem change. Norm has worked on numerous national advisory committees, including the Interagency Taskforce on the Ecological Effects of the 1988 Yellowstone Fire, the Committee on Environmental Issues in Pacific Northwest Forest Management, the Ecological Society of America Committee on the Scientific Basis for Ecosystem Management, and the U.S. Nuclear Waste Technical Review Board. He has served on the boards of The Conservation Fund, Resources for the Future, the Environmental Defense Fund, and the North Carolina Nature Conservancy. He is a fellow in the American Association for the Advancement of Science and past president of the Ecological Society of America.

Undergraduate education, especially at the introductory level, has been an important part of Norm's career at Duke. He has been honored twice by the university with awards for distinguished undergraduate teaching. He was instrumental in the development of Duke's undergraduate program in environmental science and policy, and he has taught the introductory course for this program for over 15 years. This book is very much a product of these efforts and Norm's passion for connecting students with their environment.

About Our Sustainability Initiatives

Pearson recognizes the environmental challenges facing this planet, as well as acknowledges our responsibility in making a difference. This book is carefully crafted to minimize environmental impact. The materials used to manufacture this book originated from sources committed to responsible forestry practices. The paper is Forest Stewardship Council™ (FSC®) certified and comes from facilities that minimize waste, energy consumption, and the use of harmful chemicals.

Along with developing and exploring digital solutions to our market's needs, Pearson has a strong commitment to achieving carbon-neutrality. As of 2009, Pearson became the first carbon- and climate-neutral publishing company. Since then, Pearson remains strongly committed to measuring, reducing, and offsetting our carbon footprint.

The future holds great promise for reducing our impact on Earth's environment, and Pearson is proud to be leading the way. We strive to publish the best books with the most up-to-date and accurate content, and to do so in ways that minimize our impact on Earth. To learn more about our initiatives, please visit **www.pearson.com/responsibility**.

MIX
Paper from responsible sources
FSC® C005928

Dedication

To Nicholas, Natalie, Noelle, Nicole, and all other of Earth's children; you are the inspiration for this book. May we make decisions today that ensure the future beauty, diversity, and health of the environment on which they will depend.

Preface

It has been said that change is the only constant. For billions of years, Earth's environment and the organisms that inhabit it have been constantly changing. Over tens of millennia we, our species, have constantly changed; each generation's technologies, values, and understanding of its environment have differed from those that preceded it. As a consequence of those technologies and our growing numbers, we have changed Earth's environment more than any other species living now or in the past.

You and the world around you are the current manifestation of this process of inexorable change. The health and well-being of most of Earth's people have markedly improved over the past century but our impacts on Earth's environment have increased significantly. A century ago, our global population was fewer than 2 billion; today there are 7 billion of us. What's more, each of us today uses several times more resources and generates several times more waste than our century-ago ancestors. The affects on our environment are alarming. Resources such as water and petroleum are dwindling. Air pollution and water pollution have become commonplace. Rates of extinction among Earth's species are more than 10 times higher than in pre-industrial times, and Earth's climate is warming because of human-caused changes in the chemistry of its atmosphere.

These changes threaten the health of Earth's ecosystems and the well-being of many of its people; they directly affect you. These changes are unsustainable, but they are not inevitable. Sustainability and ecosystems are important themes throughout this book. Sustainable action and change require knowledge and understanding of the ecosystems upon which we depend. Yes, they are complex, but the key elements of ecosystem function and sustainability are beautifully simple. In an increasingly urban and technology-driven world, the connections between Earth's ecosystems and our well-being may seem distant, even irrelevant. But they are at all times immediate and compelling.

I have not downplayed the significant challenges presented by the variety of environmental issues that affect our lives, because a balanced view of the challenges is needed. Naïve optimism is not likely to motivate substantial change in our actions and impacts, but neither is pessimism. I am confident that we can change and that we can change the world in directions that are truly sustainable. I am convinced you will be part of that process of change. That confidence and conviction were the motivation for writing this book; hope was the inspiration.

Features

A Focus on You

The importance of humans as agents of environmental change and the effects of those changes on human well-being are central to *The Environment and You*. This book emphasizes problem-solving and solutions so you can make more informed choices on actions to support the well-being of humans as well as the health of the planet.

- *Seeing Solutions* features highlight how individuals and groups around the world are using new approaches to solving environmental problems. Topics range from banks that promote both economic development and environmental sustainability to how rock bands, sports organizations, universities, and wedding planners are reducing their emissions of greenhouse gases.
- *Agents of Change* showcases college students and recent graduates who are taking action to produce sustainable environments and improve human well-being. These young leaders provide role models for students who are looking for opportunities to take action to influence the future.

Solid Coverage of Environmental Science

Our current understanding of environmental issues is built on a foundation of decades of careful research by generations of scientists. *The Environment and You* helps students understand not only scientific content but also the role of science in reducing uncertainties that surround environmental issues.

- Each chapter of *The Environment and You* provides numerous explanations of how scientists have found innovative ways to gather the evidence that supports current conclusions and predictions.

- *Focus on Science* features highlight the research of individual scientists and show how scientific research can be used to increase an understanding of environmental issues.

Organized for Learning

The Environment and You is organized to help students understand environmental science.

- Each lesson begins with a big idea so students always have a way to see the forest as well as the trees.
- Manageable amounts of information, organized by modules, give students manageable, complete lessons before moving on to the next topic.

Answering Real Questions

Students have many questions about what is happening in the environment today.

- We asked students around the country what they want to know, and they responded with queries such as "Is climate change the reason for increased storms?" Their questions appear in the margins of every chapter, and the author offers brief answers. These questions are located next to the text that provides a deeper understanding of the relevant issues.

Supporting All Levels of Students

Students in introductory environmental science classes have vastly different levels of science background. *The Environment and You* is designed and written to serve that diversity.

- Self-assessment: Questions at the end of every module, found in the margin, allow students to assess whether not they have truly grasped a topic before they move on. Questions at the end of each chapter are designed to encourage synthesis of concepts.
- MasteringEnvironmentalScience™: Used by over a million science students, the Mastering platform is the most effective and widely used online tutorial, homework, and assessment system for the sciences. It motivates students to come to class prepared; provides students with personalized coaching and feedback; quickly monitors and displays student results; easily captures data to demonstrate assessment outcomes; and automatically grades assignments, including Video Field Trips, 3-D BioFlix® animation activities and quizzes, Graphit! activities, and chapter reading quizzes.

Student Supplements

MasteringEnvironmentalScience™ for *Environment: The Science behind the Stories* 0-321-72147-0

New to environmental science but used by over a million science students, the Mastering platform is the most effective and widely used online

tutorial, homework, and assessment system for the sciences.

Blackboard Premium for Environment: 0-321-77582-1

Blackboard Open Access: 0-321-77630-5

Instructor Supplements

Instructor Resource Center on DVD 0-321-77628-3 with TestGen

This powerful media package is organized chapter by chapter and includes all teaching resources in one convenient location. You'll find Video Field Trips, PowerPoint presentations, Active Lecture questions to facilitate class discussions (for use with or without clickers), and an image library that includes all art and tables from the text.

Included on the IRDVD, the test bank includes hundreds of multiple-choice questions plus unique graphing and scenario-based questions to test students' critical-thinking abilities.

Instructor Guide 0-321-776291-1

This comprehensive resource provides chapter outlines, key terms, and teaching tips for lecture and classroom activities.

Blackboard Premium for Environment: 0-321-77582-1

Blackboard Open Access: 0-321-77630-5

MasteringEnvironmentalScience™ for *The Environment and You*: 0-321-77624-0

New to environmental science but used by over a million science students, the Mastering platform is the most effective and widely used online tutorial, homework, and assessment system for the sciences.

Acknowledgments

I accept all of the responsibilities of authorship for *The Environment and You*, most particularly for any mistakes or flaws. But others deserve much of the credit for its development, organization, presentation, and production. As this project evolved over the course of several years, the Benjamin Cummings publishing team and numerous environmental science colleagues provided much-needed guidance and encouragement.

I am especially grateful to my executive editor, Chalon Bridges. It was Chalon who first approached me about writing an introductory environmental science textbook and, more importantly, convinced me that I could actually do it. Many of the ideas for organization and presentation actually came from Chalon. I was continually amazed by her ability to visualize this project through the eyes of students and instructors. She has a truly remarkable ability to offer criticism that is at once honest, clear, and encouraging.

Over the course of this project, I had the benefit of working with three developmental editors. Susan Teahan guided my efforts in the early stages of this project. Among many things, she taught me the discipline of focusing on learning objectives to determine what and what not to include in the text. All of the chapters benefited from Melissa Parkin's editorial advice. I am thankful for her keen sense of organization and her constant encouragement. Julia Osborne provided editing and coaching aimed at making this text as accessible and readable as possible. She has a wonderful ability to transform writing that can be complex and technical into an inviting narrative, and she has an amazing eye for detail.

Editor-in-Chief Beth Wilbur and Director of Development Deborah Gale supported this project at every stage. In addition, we would not have been able to publish this project without the support from Frank Ruggirello and Michael Young, editorial directors, as well as Presidents Paul Corey and Linda Davis. Thank you for taking a risk on this project and for your ongoing collective leadership in science education.

This book is the result of a collaboration with our colleagues at Dorling Kindersley Education. Sophie Mitchell helped to birth the vision for this project and assembled a winning team to create it. Stuart Jackman provided the design architecture for the project Richard Czapnik, Anthony Limerick, and Ian Midson helped to craft every single page in the text.

Producing a book where text and art are created, designed, and arranged in tandem requires a highly collaborative approach to publishing. We are grateful to our production colleagues for overseeing and orchestrating this effort. Corinne Benson was the founding brain behind our processes and procedures while Gina Cheselka and Ed Thomas helped to revise the plan and successfully carry the project to the finish line. Jenna Gray at PreMediaGlobal worked closely with our in-house production team. Jay McElroy and Elisheva Marcus provided guidance and ideas about nearly every figure appearing in the book.

Project Editor Leata Holloway kept the book moving and on schedule and worked to complete the vision for the book. I would like to give special recognition to Rachel Brickner, Assistant Editor for her leadership and persistence in the development of the Agents of Change feature, for which she spent many hours researching on-campus projects and conducting interviews.

Special thanks to Lee Ann Doctor for overseeing all details on the media, including the new Video Field Trips, and Alyissa Dzaugis for her work on MasteringEnvironmentalScience. Ed Zalisko carefully crafted the reading quiz questions and wrote wrong-answer feedback so that students can now arrive to their lectures better prepared.

Brady Golden, Associate Editor, did a wonderful job overseeing the Instructor's Guide, Test Bank, and PowerPoint presentations. Nora Lally-Graves provided a careful eye editing the art for the PowerPoint presentations so they will work well in lectures. David Serrano authored content for the PowerPoint Lecture presentations and Clicker questions, and Christy Bazan wrote the Test Bank.

After many years creating the book, there comes a time to pass the torch to marketing and sales. We are grateful to Lauren Rodgers for her enthusiasm in promoting this project and to Lauren Harp for her guidance. Our market development colleagues, Brooke Suchomel, Michelle Cadden, and Cassandra Cummings helped orchestrate all class tests. Jane Campbell, Lillian Carr, and Jessica Perry helped to create all marketing pieces.

I owe special thanks to Terrence Bensel, Brian Bovard, Robert Kingsolver, and Lester Rowntree for their contributions to chapters on climate change, biodiversity, agriculture, energy, and waste management. They contributed a variety of ideas that have shaped the organization and content of this book. Their detailed outlines provided road maps through sometimes unfamiliar territory, and many elements from their drafts of several of these chapters are part of the final product.

I owe much to the students in my introductory environmental science and policy courses over the past 10 years. They helped to shape the spirit and content of this text. Over the past 2 years, they have been guinea pigs for each of its chapters. I invited their editorial comments, and they took my invitation seriously. The book is much the better for their input.

Over the years, I have had the benefit of working with wonderful mentors and colleagues. My undergraduate and master's advisor, Bert Tribbey, passed along much knowledge and wisdom that appears in these pages, and he has long served as my primary role model for teaching excellence. My Duke colleagues William Chameides, Deborah Gallagher, Prasad Kasibhatla, Emily Klein, Randy Kramer, Marie Lynn Miranda, Joel Meyer, Lincoln Pratson, William Schlesinger, and Dean Urban were key sources of information and constructive criticism. Kristy Manning helped me shape early versions of several chapters. Her gentle wisdom and laughter are missed by us all.

I am grateful to my family for their patience with me over the life of this project. Finally, they won't feel the need at every meeting to ask, "How's your book coming?" And if they do ask, I won't be embarrassed to answer. Most of all, I thank my wife, Portia. She has been a sounding board for so many of my ideas, a willing editor of all of this book's essays and features, and the best friend ever.

Reviewers

David A. Aborn
University of Tennessee, Chattanooga
Isoken Aighewi
University of Maryland
Saleem Ali
University of Vermont
John All
Western Kentucky University
Mary Allen
Hartwick College
Mark W. Anderson
University of Maine
Joe Arruda
Pittsburg State University
Daphne Babcock
Collin County Community College
Narinder S. Bansal
Ohlone College
Jon Barbour
University of Colorado, Denver
Morgan Barrows
Saddleback College
Christy Bazan
Illinois State University
Hans Beck
Aurora University
Peter Beck
St. Edwards University
Diane B. Beechinor
Northeast Lakeview College
Terry Bensel
Allegheny College
Leonard Bernstein
Temple University
William Berry
University of California, Berkeley
Lisa K. Bonnaeu
Metropolitan Community College
Brian Bovard
Florida Golf Coast University

Peter Busher
Boston University
Kelly Cartwright
College of Lake County
Paul Chandler
Ball State University
David Charlet
College of Southern Nevada
Marina Chiarappa-Zucca
DeAnza College
Van Christman
Brigham Young University, Idaho
Donna Cohen
Massachusetts Bay Community College
John Conoley
East Carolina University
Jessica Crowe
Valdosta State University
Jean DeSaix
University of North Carolina, Chapel Hill
Doreen Dewell
Whatcom Community College
Dr. Darren Divine
Community College of Southern Nevada
Rebecca Dodge
Midwestern State University
James English
Gardner-Webb University
JodyLee Estrada Duek
Pima Community College
Douglas Flournoy
Indian Hills Community College
Steven Frankel
Northeastern University
Jonathan Frye
McPherson College
Karen Gaines
Eastern Illinois University

Kurt Haberyan
Northwest Missouri State
Anne Hall
Emory University
Stephanie Hart
Lansing Community College
Harlan Hendricks
Columbus State University
Carol Hoban
Kennesaw State University
Kelley Hodges
Gulf Coast Community College
Tara Holmberg
Northwestern Connecticut Community College
Kathryn Hopkins
McLennan Community College
Meshagae Hunte-Brown
Drexel University
Emmanuel Iyiegbuniwe
Western Kentucky University
Tom Jurik
Iowa State University
Richard Jurin
University of Northern Colorado
Susan W. Karr
Carson-Newman College
David K. Kern
Whatcom Community College
Kevin King
Clinton Community College
Jack Kinworthy
Concordia University
Rob Kingsolver
Bellarmine University
Cindy Klevickis
James Madison University
Steven A. Kolmes
University of Portland

Ned Knight
Linfield College
Erica Kosal
North Carolina Wesleyan College
Janet Kotash
Moraine Valley Community College
Robert Kremer
University of Missouri
Diana Kropf-Gomez
Richland College
James David Kubicki
The Pennsylvania State University
Kody Kuehnl
Franklin University
Frank Kuserk
Moravian College
Troy A. Ladine
East Texas Baptist University
Elizabeth Larson-Keagy
Arizona State University
Jejung Lee
University of Missouri
Lissa M. Leege
Georgia Southern University
Kurt Leuschner
College of the Dessert
Honqi Li
Frostburg State University
Satish Mahajan
Lane College
Kenneth Mantai
State University of New York, Fredonia
Anthony Marcattilio
St. Cloud State University
Heidi Marcum
Baylor University
Allan Matthias
University of Arizona
Kamau Mbuthia
Bowling Green State University
John McClain
Temple College
Joseph McCulloch
Normandale Community College
Robert McKay
Bowling Green State University
Bram Middeldorp
Minneapolis Community and Technical College

Chris Migliaccio
Miami Dade College
Kiran Misra
Edinboro University of Pennsylvania
James Morris
University of South Carolina, Columbia
Sherri Morris
Bradley University
Eric Myers
South Suburban College
Jason Neff
University of Colorado, Boulder
Emily Nekl
High Point University
John Olson
Villanova University
Bruce Olszewski
San Jose State University
Gregory O'Mullan
Queens College
Stephen Overmann
Southeast Missouri State University
William J. Pegg
Frostburg State University
Barry Perlmutter
Community College of Southern Nevada
Shana Petermann
Minnesota State Community and Technical College
Julie Phillips
De Anza College
Frank Phillips
McNeese State University
John Pleasants
Iowa State University
Brad Reynolds
University of Tennessee, Chattanooga
Kayla Rihani
Northeastern Illinois University
Carleton Lee Rockett
Bowling Green State University
Susan Rolke
Franklin Pierce University
Deanne Roquet
Lake Superior College
Steven Rudnick
University of Massachusetts, Boston
Dork Sahagian
Lehigh University

Milton Saier
University of California, San Diego
James Salazar
Galveston College
Kimberly Schulte
Georgia Perimeter College
Michele Schutzenhofer
McKendree University
Rebecca Sears
Western State College of Colorado
David Serrano
Broward College
Garey Simpson
Kennesaw State University
Debra Socci
Seminole Community College
Ravi Srinivas
University of St. Thomas
Craig W. Steele
Edinboro University
Michelle Stevens
California State University, Sacramento
Robert Strikwerda
Indiana University, Kokomo
Keith Summerville
Drake University
Jamey Thompson
Hudson Valley Community College
Ruthanne Thompson
University of North Texas
Bradley Turner
McLennan Community College
Lina Urquidi
New Mexico State University
Sean Watts
Santa Clara University
John Weishampel
University of Central Florida
Timothy Welling
Dutchess Community College
Kelly Wessell
Tompkins Cortland Community College
James Winebrake
Rochester Institute of Technology
Chris Winslow
Bowling Green State University
Danielle M. Wirth
Des Moines Area Community College
Todd Yetter
University of the Cumberlands

Class Test and Interview Participants

Ginny Adams
University of Central Arkansas
John All
Western Kentucky University
Jeff Anglen
California State University, Fresno
Dave Armstrong
University of Colorado
Berk Ayranci
Temple University

Roy Barnes
Scottsdale Community College
Christy Bazan
Illinois State University
Sandy Bejarano
Pima College East Campus
Leonard Bernstein
Temple University
William Berry
University of California, Berkeley

Neil Blackstone
Northern Illinois University
Christopher Bloch
Texas Tech University
Gary M. Booth
Brigham Young University
James Brandle
University of Nebraska, Lincoln
Robert Bruck
North Carolina State College

George Byrns
Illinois State University

John Calloway
University of San Francisco

Frank Carver
Forsyth College

Ken Charters
Cochise Community College

Dave Charlet
Community College of Southern Nevada

LuAnn Clark
Lansing Community College

Jaimee Corbet
Paradise Valley Community College

Robert Cromer
Augusta State University

Wynn Cudmore
Chemeketa Community College

Jane Cundiff
Radford University

Lynnette Danzl-Tauer
Rock Valley College

James Diana
University of Michigan, Ann Arbor

Darren Divine
Community College of Southern Nevada

Rebecca Dodge
Midwestern State University

David Dolan
University of Wisconsin, Green Bay

Michael Draney
University of Wisconsin, Green Bay

Renee Dutreaux-Hai
California State University, Los Angeles

Johannes Feddema
University of Kansas

Richard S. Feldman
Marist College

Kevin Fermanich
University of Wisconsin, Green Bay

Linda Fitzguh
Gulf Coast College

Laurie Fladd
Trident Technical University

Chris Fox
Catonsville Community College

Katie Gerber
Santa Rosa Junior College

Thaddeus Godish
Ball State University

James Goetz
Kingsborough Community College

Robert Goodman
Citrus College

Larry Gray
Utah Valley University

Peggy Green
Broward Community College North

Joshua Grover
Ball State University

Kurt Haberyan
Northwest Missouri State

George Hagen
Palo Alto College

Nigel Hancock
Long Beach City College

Wendy Hartman
Palm Beach Community College

Kim Hatch
Long Beach City College

James Haynes
State University of New York, Brockport

Kathi Hopkins
McClennan Community College

James J. Horwitz
Palm Beach Community College

Joseph Hull
Seattle Central Community College

Carolyn Jensen
Pennsylvania State University, University Park

David Jones
North Eastern Illinois University

Susan Karr
Carson-Newman College

Leslie Kanat
Johnson State College

Julie Klejeski
Mesabi Range Community College

Janet Kotash
Moraine Valley Community College

Katherine LaCommare
Lansing Community College

John Lendvay
University of San Francisco

Paul Lorah
University of St. Thomas

Deborah Marr
Indiana University, South Bend

Allan Matthias
University of Arizona

Shelly Maxfield
Pima Community College

John McClain
Temple Junior College

Joesph McCulloch
Normandale Community College

Rachel McShane
St. Charles Community College

Steven J. Meyer
University of Wisconsin, Green Bay

Alex Mintzer
Cypress College

Jane Moore
Tarrant County Community College

James Morris
University of South Carolina, Columbia

William Muller
Temple University

Hari Pant
City University of New York, Lehman

Dr. Robert Patterson
North Carolina State University

Dan Pavuk
Bowling Green State University

Christopher Pennuto
Buffalo State University

Barry Perlmutter
Community College of Southern Nevada

Julie Phillips
De Anza College

Mai Phillips
University of Wisconsin, Milwaukee

John Pleasants
Iowa State University

Ron Pohala
Luzerne County Community College

Juan Carlos Ramirez-Darronsoro
Ball State University

Marco Restani
St. Cloud University

Brad Reynolds
University of Tennessee, Chattanooga

Howard Riessen
Buffalo State University

Shamili A. Sandiford
College of Dupage

Jodi Shann
University of Cincinnati

Loris Sherman
Somerset Community College

Brent Sipes
University of Hawaii, Manoa

Shobha Sriharan
Virginia State University

Edward Standora
Buffalo State University

Philip Stevens
Indiana University, Fort Wayne

John Suen
California State University, Fresno

Jamey Thompson
Hudson Valley Community College

Claire Todd
Pacific Lutheran University

William Trayler
California State University, Fresno

Carl N. Von Endem
Northern Illinois University

Zhi Wang
California State University, Fresno

Sharon Ward
Montgomery College

Jeff Watanabe
Ohlone College

Paul W Webb
University of Michigan, Ann Arbor

James W.C. White
University of Colorado

Deb Williams
Johnson County Community College

Christopher J. Winslow
Bowling Green State University

Don Wujek
Oakland Community College, Auburn Hills

Lori Zaikowski
Dowling College

Carol Zellmer
California State University, Fresno

Joseph Zurovchak
Statue University of New York, Orange Community College

TABLE OF CONTENTS

TABLE OF CONTENTS

IT'S ABOUT SCIENCE

The Environment and You, informed by decades of careful research by generations of scientists, is presented in a manner non-science majors understand.

Focus on Science highlights individual scientists and shows how their research is used to increase understanding of environmental issues.

FOCUS ON SCIENCE:
Measuring Long-Past Temperatures

What evidence do scientists use to determine climates of the past?

As you read about the climate changes during the last 100,000 years, did you wonder how scientists know the temperatures from so long ago? Certainly there were no weather forecasters with thermometers in the early Pleistocene.

A method for determining past temperatures was first developed by two chemists—University of Chicago Professor Harold Urey and his student Cesare Emiliani (**Figure 8.11**). In 1934 Urey had been awarded the Nobel Prize in chemistry for his pioneering work on isotopes, atoms of the same element that contain different numbers of neutrons (3.1). In 1950 Urey took up the challenge of inferring past climates. Later he called it "the toughest chemical problem I ever faced."

Urey understood that both hydrogen and oxygen have isotopes with differing masses. The different isotopes of hydrogen and oxygen combine to form water molecules of differing mass. Over 99% of water molecules are "light water" or $^1H_2^{16}O$; only a small fraction of water molecules are "heavy water," or $^2H_2^{18}O$. Because it is lighter, $^1H_2^{16}O$ evaporates more readily than $^2H_2^{18}O$. Consequently, a greater proportion of light water molecules evaporate from ocean water than do heavy water molecules. This reduces the ratio of light water molecules to heavy water molecules in ocean water. As temperature increases, the proportion of heavy water molecules evaporating from the ocean surface also increases. Careful analysis shows that the isotope ratios of ocean water change with temperature in a very predictable way.

Urey reasoned that the isotope ratios of ocean water should be reflected in the isotope ratios found in the bodies of organisms living in the ocean. These ratios would be preserved in the fossils of these organisms found in ocean sediments.

▲ **Figure 8.11 Teamwork**
Harold Urey (left) was fascinated by the challenge of determining the climate at different times in the past from the relative amounts of isotopes of hydrogen and oxygen in ocean sediment. But it was Urey's student, Cesare Emiliani (center), who worked out the actual techniques for estimating past atmospheric temperatures from sediment isotopes.

By measuring isotope ratios in the sediment, it would be possible to infer the temperature of the ocean water when the organisms were alive.

It's one thing to have a great idea and quite another to make it work. For this, Emiliani deserves the credit. First, Emiliani perfected methods for measuring isotope ratios in the fossil shells of microscopic organisms. Then he applied those methods to bits of sediment taken from the bottom of the ocean. Using established relationships between isotope ratios and water temperature, he was able to deduce temperature changes over a 300,000-year period.

In the years since Urey and Emiliani's pioneering research, sediment isotope studies have been refined. Today they are the standard method of inferring the climates of long ago (**Figure 8.12**).

◀ **Figure 8.12 A Core Through Time**
Each year, fine sediments accumulate on the ocean floor. Variations in the amounts of different oxygen and hydrogen isotopes in those sediments are correlated with temperature changes in the atmosphere above the ocean.

Changing Populations and Ecosystems

Because climate is a key component of habitats, global warming is producing significant changes in the populations of many species and in many ecosystems.

Global warming is affecting ecosystems as plants and animals respond to altered habitats. Satellite images of Canada, New England, and the Midwest since 1980 reveal that warmer spring temperatures are causing the leaves of plants to open earlier. On the ground, scientists have observed that many species of birds are nesting and laying eggs earlier than in the past. Nevertheless, there is still a risk that spring frosts will damage young leaves and kill nestling birds.

Milder temperatures are causing migratory birds to move to the Arctic earlier and stay in high latitudes later into the fall. This delayed departure exposes some species to fierce autumn storms as they migrate southward. The migration of some species no longer coincides with the availability of food, which also varies seasonally (**Figure 8.28**).

plankton necessary to support juvenile cod. Off the coast of California, populations of many seabirds are approaching collapse. It appears that warmer waters have altered populations of microscopic plankton. This, in turn, has diminished the supply of small fish upon which the seabirds feed.

Ocean warming appears to be a major factor in the deterioration of many coral reefs. The warmer conditions contribute to a phenomenon called "bleaching," which is caused by a breakdown in the relationship between coral animals and their symbiotic algae (11.5).

All forecasts suggest accelerated warming in the decades ahead. Continued loss of snow and ice, rising sea levels, and changes to patterns of rainfall seem inevitable, as do the effects of these changes on plant

1. Describe specific climate changes associated with global warming in three different regions in the world.

Ecosystems and Sustainability Themes woven throughout the entire text offer you a cohesive connecting thread in this broad-based discipline.

Business as Usual. The *Business as Usual* scenario occurs if we take no actions to curb greenhouse emissions. At the present rate of increase, in 50 years CO_2e emissions will more than double, to about 58 Pg per year. At this rate of increase, the amount of greenhouse gases in Earth's atmosphere would eventually triple, to over 1,800 Pg. Climate scientists are in complete agreement that these levels of CO_2e would have catastrophic consequences. Average annual global temperatures would increase by 8 °–10 °C, and sea level would rise more than 1 m (3.3 ft). Such changes would likely be devastating for humans and the ecosystems on which they depend.

Sustainable World. If over the next 50 years we are able to stabilize emissions at their current level of 28 Pg per year, atmospheric CO_2e concentration will not rise above 1,200 Pg. At this concentration, some amount of climate change will be inevitable, but scientists predict that sustainable adaptation to this amount of change is quite possible.

A Portfolio of Strategies. Ecologist Stephen Pacala and engineer Robert Socolow argue that

stabilization wedges into five categories: efficiency and conservation, fossil fuel use, renewable energy, nuclear energy, and biostorage.

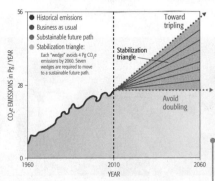

◀ **Figure 8.34 Getting to a Sustainable Future**
In this graph, the difference in accumulated emissions between *Business as Usual* and *Sustainable World* scenarios is indicated by the yellow stabilization triangle, which represents an increment of 28 Pg CO_2e/year by 2060.
Source: Pacala, S., and R. Socolow, 2004. Stabilization wedges: Solving the climate problem for the next 50 years with current technologies. *Science* 305: 968–972.

Citations are provided so you can trace the source of information.

IT'S ABOUT SOLUTIONS

An emphasis on problem solving and scalable solutions empowers you to make more informed choices to support the well-being of humans and the planet.

SEEING SOLUTIONS: Becoming Carbon Neutral

They are all working to reduce global warming by becoming carbon neutral or finding ways to reduce their CO_2e emissions to zero.

The first step toward becoming carbon neutral requires knowing your "carbon footprint"—the amount of CO_2 or other greenhouse gases your actions produce. On the Internet you can find numerous carbon counters that will do the math for you. Generally, a college student who lives in the dorm and rides a bike is responsible for generating about 7 tons of carbon per year. That contrasts with the 20 tons generated by a person living in a 3-bedroom house, driving a car 10,000 miles, and taking several short airplane trips each year. A more luxurious lifestyle generates even more carbon emissions. The Academy of Motion Pictures, for example, assumes that a "celebrity lifestyle" emits 30 tons of CO_2 per year. To measure the total footprint of the Oscars, they sum this value for each individual participating in the ceremonies.

The next step toward carbon neutrality is to take action to reduce the size of your carbon footprint. For example, Middlebury College in Vermont aims to become carbon neutral by the year 2016. To do this, they are making their old and new buildings as energy efficient as possible. A new wood biofuel plant will meet all of the campus's energy needs. As one student

leader commented, "not only are we neutralizing carbon emissions and contributing to the sustainability of our ecosystem but we're shifting the way our generation thinks about our energy use."

It is, however, difficult for individuals or organizations to take direct actions to reduce their carbon footprint to zero. The purchase of carbon offsets provides a way to complete the job. The Rolling Stones buy offsets to balance the carbon emissions produced by a concert tour. The National Football League buys them to offset the carbon footprint of the Super Bowl. The World Bank and Google purchase them to compensate for the CO_2e emissions from their daily activities. Even your wedding could be carbon neutral, if you have your guests buy offsets (**Figure 8.40**).

Where do the carbon offsets come from? Many nonprofit and for-profit organizations create carbon offsets by planting trees or developing sources of renewable energy. These offsets are then sold to individuals and organizations wishing to reduce their carbon footprint. Carbon offsets are not the total solution to global warming, nor are they a replacement for actions that individuals and organizations ought to take to reduce their carbon footprints. Carbon offsets do provide a voluntary way for each of us to compensate for unavoidable activities that influence greenhouse gas emissions.

What do the Academy of Motion Pictures, Middlebury College, the Rolling Stones, the Super Bowl, Google, the World Bank, and your wedding have in common?

Seeing Solutions highlights how individuals and groups around the world are using new approaches to solve environmental problems.

▲ **Figure 8.40 Becoming Carbon Neutral**
Buy a ticket to a Rolling Stones concert or, perhaps, plan a carbon-neutral wedding, and you could also be buying offsets to CO_2e emissions.

223

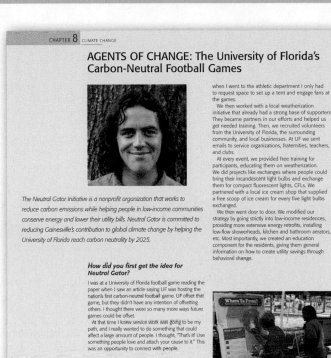

AGENTS OF CHANGE: The University of Florida's Carbon-Neutral Football Games

The Neutral Gator Initiative is a nonprofit organization that works to reduce carbon emissions while helping people in low-income communities conserve energy and lower their utility bills. Neutral Gator is committed to reducing Gainesville's contribution to global climate change by helping the University of Florida reach carbon neutrality by 2025.

How did you first get the idea for Neutral Gator?

I was at a University of Florida football game reading the paper when I saw an article saying UF was hosting the nation's first carbon-neutral football game. UF offset that game, but they didn't have any intention of offsetting others. I thought there were so many more ways future games could be offset.

At that time I knew service work was going to be my path, and I really wanted to do something that could affect a large amount of people. I thought, "That's it! Use something people love and attach your cause to it." This was an opportunity to connect with people.

What steps did you take to get started?

First, we figured out the games' imprint and what it would take to offset a season. I didn't ask permission of the athletics department. Instead, I focused on self-funding the program. That way, when I went to the athletic department I only had to request space to set up a tent and engage fans at the games.

We then worked with a local weatherization initiative that already had a strong base of supporters. They became partners in our efforts and helped us get needed training. Then, we recruited volunteers from the University of Florida, the surrounding community, and local businesses. At UF we sent emails to service organizations, fraternities, teachers, and clubs.

At every event, we provided free training for participants, educating them on weatherization. We did projects like exchanges where people could bring their incandescent light bulbs and exchange them for compact fluorescent lights, CFLs. We partnered with a local ice cream shop that supplied a free scoop of ice cream for every five light bulbs exchanged.

We then went door to door. We modified our strategy by going strictly into low-income residences, providing more extensive energy retrofits, installing low-flow showerheads, kitchen and bathroom aerators, etc. Most importantly, we created an education component for the residents, giving them general information on how to create utility savings through behavioral change.

230

Agents of Change showcases college students and recent graduates who are taking action to develop sustainable environments and improve human well-being.

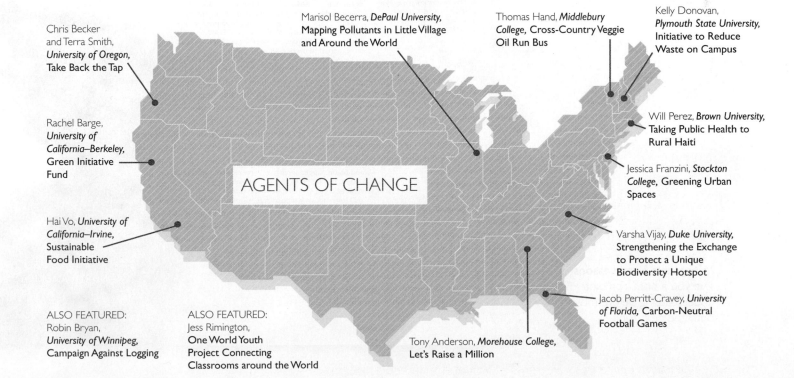

Chris Becker and Terra Smith, *University of Oregon,* Take Back the Tap

Marisol Becerra, *DePaul University,* Mapping Pollutants in Little Village and Around the World

Thomas Hand, *Middlebury College,* Cross-Country Veggie Oil Run Bus

Kelly Donovan, *Plymouth State University,* Initiative to Reduce Waste on Campus

Rachel Barge, *University of California–Berkeley,* Green Initiative Fund

Will Perez, *Brown University,* Taking Public Health to Rural Haiti

Jessica Franzini, *Stockton College,* Greening Urban Spaces

AGENTS OF CHANGE

Hai Vo, *University of California–Irvine,* Sustainable Food Initiative

Varsha Vijay, *Duke University,* Strengthening the Exchange to Protect a Unique Biodiversity Hotspot

Jacob Perritt-Cravey, *University of Florida,* Carbon-Neutral Football Games

ALSO FEATURED:
Robin Bryan,
University of Winnipeg,
Campaign Against Logging

ALSO FEATURED:
Jess Rimington,
One World Youth
Project Connecting
Classrooms around the World

Tony Anderson, *Morehouse College,* Let's Raise a Million

IT'S ABOUT YOU

The Environment and You makes navigating, focusing, and learning easier for you.

Big Idea Summaries start each lesson so you will always have a way to see the big picture as well as the details for each module and topic.

Answers to Student Questions about what is happening in the environment today are posed and answered within the margins of the text.

Manageable-sized Lessons give you a brief, complete understanding before moving onto the next topic.

8.4 Consequences of Global Warming

The changes caused by global warming vary from region to region. In some places, increasing temperatures have been accompanied by higher rainfall amounts. In others, they have brought drought. Winters have become milder and shorter in Earth's middle latitudes, and dry seasons have grown longer in some parts of the tropics. Glaciers and ice sheets are melting worldwide. Warming is causing sea levels to rise. Taken together, these changes are having a significant impact on the flora and fauna of many ecosystems.

Drier and Wetter

Global warming is producing wetter conditions in some places and drought in others.

Is climate change the reason for increased storms and global disturbances?
Ciara Tyce, Georgia Southern University

Norm: *Climate scientists are not in agreement about whether global warming is producing more storms or extreme weather events such as summer heat waves. However, climate models indicate that such changes are likely with continued warming.*

The effects of rising temperatures on precipitation vary geographically. Rainfall has increased significantly in eastern North and South America, as well as in most parts of Europe and Asia. In contrast, sub-Saharan Africa, the Mediterranean region, and western North America have been drier. Since 1970, longer and more intense droughts—as measured by decreased precipitation and higher temperatures—have affected wide areas of the tropics and subtropics.

In regions in which rain is highly seasonal, such as sub-Saharan Africa, global warming appears to be changing the length of wet and dry periods. This is a matter of special concern because food production depends on the length of wet seasons. Based on current trends and climate models, growing seasons are expected to become shorter over most of sub-Saharan Africa, with the exception of lands very near the equator.

Regional trends are complicated by variation in weather events. For example, there has been a worldwide increase in the frequency of rainstorms that result in flooding, even in areas where total annual rainfall has declined. In the aftermath of Hurricane Katrina and other major storms, there has been much debate about whether global warming is producing more intense hurricanes and typhoons. Ocean temperature is an important factor in the development of tropical storms and hurricanes, and ocean temperatures have increased between 0.25 °C and 0.5 °C (0.45 °–0.9 °F) over the past century. Some studies have found that warmer seas are associated with a large increase in the number and strength of tropical storms. Other studies, however, indicate that changes in observational techniques and instrumentation can account for these increases.

There is evidence that global warming is influencing drought cycles. For example, the El Niño/La Niña/Southern Oscillation is caused by changes in the temperature of surface waters in the equatorial Pacific Ocean. When waters off the west coast of South and Central America are cold, drought is much more common in the southwestern United States. Some climatologists think that since 1970 the length and intensity of El Niño and La Niña events have been outside the range of natural variability. Although climate models predict that such changes will occur, most scientists feel that it is not clear that they are actually underway.

RUSSIAN FEDERATION

PACIFIC OCEAN

NORTH AMERICA

Melting Glaciers and Ice Sheets

Over 80% of Earth's glaciers are retreating.

Nearly 75% of Earth's fresh water is contained in snow pack and glaciers, the ice sheets of Greenland and Antarctica, and Arctic sea ice. The total amount of this frozen water depends on the rate of snowfall and ice formation relative to the rate at which it melts. In most places today, the rate of melting exceeds the rate of ice formation. Snowpack in many mountain regions has been shrinking, and over 80% of Earth's mountain glaciers are retreating.

Since satellite observations began in 1978, Arctic sea ice has shrunk by about 3% per decade (**Figure 8.24**). As you would expect, rates of shrinkage are higher in the summer than in the dark winter, when ice rebuilds. Even so, the total shrinkage is so great that scientists are forecasting that the Arctic Ocean could be completely ice free during summer in 20 years or less. The overall mass of ice on Antarctica is also decreasing (**Figure 8.25**).

For decades, scientists have noted that Greenland's ice sheet is shrinking. Recent data indicate that melting is far more rapid than previously thought because of the prevalence of streams and rivers beneath the ice.

The glacial retreat has exposed new land along Greenland's coast. This exposure, along with longer growing seasons, has allowed farmers to plant crops that have not been grown in Greenland since the Medieval Viking period.

Greenland's small community of farmers may be benefiting from the effects of a warmer climate, but globally the loss of snow and ice is a double whammy. First, snow and ice reflect about 35% of incoming solar radiation, compared to just 10% from forests or ocean waters. A decrease in the amount of land covered by snow and ice accelerates global warming, since solar radiation is absorbed instead of being reflected back into space. Second, water melting from ice sheets and glaciers eventually makes its way to the ocean, where it adds to the rising sea level. Meltwater also has the potential to alter ocean currents, much as it did during the Big Freeze 12,000 years ago.

▲ **Figure 8.25 Antarctic Ice Shelf Collapses**
Over a 30-day period in 2002, warming caused an ice shelf the size of the state of Rhode Island (light blue) to collapse into the Antarctic Ocean.

1979–2000 median minimum

GREENLAND

ATLANTIC OCEAN

Northwest passage

◄ **Figure 8.24 A Northwest Passage at Last**
Over the past several decades, the northern polar ice cap has been steadily shrinking. Since 1965, global warming has produced a steady decline in the extent of summer terrestrial snow cover (blue on the chart below) and sea ice (red) in the Northern Hemisphere. Note the open water from Greenland across the coast of North America.

Data from: Post, E., et al., 2009. Ecological dynamics across the Arctic associated with recent climate change. *Science* 325: 1355–1358.

Navigate easily to any module using the colored tabs at the top corner of each page.

Designed in partnership with DK Publishing, an imprint known for highly visual and award winning design, this text will inspire and engage you in learning through both images and words.

MasteringEnvironmentalScience®

FOR THE STUDENT

MasteringEnvironmentalScience is designed with a single purpose: to help you reach the moment of understanding. The Mastering online homework and tutoring system delivers self-paced tutorials that provide you with individualized coaching set to your professor's course objectives. MasteringEnvironmentalScience helps you arrive prepared for lecture and lab and offers additional study tools for you to use wherever you are!

Video Field Trips give you a fascinating behind-the-scenes experience of a prescribed fire burn, solar energy, coal-fired power plants, wastewater treatment facilities, landfills, farms, and more.

eWaste

Carbon-neutral Campus Events

BioFlix activities are highly interactive and use 3D animations.

Topics include:
- Cellular Respiration
- Photosynthesis
- Mechanisms of Evolution
- Carbon Cycle
- Population Ecology

Current Events Quizzes connect recent news articles to course topics.

The Pearson eText gives you access to the text whenever and wherever you can access the Internet. The eText pages look exactly like the printed text, and include powerful interactive and customization functions.

Additional Study Tools on MasteringEnvironmentalScience:

- **Reading Quizzes** help you stay on track with your reading assignments and test your true understanding of the content.

- **Self Study Area** offers a 24/7 study resource.

- **GraphIt!** activities help you work with and interpret graphs and real data.

MasteringEnvironmentalScience®

www.masteringenvironmentalscience.com

FOR THE INSTRUCTOR

The Mastering platform is the most effective and widely used online tutorial, homework, and assessment system for the sciences. It was developed by scientists for science students and instructors and has a proven history with more than 10 years of student use and 99.8% server reliability.

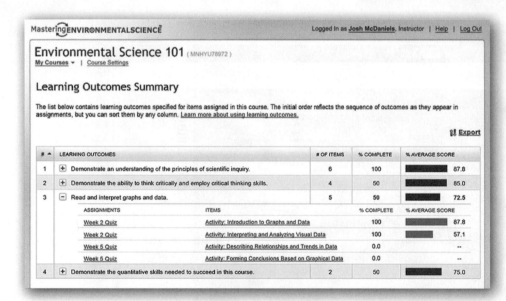

Learning Outcomes track student performance against your learning outcomes:

- Add your own or use the publisher-provided learning outcomes to track student performance and report it to your administration.

- View class performance against the specified learning outcomes.

- Export results to a spreadsheet that you can further customize and/or share with your chair, dean, and administrator or accreditation board.

- Mastering offers a data-supported measure to quantify students' learning gains and to share those results quickly and easily.

The **Gradebook** provides easy-to-interpret insights into student performance.

- Every assignment is automatically graded.

- At a glance, shades of red highlight vulnerable students and challenging assignments.

One-click **Diagnostics** show where students are struggling or progressing.

With a single click, **Student Performance Data** provides at-a-glance statistics on your class as well as national results.

Learn how students are doing compared to a national average and which topics they are struggling with.

1

Environment, Sustainability, and Science

Earth Day, Hope Day

*Can we
alter current
environmental
trends?*

Each year on April 22, people around the world gather in public parks and college amphitheaters to celebrate Earth Day. Events may include music and projects to encourage recycling or clean up roadsides and river corridors. At nearly every gathering, speakers warn that human actions threaten Earth's life-support systems. On our current course, they say, the ecosystems upon which we depend cannot be sustained. Listen to some of these speeches and you are likely to hear statements such as these.

There is a biodiversity crisis! Species of plants and animals are rapidly disappearing (**Figure 1.1**). The loss of some species is due to the destruction of their habitat. Other species, such as many fishes, are threatened by overharvesting. Still others are threatened by pollution or changing climate. If current trends continue, some biologists predict that a century from now as many as one-third of all Earth's plant and animal species will be extinct. This loss of species will affect everyone, not just birdwatchers and nature lovers. Diversity of species is a key factor in the ability of ecosystems to provide the resources that we humans need to survive.

Our lives depend on it! Over the next 24 hours, you will breathe in and out between 15,000 and 30,000 times. Even in a rural town, your lungs will be exposed to pollutants, including smoke particles and hundreds of potentially toxic chemicals. If you live in a large city, your exposure will be 10 to 20 times higher. The World Health Organization estimates that each year air pollution kills 2.4 million people (**Figure 1.2**).

Earth's atmosphere is getting hotter! The carbon dioxide content of our atmosphere has risen by over 30%, causing the average temperature of Earth's atmosphere to warm by almost 1 °C (1.8 °F). The increase in atmospheric carbon dioxide is a direct consequence of human activities—the burning of coal, oil, and natural gas and the cutting and burning of forests. This warming threatens to melt glaciers, raise sea levels, and alter ecosystems on land and in the sea (**Figure 1.3**). As we continue to remove forests and burn larger and larger amounts of fossil fuels, the problem is getting worse.

▲ **Figure 1.1 Long Gone**
Once common throughout the southeastern United States, the Carolina parakeet is now seen only as museum specimens such as these. It is one of hundreds of species of birds that have disappeared completely from Earth's ecosystems in the past 200 years.

▲ Figure 1.2 **Air Pollution in a Big City**
On a day like this, the average Shanghai citizen will inhale almost a tablespoon of toxic smoke particles.

◄ Figure 1.3 **A Global Meltdown?**
About 80% of Earth's glaciers are retreating, and their meltwater is adding to the volume of the oceans.

We have met the enemy and it is us! Over the past 75 years, the human population has increased almost threefold, and our use of Earth's natural resources has increased nearly 20-fold (**Figure 1.4**). Through agriculture, forestry, fishing, and land development, humans currently use nearly one-third of the total energy provided by all of Earth's green plants. More than half of the fresh water flowing across Earth's surface is diverted by dams and canals for human use. The world's wealthiest people use a disproportionately large amount of these resources. If all of the 7 billion people on Earth today used as many resources as does an average citizen of the United States, we would need at least five planet Earths to support them. And by 2050, the global population is predicted to reach 9 billion.

There are those who dismiss Earth Day events as all doom and gloom. Some people deny there are any problems, even in the face of strong evidence. Some people acknowledge the threats but feel that pessimism encourages complacency: "If you're booking passage on the *Titanic*, why not go first class?"

There are many reasons to be hopeful! We are confronted with urgent environmental challenges. Naïve optimism—an exuberant expectation that everything will be all right—is hardly an appropriate response to these challenges. Still, we have many reasons to be *hopeful*. National laws and international treaties now protect endangered species. The money and effort spent to halt habitat loss and restore endangered species is increasing every year. In many places, air pollution and water pollution are being cleaned up; new regulations and technologies are improving air quality in some of the world's largest cities (**Figure 1.5**). It is now internationally recognized that global warming presents a significant challenge, and the majority of the world's countries are implementing strategies to reverse it.

Yes, human populations continue to grow, but the *rate* at which our numbers are increasing has dropped by over 40% in the past 40 years. There is abundant evidence that conservation and increased reliance on renewable sources of energy can significantly lessen the impact each of us has on Earth's ecosystems—without reducing the quality of our lives.

What will we see on Earth Day 2050? We have reasons to be hopeful that the negative environmental trends witnessed over the past 40 years will be slowed or even reversed. As the Czech poet and leader, Vaclav Havel, said, "hope is not the conviction that something will [necessarily] turn out well, but the certainty that something makes sense regardless of how it turns out." Hope motivates action—not because of the certainty of success, but because of the certainty that inaction will produce failure.

Successful actions will require a greater understanding of the basic principles that govern Earth's ecosystems.

▲ **Figure 1.4 Human Population Growth**
Earth's human population has grown by nearly 600% since 1900. Today, each person uses much more energy and many more resources than a century ago.

We can then use those principles to make wise decisions and to guide the policies and actions that will ensure human well-being from generation to generation. Motivated by hope, we can achieve these goals. This is the central theme of this chapter and, indeed, of this entire book.

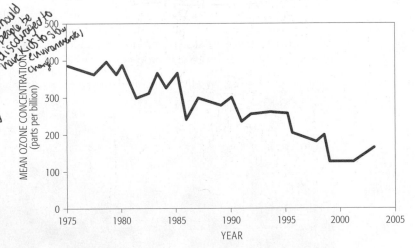

▲ **Figure 1.5 Breathing Easier**
Over the past 30 years in cities such as Los Angeles, regulations and improved auto emissions technology have produced a steady decline in air pollutants such as ozone and have improved air quality.
Source: *U.S Environmental Protection Agency*

1.1 Environment and Sustainability

You may use the word *environment* to describe where you are and everything around you. Scientists have a more specific definition: the **environment** is all of the physical, chemical, and biological factors and processes that determine the growth and survival of an organism or a community of organisms. The long list of all the factors that make up your environment would include the gases in the air you breathe and the many life forms that nourish and are nourished by you. **Environmental science** studies all aspects of the environment. **Ecology** is the branch of environmental science that focuses on the abundance and distribution of organisms in relation to their environment. Earth's environments have sustained living organisms for at least 3.8 billion years; they have sustained members of our own species for well over 100,000 years. Throughout this time, Earth's environments and the communities of organisms that depend on them have been constantly changing.

In Earth's long history, no organism has had a greater effect on the environment than have humans. Our ability to appropriate Earth's resources has been a major factor in the rapid growth in our numbers. Over the past century, we have come to understand that our actions have significant consequences for the well-being of the community of all living things and for ourselves in particular. This understanding is the foundation for determining human actions and behaviors that are sustainable.

Defining Sustainable Actions

Our understanding of sustainable behavior has changed through time.

The concept of sustainability is central to environmental science. But what does it mean to be sustainable? Over the last 150 years, we have viewed this concept in different ways.

A century and a half ago, the human use of resources, such as wildlife and fisheries, was determined largely by our needs or perceived needs. Earth's forest resources appeared to be inexhaustible, and wildlife was abundant (**Figure 1.6**). Rivers, lakes, and coastal waters were teeming with fish. We were fully confident of the environment's capacity to produce an abundance of resources and to absorb and process our wastes.

And why not? The world's population was only about 1 billion people, and the population of the United States was less than 75 million. (Today, nearly 7 billion people inhabit this planet; more than 300 million live in the United States.)

As human populations and their demand for resources grew, supplies of resources began to dwindle. Between 1920 and 1940, the number of scientific studies of the environment increased greatly. With new knowledge, resource managers began to appreciate the need to align the demand for resources with their supply.

◀ Figure 1.6 **The Frontier**
This painting by Albert Bierstadt of seemingly unending wilderness is typical of many 19th century depictions of American landscapes.

▲ Figure 1.7 Too Many Uses?
Rapid population growth in the United States following World War II placed increasing demands on public lands for timber resources and for other ecosystem services such as grazing, recreation, and the protection of water supplies.

In forestry, this included strategies to plant and regrow trees after harvest and to protect them from the threats of pests and fire. In fisheries, this meant establishing catch limits and artificially stocking lakes and streams. These actions helped to sustain the supply of resources.

By 1950, there were 2.5 billion people in the world and 152 million in the United States. As the demand for resources increased, conflicts skyrocketed. In the United States, for example, growth in housing increased the demand for wood from national forests. At the same time, there was increased public interest in using those forests for recreation, for supporting wildlife, and for protecting water supplies. People argued over which use of forest resources should be given priority. To be sustainable, management policies had to recognize the different demands on the environment and its resources (**Figure 1.7**). Policies also had to address the conflicts among humans who valued those resources differently.

For example, in the management of U.S. national forests strongly held values associated with commercial timber management, the provision of clean water, and the conservation of species have often been in conflict.

In its 1987 report, *Our Common Future*, the United Nations Commission on Sustainability declared, "at its most basic level, sustainability means meeting the needs of the present without compromising the ability of future generations to meet their own needs" (**Figure 1.8**). This declaration added a critical element of time to our concept of sustainability—sustainable management has its eye on the needs of the future.

The UN Commission on Sustainability was concerned with maintaining **human well-being**, a multifaceted concept that includes life's basic necessities, such as food and shelter, as well as good health, social stability, and personal freedom. Their report noted that the factors necessary to ensure human well-being may change from one generation to the next. Thus, sustainability is not a process of maintaining the status quo. Instead, the commission argued that sustainability is maintaining the ability to accommodate three important sources of change.

▲ Figure 1.8 Sustainability as a Commitment to the Future
The 1987 UN Commission on Sustainability was chaired by Norwegian Prime Minister Gro Brundtland.

7

1. The world is changing.

Earth's environments undergo constant change, sometimes in regular daily and seasonal cycles and sometimes in more complicated patterns. Environmental change is both inevitable and essential.

2. We are changing.

Successive generations of humans have developed and used ever-changing technologies to extract and use resources. Humans also pass their knowledge about their environment and its resources from one generation to the next. As a consequence, our needs for ecosystem goods and services are constantly changing. The value we place on these goods and services changes, too.

3. We are changing the world.

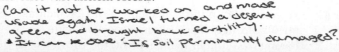

No other organism has ever shaped its environment to the extent that we have. Our use of technologies, such as fire and agriculture, has allowed our numbers to increase and has changed the world. Our increasing numbers have extended our influence (**Figure 1.9**). Our increasing use of technology has had even greater consequences.

Today we are at a unique moment in history. At no previous time has Earth supported so many humans. At no previous time has overall human well-being been better. We can point with gratitude to actions by previous generations that have helped us meet our current needs, such as the development of agricultural technologies and the establishment of national forests and parks. But we can also identify actions by our ancestors that have impoverished our world. In many regions, poor agricultural practices have permanently diminished soil fertility. Overexploitation has left us with mere snippets of the once vast ancient forests.

Although average well-being among humans is high, variation in well-being is high as well. In developed countries such as the United States, clean water is taken for granted and obesity is all too common. Yet, more than one-fifth of Earth's people lack access to clean water and suffer from malnutrition. This disparity has led the United Nations and other world leaders to broaden the definition of sustainability to include the concept of equity. In this context, **sustainability** means meeting the needs of the present *in an equitable and fair fashion* without compromising the ability of future generations to meet their own needs. This is the definition that we will use throughout this text. (It should be noted, however, that scholars and decision makers are not in agreement on what constitutes "equitable and fair.")

Does this definition of sustainability seem too focused on humans? If so, remember that Earth's environments were self-sustaining for billions of years before our time. If we were to disappear tomorrow, they would eventually recover from our impacts. Sustainability is indeed about our actions and their consequences for human well-being. We shall see throughout this book that our present and future well-being depends heavily on how we treat all of Earth's life and environments.

There is much we can learn from the past to help ensure a sustainable world for the next generation. Yet in many important respects, that generation will be quite different from all earlier generations. By all indications, it will include nearly 9 billion people. Those people will use technologies and have expectations that we can hardly imagine. If the management of our resources today is to be truly sustainable, we must be very generous in our projection of future needs. The ecosystem concept described next provides a foundation for defining sustainable behavior and actions.

QUESTIONS

1. Describe how human understanding of sustainable actions has changed over the past century.

2. What kinds of change must be included in our understanding of sustainability?

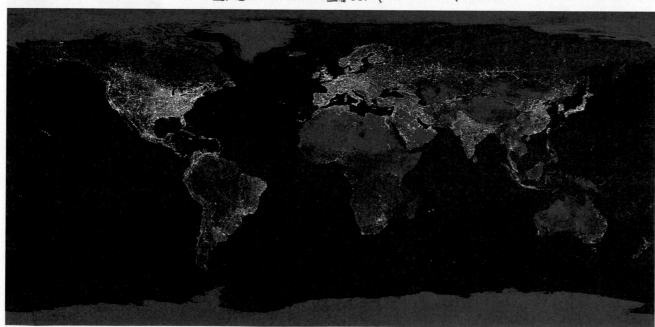

◄ **Figure 1.9 A World of Change**

This is a composite satellite image of Earth at night. Imagine how it might have looked a century ago. How will it change over the next century? The changes in these lights represent changes in Earth's ecosystems, in human values and technologies, and in our environmental impacts.

1.2 Ecosystems

We use the ecosystem concept as a way to organize all of the factors and processes that make up our environment. An **ecosystem** includes all of the organisms and their physical and chemical environment within a specific area. The flow of energy and matter through ecosystems influences the distribution and abundance of organisms within them. Ecosystems provide resources and services that are essential to our well-being. Today, human activities influence nearly all of Earth's ecosystems, often in ways that diminish the benefits they provide and threaten our well-being.

Ecosystem Function and Integrity

The ecosystem concept combines living organisms and their environment.

An ecosystem is the combination of a community of organisms and its physical and chemical environment, functioning as an integrated ecological unit. A forest and a lake are familiar examples of ecosystems.

Ecosystems include biota, or living organisms, and their abiotic, or nonliving, environment. The abiotic characteristics of an ecosystem, such as its climate, determine the distribution and abundance of the organisms that live there. At the same time, the actions of the organisms living in an ecosystem change and shape their environment. For example, trees need deep soil to grow, and as they grow, their decaying leaves increase soil fertility and their root systems penetrate rock fractures and accelerate soil formation.

Ecosystems are connected by the flow of energy and matter. Energy and matter are transformed as they move between living organisms and the abiotic environment. Ecologists refer to the flow of matter and energy and the processes influencing the distribution and abundance of organisms as **ecosystem functions**.

Ecosystem integration. An ecosystem is an *integrated system* made up of living and nonliving parts and the processes that connect them (**Figure 1.10**). **Ecosystem integrity** refers to the web of interactions that regulate ecosystem functions. Ecologists often compare the functional integrity of

an ecosystem to that of an individual organism. For example, the tissues and organs of your body work together to carry out the functions necessary for life, such as breathing and digesting food. Similarly, the microbes, plants, and animals in an ecosystem interact with each other and their environment in ways that determine the overall movement of matter and energy through an ecosystem.

Like individual organisms, ecosystems have the capacity to adjust to disturbances and changes in their environment. This capacity depends upon their integrated nature. Severe disturbances, such as hurricanes and fires, may greatly alter the species living in an ecosystem and the way the ecosystem functions. The ecosystem can then undergo a process of change and recovery that ecologists liken to the process of growth and development of an individual organism.

In one very important way, however, the ecosystem-as-organism metaphor is not accurate. The boundaries of a single organism are quite distinct. For example, your skin forms a well-defined boundary between you and your environment. The boundaries of ecosystems are not so clearly delineated. Instead, they are defined by those who study or manage them; often they are quite artificial.

Ecologists study ecosystems of many sizes. The unit of study may be as large as a forest or river basin, an entire continent, or even the entire globe.

▶ **Figure 1.10 An Integrated System**
Ecosystems comprise organisms and their nonliving environment, and they are dependent on inflows and outflows of matter and energy.

Inflow: Sunlight and materials like carbon, nitrogen, and water.

Transformation and exchange of matter and energy between biota and abiotic environment.

Outflow: Heat and materials like carbon, nitrogen, and water

However, the ecosystem concept is equally applicable at much smaller scales, such as a cup of sour milk or a rotting log (**Figure 1.11**).

✳ Ecosystems are often studied and described in the absence of human influences. Ecologists interested in the fundamental workings of ecosystems may choose to study regions of wilderness where human influences are minimal. Today, however, human influences are nearly everywhere.

Over the past century, our population has grown from just over 1 billion to nearly 7 billion people. Changes in culture and technology, such as forms of transportation, energy use, and urban development, have greatly magnified our demand for resources. As a consequence, our impacts on Earth's ecosystems have grown by far more than the sevenfold increase in our numbers.

◀ **Figure 1.11 Ecosystem Scale**
Ecosystems can be defined at a variety of scales, from a rotting log Ⓐ, the Great Lakes Ⓑ, to Earth Ⓒ in its entirety. At every scale they are open to inflows and outflows of matter and energy.

Ecosystem Services

Ecosystems provide the resources and processes upon which humans depend.

Human well-being depends on **ecosystem services**, the multitude of resources and processes that ecosystems supply to humans. Ecosystem services can be classified into four categories: provisioning, regulating, cultural, and supporting services.

Provisioning services supply us with resources, such as food, water, and the air we breathe. We humans are notable for our ability to modify—often simplify—ecosystems in order to increase their provisioning services. For example, an agricultural field of corn provides a large amount of food to humans but supports far fewer species than the grassland that once grew there.

Regulating services are the ways that ecosystems control important conditions and processes, such as climate, the flow of water, and the absorption of pollutants. These services are a consequence of the integrated, self-regulating nature of ecosystems.

Cultural services are the spiritual and recreational benefits that ecosystems provide. If you have visited the Grand Canyon, Yosemite Valley, or simply taken a quiet walk in the woods near your home, you have benefited from cultural services.

Supporting services are the basic ecosystem processes, such as nutrient cycles and soil formation, that are needed to maintain other services. For example, bees and other pollinating insects provide a supporting service by ensuring that flowers are fertilized and fruits form.

Like all living things, we humans rely on ecosystems for survival as well as shape ecosystems through our actions. But unlike other organisms, human interactions with the environment are mediated by beliefs, knowledge, technologies, and institutions that are passed from one generation to the next.

For example, rather than relying on the slow process of biological evolution to adapt to new environments, we dress appropriately for changing weather and invent new technologies to heat our homes. We then pass this knowledge on to our children. Our success as a species, as reflected by our growing numbers, is largely attributable to these unique features (**Figure 1.12**).

[Handwritten margin note: Our uniqueness. Our ability to feel and use our conscience make decisions and pass on valuable info. we aren't that species that evolved enough to pull itself ahead. We are uniquely different and purposefully so.]

In 2006, The United Nations Environment Programme published its *Millennium Ecosystem Assessment*. The work of hundreds of scientists and decision makers, this document evaluated the global impacts of humans on ecosystems and the effects of ecosystem change on human well-being. Three important themes emerged from this assessment. First, many of Earth's ecosystem services are being degraded. Second, there is growing evidence that many ecosystems are becoming more fragile and prone to disturbance. Third, these changes threaten the well-being of all humans, but especially the poor. The report also notes that the effective management of ecosystem services can be a dominant factor in the reduction of poverty and international conflict. The interconnections between ecosystems and human well-being are a major theme of this book.

[Handwritten margin note: Is this true?]

▼ **Figure 1.12 Humans and Ecosystems**
Human well-being is dependent on the services provided by ecosystems. Human activities produce significant changes in the ecosystem functions that deliver those services.

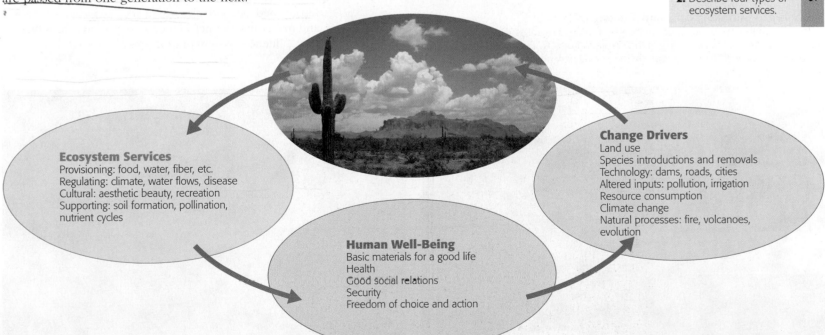

Ecosystem Services
Provisioning: food, water, fiber, etc.
Regulating: climate, water flows, disease
Cultural: aesthetic beauty, recreation
Supporting: soil formation, pollination, nutrient cycles

Human Well-Being
Basic materials for a good life
Health
Good social relations
Security
Freedom of choice and action

Change Drivers
Land use
Species introductions and removals
Technology: dams, roads, cities
Altered inputs: pollution, irrigation
Resource consumption
Climate change
Natural processes: fire, volcanoes, evolution

QUESTIONS

1. Differentiate between ecosystem functions and ecosystem services.

2. Describe four types of ecosystem services.

1.3 Principles of Ecosystem Function

Ecosystems are integrated systems that can be understood in terms of fundamental scientific principles. Although awesomely complex, ecosystems are not haphazard. Even though living things are continuously consuming and replenishing oxygen, the amount of oxygen in our air remains remarkably constant. The mix of life-forms that surrounds us changes from place to place and season to season in complicated but dependable ways. The organization and function of ecosystems depends on four fundamental principles.

1. Matter and energy are neither created nor destroyed.
2. Ecosystems are always open to gains and losses of matter and energy.
3. Ecosystem processes are self-regulated by interactions among their living and nonliving components.
4. Ecosystem change is inevitable and essential.

Conservation of Matter and Energy

Something cannot be created from nothing, and everything goes somewhere.

Energy and matter can be neither created nor destroyed, although in the cores of stars and nuclear reactors matter can be converted into energy. This is known in physics as the law of energy and mass conservation. If you could capture the ash, smoke, and gases released from burning 1 g of coal, you would find that they weighed exactly 1 g. You would also find that the light and heat energy emitted by the fire, plus the energy left in the ash and smoke, precisely equal the chemical energy that was stored in the unburned lump of coal. Something—coal—has been used to generate something else—ash, smoke, gases, heat, and light: But no matter or energy has been lost.

The conservation of energy and mass applies to ecosystems, too. By measuring the amounts of matter in the different parts of an ecosystem and the rates at which this matter flows through it, we can characterize ecosystem functions and make predictions about their future behavior. This accounting of amounts and rates is a fundamental tool in ecosystem science and management.

In environmental science, the law of energy and mass conservation is a powerful reminder of two fundamental truths. First, *something cannot be created from nothing*. Energy and matter can be converted from one form to another, but the total amount of any resource is finite. A resource can be exhausted if it is used faster than it is replenished. Second, *everything goes somewhere*. Energy and matter that we put into ecosystems have fates that may influence ecosystem functions (**Figure 1.13**).

▼ **Figure 1.13 Shrinking Lake and Stinking Lake**
Something cannot be created from nothing—Lake Mead near Las Vegas, Nevada, is gradually shrinking as water use exceeds water supply in the Colorado River Ⓐ. Everything goes someplace—China struggles to cope with the numerous chemicals that now pollute Lake Tai, one of its largest lakes Ⓑ.

Ecosystems Are Open

Because flows of matter and energy influence ecosystem functions, context matters.

Ecologists often assign arbitrary boundaries to ecosystems based on factors such as type of soil or kind of vegetation. Regardless of the boundaries we choose for them, ecosystems do not have distinct boundaries. Ecosystems are open. Matter and energy can flow into them from their surroundings, and they can also move out. For example, sunlight and rain can enter an ecosystem, and heat and streams can flow out. These gains and losses of matter and energy influence ecosystem functions. An important consequence of this openness is that *context matters*: Functions in an ecosystem are inevitably influenced by the nature of the ecosystems that surround it.

Scientists set the size and boundaries of the ecosystems they study in such a way as to measure inputs and outputs of matter or energy most easily or to monitor key processes within them. For example, ecologists often study the drainage basins of rivers. The boundaries of drainage basins are set by the mountains, hills, and valleys that determine where the water flows. Ecologists can determine the amount of water coming into a drainage basin by measuring rainfall. They can measure the amount leaving by monitoring the flow at the mouth of the river that drains the basin.

To manage ecosystem processes effectively it would seem logical to do so according to the scale and boundaries by which we can most easily monitor and sustain them. Unfortunately, our political systems often make this difficult to achieve. For example, rivers frequently define the boundaries between counties, states, and countries. These political entities are responsible for managing the water flowing through them. But rivers also divide drainage basins. As a consequence, multiple counties, states, and countries influence and manage water quality and flow (**Figure 1.14**). This often leads to mismanagement and conflicts among jurisdictions.

Similarly, consider how often straight lines form the boundaries between counties, states, and countries, such as much of the border between the United States and Canada. Such lines have little to do with the movement of the organisms, matter, and energy that shape our environment.

▶ Figure 1.14 **An Arbitrary Boundary**
The Rio Grande separates Mexico from the United States along nearly 70% of the border between the two countries. In this satellite photo the river flows between El Paso, TX, on the right and Ciudad Juarez on the left: both cities depend on its resources, contribute pollution to its waters, and argue over its management.

Ecosystem Stability

Ecosystem processes are regulated by interactions among biotic and abiotic components.

Ecosystems are constantly changing. Yet key features, such as the total number of species and the processes involving the transformations of matter and energy, remain stable. The stability of ecosystem features and functions derives from the complex interactions among their diverse array of biotic and abiotic components.

Dynamic homeostasis is the process by which systems adjust to changes in ways that minimize how much features or processes vary from their normal values. Your ability to maintain your body temperature around 98.6 °F (37 °C) is an example of dynamic homeostasis. If your temperature cools, even slightly, you put on a sweater or coat to warm yourself. If your temperature rises above this value, you are prone to start shedding layers of clothing.

Examples of dynamic homeostasis abound in ecosystems. Over time, the sizes of plant and animal populations tend to fluctuate above and below average values. The amounts of the chemical elements nitrogen and phosphorus found in lakes or soils vary from time to time, but they maintain long-term average values.

Dynamic homeostasis occurs because of **feedback**, the ability of a system to adjust based on changes in the system itself. Homeostatic systems "talk and listen to themselves." Again, consider the regulation of your body temperature. A part of your brain called the *hypothalamus* is very sensitive to changes in your body temperature. If your blood temperature drops below 98.6 °F, your hypothalamus sends an "I feel cold" signal to your consciousness. It also sends signals that stimulate involuntary responses, such as shivering or increased metabolism. If your blood temperature rises above 98.6 °F, your conscious brain receives the "I feel warm" signal from the hypothalamus.

It also stimulates involuntary responses that cool your body, such as sweating.

The regulation of temperature in your body is an example of a negative feedback system. **Negative feedback** occurs when directional change in a process alters the system in a manner that reverses the direction of that change. A change in your body temperature causes your body to respond in ways that help your body return to its normal temperature.

Many ecosystem processes are regulated by negative feedback in the interactions between their living and nonliving components. In a pond ecosystem, for example, the rapid growth of algae may deplete the supply of an element in the water that is essential for growth. Consequently, the growth of the algae slows and the supply of that element is again stabilized (**Figure 1.15**). Negative feedback systems tend to stabilize ecosystems.

Positive feedback occurs when the directional change in a process alters the ecosystem so as to reinforce that change. For example, because tree roots hold soil and limit erosion, the loss of forests on steep mountain slopes accelerates soil erosion. This erosion leads to the loss of even more forest. If not counteracted, positive feedback can destabilize ecosystems (**Figure 1.16**).

Ecosystem diversity and complexity matter.

The dynamic homeostasis of ecosystem functions derives in part from the diversity of species and the complexity of their interactions. Such diversity and complexity provide multiple alternative paths for the movement of matter and energy within an ecosystem. Human actions that diminish diversity often diminish ecosystem stability. This is explored in more detail later in this chapter.

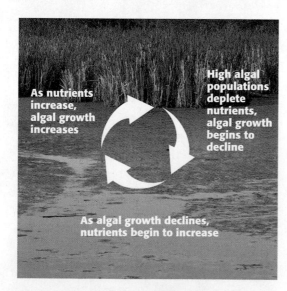

As nutrients increase, algal growth increases

High algal populations deplete nutrients, algal growth begins to decline

As algal growth declines, nutrients begin to increase

▲ **Figure 1.15 Negative Feedback**
The growth of algae and the availability of nutrients that control their growth provide an example of negative feedback in which each change is reversed by the other. Increased nutrients increase algal growth; this then diminishes nutrients, and subsequently diminishes algal growth.

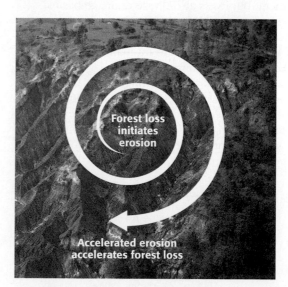

Forest loss initiates erosion

Accelerated erosion accelerates forest loss

▲ **Figure 1.16 Positive Feedback**
Deforestation and erosion can generate positive feedback. On this hillside in Guatemala, loss of forest stimulates erosion, which then results in loss of more forest and further erosion. One change is reinforced by the other.

Ecosystem Change

Change is both inevitable and essential.

Dynamic homeostasis implies change. Ecosystems are constantly changing in response to actions from outside as well as to the processes operating within them. Processes of change in ecosystems have played a significant role in the evolution of the organisms that live in them. As a consequence, species have a variety of characteristics that permit them to adjust to change.

Some changes, such as seasonal changes in temperature, are so regular and predictable that we do not consider them to be disturbances. It is the departures from these regular changes, such as abnormally cold summers or extended winter weather, that we call disturbances. Other disturbances include storms, floods, and fire. Though less predictable, such disturbances are inevitable, and ecosystems must adjust to them.

Ecosystem change in response to disturbance is often homeostatic, controlled by feedbacks that produce regular and predictable patterns. For example, a hurricane may fell many of the trees in a forest (**Figure 1.17**). This creates an opportunity for new trees to grow. If the climate remains relatively constant, these new trees will produce a very similar forest in 100 to 200 years.

Sometimes processes in ecosystems change because of long-term changes in other processes. For example, a gradual change in the climate of a region can produce new patterns of ecosystem change. Between 5,000 and 10,000 years ago, moist tropical forest grew across the Sahel of Africa, the region south of the modern Sahara Desert. When these tropical forests were disturbed by storms or other events, similar forests soon grew in their place. About 4,000 years ago, the climate of this region became hotter and drier. When tropical forests were disturbed, they were replaced by grasslands and savannas, not forests. The hotter climate had caused a new pattern of changes in the ecosystem.

Over the past 10,000 years, humans have become increasingly important agents of change in most of Earth's ecosystems. Whether or not our actions are sustainable depends on the degree to which we recognize and apply the principles of ecosystem function.

▼ **Figure 1.17 Ecosystem Disturbance**
This diverse tropical forest ecosystem in the Silver Glen Springs palmetto jungle in the Ocala National Forest, Florida, was disturbed by Hurricane Frances in 2004. It will now undergo change that will result in the regeneration of forest in 100 years.

QUESTIONS

1. What important environmental principles derive from the law of conservation of mass and energy?

2. What is meant by the phrase "ecosystems are always open"?

3. What is the difference between negative and positive feedback?

15

1.4 Acting Sustainably

Ensuring the well-being of present and future generations is a worthy, motivating vision, but how can we know which of our actions are truly sustainable? For this, we must refer to our understanding of ecosystems. An action is sustainable if it does not impair the functions of ecosystems or their ability to deliver the services upon which we depend. The four fundamental principles of ecosystems are guideposts for assessing whether an action is sustainable.

Managing Resources

To be sustainable, actions must conform to the law of mass and energy conservation.

For at least three centuries, scientists have agreed about the unfailing truth of the law of conservation of mass and energy. However, given our overexploitation of natural resources and our dumping of toxic waste into ecosystems, one might conclude that great numbers of us do not share that agreement.

The resources that we take from ecosystems often can be replenished by natural processes or from outside sources. For example, the water taken from the Great Lakes is replenished by rivers that flow into them. Those rivers, in turn, are fed by rain falling in their headwaters. The fish harvested from the Great Lakes are replenished by a complex array of biotic and abiotic interactions that support fish reproduction and population growth. Resources such as water and fish are said to be renewable, but they can be renewed only if we do not use them more quickly than they can be replenished.

Similarly, ecosystems can transform many waste materials and pollutants into less obnoxious or toxic forms. But ecosystems can carry out this function only if the rate of waste flowing in does not exceed the rate at which the ecosystem can transform it. In other words, the law of mass and energy conservation must be followed.

A resource is **nonrenewable** if its amount in an ecosystem declines with virtually any level of use. In the scale of human history, oil is a nonrenewable resource. The world's oil reserves were produced by geologic processes operating over millions of years. Today we are using these reserves about 1 million times faster than they can be replenished (**Figure 1.18**). Exploitation of nonrenewable resources defies the law of mass and energy conservation.

Sustainable use of resources requires two things: an understanding of their rate of renewal and the ability to manage the rate of their use. For example, many of the factors affecting the flow of water into the Great Lakes are understood, but the causes of decade-to-decade changes in rainfall in the Great Lakes basin are not. The ability to manage the use of water in the basin is complicated by the rapid increase in the number of people living in the region and disputes among states and between the United States and Canada.

▼ Figure 1.18 **A Nonrenewable Resource**
It took Earth's ecosystems tens of thousands of years to produce the volume of petroleum consumed by the world's billion plus automobiles and trucks in a single day.

Understanding Boundaries

To be sustainable, actions must acknowledge the importance of boundaries.

Sustainable environmental management requires an understanding of the spatial scales and landscape features that influence ecosystem functions and services. Whether it is a backyard, a national park, or an entire country, arbitrary scales and boundaries limit our ability to monitor and manage ecosystem processes. We cannot manage the quality of water in the lower reaches of a river, for example, if arbitrary boundaries exclude management options upstream.

The scale and boundaries appropriate for managing one process may not align with those appropriate for managing another process or service.

For example, river basins are useful units for managing the flow of water, chemicals in the water, and the organisms that live in the water. However, river basins are usually irrelevant to the management of air pollution or populations of large migratory animals.

Sustainable management also requires communication and collaboration across the boundaries that define differences in ownership, management purpose, and nationality. Unfortunately, these human boundaries often represent some of the most significant challenges to sustainable management.

Maintaining Balance and Integration

To be sustainable, actions must maintain the homeostatic capacity of ecosystems.

Although our knowledge of ecosystems is far from complete, we must acknowledge that their complexity is key to their integrated character. The many living and nonliving components of an ecosystem play important roles in its functions and the services it provides.

The connection between ecosystem complexity and ecosystem services is similar to the complexity of an automobile. An automobile is made up of thousands of individual parts. To a greater or lesser extent, each part contributes to the automobile's overall function and to the services it provides. The key service you expect from an automobile is to move you from one place to another. Less important services—but ones that might influence which car you purchase—include comfort, climate control, appearance, entertainment, and even navigation. As you drive along, some parts of the car are in use and others are not. If it is summer, you are unlikely to be using the heater. You don't use the brakes until you decide to slow down or stop. Some parts, such as the spare tire and emergency brake, are used only when another part fails.

Now, imagine that you are asked to remove parts from your car sequentially, while trying to maintain its most important functions (**Figure 1.19**). If you have complete knowledge of the relationship between your car's parts and its workings, you can make wise decisions about which services to lose first. For example, you would remove cosmetic features, such as hubcaps and hood ornaments before giving up a piston or fuel pump. If you were willing to tolerate risk, you might give up the spare tire or emergency brake, since they are only important if other parts fail.

People with a limited understanding of automobiles would approach this disassembly process in a much more random fashion. Unable to associate parts to functions, they would soon disable their cars. They would have a limited ability to set priorities for particular functions or services.

They would not understand the risks they might incur by the removal of parts that do not immediately impair a car's function, such as the spare tire or warning lights.

With regard to the manipulation, simplification, and management of ecosystems, we are like children aspiring to be auto mechanics. We can name only a few of the "pieces" in most ecosystems, and we have an even more limited understanding of their importance to particular ecosystem functions and services. We frequently focus on only the most obvious parts and processes or those of most immediate importance to us. Too often we give little thought to the ecosystem equivalents of spare tires, emergency brakes, and warning lights.

▼ **Figure 1.19 What's Essential?**
Are all these parts necessary for the functioning of your car?

awesome analogy

Embracing Change

To be sustainable, actions must not ignore the inevitability of change or interfere with ecosystems' capacity to change.

Sustainable environmental management demands that we do more than accept that change is inevitable. It requires that we understand change and explicitly manage it. Past experience and scientific study have taught us much about how to do this successfully. Three lessons are particularly important.

Beware of attempts to halt or alter the tempo of natural change in ecosystems.

Over and over, human attempts to slow or stop normal environmental change have produced surprising and unsustainable consequences. For example, natural factors such as currents and waves are continuously changing the ocean shoreline. Human developers have tried various tactics to halt these changes. Instead of preventing change, these tactics have often damaged coastal ecosystems (**Figure 1.20**).

Many of our most significant environmental challenges are associated with our limited ability to accommodate inevitable change. In arid regions, plans for managing and allocating water often fail to include contingencies for periods of drought, resulting in much human conflict and hardship. Because agricultural and urban ecosystems lack the homeostatic features of their natural counterparts, they are highly vulnerable to the effects of natural disturbances such as earthquakes and hurricanes.

Copy nature.
For many centuries, humans have been among the most powerful agents of change on Earth. We have altered landscapes, modified the flow of rivers, created and drained lakes, and even changed the chemical composition of air. Indeed, it appears that our activities are now warming Earth's climate. But if change is normal in ecosystems, why should we worry about human-caused change? We must be concerned because the character and tempo of much of human-caused change has no precedent in the evolution of Earth's ecosystems.

Over the past century, we have added hundreds of novel chemicals to the ecosystems. These chemicals, which are not easily broken down by natural processes, are toxic to many organisms. For example, the manufacture of paper and some herbicides and the incineration of commercial and municipal wastes have released large quantities of toxic chemicals called *dioxins* into our environment. In recent years, environmental regulations in the United States have reduced the release of dioxins by 90%. However, because dioxins decompose very slowly, high levels of them remain in many ecosystems.

In other cases, humans have altered the patterns of natural change in ecosystems. For instance, dams that have been installed on rivers to generate electricity also modify the rate of water flow. This causes significant changes in downstream ecosystems, which are adapted to the rates of flow and patterns of flooding that occurred before the dams were built.

▲ Figure 1.20 **Defying Inevitable Change**
Attempts to alter the natural movement of sand along coastal beaches have often produced accelerated erosion such as that along the coast of North Carolina where houses are literally falling into the water.

Be alert for thresholds of change, or tipping points.
Some human actions can push ecosystems beyond the limits of their normal range of change. This may alter the ecosystems so much that they cannot easily return to their previous state. The point at which an ecosystem changes from one state of homeostasis to another is called a threshold of change, or *tipping point*.

Much of the woodlands in the Sahel region of Africa are at a tipping point. The trees in these woodlands are frequently cut down for fuelwood or to make way for agricultural development. When cut on a small scale, the trees grow back and the woodlands are restored. But when large numbers of trees are repeatedly cut down, the climate becomes drier and soils rapidly erode. When these lands are abandoned, they become deserts rather than woodlands. A threshold of change has been crossed. Human abuse of the land has changed the environment so much that the ecosystem can no longer return to its previous state.

Today most scientists agree that human activities are contributing to changes in the global climate. With this consensus comes concern that climate change may cause many of Earth's ecosystems to change in ways that cannot be reversed.

1.5 Uncertainty, Science, and Systems Thinking

Uncertainty abounds. Part of this uncertainty is the consequence of ignorance—our understanding of Earth's ecosystems is very far from complete. Uncertainty is also a consequence of the complexity of ecosystems, which can produce unpredictable changes.

It's all right to be uncertain; in fact, it is inevitable. **Science** is a process that poses and answers questions objectively in order to increase knowledge and lessen uncertainty. Science depends on careful observation and experimentation to produce results that can be duplicated by others. The results and conclusions of scientific research are open to review and revision by other scientists. New knowledge gained by scientific study is most likely to produce sustainable actions and behaviors when it is applied in ways that recognize the integrated nature of ecosystems.

Uncertainty

There is much we do not understand. Why?

It is an understatement to say that there is much that we do not know about ecosystems. Two factors—ignorance and complexity—are especially important sources of uncertainty in our understanding of the environment.

Ignorance. We would be able to meet many environmental challenges more effectively if we simply knew more. Reducing ignorance is the basic justification for most environmental research.

Ignorance presents two different challenges. Sometimes the extent of our ignorance is so overwhelming that it is difficult to know where to start or how to set research priorities. At other times, we don't know what we don't know. Our ignorance leads us to think that we know more than we do, so we fail to ask the right research questions. In his book, *The Discoverers*, Daniel Boorstin wrote, "The great obstacle to discovering the shape of the earth, the continents and the oceans was not ignorance, but the illusion of knowledge."

Complexity. Although their parts may interact in a predictable manner, complex systems often behave in ways that are practically impossible to predict. It is not that such behavior is random; rather, it is because miniscule, too small to measure differences in the individual parts of a complex system can lead to very large differences in that system later on. Because ecosystems are so complex, it is virtually impossible for scientists and resource managers to measure and understand all of their interacting components.

The unpredictable behavior produced by the complexity of ecosystems limits our ability to forecast changes. This is similar to the way that the complexity of weather systems limits our ability to predict the weather. Meteorologists can make reasonably accurate predictions about the weather for tomorrow or the next day.

The accuracy of weather forecasts diminishes rapidly the further we look into the future. This is because very small variations in the components of weather systems on one day produce significant differences in the future (**Figure 1.21**).

Uncertainty regarding environmental challenges is often the basis for inaction and conflict. Most famously, this has been the case in the way that various organizations and governments have responded to the challenge of global warming. In recent years, scientific research has shown conclusively that Earth's lower atmosphere is warming, but questions about the magnitude of this change persist. Uncertainties also remain as to whether human activities are responsible for global warming and what actions would be most effective in mitigating this process. These uncertainties even cause some people to argue that we should not take any action at all.

▼ **Figure 1.21 Complexity Yields Uncertainty**
Data on maps such as this allow meteorologists to predict the weather a few days in advance. However, weather predictions for the next month or year are much less certain because of the complexity of Earth's atmospheric systems.

Reducing Uncertainty with Science

Science asks questions in a fair and unbiased fashion.

The scientific revolution of the past several centuries has greatly expanded human knowledge and resolved much uncertainty about our world. We are often awed by the technical detail and complicated instrumentation of science. Yet at its core, science is simply the process of asking questions in a way that is as objective, or unbiased, as possible (**Figure 1.22**).

To find an answer to a question, scientists propose **hypotheses**, or alternative answers that can be tested by careful observation or experimentation. For example, scientists might ask why the population of a particular animal is declining. Hypotheses might be that the animals are suffering from a disease, that too many are being killed by hunters, or that their food supply is diminishing. Scientists might test the hypothesis that food supply is important by monitoring the availability of the animal's food. They might perform an experiment in which they provide extra food to half of the population and none to the other half, then compare the rates of growth of the two groups.

The scientific process follows very careful rules regarding hypotheses and the strategies to test them. Most important, a scientific hypothesis must be *falsifiable* by observation or experimentation. This means that it is possible to devise an experiment or observation that could prove the hypothesis wrong. Hypotheses based on an immeasurable "force" or on divine intervention are simply not open to scientific inquiry (**Figure 1.23**).

Scientists take care to make precise observations and measurements that will not produce biased results. Experiments generally include one or several treatments relevant to the hypotheses, the results of which are then compared to an untreated control. Furthermore, experiments and observations are replicated and results are compared to ensure consistency. Finally, scientists subject their results and interpretations to review by other experts before they communicate them in scientific journals.

Although scientists may come to consensus regarding the answer to a particular question, no idea, theory, or hypothesis is immune from further question or revision. No uncertainty ever becomes an absolute certainty. *Can always be questioned!*

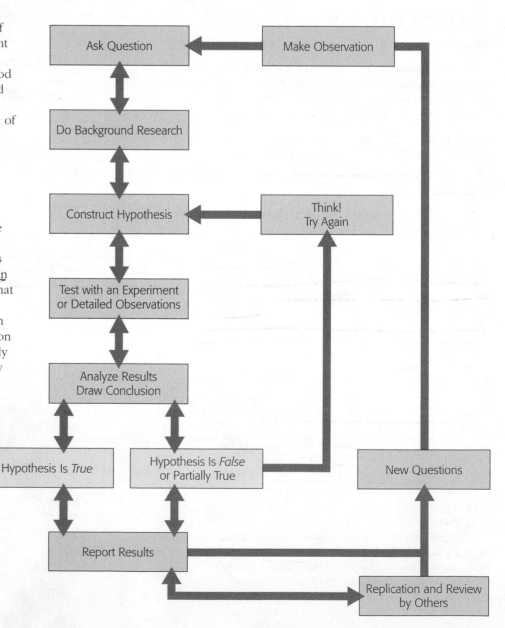

► Figure 1.22 **The Scientific Method**
Science is a systematic method for asking question in an unbiased way. Because results are subject to review and replication by others, science ensures that mistakes are corrected. Inevitably, the process also generates new and interesting questions.

A set of experiments and observations may lead scientists to agree that a particular hypothesis answers a question or that a specific model explains a phenomenon. However, further study may determine that the model is naïve or misguided and replace it with another, very different model.

As much as they aspire to objectivity, scientists are human. As humans, they may become overly protective of their ideas. Occasionally, they are guilty of advocating for particular answers to questions because they align with closely held personal beliefs or values. However, science thrives by falsifying hypotheses. In so doing, it is self-correcting.

Although scientists often have strong opinions, it is not the role of science to make decisions regarding environmental policies or the management of resources. Rather, science provides a context for assessing and resolving uncertainties regarding specific policy or management alternatives.

▲ **Figure 1.23 It Must be Falsifiable**
In 1633, Galileo defended his hypothesis that Earth and the other planets circled the sun before the Inquisition in Rome based on careful observations rather than religious beliefs.

Systems Thinking

Just because we know things doesn't mean we understand how they fit together.

Sustainable management does not require us to understand every piece and process in an ecosystem (although the more we know the better). It does, however, require systems thinking. Systems thinking recognizes the connections among the pieces of a larger, integrated system. In contrast, piece thinking focuses on the individual parts of a system, giving little attention to their association with other parts.

Returning to our car metaphor, a piece thinker interested in improving the efficiency of the braking process of a car might focus on the materials that make up the brake pads. Systems thinkers would recognize that the ability to stop is influenced by many interacting features, including brake rotors, tires, and car weight. Thinking across boundaries, systems thinkers would recognize the importance of driver reaction time and road conditions. They would also realize that changes that improve braking efficiency might affect the behavior of other parts of the car.

Many unsustainable actions arise as a consequence of piece thinking. For example, overexploitation of resources such as fishes or forest trees is often a consequence of neglecting the complex ecosystems of which fishes and trees are part. The air pollution we call smog is a result of piece thinking. City planners focused on transportation and industrial development but ignored the ecosystems of their cities, including chemical transformations in the atmosphere and their effects on human health.

Systems thinkers recognize that the parts of systems are actually systems too, with behavior that depends on complex linkages among their parts. And finally, they recognize that all systems (except, perhaps, the universe) are parts of even larger wholes.

QUESTIONS

1. Describe two important sources of uncertainty in our understanding of ecosystems.
2. Scientific hypotheses must be falsifiable. What is meant by this statement?
3. How does systems thinking differ from piece thinking?

1.6 Sustainability Science

In the end, whether or not we choose sustainable actions depends on our understanding of the connections between ecosystems and our well-being. The emerging field of **sustainability science** aims to understand the interactions between ecological systems and social systems, with a particular focus on long-term changes in global systems such as climate and the world economy. Sustainability science is also interested in how stresses, such as pollution and human conflict, affect the homeostasis of ecosystems and social systems. To be successful, sustainability science must embrace a wide range of outlooks and engage scholars from both the natural and social sciences. It is not a laboratory science; rather, it is about the actions we take in the real world. Therefore, scientific exploration and practical application must occur simultaneously. Seven questions are particularly important to the development of sustainability science.

Ecosystem—Social System Connections

How can the interactions between ecosystems and social systems be incorporated into our understanding of sustainable human development?

Throughout history, we humans have modified ecosystems to increase the provision of the resources we need or want, such as food and water. Increasing these resources has often diminished other ecosystem functions or services. In some cases, human actions have lessened the ability of ecosystems to provide essential services (Figure 1.24).

The effects of human activities may not be immediately obvious because the time lag between a particular action and its impact on ecosystem functions may be quite long. For example, social conditions in one region may encourage people to cut down woodlands. Over time, this deforestation reduces rainfall, so communities downstream receive less water. Downstream communities may respond by diverting water to meet their immediate needs, thereby diminishing the long-term capacity of the ecosystem to sustain water flows.

The interactions between social systems and ecosystems extend from the microscopic, such as those that occur among microscopic plankton in a drop of water, to the global, such as those that occur among Earth's oceans, land, and atmosphere. Research to understand the physical and biological processes that influence ecosystems from local to global scales must be integrated with studies of the social, political, and economic processes that drive human actions across these scales.

Long-Term Trends

How are long-term trends in environment and development reshaping the interactions between ecosystems and social systems?

In the past two centuries, humans have achieved a dominion over Earth's ecosystems that is unmatched by any other organism in the 3.8 billion-year history of life. Nearly 7 billion of us exploit Earth's ecosystems, using technologies that could not have been imagined a century ago. We are apt to see this as a "natural" consequence of our unique human traits such as intelligence, communication, and learning. Yet, there is no precedent that assures us that our situation is sustainable.

Changes will continue in our numbers, in our technologies, in our social systems, and in the ecosystems upon which we depend. Patterns of ecosystem change from local to global scales must be better understood in the context of the increasing number of people and the technologies they employ. The world a generation from now will be unlike any in the past. Commitment to environmentally sustainable development is a commitment to the well-being of future generations.

▼ **Figure 1.24 Human-dominated Ecosystems** Humans have planted agricultural fields with single crop species on hundreds of millions of acres that once supported diverse and complex prairie ecosystems. These fields provide food that is essential to the growing numbers of us. But they require significant subsidies of fertilizer and water and are more prone to erosion and susceptible to drought than the ecosystems they replaced.

Ecosystem–Social System Stability

What factors threaten the stability of interactions between ecosystems and social systems?

Human-dominated ecosystems, such as cities and agricultural fields, lack the homeostatic features of their more complex, natural counterparts. When we manage ecosystems to increase their provisioning services, we often sacrifice their regulating or supporting services. As a consequence, they are more vulnerable to disturbances and the effects of long-term change.

In other words, ecosystems simplified by humans are more susceptible to drastic changes. A disaster such as a hurricane is often called an "act of nature" or an "act of God." In 2005, Hurricane Katrina led to severe flooding in New Orleans. However, most of the *natural* ecosystems of southern Louisiana were quite resilient. It was the limited resilience of our *engineered* environment that caused most of the destruction and human misery (**Figure 1.25**).

Human population growth and development have increased the number of people living in semi-arid and arid environments. Too often, water management in these areas is based on average water conditions and is not designed to accommodate the inevitable variation in climate. For example, the policies for managing California's system of dams and canals were designed for the average rainfall and snowfall conditions of the 1950s and a population that was much smaller than it is today. Since the 1950s, residential, industrial, and agricultural demands for water have tripled. During dry years, there are severe water shortages in this region. The prospect of even warmer and drier conditions in the future is daunting.

Scientists are learning more about the connections between ecosystem homeostasis and ecosystem complexity. We also need to learn more about how our social systems influence such connections and how our social systems can support them.

◄ **Figure 1.25 A Natural Disaster?**
The impacts of Hurricane Katrina on New Orleans were greatly increased by the loss of wetlands in the regions surrounding the city. These complex ecosystems would have buffered many of the hydrological impacts of the storm.

Thresholds of Change

Can we define reliable thresholds beyond which ecosystems and social systems incur significant risk of degradation?

Global warming is one example of long-term directional change in the environment that has been caused by humans. Other examples include the introduction of non-native species, land development, and changes in water quantity and quality. Within a particular range, ecosystems are resilient to change. Outside that range, ecosystems may undergo significant change toward a new stable structure. At that point, it may be difficult or impossible to restore their original structure and function.

We know that such thresholds exist, but we are not yet able to predict their exact ranges. For example, humans have long depended on arid grasslands to support grazing animals.

As long as grazing pressure is moderate, such ecosystems are relatively stable and resilient. When rainfall falls below a certain amount, however, such ecosystems are no longer stable, and grasslands quickly become deserts (**Figure 1.26**). At present, we do not know the threshold of rainfall at which this important transition is likely to occur.

▼ **Figure 1.26 Thresholds of Change**
A shepherd searches for grass for his goats in Niger, Africa. As climates have become warmer and drier, overgrazing has converted grassy woodlands to desert.

Human Incentives

What incentives to human behavior will most effectively produce sustainable interactions between ecosystems and social systems?

In general, human actions are influenced by laws and regulations, economic incentives, knowledge, and belief systems. How these factors interact to affect human behavior is not well understood. On a global level, laws and regulations often have limited effects.

Long ago, incentives that influenced human actions were local; their effects on ecosystems were mostly local, too. Today, many human actions are driven by national and global incentives. Globalization of the world's economy makes it difficult to identify incentives for sustainable actions. Incentives that encourage sustainable land stewardship in Bangladesh, where farmers grow their own food, are very different from those for farmers in Kansas who grow food for global markets.

For some people, an increased knowledge of the unsustainable nature of actions such as the destruction of forests or increased consumption of fossil fuels is an incentive for action. For others, it is not. How differences in ethical systems and religious beliefs influence such actions is not at all clear.

The aspirations of the world's poorest people will be especially important in the years ahead. Improvement in their well-being is essential to reducing conflict among peoples and nations. Identifying incentives for actions that diminish poverty in a sustainable fashion is a particularly important challenge for sustainability science.

Monitoring Our Progress

How can systems for monitoring and reporting provide more useful guidance for sustainable actions?

Our modern understanding of human physiology has helped us develop a very sophisticated system for monitoring, restoring, and maintaining human health. For example, if your body temperature, as measured by an oral thermometer, rises much above 99 °F, you know that you have an infection. Such knowledge might cause you to rest and drink lots of liquids. If your condition worsens, a doctor might make additional measurements to determine the exact cause of the infection and prescribe an antibiotic. Unfortunately, we have not yet developed a system for monitoring the health or sustainability of the interactions between human social systems and ecosystems.

The challenge is a bit like the structure of an automobile's dashboard. We need indicators to show the sustainability of our ecosystems and social systems, similar to the gauges on a dashboard—the speedometer, oil pressure gauge, and fuel gauge. Local, national, and global communities must agree on what levels or amounts of these indicators are acceptable. For example, similar to setting a speed limit, we must determine what levels of chemicals such as arsenic or mercury can be tolerated in drinking water. We also need to develop protocols for the actions we will take when we approach critical values of these indicators. Finally, we need ways to ensure that these actions are taken, much as we enforce speed limits.

Integration of Learning and Action

How can the relatively independent activities of research, planning, monitoring, assessment, and decision making be better integrated?

Success in sustainability science will require changes in many of our institutions. Scholarly institutions such as universities are organized by disciplines, for example, the natural sciences, mathematics, economics, political sciences, and the arts. The challenges of sustainable development cannot be met by the methodologies of any one of these disciplines alone. Instead, it requires interdisciplinary, problem-oriented learning (**Figure 1.27**).

Many of the challenges of sustainability are urgent. But the time between the acquisition of new knowledge and its application to real-world problems is often long. It can be accelerated by more effective communication among scholars and those who will apply their scholarship. Research is needed to develop the management systems that unite scientific exploration and application.

▶ **Figure 1.27 Integrated Learning?** How is learning and research that is important to sustainability distributed among schools and departments at your college or university?

QUESTIONS

1. Give an example of an ecosystem–social system interaction.
2. What is meant by "thresholds of change"?
3. Why is monitoring important for sustainability?

AGENTS OF CHANGE: The University of California, Berkeley's Green Initiative Fund

Rachel Barge is a graduate of the University of California, Berkeley, where she helped create The Green Initiative Fund (TGIF), which provides approximately $250,000 a year to campus projects that make UC Berkeley more environmentally sustainable. Students began by paying $5 per semester into the fund, which then provided grants students could apply for to champion projects such as energy retrofits, water conservation, and "green" student internships. After graduating, Rachel began her own consulting firm to spread the green fund model to other campuses, raising more than $16,000,000 in the process.

How did you first get the idea for The Green Initiative Fund?

When I arrived at UC Berkeley as a college freshman, the university had just released its first-ever "Campus Sustainability Assessment," a comprehensive analysis of the university's performance in the areas of energy, water, buildings, food, transportation, and purchasing. Although the assessment was a great first step, I soon realized that there was no plan to *implement* most of the great ideas contained in the document. Why? The funding available for such projects was minimal, not even scraping the surface of what needed to be done.

A group of fellow students and I thought we could tackle this problem head-on by raising funds ourselves while also pressuring the administration to put its own money forward. We learned that a green fee program run by the students of UC Santa Barbara was generating more than $150,000 per year for campus sustainability efforts. We thought a similar student fee would be a great way to unleash a significant amount of capital for green projects on our campus.

If anything convinced us of the need to establish this fund, it was learning that our campus had a carbon footprint that was larger than that of the entire nation of Cambodia, a nation of 13 million people. It was an outrageous fact that served as our call to action.

What steps did you take to get started?

About eight students and I formed a core team dedicated to passing the green fee in the spring campus elections. We needed to get the majority of the students to vote in favor of the initiative, which was no easy task. We learned everything about green fees: how to develop their legal structure, how to propose them on the ballot, how to explain their value to fellow students through social media and videos, and how to determine what kinds of projects the fees could pay for.

After nine months of hard work, 69% of the students voted to pass TGIF—we had won our green fee campaign. The university now has approximately $250,000 per year for sustainability initiatives, which is $2,500,000 over 10 years!

What are the biggest challenges you've faced with the project, and how did you solve them?

A fellow student challenged our campaign in court and tried to have The Green Initiative Fund removed from the election ballot. Our opponent found a few small errors in the way we'd set up our campaign. We went to court and fought back hard, winning our legal case and prevailing in the election.

How has your work affected other people and the environment?

It's fun to see the effects of TGIF growing every year. A favorite project of mine is a student-run organic café in Berkeley that supplies students with fresh, local, organic food and serves as a hub for food education and activism. The best part of creating green funds is that you never know what amazing ideas students will have about how to make the campus a better, more sustainable place.

What advice would you give to students who want to replicate your model on their campuses?

Here's my advice for creating of a green fund on your campus.

- *Do your research.* Figure out how your campus elections work, how student fees work, and what fees have been created in the past. Every campus is different.

- *Get a core team.* You can't run this campaign alone, so make sure you've got a core team of five to eight people and lots of volunteers for election time.
- *In marketing, focus on the little thing—the size of the fee.* It's only $5, and most students spend that much just to buy breakfast. Isn't $5 a semester a small price to pay for a green campus?
- *Have fun.* These campaigns are your chance to leave a lasting legacy of sustainability on your campus, but make sure to enjoy the process—it goes by fast!

I hope people can learn from my experience: If you have a great idea, go out there and ask people for money to launch your initiative into the world! I had no idea that was possible when I was graduating, but a mentor told me, "You can create a job out of this thing you loved doing in your free time. Put together a proposal and start asking people for money!" It's the best advice I've ever received.

ONE BURRITO HELPS SAVE THE WORLD

For just $5 / semester, TGIF will make our campus more sustainable. That's less than the cost of your lunch.

27

Summary

Earth's environments comprise all of the physical, chemical, and biological factors that determine the growth and survival of organisms. Humans have altered Earth's environments more than any other single species. These alterations affect our own well-being—our access to food and shelter, health, social stability, and personal freedom. Our actions will be sustainable if they meet the needs of the present in an equitable and fair fashion without compromising the ability of future generations to meet their needs. Principles of ecosystem function provide basic guidelines for sustainable action. These include a keen awareness of the limits posed by the law of conservation of mass and energy and the openness of ecosystems, conservation of ecosystem complexity and the homeostasis it provides, and the ability to accommodate change. The sustainability of our actions depends especially on our understanding of the connections between ecosystems and human well-being. Studies of ecosystem–social system interactions and stability, long-term trends, thresholds of change, and human incentives for sustainable action—will be key to that understanding.

1.1 Environment and Sustainability

- The environment includes all of the physical, chemical, and biological factors and processes that influence the growth and survival of organisms.
- Sustaining human well-being and the well-being of other organisms depends on human actions and behavior that meet the needs of the present in an equitable and fair fashion without compromising the ability of future generations to meet their needs.

KEY TERMS

environment, environmental science, ecology, human well-being, sustainability

QUESTIONS

1. How are notions of equity and justice embedded in our current concept of sustainability?

2. Most decision makers agree that equitable treatment and justice among humans must be part of any definition of sustainability. Why is this important?

1.2 Ecosystems

- An ecosystem is a complex community of organisms (biota) and its physical and chemical (abiotic) environment, functioning as an integrated ecological unit.
- The flows of matter and energy and the processes that influence the growth and survival of organisms are called *ecosystem functions*.

KEY TERMS

ecosystem, biota, abiotic, ecosystem functions, ecosystem integrity, ecosystem services, provisioning services, regulating services, cultural services, supporting services

QUESTIONS

1. Ecologists often liken the functioning of an ecosystem to that of an organism. In what ways does this comparison make sense and in what ways does it not?

2. Over the past 24 hours, you have benefited from many ecosystem services. Name five specific examples and the categories to which they belong.

Inflow: Sunlight and materials like carbon, nitrogen, and water.

Transformation and exchange of matter and energy between biota and abiotic environment.

Outflow: Heat and materials like carbon, nitrogen, and water

1.3 Principles of Ecosystem Function

- Although they are complex, ecosystems are highly organized and function in very predictable ways.
- All ecosystems are open to inflows and outflows of matter and energy.
- Ecosystem processes and functions are self-regulated by dynamic homeostasis.
- Ecosystem change is inevitable and essential.

KEY TERMS

dynamic homeostasis, positive feedback, negative feedback

QUESTIONS

1. The boundaries between counties and states are often defined by straight lines. Why might this be a problem for the management of ecosystems?

2. Explain why positive feedbacks often destabilize ecosystems.

1.4 Acting Sustainably

- Human actions are likely to be unsustainable if they ignore the limited supply of resources, reduce the homeostatic capacity of ecosystems, or interfere with the capacity of ecosystems to change.

KEY TERM

nonrenewable

QUESTIONS

1. The law of conservation of mass and energy is universally accepted. Nevertheless, some human actions indicate the opposite belief. Give two examples of these actions.

2. Much of human management involves replacing complex and diverse ecosystems, such as a forest, with much simpler ecosystems, such as a cornfield. Why are these simpler systems more prone to disturbance?

3. If change is inevitable and normal in ecosystems, why should we be concerned about the changes wrought by humans?

1.5 Uncertainty, Science, and Systems Thinking

- Acknowledging that there is much that we do not understand about the organization and functioning of ecosystems is an important prerequisite to sustainable action.

- Sustainable management requires systems thinking, the recognition of the linkages among the parts in the context of a larger, integrated ecosystem.

KEY TERMS

science, hypotheses

QUESTIONS

1. Why is it important to recognize that our knowledge about most environmental issues is incomplete and might change with additional information?

2. Some environmentalists believe that all life is guided by a spiritual force. What role might science have in verifying or refuting that belief?

3. Disagreement among scientists is critical to the scientific process. Why?

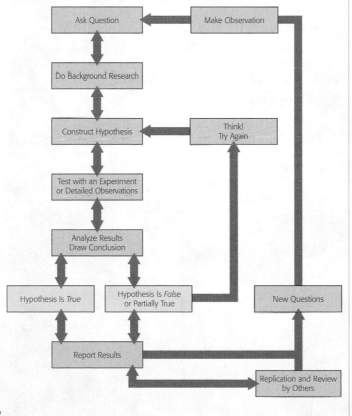

1.6 Sustainability Science

- Sustainability science is an emerging area of study that focuses on the interactions between ecosystems and human social systems, with a special focus on long-term changes in environment and human populations and society.

KEY TERM

sustainability science

QUESTIONS

1. Research aimed at sustainable solutions to environmental problems must involve many academic disciplines. Using a particular environmental challenge, explain why this is true.

2. Natural disturbances such as hurricanes or droughts may have unexpected consequences in human-dominated ecosystems. Briefly explain why.

Mastering ENVIRONMENTALSCIENCE

Go to **www.masteringenvironmentalscience.com** for homework assignments, practice quizzes, Pearson eText, and more.

2

Environmental Ethics, Economics, and Policy

Dam-nation!

Is it right to destroy a wilderness ecosystem in order to benefit human communities?

In winter, deep snow covers Tuolumne Meadows, high in Yosemite National Park. In spring and early summer, the snow melts and the water eventually flows into the Tuolumne River. The water is then captured in the Hetch Hetchy Reservoir and carried by pipeline to San Francisco's public water system, 167 miles away. Since 1934, most of the water consumed by the millions of San Francisco residents has made this journey.

Hetch Hetchy provides the people of San Francisco with exceptionally pure water at a fraction of the cost paid by people in other California cities. Yet in 2004 Governor Arnold Schwarzenegger instructed state agencies to undertake a comprehensive study of the costs and benefits of removing the 312-foot-high O'Shaughnessy Dam that backs up Hetch Hetchy Reservoir and restoring the ecosystem it submerged more than 80 years before. In addition to the cost of replacing the water that the reservoir currently provides, the costs for removing the dam and restoring the original ecosystem are estimated to be in the billions of dollars.

Why would the leader of a state that is notorious for its chronic water shortages even contemplate such an action? What benefits could possibly compensate for these costs? The answers to these questions are found in the history of the Hetch Hetchy project and the complex ethical, economic, and policy dilemmas it represents.

Like Yosemite Valley, Hetch Hetchy Valley is bounded by awesome cliffs and granite domes (**Figure 2.1**). Before the dam was built, lush meadows along the meandering Tuolumne River supported abundant wildlife. For thousands of years, the valley was treasured by all who visited, especially the Paiutes, the indigenous people who hunted and camped here.

As early as 1895, engineers were arguing that the narrow opening and high granite walls of the valley provided an ideal setting for a dam and reservoir. The reservoir would capture the flush of water from melting snows, providing a reliable year-round supply of water. San Francisco's population was growing rapidly, and its mayor, James Duval Phelan, perceived that water supply was the primary limitation for continued growth.

▲ Figure 2.1 **Before the Dam**
Hetch Hetchy Valley as painted by Albert Bierstadt in 1866.

Beginning in 1903 and continuing through 1905, Phelan and his proponents repeatedly applied to the U.S. Department of Interior for permits to proceed with the project.

Legendary wilderness advocate John Muir and the members of his newly founded Sierra Club opposed the project. They argued that it was contrary to the mission of the newly formed Yosemite National Park and would defile one of nature's wonders. In late 1905, two years after a much publicized visit to Yosemite with John Muir, President Theodore Roosevelt rejected Phelan's application (**Figure 2.2**). San Francisco formally abandoned its pursuit of Hetch Hetchy water in January 1906.

But at 5:16 A.M. on April 18, 1906, a great earthquake struck San Francisco. The earthquake and the fire that followed demolished the city. The rebuilding of the city soon reopened the Hetch Hetchy water debate because by then the city had to pay exorbitant prices to buy water from surrounding communities. The political pressures for developing Hetch Hetchy grew rapidly.

Gifford Pinchot, the chief of the newly created Forest Service and a powerful voice in the Roosevelt administration, was a strong supporter of the project. Pinchot wrote, "I fully sympathize with the desire . . . to protect the Yosemite National Park, but I believe that the highest possible use which could be made of it would be to supply pure water to a great center of population."

▲ Figure 2.2 **A Memorable Meeting**
President Theodore Roosevelt and John Muir at Glacier Point, Yosemite, in 1903. At this famous meeting, Muir thought he had convinced Roosevelt to preserve Hetch Hetchy Valley.

Muir, on the other hand, passionately fought the project on the principle that it was simply wrong to destroy a beautiful wilderness, regardless of its potential human value.

The voices of the Paiutes, who had lived in Hetch Hetchy for centuries, were not heard in the din of this public debate. Although the development of Hetch Hetchy would cost these people dearly, they were not considered at all. No matter how the valley was developed—with a dam and reservoir or with roads, hotels, and public camps—their access to the valley would be lost forever.

In 1908, Roosevelt succumbed to pressure from Pinchot and others and initiated the permitting process for the dam. But in 1909, Roosevelt's successor, President William Howard Taft, was persuaded by Muir's arguments and put the project on hold. Debate and political wrangling over the project continued until 1913, when the new president, Woodrow Wilson, insisted on its approval. In that same year, Congress passed legislation affirming its support for the project.

Construction of the O'Shaughnessy Dam was completed and the valley filled with water in July 1923 (**Figure 2.3**). It took 11 more years to complete the aqueduct and water distribution system that would deliver Hetch Hetchy water to thirsty San Franciscans. Few doubt that the city received great benefit from this project. But as California has become more urban, concerns over the continued loss of wilderness have grown. Many influential individuals and organizations now believe that the benefits of a wild Hetch Hetchy Valley exceed the costs of its restoration and replacing its water. Those powerful voices motivated Governor Schwarzenegger's action in 2004.

Was it ethical to submerge a beautiful wilderness ecosystem? Do the social and economic benefits of the Hetch Hetchy project exceed its environmental costs? Was the policy that produced this outcome just and equitable? These and similar questions are inherent in nearly all actions affecting our environment, and they are the focus of this chapter.

▼ **Figure 2.3 To Be Restored?** Hetch Hetchy Valley today with the reservoir that supplies water to the city of San Francisco.

2.1 Changing Views of Humans and Nature

What is your connection to the environment? Human attitudes toward the environment are shaped by a variety of factors, including mode of living, cultural history, religious and political beliefs, and knowledge. Connections between humans and the environment were probably obvious to primitive hunter-gatherer communities. But as communities became more agricultural and then developed industrial technologies, the connections became less obvious. During the Industrial Revolution, new knowledge, population growth, and growing environmental impacts raised concerns about the relationship of humans and the environment. Debates raged about whether nature had value in its own right or was valuable only in its service to the needs of humankind. Today, our view of the relationship of humans and the environment continues to evolve in response to the effects of human actions and their consequences for different groups of people.

Pre-Industrial Views

Our view of our relationship to nature changed as our dependence on it became less immediate and obvious.

The environment is a central feature in the spiritual traditions of many primitive cultures. For example, many Native American traditions assign sacred status to Mother Earth and Father Sky, whose union they assert produced all life. **Animism,** the belief that living and nonliving objects possess a spirit or a soul, is a common feature in the religions of many indigenous people (**Figure 2.4**). For hunter-gatherers, humanity and nature are unified.

The domestication of plants and animals allowed humans to alter ecosystems for their benefit. For example, farmers in ancient Mesopotamia replaced native grasslands with fields of domesticated wheat. By doing this, they were able to direct the productivity of these landscapes to the support of human populations. Weeding out competing plants and irrigation allowed farmers to boost crop production even more (13.2). As agricultural societies gradually became more complex, their citizens probably became less aware of their dependence on the natural environment.

The monotheistic religions of Judaism, Christianity, and Islam arose during this period of agricultural development. A key feature of these religious traditions is the centrality of humankind among all creation, as evidenced by man having been created in God's image. In Genesis, for example, man is directed to subdue the earth and is given dominion over "every living thing that moves upon the earth."

◄ **Figure 2.4 A Sacred Giraffe**
Archaeologists believe that paintings such as these done about 6,000 years ago by hunter-gatherers in the Tassili n'Ajjer Caves, Algeria, reflect the reverence of these people for the animals they depended on.

▲ **Figure 2.5 Sacred Site**
Shinto worshipers enter the Miyajima Temple through the red sea gate first erected in A.D. 1168. The temple is dedicated to the daughters of the Shinto deity of seas and storms and to the brother of the great sun deity. The gate symbolizes the connection between sea and land.

Even so, the relationship of humans to nature is described in different ways in different parts of the Hebrew Scriptures, reflecting historical changes in culture and knowledge. Other Judeo-Christian and Islamic writings place greater emphasis on human stewardship of resources.

Eastern religions such as Hinduism, Buddhism, and Shintoism also originated during the time of agricultural development. In general, these religions emphasize the importance of harmony between human actions and nature, and they view humans as inseparable from nature (**Figure 2.5**).

The environment and organisms were common themes in the writings of ancient Greek scholars. None of these scholars had a greater influence on future generations than did Aristotle (384–322 B.C.). Nearly half of his more than 400 works focused on the nature of living things and mankind's relationship to them. He saw the elements of the environment arrayed along a natural "ladder of being" (*scala naturae*) from nonliving matter to plants, insects, and higher animals. On Aristotle's ladder, humans occupy the rung immediately beneath God. For the next two millennia, Aristotle's human-centered view of nature prevailed among Western thinkers.

Time of Agricultural development

35

The Enlightenment and Industrial Revolution

New knowledge and ever-growing human impacts on the environment catalyzed new views of the relationship of man and nature.

By 1700, the scientific method was in wide use, and the Enlightenment was in full bloom. It was widely accepted that Earth was not at the center of the universe, opening up speculation about the position of humankind in the cosmos. Over the next century, philosophers such as Voltaire (1694–1778) would openly question humanity's special relationship to God. In 1809, Jean-Baptiste Lamarck proposed that the human species was simply one of many products of a process of evolution. Lamarck's proposal received little acceptance until 1859, when Charles Darwin's theory of evolution by means of natural selection provided a clear explanation for evolutionary change (4.5). Throughout this time, the fields of botany and zoology were generating an increased understanding of the connections between organisms and their environment.

Between 1750 and 1850, science spawned technologies such as steam-driven machines, giving birth to the Industrial Revolution. The human impact on the environment grew enormously. Increased commerce and food production generally improved human well-being. But, with the growth of urban industrial centers, fewer people had direct contact with natural environments (**Figure 2.6**).

As industrialization increased, some people began to look back to the natural world for inspiration. In 1836, Ralph Waldo Emerson published his short book, *Nature*. Emerson argued that the obsession of industrial societies with material goods broke the connection between humans and nature. Furthermore, science provides limited insight into nature; humans need to immerse themselves in wild nature if they wish to reunite with its transcendent truths.

The publication of *Nature* is often seen as the origin of the American transcendental movement, which included writer-philosophers such as Nathaniel Hawthorne, Walt Whitman, and Henry David Thoreau. For environmentalists, Thoreau's *Walden* is the most enduring of all transcendentalist writings. Many consider Thoreau to be among the founders of the discipline of ecology (**Figure 2.7**).

Less literary and more scientific, George Perkins Marsh's 1864 book, *Man and Nature*, is viewed by many as the genesis of the modern environmental movement. Marsh argued that nature is essentially "harmonious," but humans have separated themselves from nature and thereby violated its harmonies to the detriment of both nature and humans. Our survival, he argued, requires that we reconcile our activities with the inescapable laws of the natural world.

▼ **Figure 2.7 Walden's Inspiration**
Walden Pond as it appears today. In Thoreau's words, "I went to the woods [to] . . . see if I could not learn what it had to teach and not, when I came to die, discover that I had not lived."

▲ **Figure 2.6 Urbanization in the Industrial Age**
As more and more people moved into cities, there were fewer opportunities for people to connect with the natural world.

Conservation vs. Preservation

What is the value of nature to humans?

Yellowstone—the world's first national park—was dedicated in 1872. As the government created other parks and forest reserves, their exact mission and value to society was hotly debated. The naturalist John Muir, who was strongly influenced by transcendentalist writings and his own experiences in the Sierra Nevada wilderness, advocated the **preservationist** view that parks and public lands should *preserve* wild nature in its pristine state. Preservationists believed that humans should have access to wilderness parks for their inspiration and beauty, but that parks should be protected from consumptive uses such as logging or water diversion. To further these goals, Muir and other preservationists founded one of the earliest nongovernmental environmental advocacy organizations, The Sierra Club.

The leading spokesman for the opposing view was Gifford Pinchot. Pinchot had studied forestry in Europe and was an ardent critic of the destructive forestry practices widely used at that time. Pinchot articulated the **conservationist** view that public resources should be used and managed in a sustainable fashion to provide the greatest benefit to the greatest number of people. Conservationists valued nature for the goods and services it provided human beings. Preservationists, in contrast, argued that nature had value in its own right and therefore deserved protection.

Muir and Pinchot, and their respective viewpoints, were at the center of the debate over the Hetch Hetchy water project. To this day, the divide between conservationist and preservationist views continues to define debate on the management of public lands.

The Modern Era

Current views of the relationship of humans and the environment have been shaped by landmark books and ever-increasing human influence.

Growing up in the midst of the conservation-preservation debates, ecologist and environmentalist Aldo Leopold argued for both the sustainable human use of natural resources and the preservation of wilderness. Leopold thought that wilderness provides resources that are essential to civilization. He warned that human impacts on ecosystems differ significantly from natural disturbances. They are, therefore, likely to have consequences that are quite different from those that were foreseen.

One such unforeseen consequence resulted from the widespread use of the pesticide DDT. After World War II, this pesticide was sprayed on wetlands across the United States to kill disease-carrying mosquitoes. In addition to killing mosquitoes, DDT destroyed the birds that fed upon them as well as the predators of those birds. In 1962, the American public was alerted to this threat by marine biologist Rachel Carson. Her widely read book, *Silent Spring*, pointed out the danger that pesticides posed to natural ecosystems and, ultimately, to humans.

Six years later, Paul Ehrlich's book, *The Population Bomb*, told the public how the rapid growth of the human population was threatening the environment and human survival. Should trends in population growth continue, Ehrlich warned, our demands would soon exceed Earth's ability to supply us with food and other essential resources.

Today, environmental trends, debates, and warnings dominate news stories and political agendas around the world (**Figure 2.8**). Two issues are of particular concern—the widespread extinction of plant and animal species and the warming of Earth's atmosphere. In numerous books and articles, distinguished Harvard ecologist Edward O. Wilson has documented the loss of hundreds of plant and animal species, as well as the threats that these losses pose to our sustainability. In 2007, the Intergovernmental Panel on Climate Change and former Vice President Al Gore received the Nobel Peace Prize for their advocacy for action to respond to the threats posed by global warming. Their work has helped shape public understanding of the importance of climate change and the need for action to minimize its effects.

In recent years, theologians representing several different faiths have placed increased emphasis on environmental stewardship. Organizations such as Interfaith Power and Light bring together Jews, evangelical and non-evangelical Christians, and Muslims with common interests in environmental sustainability. These groups advocate care for the environment both as a matter of respect for "God's creation" and as a responsibility to humankind.

According to our history, our view of our relationship to the environment has come full circle. As hunter-gatherers, our dependence on nature was clear. Agriculture, industry, and the development of complex urban societies made our connection to nature less immediate and therefore less obvious. Perhaps this encouraged a sense of dominion over nature. Today, as our numbers pass the 7 billion mark, we do not doubt our ability to dominate Earth's ecosystems, yet we also have no doubt about our dependence on those ecosystems.

▲ **Figure 2.8 Headlines**
Dramatic changes in media coverage and what those changes mean regarding our view of the environment.

QUESTIONS

1. Describe changes in human attitudes about the relationship between humankind and nature that is reflected in monotheistic traditions.

2. How did transcendental writers, such as Emerson and Thoreau, view nature and our role in it?

3. Describe three major ideas or viewpoints that influence current attitudes toward the environment.

37

2.2 Environmental Ethics

The role of humans in the environment has been discussed since ancient times. Yet only in recent decades have we attempted to define appropriate human behavior in the context of the environment. Ethics is the branch of philosophy that explores ways to distinguish between right and wrong conduct or to define moral living. **Environmental ethics** studies the moral relationship of humans to the environment and its nonhuman contents. Systems of environmental ethics vary regarding who or what has value. Nonenvironmental issues, especially fairness and justice between genders or among ethnic groups or races, also influence the determination of whether an action is right or moral.

Doing the Right Thing

Different standards for right, or moral, behavior define three ethical traditions.

What makes an action right or wrong? Ethicists have long argued over how to answer this question. The Greek philosopher Plato proposed that *motivation* should be the basis for determining whether an action was good. In his **virtue ethics** system, an action is right if it is motivated by virtues, such as kindness, honesty, loyalty, and justice. Virtue ethics has played a significant role in the development of general ethical frameworks, but because it refers to the rightness of the actions of people toward other people, its language is necessarily anthropocentric. For example, according to virtue ethics, the neglect of the concerns of the Paiutes in the process that led to building the Hetch Hetchy Reservoir would be considered unkind and, therefore, wrong.

Consequence-based ethics emphasizes the importance of *outcomes*. Right and wrong are defined in terms of pleasure or pain, benefit or harm, and satisfaction or dissatisfaction. An action is right depending on whether or not it delivers pleasure, benefit, or satisfaction. **Utilitarianism** is an example of consequence-based ethics that defines right actions as those that deliver the greatest good to the greatest number. Gifford Pinchot's conservationist approach to environmental decisions was clearly utilitarian. According to consequence-based ethics, building the Hetch Hetchy Reservoir was right because it benefited a far greater number of people than did the preservationist alternative. Note that this same action would be judged unethical if it benefited fewer people than it harmed.

Eighteenth-century philosopher Immanuel Kant advocated **duty-based ethics** that the rightness or wrongness of actions should be determined by a set of rules or laws.

In duty-based ethics, a wrong action is wrong regardless of its outcome. For example, Kant suggested that it is always wrong to lie, even if lying has a positive outcome. Muir's preservationist view of Hetch Hetchy was grounded in duty-based ethics, arguing that destroying a wild place of natural beauty is wrong regardless of how it might benefit humans.

Duty-based ethics presumes an accepted framework, such as the Ten Commandments, for defining whether actions are right. Most such frameworks have been articulated in the context of religious traditions and, like virtue ethics, deal primarily with the actions of humans toward other humans. How do we know what is ethical when it comes to the environment (**Figure 2.9**)?

▼ **Figure 2.9 Doing The Right Thing**
These protestors are convinced that human actions that might be altering Earth's climate are ethically wrong.

Who or What Matters?

Environmental ethics differs regarding who or what has value.

In any ethical framework, we must identify *who* or *what* is to be valued. In consequence-based ethics, we must specify who or what should receive the greatest benefit. In duty-based ethics, we must define to whom or what we have a duty. For example, does the commandment "thou shall not kill" extend to animals?

Systems of environmental ethics also require us to consider whether the value is intrinsic or instrumental. A person, organism, or object valued as an end unto itself is said to have **intrinsic value**. A thing that is valued as a means to some other end has **instrumental value**. Most ethicists agree that every human life has intrinsic value. Some ethicists argue that other living things have value only as a means to support the well-being of humans. Thus, their value is instrumental.

Anthropocentric ethics assigns intrinsic value only to humans; it defines right actions in terms of outcomes for human beings. Other organisms and objects have instrumental value because they contribute to human well-being. The conservationist view of environmental management is certainly anthropocentric. There are also situations for which the preservationist view could be considered anthropocentric. Things that are not essential to our well-being, such as a place of beauty (Yosemite) or a charismatic animal (panda bear), acquire instrumental value because humans happen to care about them.

Biocentric ethics argues that the value of other living things is equal to the value of humans. Biocentric ethics extends intrinsic value to *individual* organisms beyond human beings; the organisms do not need to benefit humans in order to have value.

There are two distinct schools of biocentric ethics. Some biocentrists argue that for a thing to have intrinsic value, it must be able to experience pleasure or satisfaction. Therefore, it must possess qualities such as self-perception, desires, memory, and a sense of future. In this view, animals such as dogs, chimpanzees, and porpoises have intrinsic value, but plants and animals lacking a complex nervous system do not. This biocentric ethic motivates many in the animal rights movement.

Other biocentric ethicists argue that any individual organism that is the product of natural evolution has intrinsic value. Therefore, an amoeba or a jellyfish has the same value as a porpoise or a chimpanzee (**Figure 2.10**).

Ecocentric ethics places value on communities of organisms and ecosystems. Ecocentric ethicists believe that *collections* of organisms or critical features in the environment have intrinsic value. Aldo Leopold's assertion that "a thing is right when it tends to preserve the integrity, stability and beauty of the biotic community" is an example of ecocentrism. An ecocentric ethicist might argue that, because hunting individual animals improves the health of that species' populations by removing diseased animals, it is a right action.

On the other hand, many biocentrists argue that hunting is unethical because it violates the intrinsic value of the individual organism.

The most expansive form of ecocentrism asserts that intrinsic value derives directly from naturalness. Philosopher Andrew Brennan, for example, argues that a natural entity such as a river has intrinsic value and is not a "mere instrument." Nature does not exist in order to meet our particular needs. Instead, it deserves moral respect in its own right. Because we are the air we breathe, the water we drink, and the food we eat, all elements of the environment have equal intrinsic value. This concept is the basis for what is called the **deep ecology movement**.

▼ Figure 2.10 **What Has Intrinsic Value?**
To which of these organisms do you assign intrinsic value? What is the basis for your decision?

Ecofeminism and Environmental Justice

The rightness of environmental actions may be influenced by nonenvironmental considerations.

In her 1974 book, *Le Feminisme ou la Morte* (*Feminism or Death*), Françoise d'Eaubonne argued that exploitation and abuse of both the environment and women derive from male domination. She articulated an environmental ethic, **ecofeminism**, that fostered what she saw as feminine concerns for interrelationships among humans, nonhuman life, and the environment.

Many ecofeminists point out that women often suffer disproportionately from human destruction of the environment (**Figure 2.11**). Several ecofeminist writers call attention to the sexism in phrases such as "man against nature" and "Mother Nature." Most ecofeminists believe that we should conserve our living and nonliving environment whether it is valuable to humans or not. In this respect, they have much in common with the deep ecology movement.

Many actions in the environment, such as cutting a forest or locating a waste facility, have different consequences for different people. An action that is considered right according to biocentric or ecocentric criteria may have adverse consequences for some people. With respect to human equity or justice, many would consider such an action to be unethical.

QUESTIONS

1. What is the basis for judging right from wrong in virtue, consequence-based, and duty-based ethics?

2. What has intrinsic value in anthropocentric, biocentric, and ecocentric ethical frameworks?

3. What is meant by the phrase *environmental justice*?

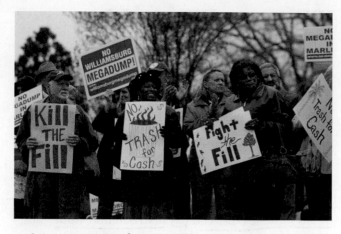

▲ **Figure 2.12 Protesting a Dump**
The environmental justice movement had its beginnings in Warrenton, North Carolina, a poor rural community that was targeted in 1982 as the site for a landfill that was to accept toxic waste dumped along the state's roadsides.

Is that due to injustice or b/c it's a cheaper area to live and people move there? (look into)

In the United States, for instance, toxic dumps and landfills have historically been located near poor communities, which often have a large racial or ethnic minority population (**Figure 2.12**). In some cases, this placement of waste facilities has been a consequence of direct racial discrimination. In other cases, sites have been selected because land near poor communities is cheap. In either situation, poor communities must bear a large share of the costs and risks associated with such facilities. Situations such as this have given rise to the **environmental justice** movement, which seeks to ensure that in the management of natural resources and the environment people are treated fairly regardless of race, gender, or economic status.

Environmental justice is a major consideration in a variety of international negotiations related to global issues. Some ethicists are concerned that the actions of many environmentalists disproportionately benefit the health and affluence of people in developed countries. For example, industrial development in poor countries is often accompanied by increased air and water pollution. Wealthy countries often step in to help control these pollutants and their negative effects on the environment. Unfortunately, many citizens of the developing countries view these efforts to set limits on pollution as unjust because such limits thwart their attempts to grow economically and reduce poverty. Justice, they feel, would be for the wealthy nations to provide economic compensation for the costs of conservation.

In summary, there is no single environmental ethic. The various ethical frameworks focus on different motivating values, such as justice, consequences for humans and ecosystems, and governing principles. Systems of environmental ethics also differ with respect to who or what in the environment has intrinsic value.

▼ **Figure 2.11 Gender Inequality**
In developing countries, land degradation may have particularly negative impacts on women because they are often responsible for tending crops and gathering fuelwood.

2.3 The Environment and the Marketplace

Many activities that affect the environment are associated with buying physical goods, such as food and clothing, and intangible services, such as transportation on airplanes or buses. An **economic system** is made up of the institutions and interactions in a society that influence the production, distribution, and consumption of goods and services. In a market economy, supply and demand determine the price and level of production for goods and services. The price that consumers are willing to pay depends on their perceived need for a good or service and the time they will receive it. For several reasons, markets may produce incentives for production or consumption that cannot be sustained without depleting natural resources or damaging the environment.

Economic Systems

Economies vary in their complexity and the extent of government control.

The most basic economic system is a **subsistence economy**, in which a society meets its needs from its environment without accumulating wealth. Individuals within a subsistence economy may barter to exchange goods and services, but there is no currency. Because people in subsistence economies do not accumulate a surplus of goods, their well-being depends on the continual renewal of resources from their environment. Societies of hunter-gatherers, fishermen, or farmers who live on what they catch or grow are examples of subsistence economies (**Figure 2.13**).

In **market economies**, the production and consumption of goods and services take place in markets guided by prices based on a system of currency. Individuals and businesses decide on their own what to produce or purchase. Producers are free to determine the price and amount of what they produce. Consumers are free to select what they purchase from among different producers. In a **free market economy**, the government does not influence the marketplace with subsidies, taxation, or regulation.

▶ **Figure 2.13 A Subsistence Economy**
The women at this market in La Paz, Bolivia, often trade one commodity for another.

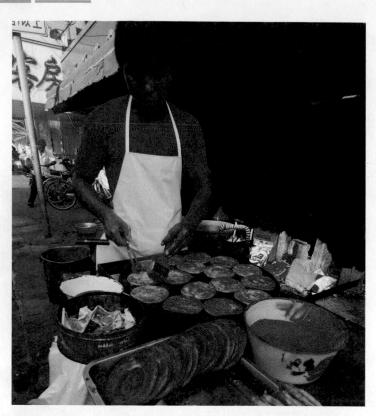

In a **planned economy**, the government regulates the prices of goods and services and the level of production. A government may simply set production, prices, and salaries by decree. More commonly, planned economies are controlled through a combination of regulation, incentives, subsidies, grants, and taxation.

Today, no country in the world has only one economic system. Although every country has a monetary system, subsistence economies in which people grow or barter for the things they need are found everywhere. In rural communities, for example, farmers may trade services, such as planting or harvesting, with no exchange of money. The United States is often said to have a free market economy, but many sectors of its economy are influenced by government regulations, subsidies, and taxation. China, North Korea, and Cuba claim to have planned economies, but in each of these countries much production and consumption takes place in "black markets" that operate outside the realm of government control (**Figure 2.14**).

◀ **Figure 2.14 It's Not All Planned**
Much of China's economy is planned, but this Shanghai street vendor openly markets his oven rolls.

Supply and Demand

The availability and price of a good or service is determined by the interaction of supply and demand.

Our understanding of what drives supply and demand in an economy has changed over the centuries. In 1776, Scottish philosopher Adam Smith proposed what he believed to be the basic principles motivating the production and consumption of goods and services in a society.

Smith noted that when a commodity was in short supply, its price increased. As a result, the producer had a greater **profit**, the difference between the cost to produce a commodity and its price in the marketplace. Smith argued that an increase in price and profit creates incentives for others to produce the same commodity. Supply increases. With a greater supply, competition among producers motivates them to lower prices. If prices for the commodity drop to near the cost of its production, incentives for production decline and supply decreases. Smith referred to this interaction among supply, price, and competition as the "invisible hand" of the economy.

Nowadays, economists accept Smith's assertion that increases in the price of a commodity will increase its supply. But they note that the price and abundance of commodities in a free market are also influenced by consumer demand—the amount of a commodity sought by a community. Consumer demand increases as goods and services become cheaper. **Neoclassical economic theory** focuses on the determination of the price and production of goods and services through the interaction of supply and demand (**Figure 2.15**).

Neoclassical economic theory is based on two key assumptions about human behavior. It assumes that those who produce and sell goods and services act to maximize profit.

HUGE

▼ **Figure 2.15 Supply and Demand**
As the price of a good or service declines, buyer demand increases, but incentives for producers or sellers decrease. Neoclassical economic theory suggests that this interaction results in an equilibrium between price and production.

SUPPLY AND DEMAND

Equally important, it assumes that consumers choose among alternative goods and services based on the satisfaction or benefit they expect to receive from them. In other words, neoclassical economists argue that consumers invariably act to maximize their satisfaction or benefit relative to cost.

Supply and demand affect the abundance of a commodity. Consider, for example, the availability of fish in a seaside village. If the price of fish drops, the demand for fish—the amount of fish that the villagers will seek to consume—will increase. If the price of fish goes up, fishermen will have a greater incentive to catch more fish and increase the supply to the market.

Thus the price and availability of fish are determined by the interaction of supply and demand.

This example assumes that the cost of catching each fish is always the same. However, there may be **economies of scale**, in which the cost per unit of a good or service declines as the level of production increases. For fishermen, the costs of running a boat are fixed, so as more fish are caught, the cost of catching each individual fish declines and profit increases. If, however, fish stocks are depleted by overfishing, the effort and cost required to catch fish may increase as each successive fish is caught.

Economic Value

The economic value of a good or service is determined both by consumers' need for it and the time when they receive it.

Economists define the **economic value** of a commodity as the price that consumers are willing to pay for it. That price is largely determined by the satisfaction or benefit that a consumer perceives in a commodity. Thus the economic value of a bottle of water, a piece of wood, or a barrel of oil is not determined by the costs of their production; rather, it is set by what consumers are willing to pay.

The benefit of a good or service—and, therefore, its economic value—is not a fixed quantity. Instead, it depends on what the consumer already possesses and when the consumer will actually receive the good or service.

In general, the satisfaction or benefit provided by a commodity declines as a consumer acquires more of it. For example, the benefit you derive from buying a car may be very high. The benefit you derive from purchasing a second car is much lower. In general, the benefit derived from a commodity is tightly coupled to the consumer's sense of need. The benefit of a glass of water is far greater to a poor soul wandering in a sun-parched desert than to someone blessed with abundant water (**Figure 2.16**).

The timing of the delivery of a good or service also affects the perception of its benefit. It is human nature to want something sooner rather than later. Its benefit diminishes if we are not to receive it immediately. Likewise, the economic value of a good or service—a consumer's willingness to pay for it—declines as the time until the consumer will receive it becomes longer. The owner of a fish market may be willing to pay a fisherman $10 for 10 pounds of fish delivered today. But the owner would not be willing to pay $10 for 10 pounds of fish that the fisherman promises to deliver a year from now. The rate at which economic value declines with time is called the **discount rate**.

Discount rates are affected by opportunity cost, risk, and consumer need. The **opportunity cost** of buying a thing is equal to the economic value of the best alternative use of that money. For example, the fishmonger who was

asked to pay $10 for fish to be delivered a year from now might be able to invest that $10 in the stock market and earn an additional $3. The opportunity cost for buying the fish in advance would be $3.

Our fishmonger's willingness to pay will be further diminished by risks, or uncertainties about the future. For example, illness or equipment problems may affect the fisherman's ability to deliver fish in the future. Or the actual price of fish a year from now might be lower than the price today.

Finally, the economic value of fish to be delivered at some future time depends heavily on the fishmonger's current needs. If she is desperate to have fish for her shop immediately, she is unlikely to spend $10 for fish that will not be delivered until a year from now. To her, that fish is of little value. In other words, she is likely to assign a very high discount rate to fish delivered in the future.

▼ **Figure 2.16 Abundance Influences Benefit**
In general, the satisfaction or benefit that a consumer or buyer derives from a unit of a commodity such as water decreases as it becomes more abundant.

AVAILABILITY OF ABUNDANCE

Market Complications

Markets are never perfect, and prices often do not reflect all costs.

Neoclassical economics assumes that the rational decisions of producers and consumers serve to maximize profits and consumer satisfaction. But in the complicated markets of the real world, there are often other outcomes. Free markets sometimes produce incentives for actions that deplete resources or damage the environment. There are three reasons why this is so.

Externalized Costs and Benefits. Buyers and sellers rarely bear all of the costs and benefits associated with a product. This is because the production and consumption of products often influence ecosystems or people who are not part of the transaction. **Externalities** are costs and benefits associated with the production of a commodity that affect people other than buyers and sellers. Externalities can be positive or negative.

Positive externalities arise when a third party benefits from an economic transaction between others. For example, a timber company may maintain a healthy forest in order to maximize profits when the timber is harvested. People who don't buy or sell timber can enjoy the recreation, wildlife, and beauty that the healthy forest provides. These benefits are positive externalities.

Negative externalities impart costs to individuals who are neither buyers nor sellers. For example, the timber company might harvest its forest in a manner that pollutes the streams or diminishes the property value of nearby homes. Here, people who are not involved in the buying and selling of the timber nevertheless pay a cost.

Externalized costs often present significant challenges to human well-being and to sustainable management. For example, coal-fired power plants release air pollution that is responsible for billions of dollars in health care costs that are paid by individuals and society. Those costs are not included in the price of coal, nor are they factored into the price of electricity. As a consequence, the marketplace provides no incentives to reduce this negative externality.

Unknown Costs and Benefits. The satisfaction or sense of benefit we realize from any commodity depends in part on what we know about its actual costs and benefits. Yet it is nearly impossible for consumers to have a full understanding of all of the costs and benefits associated with a good or service.

Fifty years ago, for example, ignorance of health risks contributed to the high frequency of cigarette and cigar use among Americans. The perception of the benefits of smoking was enhanced by ad campaigns and Hollywood films, which made smoking appear glamorous and sophisticated (**Figure 2.17**). Growth in our understanding of the health risks associated with smoking has greatly lessened the perceived benefit of cigarettes. In addition, stiff court fines to tobacco companies and increased tobacco taxes have ensured that some of the health care costs of smoking are now captured in the price of a cigarette.

Limited Resources Economic models often treat individual commodities as if they are infinitely available or easily interchangeable with other goods and services. This is rarely the case with natural resources and ecosystem services. For example, the Food and Agriculture Organization of the United Nations estimates that over the last four decades nearly 80% of the world's marine fish stocks have been fully exploited or overexploited. During this time, human demand for fish has tripled. Many argue that economic markets have encouraged this exploitation.

Neoclassical economic theory is a powerful framework for understanding human behavior related to the production and consumption of ecosystem goods and services. However, because most resources are finite and their costs and benefits are often externalized or not fully known to consumers, markets can result in depletion of resources or environmental damage. The depletion or overexploitation of a natural resource imparts an externalized cost to future generations.

▼ Figure 2.17 Risky Behavior
Fifty years ago, magazine ads such as this contributed to the poor understanding of the risks of cigarette smoking among consumers.

2.4 Valuing Ecosystems

Ecosystem services, such as clean water and beautiful scenery, are not usually bought and sold, but they do have economic value. Their value can be measured by assessing people's willingness to pay for actions that conserve them. Their value can also be assessed by estimating the costs that people must pay if the services are lost.

Economic Valuation of Ecosystem Services

Willingness to pay can be used to measure the relative economic value of ecosystem services.

What is the value of unpolluted air or the flood control provided by a forest in a floodplain? Because ecosystem services are not typically bought and sold in markets, how can we assign an economic value to them?

One way to measure their value is by assessing people's willingness to pay the costs of conserving them. Such assessments do not determine the total value of an ecosystem's services. Rather, they determine the **marginal value**, the difference in people's willingness to pay for one action compared to an alternative.

For example, the marginal value of removing the O'Shaughnessy Dam and restoring Hetch Hetchy could be assessed by asking citizens of San Francisco how much they would be willing to pay for their water if this alternative were pursued. The difference between how much they currently pay and how much they would be willing to pay if the dam were removed is the marginal value of restoring the ecosystem. Notice that the emphasis of marginal valuation is on benefits that are added or lost to humans, not to ecosystems.

What is the economic value of inspiring landscapes and recreational opportunities, such as those provided by Yosemite Valley (or, perhaps, a restored Hetch Hetchy Valley)? **Travel-cost valuation** is the amount of money that people are willing to pay for transportation and lodging to visit such places. Travel-cost estimates may also include the cost of building and maintaining the roads and infrastructure for these destinations.

This valuation technique has largely been applied to cultural, aesthetic, and recreational services.

In **hedonic valuation**, economic value is determined by the difference in the market price of real estate that is affected by different environmental alternatives. For example, the hedonic value of a public park can be estimated by the difference between the price that people are willing to pay for property near the park compared to the price they will pay for property elsewhere. Similarly, some of the economic costs of a municipal waste facility might be measured by its effect on the value of houses nearby compared to the value of similar houses located far away from the facility (**Figure 2.18**).

Estimating the value of an ecosystem service by surveying people's willingness to pay is called **contingent valuation**. For example, to determine the economic value of an endangered species, respondents might be asked to provide a statement of what they would be willing to pay for actions aimed at its conservation.

Can you put a monetary value on clean air or a beautiful forest? Aren't they priceless?

Christopher Shad, Cypress College

Norm: *They are indeed priceless, and we cannot determine their absolute value. We can, however, assess what people are willing to pay for actions that will improve air quality or preserve more forest land. We can also measure economic costs for diminished air quality or the loss of forest cover.*

▼ **Figure 2.18 Hedonic Value**
You might be willing to pay a premium for a home that is located in a natural setting, such as near a lake or a park. If you knew that a sewage treatment facility was located nearby, would that influence the price you would be willing to pay? Yes's

The value of the proposed restoration of Hetch Hetchy might be estimated by surveying the willingness of citizens to contribute to its restoration, either directly or through increased taxes.

Contingent valuation has been applied to a wide variety of situations; however, the uncertainties associated with this technique make its use controversial. Contingent valuation assumes that respondents' expression of their willingness to pay is an accurate expression of what they would actually pay. This assumption may be untrue. Many studies have shown that expressions of willingness to pay are heavily influenced by how much respondents know about particular ecosystem services and alternative actions. We are likely to assign greater value to things we understand than to those we do not.

Ecological Valuation

The value of ecosystem services can be measured by the cost of their possible loss.

In a book called *Natural Capital,* Paul Hawkins, Amory Lovins, and Hunter Lovins argue that all of Earth's resources should be viewed as economic capital. They define **natural capital** as all of Earth's resources that are necessary to produce the ecosystem services on which we depend. Because human actions can improve or degrade these resources, our actions yield consequences that have true monetary value.

The natural capital concept is the basis for **ecological valuation** of ecosystem services. Rather than focusing on willingness to pay, economists determine ecological value by the potential cost of the loss or degradation of an ecosystem service. For example, the ecological value of a forest on a mountainside could be determined by estimating the costs resulting from the floods and landslides that would occur if the forest were cut down.

Often, the costs or benefits of environmental actions such as deforestation or reforestation are not realized until long after the actions are taken. For this reason, proponents of ecological valuation argue that this approach aligns well with environmental sustainability goals that respect the rights of future generations.

In 1997, a team of scientists headed by University of Vermont professor Robert Costanza used ecological valuation techniques to estimate the annual value of all Earth's ecosystem services. That value was $33 trillion (**Figure 2.19**). This represents the costs that would be inflicted if these services were suddenly withdrawn. Their valuation demonstrates the significant economic consequences of the degradation of ecosystem services.

Some traditional economists rightly observed that the valuation by Constanza's team did not reflect economic value in the sense of people's willingness to pay. Given that the loss of all ecosystem services would mean our certain death, we humans would likely be willing to pay whatever was necessary to maintain those services. In other words, *the total value of Earth's ecosystems is priceless.*

▶ **Figure 2.19 Ecological Value**
A global map of the value of ecosystem services in dollars per hectare (2.47 acres) of land based on the costs to humans if such services were lost. Ecosystem services in forested regions and along rivers in tropical and temperate regions are especially valuable compared with those in large expanses of desert, such as in northern Africa and central Australia.

Ecological Value

100   1,000   10,000
US$ ha⁻¹ yr⁻¹

FOCUS ON SCIENCE: Calculating Ecological Value

Taylor Ricketts is Director of Conservation Science at the World Wildlife Fund. Much of his research has focused on the economic value of wild places that are not directly exploited by humans (**Figure 2.20**). Ricketts has been especially interested in how economic forces can be harnessed to help conserve biodiversity. He could see that natural ecosystems are immensely beneficial to society. But because those benefits are not easily quantified, they are often unappreciated.

More specifically, Ricketts asked the question, What is the economic value of intact tropical forests to coffee plantations? He investigated coffee because it is one of the world's most valuable commodities and it is most often grown in regions where diverse tropical forests are threatened by deforestation.

To produce beans, coffee flowers must be pollinated by insects. Ricketts hypothesized that wild bees that live in nearby tropical forests provide this important ecosystem service (**Figure 2.21**). If those forests were removed, he predicted, the number of wild bees visiting coffee flowers would decline, as would the production of coffee beans. The economic value of the pollination provided by the forest could then be calculated by the value of the lost coffee production.

At a very large coffee plantation in Costa Rica, Ricketts established study sites at varying distances from patches of native forest. At study sites within 1 km (0.66 mile) of the forests, he found large numbers of many different bee species visiting the coffee flowers. The number of bees visiting coffee flowers located farther from the forests was about 21% lower. This difference translated to a monetary loss of over $125 per hectare (over $50 per acre). Rickets concluded that if the forests near this plantation were cut down, the loss of pollinators would cause economic losses of more than $60,000 each year.

Ricketts found that each hectare of forest was providing pollinating services to the coffee grower worth $385 each year. Costa Rica has a unique Environmental Service Payments Program through which the government provides $42 per hectare annually to landowners who conserve their forests. This compensation, Ricketts noted, is only 11% of the value of a single ecosystem service, pollination.

Ricketts, T.H., G.C. Daily, P.R. Ehrlich and C.D. Michener. 2004. Economic value of tropical forest to coffee production. Proceedings of the National Academy of Sciences 101:12579–12582.

What is the value of a diverse tropical forest to a coffee farmer?

▲ **Figure 2.20 Assessing the Economic Value of Pollination**
Dr. Taylor Ricketts sampling ripe coffee beans during his Costa Rican studies.

▼ **Figure 2.21 A Key Ecosystem Service**
A wild honey bee pollinating coffee flowers.

1. Describe three approaches to estimating the economic value of the ecosystem services provided by conserving a park.

2. How does ecological valuation differ from traditional economic valuation of an ecosystem service?

3. Why does gross domestic product often overestimate the wealth of countries?

Measuring the Wealth of Nations

Measures of national wealth may externalize the costs of degraded ecosystems and their capacity to generate future wealth.

Economists often estimate the wealth of a nation by its per capita **gross domestic product** (**GDP**). GDP is the total value of goods and services produced by the citizens of a country divided by its population size. GDP is a measure of a country's economic standing. Governments and international organizations also use GDP as a basis for assigning financial aid and making loans to nations.

A number of economists have noted that certain actions increase a country's GDP but reduce its human and natural resources, thereby reducing the country's capacity to generate future wealth. In the short term, overharvesting forests or fisheries and mining practices that pollute the ground generate income and add to GDP. But in the long term, these practices diminish the wealth of the country (**Figure 2.22**). This situation is similar to a shopkeeper who sells his entire inventory and does not carry out routine maintenance. The shopkeeper will make money in the short term, but his capacity to do so in the future is greatly diminished.

To account for the long-term effects of actions, some economists have suggested an alternative measure of national wealth—the **genuine progress indicator** (**GPI**). GPI is the GDP plus or minus the economic value of enhancements or degradations to the environment. Most economists agree with the concept of GPI, but there is considerable disagreement on the appropriate methods for measuring it. The World Bank, which provides loans and grants to many of the world's poorest nations, has recently adopted the GPI concept to measure the sustainability of a country's human and environmental resources. Actions such as the destruction of forests and the neglect of waste management systems diminish a country's GPI. Reforestation and investment in health care improve sustainability and increase a country's GPI.

In summary, although many ecosystem services are not typically bought or sold in the marketplace, they do have economic value. That value can be assessed by measuring people's willingness to pay for those services or by calculating the costs that are incurred when the services are impaired. A complete accounting for the wealth of a country considers both the positive and negative impacts of its actions on its human and natural resources.

◀ **Figure 2.22 Increasing GDP, Diminishing GPI**

This logging operation in Indonesia increases the country's GDP. Because it was done in an abusive fashion that reduces future productivity, it diminishes its GPI, or capacity to generate wealth in the future.

2.5 Environmental Policy: Deciding and Acting

Policies are principles that guide governments and other institutions in setting goals, making decisions, and carrying out activities. **Environmental policies** guide decisions and actions that influence environmental conditions or processes. Policy goals may be achieved through regulation, incentives, partnerships, and volunteerism. The issues that influence environmental policies typically fall into one of eight categories.

The Policy Cycle

Environmental policy evolves in a cycle and guides several kinds of actions.

Governments typically develop and revise policies through a series of steps called the **policy cycle** (**Figure 2.23**). The cycle begins with the definition of the problem to be solved. The next step is setting the agenda, the process of determining who will deal with a particular issue and when.

Policy is then formulated through a process of public hearings, debates, and lobbying, in which interested parties present their positions on the issues. Policy formulation culminates in decision making, in which a law may be passed or a management rule developed.

Next, individuals or agencies implement the policy, or carry out the new activities and enforce the laws and rules. The implementation process often requires additional interpretation as the general directions in a law or rule are adjusted to fit the unique situations in the real world. Finally, the laws and the policies that produced them are assessed in a complex political process that includes the news media and the public, as well as elected officials (**Figure 2.24**). As a consequence of such assessments, policies may be revised or terminated.

▼ Figure 2.23 **The Policy Cycle**
Policy is developed, implemented, and amended in a series of steps.

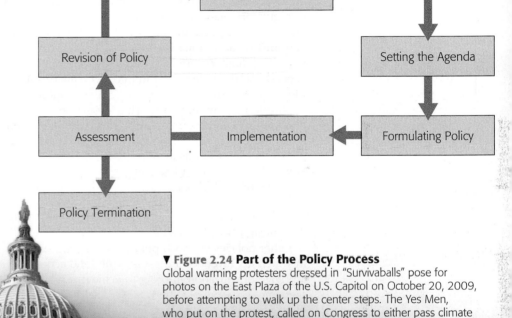

Problem Definition

Setting the Agenda

Formulating Policy

Implementation

Assessment

Revision of Policy

Policy Termination

▼ Figure 2.24 **Part of the Policy Process**
Global warming protesters dressed in "Survivaballs" pose for photos on the East Plaza of the U.S. Capitol on October 20, 2009, before attempting to walk up the center steps. The Yes Men, who put on the protest, called on Congress to either pass climate change legislation or purchase survival suits for everyone.

Environmental policies may use different methods for guiding actions. **Regulatory mandates** set legal standards for actions. For example, in 1975 the U.S. Congress enacted Corporate Average Fuel Economy (CAFE) regulations, which required automobile companies to meet standards for fuel economy set by the National Highway Traffic Safety Administration and verified by the U.S. Environmental Protection Agency. Other policies are implemented by means of **incentives**, which encourage action by offering something appealing. The Energy Policy Act of 2005, for example, offered consumers tax credits for buying fuel-efficient hybrid vehicles. People were not required to do this, but the tax credit was an important incentive to many.

Market-based policies use economic markets to guide actions. Such policies usually use a combination of regulatory mandates and incentives in the form of credits or rights that are bought and sold. For example, **cap and trade** is a market-based policy in which a regulatory limit, or cap, is set on an action; rights to exceed that limit are traded in markets. The 1990 U.S. Clean Air Act Amendment has been a very successful application of cap and trade. This legislation sets a limit on the total amount of sulfur that can be emitted by all the electrical utilities in the country. From this, performance standards are calculated for each facility. Power plants that emit less sulfur than the standard earn sulfur credits, which can be purchased by power plants whose emissions exceed the standards. These sulfur credits are bought and sold in public markets such as the Chicago Board of Trade.

Other policies establish public-private partnerships, in which tax revenues are used to fund the actions of a private organization in order to achieve a particular outcome. For example, the U.S. National Park Service often wishes to acquire land for conservation but depends upon the slow process of congressional appropriation for funds. Because opportunities for land purchase often require quick action, the Park Service works in partnership with nonprofit organizations such as The Nature Conservancy, which uses its own resources to acquire land. When Congress finally provides funding, the nonprofit organizations resell the land to the Park Service. Private organizations are generally required to resell at the price they paid, but they receive interest on their investment and a commission fee.

Some policies promote **volunteerism**, work performed freely on behalf of a community (**Figure 2.25**). For example, the Smokey Bear and Woodsy Owl campaigns were developed by the U.S. Forest Service to encourage voluntary behavior that would control destructive wildfires and diminish littering.

▲ **Figure 2.25 Volunteerism**
Volunteers clean up trash along the banks of a stream flowing through their community. A variety of public policies work to encourage such activities.

Policy Decision Framework

Environmental policy debates often center on one of eight issues.

A wide variety of issues can affect environmental policy. However, University of North Carolina political scientist Richard Andrews suggests that debates over environmental policy usually fall into one of the following eight categories.

1. Government versus individual control.
In the United States, we generally assume that governments should set and enforce standards for environmental hazards that affect public health and well-being, such as chemical pollutants. Decisions related to transportation, energy use, or waste disposal are frequently left to individuals. However, the amount of control given to individuals varies from region to region. For example, some counties and cities leave decisions about recycling metal cans, bottles, and newspapers to the discretion of individuals. Other communities provide recycling services but do not require citizens to use them. Still other communities impose fines on those who violate recycling laws. Who should make these decisions—government or individuals?

2. Competing public values.
Decisions about environmental policy often involve trade-offs with other public benefits, most notably economic growth. For example, policies for the management of U.S. National Forests are often influenced by competing values, such as the protection of water and endangered species versus the economic benefits of the jobs provided by the extraction of timber. At an international level, those developing policies to limit the emissions of greenhouse gases must also address concerns about the economic consequences of such policies. Whose values should have priority?

3. Uncertainty and action.

In 2000, the European Union endorsed what has come to be known as the **precautionary principle**: When there is reasonable evidence that an action or policy may place human health or the environment at risk, precautionary measures should be taken, even if that evidence is uncertain or inconclusive. In theory, the precautionary principle seems very logical. In practice, policymakers must agree on what constitutes "reasonable" risk. They must also agree on the trade-offs between different kinds of risk. For example, a policy banning the use of pesticides may lower the risk of certain adverse health effects but increase the risk of lowered crop production. How much do we need to know before we act?

4. Which level of government decides?

Many different levels of government are involved in regulating actions related to the environment. In the United States, for example, a federal agency—the U.S. Environmental Protection Agency—sets regulatory standards for air quality. Should individual states have the right to set and enforce their own standards for air quality, so long as they are at least as stringent as federal guidelines? States such as California and New York argue yes, they should be allowed to tailor standards to meet unique situations, such as air quality in major cities. Concerned with having to adjust to a gaggle of different requirements, manufacturers and utilities say no. Which level of government should make environmental decisions?

5. Which government agency has jurisdiction?

Many different governmental departments and agencies have responsibility for functions that affect the environment. For example, responsibility for controlling the flow of water in rivers resides with both the Bureau of Reclamation in the Department of the Interior and the Army Corps of Engineers in the Department of Defense. If hydroelectric power generation is involved, the Federal Energy Regulatory Commission of the Department of Energy may also share some responsibility. Shared responsibility may serve the public good if it ensures that competing values are represented in decisions over the management of resources. However, when the responsibilities of agencies overlap or are poorly defined, the result can be confusion, gridlock, or finger pointing. How should decision making be assigned to governmental departments and agencies?

6. Protection against selfish actions.

Scientist-philosopher Garrett Hardin argued that conflict between individual interests and the common good is inevitable when people compete for shared, limited resources, such as the fish in a public lake or grazing rights in a public pasture. If individual interests prevail, the common resources are likely to be overexploited. He called this form of overexploitation the **tragedy of the commons** (**Figure 2.26**). How can environmental policy be formulated to avert tragedies of the commons?

▼ **Figure 2.26 Tragedy of the Commons**
The livestock on this degraded pasture in the African country of Chad are owned by several different families. Because the land is shared with no restrictions, each family has the incentive to put as many animals as it can on the land.

One strategy for avoiding overexploitation of common resources is to assign private ownership to open-access resources. For example, livestock herders may be assigned a specific number of animals that they may graze on a common pasture. They may use their entire quota or sell a portion to others. Alternatively, common resources may be treated as the property of the government and their use regulated by law. Access to our national parks is such an example.

Other common resources are more difficult to protect because users are too numerous, geographically separate, or hold conflicting values. The clean water in a river is a common resource. A city or farmer may dump wastes into the river, but the negative effects of their actions may not be realized until far downstream.

Although most people would agree that common resources should be protected, people rarely want to take a large share of the responsibility for such actions. For example, most of us would agree on the need for facilities to manage sewage or municipal waste, but few of us wish to see such operations located near our homes. This "not in my backyard" syndrome has a major influence on a variety of environmental policies (**Figure 2.27**).

7. The best means to an end.

In many policy debates, people agree on the desired outcome but disagree about the appropriate strategies for achieving it. Should regulation be enforced with financial or civil penalties? Should incentives take the form of subsidies, tax benefits, or cap-and-trade markets? For example, advocates for policies aimed at reducing emissions of carbon dioxide to Earth's atmosphere

disagree on the best way to achieve that end. Some argue that a tax on carbon-based fossil fuels is the best way to reduce emissions. Others argue for setting an overall limit on carbon emissions and allowing individuals and companies to buy and sell the right to emit carbon above that limit. Yet others propose that any measures to reduce carbon dioxide emissions should be voluntary. What kinds of policies are best suited to achieve particular environmental outcomes?

8. Political power relationships.

Many policy complications arise as a consequence of complicated decision-making processes. Collective decisions in a democracy often require compromises, which may produce less than ideal outcomes. The role of power relationships in the U.S. Congress, such as which party holds a majority, is well-known. Power relationships can also arise because of the geographical distribution of resources. For example, fossil fuels such as coal and oil are not distributed uniformly among states and countries. This distribution influences the policy positions of these different entities; it is also a major factor in determining the power of individual states or nations to push for their own policy positions. How can power relationships among geographic regions and political constituencies be overcome to develop sound policies?

In summary, the policy cycle generates policies and laws that influence behavior toward the environment by means of regulation, incentives, markets, and volunteerism. Questions related to the role of individuals, the handling of uncertainty, and the nature of governance influence the specific form that environmental policies take.

QUESTIONS

1. Describe the steps in the policy cycle.

2. In what ways might policies guide human actions?

3. List four questions that are important to policy formulation.

▼ **Figure 2.27 Not In My Backyard**
We may all agree on the need for sewage treatment plants, but most of us would protest if one was to be located near our home.

2.6 U.S. Environmental Law and Policy

In the United States, environmental policies and laws are influenced by actions of the different branches of federal, state, and local governments. Provisions in the U.S. Constitution set a framework for the development of many policies, laws, and court decisions. U.S. environmental policy and law have changed in response to changing human values.

Governmental Functions

The three branches of the federal government and state and local governments play unique roles in setting environmental policy.

Through the decade following the Declaration of Independence in 1776, the power to legislate, or make laws, resided with each of the 13 states. This included laws related to commerce, land ownership, and individual rights. Each state also had the power to enforce and interpret its own laws. Fearing a return to centralized rule and not wishing to give up their individual powers, the states granted very few powers to the Continental Congress. As a result, there was considerable state-to-state variation in laws, as well as in the interpretation and enforcement of their laws.

In 1787, the United States Constitution was adopted by the Constitutional Convention and became the supreme law of the land. The Constitution vested primary legislative, executive, and judicial authority in the federal government and, thereby greatly increased the power of the federal government relative to the states. However, the organization of the federal government ensures checks and balances among the activities of the three branches of government (**Figure 2.28**).

All legislative functions are assigned to the Congress, consisting of the House of Representatives and the Senate. As a check on the power of the Congress, legislation is either approved or vetoed by the president. The Congress may override a presidential veto by a two-thirds majority of both houses.

The implementation and enforcement of legislation is the responsibility of the executive branch, which is headed by the president and includes the cabinet and many departments and agencies. The enforcement of particular legislation is generally assigned to one or a few agencies. Agencies then set **regulations**, which are specific rules that establish standards for performance, programs to ensure compliance, and protocols for enforcement.

Wisely, the framers of the Constitution perceived that social change and new knowledge and technology would necessitate continued re-interpretation of law. This interpretive function is vested in the judicial branch, which includes the Supreme Court and lower courts. The various decisions made by the individual courts

collectively make up **case law** and establish precedents that influence future court decisions.

A closer look at the law that governs water quality demonstrates how the three branches of the government work together to create, implement, and interpret environmental policies. In 1972, Congress passed the Federal Water Pollution Control Act, or Clean Water Act, and President Richard Nixon signed it into law. The act describes broad goals for water quality and assigns the responsibility for setting, monitoring, and enforcing standards of water quality to the Army Corps of Engineers and the U.S. Environmental Protection Agency.

Congress approves the executive budget and executive appointments

Congress passes laws that are then signed and enacted by the executive branch

LEGISLATIVE BRANCH
The Congress

EXECUTIVE BRANCH
The President

Courts rule on the constitutionality of legislation

President appoints judges with congressional approval

Congress approves judicial appointments and may remove judges by impeachment

JUDICIAL BRANCH
The Courts

Courts rule on the appropriate interpretation and enforcement of laws

▲ Figure 2.28 **Branches of Government**
The U.S. government is organized to provide checks and balances on the processes of law making, law enforcement, and law interpretation.

These responsibilities are carried out in accordance with hundreds of pages of detailed administrative rules and regulations. In the years since the passage of the Clean Water Act, the courts have ruled on specific matters of interpretation, such as whether activities affecting isolated ponds and bogs are actually covered by the act (**Figure 2.29**). In some circumstances, the judicial branch can require implementation of laws. In 2007, for example, the Supreme Court ruled that carbon dioxide (CO_2) is a pollutant and falls within the jurisdiction of the Clean Air Act. They further ruled that the U.S. Environmental Protection Agency has the responsibility to regulate CO_2 emissions from automobiles and industrial sources.

The first ten amendments to the Constitution, called the Bill of Rights, delineate the power of the government relative to individuals and states. The Tenth Amendment to the Constitution specifies that powers not delegated to the central government and not prohibited for the states be "reserved to the states respectively, or to the people." Indeed, much environmental policy is set by state and local governments. In many cases, federal legislation delegates enforcement of environmental statutes to individual states. States may also enact legislation regulating matters such as water quality or species protection, so long as their laws do not violate federal statutes.

▲ **Figure 2.29 Legislation and Interpretation**
The 1972 Clean Water Act regulates activities affecting the quality of the nation's water. However, the U.S. Supreme Court has interpreted this legislation to apply only to wetlands that drain into rivers and large lakes. It ruled that the act does not apply to actions affecting isolated wetlands such as these ponds in South Dakota.

Cities and counties also set policies and make important environmental decisions. Counties, for example, are empowered to make land use decisions, such as the location of landfills, commercial developments, and public parks. These governments also pass ordinances and set policies related to local water use and waste management.

The Constitution and Environmental Policy

Several articles and amendments to the U.S. Constitution significantly influence environmental policy.

Several articles and amendments to the Constitution have been especially relevant to the development of U.S. environmental policy. Article 1 establishes the right of the federal government to regulate foreign and interstate commerce. The federal government's right to set rules regarding transportation, the flow and quality of navigable waters, commerce affecting wildlife, mining, and transport of natural resources such as oil, coal, and electricity derive from the commerce clause of Article 1.

Article 4 includes the property clause, which enables the government "to dispose of and make needful rules and regulations respecting the territory or other property of the United States." This article is the basis for laws and regulations governing public lands, including the establishment and management of national parks, national forests, and national wildlife refuges. It also is the basis for legislation that allows the government to protect its property by setting restrictions on nearby landowners.

The Fifth Amendment of the Constitution has many provisions that influence environmental policy. Most importantly, it allows the government to take property for public use, but only if it provides just compensation. It is clear that the government must pay a landowner fair market value for land that it acquires for a road, military facility, or national park. However, if the government establishes a regulation that restricts the use of private property, thereby diminishing the value of the property, must the government compensate the property owner? This question is central to environmental policies affecting matters such as land use, regulation of pollution, and protection of endangered species. Over the years, the courts have ruled that if a legitimate public purpose is served, the government need not pay for every "diminution of value." In other words, owners are not automatically entitled to the highest value use of their property. However, recent court decisions have affirmed that the government must provide compensation if a regulation removes most or all of the value of a property.

The Fourteenth Amendment, added to the Constitution soon after the Civil War, requires that the states provide legal due process and "equal protection" to all citizens. Some see this amendment as a constitutional basis for environmental justice. They argue that this amendment requires that environmental services, such as clear air and water, must be provided equally to all citizens, regardless of race or economic status. The courts have supported this interpretation in some situations, but not in others. For example, case law requires consideration of equal protection issues if a new waste facility is located in an established community. Such considerations are not required if people knowingly choose to live near an established waste facility.

QUESTIONS

1. Describe the role of the three branches of government in setting environmental policy.
2. Describe four parts of the U.S. Constitution that influence environmental policy.

2.7 International Environmental Law and Policy

International environmental policies and laws are established by long-standing norms of behavior, treaties that are ratified by national governments, and decisions and precedents set in international courts. The United Nations, regional consortia, international financial organizations, and nongovernmental organizations all play significant roles in the development of international environmental law and policy (**Figure 2.30**). Over the past four decades, global policies and treaties have been developed to restore and conserve the environment while promoting sustainable economic development in poor countries.

Environmental Laws

Environmental law derives from established custom, formal treaties, and judicial decisions.

Sovereignty is the concept that a country may behave as it pleases within its borders as long as it does not violate international laws to which it has agreed. These international laws may be in one of three forms: customary, conventional, or judicial.

Customary international laws are accepted norms of behavior or rules that countries follow as a matter of long-standing precedent. The vast majority of the world's governments, including the United States, accept customary international law. For example, it is customary in international law that countries may not use their territory in such a way as to injure the territory of another country and that countries have a duty to warn other countries about environmental emergencies. Therefore, following the major earthquake and tsunami in 2011, Japan was obligated to notify other countries of serious failures at and releases of radioactive materials from its Fukashima Daiichi nuclear facility. The exact time when a principle of action becomes customary law is not clear-cut and is a matter of frequent disagreement among countries.

Conventional international laws are established by formal, legally binding conventions or treaties among countries. In 1940, for example, the United States and 17 other countries in the Americas ratified the Convention on Nature Protection and Wildlife Preservation in the Western Hemisphere. By this treaty, these nations agreed to protect migratory birds and to take actions to prevent the extinction of any migratory species.

Judicial international law sets standards for the actions of countries based on the decisions of international courts and tribunals. The International Court of Justice in The Hague, Netherlands, is the most famous of these courts (**Figure 2.30**). Regional tribunals, such as the European Court of Justice, have been established by treaty among cooperating nations. These courts often have limited authority on environmental issues, but their opinions carry considerable weight in the interpretation of customary and conventional law. Because so many environmental issues are influenced by global trade, the Dispute Settlement Board of the World Trade Organization is also an important source of judicial environmental law.

For several reasons, the enforcement of international environmental law is somewhat limited. First, it is often difficult to verify whether a particular country is complying with the terms of a treaty. Second, countries may enter into an agreement in good faith but lack the capacity to enforce its terms. Third, the international community has few options for punishing countries that violate a treaty.

Why are International Law treaties not fully enforced (or more stringent) in order to protect our air quality and climate?

Odalys Solares, Florida International University

Norm: *International treaties depend not only on the willingness of governments to participate, but also on their capacity to fund and enforce treaty agreements. Countries may fail on one or all of these points.*

▼ **Figure 2.30 International Justice**
Environmental disputes among countries are resolved in the International Court of Justice in Peace Palace in The Hague, Netherlands.

International Institutions

Environmental policies and laws are influenced by a variety of international institutions.

The United Nations. The most important organization shaping international environmental policy and law is the United Nations (**Figure 2.31**). The 54 member nations of the United Nations Economic and Social Council are charged with promoting international economic and social cooperation and development. As part of its charge, the Economic and Social Council oversees two agencies that play a significant role in international policy. The United Nations Development Programme (UNDP) provides countries with assistance to encourage environmentally sustainable development. The United Nations Environment Programme (UNEP) coordinates environmental programs for the United Nations and has authority over regional and global environmental issues.

Regional Consortia. Countries within different regions have formed organizations to encourage communication and cooperation on matters of particular interest. For example, the Organization of American States represents 35 countries in the Western Hemisphere. Its Department of Sustainable Development coordinates policies and programs aimed at alleviating poverty and promoting economic development in its member nations.

The European Union (EU), created to promote European unity and its influence in the world, is one of the world's strongest regional organizations. It can sign treaties on behalf of its 27 member countries; it also regulates their behavior in many realms, including the environment. The EU European Environment Agency sets standards in a number of areas, including pollution, habitat degradation, fisheries, and waste management. Because these standards apply to many items imported to EU countries, the EU exercises considerable influence on environmental management outside the region.

International Financial Institutions. **Multinational development banks** are institutions that provide financial and technical assistance to countries for economic, social, and environmental development. The largest of these banks is the World Bank, which provides long-term loans and grants to the poorest countries in the world. Regional banks such as the African Development Bank and the Inter-American Development Bank provide similar support. These banks are funded by donations from wealthier countries. In the past, development banks frequently funded projects, such as water diversions and community resettlement, that had positive effects on human well-being but negative environmental consequences. Over the past two decades, development banks have worked hard to incorporate principles of environmental sustainability into their programs.

The World Trade Organization (WTO) was created in 1995 to promote free trade by reducing obstacles to international commerce and enforcing fairness in trading practices. The objectives of sustainable development and environmental protection are included in the introduction to the agreement that created the WTO, and its policies and programs have a significant influence on environmental behavior. WTO agreements and judgments have confirmed the right of governments to protect the environment. However, the globalization of markets facilitated by the WTO may also encourage the unsustainable use of resources in countries with weak environmental laws or countries without the human and financial resources to enforce them. For example, the global market for tropical wood products has encouraged extensive, unsustainable logging in Indonesia. Although Indonesia has laws prohibiting such cutting, they are poorly enforced.

Nongovernmental organizations. Nongovernmental organizations (NGOs), legally constituted organizations in which there is no governmental participation or representation, have a significant influence on international environmental policy. Some NGOs, such as Greenpeace, advocate for national laws and international treaties to protect the environment. Other NGOs, such as World Wildlife Fund, work with governments to develop the capacity to implement and enforce laws and treaties (**Figure 2.32**).

▼ **Figure 2.31 The United Nations**
This is the UN General Assembly Hall in which the member countries consider all matters influencing international peace and security.

◄ **Figure 2.32 Nongovernmental Environmental Organizations**
The United Nations Environmental Liaison Center International coordinates activities of international NGOs. Since its creation in 1974, the number of member NGOs in the center has grown significantly.

Chart: NUMBER OF NGOs vs YEAR
1974: ~130; 1993: ~710; 2009: ~850

SEEING SOLUTIONS: The Global Environmental Facility

In the early 1980s, sustainable economic development among the world's poorest nations was a high priority for the World Commission on Environment and Development. It envisioned a new era of economic growth based on policies that enhance natural resources and protect humans and ecosystems from pollution. To realize this vision, it argued for greater cooperation among developing and developed countries with regard to many issues, including population growth, alleviation of poverty, energy management, and protection of species and ecosystems.

Following the commission's recommendations, the United Nations established the Global Environmental Facility (GEF) at the 1992 Earth Summit in Rio de Janeiro, Brazil. The GEF is a partnership among 179 countries, development banks, NGOs, and multinational corporations. The GEF uses grants and loans from member countries and development banks to complement funds from private companies. International NGOs often play a significant role in the management of GEF projects.

Since its creation, the GEF has provided a total of almost $30 billion to support nearly 2,000 projects designed to produce environmental benefits, such as the reduction of pollution, the management of water resources, and the support of biodiversity. In addition to producing environmental benefits, GEF projects must encourage sustainable economic development.

The GEF has provided nearly $1 billion to Mexico and the countries of Central America for the development of the Mesoamerican Biological Corridor (MBC). The MBC is not a single place. Rather, it is a group of programs that manage about 600 protected areas and the land surrounding them (**Figure 2.33**). The MBC unites the goals of species and ecosystem conservation with initiatives for sustainable development that benefit local people (**Figure 2.34**). Many MBC projects focus on individual countries, but a number of its grants have supported the development of preserves that span international boundaries. Nearly all of these grants have involved cooperation among governments, NGOs, and local communities.

The MBC is a key reason that nearly 20% of this region is now protected for conservation, helping to save the rich biological diversity of the area. The MBC has also bolstered tourism to every Central American country and improved the economic well-being of their citizens. Just as important, it has encouraged cooperation among countries in a region where, just 20 years ago, border disputes and civil wars were common.

Can a development bank promote both economic development and environmental sustainability?

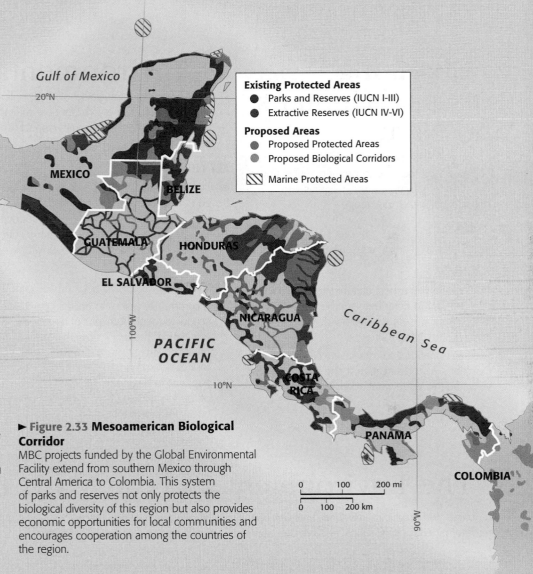

Existing Protected Areas
● Parks and Reserves (IUCN I-III)
● Extractive Reserves (IUCN IV-VI)

Proposed Areas
● Proposed Protected Areas
● Proposed Biological Corridors

▨ Marine Protected Areas

▶ **Figure 2.33 Mesoamerican Biological Corridor**
MBC projects funded by the Global Environmental Facility extend from southern Mexico through Central America to Colombia. This system of parks and reserves not only protects the biological diversity of this region but also provides economic opportunities for local communities and encourages cooperation among the countries of the region.

▶ **Figure 2.34 Environmental and Economic Sustainability**
Local naturalists guide ecotourists along the many scenic stream channels in Tortuguero National Park, Costa Rica. The GEF has provided funding for parks such as this and for naturalist training programs.

Summary

Our attitudes toward the environment are influenced by our mode of living. People who get their provisions directly from the environment usually have a good understanding of their dependence on the natural world. In more urban communities, that dependence is less obvious and often not well understood. Human beliefs about whether actions that influence the environment are right or wrong may be based on motivating virtues, consequences of actions, or rules of behavior. Beliefs are also influenced by ethical decisions about who or what has intrinsic value.

Market forces of supply and demand determine the economic value of natural resources that are harvested or extracted from the environment, such as timber and coal. However, many ecosystem services that are necessary for our well-being, such as clean air and beautiful landscapes, are not bought and sold in markets, but we value them nonetheless. Environmental policies and laws influence human actions through a variety of means. These policies are developed, implemented, enforced, and interpreted by governments and other organizations at the community, national, and international levels.

2.1 Changing Views of Humans and Nature

- Human attitudes toward the environment have changed over time as a consequence of changes in our mode of living, culture, religious and political beliefs and knowledge.

KEY TERMS

animism, preservationist, conservationist

QUESTIONS

1. How would you expect attitudes toward the environment to differ between hunter-gatherers in a remote wilderness and city dwellers in 19th-century New York?

2. Logging in national forests is very controversial. Describe how preservationist and conservationist views on this issue would likely differ.

2.2 Environmental Ethics

- Environmental ethics is a scholarly discipline that seeks to create frameworks for determining right and wrong human actions with respect to the environment.

KEY TERMS

environmental ethics, virtue ethics, consequence-based ethics, utilitarianism, duty-based ethics, intrinsic value, instrumental value, anthropocentric ethics, biocentric ethics, ecocentric ethics, deep ecology movement, ecofeminism, environmental justice

QUESTIONS

1. A community wishes to build a new elementary school to meet the needs of its growing population. The most convenient site for this new school happens to be one of the few remaining tracts of forest within the city. On what criteria would virtue ethicists, consequence-based ethicists, and duty-based ethicists decide if cutting this forest to locate the school was right?

2. In very dry years or when winters are very cold, the food supply for elk in the Yellowstone wilderness may be limited, and weak or old animals are likely to die. Some argue that park managers should provide supplemental feeding stations to prevent such deaths. What is your view on this matter, and on what ethical framework would you base it?

3. Give an example of an environmental justice issue in your community. How do you feel disputes over such matters should be resolved?

According to...	An act is ethical if...
Virtue Ethics	It is motivated by honesty, justice, or kindness.
Consequence-based Ethics	It results in satisfaction.
Duty-based Ethics	It follows a set of agreed-upon rules.

2.3 The Environment and the Marketplace

- The prices and production of goods and services are determined by their supply and the demand for them.
- Consumers' willingness to pay for a good or service is determined by the satisfaction or benefit that they feel it will bring them.

KEY TERMS

economic system, subsistence economy, market economy, free market economy, planned economy, profit, neoclassical economic theory, economies of scale, economic value, discount rate, opportunity cost, externality

QUESTIONS

1. The economic system of the United States includes aspects of subsistence, market, and planned economies. Using examples, explain this statement.

2. The cost for producing a particular good or service decreases as the amount produced increases. How is this likely to affect incentives for the producer?

3. Environmental impacts of the production and consumption of commodities are often economic externalities. Give an example of a way that economic systems can capture such external costs in the price of a commodity.

SUPPLY AND DEMAND

2.4 Valuing Ecosystems

- The economic value of human actions that either impair or enhance ecosystem services such as clean water or aesthetic beauty can be determined by the difference between people's willingness to pay for the cost of such actions and the cost of alternative actions.

KEY TERMS

marginal value, travel-cost valuation, hedonic valuation, contingent valuation, natural capital, ecological valuation, gross domestic product (GDP), genuine progress indicator (GPI)

QUESTIONS

1. Contingent valuation of economic services is very controversial among economists. Describe two reasons for this controversy.

2. Some economists argue that ecological valuation of ecosystem services does not reflect their true value in an economic system. What is their reasoning?

3. Describe two examples of actions that countries might take that increase their GDP while diminishing their GPI. Give an example of actions that would likely increase both GDP and GPI.

2.5 Environmental Policy: Deciding and Acting

- Environmental policies are plans that influence human decisions and actions relative to the environment and its natural resources. They do this through mandates, incentives, partnerships, and volunteerism.

KEY TERMS

environmental policy, policy cycle, regulatory mandate, incentive, market-based policy, cap and trade, volunteerism, precautionary principle, tragedy of the commons

QUESTIONS

1. How might variations in supply and demand influence the economic values in a market-based system to manage sulfur emissions from power plants?

2. What factors might influence a country's application of the precautionary principle to a particular environmental problem?

3. Describe two policy alternatives that might be used to minimize selfish actions that might impair a resource that is shared in common within a community.

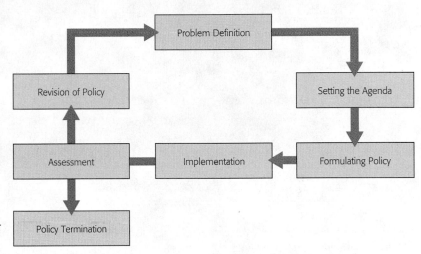

2.6 U.S. Environmental Law and Policy

- Environmental policies are cast into law by the legislative branch of government, they are implemented and enforced by the executive branch, and they are interpreted through court decisions by the judicial branch.

- The U.S. Constitution delegates many decision-making powers to governments at the state and local levels.

KEY TERMS

regulations, case law

QUESTIONS

1. In response to public concerns about an environmental issue, a law is passed. Describe the role of each of the branches of government in the development of that law and in its effects on public actions.

2. Government decisions and regulations often influence the economic value of private property. What part of the Constitution determines the government's responsibility to compensate property owners? Under what conditions do property owners have a right to compensation?

2.7 International Environmental Law and Policy

- International environmental law assumes that each nation may act as it pleases within its boundaries so long as it does not violate international laws to which it has agreed. It is shaped by the United Nations, agreements among nations, and economic and nongovernmental organizations.

KEY TERMS

sovereignty, customary international laws, conventional international laws, judicial international laws, multinational development banks

QUESTIONS

1. Describe ways in which international financial institutions influence international environmental policies.

2. How have the attitudes of individual countries regarding environmental policies been influenced by a country's level of development or wealth compared to other countries?

Customary international laws	Based on long-established precedents and accepted norms of behavior.
Conventional international laws	Based on formal and legally binding treaties.
Judicial international laws	Based on the decisions of international courts.

Mastering ENVIRONMENTALSCIENCE

Go to **www.masteringenvironmentalscience.com** for homework assignments, practice quizzes, Pearson eText, and more.

3

The Physical Science of the Environment

Searching for Life Elsewhere

Where is everybody?

The universe is truly vast! Just how immense it really is became obvious in the 1930s, when astronomers discovered that our galaxy, the Milky Way, was only 1 of more than 100 billion galaxies (**Figure 3.1**). Each of these galaxies has between 100 and 400 billion stars. A prominent astronomer of that era, Harlow Shapley, argued that somewhere in this vast universe there must be other forms of life. Even if only one star in a thousand has a solar system, and if only one in a thousand of those solar systems has a planet capable of supporting life, and if life appeared on only one in a thousand of those planets, there would still be several hundred billion planets upon which life would exist. Of course, this leaves us with the question, "Where is everybody?"

Determining whether life exists elsewhere has been a central goal for space exploration, particularly the exploration of the planet Mars. In 1975, two space probes were sent to Mars—*Viking 1* and *2*. Each of these probes had an orbiter and a lander. Photographs taken from the *Viking* orbiters revealed evidence that water had once flowed on the surface of the planet. This evidence of water was encouraging because all of the organisms we are familiar with require water to live. In other words, water is essential to life as we know it.

On July 20, 1976, *Viking 1* lander became the first spacecraft from Earth to land on Mars (**Figure 3.2**). *Viking 2* followed it two months later. Pictures taken by these landers showed an arid, boulder-strewn landscape. No signs of water or of past or present life were visible (**Figure 3.3**).

Each *Viking* lander had a robotic arm that was used to sample Martian soil. The soil samples were then

▲ **Figure 3.1 Worlds Beyond Our Own**
Over a relatively small portion of the sky, the orbiting Hubble Space Telescope captured this image of two colliding galaxies and hundreds of others in the distance.

▶ **Figure 3.2 *Viking 1* Lander**
The *Viking 1* lander was equipped with a robotic arm to sample Martian soil (upper right) and a variety of instruments to analyze that soil for signs of life.

subjected to tests designed to determine the presence or absence of life. One of these tests involved mixing the soil with a solution of nutrients similar to those that might have sustained early life on Earth. If this soil mixture released carbon dioxide, it would indicate that nutrients were being processed in a lifelike fashion. To the scientists' great surprise and elation, carbon dioxide was released in abundance!

Elation was soon replaced by confusion as the results of other tests were revealed. For example, Martian soil was heated to see if it would release any of the chemicals that are characteristic of life. None were detected. In another test, soil was exposed to a mixture of gases that might be used by living organisms. There was no evidence that any of these gases were taken up in the soil.

Some scientists argued that the carbon dioxide released from the nutrient mixture was not caused by living organisms. Rather, they proposed that intense ultraviolet (UV) light striking the surface of Mars had altered chemicals in the Martian soils, making them highly reactive with the nutrients. Other scientists hypothesized that the heating experiments were flawed; they argued that the carbon dioxide released from the nutrient solution was indeed evidence of microbes living in the Martian soils. This debate continues. It is unlikely to be resolved completely until some future Mars mission.

The Viking missions show that determining whether or not life exists on another planet is more complicated than we had hoped. To do so presumes that we understand all the physical and chemical conditions necessary for life, as well the ways that living things affect their physical environment. But everything we know about these conditions and effects has come from studies of life on Earth.

could be other ways of detection

Our current understanding allows us only to detect life as we know it on Earth. It is possible that physical and chemical conditions elsewhere in the universe may produce lifelike responses in the absence of life. It is also possible that there are other forms of life whose chemistry and behavior are totally unfamiliar to us.

In this chapter, we explore the basic principles of chemistry and energy transfer that characterize life on Earth. We compare Earth's environment with those of its nearest planetary neighbors. We also examine the physical and chemical characteristics of Earth's interior and its atmosphere and oceans. Finally, we explore how these features create climates that sustain abundant life on the one and only location in the universe where we know for certain that life exists.

▼ **Figure 3.3 Is There Life Here?**
The Martian landscape presents no obvious signs of life; however, chemical reactions in these Martian soils are thought by some to be lifelike.

3.1 Chemistry of the Environment

Matter is the tangible "stuff" of the universe. All matter is made of one or more **elements**. An element is a chemical that cannot be broken down or separated into other chemicals. The physical and chemical properties of elements are attributable to the characteristics of their most basic subunits, **atoms**. Two or more atoms may be held together by chemical bonds to form **molecules**. Water is among the most important molecules in ecosystems. It is a major constituent of all living organisms and is essential to virtually all life functions.

Atoms, the Basic Unit of Matter

The chemical properties of atoms depend on the numbers of protons, neutrons, and electrons they contain.

Nearly all matter consists of one or more of the 92 naturally occurring elements. (We say "naturally occurring" because chemists have artificially synthesized about 20 additional elements.) Chemists give each element a symbol. Hydrogen (H), carbon (C), oxygen (O), and nitrogen (N) are examples of four common elements and their symbols. These four elements play especially important roles in ecosystems.

An atom is the basic subunit of an element. Atoms have a central nucleus that contains positively charged particles called **protons** and electrically neutral particles called **neutrons** (Figure 3.4). The atoms of a particular element always have the same unique number of protons; this number is called the element's atomic number. For example, all hydrogen atoms have a single proton and an atomic number of 1. All carbon atoms have 6 protons and an atomic number of 6.

The nucleus is surrounded by one or more negatively charged particles called **electrons**. The negative charge of a single electron exactly balances the positive charge of a proton. Thus, the overall charge of an atom is neutral when the number of electrons surrounding the nucleus is equal to the atomic number.

Protons, neutrons, and electrons have **mass**, the property responsible for the gravitational attraction of all matter to all other matter. Protons and neutrons account for most of an atom's mass. The mass of each is more than 1,000 times greater than the mass of an electron.

Unlike the number of protons, the number of neutrons in the nucleus of an element can vary. Atoms of an element with different numbers of neutrons are called **isotopes**. For example, carbon atoms always have 6 protons but may have 6, 7, or 8 neutrons. These isotopes are indicated by the chemical symbol preceded by the total number of neutrons and protons. For example, ^{12}C has 6 protons and 6 neutrons; ^{14}C has 6 protons and 8 neutrons (Figure 3.5).

The various isotopes of an element generally share the same chemical properties. However, isotopes with more neutrons are slightly more massive than those with fewer neutrons. This difference in mass allows chemists to determine the relative amounts of different isotopes in a sample of a particular element.

Some isotopes have nuclei that are unstable and can break apart. Such isotopes are **radioactive**, which means they spontaneously "decay," emitting various combinations of high-energy protons, neutrons, electrons, and radiation. This decay converts them into lighter isotopes of the same or different elements. These new isotopes may be stable or radioactive.

Each radioactive isotope, or radioisotope, decays at a unique rate. That rate is measured by **half-life**, the length of time that it takes for half of a collection of atoms of a radioisotope to decay. Half-lives may be as short as tiny fractions of a second or longer than 1 billion years. For example, ^{14}C has a half-life of 5,730 years; it decays to form ^{14}N. The uranium isotope ^{235}U, which is used in nuclear power plants and weapons, has a half-life of 703 million years; it decays to lead-207 (^{207}Pb).

Nucleus

Electrons

Protons

Neutrons

Electrons

◀ **Figure 3.4 The Atom**
The nucleus of an atom is composed of positively charged protons and uncharged neutrons. Negatively charged electrons swirl in a cloud outside the nucleus.

◄ **Figure 3.5 Isotopes of an Atom**
Different isotopes of an element such as carbon have the same number of protons and electrons, but different numbers of neutrons.

^{12}C ^{13}C ^{14}C

Molecules and Ionic Compounds

Atoms are joined by chemical bonds to form molecules and ionic compounds.

Atoms can combine with other atoms to form molecules. Some molecules are made of two atoms of the same element, such as the oxygen (O_2) and nitrogen (N_2) in the atmosphere. Molecules that are made of more than one element are called **compounds**. Water (H_2O) and carbon dioxide (CO_2) are two common compounds.

The atoms in a molecule are held together by chemical bonds. The strength of a chemical bond is measured by the energy required to break it. **Covalent bonds** are strong bonds formed when atoms share electrons. For example, the two oxygen atoms in oxygen gas (O_2) are held together by a covalent bond: the electrons and electrical charge are shared equally between the atoms so the bond is nonpolar. In other molecules, atoms share electrons unequally, so part of the molecule is electrically positive and part is electrically negative. For example, in water molecules (H_2O), the oxygen atom attracts electrons a bit more than do the hydrogen atoms. The result is that the end of the molecule with oxygen has a slight negative charge and the ends with hydrogen atoms have a slight positive charge (**Figure 3.6**).

Strong bonds can also be formed when one atom transfers one of its electrons to another atom. The atom donating the electron has a net positive charge, and the atom receiving the electron has a net negative charge. Such electrically charged atoms or molecules are called **ions**, and the bonds between them are called **ionic bonds**. Compounds formed by ionic bonds are not called molecules; instead they are called ionic compounds or salts. Sodium chloride (NaCl), also known as table salt, is the most familiar example of an ionic compound (**Figure 3.7**).

▲ **Figure 3.7 Table Salt**
A sodium atom (Na) donates an electron to a chlorine atom (Cl) to form common table salt, NaCl.

Salt crystals

Dipole bonds are weaker bonds that form between atoms and molecules as a consequence of shifts in electrical charge. The strength of dipole bonds depends on the extent of the polarity, or difference in charge between the molecules. Dipole bonds affect the chemical and physical properties of polar molecules, such as water. They are also important in a variety of biological functions. For example, animals such as geckos are able to climb smooth walls because of dipole bonds between the molecules in their toes and the molecules on the surface of the wall (**Figure 3.8**).

◄ **Figure 3.8 Hanging in There**
Animals such as geckos are able to adhere to vertical surfaces because of dipole bonds between molecules in their feet and the surface.

▲ **Figure 3.6 Nonpolar and Polar Molecules**
Electrons and electrical charges are evenly distributed around the nonpolar O_2 molecule. Uneven distribution of electrons between the hydrogen and oxygen of water make it polar.

The Water Molecule
Life depends on the chemistry of water.

Every organism on Earth depends on water to live. The cells and tissues of all organisms are largely made of water, and the unique chemical properties of water are critical to their structure and function.

Because water molecules are polar, they form dipole bonds with other water molecules. The dipole bonds involve hydrogen atoms, so they are also called hydrogen bonds. Hydrogen bonds are responsible for the cohesive nature of liquid water. As liquid water heats, the hydrogen bonds weaken; at 100 °C (212 °F), virtually all of them are broken. Without hydrogen bonds, water would boil at temperatures well below its actual freezing point (0 °C or 32 °F).

Conversely, as liquid water becomes colder, more hydrogen bonds form, pulling the water molecules closer together. As a result, water becomes denser as it cools. Water reaches its maximum density at 4 °C (39 °F). Below this temperature, hydrogen bonds begin to stabilize. As water cools to 0 °C, its molecules move apart into the lattice structure of ice, which is far less dense than liquid water (**Figure 3.9**). This is why ice floats. Thus, the winter ice that floats on lakes and oceans is colder than the water underneath.

The polarity of water makes it an excellent solvent—a great many chemicals are able to dissolve in it. **Solubility** refers to the ability of a chemical to dissolve in a particular liquid. Ionic compounds such as NaCl are particularly soluble in water. In contrast, nonpolar molecules, such as those of many oils, are not very soluble in water. Virtually all of the chemical reactions that are essential to the function of living organisms depend on water's unique properties as a solvent.

Water molecules may break apart, or dissociate, into hydrogen ions (H^+) and hydroxyl ions (OH^-).

▼ **Figure 3.9 Dipole Bonds in Water**

A In liquid water, dipole bonds between water molecules are constantly forming and breaking. They become more stable as the temperature decreases.
B At 0 °C (32 °F), these dipole bonds stabilize water molecules into the crystalline structure we know as ice.

Pure water contains equal amounts of these two ions. Chemicals dissolved in water may shift the amount of H^+ ions relative to OH^- ions. The **pH scale** is a quantitative representation of the relative amounts of hydrogen and hydroxyl ions in a liquid (**Figure 3.10**). The pH of pure water is 7, which is said to be neutral. When the concentration of H^+ exceeds that of OH^-, the water is acidic and has a pH value of less than 7. When the concentration of OH^- exceeds that of H^+, the water is basic, and its pH is greater than 7. Each pH unit reflects a 10-fold change in the amount of H^+ ions relative to OH^- ions. Compared to a neutral solution, a solution of pH 6 has 10 times more H^+ ions than OH^- ions; a solution of pH 5 has 100 times more.

The pH of a solution affects many chemical reactions. Variations in the pH of solutions are especially important within organisms and ecosystems. For example, the pH of your blood is slightly alkaline at 7.4: Variations of just 0.1 pH unit can have serious health consequences. Most aquatic organisms, such as fishes and insects, are able to tolerate only very narrow pH ranges.

► **Figure 3.10 pH Scale**
In pure water, hydroxyl and hydrogen ions are equally abundant, and pH = 7. Lemon juice is acidic because hydrogen ions outnumber hydroxyl ions. The reverse is true of milk of magnesia, which is alkaline.

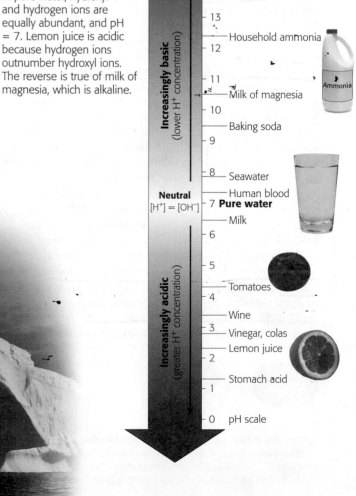

Increasingly basic (lower H^+ concentration)

- 14 1.0 M NaOH
- 13
- 13 Household ammonia
- 12
- 11
- 11 Milk of magnesia
- 10
- Baking soda
- 9
- 8 Seawater
- Human blood
- **Neutral** $[H^+] = [OH^-]$ 7 **Pure water**
- Milk
- 6
- 5
- Tomatoes
- 4
- Wine
- 3 Vinegar, colas
- Lemon juice
- 2
- Stomach acid
- 1
- 0 pH scale

Increasingly acidic (greater H^+ concentration)

3.2 The Organic Chemistry of Life

Water is essential for life. However, the primary structural and functional elements of organisms are **organic molecules**, which are made of carbon atoms covalently bonded to hydrogen and other atoms. For this reason, we say that life on Earth is carbon based. Organic molecules range in size from relatively simple hydrocarbons and sugars to large, complex molecules, such as proteins and nucleic acids. Compounds that are not made up of carbon and hydrogen are said to be **inorganic**.

Hydrocarbons and Carbohydrates

Hydrocarbons and carbohydrates are among the most abundant organic molecules.

Carbon atoms are able to bond with up to four other atoms, allowing carbon to form a great number of different compounds. **Hydrocarbons** are organic molecules composed entirely of carbon and hydrogen atoms. Methane, the primary constituent of natural gas, is the simplest hydrocarbon; its molecules have a single carbon atom and four hydrogen atoms (**Figure 3.11**). Crude oil is a mixture of hydrocarbons that range in size from a few to more than 20 carbon atoms.

Carbohydrates are organic molecules made of carbon, hydrogen, and oxygen. Carbohydrates are the fundamental building blocks of organisms. **Sugars** are carbohydrates with the general chemical formula $(CH_2O)n$, where "n" is between 3 and 7. The sugar glucose ($C_6H_{12}O_6$), or grape sugar, is a basic source of energy in most organisms and an ingredient in the synthesis of numerous other organic molecules (**Figure 3.12**). Table sugar is composed of a glucose molecule bonded to a similar sugar called fructose. Ribose ($C_5H_{10}O_5$) and deoxyribose ($C_5H_{10}O_4$) are very important 5-carbon sugars.

Lipids, or fats and oils, are organic molecules made of long chains of carbon and hydrogen atoms and a shorter region with one to several oxygen molecules (**Figure 3.13**). Lipids are the major constituents of the membranes inside and around the cells of all organisms. Fats and oils are also important forms of energy storage.

The shorter region of a lipid is polar and the longer region is not. Because of their nonpolar region, most lipids are not soluble in water. In other words, oil and water do not mix. As you know from salad dressings, oil does not dissolve in water or vinegar. Even when you shake the dressing, you can still see two separate liquids.

▼ **Figure 3.11 Hydrocarbons**
These molecules are among the array of hydrocarbons found in petroleum.

C_4H_{10}

Methane CH_3
Ethane C_2H_6
Propane C_3H_8

Butane
(straight chain)

2-Methylpropane
(branched chain)

Hydrogen
Oxygen
Carbon

▲ **Figure 3.12 Glucose**
This sugar is an important source of energy for most organisms.

Polar region
Phosphorus
Oxygen
Carbon

Nonpolar region
Hydrogen

▲ **Figure 3.13 Lipids**
The long carbon–hydrogen tails of molecules are nonpolar and do not easily in water.

Macromolecules

Many structures and functions in organisms depend on very large organic polymers.

Small organic molecules may be linked to one another to form **macromolecules**. Relatively simple organic molecules, such as carbohydrates and hydrocarbons, may be linked together to form much larger chains or networks called **polymers**. The many kinds of plastics you use every day are polymers. Polymers are also essential to the structure and function of organisms.

Polysaccharides are polymers of simple sugar molecules. **Starch** and **cellulose** are polysaccharides composed of hundreds of glucose molecules. Plants use starch to store extra sugar. Cellulose is the primary structural constituent in plant tissues; it is also one of Earth's most abundant organic molecules. The only difference between starch and cellulose is the nature of the bond between adjacent glucose molecules. However, this difference has a significant effect on the ability of organisms to break down the two kinds of molecules. Most organisms can easily break the bonds in starch, releasing glucose molecules to use for food.

The bonds between the glucose molecules in cellulose are much more resistant; only a few kinds of microorganisms are able to break those bonds.

Proteins are polymers made of nitrogen-containing organic molecules called amino acids. The 20 amino acids found in all organisms have different chemical structures, but each has a nitrogen-containing amino group (–NH$_2$) and a carboxylic acid group (–COOH). Individual proteins are composed of 100 to over 1,000 individual amino acids (**Figure 3.14**). The function of each protein is determined in part by its unique sequence of amino acids. The chemical interactions and bonds among the various amino acids in a protein chain cause it to fold into a very complex three-dimensional shape, which is important to its function.

Every organism depends on the actions of thousands of different proteins. Some proteins, such as the keratin that makes up your fingernails and hair, provide structure.

▶ **Figure 3.14 Amino Acids to Functional Proteins**
The function of each protein is determined by the specific sequence of the 20 different amino acids and by its three-dimensional structure.

Typical amino acids

H$_2$N—C—COOH
CH$_3$
Alanine

Acid group
Amine group

H$_2$N—C—COOH
CH$_2$
CH$_2$
C
O O-H
Glutamic acid

Amino acids are linked to form a protein chain

Protein chains assume a three-dimensional structure

Other proteins store energy, and still others play a role in the transport of materials within and among cells. Many proteins serve as **catalysts**, substances that promote chemical reactions without being consumed in the reaction. Proteins that serve as catalysts are called **enzymes**. Virtually every chemical reaction in an organism is catalyzed by its own unique enzyme.

Nucleic acids are polymers of chemical subunits called nucleotides. Each nucleotide is composed of a 5-carbon sugar, a phosphate group ($-PO_4$), and a nitrogenous base. **Deoxyribonucleic acid (DNA)** forms the hereditary material that is passed from generation to generation in all organisms. Each nucleotide of DNA contains the sugar deoxyribose and one of four nitrogenous bases—adenine (A), thymine (T), cytosine (C), or guanine (G). Each portion of a DNA molecule has a unique sequence of nitrogenous bases, such as ACTTGAAC. The DNA molecule usually exists as a ladderlike double strand (**Figure 3.15**). Deoxyribose and phosphate groups form the edges or backbone of this ladder, and paired nucleotides form the rungs. The nucleotides are paired in a very specific fashion: adenine (A) pairs only with thymine (T), and cytosine (C) pairs only with guanine (G). The entire ladder is twisted into the shape of a helix.

The DNA of an individual organism is made of billions of nucleotides. The hereditary, or genetic, code of each organism is carried in the unique sequence of nitrogenous bases (A, T, C, and G) along that organism's DNA. Segments of DNA called **genes** carry the code used to synthesize, or build, specific proteins. An organism's complete set of DNA is called its **genome**.

Ribonucleic acid (RNA) also plays a central role in the synthesis of proteins, using information encoded in DNA molecules. The structure of RNA is similar to that of DNA, except the nucleotides of RNA contain the sugar ribose rather than deoxyribose and the nitrogenous base uracil (U) rather than thymine (T). Furthermore, RNA typically exists as a single strand rather than a double strand.

Proteins are synthesized in a two-step process (**Figure 3.16**). In **transcription**, the DNA code for a particular gene is rewritten as a segment of RNA. This RNA then serves as the template for the assembly of a specific protein in the **translation** process. The sequence of triplets of nucleic acids in the RNA strand (AUC, GUU, or CAG, for example) determines the sequence of amino acids in the newly synthesized protein.

* RNA vs. DNA

QUESTIONS

1. How do carbohydrates differ from hydrocarbons?

2. Many lipids are not soluble in water. Why?

3. How does starch differ from cellulose? Why is this difference important?

4. What is an organism's genome?

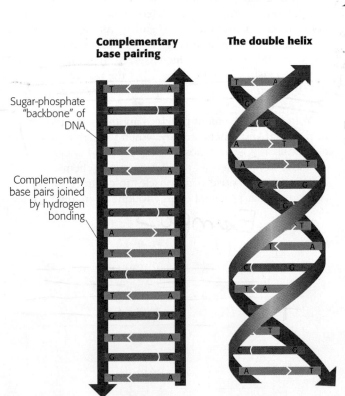

▲ **Figure 3.15 DNA Forms the Hereditary Code**
The genetic code that determines the structural and functional attributes of each organism is embedded in the sequence of nitrogenous bases along the DNA strand.

▲ **Figure 3.16 Reading the Genetic Code**
The information contained in the DNA molecule is converted into an organism's structure and function in two chemical steps, transcription and translation. The end products are the specific structural proteins and enzymes.

3.3 Energy and the Environment

Unlike matter, energy is not a tangible thing. Rather, **energy** is the capacity to do work. Energy cannot be created, neither can it be destroyed. Therefore, the amount of energy in the universe is constant. Four types of energy—the energy of motion, the energy in chemical bonds, the energy in light, and the energy contained in matter itself—are important in ecosystems. Energy can be changed from one form to another. However, energy transformations inevitably increase the disorder of the universe.

Energy

Energy is the capacity to do work.

Energy describes the ability of a system to move things, whether they are molecules, muscles, or machines. The first of Isaac Newton's laws of motion states that any object at rest will stay at rest, and any object in motion will remain in motion unless acted upon by some force. This law holds true for all objects, from atoms to planets, regardless of their size. When an object is moved from rest or from one trajectory of motion to another, physicists say that **work** has been done. Work is a force applied to an object over some distance. The greater the force or distance, the greater the amount of work.

Energy is the capacity to do work. Energy exists in one of two general forms. **Potential energy** is the energy that is stored in a system and available to do work in the future; it is sometimes called the energy of position. **Kinetic energy** is the energy of motion. For example, consider the system of an expensive ceramic vase poised on top of a somewhat tippy table. (**Figure 3.17**). The vase has potential energy that is related to its mass, its height above the floor, and the tug of gravity. As the vase falls, the potential energy in this system decreases, and the kinetic energy of the vase increases. At the instant before the vase crashes into the floor, all of the potential energy of this system has been converted into the kinetic energy of the vase.

Your body movement provides another example of potential and kinetic energy. When you are at rest, chemicals in your muscles possess potential energy for movement. When you decide to move, chemical transformations cause your muscles to contract, converting that potential energy to kinetic energy, or motion.

◄ **Figure 3.17 Look Out Below!**
A ceramic vase poised on top of a table has potential energy Ⓐ. As it falls, that potential energy is converted to kinetic energy Ⓑ and the disorder or entropy of the system clearly increases as it crashes to the floor Ⓒ.

Laws of Thermodynamics

Energy can be neither created nor destroyed, but it can be transformed.

The potential energy in a system may be transformed into kinetic energy. But all energy transformations conform to two fundamental principles, called the laws of thermodynamics.

The **first law of thermodynamics** states that the total amount of energy in the universe is constant. Energy can be transformed from one form to another, but it can be neither created nor destroyed. This is also called the law of energy conservation.

Although the total energy contained in the universe remains the same, particular systems within the universe may gain or lose energy. Once the vase crashes on the floor, the kinetic energy of its motion will be transformed into the motion of pieces of vase and the floor and the motion of the molecules in the atmosphere through which the vase fell.

The **second law of thermodynamics** states that every energy transformation increases disorder. Scientists use the word **entropy** to describe the disorder or disorganization in a system. In every energy transformation, some amount of energy is converted to entropy.

A very important consequence of this second law is that in any energy transformation, energy always tends to go from more usable, or higher quality, forms to less usable, or lower quality, forms.

The least usable form of energy is the random motion of molecules, or **heat**. Thus, much of the potential energy in the system of vase and table is transformed into the random motion of molecules and debris. Much of the chemical energy available for moving your arm ends up in the random kinetic motion of molecules in muscles and surrounding tissue. This motion of the molecules makes your body warmer.

The maintenance of life on Earth requires a continuous input of energy from the sun. This energy is necessary to transform inorganic compounds into the large, complex macromolecules found in all living things. It also takes a great deal of energy to power the growth, movement, reproduction, and other activities carried out by all organisms. At first, the order and organization of organisms and ecosystems may seem to violate the second law of thermodynamics (**Figure 3.18**). This apparent contradiction disappears when we recognize that ecosystems and their organisms only remain organized with constant input of energy from the sun.

▶ **Figure 3.18 Where Is the Entropy?**
Individual organisms and ecosystems are continuously transforming energy, but they remain highly organized. Their organization depends ultimately on the continuous input of energy from the sun, and it results in an overall increase in the disorder or entropy of the universe.

71

Ionizing radiation—photons carry energy sufficient to ionize molecules and break chemical bonds

Visible light photons can excite electrons—this is what powers photosynthesis and eyesight

| Gamma Rays | | X-Rays | | | | Ultraviolet | | Visible Light | |
| 10^{-6} nm | 10^{-5} nm | 10^{-4} nm | 10^{-3} nm | 10^{-2} nm | 10^{-1} nm | 1 nm | 10 nm | 100 nm | 1 µm | 10 µm |

▲ **Figure 3.19**
Electromagnetic Energy
Light is composed of photons whose energy is determined by their wavelength, the distance from wave crest to wave crest. The shorter a photon's wavelength, the more energy it contains. *Source: Copyright DK LTD*

Forms of Energy

Four types of energy are especially important in ecosystems.

Electromagnetic radiation. The energy of light is called **electromagnetic radiation**. This energy is transported at a remarkable speed (186,000 miles/second) as "particles" called photons, which have no mass and behave like waves. The amounts of energy carried by the different forms of electromagnetic radiation are determined by their wavelength, the distance from one wave crest to the next (**Figure 3.19**). The full range of wavelengths is called the **electromagnetic spectrum**.

Visible light—the light that you are able to see—represents a very small portion of the total electromagnetic spectrum. The different wavelengths of visible light form the various colors. Photons in this range are sufficiently energetic to elevate the energy level of electrons in organic molecules without breaking their bonds. For example, photons of visible light excite electrons in molecules in the retina of your eyes, providing a signal that your brain interprets as sight. In green plants, photons of visible light excite electrons in chlorophyll, providing the energy for photosynthesis.

Gamma rays, X-rays, and UV light have wavelengths that are shorter than those of visible light. Their photons pack energy sufficient to disrupt the bonds in many organic molecules. Thus, these forms of radiation are generally harmful to living systems.

Forms of radiation with wavelengths that are longer than visible light, such as infrared radiation, microwaves, and radio waves, carry less energy than does visible light. Nevertheless, the photons of these wavelengths carry sufficient energy to increase the kinetic energy of molecules. They can also break very weak chemical bonds.

Heat. Heat energy refers to the kinetic energy of molecules. **Temperature** is a measure of the average kinetic energy of a collection of molecules. Over the years, several temperature scales have been devised. In 1727, Daniel Gabriel Fahrenheit proposed a temperature scale in which water freezes at 32 ° and boils at 212 °. Fahrenheit based his scale on measurements of everyday phenomena, such as very cold days and human body temperature. Fifteen years later, Anders Celsius proposed a temperature scale that set 0 ° as the freezing point of water and 100 ° as the boiling point of water (**Figure 3.20**). Because it is scaled in a simple fashion against the physical property of a common and important chemical, environmental scientists prefer to express temperature in degrees Celsius (°C).

Heat can move in four ways—conduction, convection, radiation, and latent heat transfer (**Figure 3.21**). **Conduction** is the direct transfer of heat by means of the collisions of molecules.

▶ **Figure 3.20 Measuring Kinetic Energy**
Two temperature scales, Fahrenheit and Celsius, are widely used to measure the kinetic energy of collections of molecules.

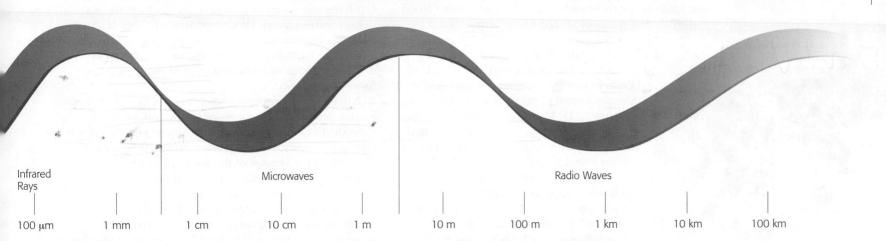

Photons with wavelengths longer than visible light carry sufficient energy to increase the kinetic energy of molecules

Infrared Rays			Microwaves				Radio Waves				
100 μm	1 mm	1 cm	10 cm	1 m		10 m	100 m	1 km	10 km	100 km	

When you touch a hot pot, heat moves from molecules on the pot to the molecules in your fingertips. **Convection** occurs because warm regions in a gas or liquid become less dense and rise, causing the gas or liquid to circulate.

All matter is constantly losing heat by **radiation**, the release of electromagnetic energy. The wavelength of the radiation that is given off decreases with increasing temperature. At 37 °C (98.6 °F), your body radiates heat primarily in the infrared portion of the spectrum, whereas a metal heated to 300 °C (570 °F) glows bright red.

Temperature is a measure of the average kinetic energy of molecules. Within a liquid, some molecules move more rapidly than others. **Latent heat transfer** occurs as the molecules with the highest kinetic energy evaporate, leaving behind molecules with lower kinetic energy and temperature. You experience latent heat transfer when you sweat and your perspiration evaporates, cooling your skin. Because oceans cover nearly 72% of Earth's surface, the total amount of latent heat transferred to the atmosphere from the oceans is quite significant.

▼ Figure 3.21 Heat Transfer Heat energy can be transferred by Ⓐ conduction, collisions among molecules; Ⓑ convection, circulation due to differences in density; Ⓒ radiation of electromagnetic energy; and Ⓓ latent heat transfer resulting from evaporation of more energetic molecules.

Ⓓ Latent heat transfer

Ⓑ Convection

Ⓒ Radiation

Ⓐ Conduction

Chemical energy: The potential energy associated with the formation or breakage of bonds between atoms is called **chemical energy**. Chemical energy is essential to life. In the process of photosynthesis, green plants use light energy to assemble carbohydrates from carbon dioxide and water (4.1). The energy in the bonds of these carbohydrates is retrieved by plants and animals to sustain all of life's functions. For example, the energy released from breaking bonds in glucose molecules is captured in the movement of proteins in muscle cells (**Figure 3.22**). Thus, the potential chemical energy in the glucose bonds is converted to the kinetic energy of the contraction of those muscles.

The formation of some chemical bonds releases energy. For example, when hydrogen gas (H_2) bonds with oxygen gas (O_2) to form water, the reaction releases so much energy that it causes an explosion. The resulting bonds between the hydrogen and oxygen atoms are very strong. The energy required to break a water molecule apart is equal to the energy that was released in its formation.

Nuclear energy: The energy contained in the structure of matter itself is called **nuclear energy**. Albert Einstein's famous equation, $E = Mc^2$, states that the energy (E) contained in matter is equal to its mass (M) times the speed of light (c = 299,792,458 meters per second) squared. Thus, even tiny amounts of mass contain enormous amounts of energy. In situations relevant to Earth's ecosystems, the transformation of mass to energy occurs as a consequence of two processes, fission and fusion.

Nuclear fission occurs when the nucleus of an atom is split, producing two or more smaller nuclei and a great deal of electromagnetic and kinetic energy. Fission changes one element into one or more other elements. The sum of the masses of the resulting products is slightly less than the mass of the original atom. The difference in the mass of the system before and after fission is equal to the electromagnetic and kinetic energy that is released. For example, neutrons striking the nuclei of atoms of the uranium isotope ^{235}U can cause the atoms to split into a variety of new atoms and free neutrons, releasing considerable energy. If the ^{235}U is sufficiently abundant, the neutrons released in this reaction may stimulate additional fission events, creating a chain reaction (**Figure 3.23**). When uncontrolled, such a chain reaction can be explosive, as with an atomic bomb.

Good Example

▲ **Figure 3.22 Chemical to Kinetic Energy**
The potential chemical energy in glucose molecules is converted to kinetic energy in the contraction of muscles.

Neutron split yet other ^{235}U atoms creating a chain reaction

$^{235}_{92}U$

Neutron (n)

Read Example

Each time an atom is split a small amount of mass is converted to a very large amount of energy

▲ **Figure 3.23 Nuclear Fission**
When a sufficient number, that is, a critical mass, of ^{235}U is bombarded by high-energy neutrons, the nuclei of these atoms split, producing two smaller atoms and releasing more high-energy neutrons.

In the core of a nuclear reactor, the chain reactions are controlled, and the energy that is released is used to boil water that is then used to generate electricity.

Nuclear fusion occurs when two atoms collide with so much energy that their nuclei fuse, forming an atom of a new element. The fusion of relatively small atoms, such as hydrogen, carbon, and oxygen, releases energy as very small amounts of mass are converted to energy. The fusion of massive atoms, such as lead or mercury, requires an input of energy.

By far the most common and most important fusion reaction in the universe occurs when two atoms of hydrogen collide to create an atom of helium (**Figure 3.24**). The kinetic energy required for this reaction to occur is equivalent to approximately 3,000,000 °C. The electromagnetic radiation and kinetic energy released from this reaction far exceed the energy contained in the colliding hydrogen atoms. This is the energy that powers most stars, including our sun. These stars are so massive that the pull of their gravity provides enough kinetic energy for nuclear fusion to occur.

▶ **Figure 3.24**
Solar Power
High temperatures and pressures within the sun's interior are sufficient to power fusion reactions between isotopes.

Hydrogen isotopes
Helium
Energy
Neutron

Energy Units

Scientists use several different units to measure energy, depending on the application.

The fundamental unit of energy is named in honor of a 19th-century scientist, James Prescott Joule. A **joule (J)** is the energy required to support a 1-kg mass against the force of gravity. Scientists and engineers also use other units of energy for particular purposes. All of these units can be converted to joules (**Figure 3.25**).

Temperature describes the average kinetic energy of the molecules in a system, but the total energy content of a system also depends on the number of molecules within it. A **calorie (cal)** is the energy required to raise the temperature of 1 g of water by 1 °C (1 cal = 4.18 J). Thus, it is sensitive to both the kinetic energy of the molecules and their amount. Biologists and ecologists frequently use calories to measure changes in the energy of organisms and ecosystems. The Calorie (with a capital *C*) that is commonly used to measure the energy content of food is equal to 1,000 cal, or 1 kcal (small-*c* calories).

The energy unit most commonly used to measure the everyday use of electricity is the **watt-hour (Wh)**. One kilowatt-hour (kWh) is equal to 1,000 Wh.

A watt-hour is the energy expended over an hour at the rate of 1 joule per second. Since there are 3,600 seconds in an hour, 1 Wh is equal to 3,600 J. Thus, a 40-watt lightbulb consumes 144,000 J in the course of an hour. In summary, energy is neither created nor destroyed; it is only transformed from one form to another. Whenever it is transformed, the entropy in the universe increases. Forms of energy include electromagnetic radiation, heat, chemical energy, and nuclear energy. The units used to measure energy include joules, calories, and watt-hours.

▼ **Figure 3.25 Energy Conversions**

	Joules (J)	Calories (c)	Watt-hours (Wh)
A joule (J) =	1	0.24	0.00028
A calorie (c) =	4.18	1	0.0012
A watt-hour (Wh) =	3,600	861	1

75

3.4 The Planet Earth in Context

The matter that would eventually form our solar system had its origins in the collapse of a vast cloud of interstellar gas. That collapse was likely triggered by the explosion of a very massive star nearby (**Figure 3.26**). Our sun formed at the center of the cloud of gases, eventually gathering up more than 99% of the cloud's mass. The remaining gases came together to form the planets and other objects that orbit the sun. The planets nearest Earth—Venus and Mars—have properties that suggest that they may once have supported life. Although it is possible that Mars may yet support life, Earth appears to be the only place in our solar system where abundant life is sustained.

▲ **Figure 3.26 Birth of Solar Systems**
As explosions of massive stars blast the Eagle nebula's gas and dust, they may be triggering the formation of planetary systems similar to our solar system.

Comparing the Planets

Earth is unique among its closest celestial neighbors.

About 4.6 billion years ago, our solar system coalesced to form a swirling disk of gases (**Figure 3.27**). Over time, gravity pulled most of the gases into the center of this disk. As the gases in the center became denser and hotter, nuclear fusion began. Our sun was born. Nuclear reactions in its core cause the sun to emit electromagnetic radiation. The sun also emits a thin but energetic stream of charged particles known as the solar wind.

Some of the gases and dust that were not consumed in the formation of the sun came together to form planets and a variety of smaller objects, such as asteroids and comets (**Figure 3.28**). The larger, planetlike objects are classified into three groups, based on their physical structure.

Planetoids are composed of ice and rock. The best-known planetoids are Pluto and its moon Charon. These objects orbit in the outer reaches of the solar system; they are so distant that from their surface the sun would look like just another bright star in the sky.

Orbiting closer to the sun are the **gas giants**—Jupiter, Saturn, Uranus, and Neptune. The gas giants are composed mainly of hydrogen and helium, with smaller amounts of other gases such as methane and ammonia.

▶ **Figure 3.27 The Infant Solar System**
Early in the evolution of the solar system, the developing sun was surrounded by a swirling disk of debris from which the plants condensed.

Pluto and Charon | Neptune | Uranus

As their name suggests, they lack a solid surface. These planets are huge. Jupiter's volume, for example, is approximately 1,000 times that of Earth. The gas giants might have become suns in their own right, but they lack the mass and gravitational pull to initiate fusion reactions.

The gas giants are notable for their numerous moons. For example, Saturn has 59 moons and Jupiter has 63. Some of these moons are only tens of kilometers across; others are larger than the planet Mercury.

The four inner planets—Mercury, Venus, Earth, and Mars—are called the **terrestrial planets**. Each has a dense metallic core surrounded by thick layers of less dense rock. Venus, Earth, and Mars are surrounded by gaseous atmospheres, although their atmospheres differ in density and chemical composition. Because of its weak gravitational pull and proximity to the sun, Mercury lacks an atmosphere.

Unlike the gas giants, the terrestrial planets have few moons. Earth's moon formed 4.5 billion years ago, when the still-forming Earth was struck by a smaller object. The collision threw off an immense amount of rocky debris, some of which coalesced to form our moon. Mars has two tiny moons, each less than 20 km in diameter. Neither Mercury nor Venus has a moon.

Astronomers have spent many long hours searching for signs of life on the menagerie of planets and moons in our solar system. A century ago, a few astronomers were convinced that intelligent life forms had built canals on Mars. Mars also appeared to undergo seasonal changes in color that might be evidence of the activities of intelligent life. The notion of intelligent life on Mars has been completely debunked by improved telescopes and evidence from the Mars space probes. However, evidence does show that Mars once had abundant water, so some form of life may have existed there in the past; it is even possible that such life may persist in some locations on Mars today.

Nevertheless, the only location in our solar system where scientists have conclusively verified the existence of life, past or present, is here on Earth. Why here? Earth's distance from the sun—93 million miles—is an important part of the answer. Mercury's orbit brings it within 32 million miles of the sun, exposing the planet to heat and high-energy radiation that would quickly destroy most hydrocarbons and carbohydrates. Jupiter's orbit is greater than 483 million miles from the sun. At that distance, Jupiter receives only about 1% of the sunlight that reaches Earth, not enough to provide the warmth and energy needed to support life as we know it. In contrast, Venus orbits the sun at 67 million miles and Mars orbits at 141 million miles, positions that allow both planets to receive amounts of sunlight that might be conducive to life. So, although distance from the sun is important, it is not sufficient to explain the abundance of life on Earth and its apparent absence elsewhere.

Scientists have concluded that several of our planet's unique features have allowed life to develop and flourish. Earth's physical structure, from its core to outer crust, has influenced the development of life. Our planet's vast oceans are a reservoir of the water necessary for life; oceans also play a central role in maintaining temperatures within livable ranges. The composition of Earth's atmosphere is quite different from that of Venus or Mars. The characteristics of our atmosphere directly influence Earth's climate and the chemistry of life. Earth's unique features are discussed in detail throughout the book.

▼ **Figure 3.28 The Planets**
The solar system includes four solid terrestrial planets, four gas giants, and many planetoids, among which Pluto is the closest and best known.

QUESTIONS

1. Describe the differences among the three groupings of planetlike objects in the solar system.
2. Why does nuclear fusion "light up" the sun, but not the gas giant planets?

Saturn   Jupiter   Mars   Earth   Venus   Mercury   Sun

3.5 Earth's Structure

During the early evolution of Earth, gravity pulled heavier materials into its center, and lighter materials moved toward its surface. As a result, Earth has three distinct layers: a dense core, a less dense mantle, and a light, rocky crust (**Figure 3.29**). The crust is divided into large segments, or plates. Over Earth's history, the movement of these plates has produced significant changes in the geographic configuration of the oceans and continents.

The Core, Mantle, and Crust

Solid Earth's three layers have very different physical properties.

At Earth's center is the **core**. The core is composed of a mixture of nickel and iron, with smaller amounts of other heavy elements. The core has two parts: a solid inner core, with a radius of 1,220 km (760 mi), and a liquid outer core that extends 2,260 km (1,400 mi) toward the surface. Both the inner and outer core are under extreme pressure—three million times greater than the atmospheric pressure at Earth's surface. The inner and outer core are also extremely hot, with temperatures in excess of 7000 °C.

Convection currents and Earth's rotation stir the molten iron of the outer core, setting up a strong magnetic field. This magnetic field orients the magnetic needle of a compass. Some birds and insects also use this magnetic field to navigate over long distances; these animals have magnetized iron crystals in their skin or specialized organs for orientation. More important to the maintenance of life on Earth, this magnetic field deflects the high-energy protons and electrons in the solar wind, causing most of these particles to travel around, rather than through, our atmosphere. Direct exposure to these particles would be lethal to most organisms.

Above the core is a layer of less dense rock called the **mantle**. The mantle, which is 2,900 km (1800 mi) thick, occupies about 70% of Earth's total volume. The rocks of the mantle are rich in the elements magnesium and silicon, with much smaller amounts of calcium and aluminum. Because they are under great pressure, the solid rocks in the mantle can slowly flow, in a plastic fashion like modeling clay. Heat from the core causes the mantle rock to circulate in huge, slow-moving convection currents. In parts of the upper mantle, rocks may be heated to a liquidlike state called **magma**.

Above the mantle is Earth's **crust**, a thin veneer of solid and relatively light rocks. The crust, which makes up less than 1% of Earth's total volume, varies in depth from 5–100 km (3–62 mi). This thin layer is the only part of Earth that directly interacts with living systems.

There are two types of crust, which differ in their chemistry and density. Continental crust makes up the continents and the areas immediately adjacent to them. The many different kinds of rocks in continental crust contain large amounts of lighter elements, especially aluminum and silicon. Ocean crust, which lies beneath most of the deep ocean, is composed of denser and more homogeneous rocks that are similar to the rocks in the mantle. Continental crust varies in thickness from 30–100 km (20–62 mi), whereas ocean crust is uniformly about 5–10 km (3–6 mi) thick.

Geologists refer to the crust and the upper reaches of the mantle that interact with it as the **lithosphere**. This is the zone of geologic activity that has shaped Earth's surface and continues to do so. Earth's hydrosphere is composed of the liquid water and ice in and on the crust and includes oceans, rivers, and glaciers.

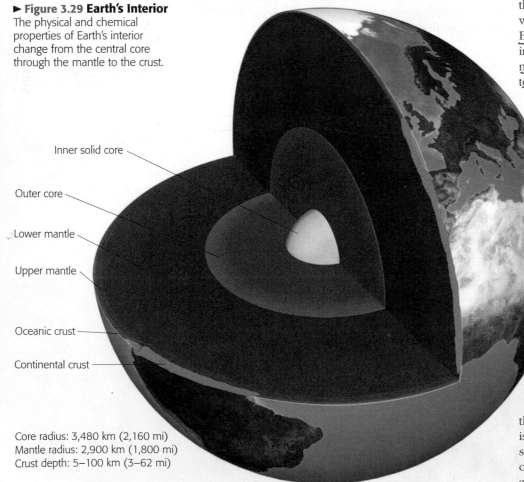

► **Figure 3.29 Earth's Interior**
The physical and chemical properties of Earth's interior change from the central core through the mantle to the crust.

Inner solid core

Outer core

Lower mantle

Upper mantle

Oceanic crust

Continental crust

Core radius: 3,480 km (2,160 mi)
Mantle radius: 2,900 km (1,800 mi)
Crust depth: 5–100 km (3–62 mi)

Building and Moving Continents

The geographic configuration and distribution of oceans and continents have changed through time.

Since the existence of relatively accurate world maps, geographers have noticed that the margins of some continents match up with those of other continents. For example, if the Atlantic Ocean separating South America and Africa were removed, these two continents would come together like matching pieces in a puzzle (**Figure 3.30**). In the 19th and early 20th centuries, geologists observed that sedimentary rocks on separate continents often hold the same kinds of fossils.

In 1911, the German scientist Alfred Wegener attempted to explain these observations by suggesting that in the distant past all of the continents were joined as one large continent. He called this ancient supercontinent Pangea. Wegener proposed that over time Pangea broke apart, and the smaller continents drifted away. However, Wegener could not propose a mechanism that would cause the continents to move. In the absence of such a mechanism, geologists largely dismissed Wegener's hypothesis for the next 50 years.

By the 1960s, evidence was beginning to accumulate that Earth's crust was indeed able to move and that its movements were a consequence of interactions between the crust and the mantle.

New techniques allowed geologists to determine when rocks had formed. These techniques revealed that the ages of rocks in the crust vary greatly. Some continental rocks are quite new, whereas others are nearly 4 billion years old. In contrast, rocks in the ocean crust are relatively young; none are more than 200 million years old. Geologists also observed that the youngest rocks in the ocean crust are found near mid-ocean ridges; the oldest rocks in the ocean crust are found next to the continents.

All of these observations have come to be understood through the theory of plate tectonics. According to this theory, Earth's crust is broken into pieces called **tectonic plates** that float on top of the mantle (**Figure 3.31**). Continents are embedded in some of these plates. The positions of the plates are not fixed. Instead, they slowly move in relation to each other. What causes tectonic plates to move? As convection currents slowly stir the plastic materials in the mantle, the plates also move. The movement of the mantle is like an enormous conveyor belt, providing the energy that moves the plates along.

The places where plates come together are called boundaries; the type of boundary depends upon how the plates are moving in relation to one another.

Fit of South America and Africa at 500m depth

▲ **Figure 3.30 A Chance Alignment?**
By the late 19th century, global maps were sufficiently exact that the alignment of the coasts of South America and Africa were quite obvious.

◄ **Figure 3.31 The Fractured and Mobile Crust**
Earth's crust is divided into 17 tectonic plates that move relative to one another.

Subduction zone
Divergent boundary
Plate movement

Plates slide past one another along **transform fault boundaries.** As plates move past one another in fits and starts, they may generate earthquakes. The San Andreas Fault is one of the best-studied transform fault boundaries (**Figure 3.32**). It separates the Pacific Plate, which includes a significant portion of south coastal California, from the western portion of the North American Plate and the rest of California.

Divergent boundaries separate plates that are moving apart. Such boundaries coincide with ocean ridges. The Mid-Atlantic Ridge is an example of a divergent boundary. It separates the North American Plate from the Eurasian Plate and the South American Plate from the African Plate. At divergent boundaries, hot magma from the mantle rises to the surface, forming new ocean crust.

Tectonic plates collide at **convergent boundaries.** Where ocean and continental plates meet, the dense ocean crust is thrust beneath the lighter continental crust in a process called subduction (**Figure 3.33**). As it is pushed into the mantle, subducted crust is heated, generating earthquakes and creating an area beneath the continent where magma is abundant.

This magma may rise to the surface, forming volcanoes. For most of its length, the west coast of South America forms convergent boundaries with other tectonic plates. The volcanic Andes Mountains are a product of subduction occurring along these boundaries.

Where continental crust collides with continental crust, the land along the boundary is thrust upward, sometimes on a grand scale. For example, the subcontinent of India (part of the Indian-Australian Plate) has been colliding with the Eurasian Plate for nearly 50 million years. Their collision produced the mountains of the Hindu Kush and the Himalayas (**Figure 3.34**). The Appalachian Mountains of the United States and Atlas Mountains of Africa were created between 240 and 300 million years ago, when southeastern North America collided with northwest Africa. These two mountain ranges once rose as high as the Himalayas, but have since been eroded to their current stature.

Mercury, Mars, and our Moon have rigid crusts that lack tectonic plates. Their surfaces bear clear evidence of their distant geologic past. That evidence includes tens of thousands of craters accumulated from impacts with meteors over hundreds of millions of years. Earth, too, has been struck by thousands of meteors, but the subduction, mountain building, and erosion associated with plate tectonics have continuously reworked Earth's surface, erasing all but a few meteor craters.

Earth's dynamic crust has been a significant factor in the development of life. The materials that make up the crust contain essential elements in proportions that are helpful for the formation of organic molecules. The continual formation and destruction of the crust has provided a continuous supply of the elements necessary to the function of living organisms and ecosystems. The moving continents have also affected the way in which individual organisms and ecosystems have evolved.

▲ Figure 3.32 **A Transform Fault**
A section of the San Andreas Fault in central California. Terrain west of the fault is moving north relative to that to the east.

Transforming boundary

Oceanic crust formed when magma cools and solidifies

Rising magma from mantle

Diverging plate boundary

Movement of oceanic plate

Magma forms as plate descends

Ocean plate is subducted beneath continental plate

▲ Figure 3.33 **Three Kinds of Plate Boundaries**
Ⓐ Ocean crust is created at divergent plate boundaries. Ⓑ Plates collide at convergent plate boundaries. Ⓒ Plates slide past one another along transform fault boundaries.

QUESTIONS

1. What is the source of Earth's magnetic field?

2. How does continental crust differ from ocean crust?

3. Describe three types of boundaries between tectonic plates.

► Figure 3.34 **When Continents Collide**
The Himalayas have been thrust up along the convergent boundary between India and Asia proper.

3.6 Earth's Atmosphere

Earth, Venus, and Mars have sufficient gravity to hold an **atmosphere**, the layer of gases above their surfaces (**Figure 3.35**). The chemical composition of Earth's atmosphere is unlike those of its planetary neighbors; over billions of years, the actions of photosynthetic organisms have decreased the concentration of carbon dioxide and increased the concentration of oxygen. On average, water accounts for about 1% of the molecules in Earth's atmosphere. However, the actual amount of water in the air varies widely from time to time and place to place. Scientists divide Earth's atmosphere into four altitudinal zones, based on changes in temperature and physical characteristics.

Composition of Gases

Earth's unique atmosphere sustains life and is a product of life.

The atmospheres of both Venus and Mars are composed of over 96% carbon dioxide (CO_2), with about 3% nitrogen (N_2) and traces of other gases, such as argon (Ar). In contrast, Earth's atmosphere contains 78% nitrogen and 21% oxygen (O_2); CO_2 accounts for a mere 0.039%. Because many organisms depend on oxygen for survival, it is tempting to attribute the presence of life on Earth to the abundance of oxygen in its atmosphere. But it is more correct to say that Earth's unique oxygen-rich atmosphere is a consequence of the presence of life.

It appears that 3 billion years ago the composition of Earth's atmosphere was similar to that of Venus and Mars. When photosynthetic microbes arose, they used atmospheric CO_2 to manufacture carbohydrates, releasing O_2 as a by-product. As a result, the concentration of oxygen began to rise, and the concentration of CO_2 began to decline. Over billions of years, photosynthetic organisms slowly altered the composition of the atmosphere. Thus, we can thank photosynthesis for the oxygen-rich air that we breathe.

Billions of yrs. ago... How Do we know!!!..

B Mars

▼ Figure 3.35 **Planetary Atmospheres**
A Earth's atmosphere is rich in nitrogen (N_2) and oxygen (O_2), and the clouds indicate lots of water vapor. **B** The Martian atmosphere is nearly 97% carbon dioxide (CO_2), with virtually no water vapor.

A Earth

Layers of the Atmosphere

The four layers of Earth's atmosphere differ from one another chemically and physically.

Earth's atmosphere is approximately 480 km (300 mi) deep. The air at Earth's surface is compressed by the weight of the great column of air above it, affecting the density of atmospheric gases (i.e., the number of molecules per unit volume). The force caused by the pull of gravity on a column of air is **atmospheric pressure**. At sea level, atmospheric pressure is 1,013 millibars (mb), or 14.7 pounds per square inch. With increasing altitude, the weight of the air above decreases, and so does the atmospheric pressure. At an altitude of 5.5 km (18,000 ft), atmospheric pressure is about one-half (500 mb) that at sea level. An airliner at 12.2 km (40,000 ft) is flying through air with an atmospheric pressure of less than 170 mb.

Scientists divide the atmosphere into four layers based on differences in temperature and chemical properties (**Figure 3.36**). The lowest layer, the **troposphere**, provides the air that we breathe. Most of our planet's weather occurs in the troposphere. Air at low altitudes is warmed by heat radiating from Earth's surface. As the air near the ground becomes warmer, it expands, which causes it to rise. As it rises, it cools, and eventually sinks back to the surface. This process of thermal convection keeps the air in the troposphere well mixed. Thus, pollutant gases, such as sulfur dioxide from smokestacks, are quickly dispersed through the troposphere.

If you could travel up through the troposphere, you would find that the air temperature decreases as the altitude increases. This decrease occurs at a rate of 6 °C every km (3.5 °F/1,000 ft). At an altitude of approximately 15 km (9 mi), air reaches a minimum temperature of about −52 °C (−62 °F), and temperature change ceases. You have reached the tropopause, the upper boundary of the troposphere.

Above the tropopause is the **stratosphere**, which extends from 15 to about 48 km (9–30 mi) above Earth's surface. Moving upward through the stratosphere, temperature gradually increases. The upper boundary of the stratosphere is the stratopause, where the temperature reaches a maximum of approximately −30 °C (27 °F)

The chemical composition of the stratosphere is very similar to that of the troposphere, except the stratosphere contains comparatively large amounts of oxygen in the form of ozone (O_3). The concentration of ozone is highest at altitudes between 15 and 35 km (9–22 mi), often called the **ozone layer**. Ozone is particularly effective in absorbing and scattering the abundant UV light that strikes Earth's upper atmosphere. The high amount of energy contained in UV photons can damage many organic molecules, including DNA. Thus, the ozone layer provides important protection to life on Earth.

Unlike the air in the troposphere, the air in the stratosphere undergoes very little convectional mixing. So, when chemical pollutants get into the stratosphere,

▲ **Figure 3.36 Combined Atmosphere**
Ⓐ At sea level, atmospheric pressure is 1,013 millibars (mb). With increasing altitude above Earth's surface, the weight of air above, and therefore the atmospheric pressure, diminishes. The atmosphere divides into four unequal layers based on temperature. Ⓑ The temperature and chemical properties of the atmosphere change from the troposphere through the stratosphere, mesosphere, and thermosphere. *Source: Copyright DK LTD*

they are not rapidly dispersed. Rather, they tend to remain there for long periods of time.

Above the stratopause is the **mesosphere**, where the air temperature begins to drop again, reaching lows of −173 °C (−279 °F) at about 90 km (55 mi). Above this is the **thermosphere**, which extends about 480 km (300 mi) into space. The density of gas molecules in the thermosphere is very low indeed. At the boundary between the mesosphere and thermosphere, atmospheric pressure is only 0.0003% of that at sea level. Above about 150 km (95 mi), the gases are so thin that they create virtually no friction with spacecraft. The International Space Station orbits in the thermosphere at an altitude between 320 and 380 km (200–235 mi).

As the name thermosphere suggests, the gases in this layer are heated by direct solar radiation to very hot temperatures, exceeding 2000 °C (3600 °F). However, these superheated molecules are so far apart that when astronauts spacewalk in the thermosphere, they do not feel warm. Many gases in the thermosphere are ions, which have electrical charge. These ions interact with the solar wind and Earth's magnetic field to produce the auroras, or northern and southern lights (Figure 3.37).

▲ **Figure 3.37 The Auroras** In polar regions, ions from the thermosphere interact with Earth's magnetic field and glow.

Water in the Atmosphere
Water content of the atmosphere varies in a predictable fashion.

The amount of water in the atmosphere varies greatly from place to place and from time to time. On average, water vapor (H_2O) makes up about 1% of the molecules in Earth's atmosphere. At sea level, the **vapor pressure** of water, its relative contribution to total atmospheric pressure, is about 10 mb, or 1% of 1,013 mb. However, the vapor pressure of water in air ranges from as little as 0.01 mb to over 40 mb.

At any particular temperature, the air can hold only so much water. That amount is the **saturation vapor pressure** for that temperature. Above the saturation vapor pressure, water condenses as into a liquid, forming droplets of fog or rain. Saturation vapor pressure—the amount of water the air can potentially hold—increases as the temperature increases (**Figure 3.38**).

Relative humidity is a measure of the extent to which air is saturated, expressed as a percentage. The relative humidity of a mass of air rises and falls with its temperature (**Figure 3.39**). Changes in relative humidity explain the condensation that forms on the outside of a pitcher of iced tea on a hot, humid day. As air circulates near the pitcher, it is cooled. This causes its saturation vapor pressure to drop and its relative humidity to increase. When the air immediately adjacent to the pitcher reaches 100% humidity, droplets of water condense on the glass.

As an air mass becomes cooler, it eventually reaches its **dew point**, the temperature at which the relative humidity is 100%. If the temperature drops to the dew point overnight, water condenses on surfaces, such as your lawn or car. As the temperature warms in the morning, the relative humidity drops and the dew evaporates.

Relative humidity affects the rate at which water evaporates, or changes from a liquid to water vapor. As humidity increases, evaporation occurs more slowly. For example, on a day with low humidity, your perspiration evaporates quickly, cooling your body. On a humid day, your perspiration evaporates more slowly. That's why hot, humid days can seem more uncomfortable than hot, dry days. Similarly, the loss of water from the surfaces of the ocean, lakes, and plant leaves slows down as relative humidity rises.

These transformations of water are especially important in the troposphere, where they influence variations in rainfall and the evaporation across Earth's surface, thereby playing an important role in determining climate. Because of the importance of water in all ecosystems, variations in the amount of water in the atmosphere are of particular interest to environmental scientists.

QUESTIONS

1. The small amount of CO_2 in Earth's atmosphere is evidence of life. Explain.
2. Describe how temperature changes through each of Earth's four atmospheric layers.
3. What is the relationship between air temperature, saturation vapor pressure, and dew point?

▼ **Figure 3.39 Temperature and Relative Humidity** Daily change in temperature and relative humidity at Sacramento, California, over 3 days in July 2009. *Source: U.S. National Oceanographic and Atmospheric Administration, National Weather Service 2009.*

◄ **Figure 3.38 Saturation Pressure** Water vapor pressure is the amount of atmospheric pressure in millibars that is due to water molecules. The saturation vapor pressure, the amount of water that air can hold, increases with increasing temperature.

3.7 Earth's Energy Budget, Weather, and Climate

On average, the amount of energy that our Earth receives from the sun is equal to the amount that it reflects or radiates back into space. However, at different times, energy gain and loss in particular locations may not be in balance. These regional differences produce differences in the temperature of land and water surfaces and the air above them, which in turn affect conditions of weather and climate, such as temperature, relative humidity, and rainfall.

Earth's Energy Budget

Earth's energy budget is balanced overall.

Every minute, solar radiation equivalent to 25×10^{17} calories strikes Earth's upper atmosphere. The majority of solar radiation arrives as visible light, although there is also considerable UV radiation (**Figure 3.40**). We know from the first law of thermodynamics that energy cannot be created or destroyed (3.3). So where does all this energy go?

The fate of this energy is described by an **energy budget**, a system of accounting that measures all the energy entering and leaving Earth (**Figure 3.41**). Of the total radiation striking Earth, 30% is reflected back into space by the upper atmosphere, clouds, and Earth's surface. Clouds are especially important, accounting for two-thirds of the reflected energy. The remaining 70% of solar energy is absorbed by the atmosphere, land, and water, increasing the kinetic energy of the molecules in these substances and making them warmer. Heat gained in this fashion is eventually radiated back into space as infrared radiation.

▲ **Figure 3.40 The Solar Spectrum**
In addition to visible light, indicated by the rainbow of colors, the solar energy striking the outer reaches of Earth's atmosphere contains significant amounts of UV and infrared radiation.

Reflected by atmosphere 6%

Reflected by clouds 20%

Reflected from Earth's surface 4%

Incoming solar energy 100%

Absorbed by atmosphere 16%

Absorbed by land and oceans 51%

▶ **Figure 3.41 Earth's Energy Budget**
Over the entire globe and throughout the year, the inflow of energy from the sun (yellow) equals the amount of energy radiated and reflected to outer space (red).

Over the course of a year, the energy budget of the entire Earth is balanced—the total amount of energy coming in is equal to the total energy that is reflected or radiated back into space. However, at any particular place or time, the energy budget may be out of balance. The most familiar example of variation in energy balance is the daily temperature cycle. During the daytime, sunlight warms the atmosphere, land, and water, causing a net heat gain. At night, that heat is lost to space through radiation. Differences in the energy budget from region to region are the primary reason for the variety of climates around the world.

Weather and Climate

Local and regional variation in temperatures produces weather and climate.

Climate refers to atmospheric conditions such as temperature, humidity, and rainfall that exist over large regions and relatively long periods of time. In contrast, **weather** refers to short-term variations in local atmospheric conditions. When we say that Los Angeles will experience a cool, foggy morning and hot afternoon, we're talking about weather. When we say that Southern California has mild, moist winters and hot, dry summers, we are describing the climate of this region.

Changes in weather occur as a consequence of changes in the temperature and pressure of air masses. For example, on a sunny summer day, humid air becomes warmer. When this happens, the air expands; because it is less dense, it rises. As the air rises in altitude, it begins to cool. When the air mass cools to its dew point, water condenses, forming a cloud and perhaps a rain shower (**Figure 3.42**). As air cools, it becomes denser, until it eventually begins to sink. As it sinks, it is warmed and its relative humidity decreases.

In areas where air is rising, atmospheric pressure tends to fall. Thus, a drop in air pressure is often associated with an increased likelihood of rain. Conversely, where air is descending, atmospheric pressure rises, usually indicating dry conditions.

Local variations in heat gain and loss produce large areas of relatively warm or cold air. Such air masses may meet along boundaries called fronts. A cold front is a boundary where a mass of cold air is replacing a mass of warm air. Along a warm front, warm air is replacing cold air. In both cases, warm air is pushed up over denser cold air. If the warm air contains sufficient moisture, it may produce rain as it is cooled.

▲ **Figure 3.42 Thunderhead**
Air warmed over the ocean surface rises in this cloud. As it rises, it cools to the dew point, producing a rain shower.

What role does our ocean play in the weather and weather patterns?

Christopher Shad, Cypress College

Norm *Oceans affect nearly every aspect of our weather. They influence rainfall patterns because they are the source of most of the water in the atmosphere. Because they absorb solar radiation and circulate heat, they influence temperature patterns worldwide.*

Absorbed by clouds 3%

Conduction and rising air 7%

Carried to clouds and atmosphere by latent heat in water vapor 23%

Radiated to space from clouds and atmosphere 64%

Radiated directly to space from Earth 6%

Radiation absorbed by atmosphere 15%

Wind Cells

Heat from the equator is transferred toward the poles by convection cells in the troposphere.

Because Earth's surface is curved, the amount of sunlight reaching its surface varies with latitude. The amount of radiation striking a square meter of land is highest at the equator and diminishes toward the poles (Figure 3.43). The uneven distribution of solar radiation results in significant differences in air temperature and pressure around the globe. These differences cause air in the troposphere to circulate in large-scale convection currents, called wind cells. The climates of the world are greatly influenced by these wind cells.

At the equator, Earth's surface and the atmosphere immediately above it experience the greatest daily heating. This produces a band of rising air around the equator called the **intertropical convergence zone** (Figure 3.44). As this warm, humid air rises, it cools, and water condenses, forming rain. Some of the wettest tropical rainforests are found in this zone. The air then diverges northward or southward, forming two convection cells, called **Hadley cells**, on either side of the equator. At about 30 to 35 degrees latitude, the air is cooled and descends.

This produces a zone of low relative humidity; many of the world's driest deserts, such as the Sahara, are found in this zone. Similar convection cells form in each hemisphere at latitudes between 30 and 60 degrees and 60 and 90 degrees; these are the **Ferrel cells** and **polar cells**, respectively.

The air circulating in the wind cells interacts with Earth's rotation to cause general wind patterns. As Earth spins on its axis, areas near the equator spin faster than areas near the poles. If you were standing on the equator, you would travel 40,000 km (24,900 mi), Earth's circumference, in a day. If you moved toward the pole, the distance you traveled in a day would decrease. Once you reached the pole, you would not travel any distance, but simply spin in place. Because the ground is moving at different speeds, air moving over it in a straight path toward the equator appears to be deflected from east to west. Air moving toward the poles appears to be deflected to the east.

▼ **Figure 3.44 Wind Cells**
Uneven heating of Earth creates atmospheric convection in the troposphere that creates three distinct wind cells in each hemisphere.

Polar cells
Circumpolar zone
Ferrel cells
Temperate zone
60°
Hadley cells
Tropical zone
25°
Direction of Earth's rotation
0° Equator
Intertropical convergence zone

North Pole

Low-angle incoming sunlight

B

A

Equator

Sunlight directly overhead

More sunlight per unit of surface area

B

Low-angle incoming sunlight

Less sunlight per unit of surface area

South Pole

▲ **Figure 3.43 Solar Angle**
The angle at which the sun's radiation strikes Earth increases, and the amount of sunlight received per unit area decreases from the equator to the poles. Solar energy at the equatorial regions is concentrated in a smaller surface area. Solar energy at the polar regions is spread out over a larger surface area.

The apparent change in wind direction due to Earth's rotation is called the **Coriolis effect**.

Because the air near the ground in the Hadley and polar cells is moving toward the equator, the winds seem to be deflected from east to west. Winds that blow from the east are said to be easterly. Thus, regions located within the Hadley and polar cells have prevailing easterly winds. In contrast, the air near the ground in the Ferrel cells is moving toward the poles, so it appears to be deflected from west to east, producing a westerly wind flow. Locations within the Ferrel cells experience prevailing westerly winds.

In zones where air is ascending or descending, there is no Coriolis effect and no prevailing winds. Winds in these zones are often weak and unpredictable. In the days of sailing ships, the lack of predictable winds in the intertropical convergence zone presented a significant obstacle. Ships attempting to cross the equator were often at a standstill. This region is often referred to as the doldrums.

The Seasons

Wind cells shift north and south over a year, influencing patterns of seasonal change.

Seasons occur because Earth is tilted on its axis by 23.5 degrees. Each hemisphere is tilted toward the sun during its summer and away from the sun during its winter (**Figure 3.45**). Solar energy per unit area is highest in the summer and lowest in the winter.

Seasonal changes in the amounts of radiation reaching different parts of the globe cause the positions of the wind cells to shift north and south. This, in turn, influences the nature of seasonal climate change. Los Angeles, California, for example, is located at about 35° north latitude. During the Northern Hemisphere winter, the city is influenced by the westerly winds of the Ferrel cell, which bring storms and rainfall from off the Pacific Ocean. In summer, the wind cells shift northward. Then the zone of descending air between the Hadley and Ferrel cells is located over Los Angeles, and little rain falls.

Large bodies of water moderate the effects of seasonal changes. Land masses gain and lose heat much faster than do bodies of water. Thus, during summers, land areas tend to gain more heat than the adjacent ocean, and locations in the center of continents become much hotter than locations near the coast. Large areas of hot rising air and low pressure develop over the middle of continents, such as North America and Asia. This low pressure pulls in moist air from adjacent oceans, creating conditions favorable for summer thunderstorms. In the winter, land cools much more rapidly than do adjacent waters. As a result, the coldest winter temperatures are experienced in locales at the center of continents.

Land-sea differences are also influenced by ocean currents. In general, ocean surface circulation tends to move relatively warm water along the east coasts of continents and cooler water along west coasts. As the temperature of the ocean surface increases, so does the amount of evaporation to the atmosphere above it. This, in turn, influences the amount of water vapor in air moving onto land. As a result, west coast locations are often drier than east coast locations at the same latitude.

March equinox: equator faces sun directly; neither pole tilts toward sun; all regions on Earth experience 12 hours of daylight and 12 hours of darkness

March

December

June

September

Summer solstice: Northern Hemisphere tilts toward sun; summer begins in Northern Hemisphere; winter begins in Southern Hemisphere

September equinox: Equator faces sun directly; neither pole tilts toward sun; all regions on Earth experience 12 hours of daylight and 12 hours of darkness

23.5°

Winter solstice: Northern Hemisphere tilts away from sun; winter begins in Northern Hemisphere; summer begins in Southern Hemisphere

▲ **Figure 3.45 The Seasons**
Through the year, the area of Earth receiving the greatest amount of solar radiation shifts, causing corresponding shifts in wind cells and weather that produce seasonal change.

For example, rainfall in the U.S. West Coast city of San Francisco is approximately 50 cm (20 in) per year. Washington, D.C., which sits at nearly the same latitude on the East Coast, receives about 110 cm (42 in) of rain per year.

In summary, the annual variation in temperature and rainfall at a location is influenced by the seasonal movement of wind cells above it and its position on a continent. Locations near cold ocean currents tend to be drier than locations near warm ocean currents.

QUESTIONS

1. What features of Earth's atmosphere contribute to the energy that it reflects and radiates back to space?

2. What is the difference between climate and weather?

3. Why is rainfall often associated with cold and warm fronts?

4. The intertropical convergence zone shifts north and south throughout the year. Why?

Summary

Earth comprises a single large ecosystem. A variety of Earth's features are responsible for the abundant life that occurs here, but apparently nowhere else in our solar system. Earth contains the elements necessary for the structure and function of organisms and ecosystems. Unlike other planets in our solar system, Earth has large amounts of water necessary to sustain life. It receives solar energy in amounts sufficient to power the transformation of inorganic molecules to complex organic compounds and to sustain a variety of climates that are favorable to life. Earth's liquid core generates the magnetic field that surrounds it and deflects particles in the solar wind that would otherwise destroy life. Earth's atmosphere filters most of the UV light in the solar spectrum. Its mobile crust has reshaped and relocated continents, thereby influencing the evolution of Earth's life and ensuring continuous provision of elements necessary to sustain ecosystems. Earth is indeed a wonderful place to live!

3.1 Chemistry of the Environment

- Elements are substances that cannot be broken down or separated into other chemicals. Oxygen, hydrogen, and carbon are 3 of the 92 naturally occurring elements. Molecules are composed of two or more atoms of the same or different elements.

- The unique properties of water are essential to the structure and function of organisms and ecosystems.

KEY TERMS

element, atom, molecule, proton, neutron, electron, mass, isotope, radioactive, half-life, compound, covalent bond, ion, ionic bond, dipole bond, solubility, pH scale

QUESTIONS

1. Hydrogen is the simplest element; it has an atomic number of 1. There are three hydrogen isotopes: ^{1}H, ^{2}H, and ^{3}H. Explain how these isotopes differ with respect to their number of protons, neutrons, and electrons.

2. ^{3}H, the hydrogen isotope called tritium, has a half-life of 12.32 years. What does this mean?

3. Based on the mass of its molecules, water should boil at temperatures well below its freezing point. What accounts for its high boiling point?

- 14 — 1.0 M NaOH
- 13
- — Household ammonia
- 12
- 11
- — Milk of magnesia
- 10
- — Baking soda
- 9
- 8 — Seawater

Neutral
$[H^+] = [OH^-]$
- — Human blood
- 7 **Pure water**
- — Milk
- 6
- 5
- — Tomatoes
- 4
- — Wine
- 3 — Vinegar, colas
- — Lemon juice
- 2
- — Stomach acid
- 1
- 0   pH scale

Increasingly basic (lower H$^+$ concentration)
Increasingly acidic (greater H$^+$ concentration)

3.2 The Organic Chemistry of Life

- Organic molecules such as carbohydrates, lipids, proteins, and nucleic acids are essential to the functions of organisms.

KEY TERMS

organic molecule, inorganic compound, hydrocarbon, carbohydrate, sugar, lipid, macromolecule, polymer, polysaccharide, starch, cellulose, protein, catalyst, enzyme, nucleic acid, deoxyribonucleic acid (DNA), gene, genome, ribonucleic acid (RNA), transcription, translation

QUESTIONS

1. Describe three different kinds of carbohydrates.

2. The proteins of all organisms are composed of the same 20 amino acids, but there are thousands of different kinds of proteins. What makes the proteins different from one another?

3. What is the general relationship between a gene and the protein that it produces?

4. Describe an important role of RNA in the conversion of the DNA code to protein.

3.3 Energy and the Environment

- Energy describes the capacity of a system to do work. Potential energy is stored in a system; kinetic energy is the energy of motion itself.

- Energy can neither be created nor destroyed, but it can be transformed from one form to another. When it is transformed, the entropy of the universe increases.

KEY TERMS

energy, work, potential energy, kinetic energy, first law of thermodynamics, second law of thermodynamics, entropy, heat, electromagnetic radiation, electromagnetic spectrum, temperature, conduction, convection, radiation, latent heat transfer, chemical energy, nuclear energy, nuclear fission, nuclear fusion, joule, calorie, watt-hour (Wh)

QUESTIONS

1. You move a book from your desk to a high shelf. Describe the change that occurs in this system in terms of kinetic and potential energy and work.

2. The potential energy represented in the book poised above your desk is considerably less than the energy you expended to put the book up there. What happened to the rest of the energy?

3. Photons of gamma rays, X-rays, and UV light are damaging to living systems. Why?

4. Cats lick themselves to cool off on a hot day. Explain how this works.

5. Gas atoms in the atmosphere are constantly colliding, but these collisions do not cause nuclear fusion to occur. Why not? What would be required to initiate fusion?

D Latent heat transfer
B Convection
C Radiation
A Conduction

3.4 The Planet Earth in Context

- Our solar system formed approximately 4.6 billion years ago; it comprises the sun, eight planets, and thousands of planetoids.
- Only Earth has been found to have an environment that can support life as we know it.

KEY TERMS

planetoids, gas giant, terrestrial planet

QUESTIONS

1. Describe the reasons why scientists think it is unlikely that life will be found on Jupiter, Pluto, or Mercury.
2. What features of Mars and Venus make them favorable for the development of life? What features would make it difficult for these planets to support life as we know it?

3.6 Earth's Atmosphere

- Earth's atmosphere is composed mostly of nitrogen (N_2) and oxygen (O_2), with smaller amounts of other gases. It is divided into four distinct layers.
- Vapor pressure is the contribution of water vapor to total atmospheric pressure.

KEY TERMS

atmosphere, atmospheric pressure, troposphere, stratosphere, ozone layer, mesosphere, thermosphere, vapor pressure, saturation vapor pressure, relative humidity, dew point

QUESTIONS

1. Life does not exist on Earth because of its unique atmosphere; rather, its unique atmosphere is a consequence of life. Explain this statement.
2. Why does atmospheric pressure decline with increasing altitude?
3. Gases in the troposphere are constantly mixed, but those in the stratosphere are not as well mixed. Why?
4. On a hot, humid day, water may condense on the outside of windows of an air-conditioned building. Explain this phenomenon using the terms vapor pressure, saturation vapor pressure, relative humidity, and dew point.

3.5 Earth's Structure

- The solid portion of Earth is divided into a dense inner core, a mantle composed of less dense rock, and an even less dense thin outer crust.
- The location and configuration of Earth's continents have changed considerably over its history.

KEY TERMS

core, mantle, magma, crust, lithosphere, tectonic plate, transform fault boundary, divergent boundary, convergent boundary

QUESTIONS

1. Although Earth's magnetic field is generated deep in its core, it has important consequences for life on the surface. Explain.
2. Some continental rocks are nearly 4 billion years old, but nowhere is ocean crust older than 200 million years. Why is this so?
3. What is the source of the energy that moves Earth's tectonic plates?
4. What does the cratered surface of our Moon tell us about tectonic activity there?

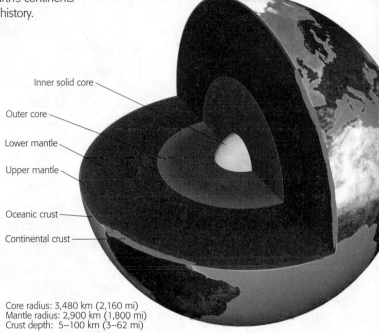

Inner solid core
Outer core
Lower mantle
Upper mantle
Oceanic crust
Continental crust

Core radius: 3,480 km (2,160 mi)
Mantle radius: 2,900 km (1,800 mi)
Crust depth: 5–100 km (3–62 mi)

3.7 Earth's Energy Budget, Weather, and Climate

- Over a year, Earth's energy budget is balanced; the amount of energy that it receives from the sun equals the amount that is reflected or radiated back into space.
- Climate refers to atmospheric conditions such as temperature, humidity, and rainfall over large regions and long times. Weather refers to short-term variation in those conditions.

KEY TERMS

energy budget, climate, weather, intertropical convergence zone, Hadley cell, Ferrel cell, polar cell, Coriolis effect

QUESTIONS

1. In particular places and at certain times of the year, the energy budget may not be balanced. Describe three factors that can produce such imbalances.
2. Many of the world's driest deserts are located at about 30–35° latitude. Why?
3. The "solar equator" is defined as the line around the Earth along which the sun's radiation is most direct. How does this change relative to the actual equator through the year?
4. How does the temperature of the ocean affect patterns of rainfall on adjacent lands?

Polar cells
Circumpolar zone
Ferrel cells
Temperate zone
Hadley cells
Tropical zone
Direction of Earth's rotation
Intertropical convergence zone
0° Equator

Mastering ENVIRONMENTALSCIENCE®

Go to **www.masteringenvironmentalscience.com** for homework assignments, practice quizzes, Pearson eText, and more.

4

Organism and Population
Ecology and Evolution

91

Genetic Change and Population Growth— Fact and Fiction

Could mutant organisms take over the world?

Mutant creatures run amok is the theme of the 1972 horror film *Night of the Lepus* (**Figure 4.1**). In it, scientists experimenting with a laboratory rabbit cause a mutation that interrupts its breeding cycle. And wouldn't you know it? The bunny gets away! Much to everyone's horror, the bunny quickly grows to a monstrous size and begins a reproductive frenzy. Matters only get worse when people discover that this mutant prefers the taste of human flesh to carrots!

Fortunately, few of us lie awake at night in fear of ravenous rabbits. But a few pieces of this story have some basis in reality.

► **Figure 4.1 Horrors Indeed**
Genetic change and explosive population growth are central themes for many horror flicks. *Lepus* is the scientific name for the subgroup of rabbits called hares.

▼ **Figure 4.2 Population Explosions**
Within 10 years of the introduction of a few rabbits to Australia, their populations exceeded millions, devastating the resources upon which they fed.

(Be assured, a humongous, human-hungry hare is not one of those pieces.)

Consider, for example, the explosive growth of the population of European rabbits in Australia. When European settlers arrived, there were no rabbits in Australia. Rabbits were first introduced in 1859 for hunting purposes. With abundant food and few natural predators, their numbers quickly increased. In just a couple years, there were thousands of rabbits. By 1870, hunters were killing 2 million rabbits each year, but with no apparent effect on their population. In 1950, Australia's rabbit population was estimated to be 600 million. These rabbits were eating everything they could find, destroying native plants and leaving little for the native animals to eat (**Figure 4.2**). To try to control their numbers, the government introduced rabbits infected with a deadly viral disease. The virus spread rapidly, reducing the population by about 60%.

The threat of genetic changes depicted in *Night of the Lepus* is overblown. But the remarkable advances in the field of genetic engineering since the movie's release lend a measure of credibility to the dangers caused by releasing genetically altered species into the environment. The growth and reproduction of individual organisms are determined by genes encoded in their DNA. More often than not, changes in the DNA code are harmful to the organisms that possess them. For example, ecologists in the Great Lakes region have recently found a large number of frogs with reproductive abnormalities and serious deformities, such as extra legs and eyes (**Figure 4.3**). Some of these changes appear to be directly linked to changes in the frogs' DNA caused by increasing amounts of pesticides in their environment.

Genetic changes in individuals within a population do sometimes proliferate with significant environmental consequences. For example, the insecticide DDT was first used in 1941 to control populations of the mosquitoes that spread malaria and typhoid; later it was used to control agricultural pests. It was noted at the time that very small numbers of these insects were more resistant to DDT than other members of their populations. Over the next two decades, the relative abundance of insects that were resistant to DDT increased dramatically. To learn the fate of the mythical mutant bunnies and their would-be prey, you will need to see the movie. In this chapter, we will focus on biological processes that take place in real organisms and real populations. We will investigate how organisms acquire energy and how that energy supports their growth and reproduction. We will then explore the factors that accelerate and limit the growth of populations. We will discuss the origin of inherited changes, their propagation in populations, and their role in the evolution of new species. Finally, we will consider how evolutionary change shapes our understanding and classification of the diversity of life on Earth.

▶ Figure 4.3 **Environmental Factors Can Alter DNA**
Environmental factors such as ultraviolet light or chemical pollutants can cause changes in an organism's DNA that produce abnormalities. Scientists suspect that insecticides may be responsible for abnormal limb growth observed in frog populations in the Lake States.

4.1 The Cell—The Fundamental Unit of Life

The cell is the basic unit of structure and function of all life. The cells of plants and some bacteria are able to use light energy to synthesize high-energy carbohydrates from carbon dioxide and water. Other kinds of bacteria are able to manufacture carbohydrates by using energy released from inorganic chemical reactions. The cells of all organisms—microorganisms, fungi, plants, and animals—depend on carbohydrates to power their functions, including cell growth and division.

Cell Structure

Cell structure is the basis for distinguishing organisms into two major groups.

The most ancient evolutionary relationships among organisms can be inferred from the structure of their cells. Biologists use cell structure to classify organisms into one of two groups—prokaryotes and eukaryotes (**Figure 4.4**). Both groups have cells that are made up of a material called **cytoplasm**. Around the cytoplasm is a plasma membrane that regulates materials flowing in and out of the cell.

Prokaryotes, which include bacteria and their relatives, are thought to resemble the earliest cellular life on Earth. Compared to the cells of eukaryotes, the cells of prokaryotes are simple. Their cells lack a membrane-enclosed nucleus, and their DNA occurs as a single chromosome in the cytoplasm. There is great diversity among the prokaryotes, with species that perform a multitude of functions in ecosystems, including photosynthesis, the decomposition of wastes, and the cycling of essential elements, such as nitrogen and sulfur.

Eukaryotes vary in complexity from single-celled algae and protozoa to plants, fungi, and animals that are made of millions of cells. The cytoplasm of eukaryotes contains a number of membrane-enclosed **organelles** ("little organs") that perform specialized functions. The most prominent eukaryotic organelle is the nucleus. The nucleus contains most of the cell's DNA, which is organized into several distinct chromosomes. All eukaryotes have mitochondria, organelles that process oxygen and food. The cells of plants and many algae have chloroplasts, organelles that use sunlight, carbon dioxide, and water to produce carbohydrates and oxygen.

▼ **Figure 4.4 Eukaryotic and Prokaryotic Cells**
All cells are bounded by a cellular membrane. Eukaryote cells contain a nucleus and several different membrane-enclosed organelles. Prokaryote cells lack organelles.

Prokaryotic cell

Cytoplasm
Plasma membrane
Cell wall
Nucleoid region (DNA)

Animal Eukaryotic cell | Plant Eukaryotic cell

Nucleus
Plasma membrane
Mitochondrion
Cytoplasm

Mitochondrion
Nucleus
Cell wall
Chloroplast
Plasma membrane

Chemical Functions

The cell is the site of life's most important chemical functions.

Plants, some eukaryotic algae, and prokaryotic blue-green algae called *cyanobacteria* nourish themselves through the cellular process of **photosynthesis**. Photosynthesis requires sunlight, carbon dioxide (CO_2), and water (H_2O). Plants and algae absorb CO_2 from the atmosphere and extract H_2O from soil, ocean water, or fresh water.

The chloroplasts in plant cells contain pigments, such as chlorophyll, that are able to capture the energy of light (3.3). Similar pigments are found in the cytoplasm of cyanobacteria. When sunlight strikes these pigments, they absorb photons of light, increasing the energy level of electrons in the pigments (3.2). Through a series of chemical reactions, the energy from these electrons is used to combine CO_2 and H_2O to form carbohydrate molecules, such as glucose ($C_6H_{12}O_6$) (**Figure 4.5**).

Carbohydrates power the cellular processes of plants and algae and allow these organisms to grow, as well as all the organisms that feed on them. Photosynthesis also produces O_2. Indeed, all of the O_2 in Earth's atmosphere is a by-product of photosynthesis (3.6).

In some environments where there is no light, certain kinds of bacteria obtain food through the process of **chemosynthesis** (**Figure 4.6**). For example, in the dark depths of the ocean, chemosynthetic bacteria take in CO_2, O_2, and hydrogen sulfide (H_2S).

Through a series of chemical reactions, bacteria convert those molecules to carbohydrates, sulfur (S), and H_2O. The conversion of H_2S to S provides the energy for this process. In such dark environments, chemosynthesis produces the carbohydrates used for the growth and nourishment of all the organisms in the ecosystem.

In **cellular respiration**, the energy in carbohydrate molecules is retrieved and used to carry out cell functions and facilitate growth (**Figure 4.7**). Cellular respiration takes place in the cytoplasm of prokaryotes and in the mitochondria of eukaryotes. In cellular respiration, cells break apart carbohydrate molecules in a sequence of chemical reactions that produces energy, H_2O, and CO_2. The formation of new molecules in these chemical reactions requires existing molecules to either take on electrons or lose them. This movement of electrons provides the chemical energy for all other cell functions. In the final step of cellular respiration, the O_2 molecules pick up electrons and hydrogen ions to produce H_2O molecules.

Nearly all organisms use the energy from cellular respiration to power their physical, behavioral, and reproductive functions—all of their life processes. Note that plants as well as animals carry out cellular respiration. The many steps in cellular respiration also generate a variety of different chemicals for building the macromolecules—proteins, starch, cellulose, nucleic acids, and lipids—that are essential for growth and reproduction (3.2). These complex macromolecules may be converted back to simple carbohydrates for use in cellular respiration.

Cellular respiration requires oxygen. But in some environments there is little or no oxygen; such environments are said to be *anaerobic*. Because they cannot carry out respiration, most plants and animals are unable to survive in anaerobic environments.

▶ **Figure 4.5 Photosynthesis**
The plant takes in CO_2 from the air and light energy from the sun. CO_2 and light energy react with the H_2O in the chloroplast of the plant cell. The reaction produces carbohydrate.

Many bacteria and fungi, however, are able to survive and grow by using **anaerobic respiration**. In this process, carbohydrates are partially broken down to produce carbon dioxide and smaller carbohydrate molecules, such as methanol or ethanol (**Figure 4.8**). Because the breakdown of carbohydrate molecules is incomplete, anaerobic respiration of carbohydrates yields only about 10–20% of the energy that is released in cellular respiration. The fact that methanol and ethanol can be used as engine fuels is evidence that anaerobic respiration is unable to take advantage of all of the potential chemical energy in carbohydrate molecules.

In summary, nearly all the chemical processes that are necessary to support life take place in cells. In all organisms, the energy that is required for these processes is generated by the stepwise breakdown of carbohydrates in cellular respiration. Plants, some algae, and cyanobacteria are able to manufacture carbohydrates from CO_2 and water using light energy (photosynthesis); some bacteria derive carbohydrates from inorganic chemical reactions (chemosynthesis). Animals obtain carbohydrates by eating plants or other animals that eat plants.

Plant

Plant cell

O_2

$C_6H_{12}O_6$
(carbohydrate)

H_2O

CO_2

QUESTIONS

1. Describe two differences between the cells of prokaryotes and eukaryotes.

2. Compare the processes of chemosynthesis and photosynthesis.

3. Give two reasons why cellular respiration is essential for all life.

H_2S (hydrogen sulfide)

CO_2

O_2

Hydrothermal vent

$C_6H_{12}O_6$ (carbohydrate)

H_2O

S

Chemosynthetic bacteria

▲ **Figure 4.6 Chemosynthesis**
Bacteria take CO_2 and O_2 from the water and H_2S from the vents. Chemosynthesis produces glucose ($C_6H_{12}O_6$) that is used as energy for the bacteria and water (H_2O) and sulfur (S) that are released into the seawater.

Plant cell

Animal cell

O_2

$C_6H_{12}O_6$
(from photosynthesis in plants or from digestion in animals)

Mitochondrion

CO_2

H_2O

Energy

Movement of electrons

▲ **Figure 4.7 Cellular Respiration**
In cellular respiration, carbohydrate molecules such as glucose are broken down to produce CO_2 and H_2O and release energy that is required for organism growth and reproduction.

Yeast fermenting beer in a vat

$C_6H_{12}O_6$ (carbo-hydrate)

CO_2

H_2O

Ethanol

Some energy

▲ **Figure 4.8 Anaerobic Respiration**
The yeast takes in carbohydrate ($C_6H_{12}O_6$). Anaerobic respiration produces energy for the yeast. Carbon dioxide (CO_2), water (H_2O), and ethanol are by-products that are released.

4.2 The Growth and Reproduction of Organisms

The growth and reproduction of organisms ultimately depends on cell growth and division. To grow and reproduce, organisms need the energy and materials stored in carbohydrates. Reproduction enables an organism to multiply and pass along its genes to the next generation. Reproduction may be asexual or sexual. **Asexual reproduction** is accomplished by simple cell division and produces offspring that are genetically identical to their single parent. **Sexual reproduction** involves mating between two parents and produces offspring that are genetically unique.

Cell Division and Differentiation

An organism grows through the division and differentiation of its cells.

For an organism to grow and develop, it needs a steady supply of carbohydrates. An organism may obtain those carbohydrates through the process of photosynthesis or chemosynthesis, by eating other organisms, or by ingesting the wastes of other organisms. Cells grow larger as they acquire carbohydrates and manufacture the macromolecules that make up cell organelles.

When cells are mature, they may divide to form new cells. The process of cell division ensures that each newly created cell receives a complete set of organelles and genes (3.2).

Before cell division, the DNA in the parent cell is precisely duplicated, so each new cell inherits an exact copy of its parent cell's DNA (**Figure 4.9**).

In complex, multicellular plants and animals, cells differentiate into tissues, such as skin and muscle, that perform specialized functions. All of the cells in an organism's tissues contain exactly the same DNA. Tissues differ from one another because the cells in each tissue use information that is activated from different portions of the organism's DNA.

(A) (B) (C) (D)

▲ **Figure 4.9 Cell Division**
During cell division, the cell's DNA is duplicated and condenses into distinct chromosomes Ⓐ and Ⓑ. A complete set of chromosomes and DNA then migrates to each of the opposite poles of the cell Ⓒ. In the final stage, a new cell membrane forms between the segregated chromosomes creating two new cells Ⓓ. Each new cell has DNA that is identical to the parent cell.

Sexual and Asexual Reproduction

The mode of reproduction affects the amount of variation among offspring.

Reproduction allows organisms to multiply, producing offspring that carry some or all of their genes. Some organisms reproduce sexually, and others reproduce asexually.

In asexual reproduction, a single parent produces offspring that are exact replicas, or clones, of their parent. In other words, the DNA of the offspring is identical to that of their sole parent (**Figure 4.10**). Most single-celled organisms and organisms with only a few cells reproduce asexually through simple cell division. For example, a single-celled amoeba reproduces by dividing into two amoebas. In more complex organisms, cells may divide to form buds that yield clones. Some plants reproduce asexually by means of runners, bulbs, or tubers.

Most plants and animals multiply by sexual reproduction, in which males and females of the same species mate to produce genetically unique offspring. These offspring possess DNA contributed by both parents. Sexual reproduction is preceded by a form of cell division that produces **gametes**, cells that are specialized for reproduction. Female gametes are called eggs, and male gametes are called sperm. Each gamete carries one complete copy of DNA. However, the DNA of every gamete is unique because it includes portions of each of its parent cell's two copies of DNA.

The union of an egg and a sperm produces a new, genetically unique cell called a **zygote**.

Fully developed hydra

Developing bud

Developing cell

Fully developed yeast cell

Scar from previous bud

Runner connecting budding offspring to parent strawberry plant

Budding offspring

Runner

Budding offspring

ⒸC

▲ **Figure 4.10 Asexual Reproduction**
This process produces offspring that are identical to one another. In simple organisms like yeast, Ⓐ this may involve simple cell division. Asexual reproduction occurs in some animals such as the sea anemone Ⓑ and in many plants, such as when the runners of a strawberry plant Ⓒ take root and separate from their parent.

The zygote contains two copies of DNA, one contributed by each gamete. The zygote then undergoes simple cell division and differentiation to produce a new, genetically unique organism that shares some characteristics with each of its parents (**Figure 4.11**).

QUESTIONS

1. How does the DNA of a gamete compare to the DNA in the cells of the organism that produced it?
2. How does the DNA of a zygote compare to the DNA of the cells of its mother and father?

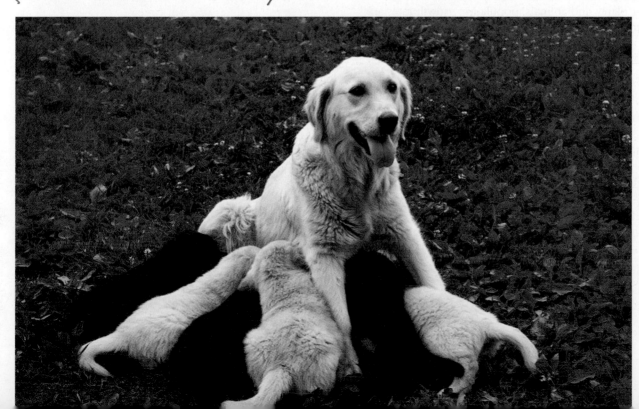

◄ **Figure 4.11 Sexual Reproduction Produces Diverse Offspring**
The puppies in this litter share the same parents, but each developed from a separate zygote. Therefore, they are genetically related but also genetically distinct from one another.

97

4.3 The Growth of Populations

If the organisms in a population are able to successfully grow and reproduce, the size of the population will increase. In a new, resource-rich environment, the growth of a population is usually exponential rather than arithmetic. A population's growth rate is determined by the rates of birth, death, and migration and how these factors vary over the lifetimes of individuals within the population.

Birth, Death, and Migration

Population growth is determined by the rate of birth, death, and migration.

Soon after a few individual organisms move into an environment that is rich in resources, their population will start to grow. In the first few weeks or months, the number of new individuals added to the population is small. But as the size of the population increases, so does the number of reproducing individuals. As a result, the growth of the population accelerates. When the growth of a population accelerates with each generation, the pattern of growth is said to be exponential. In **exponential growth**, the number of new individuals added to a population in each generation is a multiple of the number present in the previous generation. This is in contrast to **arithmetic growth**, in which the number of new individuals added at each generation is constant.

The multiple by which an exponentially growing population increases is called the **population growth rate**. Population growth rate is typically expressed as the percentage of change over a specific time interval. For example, if a population of 100 individuals grows to 110 individuals over a year, its growth rate is 10% per year. For rapidly reproducing organisms such as bacteria, population growth rates may be calculated for times as brief as an hour. For larger plants and animals, population growth rates are generally calculated for a year. Population growth rate may also be expressed as the **doubling time**, the length of time required for a population to double in size.

The doubling time of a population is approximately equal to 70 divided by the population growth rate (**Figure 4.12**).

A population's growth rate is a consequence of patterns of birth, death, and migration. Reproduction adds new individuals to the population. The **birth rate** is the number of births in the population per unit of time expressed as a percentage of the population size. Death subtracts individuals from the population. The **death rate** is the number of individuals dying per unit of time expressed as a percentage of the population size. Death rate is also referred to as **mortality rate**.

Migration describes the movement of organisms into and out of a population. The number of organisms moving into an area is calculated in terms of the **immigration rate**, the number of individuals entering the population per unit of time as a percentage of population size. **Emigration rate** is the number of organisms moving out of an area, calculated as a percentage of population size.

Thus, many factors affect the growth of a population. The total population growth rate is calculated as the birth rate plus the immigration rate minus the death rate and the emigration rate.

$$\text{Population Growth Rate} = (\text{Birth Rate} + \text{Immigration Rate}) - (\text{Death Rate} + \text{Emigration Rate})$$

▶ Figure 4.12 **Population Growth Rate and Body Size** Population growth rates are expressed here as percent change per year and doubling time in years. In general, growth rates are low and doubling times quite long for large compared to very small organisms.

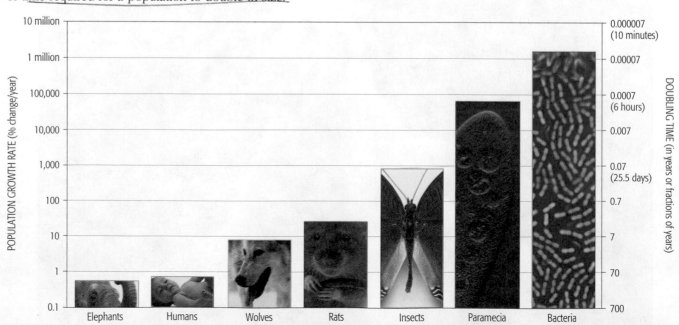

Exponential Population Growth: A Case Study

The introduction of European starlings to America provides a case study for exponential population growth.

In 1890, 60 European starlings were released in New York City for the dubious purpose of introducing to the United States all of the birds mentioned in the works of William Shakespeare. Within a decade, tens of thousands of the birds were foraging in the lawns of the city's parks. By 1920, the total population of European starlings had reached the millions, and the birds were common in all of the New England states. By 1950, hundreds of millions of European starlings were distributed across much of North America (**Figure 4.13**). Rapid, exponential growth in populations like that of the European starling is common when a species is introduced to an unexploited habitat with abundant resources.

Exponential growth can be illustrated by calculating the hypothetical growth of the population of starlings in New York City (**Figure 4.14**). A breeding pair of European starlings typically produces five offspring each year. Assuming that the original population of 60 birds was equally divided between males and females, by 1891 the population of starlings would have increased by 150 to total 210 birds.

30 females × 5 offspring = 150 offspring
150 offspring + 60 original birds = 210 starlings

Making the same assumptions for each successive year, the number of new European starlings added to the population each year would be equal to the number of breeding pairs multiplied by 5. By 1910, this hypothetical European starling population in New York would have far exceeded 4 trillion.

Although starlings are very common in New York, their numbers are, thankfully, only a very, very small fraction of this hypothetical prediction. Why are there so many fewer starlings in the real population than in our hypothetical calculation? Our calculation did not consider the rate at which starlings die. In reality, most starlings die during their first year, before they have the opportunity to reproduce. Fewer than 20% of starlings survive beyond two years. Furthermore, each year during this time interval many birds dispersed to regions outside New York City.

▲ Figure 4.13 Starlings Invade North America
The continental population of millions of starlings originated from a source population of 60 individuals released in 1890 in New York's Central Park.

1895
31,513 birds
15,756 breeding pairs
78,782 offspring

1894
9,003 birds
4,501 breeding pairs
22,509 offspring

1893
2,572 birds
1,286 breeding pairs
6,431 offspring

1892
735 birds
367 breeding pairs
1,837 offspring

1891
210 birds
105 breeding pairs
525 offspring

1890
60 birds
30 breeding pairs
150 offspring

◄ Figure 4.14 If Population Growth Were Driven by Birth Rate Only
In just 20 years, a hypothetical population of 30 male and 30 female starlings will increase to over 4.5 trillion birds. This example assumes that each breeding pair produces five offspring each year and that no deaths occur.

Survivorship and Fertility

The likelihood that an organism will survive and reproduce changes over its lifetime.

For European starlings, the chances of dying are relatively constant throughout their life—each year about 60% of the birds in the population die, regardless of their age. By comparison, the death rate for bald eagles is about 35% in the first year of life and less than 10% each year thereafter. Population biologists describe the probability of an organism dying during a particular time interval in terms of **survivorship**. Survivorship curves are constructed by plotting the percentage of individuals in a population that survive to each age (**Figure 4.15**).

Survivorship curves vary widely among species, but the ecological factors that determine survival result in three general patterns of survivorship. Organisms with *type I survivorship* are most likely to die of old age. Type I survivorship is common among large animals and predators that have few enemies. For organisms with *type II survivorship*, the probability of dying is the same at every age. Type II survivorship is observed in organisms such as the European starling, where predators and diseases affect all individuals in the population equally, regardless of their age. In organisms with *type III survivorship*, the very young have the greatest probability of dying because they are most easily taken by predators or are most vulnerable to disease. As organisms with Type III survivorship mature, they become more and more likely to survive to old age. Type III survivorship is by far the most common pattern of survivorship observed in nature.

The rate of reproduction, or **fertility rate**, also changes with age. For most species, there is an interval between birth and the age of first reproduction. European starlings, for example, do not begin to reproduce until they are at least one year old; bald eagles do not reproduce until they are at least three years old. Following first reproduction, the **age-specific fertility rate** describes the number of offspring produced by females during a particular range of ages. The age-specific fertility rate often increases to an optimum reproductive age and decreases thereafter. In some cases, such as in humans, there is an age beyond which reproduction ceases altogether.

The sum of age-specific fertility rates across all ages is the **total fertility rate**. The total fertility rate is the *potential* number of offspring that an average female in a population can produce if she survives to old age. For a population to grow, total fertility rate must exceed 2.0—the number of offspring needed to replace a female and her mate. If some females die before reaching old age, then the actual average number of births per female is less than the total fertility rate. The difference between the actual average number of offspring per female and total fertility rate is determined by differences in survivorship (**Figure 4.16**).

In general, population growth rate increases as total fertility rate increases. However, population growth rate is also influenced by **generation time**. The generation time of a population is the average difference in age between mothers and their offspring. Among populations with the same total fertility rates and survivorship, those with the shortest generation times will have the highest population growth rates.

QUESTIONS

1. Why is the growth of biological populations typically exponential rather than arithmetic?

2. What factors determine population growth rate?

3. What sorts of animals would you expect to have type III survivorship? Which would you expect to have type I survivorship?

4. Why is the actual average number of offspring born to females less than the total fertility rate?

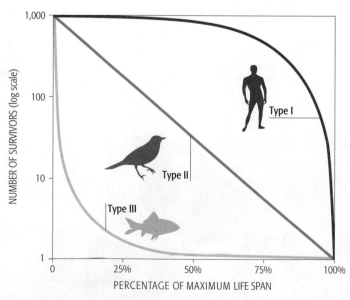

▲ **Figure 4.15 Patterns of Survival**
The risk of death is highest among old individuals in populations with type I survivorship. It is highest among young individuals in populations with type III survivorship. The risk of death is constant, regardless of age, in populations with type II survivorship.

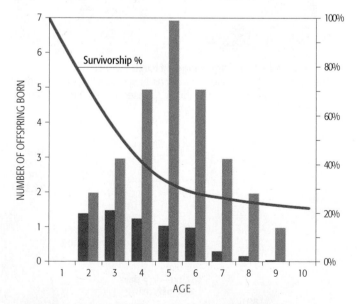

▲ **Figure 4.16 Potential and Actual Fertility Changes with Age**
The green bars represent the average number of births to females in each age class for a hypothetical population. Added together, values of these bars equal the population's total fertility rate. The orange bars represent the extent to which age class actually contributes to the overall birth rate, taking into account survivorship (orange line).

FOCUS ON SCIENCE: Dinosaur Survivorship

Is it possible to study the population dynamics of animals that have been extinct for over 63 million years? Such a task might appear hopeless, but Gregory Erickson of Florida State University does just that (**Figure 4.17**). To study long-gone populations, Erickson combines his expertise in the development and evolution of vertebrates with his lifelong fascination with dinosaurs.

◄ **Figure 4.17 Gregory Erickson** Florida State University scientist Gregory Erickson has coupled his interest in patterns of bone development in reptiles with his lifelong fascination with the ecology of dinosaurs.

Recently, Erikson investigated the fossils of a large carnivorous dinosaur, *Albertosaurus sarcophagus*. (**Figure 4.18**) (*Albertosaurus* was a close relative of *Tyrannosaurus rex*, made infamous in the *Jurassic Park* movies.) Twenty-two specimens of *Albertosaurus* were unearthed in the Barnum Brown Bone Bed, an aggregation of dinosaur fossils discovered in Alberta, Canada. Independent analysis of the Red Deer Bone Bed indicated that the dinosaurs found there had died over a period of months to years; they were not killed all at once in some sort of catastrophe.

Erickson thought that the *Albertosaurus* fossils might reveal the survivorship of these ancient carnivores. In earlier studies of large lizards and crocodiles, Erickson and other scientists had established that the lines seen in the cross-sections of bones represent yearly increments of growth. The number of lines in a bone shows the age of the animal (**Figure 4.19**). Erickson reasoned that the lines

▼ **Figure 4.18** *Albertosaurus sarcophagus* This close relative of *Tyrannosaurus* stalked prey in the forests of North America 65–83 million years ago. At 2.5 tons and nearly 25 feet from head to tail, it was one of the largest tyrannosaurs.

of growth in the *Albertosaurus* fossils could also be used to establish the age at which each dinosaur died.

By counting lines in the thighbones of each of the 22 *Albertosaurus* specimens, Erickson found that the age of the dinosaurs ranged from 2 to 28 years. Specimens younger than 2 years are not well preserved in the fossil record and so were unavailable to Erickson. Instead, he referred to other studies of large predators, which indicated a 50–80% death rate during the first years of life. For purposes of comparison, Erickson assumed that 60% of *Albertosaurus* young died before age 2. He then used the data he had amassed to construct a survivorship curve, estimating the percentage of individuals in the population surviving to various ages (**Figure 4.20**).

The results of Erickson's study reveal remarkably high survivorship: 70% of the *Albertosaurus* that were alive at age 2 survived to age 13. This pattern of survivorship is very similar to the type 1 survivorship of existing large animals, such as elephants and rhinoceroses. Erickson attributed the high survivorship of *Albertosaurus* to the fact that by age 2, they were already considerably larger than any other predators that might feed on them.

Erickson found that beyond age 13, the rate of survivorship began to decrease. This decreased survivorship corresponds to the time at which the dinosaurs approached maximum size and are thought to have become sexually mature. At that age, injuries and stress associated with mating and reproduction probably caused mortality to increase. The bones of the oldest (28 years old) specimen show clear signs of deterioration related to old age.

Erickson has recently expanded his analyses to include other members of the tyrannosaur family, including *Tyrannosaurus rex*. His other studies include analyses of bone development and growth rates in dinosaurs and birds and the mechanics of dinosaur and crocodile feeding.

Were the growth and survival of large dinosaurs similar to those of large animals that are alive today?

Younger ◄─────► Older

▲ **Figure 4.19 Aging a Dinosaur** Each line in this section of *Albertosaurus* thighbone fossil marks a year in the dinosaur's life and indicates that its growth slowed notably as the animal aged.

▲ **Figure 4.20** *Albertosaurus* **Survivorship** Gregory Erickson used the age of 22 specimens from the Red Deer Bone Bed to construct this survivorship curve for *Albertosaurus*. It is shown here in comparison with survivorship curves of several animals alive today.

Source: Erickson, G.M, P. J. Currie, B. D. Inouye, A. A. Winn. 2006. Tyrannosaur Life Tables: An Example of Nonavian Dinosaur Population Biology. *Science* 313:213–217.

4.4 Limits on Population Growth

As a population grows, the amount of unexploited resources available to it decreases. Consequently, competition for those resources increases. As the rate at which a population uses resources approaches the environment's capacity to supply those resources, birth rates decline, death rates increase, and the overall rate of population growth declines. When the rate at which a population uses resources equals the rate at which those resources are supplied, the birth rate equals the death rate, and the population growth rate becomes zero. The population size at which this occurs is called the **carrying capacity**. Thus, the rate at which resources are supplied limits the growth of a population. The growth of a population may also be limited by conditions other than its supply of resources, such as temperature or toxins in the environment. An organism's range of tolerance for many environmental factors determines where it can live.

Environmental Resistance Limits Growth

At carrying capacity, the birth rate equals the death rate and the population growth rate stalls at zero.

Population biologists have learned much from monitoring the populations of microbes, flour beetles, fruit flies, and other organisms in the laboratory. In controlled environments, essential resources can be supplied to a population at a constant rate. Factors affecting population growth, including birth rates and death rates, can be monitored precisely.

In the laboratory, changes in population size follow a predictable and repeatable S-shaped trajectory called a **logistic growth curve (Figure 4.21)**. In the beginning of logistic growth, a population grows rapidly. But as resources become increasingly limited, population growth slows until it eventually reaches the carrying capacity of its environment. At carrying capacity, the birth rate equals the death rate, and the population growth rate stalls at 0.0%. Also at carrying capacity, the rate at which resources are supplied to the population equals the rate at which they are used.

Consider an experiment in which a few bacteria are placed in a flask containing carbohydrates and all the other nutrients the bacteria need to grow; those resources are replenished at a constant rate. Initially, the rate at which the nutrients are supplied far exceeds the rate at which the small population can consume them. Conditions in the flask are favorable for the bacteria to grow and reproduce rapidly. In such ideal conditions, the population growth rate will be maximal and the size of the population will increase exponentially.

As the population of bacteria increases, the rate at which nutrients are consumed increases in direct proportion to the number of bacterial cells. Although the supply of nutrients remains constant, more bacterial cells are using them. The amount available to each individual bacterium declines, and competition among the bacteria for those resources increases. This intraspecific competition causes a decline in the rate of population growth and an increase in mortality.

Eventually, the rate of population growth begins to decline. If the population of bacteria continues to grow, then the rate of resource use will eventually equal the rate at which the resources are supplied. Populations of bacteria usually respond quickly to changes in the supply of resources. As resources diminish, reproduction slows down and bacteria begin to die. At this point, the population growth rate approaches zero, and the population stops growing.

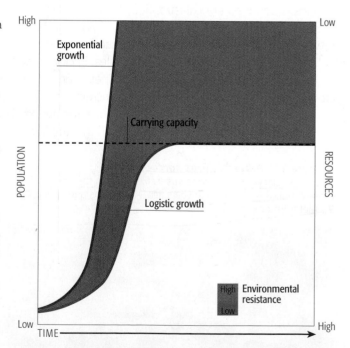

▲ Figure 4.21 **Logistic Growth**
As population size increases, resource limitations eventually slow population growth to zero, producing a logistic growth curve. The carrying capacity is the population size at which resource supply equals population resource use, and the population growth rate is zero.

Alternative Patterns of Population Growth

Birth and death rates may not respond immediately to changes in resource supply.

The smooth logistic growth curve shown by laboratory bacteria is typical of organisms with short generation times and simple life cycles. The birth rates and death rates of such organisms are tightly coupled to the supply of resources and change immediately in response to changes in that supply (**Figure 4.22A**).

In organisms that have extended or complex life cycles, the effects of limited resources on birth rate and death rate may be delayed. In such cases, populations may grow beyond their carrying capacity. Subsequent adjustment of birth rate and death rate to limited resources may cause the populations to fluctuate above and below carrying capacity. In some cases, populations may so far exceed their carrying capacities that resources become very limited. This eventually leads to a population "crash," or negative population growth, or even the disappearance of the population.

When populations exceed their carrying capacity, their use of resources exceeds the rate at which resources are supplied, and resources become scarce. The consequences of scarce resources depend on two factors: the speed with which the population growth rate responds to the diminished supply, and the rate at which the depleted resources are able to recover relative to the continued demands of the population.

Many birds and mammals have population growth rates that respond to diminished resources relatively quickly and depend upon resources that are renewed rapidly. When a population of these organisms overshoots its carrying capacity, the death rate soon exceeds the birth rate, the population growth rate becomes negative, and the population declines (**Figure 4.22B**). When the population declines, the demand for resources eases, and the supply of resources increases. As resources are renewed, the population's growth rate becomes positive, and the population begins to grow again. In such a scenario, the size of the population may oscillate above and below its carrying capacity.

If a population's response to diminishing resources is excessively delayed, or if the rate at which resources are supplied is much slower than the rate at which they are used, the oscillation in the population size becomes more extreme. Indeed, if resource use continues to outstrip resource supply, the population may "crash" and disappear (**Figure 4.22C**).

In nature, crashlike responses are rather common. Plants and animals often invade a new habitat, such as an open field or patch of forest, and increase rapidly by taking advantage of the area's resources. If those resources are supplied at a slow rate or are not renewable, the populations living in that area cannot sustain themselves and soon die out or move away.

▲ **Figure 4.22 Alternative Patterns of Population Growth**
The logistic graphs here depict hypothetical situations involving resource supply and population growth. Curve Ⓐ represents very tight coupling between population growth rate and resource supply whereas curve Ⓑ indicates a population growth rate that is delayed in response to limited resources but nevertheless recovers when resource supply is renewed. Curve Ⓒ illustrates that the delay between population growth rate response and resource supply is so significant that resource supply cannot recover and the population crashes.

Other Limits on Population Growth

Population growth may be influenced by factors other than resource use and supply.

In natural ecosystems, population changes are often erratic and don't seem to follow the growth predicted by theory or observed in the laboratory. Erratic changes occur because populations are influenced by multiple factors, some of which are not related to resources. In addition, the factors that affect population growth most significantly may change from time to time.

The population growth of a species may be limited by different factors at different seasons of the year. For example, during the breeding season, populations of many migratory songbirds are limited by the availability of nesting sites in temperate forests. During the winter, their populations are limited by the amount of food available in tropical forests.

Temperature is a particularly important factor influencing the growth rate, birth rate, and death rate of all populations. For all living things, there is a minimum temperature below which growth cannot occur and death may be likely. Above the minimum temperature, growth rates generally increase as temperature increases. At an optimum temperature, the population grows most rapidly. Above the optimum temperature, growth rates decline. At some maximum temperature, growth ceases. The span between the minimum and maximum temperatures defines an organism's **range of tolerance**.

An organism's range of tolerance is closely related to the temperatures in the environment that it inhabits.

For example, the range of tolerance for rainbow trout is a water temperature of approximately 10 °–21 °C (50 °–70 °F), which closely matches the range of temperatures in the streams where it lives (**Figure 4.23**). The desert pupfish thrives in spring-fed desert pools that vary widely in temperature; the desert pupfish's range of tolerance is very large, 3 °–40 °C (38 °–104 °F). In contrast, the ice fish lives in waters near Antarctic glaciers, where temperatures are very cold and do not vary more than a degree or two. The ice fish has a very narrow range of tolerance, between −2 ° and −1 °C (28 and 30 °F).

Populations are often sensitive to the presence of toxic chemicals in their environment. Even small amounts of toxins such as cyanide or mercury may limit a population's growth. Sometimes, the organisms themselves produce retarding chemicals. For example, yeast populations may produce ethyl alcohol, which increases in concentration as their populations grow. Depending on the particular genetic strain of yeast, alcohol concentrations between 6% and 14% define the upper limit of their range of tolerance.

Organisms may even have a range of tolerance for resources that are essential to their survival. Plant growth, for example, is often controlled by limited supplies of nitrogen. However, too much nitrogen, such as too much fertilizer applied to a crop, can retard growth or even cause death.

▲ **Figure 4.23 Range of Tolerance Varies Among Species**
In general, organism functions and population growth increase with temperature to some optimal level, above which functions rapidly decline. Temperature ranges of tolerance and temperature optima for different species are related to the environments in which they are found.

Habitat and Ecological Niche

An organism's range of tolerance for many environmental factors determines its habitat and niche.

At any particular time, each organism must respond to a wide variety of environmental factors. In addition, an organism's range of tolerance for a particular environmental factor is often influenced by the status of other factors. Consider, for example, the interaction of temperature and water. As temperature increases, evaporation also increases. As a result, most terrestrial organisms generally require a greater supply of water when it is hot. These organisms can tolerate higher temperatures when water is plentiful than when it is scarce.

A **habitat** is the complex environment in which an organism is found and upon which it depends for survival. An organism's habitat includes nonliving environmental elements, such as temperature, humidity, and soil, as well as living elements, such as insects and foliage. The size of an organism's habitat depends on the size of the organism and its ecological needs. A salamander may spend its entire life foraging on worms and insects within a few square meters of the forest floor, whereas a mountain lion may hunt for deer over hundreds of square miles of forest, grassland, and shrubland.

Ecologists often describe an organism's habitat as where the organism lives. They call the various activities that define that organism's role in an ecosystem its **ecological niche**. An ecological niche is complex, encompassing every interaction between an organism and its ecosystem—what the organism eats, what eats the organism, and the effects of the organism's activities on the flow of energy and matter through the ecosystem. An organism's habitat and ecological niche may change at different times of the year or throughout its lifetime (**Figure 4.24** and **Figure 4.25**). These topics are explored in detail in Chapter 6.

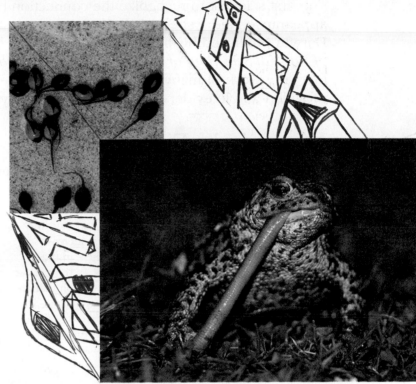

▲ **Figure 4.24 Habitat and Ecological Niche Can Change as Organisms Mature**
The habitat of American toad tadpoles is temporary ponds. The tadpoles feed on plant material and, therefore, occupy the ecological niche of herbivore. The mature toads' habitat is the terrestrial forest floor where their ecological niche is that of a carnivore, feeding on worms and insects.

QUESTIONS

1. Why does environmental resistance increase as population size increases?

2. What factors determine how well a population's growth will match the logistic growth curve?

3. A population's response to a resource or limiting factor may vary from time to time. Why?

▲ **Figure 4.25 Habitat and Ecological Niche Can Change with Seasons**
The habitat and ecological niche of the Arctic jaeger changes during the year. In the winter, it survives by stealing fish from terns and gulls in tropical waters. During the summer, it nests in the Arctic tundra where it feeds on lemmings.

4.5 Evolution and Natural Selection

The first scientist to recognize the connection between inheritance, variations in behavior and appearance, and a species' ability to persist in an environment was Charles Darwin. In 1859, Darwin published his theory of evolution in a book called *On the Origin of Species by Means of Natural Selection, or the Preservation of Favored Races in the Struggle for Life*. Since his time, the work of many other scientists and the development of the science of genetics have provided ample evidence to support his theory.

Darwin's Finches

Species evolve when populations move into new habitats and discover new ecological niches.

Darwin postulated that the diversity of life is a consequence of **adaptations**, inherited structures, functions, or behaviors that help organisms survive and reproduce. Individuals with adaptations that increase their reproductive success compared to others in the population are said to have greater **fitness**. In the process of **natural selection**, individuals in a population that are most fit survive and leave more offspring, causing their adaptations to become more common. Individuals that are less fit leave fewer offspring, so their features become less common or cease to exist.

Darwin based his theory on a diverse array of observations made over many years. But it was the remarkable diversity of finches on the Galápagos Islands—species found nowhere else—that particularly inspired Darwin's thinking (**Figure 4.26**).

Species belonging to the finch family occur on every continent except Australia. Most finches eat seeds, a diet for which their characteristic thick conical beaks are well adapted. But the 13 species of finches that live on the Galápagos occupy a variety of ecological niches, some of which are decidedly un-finchlike (**Figure 4.27**). For example, the woodpecker finch (*Cactospiza pallidus*)

▼ **Figure 4.26 The Galápagos Islands Inspire Charles Darwin**
About 600 miles (1,000 km) west of South America, volcanic activity began producing the Galápagos Islands approximately 5 million years ago. Their isolation from the mainland, austere habitats, and unique flora and fauna catalyzed Charles Darwin's ideas on evolution by means of natural selection.

▲ **Figure 4.27 Galápagos Finch Adaptive Radiation**
The 13 species of finches on the Galápagos Islands evolved from a common ancestral finch that migrated to the Galápagos Islands, perhaps as recently as 10,000 years ago.

▲ **Figure 4.28 A Finch Occupies a Woodpecker's Ecological Niche**
The Galápagos woodpecker finch feeds on insects that it prods from crevices in tree trunks and cactus pads using a sharp stick or cactus spine.

uses small sticks or cactus spines to prod insects from the crevices and holes in trees (**Figure 4.28**). The warbler finch (*Certhidea olivacea*) uses its comparatively fine beak to feed on small insects on the ground. Perhaps most peculiar of all is the sharp-billed ground finch (*Geospiza difficilis*), which feeds on the eggs of large seabirds called boobies that nest on the islands.

Because the Galápagos Islands are isolated from South America, Darwin reasoned that relatively few birds from the mainland had successfully migrated to the islands. Even so, he supposed that at some time in the distant past a few finches had found their way to the Galápagos. The finches that made it to the islands found an array of habitats quite different from those on the mainland.

Darwin recognized that individual organisms within each species varied in particular features, such as in their physical structure or behavior. Formal principles of genetics and the genetic basis of mutations were not established until long after Darwin's death. But Darwin did understand that new variations in features appear at random. Most important, Darwin knew that these variations were inherited from one generation to the next.

Darwin reasoned that finches possessing a particular bill shape and feeding behavior were more able to access food than finches with different bill shapes and feeding behaviors. Furthermore, he noted, finches that were better fed were more likely to survive and reproduce healthy offspring. Darwin reasoned that finches with bill shapes and feeding behaviors that were more conducive to accessing food would become more abundant in the population than finches with bill shapes and feeding behaviors that were incompatible with finding food.

Based on these observations and this reasoning, Darwin proposed that the different finch species of the Galápagos Islands gradually evolved into new ecological niches through natural selection. Evolutionary biologists refer to the process by which several species evolve from a single ancestor to occupy new ecological niches as **adaptive radiation**.

Finch Studies Continue

Long-term studies of Darwin's finches demonstrate that biological evolution can occur rapidly.

Rosemary Grant and Peter Grant of Princeton University have been studying the Galápagos Islands finches since 1975. The Grants have captured, weighed, measured, and placed leg bands on every finch on the island called Daphne. The data they have gathered include careful measurements of body size, beak size and thickness, and feeding behavior.

Between 1977 and 1978, the Galápagos experienced a severe drought, and the Grants recorded the death of thousands of birds. The birds had starved, apparently because their food source, seeds, was in limited supply. Curiously, the deaths did not occur randomly throughout all bird populations. The vast majority of birds that died belonged to certain species, such as the medium ground finch (*Geospiza fortis*).

Normally, medium ground finches are abundant on Daphne. During the drought, however, their population declined from over 1,400 to about 200 birds. The medium ground finch has a comparatively large beak and feeds on a wide range of seeds; it usually avoids very large, tough seeds, such as those of the widespread, non-native Jamaican fever plant. But during the drought, the only seeds that were available in large numbers were the large tough seeds of the fever plant. Because cracking large seeds requires large jaw muscles, medium ground finches with small beaks and bodies were unable to crack them. Without food, the small-beaked birds died. Only medium ground finches with the largest beaks and bodies were able to survive the drought (**Figure 4.29**). With natural selection favoring large beaks and bodies, the medium ground finch population underwent significant evolution in both beak and body size within just a few generations.

During the 1980s, rainfall in the Galápagos was plentiful, resulting in abundant and diverse plant growth and the production of a wide variety of seeds. Bountiful food resulted in very successful reproduction among the medium ground finches, and their population flourished. Measurements of birds since then show a return to the predrought range of variation in beak and body size. The Grant's studies suggest that when smaller seeds are abundant, larger beak and body size may be disadvantageous.

What is a genetic bottleneck?
Natalie Hollabaugh, Central Michigan University

Norm: *A genetic bottleneck occurs when the size of a population becomes so small that significant amounts of genetic variation are lost.*

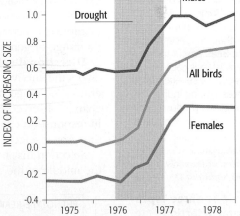

▲ **Figure 4.29 Rapid Evolution**
Between 1976 and 1977, drought on Daphne resulted in very limited seed production by herbs, with the exception of the large-seeded fever plant. During this period, the population of medium ground finches crashed. Those finches with larger beaks were able to survive on the fever plant seeds, and average beak size significantly increased in the population.
Source: Grant, B.R., and P.R. Grant, 1989. Natural selection in a population of Darwin's finches. *American Naturalist* 133: 377–393.

Inherited Variations

Mutation is the ultimate source of newly inherited variations.

Darwin's studies and those of many other researchers have demonstrated that populations include considerable individual-to-individual variation for inherited traits. What is the source of that variation? Through time, **mutations**, or chance changes in the structure of DNA, produce new forms of inherited features. Mutations may arise because of an environmental factor that affects the replication of DNA during cell division. For example, ultraviolet light and chemical pollutants are known to induce mutations. Mutations affect DNA in a random fashion; only mutations that are carried in gametes can be passed to the next generation (**Figure 4.30**).

Mutation is the source of variation upon which natural selection may act. Most mutations are detrimental and are quickly removed from the population by natural selection. A few mutations have little or no effect on fitness. These may remain in the population and contribute to the range of variation that exists for a feature.

▲ Figure 4.30 Mutation Is Random
Albino deer with white coats appear occasionally in populations of otherwise normal deer. Albinism is a consequence of mutation in one of several genes that determine the color of the animal's fur.

For example, a mutation might result in a different plumage color or a slightly larger beak in individual members of a bird species. Changes in the environment might result in natural selection on such variation.

Natural Selection

There are three types of natural selection.

As the beaks of Galápagos finches illustrate, there is a range of variation for any inherited feature (**Figure 4.31**). Some genetic variations make an individual organism more fit for its environment, whereas other genetic variations render an individual less fit. Because they can determine an individual's fitness, the range of genetic variations can be directly correlated to natural selection. Evolutionary biologists identify three types of natural selection—directional selection, stabilizing selection, and disruptive selection—based on the relative fitness across the range of genetic variation (**Figure 4.32**).

Directional selection refers to natural selection that favors the survival of individual organisms at one extreme end of the range of variation within a population. The change observed in the average size of finch beaks in response to changes in seed size is an example of directional selection.

Selection that favors traits that fall somewhere in the middle of the range of variation within a population is called **stabilizing selection**. For example, in most species, extremes of body size—very large and very small individuals—are less fit than average-sized individuals. Studies of the effects of major storms and extremely cold weather on bird populations show that very large and very small individuals suffer higher mortality rates than average-sized individuals. Stabilizing selection is thought to be important for most inherited traits.

Disruptive selection refers to situations in which individuals with genetic traits that fall at the extreme ends of the range are more likely to survive than individuals with average traits. Disruptive selection can occur when a population confronts contrasting environments. For example, if the Galápagos drought had resulted in the survival of plants that produced very small seeds in addition to the large seeds of the Jamaican fever plant, finches with small beaks and finches with large beaks would have been more likely to survive than finches with average-sized beaks.

▲ Figure 4.31 Inherited Features Vary
Bird beak size is an important attribute that influences feeding behavior. Beak size in the Galápagos Islands cactus finch is influenced by several genes. Differences among individuals for these genes produce such variation.

Source: Grant, B.R. and P.R. Grant. 1989. Natural selection in a population of Darwin's finches. *American Naturalist* 133:377–393

▶ Figure 4.32 Three Types of Natural Selection
Stabilizing, directional, and disruptive selection differ with respect to which phenotypes in a population are most fit.

Stabilizing selection
Removes extreme phenotypes—birds with very small and very large beaks—from the population

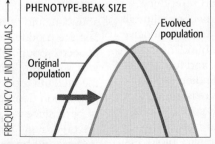

Directional selection
Removes one particular extreme phenotype—here birds with small beaks—from the population

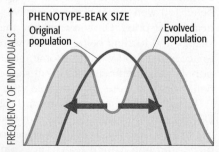

Disruptive selection
Removes medium-sized phenotypes—birds with medium-sized beaks—from the population

Genetic Change

Chance events produce genetic change in small populations.

Populations of the common Asian ladybird beetle include individuals whose wing covers display a remarkable array of colors and spot patterns. This variation is the consequence of inherited variation in a single gene. In places with small populations of Asian ladybird beetles, researchers have noted that the frequency of colors varies significantly from time to time, with some colors becoming very abundant and other colors disappearing altogether. Researchers have concluded that these changes in the frequency of colors are caused by random events.

Random events may cause the frequency of a trait to increase or decrease simply by chance. Small populations are more susceptible to the effects of random events than large populations. In large populations, the effects of chance events are too small to detect, but as population size decreases, the likelihood that a chance event may cause a noticeable change in a population increases. A change in the frequency of an inherited trait in a population that is brought about by a chance event is called **genetic drift** (**Figure 4.33**).

The relationship between population size and genetic drift can be understood by comparing two different hypothetical ladybird beetle populations. If a predator were to eat 50 ladybird beetles at random from a population, as a matter of chance the predator might end up eating more beetles of one color than beetles of other colors. In a population of thousands of beetles, this event would have little effect on the overall frequency of the genes for any particular color. But in a population of only 100 beetles, the effect of the random acts of a predator would be significant. For example, if 30 of the beetles in that small population were yellow with no spots, their frequency would be 30%. If the predator ate 50 beetles, 25 of which happened to be yellow, the surviving population of 50 beetles would include only 5 yellow individuals. The frequency of yellow beetles would then be only 10% (5/50 × 100).

Chance can play a particularly important role in determining the genetic makeup of a newly established population. A few individuals from a large population migrating to a new habitat are unlikely to have exactly the same frequencies of characteristics as those found in their parent population. Therefore, the new population they found will be genetically different. Genetic change resulting from the immigration of a small subset of a population is called the *founder effect*.

Some of the best-documented examples of the founder effect are in human populations. Huntington's disease is a severe neurological disorder caused by a variation in a gene that affects the development of brain cells. The disease occurs at an unusually high frequency among the Afrikaner people of South Africa, who

are descended from a rather small number of Dutch immigrants that colonized the region in the 18th century; that original population happened to carry an unusually high frequency of the gene for Huntington's disease.

In summary, mutation generates new variations for inherited traits, and sexual reproduction maintains such variations in populations. Evolution occurs when the relative abundance of genetically inherited traits within a population changes. Natural selection is a form of evolution in which factors in the environment favor the survival and reproduction of individuals with particular inherited traits. Genetic drift is a form of evolution that occurs in small populations as a consequence of random events.

QUESTIONS

1. Based on Darwin's study of the finches on the Galápagos Islands, describe the process of natural selection.
2. Compare directional, stabilizing, and disruptive selection with respect to their effects on inherited variation for a feature.
3. Why doesn't genetic drift occur in large populations?

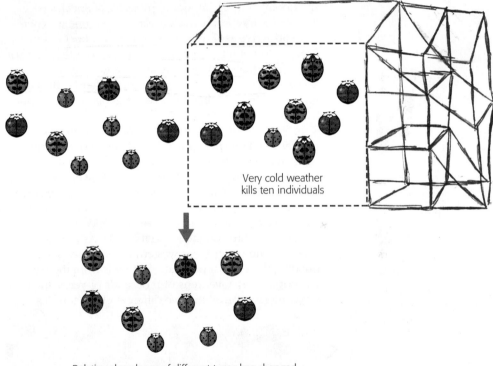

Very cold weather kills ten individuals

Relative abundance of different types has changed

Subsequent generations

Figure 4.33 Genetic Drift
The relationship of population size to genetic drift can be understood by comparing the effects of a hypothetical predator on the frequency of color types in a small ladybird beetle population.

4.6 The Evolution of Species

A **species** is a group of similar organisms that can potentially mate and produce fertile offspring. Organisms belonging to the same species but different populations can potentially interbreed, whereas organisms belonging to different species cannot. When there are barriers to interbreeding between separate populations, the two populations may undergo evolutionary changes that allow them to occupy separate ecological niches. This is the basis for the evolution of new species.

Preventing Interbreeding

Natural selection has resulted in a variety of mechanisms to prevent interbreeding.

In general, each species has a unique set of genetic traits that are not shared with other species. For example, the warbler finch differs from other Galápagos finches in having a fine beak and small body. From an evolutionary standpoint, a more important property of species is that they are reproductively isolated from one another. **Reproductive isolation** means that members of one species do not interbreed with members of other species, even those that are closely related. Closely related species are often isolated from one another by **breeding barriers.** Breeding barriers may be geographic, temporal, behavioral, or structural.

In some cases, *geographic isolation* prevents interbreeding. When species that have been geographically separated come into contact, they may be capable of producing viable offspring. The spotted owl of the northwestern United States and the barred owl of the eastern United States are thought to have evolved from a common ancestor that lived in forests that once extended across the entire continent (**Figure 4.34**). As the climate changed, the eastern and western forests were separated, isolating the eastern population of owls from the western population. Over tens of thousands of years, the two populations evolved to have different characteristics.

In recent years, barred owls have begun migrating into northwest forests, where they occasionally mate with spotted owls and produce fertile offspring.

Where the geographic ranges of closely related species overlap, unions between species are more likely to occur. Such interbreeding usually results in offspring that are sterile or in some way poorly adapted to their environment. Inherited traits that prevent interbreeding between such species will, therefore, increase their fitness and will be subject to natural selection. Over time, natural selection will favor the development of breeding barriers. Several different kinds of breeding barriers have evolved in plants and animals.

Species may be isolated from one another by reproducing at different times. This is called *temporal isolation* (**Figure 4.35**). For example, the green frog, the wood frog, and the southern leopard frog inhabit much of the southeastern United States. These three species do not interbreed because their peak breeding periods do not occur at the same time. Their different breeding times are a consequence of each species responding differently to seasonal changes in temperature.

Species may also be separated from one another by differences in behavior. This is called *behavioral isolation*.

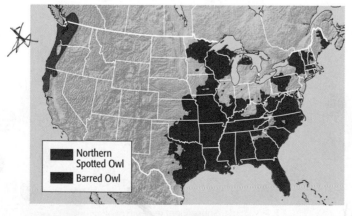

▲ **Figure 4.34 Geographic Reproductive Isolation**
The northern spotted owl inhabits Pacific Northwest forests, and the barred owl is found in forests of the eastern United States. They are reproductively isolated by means of geographic separation. Recent introductions of barred owls to the Pacific Northwest have resulted in successful interbreeding.

Northern Spotted Owl
Barred Owl

GREEN FROG
SOUTHERN LEOPARD FROG
WOOD FROG

FEB   MARCH   APRIL   MAY   JUNE

▲ **Figure 4.35 Temporal Reproductive Isolation**
Different breeding times reproductively isolate the wood frog (*Rana sylvatica*), green frog (*R. clamitans*), and southern leopard frog (*R. sphenocephala*) from one another where they share habitat in the southeastern United States.

For many animals, mating displays and courtship rituals play a significant role in the choice of mates and can be effective barriers to breeding between species. Such reproductive isolation is especially common among birds that select mates based on mating calls or dances. For example, the mating ritual of the Galápagos Islands' blue-footed boobies includes a complex dance performed first by the male and then by both sexes (**Figure 4.36**). Neither male nor female boobies will mate with individuals that don't know the dance.

Differences in reproductive structures effectively isolate many closely related species from one another, resulting in *structural isolation*. For example, the shape and color of flowers determines what pollinators are attracted to a plant and, ultimately, whether visiting pollinators will carry the flower's pollen to other flowers for successful pollination. Plants with dark-colored, tubular flowers are especially attractive to hummingbirds, which have long beaks and tongues that are able to reach the nectar at the base of the flowers (**Figure 4.37**). Yellow and blue flowers with open petals are more typically visited by insects.

Reproduction isolation prevents mating between species. However, on the rare occasions when different species do mate, the genetic differences between them are usually great enough that nonviable, sterile, or partially sterile offspring are produced. For example, no one questions that tigers (*Panthera tigris*) and lions (*Panthera leo*) are different species. However, on rare occasions where their ranges overlap, they may interbreed. The characteristics of their hybrid offspring depend on the gender of the respective parents. If the father is a lion and the mother a tiger, the offspring are *ligers*, and they resemble a very large lion, but with stripes (**Figure 4.38**). If dad is a tiger and mom is a lion, the offspring are more like a tiger and are called *tigons*.

Male ligers and tigons are completely sterile, whereas their female counterparts are fertile.

In general, interbreeding between genetically distinct populations that are adapted to specific environments produces offspring that are less fit than either of their parents. Therefore, natural selection is likely to favor the evolution of traits or behaviors—breeding barriers—that prevent such mating. This is the primary means by which new species evolve.

QUESTIONS

1. Two populations of birds may look alike but be considered separate species. On what basis do scientists make such designations?

2. Describe three different kinds of breeding barriers.

▲ **Figure 4.37 Mechanical Reproductive Isolation**
These two closely related species of columbine (*Aquilegia*) are found along streams in the Sierra Nevada. The scarlet columbine (above) grows in sunny habitats, and its red flowers attract hummingbirds. The Sierra columbine (below) grows in shadier locations, and its tubular, cream-colored flowers are visited by large hawk moths.

▲ **Figure 4.36 Behavioral Reproductive Isolation**
Blue-footed boobies are known for their peculiar mating dances by which males and females recognize each other as members of the same species.

► **Figure 4.38 The Hybrid Liger**
Distinct species may sometimes interbreed. For example, if a male lion breeds with a female tiger, their offspring, ligers, share features of both species. Ligers, however, have much reduced fertility.

Birds

FEATHERS

4.7 The Hierarchy of Life

Finches, frogs, and all Earth's life are the products of at least 3.8 billion years of biological evolution. To understand how biological evolution occurs, scientists consider how species adapt to changing conditions, how they interact with their environment, and how they compare physically, behaviorally, and genetically to other forms of life. **Taxonomy**, the classification of organisms through description, identification, and naming, provides a framework for such comparisons.

Evolutionary Map

A hierarchical system of classification reflects the evolutionary relationships among groups.

In taxonomy, each kind of organism is assigned a two-part Latin name. The first part of the name, the **genus**, identifies the organism with its closest relatives. The second part of the name is unique for each species; it may describe a unique characteristic of the species, where the species lives, or the name of a scientist who discovered the species. The two-part Latin name for the white-tailed jackrabbit is *Lepus townsendii*. The genus *Lepus* includes 28 species of large hares whose young are born with their eyes open and a full coat of fur. The species *townsendii*, named in honor of 19th-century naturalist John Kirk Townsend, is distinguished from other members of this genus by its very long ears and hind legs. Its long legs allow this jackrabbit to run as fast as 50 miles per hour.

Taxonomy is a system of hierarchical categories or taxa (singular, taxon), in which each level of classification includes all the levels beneath it (**Figure 4.39**). The lowest level of classification is the species. Closely related species are grouped together in a genus. Closely related genera are grouped into *families*, and families are grouped into *orders*. Groupings continue in this manner up the taxonomic hierarchy to encompass *classes*, *phyla* (or *divisions* for plants), *kingdoms*, and *domains*. At each level of the taxonomic hierarchy, organisms classified into a group are more similar to one another than to organisms belonging to other groups at the same level.

Kingdom	Animala
A kingdom is an overall category containing organisms that share fundamental structures and functions.	The kingdom Animala contains multicellular organisms that obtain energy by eating food. Most have muscles and nerves.
Phylum (Animals) or Division (Plants)	Chordata
A phylum or division is a major subdivision of a kingdom containing one or more classes and their subgroups.	The phylum Chordata includes the vertebrates. Among other shared features they have a nerve cord extending along their backs.
Class	Mammalia
A class is a major subdivision of a phylum or division containing one or more orders and their subgroups.	The class of "warm blooded" vertebrates that have hair or fur and have females that produce milk to nourish their young.
Order	Lagomorpha
An order is a major subdivision of a class containing one or more families and their subgroups.	Rodent-like mammals with four instead of two incisor teeth in the upper jaw.
Family	Leoporidae
A family is a major subdivision of an order containing one or more genera and their subgroups.	This family includes rabbits and hares having short furry tails and elongated ears and hind legs.
Genus	*Lepus*
A genus is a major subdivision of a family containing one or more species.	Rabbit relatives whose young are born with eyes open and a full coat of fur.
Species	*Lepus townsendii*
A species is a group of similar individuals that are able to interbreed with one another in the wild.	The whitefooted jackrabbit is noted for its long ears and well-developed legs.

▲ **Figure 4.39 Classifying *Lepus***
The hierarchical system of classification defines taxa based on shared characteristics that are inherited from common ancestors.

Bacteria

Archaea

Protists

Plants

Fungi

NERVES, MUSCLE MOBILITY

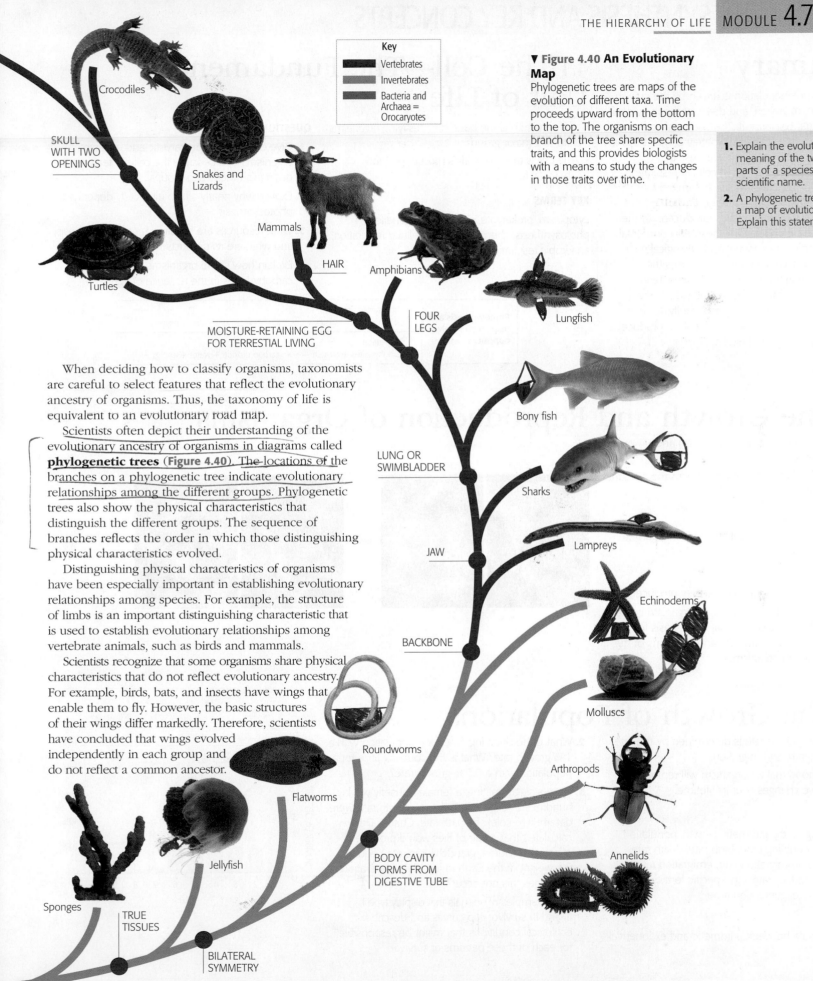

Key
- Vertebrates
- Invertebrates
- Bacteria and Archaea = Orocaryotes

Crocodiles

SKULL WITH TWO OPENINGS

Snakes and Lizards

Mammals

Turtles

HAIR

Amphibians

MOISTURE-RETAINING EGG FOR TERRESTIAL LIVING

FOUR LEGS

Lungfish

Bony fish

LUNG OR SWIMBLADDER

Sharks

JAW

Lampreys

Echinoderms

BACKBONE

Molluscs

Roundworms

Arthropods

Flatworms

Jellyfish

BODY CAVITY FORMS FROM DIGESTIVE TUBE

Annelids

Sponges

TRUE TISSUES

BILATERAL SYMMETRY

▼ **Figure 4.40 An Evolutionary Map**
Phylogenetic trees are maps of the evolution of different taxa. Time proceeds upward from the bottom to the top. The organisms on each branch of the tree share specific traits, and this provides biologists with a means to study the changes in those traits over time.

QUESTIONS

1. Explain the evolutionary meaning of the two parts of a species' scientific name.
2. A phylogenetic tree is a map of evolution. Explain this statement.

When deciding how to classify organisms, taxonomists are careful to select features that reflect the evolutionary ancestry of organisms. Thus, the taxonomy of life is equivalent to an evolutionary road map.

Scientists often depict their understanding of the evolutionary ancestry of organisms in diagrams called **phylogenetic trees (Figure 4.40)**. The locations of the branches on a phylogenetic tree indicate evolutionary relationships among the different groups. Phylogenetic trees also show the physical characteristics that distinguish the different groups. The sequence of branches reflects the order in which those distinguishing physical characteristics evolved.

Distinguishing physical characteristics of organisms have been especially important in establishing evolutionary relationships among species. For example, the structure of limbs is an important distinguishing characteristic that is used to establish evolutionary relationships among vertebrate animals, such as birds and mammals.

Scientists recognize that some organisms share physical characteristics that do not reflect evolutionary ancestry. For example, birds, bats, and insects have wings that enable them to fly. However, the basic structures of their wings differ markedly. Therefore, scientists have concluded that wings evolved independently in each group and do not reflect a common ancestor.

Summary

The growth of a population is driven by the growth, reproduction, migration, and death of the individual organisms that comprise it. The chemical energy needed for these activities ultimately comes from photosynthesis, and organisms access that chemical energy through the process of cellular respiration. The growth of populations may be limited by resources and other factors in the environment that influence the growth and reproduction of the organisms in the population. These environmental factors determine the habitat and ecological niche of each species. Genetic variation causes the individuals within a population to have different traits, which determine whether individuals are able to grow and reproduce successfully. Natural selection acts on that genetic variation to produce evolutionary change and new species.

4.1 The Cell—The Fundamental Unit of Life

- Cell structure is the basis for classifying organisms into two major groups.
- The cell is the site of life's most important chemical functions.

KEY TERMS

cytoplasm, prokaryote, eukaryote, organelles, photosynthesis, chemosynthesis, cellular respiration, anaerobic respiration

QUESTIONS

1. Give examples of prokaryotic and eukaryotic organisms and describe the cellular features that distinguish these two groups.

2. Explain why nearly all life ultimately depends on photosynthesis.

3. What resources are required for photosynthesis, and what are its products?

4. Explain how some organisms produce carbohydrates in the absence of light.

Photosynthesis and respiration involve opposite reactions	Photosynthesis carbon dioxide + water + energy ⟶ carbohydrates + oxygen
	Respiration carbohydrates + oxygen ⟶ carbon dioxide + water + energy

4.2 The Growth and Reproduction of Organisms

- An organism grows through division and differentiation of its cells.
- The mode of reproduction influences the amount of variation among offspring.

KEY TERMS

asexual reproduction, sexual reproduction, gamete, zygote

QUESTIONS

1. What are clones and how are they produced?

2. All of the cells that make up an organism contain exactly the same DNA. What is the basis for differences in the structure and functions of the cells within an organism?

Ⓐ   Ⓑ   Ⓒ   Ⓓ

4.3 The Growth of Populations

- Population growth rate is determined by the rates of birth, death, and migration.
- The likelihood that an organism will reproduce and survive changes over its lifetime.

KEY TERMS

exponential growth, arithmetic growth, population growth rate, doubling time, birth rate, death rate, mortality rate, immigration rate, emigration rate, survivorship, fertility rate, age-specific fertility rate, total fertility rate, generation time

QUESTIONS

1. Differentiate between arithmetic and exponential growth.

2. What is the doubling time for a population with a 1% growth rate? What is the doubling time for a population with a 0.5% growth rate?

3. From a single mating, a female housefly will lay hundreds of eggs. If population growth rate were determined solely by birth rate, Charles Darwin calculated that a pair of flies would produce offspring (children, grandchildren, etc.) equal to the weight of the Earth in just eight generations. Why does this not occur?

4. Give examples of animals that display type I, II, and III survivorship curves and describe ecological conditions that might be responsible for each of these patterns of survival.

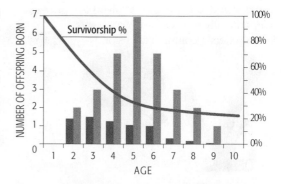

4.4 Limits on Population Growth

- At carrying capacity, the birth rate equals the death rate and the population growth rate stalls at zero.

KEY TERMS

carrying capacity, logistic growth curve, range of tolerance, habitat, ecological niche

QUESTIONS

1. As the size of a population nears carrying capacity, how does its use of resources change in relation to the supply of resources?

2. Populations may grow beyond their carrying capacity. Describe and explain two possible consequences of such growth in regard to resource use and resource supply.

3. Explain why an organism's range of tolerance for one factor may depend on the status of other factors in its environment.

4. Using a specific example, differentiate between an organism's habitat and its ecological niche.

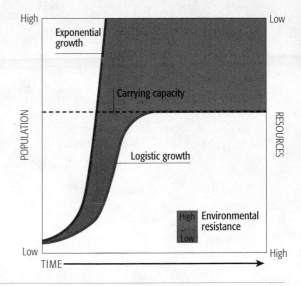

4.5 Evolution and Natural Selection

- Species evolve when populations move into new habitats and discover new ecological niches.

KEY TERMS

adaptation, fitness, natural selection, adaptive radiation, mutations, directional selection, stabilizing selection, disruptive selection, genetic drift

QUESTIONS

1. Natural selection is sometimes described as survival of the fittest. In addition to survival, what characteristic of an individual organism determines its fitness?

2. As insecticides have become more abundant in the environment, so has the abundance of insects with inherited resistance to them. Explain the likely basis for this change in the relative abundance of insecticide-resistant insects.

3. Explain why genetic drift often occurs in populations recently founded by a few immigrant individuals.

4.6 The Evolution of Species

- Natural selection has resulted in a variety of mechanisms that prevent interbreeding.

KEY TERMS

species, reproductive isolation, breeding barrier

Temperal isolation	Differences in mating season
Behavioral isolation	Differences in mating calls or dances
Structural isolation	Differences in reproductive anatomy

QUESTIONS

1. Why does natural selection favor the evolution of breeding barriers between species?

2. Describe how two new species might evolve from a parent species, even though they live in the same geographic area.

4.7 The Hierarchy of Life

- A hierarchical system of classification is an evolutionary map.

KEY TERMS

taxonomy, genus, phylogenetic tree

QUESTIONS

1. The hierarchical system of classification is intended to be a map of evolution. Explain this statement.

2. What is implied about species that occur on the same branch of a phylogenetic tree?

Go to **www.masteringenvironmentalscience.com** for homework assignments, practice quizzes, Pearson eText, and more.

115

5

Human Population Growth

Human Population Growth— What's the Problem?

Can we determine how many people Earth can support?

At some moment during the autumn of 2011, the global human population passed the 7 billion mark (**Figure 5.1**). We don't know the exact day, much less the hour or minute; United Nations census data and estimates of world population growth are just not that precise. Nevertheless, many noted this milestone in human history.

Some people feel that the rapid growth in our numbers is the "mother of all environmental problems." And they have a point. Our numbers have doubled twice over the past century. If you are 20 years old, Earth's population has grown by nearly 30% (1.6 billion) since your birth. Many argue that the growing number of humans is the primary factor driving our increasing impact on our planet and its life-support systems.

▲ **Figure 5.1 It's Getting Crowded**
In less than 150 years, Earth's human population has increased sevenfold, and will soon top 7 billion.

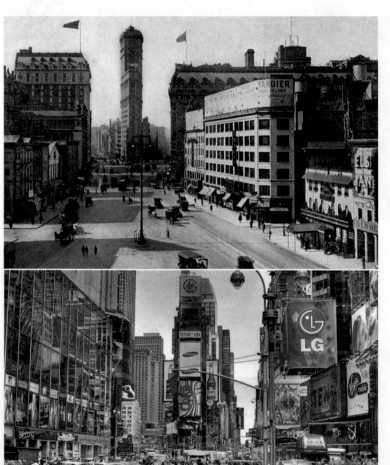

From other commentators you will hear, "It's not that simple." Yes, the growth in our numbers is a concern, but the growth in our consumption is even more troubling. As a measure of the change in our consumption, they point to a milestone related to energy use. Sometime between 2007 and 2009 the amount of energy that the average person uses per year rose above 80 billion joules, which is equivalent to the energy in about 615 gallons of gasoline. This milestone represents an increase in personal energy use of more than 40% over the past century.

Even more troubling is the change in the sources of energy we use. In the early 1900s, nearly half of our energy came from renewable sources, such as wind, water, and wood. The remainder came from fossil fuels, mostly coal. Today, over 80% of the energy used by humans comes from fossil fuels (**Figure 5.2**). Not only is the supply of fossil fuels limited, but using them pollutes the air

◄ **Figure 5.2 We're Consuming More**
The increase in human energy consumption is clear from these scenes of Times Square in New York City in 1910 (top) and 2010 (bottom).

we breathe and the water we drink. It is likely that the average human being's impact on Earth's resources and ecosystems has more than doubled in the past 100 years.

Population growth rates and resource consumption vary widely among the world's countries, largely in relation to their wealth. Population growth rates in some poor African countries are two to three times higher than those in wealthier European or North American countries. With resource consumption, the pattern is reversed. The average American uses at least 10 times more energy and materials than the average person from Zimbabwe or Zambia.

This geographic variation is also correlated with human well-being, as shown by measures such as nutrition, disease, and childhood death rates. Well-being is highest where population growth rates are low and consumption is high. It is lowest where population growth rates are high and consumption is low (**Figure 5.3**).

It's not surprising that these geographical differences influence perspectives on the debate about which is more important, growth in numbers or growth in consumption. In developed countries, many argue that control of population growth is the key to improved well-being in poor countries. In contrast, many in the developing world argue that excessive consumption in wealthy nations prevents poorer nations from improving their well-being. These alternative views underlie tensions among many nations, although the solutions to the issues are not mutually exclusive.

How many people can the world support? This question has been at the center of much research and argument about human population change. The truth is that no one knows that number. Many feel that we are fast approaching it. Some feel we've already passed it. What are the factors that have influenced human population change through time? What factors account for the differences that currently exist among countries? How can we use this information to predict future population trends? We can quantify the changes in our population; can we do the same for our impact on resources and ecosystems? How can we slow growth in both numbers and consumption to ensure a sustainable future? These are the central questions for this chapter.

▼ **Figure 5.3 Population, Consumption, and Well-Being**
Population growth rates are high in poor areas, such as this slum in Addis Ababa, Ethiopia, whereas resource consumption and well-being are low.

5.1 The History of Human Population Growth

Over the past 100,000 years, the human population has increased enormously. The earliest humans lived in Africa. As humans moved into new regions, new resources became available and the number of people increased. Eventually, humans migrated throughout the world (**Figure 5.4**). A number of developments have made it possible for people to live in so many different places on Earth. During the past 10,000 years, innovations in agriculture have increased the food supply, allowing the size and density of human populations to increase. Over the past 300 years, some regions have experienced advances in industry and economic progress that have led to decreasing birth rates, death rates, and rates of population growth (**Figure 5.5**).

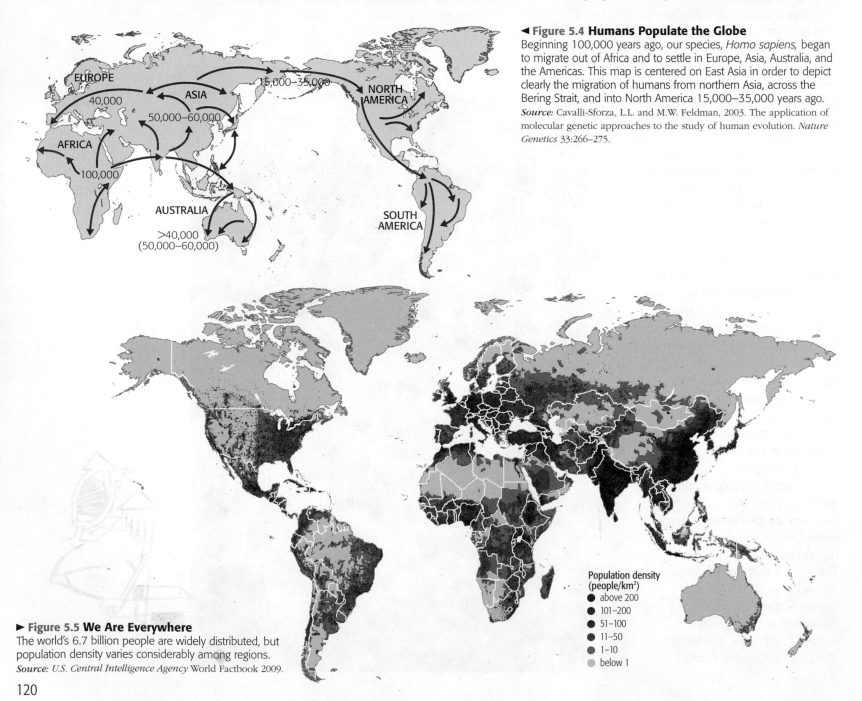

◄ **Figure 5.4 Humans Populate the Globe**
Beginning 100,000 years ago, our species, *Homo sapiens,* began to migrate out of Africa and to settle in Europe, Asia, Australia, and the Americas. This map is centered on East Asia in order to depict clearly the migration of humans from northern Asia, across the Bering Strait, and into North America 15,000–35,000 years ago.
Source: Cavalli-Sforza, L.L. and M.W. Feldman, 2003. The application of molecular genetic approaches to the study of human evolution. *Nature Genetics* 33:266–275.

► **Figure 5.5 We Are Everywhere**
The world's 6.7 billion people are widely distributed, but population density varies considerably among regions.
Source: *U.S. Central Intelligence Agency* World Factbook 2009.

Population density (people/km²)
- above 200
- 101–200
- 51–100
- 11–50
- 1–10
- below 1

Three Periods of Growth

Human populations experienced three periods of growth.

People who study the size, density, and distribution of human populations are called **demographers**. Demographers identify three periods of human population growth—the pre-agricultural period, the agricultural period, and the industrial period (**Figure 5.6**). The transitions between these periods occurred gradually, over decades or centuries. Furthermore, these transitions occurred at various times in different parts of the world.

Pre-agricultural period. The pre-agricultural period, which extended over more than 100,000 years, coincided with the development of human culture. The advancement of skills such as tool making, as well as the ability to pass those skills on to successive generations, allowed humans to adapt quickly to new environments. The rate of global population growth during the pre-agricultural period depended on the speed at which humans spread into new regions. It took tens of thousands of years for the human population to double in size. At the end of the pre-agricultural period, the global human population numbered between 5 and 10 million.

Agricultural period. The agricultural period began about 10,000 years ago, with the domestication of plants and animals and the development of farming. Breeding and growing livestock and nutrient-rich crops increased the supply of food. New technologies such as irrigation and plowing helped to increase the size of harvests. With more food available, human populations continued to increase. The time required for human populations to double shortened to less than 1,000 years. By the end of the agricultural period, the human population had grown to more than 500 million.

Industrial period. With the advent of the Industrial Revolution around 1800, the industrial period of population growth took hold. Technologies powered by fossil fuels increased food production and allowed agriculture to expand onto land that previously was unfit for farming. With the industrial period came vastly improved sanitation and huge advances in medicine. The rate of population growth accelerated. We are, of course, still in the industrial period, and our population currently exceeds 6.7 billion.

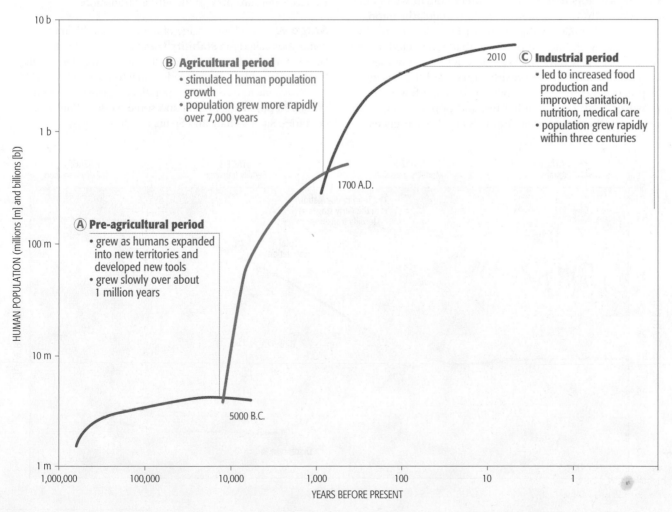

◄ **Figure 5.6 Three Periods of Human Population Growth**
The first period of population growth was caused by global migration Ⓐ. The second period was initiated by the introduction of agriculture Ⓑ, and the third commenced with the Industrial Revolution Ⓒ. Note that both the time and population scales on the graph are exponential.

121

Demographic Transition Model

Demographic transition model describes four stages of change in human population growth.

In 1929, the American demographer Warren Thompson published a paper titled simply "Population." In it, Thompson described trends of population growth found in countries that were in the industrial period. He noted that during the 200 years in which Europe's economy transformed from an agricultural to an urban-industrial base, birth rates on the continent declined by at least 50%. During that same time, the average life expectancy of Europeans doubled, from approximately 35 years to nearly 70 years. But in countries in which industrial development had been slow to progress, birth and death rates remained high. Thompson concluded that economic development improves overall well-being. This improvement results in fewer infant deaths and increased life expectancy. More surprisingly, Thompson found that economic development causes birth rates to decrease. Thompson's findings have subsequently been verified and translated into the **demographic transition model** (**Figure 5.7**).

Stage 1. The first stage of the demographic transition model is the period prior to economic development. During this stage, human populations are limited by a low availability of food and a high prevalence of disease—factors that negatively affect human well-being. Consequently, death rates are high, particularly among the young. Families compensate for high death rates by having many children, a behavior that is reflected in high birth rates. Over time spans of decades and centuries, birth and death rates are nearly equal, and there is zero population growth. Within periods of just a few years, however, birth rates, death rates, and population growth rates may fluctuate considerably. Natural occurrences

such as drought or plague, for example, may cause a population to temporarily experience low birth rates and high death rates. One hundred years ago, many of the world's poorest countries were in stage 1.

Stage 2. The second stage of the demographic transition is called the **mortality transition**. In this stage, improved economic conditions relieve food shortages, produce better living conditions and health care, and expand access to education. The death rate decreases while the birth rate stays high or even increases. As a result, there is rapid, sustained population growth. Today, very poor countries such as Afghanistan and Niger are in this stage.

Stage 3. In the third stage of the demographic transition, the **fertility transition**, continued economic development produces social and cultural changes that lead to lower birth rates. People may delay starting families, which lowers total fertility rates. Families may limit their number of children by using contraceptives or other means of family planning. In the fertility transition, the death rate remains low and the birth rate decreases, so population growth slows. Most of the world's countries, including the United States, are now in the fertility transition.

Stage 4. The fourth stage of the demographic transition, sometimes called the **stability transition**, is characterized by a low birth rate and a low death rate. During the stability transition, the birth rate is nearly equal to the death rate. This stage is characterized by zero population growth or even a decline in population. Japan and some western European countries, such as Italy and Spain, are now in stage 4.

▶ Figure 5.7 **Demographic Transitions**
In this hypothetical representation, the four stages of the demographic transition are characterized by distinct changes in population birth rates and death rates. The total time span for all four stages exceeds 200 years.

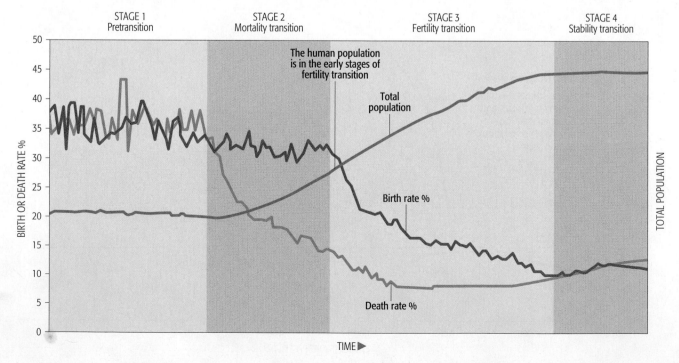

Latin America

Latin America offers a case study on demographic transition.

Over the past four centuries, the population of the Latin American region has generally followed the four stages of the demographic transition model (**Figure 5.8**). However, the region includes 46 separate countries, each with its own geography, history, and circumstances of economic development. As a result, the process of demographic transition varies considerably from country to country.

Stage 1. During the time of European colonization, from 1600 to 1900, Latin America was in the first stage of the demographic transition. The vast majority of families lived on farms. Most people were very poor, and many suffered from malnutrition. Diseases such as yellow fever and malaria were widespread. Reliable population data for the region during this time period are not available, but life expectancy in Latin America was generally less than 30 years; in some countries, it was as low as 25 years.

Stage 2. The mortality transition in Latin America probably began around 1900. As industrialization progressed, more and more people moved into cities. Data from 1950 to 1965 indicate that the death rate declined rapidly, from 1.6% to 1.2%. During that same 15-year period, birth rates declined only slightly, from 4.0% to 3.9%. Consequently, the population growth rate increased from 2.6% to 2.8%.

Stage 3. By 1970, the decline in birth rates had accelerated, and the population moved into the fertility transition. The well-being of Latin Americans improved dramatically, with an increase in personal wealth, access to clean water, a lower incidence of infectious disease,

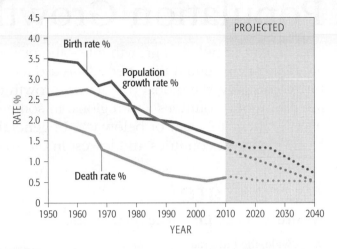

◄ **Figure 5.8 Latin America in Transition**
In 1950, Latin America was in the late stages of the mortality transition. The region's fertility transition began between 1960 and 1970. The United Nations Population Program projections indicate that Latin America will begin the stability transition between 2020 and 2030.
Source: U.S. Central Intelligence Agency World Factbook 2009.

and a rise in literacy (**Figure 5.9**). By 2010, the birth rate was only 1.73%, and the population growth rate declined to 1.0%. From 1965 to 2010, the median age of Latin Americans increased from 18 to 28 years.

Stage 4. If the current rate of development continues, the United Nations projects that Latin America will enter the stability transition near the year 2050. The birth rate is expected to continue to decline, dropping as low as 1.1%. With fewer children and greater life expectancy, the median age of Latin Americans will increase to 41 years. As the population ages, the death rate is expected to gradually increase to 0.9%. The population growth rate in Latin America will then be a mere 0.22%.

▼ **Figure 5.9 Latin America Then and Now**
As these pictures of Mexico spanning the last 100 years show, the past century has produced economic development and improved human well-being across parts of Latin America.

QUESTIONS

1. What factors contributed to the rapid population growth during the industrial period of human history?

2. During the first stage of the demographic transition, birth and death rates may differ widely from year to year but are nearly equal when calculated over decades. Explain.

3. Why does population growth rate increase at the beginning of the mortality transition?

4. During the stability transition stage, death rates may increase. Explain why.

5.2 Global Variation in Human Population Growth

The human population growth rate—the percent change in the number of people living in an area over a certain time period—is a direct consequence of births, deaths, and migration (4.3). Currently, the global population growth rate is approximately 1.11%, but this rate varies greatly among countries and regions. In some regions, the population growth rate is over 3%; in others it is near or below 0%. In general, the population growth rate is highest in the world's poorest countries and lowest in the world's wealthiest countries.

Birth Rate

Wealth and age influence birth rates.

Worldwide, the birth rate—the number of babies born each year expressed as a percent of the population—is decreasing. Between 1950 and 2010, it dropped from 3.75% to 1.94% (4.3). There is, however, considerable variation in birth rates among regions and countries.

Birth rates are highly correlated with the relative wealth of countries, as measured by per capita gross domestic product (GDP) (2.4). In the world's poorest countries, birth rates are relatively high, between 3.0% and 5.0%. In the wealthiest countries, birth rates are lower, between 0.8% and 1.5% (**Figure 5.10**). For example, in Ethiopia the birth rate is 4.34%, whereas in the United States it is 1.38%. Because of this difference in birth rates, the number of children born in Ethiopia in 2010 was nearly two-thirds the number born in the United States, even though the total population of Ethiopia is only about one-fourth that of the United States.

The difference between the birth rates of poor and wealthy countries can be largely attributed to differences in the age at which women bear children.

What number of children per family would produce zero population growth?

Brandon MacDonald, Citrus Community College

Norm: *The exact number depends on age-specific mortality rates. As childhood mortality rates diminish, the number of children per family at zero net population growth approaches two.*

Birth rates (%)
- above 5
- 4.1–5.0
- 3.1–4.0
- 2.1–3.0
- 1.0–2.0
- below 1

GDP per capita (U.S. dollars)
- above $30,000
- $20,001–$30,000
- $10,001–$20,000
- $5,001–$10,000
- $2,001–$5,000
- below $2,000

▲ **Figure 5.10 Birth Rate and Wealth**
These maps compare birth rates and wealth in countries across the globe. In map Ⓐ, notice that the United States has a low birth rate and the countries of sub-Saharan Africa have very high birth rates. Map Ⓑ shows the opposite trend for wealth—GDP is very high in the United States and very low in sub-Saharan Africa.

(A)

(B)

▲ **Figure 5.11 Comparing Birth Rates by Age**
During the years 2000–2004, the birth rates of women in the 10 poorest countries (blue) were significantly higher across all age groups than the birth rates of women in the 10 wealthiest nations (red). The difference is most striking for women between the ages of 15 and 19, where the birth rate is almost 9 times greater in poor nations than in wealthy ones.
Data from: United Nations World Population Prospects: The 2010 Revision Population Database.

Women in poorer countries typically have children at a younger age than do women in wealthier nations. **Age-specific birth rates** tell the number of children born in a year per 1,000 women within defined age groups (**Figure 5.11**). Between 2005 and 2010, for example, the birth rate for the 15- through 19-year age range in Ethiopia was 20.8, whereas in the United States the birth rate for that age range was 7.2. Note that age-specific *birth* rates describe age-related variation in birth rates for an entire population. In contrast, age-specific *fertility* rates describe the number of children born each year to an average female within an age class (4.3).

Higher age-specific birth rates among younger women shorten the time between successive generations. Shorter generation times lead to more children growing up and having children, causing a higher population growth rate. The effect of high birth rates among the young is magnified in poor countries, where there are typically more women in younger age classes. In Ethiopia in 2010, 15- to 19-year-old women outnumbered 45- to 49-year-old women by a ratio of 3.2 to 1. In the United States at that time, 45- to 49-year-old women slightly outnumbered 15- to 19-year-old women.

In Ethiopia, women who survive their full childbearing period will bear an average of 6.1 children. Therefore, Ethiopia has a total fertility rate of 6.1. American women who live through their childbearing years will have an average of 2.1 children, resulting in a total fertility rate of 2.1 (4.3). In the world's poorest countries, total fertility rates range from 5.0 to over 7.0, compared to a range of 1.3 to 2.1 in the world's 10 wealthiest nations (**Figure 5.12**).

Total fertility rates are calculated under the improbable assumption that all women within a population live through their reproductive years (ages 15–49). In wealthy countries, this is close to true. But it is not true in poor countries, where the high incidence of infectious diseases and limited access to medical care cause much higher death rates. In studies completed in 2005, for example, fewer than 3% of American women who lived to age 15 died before reaching 50. In contrast, over 18% of Ethiopian women within that age range died, often from complications associated with pregnancy or childbirth. Thus, although the total fertility rate for Ethiopia was 6.1, the average number of children born to women between ages 15 and 50 was actually only 4.8.

WORLD'S WEALTHIEST NATIONS

WORLD'S POOREST NATIONS

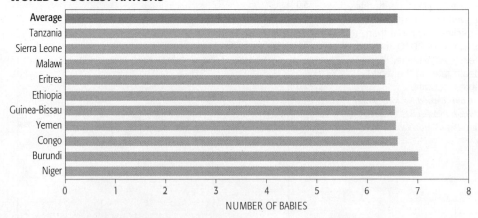

▲ **Figure 5.12 Total Fertility Rate and Wealth**
Total fertility rates among the world's poorest countries are approximately three times higher than among the world's wealthiest countries.
Data from: United Nations World Population Prospects: The 2010 Revision Population Database.

125

Death Rate

Wealth and age also influence death rates.

The death rate is the percent of individuals in a population that die each year (4.3). Like birth rates, death rates are highly correlated with the level of economic development. Over the past 35 years, the worldwide death rate has declined from 1.95% to its current rate of 0.85%. The decline in death rate corresponds to improved nutrition, sanitation, and health care. The healthier we are, the lower our death rate. Because conditions affecting human survival in the world's poorest countries remain well below the standards of developed countries, death rates in the world's poorest countries are nearly twice those of the wealthiest countries (**Figure 5.13**).

A significant component of the higher death rates in poor countries is infant mortality. **Infant mortality rate** is the percent of infants within a population who die before age 1. Infant mortality is heavily influenced by factors such as sanitation, water quality, and the availability of health care before and after birth. Babies born in poorer countries are particularly vulnerable to disease and malnutrition during their first year of life. As a result, infant mortality rates in the world's poorest countries are 18 times higher than in the wealthiest countries (**Figure 5.14**). Of the babies born in Ethiopia in 2010, 7.1% died before age 1;

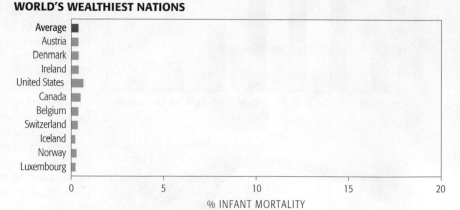

WORLD'S WEALTHIEST NATIONS

WORLD'S POOREST NATIONS

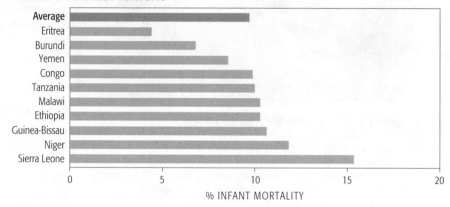

▲ **Figure 5.14 Infant Mortality and Wealth**
Infant mortality rates in the world's 10 poorest countries are 10 to 20 times higher than in the 10 wealthiest countries. The variation in these rates among poor countries reflects significant differences in nutrition, sanitation, and access to health care.
***Data from:** United Nations World Population Prospects: The 2010 Revision Population Database.*

► **Figure 5.13 Death Rates Compared in Poor and Wealthy Countries**
In the last 60 years, death rates in the world's poorest countries have declined by more than 40%. But as of 2010, they were still nearly 70% higher than the death rates in the world's wealthiest countries.
***Data from:** United Nations World Population Prospects: The 2010 Revision Population Database.*

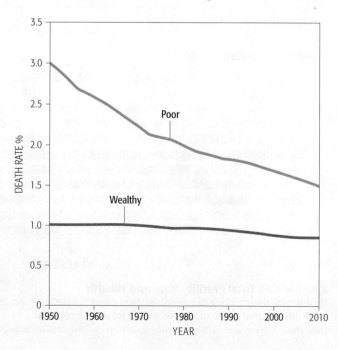

in the United States, only 0.6% of the babies died before their first birthday.

Death rates that deal specifically with age can be expressed in terms of survivorship, the number of years that a specific group of individuals born into a population lives (4.3). Like birth and death rates, patterns of survivorship differ between wealthy and poor countries. For example, in the African country of Niger, only about 60% of the population survives to age 25, compared to 97% in the United States (**Figure 5.15**). Nutrition, sanitation, and access to health care account for much of this difference.

In some of the world's wealthiest countries, increasing survivorship has lead to a slight increase in death rates. For example, during the last 25 years, the death rate in Canada has risen from 0.69% to 0.74%. The higher death rates in wealthy countries are a direct consequence of the increasing average age of individuals within their populations—older people are more apt to die than younger people.

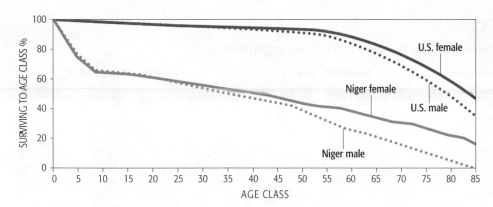

◄ **Figure 5.15 United States and Niger**
Survivorship curves show the percentage of people living to a particular age. These curves are for men and women in the United States and Niger. In 2003, over 98% of children born in the United States lived to be at least 5 years old, compared to only 66% in Niger. In both countries, differences in survivorship between males and females are small up to ages 40–50. After age 50, women have considerably better survivorship.

Data from: United Nations World Population Prospects: The 2010 Revision Population Database.

The survivorship of men and women may differ significantly. Adolescent women often have somewhat higher survivorship than adolescent men, owing to the higher propensity of young men to be involved in accidents and to exhibit violent behavior. In some developing countries, however, poor health care associated with childbearing results in lower survivorship among young women.

Nearly everywhere in the world, the survivorship of women beyond the age of 40 exceeds that of men. The reasons why women live longer than men are not well understood. Men appear to be more susceptible to heart disease, the most significant cause of death in developed countries. Men are also more likely to die in violent situations and to indulge in risky behaviors, such as smoking and drug use, which increases their chances of earlier death.

Life expectancy at birth, the average age to which a baby born at a given time will live, provides a summary of differences in survivorship. As with survivorship, life expectancy is generally higher for women than for men. In 2010, for example, the life expectancy of a baby girl born in the United States was 82.1 years, and that of a baby boy was only 77.7 years.

From a global perspective, life expectancy has increased dramatically over the past five decades, from 45.6 years to 68.9 years. This increase in life expectancy is largely a result of improvements in the production and distribution of food, cleaner water, and advances in medicine and health care. Sadly, the differences in life expectancy between poor and wealthy countries that existed in 1950 persist today (**Figure 5.16**). These differences reflect significant variations in human well-being among countries.

▲ **Figure 5.16 Life Expectancy and Wealth**
Over the past 50 years, life expectancy has increased in all countries. However, an average baby born in one of the world's 10 poorest countries will live fewer than 50 years. In contrast, a baby born in one of the world's 10 wealthiest countries has a life expectancy of nearly 80 years.

Data from: United Nations World Population Prospects: The 2010 Revision Population Database.

Age Structure

A population's age structure is a key factor in its future growth.

Survivorship and age-specific birth rates determine a population's age structure, the distribution of a population based on age classes. **Age-structure pyramids**, also known as *age-sex pyramids* or *population pyramids,* illustrate graphically how populations are apportioned according to age and gender.

In regions where age-specific birth rates are high and survivorship is low, populations have a large proportion of young people. In regions with low age-specific birth rates and high survivorship, age-structures are more evenly distributed. In 2010, the median age among Ethiopians was 17.2 years, meaning that half of the population was older than 17.2 years and half was younger. In the United States, the median age was 38.1 years, more than twice the median age of the Ethiopian population. Half of all Ethiopians were younger than 17 years, compared to only about one fifth of Americans (**Figure 5.17**).

The age structure of a population indicates present population growth rates; it also forecasts future growth rates. Where the median age is young and a large proportion of the population is of reproductive age, overall population growth rates are likely to be high. This is because young people have comparatively high birth rates and low death rates (**Figure 5.18**).

▲ **Figure 5.17 Population Age Structures**
In 2010, the median age in Ethiopia was less than half that of the United States. A much larger proportion of Ethiopia's population falls into the reproductive age range of 15–49 years.
Data from: *United Nations World Population Prospects: The 2010 Revision Population Database.*

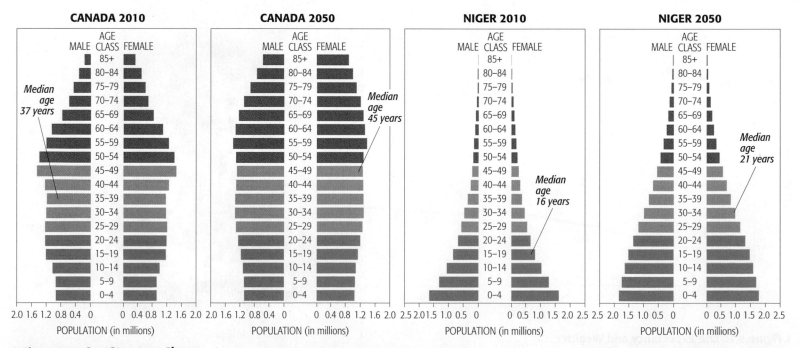

▲ **Figure 5.18 Age Structure Change**
The current population age structures of Niger and Canada reflect significant differences in their age-dependent fertility rates and survivorship. These differences are projected to result in very different population sizes and age structures by the year 2050.
Data from: *United Nations World Population Prospects: The 2010 Revision Population Database.*

Migration

The relocation of humans is influencing human population growth in many regions.

At any given time, some people are moving into a country (immigrating) and others are moving out (emigrating). Demographers calculate the **net migration rate** as the difference between immigration and emigration per 1,000 individuals in the population. Net migration rates can have a very significant effect on population growth rates. If a population increases because large numbers of people move in, that population has a *positive* net migration rate. If a population decreases because many people move out, it has a *negative* net migration rate. In general, the wealthiest countries are experiencing positive net migration, and the world's poorest countries vary between positive and negative net migration (**Figure 5.19**).

Among the world's wealthiest nations, positive net migration rates account for 65% of total population growth. In the United States, the rate of population growth in 2010 was close to 1%, and the net migration rate was 0.43%. This rate indicates that immigration accounted for 43% of the total population growth. In Austria, the net migration rate was 0.18%, and the total population growth rate was only 0.06%. The rate of immigration into Austria

actually exceeded its population growth rate. Were it not for immigration, Austria's population would actually have decreased by 0.12%.

Net migration is often influenced by situations in neighboring countries. **Push factors** are conditions that force people to emigrate. Push factors include epidemics, drought, famine, war, religious or political persecution, and crowded conditions (**Figure 5.20**). Conditions that encourage people to immigrate into a country are called **pull factors**. Pull factors include opportunities for employment and education, higher living standards, and religious or political freedom.

Migration into poor countries is often due to push factors, such as refugees escaping war in nearby countries. In 2002, both Eritrea and Sierra Leone—two of the world's poorest nations—were surrounded by war-torn countries. That year, the net migration rates into Eritrea and Sierra Leone were three to four times greater than the net migration rate into the United States. Immigration into poor countries complicates and often thwarts efforts for economic and social development.

In the wealthiest countries, aging populations and low rates of population growth have resulted in dwindling workforces. Immigration often provides the additional labor needed to maintain economic output. However, high rates of immigration can lead to political conflicts. In the United States, for example, the cost of meeting immigrants' needs for education, health care, and other public services is frequently a source of contention.

▲ **Figure 5.20 Emigration Push Factors**
Disease, famine, war, and persecution continue to push people to migrate from their home countries. (The refugees pictured here have fled drought and famine in Afghanistan.)

WORLD'S WEALTHIEST NATIONS

NET MIGRATION RATE %

WORLD'S POOREST NATIONS

NET MIGRATION RATE %

▲ **Figure 5.19 Net Migration Rates and Wealth**
The "pull" of favorable economic and social conditions produces positive net migration rates into wealthy nations. Patterns of migration for poor nations are more complex, being influenced by internal "push" factors as well as conditions in neighboring countries.
Data from: United Nations World Population Prospects: The 2010 Revision Population Database.

QUESTIONS

1. Give two reasons why high fertility rates among young women have a greater impact on population growth rate than equally high rates among older women.

2. Demographers expect death rates to increase in many wealthy countries in the years ahead. Why?

3. Net migration is positive in some poor countries and negative in others. What factors contribute to this difference?

5.3 Predicting Human Population Growth

Predictions about human population growth are usually based on the logistic population model and assumptions about human values and behaviors (4.4). The English philosopher Thomas R. Malthus was the first to predict limits on the exponential growth of human populations. He also identified correlations among human well-being, culture, and population growth. Since Malthus's time, those predicting human population growth have considered human well-being and culture. However, their predictions have varied considerably because of the quality of the demographic data available and changing assumptions about human values and behavior.

Thomas Malthus

Malthus recognized the connection between human well-being and human population growth.

Working in the late 1700s, Malthus explored the effects of limited resources on human populations by comparing data on population growth in the American colonies and on the European continent. Malthus found that in the American colonies, where resources were abundant, human populations were doubling every 25 years (4.3). In Europe, where resources were much more limited, doubling times were considerably longer (**Figure 5.21**).

Malthus noted that human populations grow exponentially, but the resources upon which humans depend typically do not. In fact, Malthus predicted that exponential human population growth would lead to widespread poverty and diminished human well-being, or what he called "misery and vice." By *misery*, Malthus was referring to the effects of starvation and its associated suffering on human survival and reproduction.

In Malthus's view, the availability of resources, or what he described as "the prospects for human happiness," related directly to the decisions people made about reproduction. If resources were limited, Malthus observed, people were apt to marry late in life or not at all. Furthermore, couples took measures to avoid conception to control the number of children they had. Reflecting the religious and social attitudes of his time, Malthus saw the avoidance of marriage and childbearing as *vices*.

▼ **Figure 5.21 18th Century Contrasts in Human Well-Being** Malthus observed that limited resources produced widespread poverty and lower rates of population growth in Europe (right) compared to the American colonies (below) in the 1700s.

Population Growth Forecasts

Forecasts for human population growth rates are based on past growth and assumptions made about future human values and behavior.

Statisticians Raymond Pearl and Sophia Gould were among the first to use logistic growth models to predict human population change (4.4). In their 1936 paper, "World Population Growth," Pearl and Gould argued that the rate of global human population growth was slowing. Applying the logistic model to the very limited data that were available in the 1930s, they estimated the world's carrying capacity to be 2.6 billion people (4.4). Assuming that the birth rate and death rate trends of their time would continue, Pearl and Gould predicted that the world population would reach its carrying capacity in 2100.

Pearl and Gould based their predictions on data gathered during the period between World Wars I and II. This was a time of widespread economic depression. Demographers of their time assumed that the worldwide poverty would continue. But in fact, the two decades following World War II were characterized by economic progress, high birth rates, and dropping death rates. As a result, population growth accelerated rapidly, and the world population surpassed Pearl and Gould's estimated carrying capacity in 1953.

In 1968, Paul R. Ehrlich's book, *The Population Bomb*, quickly became a best-seller. Ehrlich warned that if the human population growth rate continued to accelerate,

the world's exploding population would result in "a dying planet." Four years later, environmental scientists Donella Meadows and Dennis Meadows, working with a team of scholars, published *The Limits to Growth*. This highly influential book forecast population growth that far exceeded the Pearl and Gould predictions (**Figure 5.22**). Meadows and Meadows predicted that the human population would surpass 7.4 billion in the year 2000. They also suggested that as many as 15 billion people might be competing for Earth's limited resources by the year 2050.

The forecasts made by Meadows and Meadows were based on the assumption that the birth rates and death rates of the 1960s would continue. Around 1970, however, the global population growth rate began to decrease. People in many developing countries began having children later in life, a behavior that significantly slowed the rate at which new individuals were added to the global population. Instead of the 7.4 billion people that Meadows and Meadows had predicted, the world population tipped just over 6 billion in 2000.

Demographers such as Pearl and Gould, Ehrlich, and Meadows and Meadows could not foresee how human values and behavior would change with economic development. In later publications, Ehrlich and Meadows and Meadows adjusted their forecasts using updated data and new assumptions about future human values and behavior.

The latest United Nations forecast on population growth is that the global growth rate will decrease to about 0.4%, and the world population will reach approximately 9.3 billion by the year 2050 (**Figure 5.23**). Like past predictions, current predictions of moderate population growth are based on assumptions about human values and behaviors. Not only do human values and behavior affect birth and death rates in a population, they also influence the ways in which people use resources.

QUESTIONS

1. Why did Malthus expect population growth in the 18th century to be faster in the American colonies than in Europe?

2. Forecasts of human population growth depend on assumptions about human values and behavior. Explain.

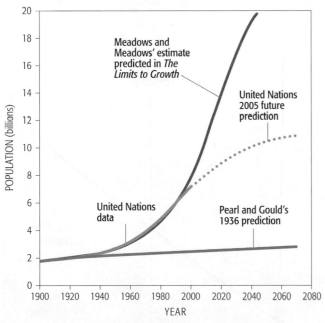

▲ **Figure 5.22 Contrasting Forecasts for Human Population Change**
The forecasts of world population growth made by Pearl and Gould in 1936 (green line) and Meadows and Meadows in 1972 (red line) were based on vastly different assumptions about factors affecting future birth and death rates. The blue line shows actual census data up to 2010; between 2010 and 2050, the blue line shows the current forecast made by the United Nations Developmental Program.

Data from: United Nations World Population Prospects: The 2010 Revision Population Database.

◀ **Figure 5.23 Declining Population Growth Rates**
Worldwide, population growth rates have been decreasing since 1970. The United Nations Department of Economic and Social Affairs forecast is for that trend to continue through the middle of this century.

Data from: United Nations World Population Prospects: The 2010 Revision Population Database.

5.4 Resource Use and Population Sustainability

According to the current United Nations forecast, the world population will reach 9 billion people in approximately four decades. Can Earth sustain that many people? Whether or not a human population is sustainable depends not only on its size but also on the demands that individuals in the population make on their environment's resources (1.3). The resources available within countries vary widely, as do the demands for resources made by the citizens of those countries. Generally, the more affluent and technologically advanced a population, the greater its impact on the ecosystems upon which it depends.

Sustainability versus Carrying Capacity

A human population is unsustainable long before it nears its carrying capacity.

How many people can Earth support in a sustainable fashion? Asking such a question is often thought to be the same as asking, What is Earth's carrying capacity? Yet the two questions are not the same.

Carrying capacity refers to the point at which competition for limited resources causes birth rates and death rates to reach equal numbers, resulting in zero population growth. Theoretically, a population can be maintained at carrying capacity as long as its supply of resources remains constant. But at carrying capacity, competition for resources is very intense. The lower birth rates and higher death rates characteristic of populations nearing their carrying capacity indicate that stresses on individuals are extreme (**Figure 5.24**). The well-being of individuals within a population at carrying capacity is considerably worse than that experienced by individuals in a population that is well below its carrying capacity.

As a human population approaches its carrying capacity, suffering increases. Because human well-being requires more than mere survival, the sustainability of a human population must be below carrying capacity. In fact, sustainability means that enough natural resources are available to guarantee a sufficient level of well-being for generations to come (1.2).

Thus, two things must be established to answer the question, How many people can Earth support in a sustainable fashion? First, we must agree on a standard of human well-being that is sustainable. Second, we must establish the capacity of Earth's ecosystems to provide goods and services to meet that standard.

▲ **Figure 5.24 Carrying Capacity and Well-Being**
As a population's size nears carrying capacity, the quality of life of its members, as measured by decreasing birth rates and increasing death rates, declines.

Ecological footprint of
average Haitian citizen

Ecological footprint of
average U.S. citizen

◀ **Figure 5.25 Ecological Footprints**
An ecological footprint represents the amount of land needed to supply the demand for ecological goods and services. The ecological footprint of an average U.S. citizen is much larger than that of an average Haitian.

Human Resource Use

Ecosystem goods and services determine the potential growth of human populations.

The impact of a human population on an environment depends on the total number of people living there and on the consumption habits of each individual within that population. If individuals in a population consume increasing amounts of an ecosystem's goods and services, their collective impact on the environment increases, even if the size of the population does not change. Because human consumption varies so dramatically across the globe and over time, estimates of the number of humans that any given environment can sustain are highly variable.

The extent to which humans consume ecosystem goods and services is often described in terms of an **ecological footprint** (**Figure 5.25**). An ecological footprint is an estimate of the area of land needed to supply human demands for resources. Such demands include shelter, energy, food, and land to absorb wastes. For example, the construction of a house requires timber that grew on a certain amount of land. Housing construction also depends upon the land used to build factories to make concrete for foundations, roofing materials, and many other supplies. In a similar respect, the fossil fuels used to heat homes and to power cars are extracted from land and processed in refineries built on land. Even the carbon dioxide emissions from burning those fossil fuels can be converted into the acres of forest required to absorb those emissions.

Sometimes ecological footprints are calculated as the total land area needed to sustain an entire

POOR COUNTRIES

DEVELOPING COUNTRIES

WEALTHY COUNTRIES

▲ **Figure 5.26 Footprint Components and Wealth**
As countries become more affluent, their footprints are increasingly determined by the extraction and burning of fossil fuels and the generation of electric power.
Data from: Venetoulis, J., et al., 2004, Ecological footprint of nations. *Oakland CA: Redefining Progress.*

Key
● Fossil fuels
● Cropland
● Forests
● Fisheries
● Built
● Nuclear and hydropower

population. More often, ecological footprints are expressed as the land area needed to sustain an average person within a population, or the total ecological footprint divided by the population size. The size of ecological footprints varies greatly from country to country. As a country becomes more affluent, its ecological footprint is increasingly defined by its use of fossil fuels and nuclear and hydrologic power (**Figure 5.26**).

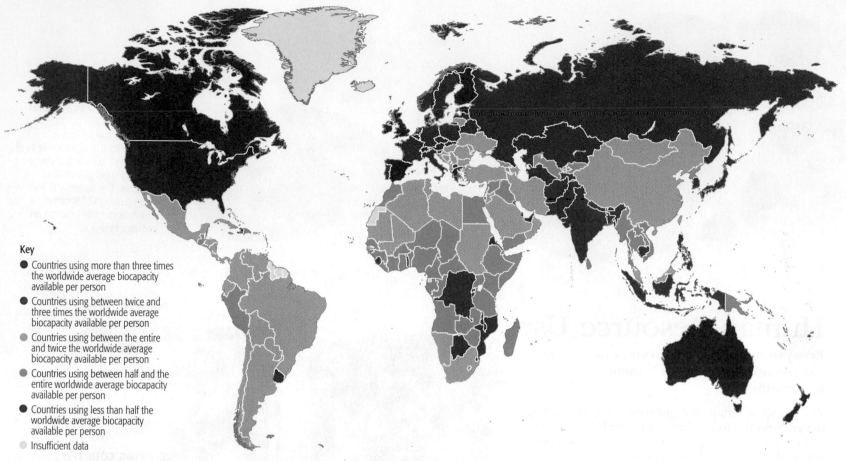

Key

● Countries using more than three times the worldwide average biocapacity available per person

● Countries using between twice and three times the worldwide average biocapacity available per person

● Countries using between the entire and twice the worldwide average biocapacity available per person

● Countries using between half and the entire worldwide average biocapacity available per person

● Countries using less than half the worldwide average biocapacity available per person

● Insufficient data

▲ **Figure 5.27 Ecological Footprints and Biocapacity** This map compares the ecological footprint of the average citizens of each country to the average worldwide biocapacity available to each person. In the wealthy nations, the average footprint exceeds worldwide biocapacity; in poor nations, the average footprint is less than worldwide biocapacity. *Source: Global Footprint Network, www.footprintnetwork.org.*

The ecological footprint of a human population is often compared to its **biocapacity** (**Figure 5.27**). Biocapacity is a measure of the area and quality of land available to supply a population with resources. Countries with small populations relative to their total land area, such as Australia, Brazil, Iceland, have biocapacities that are larger than their ecological footprints.

A population exceeds its biocapacity when it draws a portion of its ecosystem goods and services from outside its geographical boundaries. For example, Haiti has the relatively small ecological footprint of 1.2 acres per person. However, deforestation and environmental degradation have left Haiti with a biocapacity of only 0.7 acres per person; thus, its ecological footprint exceeds its biocapacity. As a result, the people of Haiti must depend on ecosystem goods and services imported from other countries.

The average American has a very large ecological footprint, with each person requiring about 23.4 acres of land to supply food, housing, transportation, and other needs. The biocapacity to meet these needs within the United States is 12 acres per person. The ecological footprint of Americans extends well beyond the borders of the United States. It can only be accommodated by appropriating the biocapacity of other parts of the world. Countries whose ecological footprints exceed biocapacity are sometimes called *ecological debtors*.

The great majority of the world's countries use more resources than are available within their borders. Some estimates suggest that the combined ecological footprint of the world's countries—each living its current lifestyle—already exceeds global biocapacity by approximately 40%. If all countries adopted the lifestyle of the average North American, the world would exceed its biocapacity five times over. Put another way, it would take five planet Earths to sustain the planet's 6.9 billion people according to the North American lifestyle. Between 1991 and 2001, the world population increased by 15%, and the ecological footprint of the average world citizen shrank by 2%. Because of the high population growth during that decade, the biocapacity available per individual declined considerably. In that time, the average ecological footprint in the world's 50 wealthiest countries increased by 8%, while that of the 50 poorest countries declined by 11%. In other words, the average person in a wealthy country is making ever-greater demands on Earth's resources, while the average person in a poor country is consuming much, much less. As a result, the gap in resource use between wealthy and poor people is growing wider.

Affluence and Technology

Affluence and technology impact the environment at a global scale.

The different sizes of the ecological footprints of rich and poor countries illustrate the central role of affluence in the rate of consumption of resources. A great deal of modern consumption is related to the use of technology, such as appliances for cooking and cleaning, cars, and computers. The consumption of resources and the use of technology impact the global environment.

Scientists Paul Ehrlich and John Holdren have proposed that human impact is determined by population size, affluence, and technology. They have summarized this relationship in the **IPAT equation**: $I = P \times A \times T$ (**Figure 5.28**). In this equation, I represents environmental impact (loss of resources and degradation of ecosystems). P represents population (including size, growth, and distribution). A represents affluence (consumption per individual), and T represents technology (things that demand resources and energy). Although some technologies improve the environment, the T in the IPAT equation represents the technologies that are destructive to the environment. The IPAT equation is a conceptual model; it is not designed to calculate numerical values.

The IPAT equation emphasizes that the size of a population is not the only factor to consider when determining its impact on the environment. Consumption of ecosystem goods and services and use of technology are also key factors. For the sake of demonstration, you can calculate the environmental impact of an imaginary population by plugging simple numbers into the IPAT equation: $P = 2$, $A = 2$, $T = 2$. The equation becomes $I = 2 \times 2 \times 2$, or $I = 8$. If the population size (P) doubles to 4, its impact (I) also doubles: $I = 4 \times 2 \times 2 = 16$. Note that the impact on the environment would be the same if the size of the population (P) remained constant but its use of technology (T) doubled: $I = 2 \times 2 \times 4 = 16$.

Of course, the interaction of the factors is not as simple as the IPAT equation may suggest. A change in one factor invariably causes another factor to change. For example, increases in population may lead to increases in technology. Increases in affluence also tend to increase technology.

Much of the impact of affluence and technology is associated with the use of energy to power cars and generate electricity. Although people living in developed countries make up only about 16% of the world population, they consume approximately 75% of the total energy used each year. Ehrlich and Holdren estimate that the average citizen of a developed country has 7.5 times more impact on Earth's ecosystems than the average inhabitant of a developing country. Based on energy use alone, Ehrlich and Holdren suggest that the environmental impact of an average American is nearly 30 times greater than that of a citizen of Haiti or Niger.

When do we need to "worry" about population growth?

Natalie Hollabaugh, Central Michigan University

Norm: *For the foreseeable future, we should constantly monitor our numbers and act to limit the impact of our population's resource use on Earth's ecosystems and our own well-being.*

▼ **Figure 5.28 IPAT**
The IPAT equation proposes that human impact (**I**) is determined by population size (**P**), affluence (**A**), and technology (**T**).

Impact = **Population** × **Affluence** × **Technology**

loss of resources and degradation | including size, growth, and distribution | consumption per individual | things that demand resources

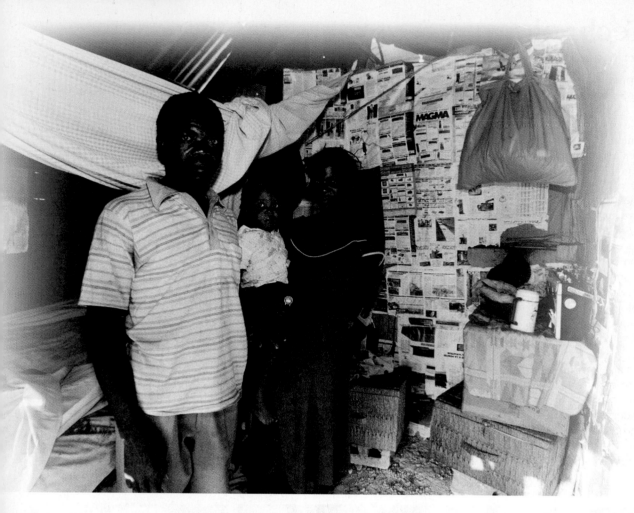

increase dramatically. For example, between 1980 and 2001, people in developing countries in the Middle East, North Africa, and Asia increased their consumption of electricity by over 370%. During those two decades, the population in the developing countries of Asia grew 1.7%, and the region led the world in pace of economic expansion.

China is the world's most populous country. It is also the second-largest consumer of energy, behind the United States. On average, an American citizen uses over 5 times more energy than a Chinese citizen. However, between 1990 and 2010, per capita energy use in China grew by over 100% compared to growth of about 12% in the United States.

Because the growth rate of the world's human population is declining, there is hope that we may be able to achieve sustainability. Yet to reach this goal, the rate of consumption will need to change, particularly in the world's richer countries (**Figure 5.30**). Sustainable development will require increased conservation of ecosystem goods and services. New technologies that help us use those goods and services more efficiently may be the answer. Greater efficiency may provide the means to diminish our collective footprint while sustaining our well-being.

The people of developing countries are striving to improve their well-being by acquiring more material goods and better access to technology (**Figure 5.29**). As the population and per capita consumption of developing countries increase, their impact on the environment will

QUESTIONS

1. Explain why a human population at carrying capacity would not be considered sustainable.

2. Although Brazil has 11% less total land area than China, its biocapacity is 77% greater. What factors may be contributing to this difference?

3. Even if the populations of many developing countries were to stabilize, their impacts on the environment are likely to continue to increase. Explain why.

Present: World population 7 billion
With its present population of 7 billion, Earth's natural resouces are supporting many millions of people at a high level of expected well-being and many more millions at a low level.

2050: World population 9 billion
If, as expected, the world's human population increases by nearly 50% by 2050, expectations of what is a sufficient level of well-being will have to decrease.

Future: World population 13 billion
If the world's population doubles to over 13 billion, there will have to be a radical shift in expectations with the major consumers modifying their behavior, attitudes, and lifestyles to minimize consumption.

▲ **Figure 5.30 Population Growth Scenarios**
The ability of Earth's environment and natural resources to help future generations meet their needs depends not only on how populations grow but also on how demands for resources change.

FOCUS ON SCIENCE: Economic Development and Measures of Human Well-Being

◀ **Figure 5.31 Sir Partha Dasgupta** Sir Partha Dasgupta is currently the Frank Ramsey Professor of Economics at Cambridge University.

Must poor, developing countries give up civil rights and political liberty in order to improve other measures of well-being?

Growing up in India, Sir Partha Dasgupta witnessed much poverty (**Figure 5.31**). That experience kindled his interest in the relationship between the high population growth in developing countries and economics, technology, and nutrition.

In his paper, "Well-being and the Extent of Its Realization in Poor Countries," Dasgupta noted that per capita GDP, infant mortality, and life expectancy are the measures most often used to describe individual wealth and human well-being (2.4). He then argued that civil and political liberties also factor into individual wealth and human well-being. Civil liberties include freedom of the press and independence of the judicial system. Political liberties include the right to take part in selecting leaders and the formulation of laws (**Figure 5.32**).

Some social scientists have argued that poor countries must forgo political and civil liberties in order to achieve rapid economic growth, improved health, and diminished population growth rates. To test this proposition, Dasgupta amassed and analyzed data from 50 of the world's poorest countries. The data sets for each country measured the change in human well-being during the decade from 1970 to 1980. The data showed changes in per capita GDP, life expectancy, infant mortality rate, and literacy rate. Dasgupta then compared the changes in wealth and well-being with independent numerical ratings of each country's system of political and civil rights, which were available from the United Nations.

Dasgupta found that wherever GDP increased, infant mortality rates decreased. However, the greatest decrease in infant mortality rates was found

where political and civil liberties were highest. Countries that ranked high with regard to political and civil liberties, such as Botswana, Sri Lanka, and Gambia, had lower infant mortality rates. Countries that had poor rankings of political and civil liberties, such as Rwanda, Uganda, and Ghana, had much higher infant mortality rates.

Dasgupta's results show that poor countries can experience all aspects of human well-being—economic progress, quality health care, increased life expectancy, and so forth—while also promoting political and civil rights. In subsequent work, Dasgupta has explored connections among economics, human population growth, food resources, and gender relations. Currently, he is working with ecologists to identify economic means to ecological sustainability.

Source: Dasgupta, P., 1990. Well-being and the extent of its realization in poor countries. The Economics Journal 100: 1–32.

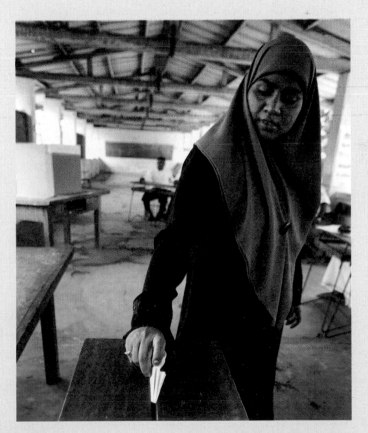

▲ **Figure 5.32 Economic Development and Civil Liberty** Many countries have experienced significant economic development and improvement in human well-being, including the right to vote and other civil liberties.

5.5 Managing Population Growth

Efforts to manage human population growth began a century before the establishment of demography as a science and the development of the demographic transition model. In the early 1800s, people who were influenced by Malthus began to advocate birth control and limitations on family size. At the advent of the 20th century, providing greater access to family planning emerged as a practical solution to the human population growth problem. As the 20th century came to a close, it was understood that economics, education, the empowerment of women, and family planning must be considered when attempting to manage human population growth.

Family Planning

Family planning advances on both national and international fronts.

In 1822, the British politician Francis Place published a book called *Ilustrations and Proofs of the Principle of Population*. This book initiated the **neo-Malthusian movement**. Like Malthus, the neo-Malthusians believed that as the human population increased, it would soon outpace the supply of food. Place and his fellow neo-Malthusians were the first to advocate contraception. (Ironically, Malthus himself was firmly against contraception as a means to manage population growth.) At first the ideas of the neo-Malthusians were dismissed, but by the end of the 19th century many scholars and some governments believed that there was a need to manage human population growth and to promote contraception.

In the first half of the 20th century, two world wars and the Great Depression consumed the attention of the United States. Consequently, national efforts to slow population growth gained only modest momentum. In 1916, Margaret Sanger succeeded in opening the first birth control clinic in the United States, where she provided poor women with access to contraception. In 1921, she founded the American Birth Control League, which later became Planned Parenthood. Despite sometimes violent opposition, scores of birth control clinics opened over the next two decades.

The period immediately following World War II produced a **baby boom** within the United States, as returning soldiers and their spouses focused on starting and growing families (**Figure 5.33**). The 1950s and 1960s were a period of international dialogue and activity on the subject of human population growth. In 1952, representatives of numerous countries, including India, the United Kingdom, and

the United States, ratified the constitution of what is now the International Planned Parenthood Federation (IPPF). A nongovernmental organization, the IPPF promoted sexual and reproductive health and the rights of individuals to make their own choices in family planning.

By the 1970s, the IPPF and other nongovernmental organizations such as Family Life and Family Health International were working to slow human population growth. They did so through promotion of universal access to contraceptives, such as birth control pills and condoms, as well as family planning, educational programs, and counseling services.

About the same time as the ratification of the IPPF constitution, the United Nations (UN) began to support international family planning organizations and to construct a network for international conferences. In these conferences, representatives from countries around the world met to negotiate solutions to population growth issues. In 1954, the first UN World Population Conference resolved to gather more complete

▶ **Figure 5.33 The Baby Boom**
The economic growth that followed World War II fostered a period of unprecedented population growth and development known as the baby boom.

information on population changes in developing countries. The conference also resolved to found regional training centers where local population issues could be addressed. Since then, UN World Population Conferences have been held every 10 years. Recent conferences have focused on the relationships between population growth, economic development, and human rights.

Today, family planning policies encompass a broad range of strategies and priorities. Access to contraceptives remains a high priority. Programs that deal with education on reproductive health, including the prevention, diagnosis, and treatment of HIV/AIDS, have also become important for governmental and nongovernmental organizations.

Development and Population

Economic development, education, and empowerment of women decrease human population growth.

The 1974 UN World Population Conference marked a milestone in international negotiations on human population. Developing nations argued that *economic development,* not expanded family planning programs, offered the most promising approach to slowing population growth and alleviating poverty. Falling into line with Thompson's demographic transition model, developing countries adopted the motto, "development is the best contraceptive."

In 1994, the international dialogue on slowing population growth shifted from economic development and the alleviation of poverty to education and the empowerment of women. That year, the International Conference on Population and Development held in Cairo, Egypt, produced the Cairo Consensus.

The Cairo Consensus emphasized improvements in health care, well-being, and rights—particularly for women—as a way to achieve the social, political, economic, and environmental stability necessary to slow population growth. Behind the Cairo Consensus was this reasoning: Once basic educational and health care needs are met, an individual is in a better position to secure employment and to make deliberate decisions about family size. Employment provides greater economic opportunity for both men and women. It also reduces the perceived need for domestic labor, which is often the motivation for having numerous children. Education

and employment also give women greater social status than the traditional roles of bearing and rearing children (**Figure 5.34**).

▲ Figure 5.34 **Education and Population Growth**
Where economic development has improved educational opportunities for women, birth rates and population growth rates have declined.

Too High or Too Low?

Some countries consider their population growth to be too high, whereas others consider theirs to be too low.

A 2003 UN report found that over three-quarters of the world's 50 least-developed countries consider their rate of population growth to be too high. Many of these countries support the provision of contraceptive methods through governmental health clinics or in collaboration with nongovernmental organizations. These nations focus their attention on opportunities to lower birth rates while also reducing infant mortality. However, efforts to lower fertility and infant mortality rates can be complicated or derailed by factors such as HIV/AIDS, civil unrest, and war.

In contrast to these developing countries, 40% of developed countries surveyed by the UN, including Japan, Spain, and Italy, consider their population growth rates to be too low. These countries are considering policies to encourage women to have more children. Developed countries are finding that low birth rates and increased life expectancy result in aging populations. In an aging population, fewer and fewer people are willing and able to work. Slower population growth and a shrinking workforce have potentially significant economic consequences for these countries.

In what ways would the government be able to recognize that the population is getting out of hand?

Megan Reinle, Northern Illinois University

Norm: *Symptoms of overpopulation or excessive resource use include increased air and water pollution and shortages of essential resources. They also include measures of decreasing human well-being, such as high infant mortality and diminished life expectancy.*

Three Approaches to Population Growth

China, Mexico, and India offer case studies on various approaches to managing human population growth.

Are policies on having children such as China's "one child per family" rule really necessary?

Jaymee Castillo, University of California, Los Angeles

Norm: *National policies on population growth are essential, but they need not be as restrictive as China's.*

As with all other aspects of human population and well-being, the strategies for managing human population growth and their results vary greatly across regions and countries. The policies of China, Mexico, and India show three different approaches to managing national population growth.

China. In the early 1970s, China began a public campaign to reduce its population growth, which culminated in 1978 with the One-Child Policy. The One-Child Policy aimed to reduce China's population growth rate from 1.2% to 0% by the year 2000. Through education, financial subsidies, and strict penalties, China encouraged couples to have only one child. China's population growth rate is currently about 0.6%, and its birth rate has dropped from 3.6% to 1.32%. It appears that the One-Child Policy is working, and the UN forecasts that China's population growth rate will hit 0% by 2030 (**Figure 5.35**). China's success in reducing its birth rate and population growth has received considerable attention. Some of that attention is focused on how the One-Child Policy may be an infringement on the civil and personal rights of the Chinese.

about the loss of civil rights, Mexico's National Family Planning Program has achieved results very similar to those of China.

India. In some ways, India's approach to population management is similar to that of Mexico, but neither India's political and economic commitment nor its progress in reducing its population growth rate has matched Mexico's. In 1966, India created a Department of Family Planning within its Ministry of Health. That program was firmly rooted in the philosophy that family planning should be voluntary and decisions regarding strategies for managing population growth should be delegated to individual states. In 1986, the Department of Family Planning began an educational program to encourage couples to postpone marriage and children. This program was developed in the context of a larger governmental campaign to increase child survival, elevate women's status and employment levels, and increase literacy in the general population.

In 2010, India's population growth rate was 1.3% and its birth rate was 2.1%. Although India's birth rates are declining in accordance with the fertility transition of the demographic transition model, its population growth rate will not drop below 1% until about 2020. Then the country's population will be nearly equal to China's projected 2020 population of 1.35 billion.

► Figure 5.35 **China's Changing Demography** Implemented in 1978, China's One-Child Policy has had a significant effect on birth rates and population growth rates. *Source: United Nations World Population Prospects: The 2010 Revision Population Database.*

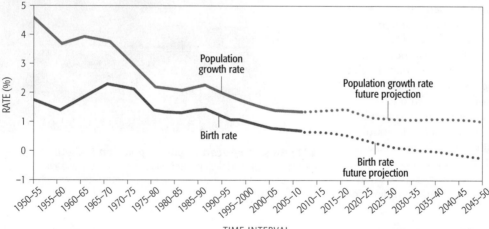

Impact of China's One-Child Policy

Population growth rate

Birth rate

Population growth rate future projection

Birth rate future projection

RATE (%)

TIME INTERVAL

QUESTIONS

1. How have the agendas of the UN World Population Conferences changed over the past 50 years?

2. Some countries are considering policies to increase their population growth rates. Which countries and why?

3. Why do many programs focused on managing human population growth encourage the education and empowerment of women?

Mexico. In 1970, Mexico's population growth rate was 3.4%. Through tax incentives for larger families, Mexico's General Population Law was actually encouraging high birth rates. Upon recognizing that the country's economic growth was not keeping pace with its rapid population growth, the Mexican government amended its General Population Law. In 1974, Mexico created its National Population Council and implemented its National Family Planning Program, which aimed to change public attitudes about family size by means of communication, education, counseling, and contraceptives (**Figure 5.36**). Between 1975 and 1985, birth rates in Mexico plummeted from 4.4% to 3.24%, and the population growth rate dropped to 2.24%. Today, Mexico's birth rate is 1.70%, and its population growth rate is 1.13%. With strong political and economic support and absent concerns

▼ Figure 5.36 **Mexico's Family Planning Strategy** Mexico's National Family Planning Program funds clinics such as this one to provide counseling on family planning and reproductive health.

SEEING SOLUTIONS: Women Deliver!

Efforts to curb population growth often attract furious debate. More often than not, these debates center on emotional issues such as immigration policy or access to abortion. Yet in many countries, population growth rates are declining as a consequence of actions that almost everyone agrees are appropriate—education, health care, and improved opportunities for women.

Since 1969, the central mission of the United Nations Fund for Population Activities (UNFPA) has been to "realize a world where every pregnancy is wanted, every birth is safe, every young person is free of HIV, and every girl and woman is treated with dignity and respect." In 2010, UNFPA Executive Director Thoraya Ahmed Obaid said, "change cannot be imposed from the outside; to be lasting it must come from within. Development should be locally owned." This is the spirit in which UNFPA works with countries and communities to reduce population growth while improving the physical and social well-being of women.

UNFPA has three core areas of focus. First, it assists governments in gathering data on the dynamics of their populations. These data are needed to generate the political will for action and to create and maintain sound policies. Second, UNFPA assists governments in delivering reproductive health care to women. This assistance is far ranging, including medical care of mothers and children before and after birth, diagnosis and treatment of infertility and sexually transmitted diseases, and family planning. Third, UNFPA encourages gender equality and the empowerment of women through programs that educate girls, create economic opportunities for women, and encourage women's participation in government.

The education of women in developing countries tends to lower the rate of population growth. In countries in which at least 48% of high school students are women, the average annual birth rate is 2.18%. In countries in which fewer than 45% of high school students are women, the average birth rate is 3.57% (**Figure 5.37**). Girls who attend high school are likely to delay marriage and childbearing until they are older. Married women with a secondary education are more likely to use contraceptives and rear smaller families.

Improved maternal education and reproductive health also diminish infant mortality. For example, a study in Egypt revealed that nearly 9% of children born to mothers with no education die before the age of 5, compared to 3.8% of children born to mothers who have completed secondary education. It may seem paradoxical, but in the long run improved infant survival results in *decreased* birth rates. This is because the loss of children is often an incentive for women to conceive more children.

Can equality, health care, and education for women reduce population growth and increase well-being?

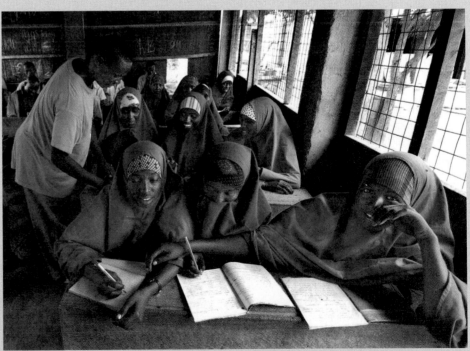

▲ Figure 5.37 **Yes, Education Matters**
Somali girls are studying geography in a Mogadishu high school. The graph compares birth rates with girls' representation in high schools (percentage of student body) for 78 countries in which per capita GDP is less than $10,000/year.

Summary

Although the global population growth rate is declining, in many of the world's poorest countries it remains high. In general, where population growth is high, human well-being is comparatively low. Economic development improves nutrition, sanitation, and health care; it also results in lower birth, death, and population growth rates. However, economic development often increases demands on Earth's resources. Our prospects for future sustainability depend on both the continued decline in the human population growth rate and the reduction of our individual demands on the environment.

5.1 The History of Human Population Growth

- Over the past 100,000 years, human populations have experienced three general periods of growth: the pre-agricultural, agricultural, and industrial periods.

- Industrial development and economic progress produce demographic transitions that involve a general decline in birth rates and death rates and an overall decline in population growth rate.

KEY TERMS

demographer, demographic transition model, mortality transition, fertility transition, stability transition

QUESTIONS

1. What factors have caused changes in human population growth rates during the past 100,000 years?

2. Describe the four stages of the demographic transition model of population change during the industrial period.

3. How do population age-structure pyramids change during a demographic transition?

4. Explain why economic development and human well-being result in continuously declining birth rates during the fertility transition stage of the demographic transition.

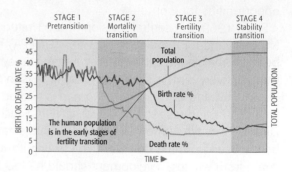

5.2 Global Variation in Human Population Growth

- Since 1970, the world's human population has continued to grow, but the population growth rate has steadily decreased.

- The magnitude of these changes varies among regions and countries and is highly correlated with wealth and well-being.

KEY TERMS

age-specific birth rate, infant mortality rate, life expectancy, age-structure pyramid, net migration rate, push factors, pull factors

QUESTIONS

1. How are birth rate, total fertility rate, and age-specific birth rate related to one another?

2. Total fertility rates overestimate the actual number of children an average woman will bear in her lifetime. Why is this so?

3. What factors account for the global decrease in infant mortality rates and increase in life expectancy over the past 50 years?

4. Two countries with identical age-specific birth rates may have very different total birth rates because they have different age-structure pyramids. Explain.

5. In Germany, the birth rate is lower than the death rate, yet the population growth rate is positive. Why is this so?

	Wealthy Countries	Poor Countries
Total Fertility Rate	Less than 2.0 children	Between 4 and 7 children
Birth Rate	Low	High
Infant Mortality Rate	Low	High
Life Expectancy	High	Low

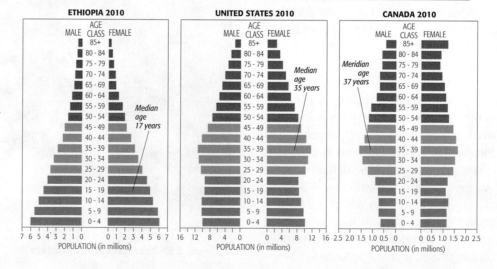

5.3 Predicting Human Population Growth

- Forecasts of future population growth are dependent on the quality of data used and the assumptions made with regard to human values and behaviors.

- Current forecasts predict that the global population growth rate will decrease and that the world population will reach approximately 9 billion by the year 2050.

QUESTIONS

1. To understand the importance of resources to population growth, Malthus chose to compare 18th century populations in the American colonies to Europe. Why?

2. Why is forecasting human population growth more difficult than forecasting population growth in other animals?

5.4 Resource Use and Population Sustainability

- The ecological footprint of a population is the area of land necessary to meet human demands for shelter, energy, food, and other resources. Biocapacity is the area of land available to meet those needs.

- Human impact on Earth's resources is determined by population size, affluence, and technology.

KEY TERMS

ecological footprint, biocapacity, IPAT equation

QUESTIONS

1. Why is the number of people that the world can support in a sustainable fashion considered to be much lower than the global carrying capacity?

2. The ecological footprint for the average person in a country may be small, but the country may still be an ecological debtor. Explain.

3. Between 1991 and 2001, the ecological footprint of the average person on Earth declined by 4%, but the total human footprint increased by 12%. Explain.

4. The world's biocapacity is estimated by some to have declined by as much as 12% between 1991 and 2001. What factors may have contributed to this decline?

5. What is the IPAT equation? How is it related to biocapacity and ecological footprint?

5.5 Managing Population Growth

- Efforts to manage human population growth began in the early 19th century and continue today with an emphasis on family planning, gender equality, and sexual and reproductive health.

KEY TERMS

neo-Malthusian movement, baby boom

QUESTIONS

1. What was the neo-Malthusian movement and why was it important?

2. Why have global conferences and initiatives on human population growth recently begun to focus on the education and empowerment of women and reproductive health?

3. Why do some argue that understanding and closing the gap between the world's poorest and wealthiest individuals, regions, and countries are critical in defining and securing sustainable population growth?

Impact = Population × Affluence × Technology
loss of resources and degradation / including size, growth, and distribution / consumption per individual / things that demand resources

Mastering ENVIRONMENTALSCIENCE

Go to **www.masteringenvironmentalscience.com** for homework assignments, practice quizzes, Pearson eText, and more.

143

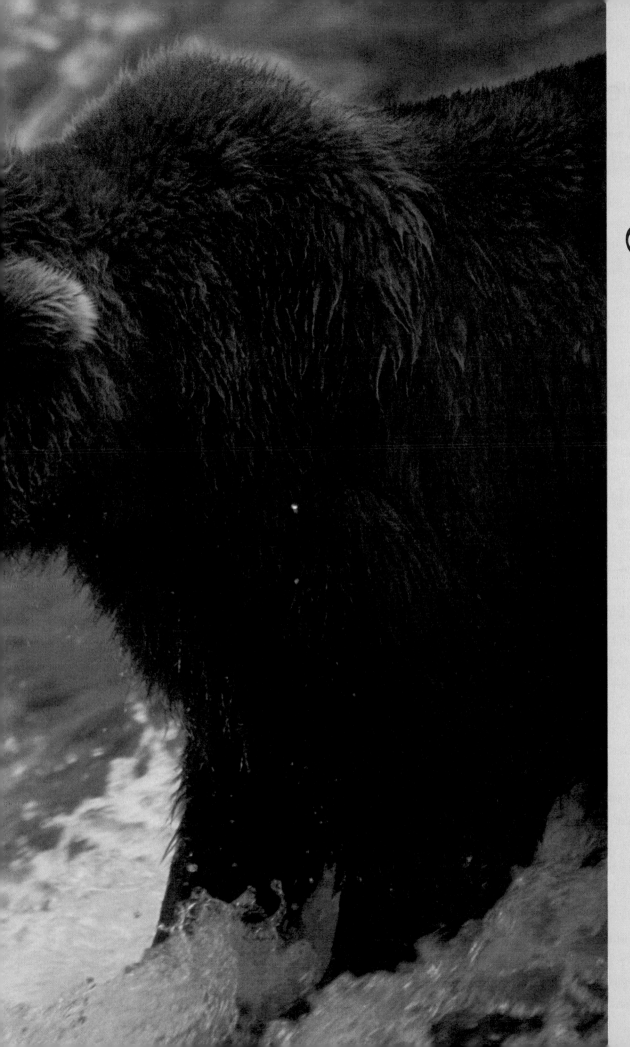

6

Ecological Communities

The Web of Life

What are the interconnections among the many species in an ecosystem?

Images of old-growth forests in the Pacific Northwest often show spotted owls perched on the limbs of towering ancient firs, cedars, and hemlocks (**Figure 6.1**). But the growth and reproduction of these majestic trees and charismatic animals depend on the activities of a myriad of organisms that are largely unseen and certainly underappreciated.

In the soil, many different species of fungi form very intimate associations with the roots of these majestic trees, benefiting both the fungi and the trees. Thin fungal strands grow through the mineral soil and organic litter, extracting nitrogen, phosphorus, and other nutrients and transferring them to the trees. For their part, the trees provide the carbohydrates necessary to support the growth of the fungi.

Fungi reproduce by means of spores. Most fungi form reproductive structures that grow above the ground. The most familiar of these reproductive structures are mushrooms; each kind of mushroom represents a different species of fungus. The spores of mushrooms are usually dispersed by the wind.

However, one group of forest fungi produces its spores inside odd-shaped structures that grow underground. These structures are known to the gourmet as truffles. The spores of truffles are dispersed by animals that feed on them, such as chipmunks and voles. To attract would-be dispersers, truffles release an array of aromatic—and often very tasty—chemicals. When animals eat the truffles, the spores pass through their digestive systems unharmed. The spores are eventually deposited in the animals' feces. In the new location, the feces provide a dose of nutrients that help the spores grow into new fungi.

Among the most important animals that disperse truffle spores in the forests of the Pacific Northwest are northern flying squirrels. These small squirrels nest in tree cavities high above the ground, but they feed on the forest floor. Several tree species depend upon the truffle-forming fungi to obtain the nutrients they need to grow; the distribution of these trees is tightly linked to the activities of flying squirrels (**Figure 6.2**). This complex connection between the trees and flying squirrels is so strong that if the forest experiences a disturbance, such as fire or logging, the flying squirrel population must be reestablished in order for the forest itself to be reestablished.

Flying squirrels form a significant part of the diet of the spotted owls, foxes, and weasels that live in the

▶ **Figure 6.1 Old-Growth Forest**
The large fir, cedar, and hemlock trees of Oregon's old-growth forests depend on fungi associated with their roots to acquire nitrogen and phosphorus.

old-growth forest. These predators prevent flying squirrels from becoming so abundant that they eat the truffles faster than they are produced. The chemical energy that powers the movements of a flying squirrel, an owl, a fox, or a weasel is provided through a complex web of feeding interactions that begins with photosynthesis in the leaves of the ancient firs, cedars, and hemlocks. Equally, the sustained growth of these trees depends on the interactions among the animals in this web.

In this chapter, we will explore the different kinds of interactions among organisms in ecological communities. These include interactions between community members that compete for shared resources, such as flying squirrels and other animals that compete for truffles. They also include interactions in which one organism is an essential resource for another, as truffles are for flying squirrels or squirrels are for owls. We will consider how the flow of energy through ecosystems influences the relative abundance of organisms. Finally, we will ask how the complex webs of interactions among organisms are affected by disturbances.

▼ **Figure 6.2 Connections in the Forest**
Truffle fungi Ⓐ are dispersed by animals such as the flying squirrels Ⓑ that feed on their reproductive structures. In their turn, flying squirrels are an important food source for several carnivorous animals, including the endangered northern spotted owl Ⓒ.

6.1 Competition for Shared Resources

The species that interact within a specific area make up an **ecological community**. Interactions among organisms that compete for shared resources are particularly important. In **intraspecific competition**, members of the same species pursue limited resources. In the forests of the Pacific Northwest, flying squirrels compete with each other for truffles, a favored food. If the number of flying squirrels increases, the demand for truffles also increases. Thus, competition among individual flying squirrels becomes more intense.

Other forest animals, such as chipmunks, also eat truffles. The interactions among different species as they compete for shared resources are called **interspecific competition**. Species may compete directly by using and depleting a shared resource. They may also compete by interfering with one another's ability to access a resource. Interspecific competition limits the distribution and population size of most species. Competing species are able to coexist because they use different portions of their shared environment.

Interspecific Competition

Interspecific competition occurs when two or more species seek the same limited resource.

Foresters were among the first to describe interspecific competition. They found that when trees of one species were removed from a forest, the remaining trees grew more rapidly. The abundance and diversity of herbs and seedlings growing on the forest floor also increased. Foresters hypothesized that this response was due to decreased competition for water and nutrients in the soil. This hypothesis was tested by digging deep trenches around 3–6 meter (10–20 feet) square plots of land; the trenches cut off all the roots growing into the plots. Conditions in the trenched plots were then compared to those in nearby, untrenched plots (the controls). In all cases, the soil in the trenched plots was found to contain more water and nutrients than the soil in the control plots. The number and growth of plants in the trenched plots were also greater than in the controls. Reduced competition for water and nutrients in the trenched plots had allowed more plants to grow (**Figure 6.3**).

In the 1930s, Russian biologist G.F. Gause devised a set of elegant laboratory experiments that provide the basis for our formal understanding of competition. Gause grew two different species of the single-celled *Paramecium*—*P. aurelia* and *P. caudatum*—separately and together (**Figure 6.4**). Populations of both species always increased more rapidly when they were grown alone (4.3). When grown together, populations of both species grew more slowly. Eventually, *P. aurelia* totally displaced *P. caudatum*.

The results of his experiments with *Paramectum* species, along with similar experiments performed on other organisms, led Gause to form this postulate: two species that directly compete for essential resources cannot coexist; one species will eventually displace the other. This postulate has come to be known as the **competitive exclusion principle**.

▼ Figure 6.3 **Forest Competition**
Forest trees compete for soil nutrients and water with the herbs and shrubs that grow beneath them.

How Competitors Coexist

If the competitive exclusion principle is true, why are ecological communities so rich in species?

An acre of tropical forest may include over 100 species of trees, all of which depend on the same soil, water, and nutrients. Freshwater lakes may have dozens of species of fish, all of which feed on the planktonic algae and animals suspended in the water. Indeed, two or more species of *Paramecium* may be found in the same lake. These and many other examples from ecological communities in nature seem to contradict Gause's principle. If two competing species cannot coexist in the laboratory, how are they able to coexist in natural settings? This question has been the basis for hundreds of ecological studies.

The distinguished ecologist G. Evelyn Hutchinson provided one of the most important explanations for the coexistence of competing organisms. He did this by refining the concept of the ecological niche (4.4). Hutchinson suggested that each species has a **fundamental niche**, the complete range of environmental conditions, such as requirements for temperature, food, and water, over which the species might possibly exist. Hutchinson noted, however, that few species actually grow and reproduce in all parts of this theoretical range.

Rather, species usually exist only where they are able to compete effectively against other species. Hutchinson used the term **realized niche** to describe the range of conditions where a species actually occurs given the constraints of competition.

Species whose fundamental niches overlap significantly are potential competitors. Hutchinson suggested that these potential competitors are able to coexist because they divide up the fundamental niche. Hutchinson called this division of resources **niche differentiation**.

Niche differentiation occurs among many different kinds of organisms. For example, five different species of warblers, small insectivorous birds, occur together in the evergreen forests of New England. During nesting season, the primary food of all the warblers is caterpillars, especially spruce budworms. Careful studies of the birds' feeding behavior reveal that each species competes most effectively in a different part of the

GROWN SEPARATELY

GROWN TOGETHER

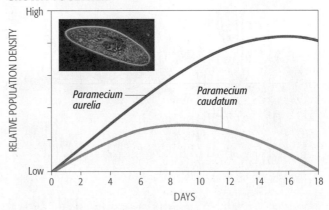

▲ **Figure 6.4 Competitive Exclusion**
The growth rates of both *Paramecium aurelia* and *P. caudatum* were not only lower when grown together than when grown in isolation, but one species actually eventually excluded the other.
Source: Gause, G.F. 1932. Experimental studies on the struggle for existence. *Journal of Experimental Biology* 9: 389–402.

► **Figure 6.5 Coexistence in Competing Warblers**
Although they feed on the same insect larvae, the diverse species of warblers in the spruce and fir forests of New England coexist because they feed in different portions of the forest canopy.
Source: MacArthur, R.H. 1958. Population ecology of some warblers of northeastern coniferous forests. *Ecology* 39: 599–619.

Cape May Warbler   Bay Breasted Warbler

Black-throated Warbler   Blackburnian Warbler   Myrtle Warbler

forest canopy, and that is where each species can be found (**Figure 6.5**).

The diverse grasses and herbs that grow in native prairies provide another example of niche differentiation. Above ground, these plants appear to be vying for the same space and resources. However, a careful mapping of root systems reveals that different species are adapted to exploiting different portions of the soil (**Figure 6.6**). In addition, some species compete most effectively when growing in bright light, whereas others compete effectively when growing in the shade of taller plants.

► **Figure 6.6 Niche Differentiation among Prairie Plants**
Many plant species coexist in grassland ecosystems, such as the tallgrass prairie of North America, because they have very different niches below ground.

Some of Gause's experiments support Hutchinson's niche differentiation hypothesis. Under any specific set of conditions—the same temperature, water availability, food source, etc.—Gause's principle holds true. But if conditions change, competition among species may produce different winners and losers. For example, if waste products are periodically removed from the growth medium, the outcome of the competition between *P. aurelia* and *P. caudatum* is reversed and *P. caudatum* wins. Thus, in a complex environment where waste materials are collected in some places and not in others, these two species could coexist.

Time is required for one species to competitively displace another, and the competitive exclusion principle presumes that environmental conditions remain constant during that time. In nature, however, environments change from season to season and from year to year, so conditions that are favorable to a particular species may not persist. Environments that are constantly changing may allow competing species to coexist.

In estuaries, where rivers flow into the sea, tides and storms are constantly changing the temperature, salinity, and oxygen content of the water. These changing conditions produce rapid changes in the populations of various species of plankton. Particular species become abundant, but soon, changing conditions favor other species. Sometimes, however, human activities around estuaries create persistent high-nutrient conditions that favor a few species to the exclusion of others.

Exploitation and Interference

Mechanisms of competition vary.

Organisms compete with one another by two general mechanisms—exploitation and interference. Imagine that two diners are served a rather meager meal. Both are very hungry, and there is not enough food to satisfy both of them. One diner might outcompete the other by simply eating more quickly. This is **exploitation competition**. Alternatively, one diner might get more food by threatening to stab his competitor with a fork, thus scaring him from the table. This is **interference competition**.

In exploitation competition, successful competitors are able to take up or utilize resources more efficiently. Therefore, the growth rate of their population per unit of limiting resource is higher. In deserts, plants compete fiercely for water. Individual plants deplete the supply of water near their roots, limiting the ability of other plants to grow nearby. The very regular spacing of desert shrubs is a consequence of exploitation competition (**Figure 6.7**).

Interference competition is common among animals that fend off would-be competitors with aggressive or territorial behavior. Such behavior is common among the various scavengers that compete for the remains of animals killed by predators (**Figure 6.8**).

Interference competition is also common among plants and microbes. These organisms compete by releasing chemicals that slow the growth of their competitors. For example, the leaves of garlic mustard release a chemical that inhibits the development of associations between fungi and the roots of trees and other herbs, thereby slowing the growth of the other plants.

▲ **Figure 6.7 Exploitation Competition**
Desert shrubs and cacti exploit water and nutrients and thereby prevent other shrubs from establishing in their vicinity. This often produces regular spacing among desert plants.

Garlic mustard was brought to North America as a culinary herb in about 1860. Soon it escaped into eastern forests, where its populations grew rapidly. In many places it is now the dominant herb, excluding other plant species. Through interference competition, garlic mustard has significantly decreased the diversity of the ecological community where it now lives.

▼ **Figure 6.8 Interference Competition**
Scavengers such as hyenas, jackals, and vultures demonstrate interference competition by aggressively competing with one another for remains of an animal recently killed by lions.

QUESTIONS

1. How does competition affect the growth rates of competing species?
2. Explain how species that depend on the same resources may coexist.
3. What is the difference between exploitation and interference competition?

151

6.2 Herbivory, Predation, and Parasitism

A key question about any ecological community is *Who is eating whom?* Feeding relationships are important in determining the abundance of most organisms. **Consumers** feed on other live organisms. **Herbivores** eat plants, and **predators** hunt and kill animals. **Parasites** live in or on other plants or animals but usually do not kill them directly. The fitness of consumers depends on adaptations that allow them to acquire and utilize their food most effectively. Because consumers directly influence the fitness of the organisms they eat, they exert strong selective pressure on their prey (4.5).

Herbivores

Herbivores feed on plant materials.

Carbohydrates and all other organic food molecules are ultimately produced by photosynthetic plants and algae. Herbivores play a crucial role in transferring the energy stored in those molecules to the rest of the ecosystem (4.1).

Herbivores use several different feeding strategies to exploit plants. Fructivores feed on fruits, and gramnivores feed on seeds. Fructivores and gramnivores often play an important role in plant dispersal. A diverse array of herbivores, particularly insects, feeds on pollen and nectar. In so doing, they obtain energy for themselves while also facilitating sexual reproduction and ensuring genetic diversity in the plants they pollinate (4.3).

Many herbivores are grazers, feeding directly on the leaves and young stems of plants. In most cases, grazing animals do not kill the plants, but they certainly do slow their growth. Grazing animals face significant feeding challenges, not the least of which is related to the basic chemistry of the plants upon which they feed. The leaves and stems of plants are supported by cellulose, a polymer of the sugar glucose (3.2). The chemical bonds that link glucose molecules together in cellulose are very resistant to the digestive enzymes of animals. Ungulates—cloven-hoofed herbivores such as cattle and goats—are able to digest cellulose because of the activity of microorganisms that live in their multichambered stomachs (**Figure 6.9**). These microbes produce an enzyme that can break down cellulose. Many other grazing animals, including horses, rodents, and insects, also depend on microbes that live in specialized areas of their digestive tracts.

Plants have evolved a variety of defenses against herbivores, including thorns, irritating hairs, and distasteful or toxic chemicals. Natural selection has favored the evolution of herbivores with adaptations that allow them to thwart, or even take advantage of, these plant defenses. This tit-for-tat evolution of prey and the species that eat them is called **coevolution**.

Coevolution can produce some very surprising results, as in the case of milkweed plants and monarch butterflies. Most species of milkweed produce and store large amounts of chemicals called alkaloids, which are toxic to most animals. Nonetheless, the caterpillars of

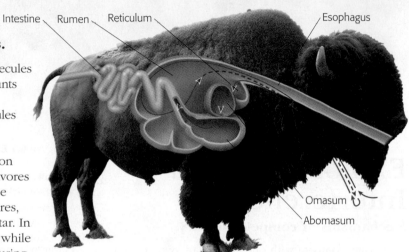

▲ **Figure 6.9 Herbivores Are Adapted to the Food They Eat** Ungulate mammals such as the American bison are able to digest plant cellulose with the help of microbes that reside in their complex, multichambered stomachs.

monarch butterflies feed almost exclusively on milkweed plants (**Figure 6.10**). Not only are these caterpillars able to tolerate the alkaloids, but they actually store these chemicals in their own tissues. The alkaloids make the monarchs toxic to birds and other predators that otherwise might eat them.

Given the wide diversity of herbivores, ecologists have wondered why the world is so green. Most ecosystems seem to have abundant plant resources that are not being eaten by herbivores. Although defense mechanisms in plants may play a role, it appears that populations of herbivores are often limited by predators, the things that eat them, rather than by the abundance of the things they eat.

▶ **Figure 6.10 Milkweed–Monarch Butterfly Coevolution**
The milkweed and the caterpillar that becomes a monarch butterfly present an example of coevolution. Just as the milkweed protects itself from most predators by producing toxic alkaloids, the monarch has adapted so it is able to feed on milkweed in spite of these toxins.

Predators

Predators hunt, kill, and consume their prey.

Predation is the process by which animals capture, kill, and consume animals of another species, their prey. Predators employ two basic feeding strategies—filter feeding and hunting.

Filter-feeding predators use webs or netlike structures to catch their prey. Spiders use webs to "filter" organisms from their environment. As marine shrimp swim through the water, they trap small organisms in the hairlike setae on their legs and transfer them to their mouths. Blue whales, Earth's largest living animals, are filter feeders (**Figure 6.11**). Blue whales and their baleen whale relatives use their comblike "teeth" to filter plankton from the water.

Hunting predators actively stalk and capture their prey. Natural selection has favored predators with keen senses of sight and smell, as well as structures such as talons, claws, and sharp teeth that allow predators to seize and kill their prey. Coevolution has played a significant role in predator-prey relationships, and prey organisms have evolved a variety of defenses against capture (**Figure 6.12**).

Populations of predators are often limited by the availability of prey. Likewise, populations of prey are often limited by their predators. Where predators are dependent on a single species of prey, interactions between the two species may result in synchronized cycles of population growth and decline (**Figure 6.13**).

Most predators feed on a number of different prey species. When a particular prey species is abundant, predators focus their attention on it. As predation depletes that species' numbers, the predators switch their attention to other, more abundant species. **Prey switching** provides predators with a steady supply of food; it also ensures that none of the prey species will be totally eliminated.

From the standpoint of an individual prey animal, the consequences of predation are far from positive. However, predation often has beneficial effects on the health of prey populations as a whole. Predators usually do not select their prey at random. Rather, they pick out the youngest, oldest, or sickest individuals. As a consequence, predation can increase the overall health of the prey population.

In many regions of the eastern United States, populations of white-tailed deer have surged, partly due to the absence of natural predators. The incidence of chronic wasting disease, a fatal neurological disease of deer, has also surged. Where predators are present, deer with chronic wasting disease are more likely to be preyed upon than healthy deer. Thus, the incidence and spread of chronic wasting disease is diminished by predation.

▼ **Figure 6.11 Giant Filter Feeders**
Baleen whales, like this blue whale, fill their expandable mouths with plankton-rich water, then keep the plankton in their mouth while forcing the water back out through their comblike upper teeth.

◀ **Figure 6.12 Defending against Predation**
Natural selection has produced adaptations among prey organisms that allow them a better chance of escaping their predators. Such adaptations include body structures that imitate nonfood parts of the environment, like this walking stick.

▲ **Figure 6.13 Predator-Prey Population Cycles**
Numbers of snowshoe hares and Canada lynx trapped by the Hudson Bay Company reveal changes in their populations between 1840 and 1940. As snowshoe hare population size grows, predation by the Canada lynx increases and hare populations then decline. This decline in hare population size may subsequently cause a decline in the lynx population, although this appears to depend on the availability of other prey.
Data from: MacLulich, D. A. (1937). 'Fluctuations in numbers of the varying hare (Lepus americanus).' Univ. Toronto Studies, Biol. Ser. no. 43: 1–136.

Parasites

Parasites depend on living hosts for nourishment.

Parasites live and feed in or on other organisms, their hosts. Usually, a parasite does not kill its host, but it does harm the host and may contribute to its eventual death. Many parasites, such as viruses and tapeworms, live inside their hosts. Other parasites, such as ticks and leeches, are external.

Some parasites have relatively simple life cycles. For example, the common human cold virus is a parasite that multiplies in host cells and is easily transmitted to the next host in droplets of saliva from a sneeze. Other parasites, such as the single-celled protist that causes malaria, have very complex life cycles that require multiple hosts and vectors. **Vectors** are organisms that carry the parasite but are unaffected by it.

Parasites multiply within individual hosts. However, scientists measure the population growth of parasites in terms of their spread, or the spread of the disease they cause, in the host population. The spread of parasitic disease is governed by the following four factors.

Abundance of hosts. As the number of individual hosts increases, the likelihood that a parasite will be transmitted also increases. For example, the devastating outbreaks of bubonic plague in Europe during the 14th century were associated with the movement of a large number of people to cities (**Figure 6.14**). The spread of the plague was enhanced by the large numbers of another host for the plague bacteria, the rats that lived in the cities.

Accessibility of hosts. The ability of parasites to locate and infect hosts is influenced by the diversity of the ecological community in which they live.

▲ Figure 6.14 **Disease Spread and Host Density**
The density of humans in medieval towns and villages is captured in the painting *The Fight between Carnival and Lent* by artist Pieter Bruegel. Although not visible here, rats harboring plague-infested fleas were even more abundant than people.

In the southeastern United States, for example, the parasitic southern pine bark beetle lays its eggs beneath the bark of pine trees. Its larvae then feed on the trees' living tissues, killing the trees in less than a year. Adult beetles locate new host trees by sensing aromatic chemicals given off by the pine. In plantations where only pines are grown, infestations of pine bark beetles spread very quickly (**Figure 6.15**). In forests with many species of trees, it is more difficult for beetles to locate new hosts, so the infestation spreads from pine to pine much more slowly.

Transmission rate of parasites. All parasites must have adaptations that allow them to move from one host to another, but rates of transmission vary. For example, two fungal diseases that have had significant effects on the forests of eastern North America—Dutch elm disease and chestnut blight—have different rates of transmission. Both of these fungi were accidentally introduced to U.S. forests on non-native nursery trees imported from Asia. (Dutch elm disease was first identified in the Netherlands, where trees were also accidentally infected by fungi from Asia; hence its name.)

The fungus that causes chestnut blight releases large numbers of wind-dispersed spores, which move from tree to tree very quickly. The blight was first introduced in 1890, and by 1930 virtually all of the adult American chestnut trees in eastern North America were dead. In contrast, the spores of the fungus that causes Dutch elm disease are transmitted by an insect parasite, the elm bark beetle. Although the Dutch elm fungus kills trees very quickly, its insect vector moves from tree to tree comparatively slowly. The disease was first observed in North America in 1928. Although it has killed many trees, it has spread much more slowly than the chestnut blight. Thus far, many elm trees have escaped infection.

Length of life of an infected host. The longer an individual host lives with a parasite, the longer it can transmit it. A host that lives a long time provides a great number of opportunities for its parasite to be transmitted to other hosts. Highly virulent parasites that quickly immobilize or kill their hosts limit the opportunities for the parasites to reproduce and spread. Thus, the negative effects of parasites on hosts with which they have coevolved over long periods are generally considerably less than those in new host-parasite interactions.

For example, the chestnut blight fungus is found in nearly all Asian chestnut trees but has very little impact on their growth or reproduction. It is likely that the fungus coevolved with Asian chestnut trees for tens of thousands of years. During that time, Asian chestnut trees developed adaptations that reduced the effects of the fungus.

It is also likely that the fungus evolved in ways that prolonged the life of its host. However, when the chestnut blight was accidentally introduced into North America, it came into contact with chestnut trees that had not evolved resistance to it. As a result, the blight rapidly killed almost all of the American chestnut trees.

Clearly, what an organism eats and what it is eaten by affect its ability to grow and reproduce. Herbivory, predation, and parasitism significantly affect the relative abundance and diversity of species in an ecological community. Coevolution produces balance and stability among populations of the eaters and the eaten.

QUESTIONS

1. Describe two factors that may prevent herbivores from devouring all of the plants upon which they feed.
2. What is prey switching?
3. What four factors influence the spread of a parasitic disease?

▼ **Figure 6.15 Disease Spread and Community Diversity**
Where pines are grown in single-species plantations Ⓐ, parasitic pine bark beetles Ⓑ can spread rapidly from host tree to host tree. However, in complex communities where pines grow interspersed with oaks Ⓒ, beetles have more difficulty locating their host.

6.3 Mutualism and Commensalism

Intimate interdependencies between species are called **symbioses**. The relationship between a parasite and its host is one form of symbiosis. A symbiotic relationship between two species in which both benefit is called **mutualism**. Mutualistic species are better able to secure resources by living together than by living separately. The symbiosis between a fungus and the roots of a tree is an example of a mutualism. In addition to the benefits they provide to the mutualistic pair, mutualisms may provide benefits to the larger ecological community. **Commensalism** refers to associations that benefit only one species, leaving the other species unaffected.

Mutualisms and Commensalisms

Mutualistic relationships benefit both interacting species; commensal relationships benefit only one species.

Pollinating systems—the relationships between flowering plants and the animals that pollinate them—are among the best-studied examples of mutualism. Plants provide pollinators with a reliable source of food, either nectar or the pollen itself. In return, pollinators ensure the breeding success and genetic diversity of the plants they pollinate. Plants often invest a large amount of energy to maintain pollinator populations. For example, in some species of milkweed, nearly 40% of photosynthetic production is dedicated to producing the nectar that attracts pollinating insects.

Ecologists describe some pollination systems as "loose," meaning that a particular plant may be visited by multiple species of pollinators, and those pollinators gather nectar and pollen from multiple plant species. In contrast, many pollination systems are very "tight" relationships between a single plant species and a single species of pollinator. There are, for example, over 900 species of fig, each of which is pollinated by a unique species of wasp in the family Agaonidae (**Figure 6.16**).

▶ **Figure 6.17 Mutual Dependency**
The bullhorn acacia is named for its large spines that provide a home for *Pseudomyrmex* ants. In addition to a home, the acacia provides food for the ants from nectar-producing glands and high-protein nodules. The ant aggressively protects the acacia from other herbivores.

▼ **Figure 6.16 Figs and Wasps**
Each species of fig depends on a unique species of tiny wasp to transfer pollen from male to female flowers. In return, the fig fruit provides habitat for wasp larvae to grow and mature. Here, mature wasps emerge from a sliced-open fig.

Thus, the species in each fig-wasp pair are totally interdependent. Clearly, the evolution of species within the fig genus has been tightly coupled to the evolution of agaonid wasps.

In some mutualisms, one partner provides food to the other partner in exchange for protection from predators or herbivores. The interdependency between acacia ants and the bull-horn acacia trees that grow in the dry tropical forests of Central America is one such mutualism (**Figure 6.17**). The bullhorn acacia gets its name from its large hollow thorns, in which acacia ants make their nests. In addition to providing a home for the ants, the acacia also provides them with two sources of food—protein-rich nodules that grow on the tips of its leaves and carbohydrate-rich nectar that is secreted from the base of its leaves.

For their part, the very aggressive acacia ants ward off the many insect herbivores common in these forests. When acacia ants were experimentally removed from acacias, other insect herbivores increased by nearly 800%, and the acacias were quickly defoliated.

Scientists hypothesize that many mutualistic relationships have developed as a consequence of coevolution between predators or parasites and their hosts. In this scenario, natural selection would favor changes in the behavior of parasites or predators that help them protect their hosts from competing parasites or predators. Natural selection would also favor traits that enhance the host's relationship with its former predator or parasite. Such traits might include the protein-rich nodules and nectar that are produced by the bullhorn acacia.

The effects of mutualism often extend to the whole ecological community. For example, fungus–plant root associations and the interdependency between many herbivores and the cellulose-digesting microbes in their guts are critical to the functioning of entire food webs.

Mutualism plays a particularly important role in the global cycling of nitrogen. Nearly 80% of Earth's atmosphere is nitrogen gas (N_2). Although plants and animals require nitrogen, they are not able to utilize, or fix, atmospheric N_2. Only certain bacteria and cyanobacteria are able to fix nitrogen. Some nitrogen-fixing bacteria form very tight mutualisms with the roots of some plants. In these relationships, the plant provides carbohydrates to the bacteria, and the bacteria provide nitrogen to the plant. Symbiotic nitrogen fixation occurs in many groups of plants but is particularly common among members of the legume family.

In commensalisms, one member of a pair of interacting species benefits and the other is unaffected. For example, hermit crabs protect their soft bodies by occupying the empty shells of marine snails (**Figure 6.18**). Since the snails have long been dead, they are not hurt by the crabs, nor do they derive any benefit from this relationship. Similarly, the branches of tropical trees provide support for a diverse array of epiphytes, such as orchids, bromeliads, and cacti. The trees are not harmed by this association, nor do they appear to receive any benefit.

Mutualistic and commensal relationships are key to the success of many species in ecological communities. These relationships are the result of coevolution; many probably arose from what were once predator-prey or parasite-host interactions.

QUESTIONS

1. Differentiate between mutualism and commensalism.
2. Describe the role of mutualism in the cycling of nitrogen in ecosystems.

▼ **Figure 6.18 Commensal Hermit Crab**
In this particular commensal relationship, the hermit crab inhabits the shell of a snail species, in this case a triton. The crab thus depends on snails for their shells, but the snails are unaffected by this relationship.

6.4 The Flow of Energy in Ecological Communities

Much of the diversity of ecological communities relates to the patterns of **energy flow**—the transfer and transformation of high-energy organic molecules—within them. These patterns are determined by who is feeding on whom. To understand the flow of energy in communities, ecologists classify organisms by feeding levels, or **trophic levels**, based on their source of food. The feeding interactions among species in a community are best described as a **food web**. The stability or sustainability of a food web may be very sensitive to the abundance of one or a few particular species.

Food Chains

The flow of energy through ecological communities is often depicted in simplified food webs or food chains.

Feeding relationships among organisms are often depicted as simplified food webs called **food chains**. In a food chain, one species or trophic level is eaten by another, which is eaten by yet another, and so on (**Figure 6.19**).

In the first, or lowest, trophic level are the primary producers. **Primary producers** transform energy from sunlight or certain inorganic chemicals into high-energy carbohydrates. Green plants, algae, and cyanobacteria, as well as chemosynthetic bacteria found in soils and deep-sea hydrothermal vents, are primary producers (4.1). Primary producers are the ultimate source of the chemical energy used by all members of ecological communities.

Above the first trophic level are those made up of consumers—organisms that feed on other living organisms. The second trophic level is composed of herbivores, such as grazing mammals and insects; animals that feed on primary producers are also called **primary consumers**. Carnivores such as lions or insect-eating birds that feed directly on herbivores make up the third trophic level; they are called secondary consumers. Tertiary consumers—organisms that feed on secondary consumers—form the fourth trophic level. You might imagine such a feeding chain extending indefinitely. But in fact, communities with more than four trophic levels are quite rare, for reasons that will soon become clear.

Much energy is consumed by **decomposers**, organisms that feed on nonliving organic matter.

4th trophic level
Tertiary consumers

3rd trophic level
Secondary consumers

2nd trophic level
Primary consumers

Energy from the sun

1st trophic level
Primary producers

Decomposers
Organisms that feed on nonliving organic matter

▲ **Figure 6.19 Energy Flow in a Grassland Ecosystem**
Ecologists organize ecological communities by feeding rank or trophic level. The arrows indicate the flow of energy among trophic levels. Dead organic matter and waste are consumed by a diverse array of decomposers.

Scavengers, such as vultures, jackals, and hyenas, feed on the dead bodies of other animals, often the remains left by predators. Termites, wood roaches, and a great variety of other organisms feed on plant litter, such as fallen leaves and dead wood. Microorganisms such as bacteria and fungi obtain their energy from very small bits of organic matter.

Energy and Biomass Pyramids

Available energy diminishes at each successive trophic level.

Organisms at each trophic level transform what they eat into **biomass energy**, the food that can be consumed by higher trophic levels. To organisms at a particular trophic level, the creatures that make up the trophic level upon which they feed might be considered "machines" that convert inaccessible energy into food energy. Thus, a second-level herbivore transforms leaves and stems into flesh that can be eaten by a third-level carnivore. However, most of the energy that is consumed by any organism is burned in respiration or lost as waste. Only a small fraction becomes biomass. (Imagine how large you would become if you converted all of your food into body mass, pound for pound!)

The fraction of energy that the organisms in one trophic level make available to the next trophic level is called **trophic level efficiency** (**Figure 6.20**). As a rule of thumb, trophic level efficiencies are assumed to be about 10%. For example, herbivores convert one-tenth of the energy available to them from plants into herbivore biomass. However, trophic level efficiencies vary a great deal and are often much lower than 10%. For example, warm-blooded carnivores use a large proportion of the energy from the food they consume to maintain their body temperature and to stalk and kill their prey. Thus, the trophic level efficiency for tertiary consumers is often only 2–3%.

TROPHIC LEVELS

▲ Figure 6.21 **The Pyramid of Energy**
In general, the biomass of organisms and the total amount of energy they store decrease from one trophic level to the next, resulting in pyramids of biomass and energy. These numbers are typical for a grassland ecosystem.

As a consequence of low trophic level efficiencies, the total energy stored by organisms in an ecosystem usually decreases significantly from one trophic level to the next. In other words, the amount of energy available to each trophic level is considerably less than the energy available to the trophic level beneath it. Both the numbers and biomass of organisms generally decrease from one trophic level to the next. This produces a pyramid of trophic level energy (**Figure 6.21**).

It is logical that the biomass at each trophic level would form such a pyramid. Throughout its life, an herbivore must consume many times its weight in plant matter, and a carnivore must consume many times its weight in prey. Food becomes increasingly scarce at each successive trophic level, and organisms that feed at higher trophic levels must expend more energy per unit of biomass to obtain their food. In most ecosystems, there is not enough energy available to support the activities of consumers beyond the third or fourth trophic level.

An exception to the pyramid pattern is found in many lakes and parts of the ocean. There the biomass of the primary producers—planktonic algae, or phytoplankton—may be less than the biomass of the animals that feed on them. This occurs because phytoplankton carry out photosynthesis, grow, and multiply very rapidly, but are constantly being grazed by herbivorous zooplankton.

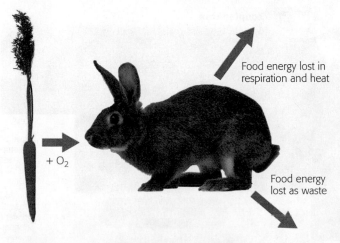

▲ Figure 6.20 **An Organism's Energy Budget**
Of the food consumed by an organism, less than 10% is converted to organism tissue and is available to the organisms that feed on it. The remainder is respired or passed as waste.

Food Web and Species Diversity

Food web complexity supports the stability of ecological communities.

The hierarchy of trophic levels is often described as a food chain. In most ecosystems, however, feeding relationships are more complex. Each trophic level usually includes many species; predators often feed on more than a single prey organism; and many species obtain their food from more than one trophic level. Because the connections between the eaters and eaten rarely form a simple chain, they are more accurately described as a food web (**Figure 6.22**).

The linkages among primary producers, herbivores, and predators form a consumer food web. Decomposing organisms are often eaten by other organisms, forming decomposer food webs. Consumer and decomposer food webs are nearly always highly interconnected (**Figure 6.23**). For example, earthworms are decomposers that feed on decaying organic matter in the soil that comes from many different microorganisms. Earthworms are often eaten by birds. These birds are, in turn, eaten by predators such as hawks. Waste organic matter from these consumers eventually reenters the decomposer food web.

The stability of a community food web is directly related to the diversity of its species and the complexity of their connections. In a relatively simple ecosystem with few interconnections, the loss of a single species may be very disruptive. But if a species disappears from a complex food web, many alternative pathways for energy transformation remain.

Why are carnivores so frequently on the list of endangered species? Is it always a result of hunting?

Whitney, Old Dominion University

Norm: *Because carnivore populations are small compared to herbivores, they are more vulnerable to habitat loss, environmental change, and hunting pressures.*

▼ **Figure 6.22 Marine Food Web**
Marine food webs are often quite complex. The thickness of arrows refers to relative amounts of energy transferred among food web members.

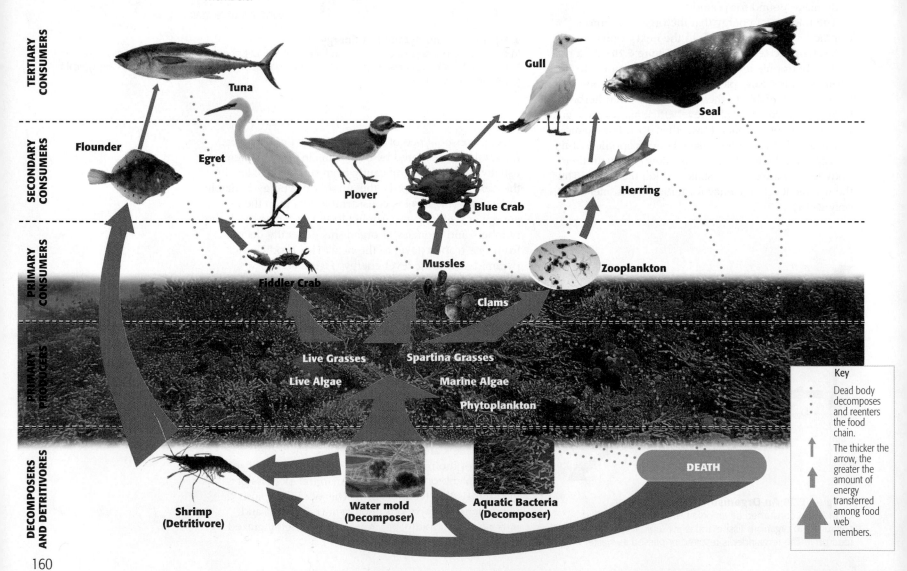

TERTIARY CONSUMERS

SECONDARY CONSUMERS

PRIMARY CONSUMERS

PRIMARY PRODUCERS

DECOMPOSERS AND DETRITIVORES

Tuna

Gull

Seal

Flounder

Egret

Plover

Blue Crab

Herring

Fiddler Crab

Mussels

Clams

Zooplankton

Live Grasses

Spartina Grasses

Live Algae

Marine Algae

Phytoplankton

Shrimp (Detritivore)

Water mold (Decomposer)

Aquatic Bacteria (Decomposer)

DEATH

Key

Dead body decomposes and reenters the food chain.

The thicker the arrow, the greater the amount of energy transferred among food web members.

▼ Figure 6.23 Terrestrial Food Web
Because plants produce large quantities of wood and cellulose that cannot be digested by herbivores, much energy in terrestrial food webs is processed by decomposing organisms. Like birds feeding on worms, decomposers may then be consumed by predators.

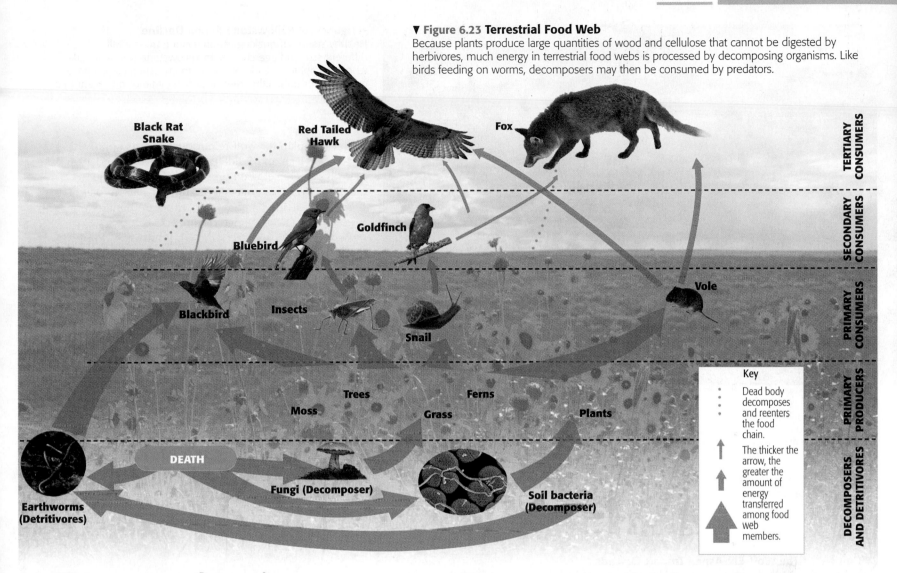

TERTIARY CONSUMERS

SECONDARY CONSUMERS

PRIMARY CONSUMERS

PRIMARY PRODUCERS

DECOMPOSERS AND DETRITIVORES

Black Rat Snake

Red Tailed Hawk

Fox

Goldfinch

Bluebird

Blackbird

Insects

Snail

Vole

Trees

Ferns

Moss

Grass

Plants

DEATH

Fungi (Decomposer)

Soil bacteria (Decomposer)

Earthworms (Detritivores)

Key

Dead body decomposes and reenters the food chain.

The thicker the arrow, the greater the amount of energy transferred among food web members.

Keystone Species

Keystone species are important in the stability of many food webs.

Keystone species, so named in reference to the wedge-shaped keystone that holds an arched structure together, play a particularly important role regarding the abundance of other species. Not surprisingly, plant species that account for a significant portion of total primary production are often keystone species.

Carnivores can also be keystone species, even though they are generally much less abundant than organisms at lower trophic levels. Carnivores often regulate the size of herbivore populations. The loss of carnivores can create a **trophic cascade**, in which the populations of herbivores increase, resulting in the overconsumption of primary producers.

Our increasing understanding of the importance of keystone carnivores is illustrated by the change in attitudes toward the management of wolves in Yellowstone National Park over the past century. In the early 1900s, managers believed that large predators such as wolves and mountain lions were a threat to deer and elk, which were considered to be

more desirable species. In 1914, managers began a program to rid Yellowstone of wolves and mountain lions. This campaign was especially effective against the wolf; the last of Yellowstone's native wolves was killed in 1924. Within a few years, the number of elk feeding on the park's North Range had more than doubled, and their numbers remained high over the succeeding decades.

Beginning in the 1980s, ecologists began to notice the absence of young trees in the many populations of quaking aspen that border the North Range. Data gathered from these populations in the 1920s revealed that aspen were then producing abundant sprouts and young trees (**Figure 6.24** on page 162). In the decade that followed, aspen reproduction ceased. On close examination, similar patterns of change were found in populations of cottonwood and willow. Amid much controversy, several researchers concluded that the decline in the reproduction of these trees was a consequence of overgrazing by the abnormally abundant elk.

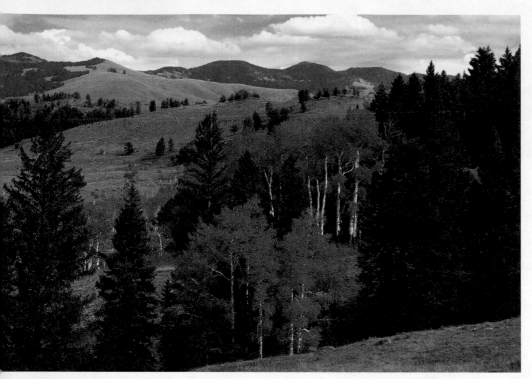

◄ **Figure 6.24 Yellowstone Aspen Decline**
Healthy stands of quaking aspen include trees in all different size and age classes. In Yellowstone National Park, where a trophic cascade occurred, many aspen stands have only older or unhealthy trees, indicating little or no recent tree reproduction.

however, a plan for wolf reintroduction was approved by the U.S. Fish and Wildlife Service. In 1995 and 1996, 31 wolves were released on the North Range. The reintroduction was quite successful. By 2008, a total of 124 wolves divided among 14 packs were hunting on Yellowstone's North Range (**Figure 6.25**).

Between 1995 and 2005, the elk population dropped from approximately 17,000 to 8,000 animals. Several studies indicate that that change was only partially due to wolves; hunting pressure and several dry years with diminished grass production also played a role. Wolf activity on the North Range is quite variable; where such activity is highest, elk populations are lowest. Where wolves are most active, young sprouts and shoots of aspen have begun to appear in the aspen stands.

Many conservation ecologists view the reestablishment of wolves on Yellowstone's North Range as a model for how to restore species that were once extinct in an area. However, this project is still in its early stages, and important challenges lie ahead. Nevertheless, it demonstrates how comparatively small numbers of predators can have a significant effect on the diversity of an ecological community through their effects on the flow of energy in complex food webs.

QUESTIONS

1. Why do ecological communities usually have a maximum of four trophic levels?

2. Why does food web complexity generally increase community stability?

Arguing that the removal of the wolf had been a colossal moral and ecological error, conservation organizations mounted campaigns to reintroduce wolves to the park. Fearing negative effects on game and livestock, most hunters and ranchers opposed the reintroduction of wolves. After years of negotiation,

▼ **Figure 6.25 The Wolf-Elk-Aspen Trophic Cascade**
In Yellowstone's North Range, where wolves have become particularly common, elk activity is diminished, causing young trees to appear in the quaking aspen groves.

FOCUS ON SCIENCE: Trophic Cascades Across Ecosystems

Tiffany Knight and her collaborators discovered a remarkable trophic cascade that connects the ecological communities of ponds to the terrestrial communities that surround them (**Figure 6.26**). Knight's team studied eight ponds in the University of Florida's Ordway-Swisher Biological Station. These ponds were similar in size and depth, but four of the ponds contained fish and four had no fish. Fish are important predators of the insect larvae that live in these ponds.

◄ **Figure 6.26 Plant Population Biologist**
Tiffany Knight of Washington University, St. Louis, Missouri, is primarily interested in factors influencing the long-term persistence of plant populations.

Knight and her collaborators predicted that the presence or absence of fish might influence insect populations. They also hypothesized that the species that interact with those insects, such as the plants that they pollinate, might be affected.

The aquatic larvae of dragonflies are eaten by several species of fish that live in some of the ponds.

▼ **Figure 6.27 Dragonflies Eat Pollinating Insects**
Dragonflies feed on a variety of flying insects, including many important pollinators. Here a dragonfly devours a small butterfly.

Adult dragonflies are voracious predators of other flying insects, including various flies, bees, and butterflies that pollinate plants growing around the ponds (**Figure 6.27**). A shrub called St. John's wort grows abundantly around all the ponds and depends on these insects for pollination (**Figure 6.28**). Knight hypothesized that a variation in the abundance of dragonflies might influence the abundance of pollinators, thereby affecting the reproductive success of plants that depend on those pollinators, such as St. John's wort. Censuses revealed that the larvae of all dragonfly species were two to five times more abundant in ponds without fish than in ponds with fish. Adult dragonflies were also much more abundant around fish-free ponds.

Knight monitored the pollinators visiting St. John's wort. She found that the visitation rates of flies, butterflies, and bees were much higher around ponds with fish and few dragonflies than around ponds with no fish and abundant dragonflies. Knight also demonstrated that seed production in St. John's wort was limited by pollen supply. Because more pollinators visited them, the flowers of shrubs growing around ponds with fish produced many more seeds than did the flowers of shrubs growing near ponds without fish.

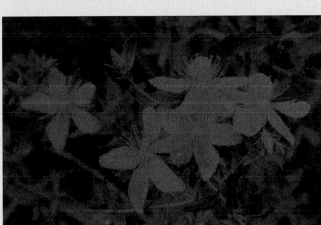

▲ **Figure 6.28 St. John's Wort**
Seed production in the St. John's wort shrub flowers depends on pollen delivered by a variety of flies, bees, and butterflies.

The work of Knight and her collaborators is among the first to demonstrate that trophic cascades can operate across the boundaries of very different ecosystems, such as between ponds and the land adjacent to them. Now at Washington University, Knight is continuing to research how interactions among organisms at different trophic levels influence processes such as plant pollination.

How could the presence or absence of fish in ponds possibly influence the production of seeds in the plants that grow around them?

6.5 **Disturbance and Community Change**

Ecological disturbances, such as fires, hurricanes, or the logging of forests, may result in the loss of many or all of the species in an ecological community. Environmental features, such as soil and woody debris, that are left behind following a disturbance are called **ecological legacies**. As species reestablish themselves, they alter the environment, often making it possible for other species to become established. This process of post-disturbance change in an ecological community is called **succession**. The nature of successional change is influenced by the severity of the disturbance and the ecological legacies that the disturbance left behind. Succession may also influence the likelihood of future disturbances, producing disturbance cycles. The specific patterns of succession vary from place to place and from time to time, depending on factors such as variations in environmental conditions, climate change, and the character of surrounding ecological communities.

Primary Succession

Primary succession depends on environmental change brought about by pioneer species.

Primary succession occurs where a disturbance has removed virtually all ecological legacies. Because it occurs slowly, often over centuries or even millennia, primary succession is difficult to study directly. Instead, ecologists infer the process of primary succession from relatively short-term studies of landscape patterns. Among the first ecologists to use this approach was William S. Cooper, who studied the recently exposed lands in Glacier Bay, Alaska.

When explorer George Vancouver first mapped Glacier Bay in 1798, a glacier that was thousands of feet thick extended nearly to the mouth of the bay. In the 200 years since then, that glacier has retreated nearly 100 km (60 miles), leaving behind newly exposed rock. Cooper wondered how ecological communities had become established on such barren sites and how they had changed over time. He reasoned that rock surfaces at the mouth of the bay had been exposed for a longer period than those closer to the glacier. Using historical maps, Cooper was able to reconstruct the exact age of the rock surface at each location.

By comparing communities at various distances from the glacier, he was able to reconstruct the overall pattern of change (**Figure 6.29**).

Primary succession in Glacier Bay begins when bare rock is first exposed. Within a couple of years, lichens

▼ Figure 6.29 **Glacier Retreat and Primary Succession**
The red lines on this map indicate the position of the glacier at different dates. In 1760, Glacier Bay was completely filled with glacial ice. Since that time, glaciers have retreated over 100 km (62 miles), gradually making rock surfaces available for primary succession.

and mosses colonize the rock (**Figure 6.30A**). Ecologists call these earliest colonists **pioneer species**. Pioneer species share three important characteristics. First, they are widely dispersed; lichens and mosses are dispersed by spores. Second, they are able to grow under very harsh and resource-poor conditions. Third, as they grow, they alter their environment in ways that allow other plants to become established. Lichens and mosses, for example, release small amounts of organic acids that accelerate the weathering of nearby rocks. This rock weathering, in combination with the organic litter produced by the lichens and mosses, results in the accumulation of small amounts of soil around these plants.

After about 10 years, sufficient soil accumulates in some locations to support the growth of herbs and a few low shrubs. Roots from these plants penetrate cracks in the rocks, enlarging them and accelerating the processes of weathering and soil accumulation. Over the next 50 years, the community of herbs and shrubs becomes more diverse and expands across the bare rock. Among these plants, the mountain avens, or *Dryas*, is especially important; symbiotic bacteria associated with its roots fix nitrogen from the air, enriching the soil around it (**Figure 6.30B**).

Within 60–100 years, birch and alder trees become established, forming a dense forest. Like *Dryas*, the alder has symbiotic nitrogen-fixing bacteria associated with its roots; these bacteria enrich the soil (**Figure 6.30C**). Eventually, the nitrogen-rich soil allows the seedlings of white spruce and other cone-bearing trees to take hold. After 100 years, saplings of these trees begin to grow through openings in the birch and alder canopy. After 250 years, these trees form a mature white spruce forest that is similar to the forests growing on sites that have been free of glacial ice for a thousand years or more (**Figure 6.30D**).

From this and similar studies of primary succession in other regions, ecologists have inferred five important principles.

1. As plant species become established and grow, they may alter the environment around them in ways that make it more habitable for other species. This process, called **facilitation**, is especially important in the early stages of succession.
2. Facilitating species, such as *Dryas* and alder, eventually disappear because they are unable to compete effectively with their successors.
3. The migration of plant species to a site depends on the proximity of seed sources and the ability of the plants to disperse their seeds. Early invading herbs and shrubs often have small wind-dispersed seeds and are abundant in nearby locations.
4. The combination of migration, facilitation, and competition results in populations of different species replacing one another over time.
5. Eventually, a community of plants that is able to reproduce generation after generation is established; this is the **climax community**. In southern Alaska, the climax community is the group of species associated with a white spruce forest.

◄ **Figure 6.30 Primary Succession**
As glaciers retreat, pioneer species such as lichens and mosses Ⓐ initiate the process of primary succession, which culminates in 200–300 years in mature spruce forest Ⓓ. Early invading species often alter the environment, making it more favorable for their successors Ⓑ and Ⓒ.

Ⓐ Pioneer mosses and lichens—rock weathers and organic matter accumulates

Ⓑ Herbs and low shrubs—diversity increases and soil develops

Ⓒ Invasion by alder, a nitrogen-fixing tree

Ⓓ Invasion by spruce and other conifers

Secondary Succession

Secondary succession is often accelerated by pre-disturbance ecological legacies.

Primary succession occurs following major disturbances that leave very few ecological legacies. **Secondary succession** is the process of change following disturbances that leave behind legacies such as soil, woody debris, or plant seeds. Disturbances that initiate secondary succession include ice and wind storms, fire, logging, and various kinds of human land use. The patterns and rate of change that such disturbances initiate are greatly influenced by their legacies.

Old-field succession, the reforestation of abandoned farmland is one of the most carefully studied examples of secondary succession. During the 1700s, much of the original forest in the southeastern United States was cleared for agriculture. That land was then farmed for a century or more. Over this period, poor farming practices led to extensive soil erosion and nutrient loss. By the time of the Civil War (1860), these lands had lost much of their original productivity. In the next 80 years, loss of soil productivity and economic hard times caused many farmers to abandon their fields. This started the process of old-field succession on millions of acres (**Figure 6.31**).

In the year immediately following abandonment, old fields are dominated by short-lived grasses and herbs, including several weeds that had previously competed with the farmers' crops. The seeds of longer-lived perennial herbs, such as aster, goldenrod, and broomsedge, also germinate in the first year (**Figure 6.32A**). During the next few years, these perennials outcompete the first-year weeds and dominate the old field. The litter they produce gradually replenishes the organic matter and nutrients in the soil.

During this time, seeds of wind-dispersed trees, such as pines, sweet gum, and tulip tree, also germinate. In 3–4 years, saplings of these trees begin to emerge above the grasses and herbs. These saplings provide perching sites for birds, which disperse the seeds of trees such as black cherry and dogwood.

In these high-light and rather nutrient-poor environments, pines generally grow faster than the other trees. After 8–10 years, the young pine trees form a closed canopy that

▲ **Figure 6.31 Abandoned Farmland Creates Opportunities for Succession**
Poor farming practices and economic hardships led to the abandonment of much badly eroded land in the southeastern United States following the Civil War. This abandonment initiated old-field succession over broad expanses of this region.

casts dense shade. Old-field grasses and herbs are unable to grow in dense shade, so they gradually disappear. In contrast, many of the shrubs and trees that grew in the old field are able to subsist beneath the pines. Soon, other shade-tolerant herbs, shrubs, and trees begin to invade. These new species are often transported by the growing populations of squirrels and other animals for which the pine forest provides suitable habitat (**Figure 6.32B**).

Pines in this region grow quickly. In just a decade or two, this process converted vast areas of the Southeast from eroded farmland to pine forest. Lands that had once produced cotton and corn became some of the nation's most important sources of timber. The rapidity of this reforestation was facilitated by the legacy of old-field soils, which were enriched by the growth and decay of pioneer grasses and herbs.

► **Figure 6.32 Old-Field Succession**
In secondary succession on abandoned farmland, the legacy of soil accelerates the process of change from old-field pioneers Ⓐ, to pine forest, Ⓑ to mature deciduous forest Ⓒ.

0–10 YEARS

Ⓐ Hardy wind-dispersed herbs and grasses dominate old fields in the first few years following abandonment.

10–15 YEARS

Ⓑ In the high light and poor soil conditions, pines are able to grow more rapidly than other trees and form an even-aged stand in 10–15 years.

150 YEARS

Ⓒ Pines are not able to reproduce in their own shade and are eventually replaced by deciduous, broad-leaved trees like oaks, maples, and hickories.

After 70 to 150 years, individual pine trees begin to die. Pines are well adapted for growing in sunny old fields, but their seedlings are unable to become established in the shade cast by mature trees. As the pines die, they are gradually replaced by broad-leaved trees that are more tolerant of shade, such as oaks, hickories, and maples. Many southeastern forests that have not been harvested for timber are currently in this stage of secondary succession. If left uncut, in about 150 years these forests will be dominated by a climax forest composed of a diverse mixture of broad-leaved deciduous trees (**Figure 6.32C**).

Because succession produces important changes in habitat, the animal communities also change. Studies in Georgia, for example, revealed significant changes in the abundance and diversity of birds during old-field succession (**Figure 6.33**). Because many of these birds transport seeds, their presence affects the composition of the plant community.

The character and rate of secondary succession depends on the ecological legacies left by the disturbance. For example, the pattern of change following logging in a 50-year-old pine forest is quite different from that in old-field succession. The soils left by logging contain considerably more organic matter and nutrients than those available in the soil of abandoned fields. Logging also leaves the seeds and saplings of many shrubs and trees. Within as few as 3 years, a diverse mixture of shrubs and trees forms a dense cover. Unless foresters use fire or herbicides to reduce their growth, broad-leaved deciduous trees soon grow over and eliminate the pines.

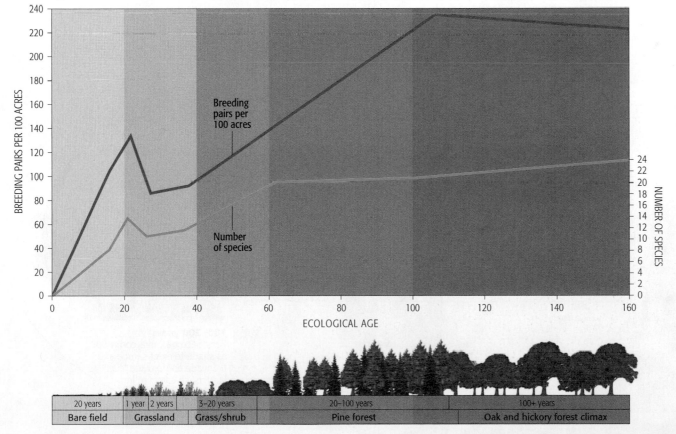

◀ **Figure 6.33 Succession and Bird Diversity**
The number of bird species and the overall abundance of birds increase during old-field succession as more bird habitats become available. For their part, birds accelerate this process by transporting seeds of many late-succession species.
Source: Johnston, D.W. and E.P. Odum. 1956. Breeding bird populations in relation to plant succession on the Georgia Piedmont. *Ecology* 37: 50–62.

Cyclic Succession

Succession may increase the likelihood of disturbance, which leads to cycles of change.

Pine cone seeds

Fire causes pine cone to open up and release seeds

▲ **Figure 6.34 Lodgepole Pine Requires Fire to Reproduce**
The seed-bearing cones of lodgepole pine remain closed until heated by fire. They then release their seeds, which are able to establish themselves in the high-light environment that follows the fire.

▼ **Figure 6.35 Lodgepole Pine Forest Fire Cycle**
As lodgepole pine forests undergo succession, wildfire becomes more likely because of changes in the amount and distribution of flammable plant material. The resulting pattern of change is called cyclic succession.

Early studies seemed to suggest that ecological communities become increasingly stable as succession proceeds. In some circumstances, ecosystems do appear to become less disturbance-prone over time. However, recent research reveals that succession in some ecosystems may actually increase the likelihood of disturbance. Where this is true, climax communities may not exist. Instead, there is a continuous cycle of disturbance and succession.

One such cycle of change occurs in the lodgepole pine forests of the mountains of western North America. More than 100 years ago, ecologists recognized that the lodgepole pine depends on fire for its survival. Many lodgepole pines produce cones that do not open when they mature. Instead, the scales of their cones are glued shut by resins that keep their seeds inside (**Figure 6.34**). The heat of forest fires melts these resins, opening the cones and releasing the seeds. Fires usually leave behind bare mineral soils; the seeds of lodgpole pines germinate best in such soils. Fires also burn away underbrush, allowing bright sunlight to reach the ground; the seedlings of the lodgepole pine require bright light to become established. Thus, without fire, the lodgepole pine cannot disperse its seeds, nor can its seeds germinate successfully.

In the years immediately following an intense fire, large numbers of lodgepole pine seedlings become established and form a dense shrublike cover. In a few decades they grow into a stand of trees that is 40–60 feet tall (**Figure 6.35**). Lighting or careless humans may start fires in these young forests. Fires typically ignite in the needles and small branches on the ground. Because there is little fuel between the forest floor and the needles and branches in the forest canopy, fires that burn through these young forests tend to remain on the ground. Such fires burn with very low intensity—not enough to damage the trees or open their cones.

After 150 years, the pines mature and individual trees begin to die. This creates openings in the canopy and opportunities for the establishment of shade-tolerant spruce and fir trees. As this process proceeds, woody debris accumulates. The distribution of fine, flammable fuels from the forest floor to the canopy becomes much more continuous. Because the branches of the spruce and fir trees are closer to the ground than those of the pines, these trees create "ladders" that can carry fire from the forest floor into the forest canopy.

Thus, as succession proceeds, the forest becomes much more prone to the very intense fires that kill trees. Intense fires create the conditions that are necessary for the reestablishment of the lodgepole pines. The length of this cycle is generally about 200–300 years, although it is affected by factors such as yearly rainfall and the occurrence of lightning.

Other kinds of disturbances can also cause cycles of change. In the high-elevation fir forests of northern New England, wind, ice, or insects can damage the canopy, exposing the trees to the prevailing winds. Eventually, these strong, icy winds kill the trees. As trees die,

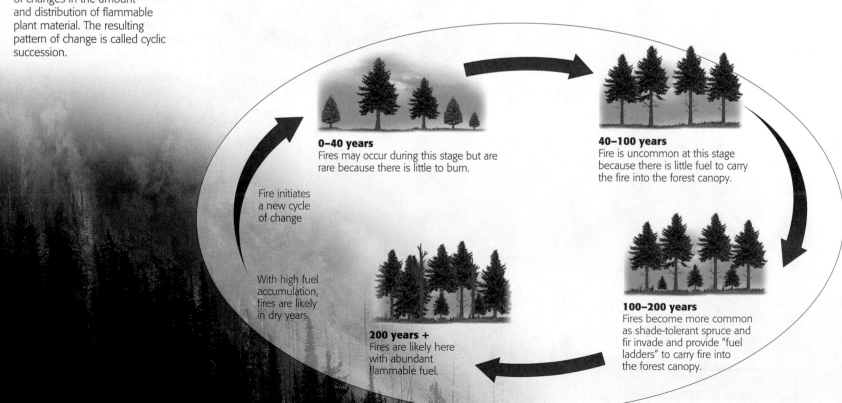

0–40 years
Fires may occur during this stage but are rare because there is little to burn.

40–100 years
Fire is uncommon at this stage because there is little fuel to carry the fire into the forest canopy.

Fire initiates a new cycle of change

With high fuel accumulation, fires are likely in dry years

200 years +
Fires are likely here with abundant flammable fuel.

100–200 years
Fires become more common as shade-tolerant spruce and fir invade and provide "fuel ladders" to carry fire into the forest canopy.

the trees immediately behind them are exposed to the same stresses, increasing the likelihood of their dying. This process initiates a "wave" of tree death that moves across a landscape at a rate of 3–10 feet per year. Succession occurs in the "wake" of such waves. The youngest trees grow near the front of the wave, with successively older trees trailing behind. At any given time, several such waves can be observed moving across many northern mountainsides.

Because they often occur in clusters, cyclic disturbances on large landscapes often produce a mosaic of patches representing different stages of succession. **Return time** is the average time between disturbances at any given place on such a landscape. Return times may be as short as a few years, as in frequently burned grasslands and savannas; many decades long, as in the chaparral of the southwestern United States; or more than a thousand years, as in tropical or temperate rainforests.

The Importance of Place

Successional change at a place is affected by the changes occurring in surrounding places.

Ecologists once viewed climax communities as the inevitable end state of succession. Within a particular climatic zone, they thought that succession would eventually produce the same climax assemblage of species, regardless of the nature of the disturbance. In southern Alaska, for example, succession would inevitably lead to a white spruce forest. In the southeastern United States, it would lead to a broad-leaved deciduous forest. Many ecologists argued that the climax community was the most stable assemblage of species possible in that environment.

Today, ecologists recognize that multiple factors can affect the process of change following a disturbance. Local variations in terrain can produce conditions that favor the establishment of different assemblages of species. In the Glacier Bay region of Alaska, for example, succession in low, moist areas leads to the formation of peat bogs and forests of black spruce rather than white spruce (**Figure 6.36**).

Succession in a particular disturbance patch is also influenced by the composition of species in communities immediately surrounding it. After 150 years, an old field surrounded by a forest of oak and hickory trees is more likely to have a large number of oaks and hickories than a similar field in the midst of a pine forest.

Recent studies in North Carolina reveal that patterns of succession in the herbs that grow in pine forests are significantly different from the patterns found in studies conducted between 1930 and 1950. Three factors appear to have caused this difference. First, changes in the landscape associated with human development have altered the patterns of seed dispersal. Second, deer populations have increased 10-fold, and deer grazing has significantly diminished the abundance of many herb species. Third, introduced non-native species, such as garlic mustard, have become more abundant in many locations.

From decade to decade and century to century, changing environments may cause important changes in the process of succession. Ecologists now understand that long-term climate change and human-caused changes, such as widespread deforestation or the introduction of non-native species, can alter the course of succession.

QUESTIONS

1. Explain why pioneer species may be critical to the recovery of communities after a disturbance.

2. Why do ecological legacies accelerate the process of succession?

3. Describe a situation in which successional change increases the likelihood of disturbance.

► Figure 6.36 **Local Environment Alters Succession**
Where drainage is poor around Glacier Bay, succession leads to a so-called muskeg bog with scattered black spruce and an understory of peat moss, which all flourish in moist environments.

Summary

Ecological communities are clearly more than a haphazard assemblage of species. Coevolution among competitors, predators and prey, parasites and hosts, and mutualistic partners has resulted in a high level of integration and interdependency among community members. This integration results in the efficient processing of energy in food webs. Organisms at each trophic level store a relatively small fraction of the energy that they take in. Therefore, the number of organisms and amount of biomass diminishes from one trophic level to the next. The integration among members of ecological communities is also central to their ability to recover from disturbance. This recovery process, called succession, is influenced by the kind of disturbance that initiates it, as well as variations in the physical environment, in the species composition of nearby communities, and in the climate.

6.1 Competition for Shared Resources

- Competition is the set of interactions among individual organisms that are pursuing a shared but limited resource.

KEY TERMS

ecological community, intraspecific competition, interspecific competition, competitive exclusion principle, fundamental niche, realized niche, niche differentiation, exploitation competition, interference competition

QUESTIONS

1. When two organisms compete, both have lower population growth rates than when they grow separately. Explain why.

2. In the Northwest old-growth forests, three mammals—northern flying squirrels, chipmunks, and voles—eat truffles. By what mechanisms might these three mammals coexist?

3. The diversity of herbs along frequently disturbed roadsides is often high. What factors might account for that diversity?

4. Describe an example of exploitation competition and an example of interference competition.

Cape May Warbler

Bay Breasted Warbler

Black-throated Warbler

Blackburnian Warbler

Myrtle Warbler

6.2 Herbivory, Predation, and Parasitism

- Consumers include herbivores, predators, and parasites; the growth of their populations may be limited by the things they eat or the things that eat them.

KEY TERMS

consumer, herbivore, coevolution, predator, parasite, vector, prey switching

QUESTIONS

1. Why is herbivory an especially challenging way to obtain food?

2. Why are predators important to the well-being of their prey as well as the entire ecological community?

3. The common cold is caused by a viral parasite. New strains of the cold virus spread rapidly, especially in the wintertime. What factors allow new strains to spread so quickly?

4. Newly introduced, non-native predators and parasites often have devastating effects on indigenous host species. Such effects are far more severe than those caused by native predators and parasites. Why might this be so?

6.3 Mutualism and Commensalism

- Symbioses are intimate interdependencies between different species; mutualism and commensalism are two kinds of symbiosis.

KEY TERMS

symbiosis, mutualism, commensalism

QUESTION

1. What role does coevolution play in the development of mutualistic relationships in ecological communities?

6.4 The Flow of Energy in Ecological Communities

- Food chains and food webs describe the path of energy flow among organisms in an ecological community.

- Within a food chain or web, organisms are assigned to trophic levels based on their food source. Green plants make up the first trophic level; herbivores make up the second trophic level; carnivores that feed on herbivores make up the third trophic level, and so on.

KEY TERMS

energy flow, trophic level, food web, food chain, primary producer, decomposer, biomass energy, trophic level efficiency, keystone species, trophic cascade, primary consumer

QUESTIONS

1. What factors might influence trophic level efficiency?

2. Sharks are considered to be important keystone species in some marine communities. Why might this be the case?

3. Why have conservationists been particularly interested in the restoration of carnivore populations?

4th trophic level
Tertiary consumers

3rd trophic level
Secondary consumers

2nd trophic level
Primary consumers

Energy from the sun

1st trophic level
Primary producers

Decomposers
Organisms that feed on nonliving organic matter

6.5 Disturbance and Community Change

- Disturbance such as fires, storms, or land clearing initiate the process of change in ecological communities called succession.

- The processes of primary, secondary, and cyclic succession are influenced by local variations in the physical environment, the composition of species in nearby ecological communities, and long-term changes in climate.

KEY TERMS

ecological disturbance, ecological legacy, succession, primary succession, pioneer species, facilitation, climax community, secondary succession, old-field succession, return time

QUESTIONS

1. Describe the roles of facilitation and competition in successional change.

2. What ecological features do pioneer species share?

3. How might successional change actually increase the likelihood of ecosystem disturbance?

4. Details of succession such as the rate of change or the sequence of species establishment often vary from location to location within a region. What factors influence such variation?

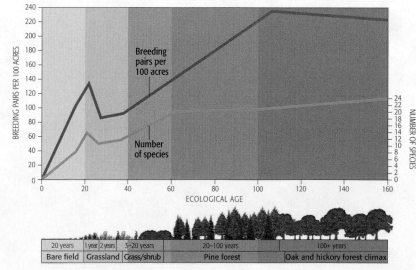

Mastering ENVIRONMENTALSCIENCE®

Go to **www.masteringenvironmentalscience.com** for homework assignments, practice quizzes, Pearson eText, and more.

Nitrogen Runs Through It

How do human activities change how nitrogen flows through ecosystems?

The Fresh River. That's what the Dutch Explorer Adriaen Block dubbed it, perhaps in contrast to the fouled waters he remembered from his native Amsterdam. It was the spring of 1614, and Block was the first European to travel up what we now call the Connecticut River (**Figure 7.1**). The river had, of course, long been known to the Native Americans. At the time of Block's visit, nearly 10,000 of these people grew maize on the river's floodplain and harvested a bounty of salmon and shad from its waters.

The waters of the Connecticut River were indeed fresh. Ninety-nine percent of rain and snow in the river's basin fell on forested land. These forests moderated both the flow and chemistry of the water as it moved to the river.

The purity of a river's water is often measured by the amount of nitrogen it contains. High nitrogen content is correlated with poor water quality. Scientists estimate that when Block arrived, the Connecticut River carried about 2.7 million kg (6 million pounds) of nitrogen to Long Island Sound each year. That's about 93.4 kg/km^2 (533 pounds/mi^2) of the watershed. Although that may seem like a lot of nitrogen, dissolved in the 16.6 trillion liters of water that flow out of the river each year, it is far less than 0.1 part per million of nitrogen.

▲ **Figure 7.1 The Connecticut River Watershed**
The Connecticut River originates in the very northern tip of New Hampshire and delivers its water to Long Island Sound, 400 miles to the south. Its tributary streams gather water from over 29,025 km^2 (11,250 mi^2) of rural and urban New England.

The drinking water in most U.S. cities has considerably higher levels of nitrogen.

By 1700, Europeans had settled along the entire length of the Connecticut River. The wildlife in its waters and on its shores sustained a thriving fishery and fur trade. Between 1750 and 1850, agriculture spread across most of its floodplain (**Figure 7.2**).

◄ **Figure 7.2 The Connecticut River circa 1800**
The View from Mt. Holyoke, a painting by Hudson River School artist Thomas Cole, shows that agriculture had replaced forest in much of the Connecticut River watershed right up to the water's edge. Forest loss diminished the ability of land to retain nitrogen, and sediment from agriculture added even more nitrogen to the river.

By 1870, much of the forest in its watershed had been cut to provide land for farms. Numerous dams and mills, especially in industrial centers such as Hartford and Springfield, captured the energy of its waters (**Figure 7.3**). Then, the Connecticut River was anything but fresh. Its waters were murky with sediments eroded from farmlands and pollution from the expanding cities along its banks. Atlantic salmon and shad had all but disappeared.

In 1970, 2.4 million people lived in the Connecticut River watershed, about 215 people for each square mile of land. Wastewater from the stormwater and sewage-treatment facilities of dozens of cities poured directly into the river. The stench from some parts of the river was so strong that several newspapers referred to it as a "landscaped sewer." The river was transporting over 13.6 million kg (30 million pounds) of nitrogen each year, over 1,000 kg of nitrogen for each km^2 of its watershed. This nitrogen contributed to frequent algal blooms and very low levels of oxygen in the waters of Long Island Sound near the river's mouth. These conditions, in turn, caused commercial shellfish and fish stocks to decline.

The increased nitrogen load was a consequence of several factors acting simultaneously and, in some cases, synergistically. Prior to European settlement, dust and rain brought nitrogen into the Connecticut watershed at a rate of about 90 kg/km^2 (500 pounds/mi^2). During the 20th century, the burning of fossil fuels for transportation and power generation greatly increased the amounts of nitrogen pollutants in the atmosphere. Today, atmospheric nitrogen falls onto the Connecticut watershed in dust and rain at nearly six times the precolonial rate. Much of this nitrogen eventually finds its way into streams.

Nitrogen fertilizers, which are widely used by farmers throughout the watershed, also contribute to the nitrogen load. Streams draining agricultural landscapes often contain two to three times more nitrogen than is found in streams coming from forests. Well over half of the river's current nitrogen load comes from the disposal of industrial and municipal waste from the numerous towns and cities along the river.

The situation was much worse before the 1972 Clean Water Act, which set new standards for water quality and mandated the cleanup of America's rivers. Since then, the amount of nitrogen released from urban wastewater facilities has been halved. Forests and streamside vegetation, which absorb nitrogen, have been restored along hundreds of miles of streams. Agricultural practices now include protocols to reduce the amount of nitrogen in the water that drains fields. Consequently, the overall nitrogen load of the Connecticut

River is declining (**Figure 7.4**). This reduction is increasing the vitality of the river's ecosystems. Populations of fishes such as Atlantic salmon and shad are now on the rise.

All of the chemical elements required by organisms are transported to a greater or lesser extent by moving water. Thus, rivers are an integral part of most element cycles. In this chapter we will explore the principles that govern how chemical elements are cycled in the biosphere. We will consider the connections of these cycles to geologic processes in the lithosphere and the movement of water in the water cycle. We will compare the global cycles of four elements that are especially important to organisms—carbon, nitrogen, phosphorus, and sulfur. We will explore the effects of human activities on these cycles. Finally, we will consider how seasonal changes and disturbances affect nutrient cycles.

▲ Figure 7.3 **The Industrial Revolution**
By the late 19th century, industrial and urban development in large cities multiplied the amount of nitrogen being carried by the Connecticut River.

▼ Figure 7.4 **Cleaning Up the River**
With the implementation of management strategies such as streamside restoration and improved wastewater treatment, nitrogen levels in the Connecticut River began to decline in the mid-1990s.

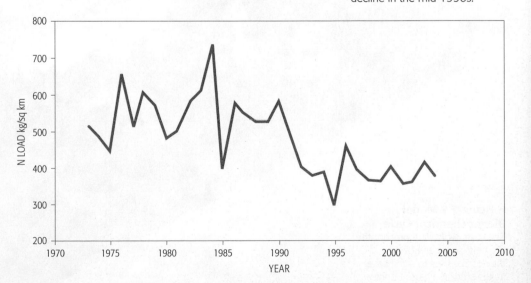

7.1 Ecosystem Ecology and Biogeochemistry

All of the organisms on Earth and the nonliving environment with which they interact make up the **biosphere**. The biosphere includes the rocks and soils that make up Earth's crust, its fresh water and oceans (the hydrosphere), and its atmosphere. A **biogeochemical cycle** describes the flow of matter through an ecosystem. Most biogeochemical cycles describe the flow of elements, such as carbon or nitrogen, but some describe the flow of compounds, such as water. Fundamental ecosystem principles are central to our understanding of biogeochemical cycles (1.3). The distribution and rate of movement of elements in the biosphere is determined by whether or not they are required by organisms, are abundant in the atmosphere, are soluble in water, and are easily immobilized in the lithosphere. The cycling of rocks, water, carbon, nitrogen, phosphorus, and sulfur are particularly important in ecosystems.

Material Pools and Fluxes

The distribution, abundance, and movement of elements in an ecosystem can be tracked by measuring element pools and fluxes.

Ecosystem scientists describe biogeochemical cycles in terms of pools and fluxes (**Figure 7.5**). **Pools** are the parts of an ecosystem in which matter may reside, such as the atmosphere or soil. **Fluxes** are the rate at which matter moves from one pool to another. By knowing the mass of an element in each pool in an ecosystem and the fluxes of the element through that ecosystem, scientists can account for changes in the abundance of that element within an ecosystem. Ecosystem scientists call this **mass-balance accounting.**

Biogeochemical cycles may be considered for Earth in its entirety or for smaller areas, such as an individual forest or watershed. Ecosystem scientists define pools at various levels of detail, depending on their research interests and management goals. In studies of the global cycling of an element such as carbon or nitrogen, scientists may define only four pools—Earth's crust, waters, atmosphere, and living organisms. More often, scientists divide ecosystems into multiple pools. Thus, a scientist interested in the cycling of an element within a lake would probably designate pools such as lake water, organic sediment, inorganic sediment, and organisms. Because many elements move through food webs, that scientist might further divide the

► **Figure 7.5 Model Biogeochemical Cycle**
Scientists typically represent the element pools as boxes, and they use arrows to signify the fluxes among different pools.

organism pool into primary producers, consumers, and decomposers.

The **capital** of a particular pool is the total amount (mass) of an element or molecule that it contains. For the entire biosphere, pool sizes are generally measured in teragrams (1 Tg = 10^{12} g) or petagrams (1 Pg = 10^{15} g). At smaller scales, such as an individual forest or a watershed, pool sizes are measured in grams (g) or kilograms (kg) per unit area, usually per square meter (m^2). The total capital of a particular element in an ecosystem is the sum of the capital in each of the individual pools that compose that ecosystem.

Element fluxes are expressed as the amount of an element that moves from one pool to another in a unit of time. Global fluxes are expressed as teragrams or petagrams per year. Fluxes at smaller scales are described as grams or kilograms per unit area per year (for example, kg/m^2/yr).

Fluxes into a pool are positive, and fluxes out are negative. The net element flux for a pool is the sum of all of the fluxes into and out of it. If the net flux of an element is positive, then the pool's capital of that element is growing. If the net flux is negative, then the pool's capital of that element is shrinking. If the net flux of an element is zero, then the capital of the element in that pool remains constant; such a pool is said to be in **equilibrium**.

Water in a bathtub is a helpful metaphor for these concepts. The flux into the tub is the rate at which water flows from the spigot. The flux out of the tub is the rate at which gravity pulls water down the drain. If you turn on the spigot full force, the flux of water into the tub may exceed the capacity of the drain to let water out. The net flux will be positive, and the tub will fill up. But if you slowly turn the spigot down, you can find a rate of flow at which the level of water in the tub remains unchanged. Although water is still flowing in and out of the tub, its net flux is zero and the water level is at equilibrium.

Residence time is the average time that an atom of an element or molecule of a compound spends in a pool. Residence time is easily calculated as the size of the equilibrium pool divided by the flux through the pool. Consider the example of the bathtub. If the volume of water at equilibrium is 5 gallons and the spigot is flowing at 1 gallon/minute, then the residence time of a molecule of water in the tub equals 5 minutes. (Note that unlike the capital of elements, quantities of water are measured by volume.) **Cycling time** is the average time that it takes an element or molecule to make its way through an entire biogeochemical cycle.

1. What is the difference between a negative and a positive flux?
2. What two factors determine the residence time of an element in a particular pool?
3. What is a nutrient element?
4. Name and describe four factors that influence the global cycling time of nutrient elements.

Nutrients

The cycles of nutrient elements are influenced by their abundance and mobility in the atmosphere, biosphere, and lithosphere.

Organisms use at least 25 nutrient elements to carry out their life functions (**Figure 7.6**). Elements that organisms require in comparatively large amounts, such as carbon, oxygen, hydrogen, nitrogen, phosphorus, and sulfur, are **macronutrients**. Macronutrients are important in the energy transformations of photosynthesis and cellular respiration and are also necessary to the structure and function of organisms (4.2). Organisms also require **micronutrients**, such as molybdenum, manganese, and boron, but only in very small amounts. These elements often play a central role in the catalytic function of enzymes. Different organisms may require different micronutrients.

Many nutrient elements are cycled as organisms eat or are eaten by other organisms. For example, producers such as plants, algae, and cyanobacteria are able to take up inorganic forms of carbon. All other organisms obtain carbon by eating either producers or organisms that eat producers. The rate at which carbon and other nutrients move through the pool of living organisms in Earth's geochemical cycles is influenced by the complexity of food webs.

Cycling times for nutrient elements are also related to the size and characteristics of their pools in Earth's atmosphere, crust, and waters. Elements that are abundant in the atmosphere, such as nitrogen, oxygen, and carbon, have short cycling times. Other elements, such as phosphorus, iron, calcium, and magnesium, have chemical forms that are easily tied up in soils and rocks; these elements flow through the biosphere slowly, and their atmospheric pools are typically small. Nutrients that are soluble in water are often abundant in Earth's waters. The cycles of soluble nutrients are greatly influenced by patterns of water movement across the land and in the ocean.

In summary, living organisms require nutrient elements to grow and reproduce. Cycles of these elements are affected by their movement through food webs and by factors that influence their residence times in Earth's atmosphere, crust, and waters. In this chapter, we will explore in detail the global cycling of four macronutrients—carbon, nitrogen, phosphorus, and sulfur.

▼ **Figure 7.6 Nutrient Elements**
What elements do you and all life require? Macronutrients are elements we need in large amounts and compose a significant part of our body's mass. Microelements are required in smaller quantities, but they play essential roles in our growth and reproduction.

Oxygen (O): 65.0%

Carbon (C): 18.5%
Hydrogen (H): 9.5%
Nitrogen (N): 3.3%

Macronutrients
Calcium (Ca): 1.5%
Phosphorus (P): 1.0%
Potassium (K): 0.4%
Sulfur (S): 0.3%
Sodium (Na): 0.2%
Chlorine (Cl): 0.2%
Magnesium (Mg): 0.1%

Microelements: less than 0.01%
Boron (B)   Molybdenum (Mo)
Chromium (Cr)   Selenium (Se)
Cobalt (Co)   Silicon (Si)
Copper (Cu)   Tin (Sn)
Fluorine (F)   Vanadium (V)
Iodine (I)   Zinc (Zn)
Iron (Fe)
Manganese (Mn)

7.2 The Rock Cycle

In the **rock cycle**, elements within Earth's crust and mantle are slowly converted from one type of rock to another (**Figure 7.7**). Geologists classify rocks into three general categories. **Igneous rocks** form as magma solidifies. **Sedimentary rocks** form when sediments, such as sand, silt, and the remains of dead organisms, become "glued together" under pressure. **Metamorphic rocks** form when great heat and pressure transform the physical and chemical properties of sedimentary or igneous rocks. The processes that change rocks include melting and solidification, weathering, and the erosion and deposition of sediments. Organisms play key roles in the weathering of rocks and the formation of sediments.

▼ **Figure 7.7 The Rock Cycle**
Over long periods of time, the matter that makes up one particular rock type can be transformed to create other types of rock. Note that this figure is not drawn to scale.

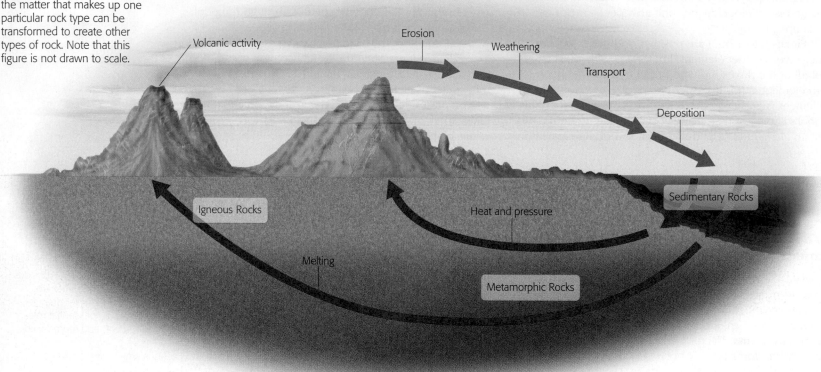

Elements Cycle in the Lithosphere

The rock cycle explains the movement of elements among different rock types.

The forces of plate tectonics in combination with weathering, erosion, and deposition cause elements to slowly move through the lithosphere. These processes also form the wide variety of rocks that make up Earth's crust (3.5).

Where temperatures and pressures in the crust or upper mantle are sufficient, all types of rocks melt, forming magma. As magma cools, it solidifies to form igneous rock. The characteristics of igneous rock depend on the chemical composition of the magma and the rate at which the magma cools. Magmas formed from material in the mantle solidify into rocks that contain large amounts of

iron and magnesium. Magmas formed from continental crust that melted in subduction zones solidify into rocks that contain large amounts of aluminum and silicon.

If magma solidifies beneath Earth's surface, it forms intrusive igneous rock, such as granite (**Figure 7.8**). Because magma cools very slowly underground, the crystals in the rock have time to grow large, giving intrusive igneous rocks a rough, granular texture. Magma that flows or is ejected onto Earth's surface forms extrusive igneous rock. Basalt, which forms from volcanic lava, is an example of an extrusive igneous rock. Because basalt cools rapidly, it has very small crystals.

A variety of physical and chemical processes on Earth's surface cause rocks to weather, or break into smaller particles. Flowing water, ice, wind, gravity, and plant roots can break rocks apart. As the roots of plants grow into crevices, they force rocks apart and release chemicals that accelerate weathering. Chemicals in decaying organic matter also accelerate rock weathering. The rock particles left by weathering make up the major part of soil. The physical and chemical properties of soils are determined in large part by the nature of the rock from which they develop.

In the process of erosion, weathered rock particles and soil are carried away by gravity, water, wind, or glacial ice. Eventually, these materials are deposited as sediment in locations such as the mouths of rivers and the bottoms of lakes and oceans. Shells, skeletal remains, and unde-composed organic matter from the biosphere also contribute to sediment. Minerals that percolate through sediment glue the particles together, forming rock. As thick layers of sediment accumulate, their weight creates pressure on the sediment layers below, accelerating the process of rock formation.

The physical and chemical characteristics of sedimentary rocks depend on the sediment from which they form (**Figure 7.9**). For example, sand forms silica-rich sandstones; silt and clay form shale. The calcium-rich shells and skeletons of marine organisms that settle on the ocean floor become limestone. Coal is a sedimentary rock formed from peat and other organic matter that has been transformed by the pressure of the layers of sediment above it.

The very high temperatures and pressures associated with tectonic processes can alter the chemical and physical properties of both sedimentary and igneous rocks, producing metamorphic rocks. The characteristics of metamorphic rocks depend on the parent rock and the processes of transformation. Under high pressure and temperature, shale becomes slate and limestone becomes marble (**Figure 7.10**).

The conditions under which a metamorphic rock forms determine its chemical structure. This, in turn, influences the nature of soils that are formed as it weathers. For example, in the presence of water and metal elements such as chromium, cobalt, manganese, and nickel, several different kinds of igneous rocks are transformed into a metamorphic rock called serpentine. On soils derived from serpentine, these metals limit plant growth and support unique ecosystems.

Over Earth's long history, plate tectonics, weathering, erosion, and deposition have slowly cycled elements through Earth's crust. Elements have been transported from the upper mantle to Earth's surface. As the continents evolved, the amounts of sedimentary and metamorphic rocks increased. Weathering of the various types of rocks produces soil and releases nutrient elements. These elements can then be taken up in the biosphere or dissolved in the waters of the hydrosphere. By weathering rocks and contributing to the formation of soils and sediments, living organisms play important roles in the rock cycle.

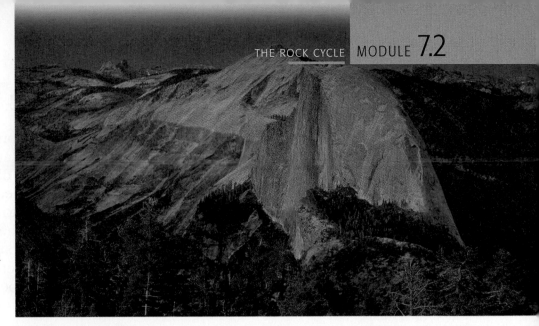

▲ **Figure 7.8 Igneous Rock**
The granite of Yosemite's Half Dome is an intrusive igneous rock that solidified from magma within Earth's crust.

▲ **Figure 7.9 Sedimentary Rock**
Sedimentary rocks, such as the sandstone and shale that make up the layers of the Grand Canyon, are formed when sediments such as sand or silt are chemically cemented into rock.

◄ **Figure 7.10 Metamorphic Rock**
The marble from which Michelangelo carved the statue of David is metamorphic rock that was formed as sedimentary limestone and was altered chemically under high temperature and pressure.

QUESTIONS
1. Name and describe the three basic kinds of rocks.
2. Describe two roles of plate tectonics in the rock cycle.

7.3 The Hydrologic Cycle

The **hydrologic cycle** describes the distribution and flux of water through Earth's biogeochemical system. Over 98% of Earth's water is found in the ocean, and 99% of the remaining water occurs in saturated rocks out of reach of plant roots. At any moment, less than 0.001% of Earth's water is in the atmosphere. Water falls from the atmosphere as rain or snow. Rainwater may then evaporate back to the atmosphere, flow across the surface into rivers and streams, or percolate through the soil to underlying rock. Snow may melt or accumulate as ice in the glaciers of high mountains, the Arctic, and Antarctic. Patterns of precipitation, evaporation, surface flow, and percolation vary widely through the seasons and among different locations. Over the entire globe, however, total annual precipitation equals total annual evaporation.

Earth's Water

Solar energy and gravity drive the hydrologic cycle.

The hydrologic cycle includes three phases of water—solid ice, liquid water, and gaseous water vapor. The energy of solar radiation causes ice to melt and liquid water to evaporate. The force of gravity moves water from the sky to the land and across the land to the sea (**Figure 7.11**).

Precipitation transfers water vapor in the atmosphere to liquid water in the hydrosphere. This flux is usually measured with a rain gauge, which shows the depth of water that falls at a particular location. Annual amounts of precipitation vary widely, from 5 to 10 cm (2–4 in.) in the driest deserts to over 6 m (20 ft) in some tropical rain forests.

The total pool of water in the atmosphere is about 13,000 km³ (3,119 mi³). That may seem like a large number, but if it were all to condense and fall at once, it would equal only about 2.5 cm (1.0 in.) of rain over Earth's entire surface (13,000 km³ divided by Earth's area, 510,072,000 km²). What truly is large is the flux of water into and out of the atmosphere— each year approximately 496,000 km³ of liquid water falls on Earth's surface, and an equal amount evaporates to the atmosphere. That's enough water to fill all of the Great Lakes 22 times each year! At any given location and at any particular moment, the rates of precipitation and evaporation are not likely to be equal. Over the entire globe, however, these rates are equal and relatively constant.

▶ **Figure 7.11 The Hydrologic Cycle**
Solar energy drives the evaporation of liquid water from the ocean and lakes as well as the circulation of the atmosphere that carries moisture-laden air over land.

Once on the ground, liquid water has three general fates—it may evaporate back to the atmosphere, flow across Earth's surface into streams and lakes and eventually to the ocean, or percolate through soil to become groundwater. Water evaporates from bodies of water, from soils, and from the leaves of plants that have absorbed water from the soil. Evaporation from leaves is called **transpiration**. Because of the large surface area of plant leaves, transpiration is a major component of evaporation in terrestrial ecosystems.

The source of water contributing to rainfall depends on the location. Nearly 100% of the rain that falls over the ocean and near coasts comes from water that evaporated from the ocean surface. In some places on land, a considerable proportion of rainfall comes from water that evaporated from terrestrial ecosystems. For example, 25–50% of the rain that falls on the Amazon rain forest region comes from water that evaporated from those forests. In many parts of the world where large areas of tropical rain forest have been cut, rainfall has declined (**Figure 7.12**).

Rainwater that falls on land may flow across the surface as **runoff**, eventually entering lakes and streams. Although some moisture evaporates along the way, gravity pulls the water in lakes and streams inexorably toward the sea. Approximately 10% of the water that evaporates from the ocean returns to it via rivers each year.

Gravity causes some water to percolate through the soil and into the rock below as **groundwater**. A layer of soil or rock that is saturated with groundwater is called an **aquifer**. The proportion of rainfall that percolates into aquifers varies widely, depending on factors such as total rainfall, vegetation cover, and soil porosity. In most locations, only 0.1–5.0% of total rainfall reaches an aquifer. Groundwater moves through rock very slowly, often at rates of less than 1 ft per year, eventually delivering water to lakes, streams, or the ocean.

The residence time of water—the average time that a water molecule resides in a particular part of the cycle—varies widely. A water molecule evaporating from the surface of the ocean has a residence time of 9.5 days in the atmosphere. If it then falls as rain and is captured as runoff, it will have a residence time of 14 days in rivers and streams. Residence time for groundwater is much longer, ranging from hundreds to tens of thousands of years. In the ocean, the residence time for an average water molecule is over 2,700 years.

Snow that falls in temperate latitudes may stay on the ground for weeks or months before melting. Snow that falls in high mountains or polar regions may accumulate as ice to form glaciers, where it may remain for a very long time. For example, ice in the lower layers of glaciers in Greenland and Antarctica fell as snow hundreds of thousands of years ago.

The movement of water significantly influences many of Earth's biogeochemical cycles. Earth's liquid water carries with it dissolved molecules and sediment, even whole ecosystems. Raindrops condense on small dust particles and pick up other chemicals in the atmosphere.

A multitude of chemicals dissolve in water as it moves through plant foliage and soils. Water flowing in streams and rivers transports vast amounts of sediment to lakes and to the ocean. Thus, the hydrologic cycle plays a significant role in the cycling of most nutrient elements.

Human impacts on the hydrologic cycle are quite significant. Humans have diverted large quantities of water from rivers and streams for agricultural, industrial, and residential uses. Deforestation and extensive areas of pavement have altered the relative amounts of evaporation, runoff, and percolation to groundwater aquifers. In many places, humans are taking water from aquifers far more rapidly than the natural rates of replenishment. Human activities are causing worldwide climates to warm; this warming is diminishing the amount of water stored in glacial ice. The ways that humans affect the water cycle are discussed in detail in Chapters 8 and 12.

▼ **Figure 7.12 Deforestation and Rainfall**
The loss of once-extensive forests has diminished the rainfall over large areas of Madagascar because less water now evaporates back to the atmosphere from this landscape.

181

7.4 The Carbon Cycle

Earth's atmosphere and lithosphere are composed mainly of the elements oxygen (45.2%) and silicon (29.5%). In comparison, carbon (C) contributes only a minuscule amount (0.032%) to the total mass of the lithosphere and atmosphere. The vast majority of this carbon is found in sedimentary rock; this carbon has cycling times of hundreds of millions of years. The remaining carbon—just 0.0005%—cycles through the biosphere quite rapidly. But of all the elements, carbon is the one most closely identified with life and the activities of ecosystems; we are composed of numerous elements, but we refer to ourselves as carbon-based life forms. Humans have significantly modified Earth's carbon cycle by the direct use of high-energy carbon from ecosystems, by the use of land, and by the burning of fossil fuels (**Figure 7.13**).

Life Drives the Carbon Cycle

Photosynthesis and cellular respiration are key steps in the carbon flux between the biosphere and the atmosphere.

The flux of carbon between the atmosphere and terrestrial and aquatic ecosystems is tightly coupled to the rate of photosynthesis and the rate at which organic carbon is either respired or burned. Photosynthetic organisms—plants, algae, and cyanobacteria—use light energy to add hydrogen ions and electrons to CO_2, producing carbohydrates and other organic molecules (4.1).

Organic forms of carbon, such as carbohydrates, hydrocarbons, lipids, and proteins, store large amounts of chemical energy; all organisms access this energy through the process of cellular respiration. In cellular respiration, hydrogen is removed from organic compounds, oxygen is added, and organic carbon is converted back to CO_2. A similar process occurs when organic compounds are burned, as in a wildfire or in the engine of your car.

The total amount of CO_2 that photosynthetic organisms convert to organic carbon each year is called **gross primary production (GPP)**. These producers use approximately half of the GPP in cellular respiration to carry out their own life functions. The remaining half is **net primary production (NPP)**. NPP is the amount of organic carbon available to all the nonphotosynthetic organisms, or consumers, in an ecosystem.

Net ecosystem production (NEP) is the amount of organic carbon left each year after subtracting the respiration of nonphotosynthetic organisms from NPP. In other words, NEP is the net flux of carbon into an ecosystem. If the total carbon in the biomass of organisms in an ecosystem remains constant, NEP is zero. If the biomass of an ecosystem increases, as it does during succession, NEP is positive. If the biomass decreases, as it does when an ecosystem is burned or degraded by human activities, NEP is negative.

▶ **Figure 7.13 The Terrestrial Carbon Cycle**
The important pools (Pg = 10^{15} g) of the carbon cycle are indicated in black and fluxes (Pg/yr) are indicated in red. The pools and fluxes enclosed by the blue dashed line are those specifically affected by human activities.

Terrestrial Carbon

Most of the carbon stored in living biomass is found in terrestrial ecosystems.

Worldwide, approximately 560 Pg of carbon is stored in biomass on land. Well over 90% of this biomass is in plants. In the absence of human activities such as deforestation, scientists assume that Earth's total biomass pool remains relatively constant. This implies that the total rate of respiration by all the organisms on Earth is equal to GPP and that global NEP is 0.

The soils of terrestrial ecosystems contain large amounts of organic carbon—approximately 1,500 Pg. (This is a very rough estimate because it is difficult to measure soil carbon, and soils are poorly studied.) Plant litter, waste, and dead organisms are constantly adding carbon to soils. Some of this carbon is in organic compounds that are quickly consumed by decomposers, such as insects, worms, fungi, and bacteria. Other organic compounds are resistant to decay and may reside in soils for hundreds or even thousands of years (**Figure 7.14**).

Because of geographical differences in the length of growing seasons and the availability of water, terrestrial NPP and biomass pools vary widely among ecosystems (**Figure 7.15**). In general, NPP and biomass pools are high in warm, moist regions and low where conditions are cold and dry. In some tropical rain forests, NPP may be as high as 1,000 g C/m^2/yr, and the carbon in biomass may exceed 15,000 g/m^2. In the driest deserts, NPP may be only 40 g C/m^2/yr, and the carbon in biomass may be less than 300 g/m^2.

◄ **Figure 7.14 Soil Carbon**
Much of this organic carbon is rapidly respired by decomposer organisms, but some is resistant to decay and may accumulate in the soil as shown here. The top layer of soil is often dark because of abundant organic carbon. In spite of the obvious color difference, amounts of organic carbon in soil are very difficult to measure.

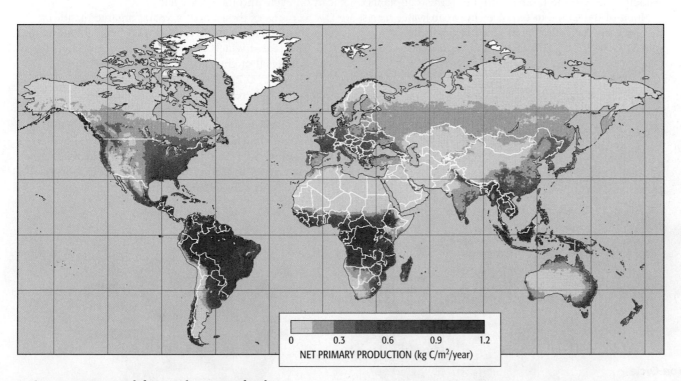

NET PRIMARY PRODUCTION (kg C/m^2/year)

▲ **Figure 7.15 Terrestrial Net Primary Production**
Net primary production in the tropical forests of warm and moist regions, such as the Amazon of South America, central Africa, and southeastern Asia, is 10–20 times that of hot, dry deserts such as those found in the southwest of North America and parts of Africa and Asia.

Aquatic and Marine Carbon

Living biomass forms a tiny fraction of total ocean carbon but has a very significant influence on the ocean carbon cycle.

Oceans, lakes, and rivers contain a very significant global pool of carbon, about 38,000 Pg. Much of this carbon is in the form of carbon dioxide, written as CO_2-C. Each year, approximately 92 Pg of CO_2-C from the atmosphere dissolves into oceans, lakes, and rivers. Of this amount, 51 Pg is consumed in NPP by phytoplankton—the photosynthetic algae and bacteria suspended in the water. This rate of productivity is remarkable because the total biomass of algae and phytoplankton in all of Earth's oceans is far less than 0.1 Pg, a minuscule fraction of total ocean carbon. Considered another way, every gram of biomass in the primary producers in the ocean contributes 510 g of NPP per year. In terrestrial ecosystems, each gram of plant biomass contributes only about 0.12 g of NPP per year (**Figure 7.16**).

What causes this difference between terrestrial and ocean ecosystems? There are two causes. First, much of the biomass of plants, the primary producers on land, is not green; rather, it is found in long-lived, nonphotosynthetic structures such as roots and wood. Second, phytoplankton grow and reproduce very rapidly but are constantly being eaten by marine herbivores. So although the NPP of phytoplankton is high, their total biomass at any moment is comparatively low. Consequently, the residence time for carbon in the biomass pool of phytoplankton is much shorter than its residence time in the biomass pool of terrestrial plants.

Much of the CO_2 that dissolves in ocean water reacts to form carbonate (CO_3^{2-}) and biocarbonate (HCO_3^-) ions. Dissolved CO_2, carbonate, and bicarbonate account

▲ **Figure 7.17 Carbon in Shells and Skeltons**
These microscopic foraminifera are among a diverse array of marine plants and animals that produce shells or skeletons containing carbon as calcium carbonate.

for more than 99% of ocean carbon. Carbonate and bicarbonate combine with calcium to form calcium carbonate ($CaCO_3$), which is the primary structural element in the shells of familiar marine animals such as clams and mussels. However, most calcium carbonate is used to form the skeletons and shells of less well-known microscopic organisms, such as foraminifera (**Figure 7.17**).

When marine organisms die, the calcium carbonate in their shells and skeletons may dissolve, and the carbonate may be converted back to CO_2. But often the remains of marine organisms sink to the ocean floor, forming layers of sediments that eventually are transformed into limestone and other sedimentary rocks. This process is an important connection between the carbon cycle and the rock cycle.

Over tens of thousands of years, organic matter in ocean sediments may be transformed into oil and natural gas. The rate at which organic carbon is buried in sediments is thought to be far less than 0.01 Pg/yr.

Each year, 90 Pg of carbon from the oceans return to the atmosphere as CO_2. This is about 2 Pg less than the amount that is absorbed each year. This 2 Pg surplus is much greater than the annual flux of carbon into sediments. Thus, marine scientists believe that the total pool of carbon in the oceans must be increasing. However, given the large size of the ocean carbon pool, this increase is very difficult to detect.

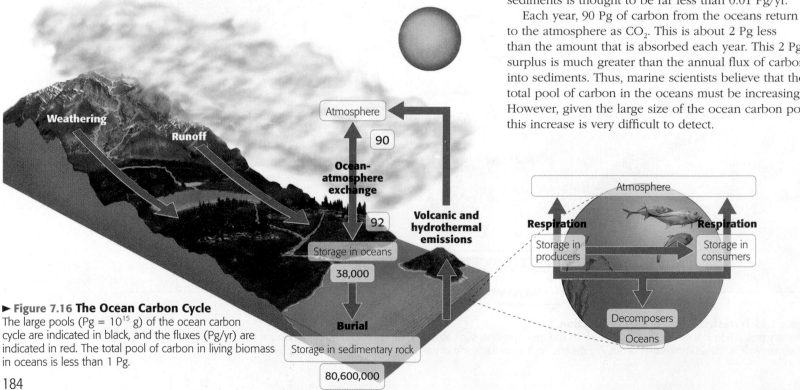

► **Figure 7.16 The Ocean Carbon Cycle**
The large pools (Pg = 10^{15} g) of the ocean carbon cycle are indicated in black, and the fluxes (Pg/yr) are indicated in red. The total pool of carbon in living biomass in oceans is less than 1 Pg.

Weathering

Runoff

Atmosphere

90

Ocean-atmosphere exchange

92

Volcanic and hydrothermal emissions

Storage in oceans

38,000

Burial

Storage in sedimentary rock

80,600,000

Atmosphere

Respiration

Respiration

Storage in producers

Storage in consumers

Decomposers

Oceans

The rates of exchange of carbon between the ocean and the atmosphere vary considerably from location to location. Among the factors influencing this variation are temperature and local factors influencing NPP. The solubility of CO_2 gas in water decreases as water temperature increases. Therefore, all other things being equal, the flux of CO_2 from the atmosphere to the ocean decreases with increasing surface ocean temperatures.

Because light can penetrate ocean water to a depth of only 50–100 meters, nearly all of the photosynthesis in the ocean takes place near the surface. Actual values of NPP vary greatly from location to location. Over most of the open ocean—more than 90% of the ocean surface—NPP averages 130 g $C/m^2/yr$. But near land where there is an influx of nutrients such as nitrogen and phosphorus, NPP increases by 200–400%. This difference in productivity has significant consequences for local rates of CO_2 exchange with the atmosphere; it also affects the composition and diversity of ocean ecosystems (**Figure 7.18**).

NET PRIMARY PRODUCTION (kg C/m^2/year)

▲ **Figure 7.18 Marine Net Primary Production**
Except where nutrient inputs are high, such as near coastlines or where nutrient-rich water from the depths is brought to the surface, net primary production over most of the ocean is less than 150 g $C/m^2/yr$.

Human Impacts

We have significantly altered Earth's carbon cycle.

Like all consumers, humans obtain their energy from organic compounds that are ultimately derived from the NPP of ecosystems. As such, we compete with all other forms of life for that NPP. The rapid increase in our numbers, particularly over the past century, is powerful evidence that we have been quite successful in that competition (5.3). Scientists estimate that today humans harvest directly or manage 32–40% of global NPP. By appropriating this large fraction of global production, our impact on the carbon cycle has been profound.

In many regions, humans have replaced natural ecosystems with farmland, cities, and suburbs. Human uses of land, such as agriculture and the harvesting of timber, have significantly diminished the amount of carbon stored in Earth's terrestrial ecosystems. For example, the biomass in each square meter of a typical tropical forest stores between 12,000 and 15,000 g of carbon. A pasture stores only about 10% of that amount. So when a tropical forest is cleared and converted to pasture, about 10,000–13,500 g C/m^2 is released into the atmosphere. Similarly, hundreds of millions of acres of grasslands, which typically store 3,000–5,000 g C/m^2, have been replaced by croplands that store only 40–70% of those amounts. Furthermore, NPP and CO_2 uptake of agricultural ecosystems is generally less than the ecosystems they replaced. Worldwide, scientists estimate that human land use adds 0.9–1.2 Pg of CO_2-C to the atmosphere each year.

The burning of fossil fuels has a particularly profound effect on Earth's carbon cycle. Since the beginning of the Industrial Revolution, the use of fossil fuels has increased exponentially (**Figure 7.19**). In 1800, the burning of coal put approximately 0.01 Pg CO_2-C into the atmosphere.

Today, the burning of coal, oil, and natural gas adds about 6 Pg CO_2-C to the atmosphere annually. Humans are burning fossil fuels thousands of times faster than they are being formed.

Between land use and the burning of fossil fuels, human activities release more than 7 Pg of CO_2-C into the atmosphere each year. Thus, the total flux of CO_2-C into the atmosphere is now greater than the flux out of the atmosphere into terrestrial and marine ecosystems. As a result, the concentration of CO_2 in the atmosphere is rising about 1% per year. And every year, human activities are releasing greater and greater amounts of CO_2. At the current rate of increase, the total flux will double to 14 Pg of CO_2-C by 2060. Scientists agree that this increase in atmospheric CO_2 is a major factor influencing climate change (8.2).

QUESTIONS

1. What is the difference between net primary production (NPP) and net ecosystem production (NEP)?

2. How does the biomass of primary producers in terrestrial ecosystems compare with the mass of primary producers in marine ecosystems?

3. What percent of Earth's NPP is directly harvested or managed by humans?

GLOBAL CO_2-C FOSSIL FUEL EMISSIONS

◄ **Figure 7.19 Fossil Fuels and the Carbon Cycle**
The flux of CO_2-C to the atmosphere from the burning of coal, oil, and natural gas has increased exponentially since the beginning of the Industrial Revolution.

185

7.5 The Nitrogen Cycle

Nitrogen is the most abundant element in Earth's atmosphere, but only small amounts of it are found in Earth's crust. Certain microorganisms are able to convert nitrogen gas (N_2) from the atmosphere to ammonia (NH_3). Plants convert inorganic forms of nitrogen into nitrogen-containing compounds that are essential to life, such as amino acids and nucleic acids. Animals obtain these compounds by eating plants or other organisms. When the wastes and remains of organisms decay, inorganic forms of nitrogen are regenerated. These inorganic forms may be taken up by plants or return to the atmosphere as nitrogen gas. Human activities such as agriculture, the synthesis of nitrogen fertilizers, and the burning of plants and fossil fuels have significantly altered the chemical composition of the atmosphere, lakes, and streams, and have doubled the rate of nitrogen exchange between the atmosphere and biosphere.

What can we do to diminish our impacts on Earth's nitrogen cycle?

Breeann Sharma, Arizona State University

Norm: *It is most important to find ways to diminish the amount of nitrogen fertilizer applied to agricultural fields and to limit nitrogen-rich runoff from those fields into streams and lakes.*

Nitrogen Transformations

Nitrogen rapidly cycles between the atmosphere and the biosphere.

Nitrogen accounts for 78% of the molecules in Earth's atmosphere but less than 0.003% of its crust. There are three reasons for this difference in abundance. First, nitrogen is not a stable part of the crystal structure of any of the minerals in rocks. Second, most molecular forms of nitrogen are volatile and tend to migrate to the atmosphere. Third, nitrogen gas does not react with other gases in the atmosphere and is very stable in the presence of solar radiation.

More than 99.9% of the nitrogen in the atmosphere is in the form of nitrogen gas (N_2). The atmosphere also contains much smaller amounts of ammonia (NH_3) and oxides of nitrogen, such as nitrous oxide (N_2O).

Nitrogen enters the biosphere through the process of **nitrogen fixation** (**Figure 7.20**). In nitrogen fixation, N_2 is converted to NH_3. A small amount of nitrogen fixation is caused by lightning. However, most nitrogen fixation is carried out by nitrogen-fixing bacteria, which convert N_2 to NH_3 and then into nitrogen-containing organic molecules, such as amino acids and nucleic acids. Some nitrogen-fixing microbes, such as many cyanobacteria, live independently in soil and water. Others coexist in symbiotic relationships with plants (6.3).

▶ **Figure 7.20 The Nitrogen Cycle**
Nitrogen pools (Tg = 10^{12} g) are indicated in black and fluxes (Tg/yr) are indicated in red. Vast amounts of nitrogen occur in the atmosphere as N_2 gas. Specialized microorganisms play especially important roles in the nitrogen cycle, including nitrogen fixation, nitrification, and denitrification. Human impacts on the nitrogen cycle are enclosed with a blue dashed line.

In soil and water, nitrogen undergoes a number of transformations that allow it to cycle through the biosphere. Ammonia dissolves in water to form ammonium (NH_4^+). Specialized bacteria in the soil then convert NH_4^+ to nitrite (NO_2^-) and nitrate (NO_3^-) in a process known as **nitrification**. Plants take up ammonium, nitrite, and nitrate through their roots and use them to manufacture nitrogen-containing organic molecules, such as amino acids and nucleic acids. Animals must obtain their nitrogen by eating plants or other animals. When decomposers break down the remains and wastes of plants and animals, ammonia (NH_3) is released into the soil.

Nitrite and nitrate are particularly soluble in water, so they are easily removed from soil and transported into streams and groundwater. Nitrogen in soil and water is returned to the atmosphere by **denitrification**. In this process, specialized bacteria convert NO_3^- to N_2 gas. The total nitrogen in Earth's organisms, soils, and waters is determined by the balance between nitrogen fixation and denitrification.

QUESTIONS

1. How does nitrogen fixation differ from denitrification?

2. What process ultimately limits the amount of nitrogen available to organisms in an ecosystem?

3. Describe three significant ways that humans affect the nitrogen cycle.

Human Impacts

Humans have nearly doubled the global rate of nitrogen fixation.

The development and expansion of agriculture over the past 10,000 years is largely responsible for the worldwide growth of human populations (5.2). However, agricultural ecosystems are limited by the availability of nitrogen in the soil. As crops grow, they take ammonium, nitrite, and nitrate from the soil, converting them into nitrogen-containing compounds that are stored in plant tissues. When crops are harvested, much of this nitrogen is removed, so only a fraction of the original nitrogen is left behind. In most areas, the amount of nitrogen lost from harvest is far greater than the rate of nitrogen fixation. As a result, each year the total amount of nitrogen in the field decreases, limiting the rate of primary production more severely.

Ultimately, the availability of nitrogen is limited by the rate of nitrogen fixation. Although they did not understand why, early farmers increased this rate by cultivating plants in the legume family, such as peas and beans. These plants maintained their productivity from year to year. We now know that legumes benefit from the large amounts of nitrogen fixed by the symbiotic bacteria that grow in their root nodules (6.3). By the 1700s, farmers in many regions had discovered that they could maintain the productivity of their fields by rotating between legume and nonlegume crops.

By increasing the amount of land devoted to growing nitrogen-fixing plants, humans have increased the global rate of nitrogen fixation by about 40% compared to pre-agricultural times. Even so, the resulting increases in crop productivity are still limited by the activities of nitrogen-fixing microbes.

The situation changed dramatically in the early part of the 20th century when chemists discovered the **Haber-Bosch process**, a nonbiological method of nitrogen fixation. Initially developed for the manufacture of explosives, the Haber-Bosch process was quickly adapted to the production of chemical fertilizers. Since 1950, the amount of fertilizer manufactured each year has increased exponentially. The Haber-Bosch manufacture of fertilizer has increased the global rate of nitrogen fixation by about 40% compared to pre-agricultural rates, and that percentage continues to rise.

Because ammonium and nitrate dissolve in water, runoff from agricultural fields eventually carries them into lakes, streams, and estuaries. Thus higher rates of nitrogen fixation ultimately lead to greater amounts of ammonium and nitrate in lakes, streams, and estuaries. Higher concentrations of nitrogen encourage the rapid growth of algae. Their growth leads to a process called **eutrophication**, in which the abundant organic matter from the algae encourages high rates of respiration by decomposing organisms. The important effects of eutrophication on aquatic and marine ecosystems are explored in Chapter 12.

Significant amounts of nitrogen are also added to the biosphere when humans burn fossil fuels, wetlands, and forests. As oil and coal burn, the organic nitrogen in them is converted to gaseous nitric oxide (NO). In the atmosphere, NO reacts with oxygen to form nitrogen dioxide (NO_2). NO_2 dissolves in rainwater to form nitric acid (HNO_3). This series of reactions is one of the major causes of acid deposition, often called acid rain (9.2).

Taken together, the increased cultivation of legumes and production of chemical fertilizers have nearly doubled the global rate of nitrogen fixation (**Figure 7.21**). As the extent of these human activities continues to increase, we can only presume that the global rate of nitrogen fixation will continue to increase as well. This change is responsible for much of the increased agricultural production that has supported the exponential increase in our population. It has also increased the amounts of nitrogen pollutants in our atmosphere and the concentrations of ammonium and nitrate in streams, lakes, and coastal ecosystems.

▼ **Figure 7.21 Nitrogen in Fertilizer**
Worldwide, nitrogen fertilizer application has increased 15-fold since 1950. Altogether, human input of nitrogen into Earth's ecosystems exceeds natural rates of nitrogen fixation.
Source: United Nations. 2007. Millennium Ecosystem Assessment.

TERAGRAMS OF NITROGEN PER YEAR

- Total human input
- Agricultural legumes
- Fertilizer and industrial issues
- Fossil fuels

Projected human input

Natural nitrogen fixation rate

RANGE OF TERRESTRIAL BACTERIAL NITROGEN FIXATION (except in agroecosystems)

300 / 250 / 200 / 150 / 100 / 50 / 0

1900 1920 1940 1960 1980 2000 2020 2040 2060

DAYS

7.6 The Phosphorus Cycle

Unlike nitrogen, phosphorus (P) is relatively abundant in Earth's crust but virtually absent from the atmosphere. Organisms require phosphorus in the form of phosphate (PO₄) to manufacture a variety of organic compounds. The supply of phosphate available to organisms is limited by the rate of weathering of sedimentary rocks. Plants and some microbes can absorb phosphate that is dissolved in soil water or the waters of streams, lakes, or the ocean. Other organisms obtain phosphate through the food web. Phosphate is an especially important limiting factor on ocean NPP. Humans mine large quantities of phosphorus for use as fertilizer, thus increasing its availability in the biosphere.

Phosphorus Supply

The supply of phosphorus is limited by the weathering of rocks.

Phosphorus accounts for about 0.13% of the mass of Earth's crust, making it the 11th most abundant element in the lithosphere. The vast majority of this phosphorus occurs in rocks as calcium phosphate, which is also known by its mineral name, apatite. Apatite is especially common in sedimentary rocks that form in some marine environments. As these rocks weather, the calcium phosphate in them dissolves, releasing phosphate ions (HPO_4^{2-}) into the soil. This very slow weathering process is the ultimate source of all the phosphorus that cycles through the biosphere (**Figure 7.22**).

Phosphorus makes up 0.5–1.0% of the tissues of organisms; this phosphorus is located in phosphate-containing organic compounds, including DNA, RNA, and the phospholipids in cell membranes. As organic matter decomposes in soil, inorganic phosphate is released and dissolves in soil water. Plants absorb this phosphate and use it to manufacture organic phosphate compounds. All other terrestrial organisms obtain organic phosphorus from the things they eat. Waste and dead organisms replenish the stores of organic phosphate in the soil, completing the terrestrial portion of the phosphorus cycle.

▶ **Figure 7.22 The Phosphorus Cycle** The black numbers (Tg = 10¹² g) indicate pool sizes and red numbers fluxes (Tg/yr) in the global phosphorus cycle. The atmospheric pool of phosphorus is very small. Human impacts are enclosed with a blue dashed line.

Each year streams and rivers carry a significant amount (21 Tg) of phosphorus to the sea. Yet only about 10% of this phosphorus is available to aquatic organisms as dissolved phosphate. The remainder is tightly bound to soil particles and other sediments that sink to the ocean bottom; these sediments are destined to become sedimentary rocks. Algae, as well as some of the small animals that feed on them, are able to take up dissolved inorganic phosphate and convert it into organic phosphates. Larger aquatic animals obtain phosphate through the food web.

The phosphorus cycle exerts considerable influence on the net primary production of the oceans. In most marine ecosystems, NPP is limited by the availability of phosphate. Because light penetrates only 50–100 m into the ocean, virtually all marine photosynthesis occurs in these surface waters. However, the phosphorus in surface waters is constantly being lost as organic sediment sinks to the ocean depths. Rivers are the only new phosphate source to the oceans, but they provide less than 1% of the amount required to maintain global marine NPP. The rest must be provided by the recycling of phosphorus among algae, consumers, and decomposers. Where streams and rivers flow into the ocean, the concentration of phosphate is higher, so NPP is usually higher along coasts and near river outflows.

In some locations, such as along the west coasts of the Americas and Africa, ocean currents bring deep ocean water to the surface. This upwelling carries large quantities of phosphate from the ocean depths. As a consequence, NPP in these areas is nearly four times that of the open ocean.

The high NPP caused by upwelling supports complex marine food webs. The higher levels of these food webs often include a diverse array of sea birds, such as gulls, terns, boobies, and puffins. The excrement of these birds, called guano, is very rich in phosphate. Over the centuries, large deposits of guano may accumulate at the nesting sites of these sea birds (**Figure 7.23**).

As marine organisms in the regions of upwelling die, their remains sink to the bottom of the ocean. Over millions of years, these sediments form sedimentary rocks that contain high levels of phosphates.

▲ **Figure 7.23 Phosphate from Guano**
Phosphate from deep-water upwelling increases net primary production and supports a diverse ecosystem. The excrement of predatory sea birds such as these puffins accumulates as guano.

Human Impacts

Humans have increased the availability of phosphorus in fresh water and ocean ecosystems.

As with nitrogen fertilizers, the application of fertilizers containing phosphate has been found to increase the productivity of many crops. The sources of these phosphate fertilizers are guano deposits and sedimentary rocks produced in areas of ocean upwelling millions of years before. Currently, about 12 Tg of phosphorus in the form of phosphate is mined from these deposits annually (**Figure 7.24**).

Human use of phosphate fertilizers and phosphate-rich detergents has significantly increased the amount of phosphate flowing in streams and rivers. Treated and untreated sewage contains large amounts of phosphate; its addition to streams also increases

the supply of phosphorus. Where the growth of algae has been limited by the availability of phosphate, this enrichment can produce eutrophication similar to that produced by excess nitrogen. Such additions increase the growth of algae and reduce the biodiversity of many stream, lake, and coastal ecosystems (see Chapter 12).

QUESTIONS

1. What is the primary source of phosphorus to organisms in ecosystems?

2. How are marine primary production, ocean upwelling, and phosphorus related to one another?

▶ **Figure 7.24 Phosphate Mining**
Today, phosphate is mined from marine sedimentary rocks that were formed millions of years ago from sediments produced in areas of ocean upwelling.

7.7 The Sulfur Cycle

Sulfur accounts for approximately 0.07% of Earth's crust. Most of this sulfur is chemically bound in rock minerals. Rock weathering and volcanic activity release forms of sulfur that can be used by organisms. The annual flux of sulfur through the atmosphere is high, but its residence time in the atmosphere is only a few days. Therefore, the atmospheric pool of sulfur is generally small. Humans have doubled the annual flux of sulfur through the biosphere. Although this has not increased NPP in most ecosystems, it has had significant environmental and health consequences.

Sulfur Transformations

The flux of sulfur through the atmosphere is very rapid.

Sulfur (S) is an important constituent of several of the amino acids that make up proteins and so is an essential element for all organisms (3.2). As rocks weather, the sulfur in them dissolves in water, forming sulfate ions (SO_4^{2-}). These ions are easily taken up by organisms. Some chemosynthetic microorganisms use the energy released when hydrogen sulfide (H_2S) is converted to elemental sulfur (S) (4.1).

The flux of sulfur through the atmosphere is approximately 182 Tg/yr (**Figure 7.25**). Considering that the total pool of sulfur in the atmosphere is generally less than 3 Tg, this flux is very large indeed. It indicates that the average residence time for sulfur in the atmosphere is only a few days. This short residence time is a consequence of the fact that the various chemical forms of gaseous sulfur are quickly converted to sulfate. Sulfate readily dissolves in water to form sulfuric acid, which is removed from the air in rainwater. Sulfate is also an important part of the dust that falls back to Earth's surface.

Seawater, decomposition, and volcanic eruptions add sulfur to the atmosphere. The largest natural input comes from minuscule droplets of seawater wafted from the ocean surface. The decay of organic matter releases gaseous hydrogen sulfide to the atmosphere. Volcanic eruptions also release sulfur into the atmosphere, mostly as small crystals of sulfur dioxide (SO_2). The annual sulfur emissions of volcanoes are generally considerably less than 10% of the total atmospheric flux.

Because the small crystals of sulfur dioxide given off by volcanoes reflect solar radiation, they have a cooling effect on the atmosphere. Major volcanic eruptions that emit large quantities of sulfur dioxide may have significant short-term effects on Earth's climate. In April of 1815, for example, an Indonesian volcano called Mt. Tambora erupted explosively. In a matter of days, it injected more than 50 Tg of sulfur in the form of SO_2 into the atmosphere. During the next 2 years, peculiar glowing sunsets and abnormally cool summers were reported around the world. In July of 1816, it actually snowed in New England!

◀ **Figure 7.25 The Sulfur Cycle**
Fluxes (Tg/yr) of sulfur are indicated by red numbers. Total pools of sulfur are not well-known. Human impacts on the sulfur cycle are enclosed with a blue dashed line.

Human Impacts

Human activities have doubled the flux of sulfur into the biosphere.

Through mining activities and the burning of fossil fuels, humans release approximately 150 Tg of sulfur from Earth's crust each year—roughly twice the amount that is released by the natural weathering of rock. Of that 150 Tg, about 90 Tg is released into the atmosphere as emissions from such sources as power plants. This atmospheric sulfur has increased the acidity of rainfall in many areas and contributed to the acidification of lakes, streams, and soils in some regions.

Human activities have also increased the flux of sulfur into rivers, lakes, and near-coastal marine ecosystems. Since the availability of sulfur rarely limits photosynthesis in aquatic or marine ecosystems, the increased level of sulfur has neither increased the productivity of these ecosystems nor caused eutrophication.

QUESTIONS

1. What is the primary source of sulfur in the atmosphere?

2. What is the approximate magnitude of the effect of human activities on the flux of sulfur through the biosphere?

FOCUS ON SCIENCE: Long-term Change in the Role of Nutrients in Net Primary Production

All nonphotosynthetic organisms depend on the net primary production (NPP) of plants and other photosynthetic organisms. In forest ecosystems, the most important factor limiting NPP is usually the amount of nitrogen in the soil, although phosphorus may be more important in some locations.

Does the role of nitrogen and phosphorus in forests change over time?

▲ **Figure 7.26 Peter M. Vitousek**

Soil is made of weathered rock, living organisms, and decaying organic matter. Phosphorus is released by the weathering of rocks, so young soils often have high levels of phosphorus. But because only nitrogen-fixing bacteria can add nitrogen to soil, young soils usually have low levels of nitrogen. Because of this, soil scientists hypothesize that in young soils nitrogen should be a more important limiting factor than phosphorus. But as soils develop over hundreds and thousands of years, the importance of phosphorus in relation to nitrogen should increase.

Stanford University's Peter M. Vitousek was interested in testing this hypothesis (**Figure 7.26**). Knowing that the soils on its various islands spanned a wide range of ages, Vitousek selected Hawaii as an ideal site to test the hypothesis (**Figure 7.27**).

Vitousek measured the nitrogen and phosphorus content of Hawaiian soils ranging from 300 years to 4.1 million years in age. As predicted, the ratio of nitrogen to phosphorus increased with soil age. This observation, however, did not firmly establish which nutrients were limiting plant growth at which time.

To evaluate nutrient limitations on plant growth, Vitousek identified similar forests of o'hia lehua trees on soils that were 300 years old, 20,000 years old, and 4.1 million years old. In each forest area, he applied various combinations of nitrogen and phosphorus fertilizer. In each treatment area he then measured tree growth for several years. He also measured tree growth in control areas to which he applied no fertilizer.

On the 300-year-old soils, only the addition of nitrogen improved tree growth. On the 20,000-year-old soils, nitrogen alone improved growth by a small amount, phosphorus alone had no effect, but the combination of nitrogen and phosphorus increased growth by more than 50%. On the 4.1 million-year-old soils, tree growth was unaffected by nitrogen alone; the addition of phosphorus alone increased growth by more than 70%.

These results confirmed Vitousek's hypothesis. They also provide a basis for understanding patterns of nitrogen and phosphorus limitations in forest ecosystems.

Source: Vitousek, P.M., 2004. *Nutrient cycling and limitation: Hawai'i as a model system.* Princeton, NJ: Princeton University Press.

▲ **Figure 7.27 Principal Islands of Hawaii**

Vitousek took advantage of the fact that the volcanic islands of the Hawaiian chain are progressively older as one travels from southeast to northwest.

7.8 The Dynamics of Biogeochemical Cycles

The pools and fluxes of nutrients within ecosystems vary over different time intervals. Seasonal variations in temperature and moisture affect patterns of plant growth, which in turn affect terrestrial nutrient cycles and nutrients in nearby aquatic ecosystems. Seasonal changes in water temperature and density may produce seasonal nutrient change in lakes. Disturbances such as logging and fire can affect nutrient cycles for years or decades. In experiments where natural successional change is prevented, nutrient loss from ecosystems may be very significant.

Seasonal Change

Biogeochemical cycles vary through the year.

Biogeochemical cycles for many elements vary considerably across the seasons. This is especially true in the deciduous forest ecosystems that dominate large areas of the eastern United States, Europe, and eastern Asia. Over the course of a year, the total amount of nitrogen and phosphorus in these forests remains relatively constant, but its distribution among various pools, such as the trees, forest floor, and soil, changes with the seasons.

Deciduous trees lose their leaves in the autumn and are dormant during the winter. In a hectare (2.47 acres) of a typical deciduous forest, about 5 tons of leaves (dry weight) fall to the ground each autumn. These leaves bring about 200 g of carbon, 5 g of nitrogen, and 3 g of phosphorus to each square meter of forest floor. Over the winter, these leaves decompose, slowly releasing the nitrogen and phosphorus they contain to the pool of nutrients in the soil. Because all plants are dormant in the winter, they are unable to absorb these nutrients.

By early spring, the pool of nitrogen and phosphorus in the soil is several times larger than in midsummer. Most of these nutrients are rapidly reabsorbed by trees as they break dormancy and begin to produce new leaves.

Large amounts of nutrients are absorbed by a diverse group of forest-floor herbs called spring ephemerals. In early spring, these herbs break dormancy and grow very quickly, absorbing nitrogen and phosphorus (**Figure 7.28**). By late spring, the leaves and stems of these plants have died back to the ground, returning that nitrogen and phosphorus to the soil.

Biogeochemical cycles in some aquatic ecosystems also undergo seasonal changes. The amount of nutrient elements carried in streams is influenced by seasonal changes in the ecosystems that surround them. In spring, for example, water percolating through the forest floor removes some nutrients. This results in a spring increase in the amount of nitrogen and phosphorus transported in streams (**Figure 7.29**). Such seasonal changes in nutrient levels contribute to seasonal variations in stream NPP.

In lakes, nutrients such as nitrogen and phosphorus are constantly being lost from surface water as sediment and organic debris settle to the bottom. In some lakes, seasonal variations in water temperature and density mix the water and redistribute these nutrients.

▲ **Figure 7.28 Spring Ephemeral Herbs**
Forest spring ephemeral herbs such as this trout lily absorb a great deal of the nitrogen in early spring. They die back in early summer, releasing their nitrogen back to the soil.

▶ **Figure 7.29 Seasonal Change**
Each point on this graph represents the average amount of nitrate-nitrogen transported each month from a forested watershed in New England. Nitrogen loss from the watershed peaks in the early spring as warmer temperatures accelerate mineralization, but uptake by plants is limited. It is especially low during the growing season when plants are actively taking up nitrogen.
Sources: Likens, G.E., F.H. Bormann, R.S. Pierce, J.S. Eaton and N.M. Johnson. 1977. *Biogeochemistry of a Forested Ecosystem.* Springer-Verlag, New York. 146 pp.

Disturbance and Succession

Plant growth following disturbance influences biogeochemical fluxes and pools.

When ecosystems are disturbed, biogeochemical pools and fluxes are often altered. For example, when a forest or grassland is cut or burned by wildfire, the pool of nutrients in plants is greatly diminished, and the flux of nutrients in and out of the remaining plants is altered. Such disturbance initiates a process of succession that restores both the plant community and biogeochemical cycles.

A long-term study carried out at the U.S. Forest Service Hubbard Brook Experimental Forest in New Hampshire provides an excellent case study for understanding such changes. The Hubbard Brook Forest contains a number of forested watersheds, each of which is monitored for nutrient inputs in rainfall and nutrient outputs in stream water and groundwater (**Figure 7.30**). Thus, scientists are able to measure the flux of nutrients through each watershed.

Late in 1966, one of these watersheds was completely logged. For 3 years afterward, herbicides were applied to prevent the regrowth of any vegetation. Prior to cutting, the annual nitrate flux in stream water from the experimental watershed was less than 2 g/m². In the year following cutting, nitrate loss was over 400 g/m² (**Figure 7.31**). Nitrogen losses remained high throughout the time that regrowth was prevented.

This increased flux of nitrate was due in part to high rates of decomposition of the organic debris left behind following cutting, as well as high rates of nitrification. Because there were no plants to absorb this nitrate, much of it was carried into the stream by runoff.

▲ Figure 7.30 Watershed Experiment
A small watershed within the U.S. Forest Service Hubbard Brook Experimental Forest was clear-cut, and forest regrowth was prevented for 3 years. Nutrient input in rainfall and dust to this watershed and a similar but uncut control watershed were monitored. Nutrient outputs were then monitored in stream water at gauging stations.

In 1970, the herbicide treatment ceased and the watershed was allowed to undergo succession. The regrowth of herbs, shrubs, and trees was rapid. As these plants grew, they began absorbing increasing amounts of nitrogen. Within 2 years, the annual flux of nitrate in stream water was equal to the annual flux before the disturbance. As the forest has recovered, nitrogen loss from the watershed has been negligible, and the nitrogen stored in forest trees continues to increase.

Some have criticized this experiment, suggesting that the herbicide treatment produced nitrogen losses far in excess of what would have been observed had the forest been allowed to regrow immediately. Even so, the treatment provided a powerful demonstration of the importance of the growth of vegetation in biogeochemical cycles. Equally, it demonstrated the role that succession plays in restoring biogeochemical fluxes and pools following disturbance.

◄ Figure 7.31 Forest Loss and Nutrient Loss
Losses of nutrient elements from the treatment watershed (dotted lines) increased significantly when it was cut, and the losses remained high during the 3 years that forest regrowth was prevented. Changes in the untreated control watershed are represented by solid lines. When forest succession was allowed to proceed, nutrient losses returned to their pretreatment levels in 2–3 years.

QUESTIONS

1. In regions with cold winters, why do nitrogen concentrations in streams often increase in the early spring?

2. Describe the pattern of change in stream nitrate concentrations immediately following the cutting of the forest surrounding the stream.

Summary

The pools and fluxes in biogeochemical cycles are shaped by interactions between the living and nonliving components of ecosystems. As the rock cycle slowly moves elements through the lithosphere, organisms play a significant role in the weathering of rocks and the formation of sediments. In the hydrologic cycle, water moves among the oceans, atmosphere, groundwater, streams and lakes. Because many elements dissolve in water, the hydrologic cycle plays an important role in the cycling of many nutrients. The cycling of nutrient elements among organisms, soils, the hydrosphere, and the atmosphere is often regulated interactions between organisms and their physical environment. Human activities such as mining and the exploitation of fossil fuels have accelerated the flux of many elements from the lithosphere to other portions of the biosphere. Human land use and the production of synthetic fertilizers have also altered fluxes among pools within the biosphere. In some cases these alterations have directly influenced ecosystems in ways that diminish the stability of biogeochemical cycles.

7.1 Ecosystem Ecology and Biochemistry

- The distribution, abundance, and movement of elements in ecosystems can be tracked by measuring element pools and fluxes.
- The cycling of nutrient elements that are required by organisms are influenced by their abundance in the atmosphere, their mobility in the lithosphere, and the complexity of food webs.

KEY TERMS

biosphere, biogeochemical cycle, pool, flux, mass-balance accounting, capital, equilibrium, residence time, cycling time, macronutrients

QUESTIONS

1. If the atmosphere contains 300 Tg of element X, and the flux into and out of the atmosphere of element X is 20 Tg/yr, what is the residence time of element X in the atmosphere?

2. If the net flux of element X into the atmosphere were suddenly to change to +5 Tg/yr, what would you expect the size of the atmospheric pool of element X to be in 10 years?

3. Why is the cycling of most elements among organisms intimately tied to food webs?

7.2 The Rock Cycle

- The rock cycle explains the movement of elements among different rock types.

KEY TERMS

rock cycle, igneous rock, sedimentary rock, metamorphic rock

QUESTIONS

1. In which parts of the rock cycle do organisms play the most significant roles?

2. Describe how rocks in each of the three major rock categories are formed.

7.3 The Hydrologic Cycle

- Solar energy and gravity drive the hydrologic cycle.

KEY TERMS

hydrologic cycle, transpiration, runoff, groundwater, aquifer

QUESTIONS

1. The residence time for water in Earth's atmosphere is about 9.5 days. How do scientists calculate that number?

2. About 10% of the water that evaporates from the ocean returns to it in rivers. How does the other 90% get back to the ocean?

3. In their studies of element cycles, scientists pay special attention to the hydrologic cycle. Why?

7.4 The Carbon Cycle

- Photosynthesis and cellular respiration are key steps in the carbon flux between the biosphere and the atmosphere.
- Human activities are increasing the flux of carbon into the atmosphere.

KEY TERMS

gross primary production (GPP), net primary production (NPP), net ecosystem production (NEP)

QUESTIONS

1. Describe a specific situation in which net ecosystem production is positive and one where it is negative.

2. The total pool of carbon in green plants is nearly 5,000 times greater in Earth's terrestrial ecosystems than the pool of carbon in the primary producers in the oceans. However, the total net primary production of Earth's marine ecosystems is nearly equal to that of its terrestrial ecosystems. Explain how this can be true.

3. What factors influence differences in NPP from one place to another in terrestrial ecosystems? How do those factors compare with the factors that influence NPP in marine ecosystems?

4. Describe how human land use is contributing to the increased flux of CO_2 to Earth's atmosphere.

5. Fossil fuels such as oil and coal are the net ecosystem production (NEP) of a distant time. Explain.

7.5 The Nitrogen Cycle

- Nitrogen rapidly cycles between the atmosphere and the biosphere.
- Humans have doubled the global rate of nitrogen fixation.

KEY TERMS

nitrogen fixation, nitrification, denitrification, Haber-Bosch process, eutrophication

QUESTIONS

1. Why is the global amount of nitrogen available to organisms ultimately limited by the rate of nitrogen fixation?

2. In ecosystems where organic matter decomposition is limited, net primary production may be reduced as well. Explain this connection.

3. What is the Haber-Bosch process and what is its relationship to increased net primary production in many streams and rivers?

4. How does the burning of oil and coal increase nitrogen availability in the biosphere?

TERAGRAMS OF NITROGEN PER YEAR

7.6 The Phosphorus Cycle

- The supply of phosphorus to ecosystems is limited by the weathering of rocks.
- Human use of fertilizers has increased the supply of phosphorus in fresh water and ocean ecosystems.

QUESTIONS

1. Only about 10% of phosphorus transported to the ocean in rivers can be used by marine algae. Why is this? What happens to the remaining 90%?

2. Phosphate is typically mined from marine sediments that were formed in regions of ocean upwelling. Why are such sediments especially rich in phosphate?

7.7 The Sulfur Cycle

- The flux of sulfur through the atmosphere is very rapid.
- Human mining activities and fertilizer use have doubled the flux of sulfur into the biosphere.

QUESTIONS

1. Why is the residence time for sulfur in the atmosphere so brief?

2. Some have suggested that we might counteract the effects of global warming by putting sulfur dioxide into the atmosphere. Why would this cause cooling? Why might this not be a good idea?

7.8 The Dynamics of Biogeochemical Cycles

- Biogeochemical cycles vary through the year.
- Plant growth following disturbances influences biogeochemical fluxes and pools.

QUESTIONS

1. How does the growth of spring herbs influence the flux of nitrogen in forest streams?

2. Why might net primary production in shallow ponds and lakes increase after windstorms?

3. Explain why the outflow of nutrient elements in streams increases significantly immediately following a disturbance such as a fire.

4. As succession proceeds, the loss of nutrients to streams quickly diminishes to very low levels. Why?

Mastering**ENVIRONMENTALSCIENCE**

Go to **www.masteringenvironmentalscience.com** for homework assignments, practice quizzes, Pearson eText, and more.

8

Climate Change

197

Arctic Meltdown

Is Earth's climate warming?

You may have heard that the Inuit—the people who live in the Arctic—have dozens of words for snow (**Figure 8.1**). That's an appealing story, but it's only a myth. What is true is that the Inuit have an incredibly close relationship to snow. The Inuit depend on the snow and ice of long, cold winters for their survival. Thick ice helps them to hunt and fish and to travel from village to village (**Figure 8.2**). Snow is the building material for their iconic temporary shelters, igloos. The permanently frozen ground beneath the tundra provides a solid foundation for their homes.

Like all people who live in extreme environments, the Inuit must pay particularly close attention to changes in their environment. Today, Inuit from Alaska, Canada, and Greenland are saying that their weather is *uggianaqtug*—behaving in an unfamiliar way. Snowfall is far less predictable. Sea ice has thinned and is now absent over longer stretches of the year. In the words of the chair of the Inuit Circumpolar Conference, "the Earth is literally melting" (**Figure 8.3**).

Earth's atmosphere is indeed warming, especially in the Arctic. Over the past 30 years, the number of days in which the average temperature is above freezing has increased significantly. Yearly average temperatures in the

▲ **Figure 8.1 Polar People**
The Inuit live on the northernmost shores of North America bordering the Arctic Sea.

Arctic have increased at nearly twice the rate as those in temperate and tropical regions.

Over 90% of Arctic glaciers are retreating at unprecedented rates. The total mass of ice floating on the Arctic Ocean is steadily diminishing. So much ice has melted that most summers ships can navigate from the north Atlantic Ocean across northern

► **Figure 8.2 Living on Ice**
A century ago, Inuit wore sealskin parkas and depended on dogsleds for transportation over long icy winters (left). Though technologies have changed (right), the Inuit still depend on the cold winter, snow, and ice for hunting and transportation.

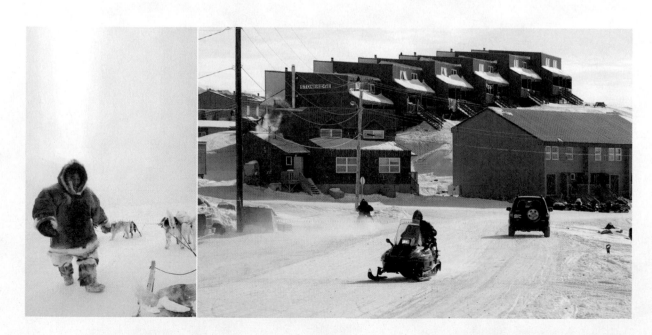

Canada and Alaska to the Pacific Ocean. The loss of sea ice has contributed to a 30% decline in the number of North American polar bears, which hunt much of their prey from ice floes. As a result, polar bears are now listed as "threatened" under the U.S. Endangered Species Act.

The warming climate is affecting entire ecosystems. The boundary between the forests of the taiga and the treeless tundra is shifting northward. Fires are becoming more common in both ecosystems (**Figure 8.4**). Warmer temperatures and longer growing seasons have changed the vegetation of the tundra, reducing the habitat of small herbivores called lemmings. These once numerous rodents are now in decline across the Arctic. Because lemmings are the primary food of

▲ **Figure 8.3 The Earth Is Melting**
The permafrost, permanently frozen ground, has melted beneath the foundation of this house in the northern Canadian town of Inuvik, causing it to sink.

many predators, their decline is jeopardizing the survival of carnivorous birds and mammals such as jaegers and Arctic foxes.

Marine and coastal ecosystems are also affected. Rivers fed by melting glaciers are carrying more water to the sea. This fresh water has lowered the salinity of estuaries and coastal waters, changing the habitats of marine plants and animals.

Several sources of data indicate that the Arctic climate has been alternating between warm and cold periods for the past million years. Throughout those cycles, ecosystems shifted and organisms adapted. Can we expect the same in the years ahead? For some organisms the answer may be yes. For others, the more likely answer is no. The current rate of temperature change in the Arctic is unprecedented. If this rate of change continues, future temperatures will be much warmer than any experienced in the past.

As the nations of the world argue about the causes of climate change and possible responses, the Arctic is being rapidly transformed. Why does change seem to be proceeding so much more rapidly in the Arctic than in temperate and topical zones? Many experts see the environmental changes in the Arctic as omens of the major changes that may soon occur worldwide. Scientists are committed to identifying actions to slow or halt global warming as well as actions that will minimize the effects of warming that are already likely to occur. What national and international policies are needed to ensure that action is taken? These challenges and questions are central themes of this chapter.

▼ **Figure 8.4 Fire on Ice**
The warmer conditions of the last several decades have caused wildfires, such as this one burning on the northern coastal plain of Alaska in 2008, to become more common in the Arctic.

8.1 Long-Term Climate Patterns

A long view—one that spans thousands or even millions of years—shows that Earth's climate is dynamic. Over hundreds of thousands of years, the global climate has varied in response to regular changes in Earth's orbit that influence the amount of sunlight it receives. During the past 10,000 years, Earth has grown warmer, but with notable ups and downs. In the past 1,000 years, the climate has warmed, cooled, and then warmed again. Long-term changes in climate have significantly influenced human evolution and history. A study of long-term climate change provides an understanding of the factors that have determined global climate in the past and will surely shape it in the future.

The Pleistocene—The Last 2 Million Years

For nearly 2 million years, Earth's climate has cycled between cold and warm periods.

You might not think so from current trends, but for the past several million years, Earth's climate has been cooling. Between 3 and 5 million years ago, a huge ice sheet in Antarctica grew until it totally covered the continent. In the Northern Hemisphere, mountain glaciers grew larger and more numerous.

About 2 million years ago, the global climate began to display cycles of colder and warmer temperatures. Each cycle of cold and warm lasted about 100,000 years. Scientists call this time of alternating cold and warm periods the **Pleistocene epoch**.

The cold period beginning 800,000 years ago brought unprecedented glaciation. Huge glaciers between 1,000 and 3,000 m (3,300 and 10,000 ft)

thick spread across vast areas of North America and Eurasia. During warm periods, the glaciers retreated or disappeared entirely. Later cold periods, or ice ages, again produced glaciers. The most recent continental glaciation lasted from about 110,000 years to 12,000 years ago. At the peak of this last ice age, glaciers covered 30% of Earth's land. In North America, glaciers reached as far south as central Illinois and Long Island, New York (**Figure 8.5**).

These cycles of changing temperatures produced significant changes in nearly all ecosystems. Fossils found in ocean sediments show major shifts in the distribution of ecosystems, both on land and in the oceans.

ARCTIC OCEAN

ATLANTIC OCEAN

PACIFIC OCEAN

PACIFIC OCEAN

► **Figure 8.5 Ice Age**
At the peak of the last glacial period 15,000–19,000 years ago, massive glaciers extended across the northern portions of Eurasia and North America. Glaciers extended as far south as central Illinois and Long Island, New York.

Maximum extent of glacial ice

During the ice ages, much of the water evaporating from the ocean fell as snow and became part of the glaciers. This reduced the volume of water in the ocean. During the coldest periods, sea level dropped by about 100 m (330 ft). The lower sea level exposed coastlines and created land bridges between places that today are separated by water (**Figure 8.6**). On the continents, water melting from glaciers produced complex river systems and expansive lakes. At the height of the last ice age, for example, Lake Agassiz covered most of what is today Manitoba, western Ontario, northern Minnesota, and eastern South Dakota.

What caused these cycles of cold glacial periods and warm interglacial periods? Changes in Earth's orbit undoubtedly played a role. Over periods of 100,000 years, Earth's motion around the sun shifts. This shifting causes regular changes in the amount of solar radiation striking different regions of Earth. By itself, however, this variation in the amount of sunlight striking Earth is not sufficient to cause major shifts in climate.

During periods of reduced radiation, snow and ice begin to accumulate on the continents. The snow and

▲ **Figure 8.6 The Bering Land Bridge**
At the peak of the last ice age, sea level was much lower, and a wide isthmus of land (in beige) across what is today the Bering Strait joined northeast Asia to northwestern Alaska.

ice reflect more of the sun's energy. As a consequence, the land absorbs less heat. The cooler land allows the ice to spread. Thus the growing glaciers speed up the cooling process by reflecting more and more radiation back into space.

Other factors may also have contributed to the spread of the glaciers. Changes in ocean currents could have altered the patterns of heat transfer around the globe. Volcanoes that spewed gases and ash into the atmosphere could have reduced the amount of sunlight reaching Earth.

Changes in the chemical composition of Earth's atmosphere played an important role in the cycle from glacial to interglacial conditions. During glacial periods, very high net primary production (NPP) in the oceans pulled great amounts of carbon dioxide (CO_2) from the atmosphere. In interglacial times, ocean NPP was as much as 50% lower, resulting in much higher atmospheric CO_2 (**Figure 8.7**). As you will learn in detail later in the chapter, CO_2 influences the amount of the sun's energy that is absorbed by our atmosphere.

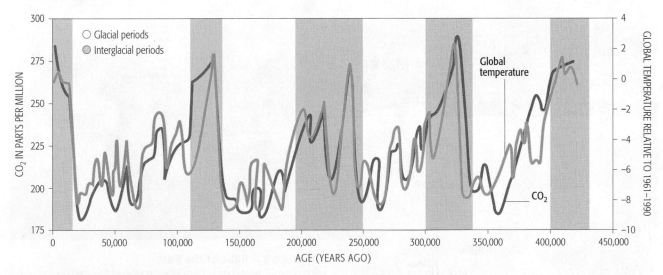

▲ **Figure 8.7 CO_2 and Glacial Cycles**
Changes in atmospheric CO_2 concentrations over the last 420,000 years can be measured by analyzing bubbles of air in Antarctic ice cores. Concentrations were low during cold glacial periods and high during warm interglacial periods (yellow bands). Nevertheless, CO_2 remained below 300 ppm throughout this time.
Data from: Petit, J.R., et al., 1999. Climate and atmospheric history of the past 420,000 years from the Vostok Ice Core, Antarctica. *Nature* 399: 429–436.

Holocene—The Last 10,000 Years

Periods of warming and cooling have produced different climate patterns in specific regions.

Earth's climate was still cold 15,000 years ago, but a warming trend was underway. Glaciers had begun to retreat, and the bare rock and sediments they left behind were undergoing rapid succession to forest (6.5). But climate transitions often occur in fits and starts. About 12,000 years ago, climates turned abruptly colder. In many places, glaciers began to advance again. This cold period, often called the Big Freeze, continued for more than a thousand years.

What caused the Big Freeze? The warming process itself. As the initial warming melted the glaciers, enormous volumes of fresh water poured into the ocean. Much of this water flowed out through the St. Lawrence River. The fresh water diluted the ocean water and interrupted the circulation of warm water from the tropics to the north Atlantic. This interruption produced much cooler conditions in polar and north temperate regions.

The last 10,000 years of the Pleistocene is called the **Holocene epoch**. During the Holocene, Earth has been in a warm interglacial period. Even so, global temperatures have varied. Climates were especially warm between 9,000 and 5,000 years ago. This period correlates closely with variations in Earth's orbit that favored solar heating. Average temperatures became a bit cooler 5,000 to 1,000 years ago.

The warming and cooling of the Holocene affected the climates of individual regions in different ways. For example, during the early Holocene warm period, the interior of North America was relatively warm and dry. These conditions allowed southern deserts to expand northward and grasslands to grow in regions that today support forest. At the same time, North Africa was comparatively moist; lakes, savanna, and forest were widespread across many parts of the Sahara region (**Figure 8.8**). Since then, North America has become moister, and the Sahara has become a vast desert.

The last 1,000 years have also witnessed significant climate variation. Climatologists call the years from A.D. 1000 to 1300 the **Medieval Warm Period**. During this time, average global temperatures may have increased by 0.5 °C, with greater warming in northern latitudes. Evidence from tree rings suggests that Iceland was 2 °–3 °C (3.6 °–5.4 °F) warmer than today (**Figure 8.9**). Vikings who settled Iceland and Greenland at this time found these lands hospitable for farming; today they are not. In western North America, prolonged drought resulted in the widespread migration of Native Americans (**Figure 8.10**). The warmer climate also led to frequent fires in western forests.

▲ **Figure 8.8 A Relic of the Past**
Lake Umm al-Maa in southern Libya is one of the few bodies of water in the nearly rainless Sahara desert. The lake is fed by groundwater that accumulated beneath the sands between 5,000 and 10,000 years ago. At that time, the region was much wetter, and Lake Umm al-Maa was surrounded by tropical savanna.

Between A.D. 1400 and 1500—just as the Renaissance was blossoming in Europe—a cold and wet period set in over the Northern Hemisphere. Once again, mountain glaciers in Europe and North America grew larger and advanced downslope. These colder conditions were almost certainly the reason that Viking settlements in Greenland were abandoned; growing seasons simply became too short to support agriculture. This cold period, called the **Little Ice Age**, continued well into the 1700s and the beginning of the Industrial Revolution.

Climate science had its beginnings during the Little Ice Age. Had those scientists understood the history of Earth's climate over the previous 2 million years, they might have thought the next ice age was imminent. In fact, the opposite was happening. The chemistry of Earth's atmosphere was undergoing changes that would initiate a new period of warming.

We humans are creatures of the Pleistocene. The first members of the genus *Homo* appeared in Africa about 2 million years ago. Climatic changes since then have had much to do with our evolution and migration around the globe. The warm, moist climates in North Africa and the eastern Mediterranean region during the early Holocene played a key role in the development of agriculture and irrigation. Climate change over the past thousand years first facilitated colonization and then forced the abandonment of lands in northern Europe and Scandinavia.

Through ice ages and warm interglacial periods, climates changed as a consequence of periodic changes in Earth's orbit, volcanic eruptions, changes in Earth's snow and ice cover, and shifts in net primary production of oceans that altered atmospheric CO_2. These factors were outside the control of humans,

▲ Figure 8.9 **A Living Climate Record**
Variations in the width of the annual rings in tree trunks reveal differences in the trees' rate of growth. In cold regions like Iceland, variations in growth are due to annual changes in temperature that affect the length of growing seasons.

A year of slow growth

A year of fast growth

but the climatic changes that they caused most certainly influenced human history. As we shall see, the activities of modern humans play a central role in the current warming trends. These changes will, with equal certainty, influence our future well-being.

▼ Figure 8.10 **Abandoned**
Dry conditions in the Medieval Warm Period forced the Anasazi peoples to abandon their farms and cliff dwellings (such as these in Mesa Verde) and migrate to moister locations in what are today New Mexico and Arizona.

QUESTIONS

1. What factors contributed to the alternating glacial and interglacial periods over the past 800,000 years?

2. What was the Big Freeze, and what was its cause?

3. Describe the general pattern of change in average global atmospheric temperature during the past 1,000 years.

FOCUS ON SCIENCE:
Measuring Long-Past Temperatures

What evidence do scientists use to determine climates of the past?

As you read about the climate changes during the last 100,000 years, did you wonder how scientists know the temperatures from so long ago? Certainly there were no weather forecasters with thermometers in the early Pleistocene.

A method for determining past temperatures was first developed by two chemists—University of Chicago Professor Harold Urey and his student Cesare Emiliani (**Figure 8.11**). In 1934 Urey had been awarded the Nobel Prize in chemistry for his pioneering work on isotopes, atoms of the same element that contain different numbers of neutrons (3.1). In 1950 Urey took up the challenge of inferring past climates. Later he called it "the toughest chemical problem I ever faced."

Urey understood that both hydrogen and oxygen have isotopes with differing masses. The different isotopes of hydrogen and oxygen combine to form water molecules of differing mass. Over 99% of water molecules are "light water" or $^1H_2^{16}O$; only a small fraction of water molecules are "heavy water," or $^2H_2^{18}O$. Because it is lighter, $^1H_2^{16}O$ evaporates more readily than $^2H_2^{18}O$. Consequently, a greater proportion of light water molecules evaporate from ocean water than do heavy water molecules. This reduces the ratio of light water molecules to heavy water molecules in ocean water. As temperature increases, the proportion of heavy water molecules evaporating from the ocean surface also increases. Careful analysis shows that the isotope ratios of ocean water change with temperature in a very predictable way.

Urey reasoned that the isotope ratios of ocean water should be reflected in the isotope ratios found in the bodies of organisms living in the ocean. These ratios would be preserved in the fossils of these organisms found in ocean sediments.

▲ **Figure 8.11 Teamwork**
Harold Urey (left) was fascinated by the challenge of determining the climate at different times in the past from the relative amounts of isotopes of hydrogen and oxygen in ocean sediment. But it was Urey's student, Cesare Emiliani (center), who worked out the actual techniques for estimating past atmospheric temperatures from sediment isotopes.

By measuring isotope ratios in the sediment, it would be possible to infer the temperature of the ocean water when the organisms were alive.

It's one thing to have a great idea and quite another to make it work. For this, Emiliani deserves the credit. First, Emiliani perfected methods for measuring isotope ratios in the fossil shells of microscopic organisms. Then he applied those methods to bits of sediment taken from the bottom of the ocean. Using established relationships between isotope ratios and water temperature, he was able to deduce temperature changes over a 300,000-year period.

In the years since Urey and Emiliani's pioneering research, sediment isotope studies have been refined. Today they are the standard method of inferring the climates of long ago (**Figure 8.12**).

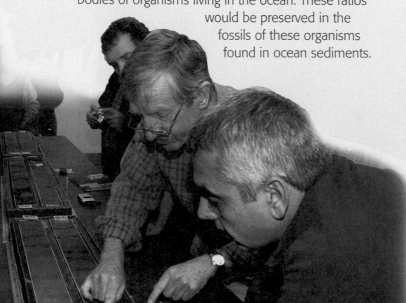

◀ **Figure 8.12 A Core Through Time**
Each year, fine sediments accumulate on the ocean floor. Variations in the amounts of different oxygen and hydrogen isotopes in those sediments are correlated with temperature changes in the atmosphere above the ocean.

8.2 Measuring Global Warming

Long-term trends in global temperature can be inferred from changes in the fossils and chemical isotopes found in sediments and glacial ice. Climate change during the past 3,000 years can be estimated from tree growth, as measured by the width of annual rings. These techniques lack the precision necessary to detect year-to-year or decade-to-decade global trends. Scientists can, however, determine differences in average annual temperatures with great precision by analyzing data gathered from thousands of locations around the globe. These analyses reveal that Earth's temperature has warmed by 1 °–2 °C since the late 1800s. This warming has been more intense in some regions than in others. Changes in ocean currents and events such as volcanic eruptions complicate the pattern of global warming.

Measuring Recent Climate Change

Since the beginning of the Industrial Revolution, Earth's atmosphere has warmed significantly.

Using computer models, scientists estimate that the average annual air temperature near Earth's surface is about 14 °C (57 °F). Given the current interest in global warming, you may be surprised to learn that scientists do not know this number with much precision. Measuring global annual surface air temperature is a daunting challenge.

It is not difficult to measure air temperature at a particular place—all that is needed is a calibrated thermometer. Such measurements are made in well-ventilated, shaded shelters to avoid the effects of the sun's direct radiation (**Figure 8.13**). Determining the average temperature of that place over a period of a year requires more persistence and care. Individual readings must be taken in a uniform manner and at carefully determined times to avoid sampling one particular time of day or one season more than another.

Estimating the average annual air temperature over the entire globe is a far greater challenge. For this, scientists need a large number of sampling stations at which measurements are taken in exactly the same way. Even so, it is nearly impossible to determine whether such a collection of samples is biased toward warm or cold regions. Given these uncertainties, how can we possibly know if the surface air temperature of our planet is changing from year to year, century to century, or millennium to millennium?

Fortunately, it isn't necessary to know the exact value of the average global surface air temperature to determine whether it is changing. Instead, scientists study patterns of temperature variation among the many locations in the world where reliable data exist. First, scientists determine a benchmark of the average annual temperature for each location over a specific time interval, such as 1960–1990. Then they compare each year's average temperature with that benchmark. The difference between each year's average temperature and the benchmark is called a **temperature anomaly**. A positive anomaly indicates a year that is warmer than the benchmark. A negative anomaly represents a year that is cooler than the benchmark.

Temperature anomalies for each year are then averaged across the thousands of locations where data are available. This average provides a measure of global temperature change.

▼ **Figure 8.13 Temperature Measurement**
Air temperature near Earth's surface is usually measured with calibrated thermometers that are protected from the direct rays of the sun in a standard shelter.

Is climate change something to be concerned about, or are variations in climate normal?

Alexcis Zare, University of Houston

Norm: *Variations in climate are normal. However, the rate at which Earth's atmosphere is warming is truly unusual, and climate scientists are concerned that warming may increase even faster if emissions of greenhouse gases continue to rise.*

Such an analysis reveals that, although the global temperature varies considerably from one year to the next, it has been gradually warming over the past 130 years (**Figure 8.14**). Warming is especially notable over the past 30 years. During the period from 2001 to 2005, the average global temperature was 0.51 °C (0.92 °F) warmer than during the period from 1971 to 1975. The 10 warmest years on record have occurred since 1999. The data also indicate that temperatures will continue to rise.

Today we are witnessing warming equivalent to what occurred in the Medieval Warm Period. The term **global warming** refers specifically to the increase in atmospheric temperature since the late 1800s. This period of global warming coincides with the Industrial Revolution.

▲ **Figure 8.14 It's Getting Warmer**
Based on data from thousands of locations, the global annual average temperature anomaly shows a steady increase in the temperature of Earth's atmosphere since 1880. The red line represents the five-year mean.
Source: U.S. National Aeronautics and Space Administration.

Causes of Natural Climate Variation

Temperature trends vary among regions and from year to year.

Scientists have been able to make extremely accurate measurements of the general increase in world temperatures known as global warming. But this trend is not uniform throughout the world. Significant differences are found from place to place and year to year.

Regional Variation. Annual temperature anomalies vary considerably from location to location (**Figure 8.15**).

In much of the Arctic, for example, temperature anomalies may exceed 2 °C (3.6 °F). These higher-than-average temperature anomalies are related to circulation patterns in the troposphere and ocean currents, which move heat from the equator toward the poles. This movement intensifies temperature change in the polar regions. It is these above-average temperature changes that are producing such rapid change in Arctic ecosystems.

► **Figure 8.15 Temperature Anomalies 2000–2009**
Temperature anomalies vary considerably from place to place. During this 10-year period, anomalies were generally higher in polar regions and in the centers of continents and lower over the oceans.
Source: U.S. National Aeronautics and Space Administration.

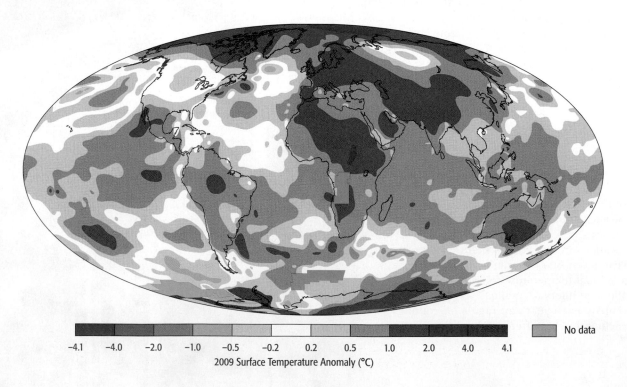

Land masses heat up more quickly than bodies of water. Because continental regions heat up more quickly than nearby oceans, temperature anomalies are generally higher over the interior of continents, such as North America and Eurasia. In other regions, such as over parts of the Southern Hemisphere oceans, temperature anomalies are actually negative.

As was the case with climate change in the Holocene, the specific effects of global warming on factors such as rainfall and length of growing season vary from location to location. In some places, rising temperatures are accompanied by increased aridity and drought. In other places, warming increases evaporation from nearby bodies of water, thereby increasing rainfall and even winter snowfall.

El Niño/Southern Oscillation.
Weather conditions vary from year to year, often as a consequence of changes in currents that affect the temperature of water near the ocean's surface. The **El Niño/Southern Oscillation (ENSO)** in the Pacific Ocean is one of the most studied examples of such variation. Over a 3- to 8-year period, surface ocean temperatures cycle between relatively cold and warm conditions. ENSO affects weather conditions around the world.

During the cold part of the ENSO cycle, ocean currents bring cold water from the Antarctic Ocean to the equator along the west coast of South America. The upwelling of deep ocean water contributes to the cold temperature of the surface water. The situation changes when water warmed in the western equatorial Pacific moves eastward, overriding the colder, denser water. Because the warm part of the cycle often begins around Christmas time, South American fishermen call it El Niño (Spanish for the baby boy), in reference to the Christ child. Given that name, scientists called the part of the cycle dominated by cold surface water La Niña (the baby girl) (**Figure 8.16**).

El Niño's warm waters feed thunderstorms and increase rainfall across the eastern Pacific Ocean and western South America. But to the east, across the Andes and in the Amazon Basin, warm, dry conditions prevail. In North America, El Niño winters are warmer than normal in the Midwest, Northeast, and Canada, but in the Southwest and Mexico, winters are cooler and wetter. In contrast, the Southwest and Mexico often experience prolonged droughts during La Niña (**Figure 8.17**). Although the effects are less dramatic, temperature and rainfall patterns in Australia, Southeast Asia, and Africa are also correlated with the ENSO cycle.

Volcanic Eruptions.
Global climates can be influenced by large volcanic eruptions. Such eruptions add tons of sulfur dioxide to the lower atmosphere. Small crystals of sulfur dioxide reflect the sun's rays and cool the atmosphere. In 1815, for example, the massive eruption of Mount Tambora, a volcano in Indonesia, resulted in two years of abnormally cold temperatures in places as distant as Europe and New England (7.7).

Over the long term, ice ages have alternated with warm interglacial periods. Climates have also undergone centuries-long fluctuations, such as the Medieval Warm Period and the Little Ice Age. Even shorter periods of fluctuation characterize the El Niño/Southern Oscillation. These fluctuations are caused by natural processes, such as shifts in Earth's orbit relative to the sun, variations in global snow and ice cover, changes in ocean currents, and volcanic eruptions.

QUESTIONS

1. The global annual temperature anomaly for 2008 was +0.44 °C. What does this mean?

2. What causes the shift from La Niña to El Niño conditions in the ocean waters near Ecuador, South America?

3. Why does the eruption of large volcanoes often result in one to two years of global cooling?

EL NIÑO OCEAN TEMPERATURES: JANUARY–MARCH 1998

LA NIÑA OCEAN TEMPERATURES: JANUARY–MARCH 1989

Sea Surface Temperatures (°C)

▲ **Figure 8.16 ENSO**
El Niño conditions are indicated by the extension of warm water across the entire equatorial Pacific Ocean. During La Niña, colder water extends along the west coast of South and Central America.

▼ **Figure 8.17 ENSO and Climate**
The graph shows rainfall in Los Angeles, expressed as a percentage of the long-term average. El Niño conditions coincide with high-rainfall years. Intermediate and La Niña years tend to be drier.
Source: U.S. National Oceanographic and Atmospheric Administration data.

Highest rainfall is in El Niño years

Lowest rainfall is in La Niña years

RAINFALL AS % OF NORMAL

La Niña ◄——— ENSO INDEX ———► El Niño

8.3 Causes of Global Warming

The temperature of Earth's atmosphere is influenced by many factors. Some, such as the amount of energy radiating from the sun and Earth's orbital position relative to the sun, are outside the realm of human influence. Other factors, such as the chemical composition of the troposphere and the reflectivity of Earth's surface, are increasingly influenced by human activities. Deforestation and the burning of fossil fuels add large quantities of carbon dioxide (CO_2) to our atmosphere. Industrial and agricultural activities release methane (CH_4) and nitrous oxide (N_2O). These three gases absorb a disproportionate amount of light energy and trap it as heat. The rapid increase in the atmospheric concentration of these three heat-trapping gases is responsible for global warming.

The Greenhouse Effect

Some gas molecules trap heat radiated from Earth's surface in the lower atmosphere.

Nitrogen (N_2) and oxygen (O_2) molecules make up about 99% of the volume of Earth's atmosphere. Together, water (H_2O) and carbon dioxide (CO_2) account for less than 0.5%. Yet, molecules of H_2O and CO_2 are thousands of times more efficient than molecules of N_2 and O_2 in capturing the infrared light that is constantly being radiated from Earth's surface (3.7). This captured light energy is transformed into heat. Thus, H_2O and CO_2 have an impact on the atmospheric temperature far in excess of their relative abundance. Were it not for the presence of these two molecules, the average temperature of Earth's troposphere would be 33 °C (59 °F) cooler. Without them, our world would be a frozen, ice-covered rock without the possibility of plant and animal life.

Although the analogy is not perfect, the trapping of heat in the troposphere is often referred to as the greenhouse effect. On a sunny winter day, the air inside a greenhouse absorbs light energy, and the temperature rises. The glass of the greenhouse prevents the warm air from mixing with the cold air outside, trapping heat in the greenhouse. In a similar fashion, there is very little mixing of gases in the troposphere with the gases in the stratosphere; this lack of mixing traps heat in the troposphere (**Figure 8.18**).

Gases that efficiently capture heat are called **greenhouse gases**. The most abundant natural greenhouse gas is water vapor. Even so, water is not considered to be adding to global warming. The concentration of water vapor in the atmosphere fluctuates widely with seasons and at different locations around the world. The effect of water vapor on the temperature of the atmosphere is complicated by water's phase transitions from solid to liquid to gas.

► Figure 8.18 **The Greenhouse Effect**
The stratosphere acts like the glass on a greenhouse, limiting the escape of heat absorbed by greenhouse gases in the troposphere.

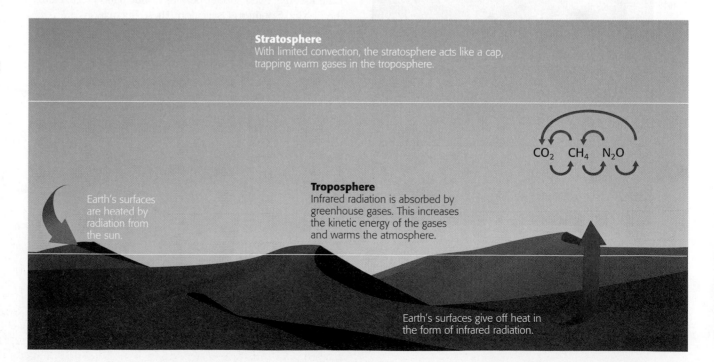

Stratosphere
With limited convection, the stratosphere acts like a cap, trapping warm gases in the troposphere.

CO_2 CH_4 N_2O

Troposphere
Infrared radiation is absorbed by greenhouse gases. This increases the kinetic energy of the gases and warms the atmosphere.

Earth's surfaces are heated by radiation from the sun.

Earth's surfaces give off heat in the form of infrared radiation.

As air becomes warmer, its ability to hold water vapor increases. Warmer air causes more water to evaporate from bodies of water and the leaves of plants. This additional evaporation might cause runaway warming, as the increasing levels of water vapor trap even more heat. Instead, the heated air rises and cools, causing the water vapor to condense into clouds. Because they reflect incoming solar radiation back into space, clouds cool Earth's surface and atmosphere.

Carbon dioxide is the second most important natural greenhouse gas. Gas bubbles in glacial ice reveal that prior to the Industrial Revolution the tropospheric concentration of CO_2 was about 285 parts per million (0.028%). At that time, CO_2 concentrations did not vary much from place to place, although they did vary by a few parts per million with the seasons.

Methane (CH_4) and nitrous oxide (N_2O) are greenhouse gases that occur naturally at comparatively low concentrations. Before the Industrial Revolution, CH_4 was found in concentrations of about 700 parts per billion (0.00007%), and nitrous oxide occurred at concentrations of 270 parts per billion (0.000027%). Methane is released into the atmosphere naturally from the decomposition of vegetation by microbes and termites and from the digestive processes of various animals, including you. Nitrous oxide is a natural by-product of the denitrification process (7.5).

Human Impacts

Human activities have increased the abundance of several greenhouse gases in the atmosphere.

Humans first began to change the chemical composition of the atmosphere thousands of years ago when they began cutting and burning forests to expand agricultural fields. This reduced the total amount of carbon stored in trees and increased the amount of CO_2 in the atmosphere (7.4). When humans domesticated cows and sheep, CH_4 was added to the atmosphere by those animals' digestive processes, or less delicately, burps and flatulence. Later, humans cut forests and burned wood to make charcoal to fuel Iron Age furnaces. Until the Industrial Revolution, however, human populations were comparatively small. As a result, *anthropogenic,* or human-caused, greenhouse gas emissions were minuscule compared to the total volume of gases in the atmosphere.

Truly significant increases in anthropogenic greenhouse gas emissions came with the burning of fossil fuels: first coal, then petroleum and natural gas. These high-energy carbon fuels were produced and stored by ecosystems eons ago. Their combustion releases greenhouse gases, especially CO_2 (7.4). Each year, the burning of fossil fuels adds about 22 petagrams, or Pg (~25 billion tons), of CO_2 to the atmosphere, as well as smaller amounts of other greenhouse gases.

Deforestation contributes to the emission of greenhouse gases, as do other land uses that diminish the amount of carbon stored in Earth's ecosystems. Scientists estimate that human land use adds 3.2–4.3 Pg CO_2 to the atmosphere each year. Fires associated with deforestation and agriculture also emit CH_4 and N_2O.

Forest fires and the burning of fuelwood and charcoal release solid particles called **black carbon** into the atmosphere. Black carbon particles absorb light and warm the atmosphere. They reduce the reflectivity of the surfaces they settle on, thereby contributing to additional warming. The black carbon deposited on snow and glacial ice is believed to be an important factor contributing to the worldwide retreat of glaciers (**Figure 8.19**).

In addition to naturally occurring greenhouse gases, humans have been releasing a variety of synthetic greenhouse gases into the atmosphere. These include chlorofluorocarbon (CFC) compounds, which are used as refrigerants, cleaning solvents, and fire retardants and in the manufacture of plastic foam products. Compared to other emissions, only small amounts of CFCs are added to the atmosphere each year. However, these gases absorb infrared light 10,000–20,000 times more efficiently than CO_2. Because of their destructive effect, the manufacture of CFCs was greatly curtailed by an international treaty, the Montreal Accord of 1987. This is covered in more detail in Chapter 9.

▼ Figure 8.19 **Black Carbon**
The burning of coal and plant biomass produces great quantities of black carbon particles, such as those on this snow. They absorb solar radiation, and their increased warmth causes the snow beneath them to melt.

The impact of the emission of different anthropogenic greenhouse gases depends on two factors: their capacity to absorb infrared light and retain heat and the length of time they stay in the atmosphere, or atmospheric residence time (7.1). These two factors are used to calculate a molecule's **global warming potential (GWP)**. GWP is a measure of an individual molecule's long-term impact on atmospheric temperature.

The GWP for different gases is calculated relative to CO_2, which is assigned a GWP of 1. An average molecule of CO_2 stays in the atmosphere for 100 years. CH_4 has a GWP of 25—it absorbs infrared photons 160 times more efficiently than CO_2, but an average CH_4 molecule resides in the atmosphere for only 14 years. N_2O has a long residence time, 120 years; its GWP is 296. CFCs absorb infrared light very efficiently and also hang around in the atmosphere for a long time; their GWPs are a whopping 1,300 to 13,000.

Prior to 1957, there were no long-term measurements of atmospheric gases. Therefore, there was no way of knowing whether anthropogenic emissions were altering our atmosphere's chemistry. That year, Charles D. Keeling of Scripps Oceanographic Institute began making systematic monthly measurements of atmospheric CO_2. Keeling chose to work at the Mauna Loa Observatory on the island of Hawaii because it was far from urban and industrial centers.

In 1957, the atmosphere contained about 312 parts per million (ppm) CO_2 (**Figure 8.20**). Keeling's data revealed clear seasonal cycles in the concentration of CO_2, which were related to the changes in the net productivity of ecosystems during summer and winter. After a few years of sampling, Keeling realized that CO_2 concentrations were increasing by 1–2 ppm each year. Measurements on Mauna Loa continue to the present. Since Keeling's first measurements, CO_2 has risen by nearly 25%. In 2010, it was 388 ppm.

More recently, atmospheric scientists monitoring other greenhouse gases have found trends that are similar to those of CO_2 (**Figure 8.21**). CH_4 in the atmosphere has more than doubled from its natural state to over 1,750 parts per billion (ppb). N_2O has increased to nearly 320 ppb.

Monitoring for CFCs began in the mid-1970s, and their concentrations increased steadily into the early 1990s. CFC levels have since stabilized as a consequence of an international agreement, the Montreal Protocol, that limits their production (9.5).

Although the issue was hotly debated in the 1980s and 1990s, today there is no doubt that global warming is a direct consequence of the increased amount of greenhouse gases in Earth's atmosphere. Nor is there any doubt that these atmospheric changes are directly linked to human activities, such as the burning of fossil fuels, deforestation, and pollution from industry and agriculture.

▼ **Figure 8.20 The Keeling Curve**
Measurements taken at Hawaii's Mauna Loa Observatory indicate that the level of CO_2 in Earth's atmosphere is increasing at a rate of 1–2 ppm per year. The inset graph shows how the concentration of CO_2 changes during a year. The orange arrows correspond to the Northern Hemisphere growing season, when net primary production is high and more CO_2 is being withdrawn from the atmosphere.

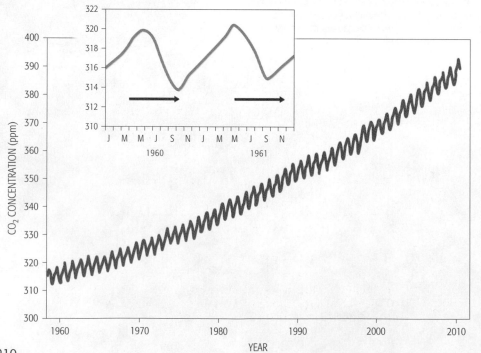

▼ **Figure 8.21 Other Greenhouse Gases**
Atmospheric concentrations of methane and nitrous oxide are increasing as a consequence of human activities such as deforestation, agriculture, and the burning of fossil fuels.
Source: NOAA, 2009. *Annual greenhouse gas index (AGGI).* http://www.esrl.noaa.gov/gmd/aggi/.

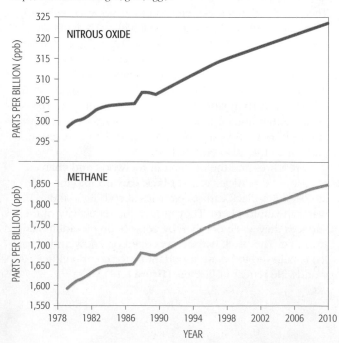

Sources of Greenhouse Gas Emissions

Heat-trapping gases are generated from a variety of human activities.

Scientists express total greenhouse gas emissions in terms of **carbon dioxide equivalents (CO_2e)**, taking into account the global warming potential of each gas. CO_2 is by far the most important, accounting for approximately 75% of warming. Increased emissions of CH_4 and N_2O are responsible for an additional 22%. Industrial pollutants such as CFCs are responsible for the remaining 3%.

Various human activities contribute differing amounts to total greenhouse gas emissions (**Figure 8.22**). Most of the electricity we use is generated by burning fossil fuels, largely coal and natural gas, so it's not surprising that electric generation produces the largest portion of emissions. Industrial processes are next in importance. These include activities such as the manufacture of concrete, which is responsible for 30–40% of all industrial emissions. Agriculture is the source of about 12.5% of greenhouse gas emissions, primarily in the form of CH_4 and N_2O from livestock and the use of fertilizer.

Greenhouse gas emissions vary widely among countries (**Figure 8.23**). In 2006, the United States and China accounted for over 40% of anthropogenic emissions. The United States has fewer than 5% of Earth's people but emits over 19 metric tons of CO_2e per person. With 18% of the world's population, China emits about 5 metric tons of CO_2e per person each year. The nations of the European Union account for about 14% of greenhouse gas emissions, about 11 metric tons of CO_2e per person. Russia, India, Japan, Indonesia, and Canada also account for a significant portion of annual CO_2e emissions.

In industrialized countries, the vast majority of greenhouse gas emissions comes from burning fossil fuels. In most other countries, deforestation and agriculture are the predominant sources. In Brazil, for example, emissions equal nearly 13 metric tons of CO_2e per year per person; 70% of this is due to deforestation. In New Zealand, greenhouse gas emissions equal nearly 15 metric tons of CO_2e per person. Over half of that is methane produced by that country's 40 million sheep.

The rate at which greenhouse gas emissions are increasing is a matter of great concern (**Figure 8.24**). Since 1990, U.S. emissions have increased at an average rate of 1.2% per year. The rate of increase in U.S. emissions is correlated with economic conditions. In good economic times, rates are above 2% per year; when the economy is down,

rates of increase are low. In European Union countries, the annual rate of growth in greenhouse emissions since 1990 has been below 0.5%. In several European countries, greenhouse emissions are shrinking.

In comparison, the rate of increase in greenhouse emissions in countries with rapidly developing economies is alarmingly high. Since 2005, China's greenhouse emissions have been increasing at an annual rate of just over 10%. Other developing countries, such as India and Indonesia, have greenhouse gas emissions that are increasing at rates above 5% per year.

These changes and the challenges they present are now acknowledged by nearly every nation. Governments have begun to develop policies to halt or reverse these trends; they are also finding ways to adapt to the changes that may be inevitable. The Montreal Protocol's success in reversing CFC concentrations provides some reason for optimism.

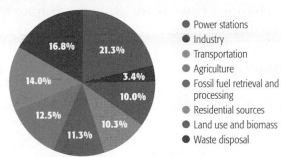

● Power stations
● Industry
● Transportation
● Agriculture
● Fossil fuel retrieval and processing
● Residential sources
● Land use and biomass
● Waste disposal

21.3%
3.4%
10.0%
10.3%
11.3%
12.5%
14.0%
16.8%

◄ **Figure 8.22 Sources of Greenhouse Gases**
The major source of anthropogenic greenhouse gas emissions is the use of fossil fuels for industry, transportation, and power generation.

QUESTIONS

1. How does the greenhouse effect get its name?

2. Name four important greenhouse gases and the human activities that are most important in their production.

3. What factors influence the global warming potential (GWP) of a particular gas?

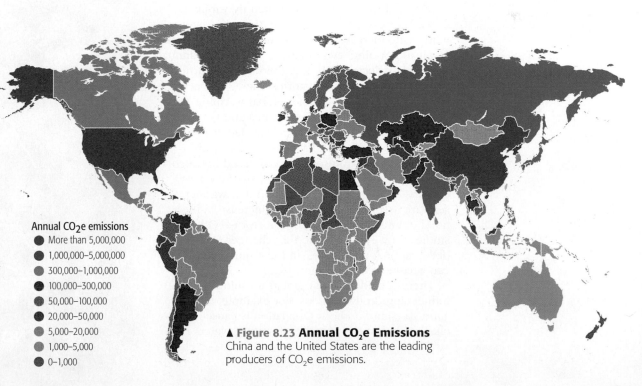

Annual CO_2e emissions
● More than 5,000,000
● 1,000,000–5,000,000
● 300,000–1,000,000
● 100,000–300,000
● 50,000–100,000
● 20,000–50,000
● 5,000–20,000
● 1,000–5,000
● 0–1,000

▲ **Figure 8.23 Annual CO_2e Emissions**
China and the United States are the leading producers of CO_2e emissions.

8.4 Consequences of Global Warming

The changes caused by global warming vary from region to region. In some places, increasing temperatures have been accompanied by higher rainfall amounts. In others, they have brought drought. Winters have become milder and shorter in Earth's middle latitudes, and dry seasons have grown longer in some parts of the tropics. Glaciers and ice sheets are melting worldwide. Warming is causing sea levels to rise. Taken together, these changes are having a significant impact on the flora and fauna of many ecosystems.

Drier and Wetter

Global warming is producing wetter conditions in some places and drought in others.

Is climate change the reason for increased storms and global disturbances?

Ciara Tyce, Georgia Southern University

Norm: *Climate scientists are not in agreement about whether global warming is producing more storms or extreme weather events such as summer heat waves. However, climate models indicate that such changes are likely with continued warming.*

The effects of rising temperatures on precipitation vary geographically. Rainfall has increased significantly in eastern North and South America, as well as in most parts of Europe and Asia. In contrast, sub-Saharan Africa, the Mediterranean region, and western North America have been drier. Since 1970, longer and more intense droughts—as measured by decreased precipitation and higher temperatures—have affected wide areas of the tropics and subtropics.

In regions in which rain is highly seasonal, such as sub-Saharan Africa, global warming appears to be changing the length of wet and dry periods. This is a matter of special concern because food production depends on the length of wet seasons. Based on current trends and climate models, growing seasons are expected to become shorter over most of sub-Saharan Africa, with the exception of lands very near the equator.

Regional trends are complicated by variation in weather events. For example, there has been a worldwide increase in the frequency of rainstorms that result in flooding, even in areas where total annual rainfall has declined. In the aftermath of Hurricane Katrina and other major storms, there has been much debate about whether global warming is producing more intense hurricanes and typhoons. Ocean temperature is an important factor in the development of tropical storms and hurricanes, and ocean temperatures have increased between 0.25 °C and 0.5 °C (0.45 °–0.9 °F) over the past century. Some studies have found that warmer seas are associated with a large increase in the number and strength of tropical storms. Other studies, however, indicate that changes in observational techniques and instrumentation can account for these increases.

There is evidence that global warming is influencing drought cycles. For example, the El Niño/La Niña/Southern Oscillation is caused by changes in the temperature of surface waters in the equatorial Pacific Ocean. When waters off the west coast of South and Central America are cold, drought is much more common in the southwestern United States. Some climatologists think that since 1970 the length and intensity of El Niño and La Niña events have been outside the range of natural variability. Although climate models predict that such changes will occur, most scientists feel that it is not clear that they are actually underway.

RUSSIAN FEDERATION

PACIFIC OCEAN

NORTH AMERICA

Melting Glaciers and Ice Sheets

Over 80% of Earth's glaciers are retreating.

Nearly 75% of Earth's fresh water is contained in snow pack and glaciers, the ice sheets of Greenland and Antarctica, and Arctic sea ice. The total amount of this frozen water depends on the rate of snowfall and ice formation relative to the rate at which it melts. In most places today, the rate of melting exceeds the rate of ice formation. Snowpack in many mountain regions has been shrinking, and over 80% of Earth's mountain glaciers are retreating.

Since satellite observations began in 1978, Arctic sea ice has shrunk by about 3% per decade (**Figure 8.24**). As you would expect, rates of shrinkage are higher in the summer than in the dark winter, when ice rebuilds. Even so, the total shrinkage is so great that scientists are forecasting that the Arctic Ocean could be completely ice free during summer in 20 years or less. The overall mass of ice on Antarctica is also decreasing (**Figure 8.25**).

For decades, scientists have noted that Greenland's ice sheet is shrinking. Recent data indicate that melting is far more rapid than previously thought because of the prevalence of streams and rivers beneath the ice.

The glacial retreat has exposed new land along Greenland's coast. This exposure, along with longer growing seasons, has allowed farmers to plant crops that have not been grown in Greenland since the Medieval Viking period.

Greenland's small community of farmers may be benefiting from the effects of a warmer climate, but globally the loss of snow and ice is a double whammy. First, snow and ice reflect about 35% of incoming solar radiation, compared to just 10% from forests or ocean waters. A decrease in the amount of land covered by snow and ice accelerates global warming, since solar radiation is absorbed instead of being reflected back into space. Second, water melting from ice sheets and glaciers eventually makes its way to the ocean, where it adds to the rising sea level. Meltwater also has the potential to alter ocean currents, much as it did during the Big Freeze 12,000 years ago.

▲ **Figure 8.25 Antarctic Ice Shelf Collapses**
Over a 30-day period in 2002, warming caused an ice shelf the size of the state of Rhode Island (light blue) to collapse into the Antarctic Ocean.

1979–2000 median minimum

GREENLAND

ATLANTIC OCEAN

Northwest passage

◄ **Figure 8.24 A Northwest Passage at Last**
Over the past several decades, the northern polar ice cap has been steadily shrinking. Since 1965, global warming has produced a steady decline in the extent of summer terrestrial snow cover (blue on the chart below) and sea ice (red) in the Northern Hemisphere. Note the open water from Greenland across the coast of North America.

Data from: Post, E., et al., 2009. Ecological dynamics across the Arctic associated with recent climate change. *Science* 325: 1355–1358.

Rising Sea Level

Thermal expansion and meltwater from glaciers are increasing ocean volume.

Over the past 50 years, sea level has been rising at an average rate of 1.8 mm (0.07 in.) per year. More recent satellite measurements suggest that the rate may be as high as 3 mm (0.12 in.) per year (**Figure 8.26**). Two factors are responsible for this change—the expansion of water as it warms and melting ice.

The ocean absorbs much of Earth's heat. Over the past century, surface ocean waters have warmed by as much as 0.5 °C (0.9 °F). This increase in temperature causes liquid water to expand slightly. In small volumes of water, the expansion is barely detectable. In the huge volume of the ocean, its effect is significant. Scientists estimate that thermal expansion accounts for as much as half of the observed rise in sea level.

The influx of water melting from mountain glaciers and ice sheets also contributes to the rising sea level. Note that the Arctic ice sheet floats on the ocean surface. Its melting has little effect on sea level, for the same reason that melting ice cubes do not cause your glass of iced tea to overflow.

The effects of rising sea level are already being felt in many coastal areas. Shorelines and barrier islands are eroding. Higher water levels and salinity are changing the distribution of some plant communities. Projections vary, but sea level may rise between 0.5 and 1.0 m (1.6–3.3 ft) by 2100. This rise could dramatically alter coastlines and cause many low-lying ocean islands to disappear (**Figure 8.27**).

► **Figure 8.26 Sea Level Rise**
Precise satellite measurements indicate that sea level is currently rising at the rate of about 30 mm (1.2 in.) every decade. Scientists fear that additional warming may cause this rate to accelerate.
Source: Cazenave, A., and R.S. Nerem, 2004. Present-day sea level change: Observations and causes. *Review of Geophysics* 42, RG3001, doi:10.1029/2003RG000139.

▼ **Figure 8.27 Islands at Risk**
The nation of Maldives in the Indian Ocean comprises a number of islands that rise barely 1 m above sea level. Within the next century, these islands could be completely inundated by the rising ocean.

Changing Populations and Ecosystems

Because climate is a key component of habitats, global warming is producing significant changes in the populations of many species and in many ecosystems.

Global warming is affecting ecosystems as plants and animals respond to altered habitats. Satellite images of Canada, New England, and the Midwest since 1980 reveal that warmer spring temperatures are causing the leaves of plants to open earlier. On the ground, scientists have observed that many species of birds are nesting and laying eggs earlier than in the past. Nevertheless, there is still a risk that spring frosts will damage young leaves and kill nestling birds.

Milder temperatures are causing migratory birds to move to the Arctic earlier and stay in high latitudes later into the fall. This delayed departure exposes some species to fierce autumn storms as they migrate southward. The migration of some species no longer coincides with the availability of food, which also varies seasonally (**Figure 8.28**).

A synthesis of more than 100 scientific studies covering 1,400 different species found that global warming is affecting the behavior of plants and animals in both land and aquatic ecosystems. In rivers and streams, many fish species are migrating earlier because of warmer waters. On land, hibernating mammals are awakening earlier than in the past as they respond to temperature clues that winter is over and spring has arrived. One study shows that marmots in the mountains of Colorado are waking up three weeks earlier than in they did in the 1970s. The consequences of these seasonal shifts are not clear in every case, but they certainly do pose problems when a behavior gets out of synch with the availability of food or other resources necessary for survival (**Figure 8.29**).

A changing climate may allow organisms that carry human diseases to expand their ranges, thus having a direct effect on human well-being. In some places, warmer and moister conditions have encouraged the spread of disease vectors such as mosquitoes. This has resulted in a marked increase in malaria and dengue fever in regions where these diseases have not been seen for over a century. It may also be important in the recent spread of the mosquito-borne West Nile virus in the United States.

While some plants and animals may be adjusting successfully to global warming, others are not. Many ocean species seem to be struggling. For example, the management plan to help North Sea codfish recover from overfishing is failing. Apparently, the waters off Europe's coast are too warm for the

▲ Figure 8.28 Bad Timing
Warmer spring temperatures across Europe cause insects to mature weeks earlier than in the past. As a result, the migration of pied flycatchers from Africa no longer matches the peak abundance of insects, so the flycatchers have insufficient food for their nestlings.

plankton necessary to support juvenile cod. Off the coast of California, populations of many seabirds are approaching collapse. It appears that warmer waters have altered populations of microscopic plankton. This, in turn, has diminished the supply of small fish upon which the seabirds feed.

Ocean warming appears to be a major factor in the deterioration of many coral reefs. The warmer conditions contribute to a phenomenon called "bleaching," which is caused by a breakdown in the relationship between coral animals and their symbiotic algae (11.5).

All forecasts suggest accelerated warming in the decades ahead. Continued loss of snow and ice, rising sea levels, and changes to patterns of rainfall seem inevitable, as do the effects of these changes on plant and animal communities. Yet, just as human actions have caused global warming, there are many human actions that can slow or even reverse these trends.

QUESTIONS

1. Describe specific climate changes associated with global warming in three different regions in the world.
2. Why does the loss of ice and snow cover accelerate global warming?
3. Describe two factors contributing to the global rise in sea level.

▼ Figure 8.29 Bad Bear!
In Yosemite National Park, black bears now emerge from hibernation earlier than in the past and before their natural food sources are available. As a consequence, they are more prone to raid picnic baskets and garbage cans.

8.5 Forecasting Global Warming

Some say that global warming is the largest experiment humans have ever undertaken. However, unlike usual scientific experiments, this experiment has only one treatment, no replication, and no control. Given the scale and complexity of climate change, how can we forecast future trends? More important, how can we predict the effects of human actions on the nature of climate change? Climate scientists use computer simulations to forecast climate change. Depending upon the assumptions made about human actions that determine greenhouse gas emissions, these computer models forecast very different futures. By taking action to reduce greenhouse gas emissions, we can avoid many adverse changes in Earth's resources and ecosystems.

Computer Simulation of Global Warming

Global circulation models use basic physical principles to forecast future climate change.

Climate scientists use computer programs called **global circulation models (GCMs)** to forecast climate change. GCMs use mathematical equations to simulate the physical processes that determine Earth's energy budget. These processes include the absorption and reflection of sunlight by the atmosphere, the heating of the land and ocean, infrared radiation given off by the land and ocean, and the circulation of heat from the equator to the poles (3.7). Where it is relevant, GCMs also include ocean circulation.

These models divide Earth's surface and the atmosphere above it into a grid of cells. GCM programs use specific equations to represent the transfer of gases and energy among individual cells. Starting conditions for parameters such as temperature, wind speed, and humidity must be determined for each cell prior to running the model. The effects of factors such as changes in greenhouse gas emissions can be varied from one model run to the next. Forecasts from these different runs can then be compared to evaluate the importance of the different factors.

How can we determine the accuracy of the forecasts produced by GCMs? First, it is important that the equations that drive these models be based on our best understanding of the physical processes they portray. For this, experimental observations are essential. The accuracy of forecasts can also be checked by "backcasting." Using past meteorological records, GCMs are evaluated by determining how well they predict past changes in weather and climate.

GCM forecasts are most accurate for large areas, such as entire continents or oceans, or the entire globe. As the area being considered gets smaller, accuracy diminishes. The forecasts of individual models differ somewhat for regions such as the southwestern United States or southern Africa. They agree even less with regard to forecasts for smaller areas, such as Botswana or the state of Arizona.

Forecasting Scenarios

Action to reduce greenhouse emissions will result in much lower future temperatures compared to no action.

Scientists vary GCM starting conditions and equations to simulate different scenarios. These scenarios differ in their assumptions about such things as economic growth, fossil fuel consumption, and government policies that affect energy use. Different scenarios of human actions result in different greenhouse gas emissions. This triggers different climate interactions, producing different climate change forecasts. Three commonly simulated scenarios are *Today's World, Business as Usual,* and *Sustainable World* (**Figure 8.30**).

▼ **Figure 8.30 GCM Scenarios**
This graph compares GCM predictions of future temperatures based on *Today's World* (yellow), *Sustainable World* (blue), and *Business as Usual* (red) scenarios. The black line represents temperature change up to 2000, and the shading indicates the uncertainty in the predictions.
Source: IPCC, 2007. *Climate change 2007: The physical basis.* New York: Cambridge University Press.

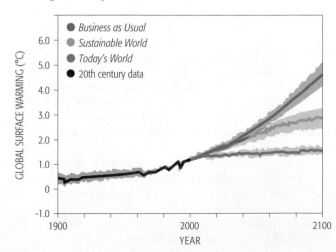

Today's World scenarios assume that anthropogenic greenhouse emissions cease immediately and that CO_2e in the atmosphere stays at current levels. This is not a realistic scenario, but it does provide an important point of reference. *Business as Usual* scenarios assume that individuals and governments will take no actions to reduce emissions and that annual CO_2e emissions will continue to increase. If this were to occur, by 2060 atmospheric CO_2 concentrations would be nearly 900 ppm, or more than three times their pre-industrial levels. *Sustainable World* scenarios assume that people and their governments will take significant actions to slow the increase in CO_2e emissions and return them to current rates within the next 50 years.

Numerous intermediate scenarios have also been modeled. Scenarios that assume little or no action to reduce greenhouse gas emissions forecast much higher future temperatures than those in which aggressive action is assumed. But even if we were to stop all emissions immediately, some warming would still occur. Climatologists call this **committed warming**, and it is about 0.1 °C per decade. *Sustainable World* scenarios predict that Earth's temperature will increase 1.8 °C (3.2 °F) by 2100. *Business as Usual* scenarios predict a temperature increase of 4.0 °C (7.2 °F) over that time.

There are, of course, uncertainties in these estimates. For example, *Sustainable World* forecasts have a 95% confidence range of 1.1 °–2.9 °C (2.0 °–5.2 °F). This means that climatologists feel there is a 95% probability (a chance of 19 in 20) that the actual future temperature will fall in this range. In *Business as Usual* forecasts, the margin of error increases considerably, to 2.4 °–6.4 °C (4.3 °–11.5 °F). The high uncertainties associated with *Business as Usual* forecasts concern climatologists because they may indicate that there will be changes in Earth's climate that are beyond our current understanding.

Changes in individual GCM cells provide a means to forecast climate change in different regions (**Figure 8.31**). Temperatures in 2100 are forecast to be much higher in polar regions and in the centers of continents. This geographic variation in warming is much greater in *Business as Usual* forecasts.

GCM forecasts show that rising temperatures will affect geographical patterns of rainfall (**Figure 8.32**). Forecasts for some regions differ, but most GCMs agree on two important points. First, rainfall will generally increase in regions at higher latitudes, such as Canada and northern Eurasia. Second, rainfall will decrease markedly in many subtropical and temperate regions that are already arid. Such places include the southwestern United States, large parts of Africa, and the Mediterranean region.

▶ **Figure 8.32 Changing Rainfall**
These maps represent GCM predictions for changes in rainfall for December–February and June–August over the next 50 years. The models predict that rainfall will increase in some regions and decrease in others.
Source: IPCC, 2007. *Climate change 2007: Synthesis report.* New York: Cambridge University Press.

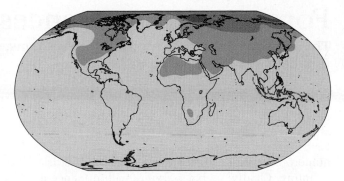

◀ **Figure 8.31 Regional Variation**
GCMs for the *Business as Usual* scenario predict that the extent of warming will vary at different locations around the world. Note that heating is greatest over continents and in polar regions.
Source: IPCC, 2007. *Climate change 2007: The physical basis.* New York: Cambridge University Press.

2020–2029

2090–2099

0.0 0.5 1.0 1.5 2.0 2.5 3.0 3.5 4.0 4.5 5.0 5.5 6.0 6.5 7.0 7.5 8.0
Regional Temperature (°C)

DECEMBER, JANUARY, FEBRUARY

JUNE, JULY, AUGUST

−20 −10 −5 0 5 10 20
Precipitation Change (%)

Forecast Consequences

The impacts on natural resources, ecosystems, and human health are high with the *Business as Usual* scenario.

The United Nations Intergovernmental Panel on Climate Change (IPCC) compiled a list of how different levels of temperature change are likely to affect key resources and human health (**Figure 8.33**). Each of these estimated impacts is based on numerous published scientific studies. The list is daunting. Clearly, global warming will influence a great many resources and processes that have direct consequences for people. However, the impact of the temperature increase forecast by the *Sustainable*

World scenario (1.8 °C) is much less than that for *Business as Usual* (4 °C).

Yes, like it or not, we are committed to some change. In some cases, the response to that change may be costly. However, we can take a variety of actions that will enable us and many of the ecosystems we depend on to adapt to committed warming. But taking action is also needed now to reduce greenhouse gas emissions to prevent us from moving into a very uncertain and unsustainable future.

GLOBAL MEAN ANNUAL TEMPERATURE INCREASE RELATIVE TO 1980–1999 (°C)

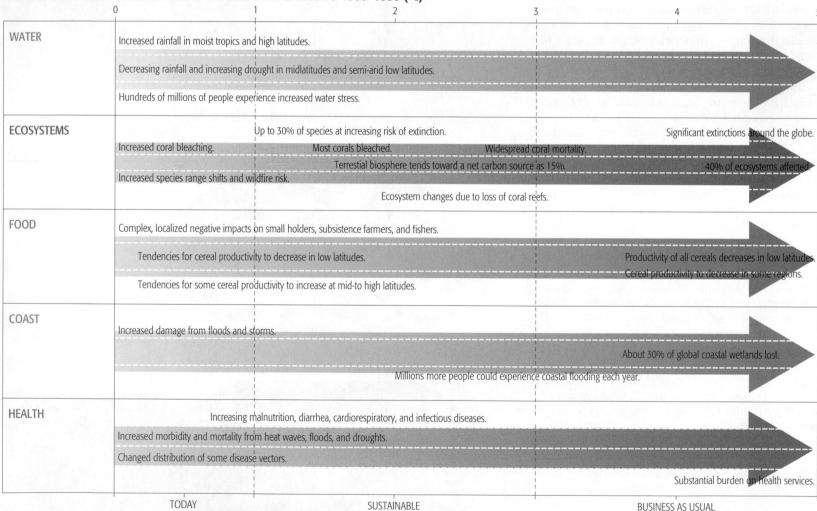

▲ Figure 8.33 **Consequences of Global Warming**
This chart shows IPCC estimates of changes that are likely to be associated with different levels of global warming. As temperatures increase, the severity of the changes and their implications for human well-being also increase.
Source: IPCC, 2007. *Climate change 2007: Synthesis report.* New York: Cambridge University Press.

8.6 Mitigating Global Warming

Emissions of anthropogenic greenhouse gases are increasing each year. If, as in the *Business as Usual* scenario, no actions are taken, annual emissions will double by 2060. Concentrations of CO_2e in the atmosphere will triple. Environmental scientists refer to actions that directly reduce an environmental threat such as CO_2e emissions as *mitigation*. Mitigation to hold annual emissions at their current level, the *Sustainable World* scenario, means that concentrations of CO_2e in the atmosphere will remain at less than twice their pre-industrial amount. Many potential actions—increased energy efficiency and conservation, changes in the use of fossil fuels, and increased use of renewable energy resources, nuclear energy, and biostorage—can help meet the challenge of moving from a *Business as Usual* world to a *Sustainable World*. By itself, no single action or technology is sufficient to meet the goal; a portfolio of actions is required.

Defining the Challenge

Business as Usual **and** ***Sustainable World*** **scenarios imply big differences in future annual greenhouse gas emissions.**

Global warming is a complex challenge. Its most direct cause is the ever-increasing amount of greenhouse gases in Earth's atmosphere. Earth's atmosphere currently contains about 800 Pg, or 388 ppm, of greenhouse gases expressed as CO_2e. Prior to the Industrial Revolution, it contained about 600 Pg, or 285 ppm. Each year human activities such as electricity generation, transportation, deforestation, and agriculture put about 28 Pg of CO_2e into the atmosphere. It is especially troubling that each year anthropogenic emissions increase by several percent. However, there is a great difference in the greenhouse gas emissions associated with the *Business as Usual* scenario and the *Sustainable World* scenario.

Business as Usual. The *Business as Usual* scenario occurs if we take no actions to curb greenhouse emissions. At the present rate of increase, in 50 years CO_2e emissions will more than double, to about 58 Pg per year. At this rate of increase, the amount of greenhouse gases in Earth's atmosphere would eventually triple, to over 1,800 Pg. Climate scientists are in complete agreement that these levels of CO_2e would have catastrophic consequences. Average annual global temperatures would increase by 8 °–10 °C, and sea level would rise more than 1 m (3.3 ft). Such changes would likely be devastating for humans and the ecosystems on which they depend.

Sustainable World. If over the next 50 years we are able to stabilize emissions at their current level of 28 Pg per year, atmospheric CO_2e concentration will not rise above 1,200 Pg. At this concentration, some amount of climate change will be inevitable, but scientists predict that sustainable adaptation to this amount of change is quite possible.

A Portfolio of Strategies. Ecologist Stephen Pacala and engineer Robert Socolow argue that it is possible to take actions that will realize the *Sustainable World* scenario and, thereby, a sustainable outcome. By itself, no single strategy, such as reforestation or generating electricity from renewable energy sources, can reduce emissions enough to meet the goal. Instead, Pacala and Socolow suggest that a *portfolio,* or combination of different activities, will produce the needed reductions.

Pacala and Socolow note that the difference between current annual emissions and those projected for 2060 in *Business as Usual* is 28 Pg. They divide this total amount of necessary reduction into seven 4-Pg "stabilization wedges" (**Figure 8.34**). They group stabilization wedges into five categories: efficiency and conservation, fossil fuel use, renewable energy, nuclear energy, and biostorage.

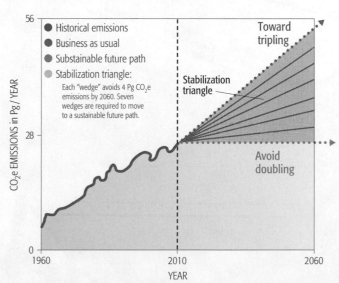

◀ **Figure 8.34 Getting to a Sustainable Future** In this graph, the difference in accumulated emissions between *Business as Usual* and *Sustainable World* scenarios is indicated by the yellow stabilization triangle, which represents an increment of 28 Pg CO_2e/year by 2060. *Source:* Pacala, S., and R. Socolow, 2004. Stabilization wedges: Solving the climate problem for the next 50 years with current technologies. *Science* 305: 968–972.

219

Efficiency and Conservation

Changes in technology and behavior can result in more efficient energy use.

By simply being more efficient in our use of energy, emissions could be reduced by as much as 16 Pg (four wedges). Pacala and Socolow propose action in four areas: transport efficiency, transport conservation, building efficiency, and efficiency in electricity production. Each of these actions would reduce the burning of fossil fuels and has the potential to form a stabilization wedge.

Transport efficiency. Today there are about 600 million automobiles in the world. Given population growth and current trends in economic development, that number could top 2 billion by 2060. A single car traveling 10,000 miles using 30 miles per gallon of gas puts 3.6 tons of CO_2e into the atmosphere each year. If fuel efficiency were doubled to 60 miles per gallon, emissions from autos would be halved. New hybrid and diesel technologies, as well as lighter construction materials, have already increased mileage significantly. There are also opportunities to improve fuel efficiency in trucks and planes.

Transport conservation. In terms of the energy used per person, trains, buses, and other forms of public transportation are far more efficient than automobiles. Investment in mass transit and urban planning to make public transportation more accessible could cut the number of miles traveled by automobiles by half (**Figure 8.35**).

Building efficiency. Residential and commercial buildings account for a significant amount of CO_2e emissions. Buildings use energy for heating and air conditioning, heating water, lighting, and electrical appliances. Improved design and use of materials could reduce energy use and emissions considerably. A 25% improvement in efficiency in all new and existing buildings over the next 50 years would produce a billion-ton wedge.

Efficiency in electricity production. Coal-burning power plants produce about 65% of the world's electricity and account for over 25% of its CO_2e emissions. Increased efficiency in the conversion of the energy in coal into electricity could add a wedge. Increased efficiency could come from technologies such as high efficiency turbines. More even distribution of the demand for electricity could also save energy.

▼ **Figure 8.35 More Efficient Transportation.**
Together, more efficient mass transit systems and improved gas mileage could reduce emissions by as much as two stabilization wedges, or 8 Pg, by 2060.

Fossil Fuel Use

Changes in the use of different kinds of fossil fuels and capture of the CO$_2$ released from their burning can reduce emissions.

Changes in the kinds of fossil fuels we burn and how we manage the CO$_2$ that such burning emits have the potential to add several stabilization wedges. For example, power plants fired by natural gas emit about half as much CO$_2$e as coal-fired plants. Replacing 1,400—about half—of the world's large coal-fired plants by natural gas–fired plants would equal one stabilization wedge. This would result in a fourfold increase in electricity generation from natural gas.

Much attention is being given to new technologies called **carbon capture and storage (CCS)**. Rather than allowing CO$_2$ to escape to the atmosphere, CCS technologies capture it and transfer it into a form that can be permanently stored underground (**Figure 8.36**). Pilot projects are testing a number of storage forms, including super cold liquid CO$_2$ and baking soda–like salts. Many technical uncertainties about CCS remain. For example, scientists disagree about the ability of underground sediments to retain CO$_2$. If CCS is found to be effective, Pacala and Socolow estimate that its widespread use could provide as many as three stabilization wedges.

▲ **Figure 8.36 Carbon Capture**
Rather than allowing CO$_2$ to escape into the atmosphere, CCS systems liquefy it. The liquid CO$_2$ is then stored underground, either in saline deposits or in the sedimentary layers from which coal and oil are extracted.

Renewable Energy

Renewable energy sources produce few CO$_2$e emissions.

Why burn fossil fuels at all if alternative energy sources will emit much less CO$_2$e? Renewable sources of energy include wind, sunlight, and biofuels.

Wind generates electricity without emitting any CO$_2$e. Today, wind energy produces less than 1% of energy worldwide. Increasing our capacity for wind generation by 30-fold will cut future emissions by 1 billion tons (**Figure 8.37**). Currently, the use of wind energy is growing very rapidly, over 30% per year. If this rate of growth continues, the goal can be reached. Wind generation at this level would, however, require a combined area larger than the state of Texas. Wind energy has also been proposed as a means of generating hydrogen that could be used to fuel future cars and trucks.

The ultimate source of most of our energy is the sun. Why not tap it directly? Using solar energy can be as simple as hanging your wet laundry out to dry. Solar energy is widely used to heat water. **Photovoltaic (PV) cells** are devices that directly convert sunlight to electrical current. At present, PV cells provide less than 0.1% of global electricity. We would need a 700-fold increase in this amount to make a wedge.

Biofuels are derived directly from plant materials, such as corn or wood. Burning them has no net effect on CO$_2$e emissions, because photosynthesis in these plants

withdraws CO$_2$ from the atmosphere. Today, the most commonly used biofuel is ethanol, or grain alcohol, which is made from corn and sugar cane. To reduce emissions by one wedge, we would need to convert one-sixth of Earth's cropland from food to biofuel production. Genetically modified crops and new technologies that allow more efficient use of crop wastes could reduce that demand.

▶ **Figure 8.37 Wind Energy**
Renewable energy sources such as this wind-powered generator offset the use of fossil fuels. Steady growth in the use of renewable technologies could account for one or more stabilization wedges.

AGENTS OF CHANGE: The University of Florida's Carbon-Neutral Football Games

The Neutral Gator Initiative is a nonprofit organization that works to reduce carbon emissions while helping people in low-income communities conserve energy and lower their utility bills. Neutral Gator is committed to reducing Gainesville's contribution to global climate change by helping the University of Florida reach carbon neutrality by 2025.

How did you first get the idea for Neutral Gator?

I was at a University of Florida football game reading the paper when I saw an article saying UF was hosting the nation's first carbon-neutral football game. UF offset that game, but they didn't have any intention of offsetting others. I thought there were so many more ways future games could be offset.

At that time I knew service work was going to be my path, and I really wanted to do something that could affect a large amount of people. I thought, "That's it! Use something people love and attach your cause to it." This was an opportunity to connect with people.

What steps did you take to get started?

First, we figured out the games' imprint and what it would take to offset a season. I didn't ask permission of the athletics department. Instead, I focused on self-funding the program. That way,

when I went to the athletic department I only had to request space to set up a tent and engage fans at the games.

We then worked with a local weatherization initiative that already had a strong base of supporters. They became partners in our efforts and helped us get needed training. Then, we recruited volunteers from the University of Florida, the surrounding community, and local businesses. At UF we sent emails to service organizations, fraternities, teachers, and clubs.

At every event, we provided free training for participants, educating them on weatherization. We did projects like exchanges where people could bring their incandescent light bulbs and exchange them for compact fluorescent lights, CFLs. We partnered with a local ice cream shop that supplied a free scoop of ice cream for every five light bulbs exchanged.

We then went door to door. We modified our strategy by going strictly into low-income residences, providing more extensive energy retrofits, installing low-flow showerheads, kitchen and bathroom aerators, etc. Most importantly, we created an education component for the residents, giving them general information on how to create utility savings through behavioral change.

What are the biggest challenges you've faced with the project, and how did you solve them?

The main challenges have been communication and implementation. We have to teach people what carbon offsets are, what a carbon footprint is, and why it's important. Then we need to get them to take impact-reducing actions. Sometimes it's having them replace their incandescent bulbs with CFLs, installing low-flow showerheads, or taking shorter showers. Those things add up. It's important to find a way for people to get involved that works for them.

How has your work affected other people and the environment?

In our first summer, our group consisted of 15 volunteers. As we continued, the amount of people grew exponentially. At our last event that summer we had an astonishing 75 people show up! All summer we did projects to educate the community and help them save money. By the end of the summer, we had offset approximately 3,000 tons of CO_2—enough to offset all the football games that season.

Since we began, the economic benefits for the low-income community have been substantial. Our weatherization efforts have retrofitted 180 low-income residences and distributed 63,000 energy-effcient light bulbs, creating a combined savings of more that $3.2 million.

What advice would you give to students who want to replicate your model on their campus?

Every community has different resources and needs. However, moving forward there are a few key steps to take:

First, know the environmental impact of what you're trying to offset. This will give you an idea of what you'll need to do to generate offsets. For example, how many trees will you need to plant? How many houses will you need to weatherize? How many light bulbs will you need to install?

Then create an education strategy. It's important to be specific about what you want people to know and do. It can be as simple as making them aware of the environmental impact of a football game, and then asking them to participate in reducing their own footprint by signing a commitment to change a personal habit or donate to your cause.

Finally, follow up with people, be consistent with your outreach, and let people know how they can learn more.

And you can always find us at:
http://www.neutralgator.org
for support along the way!

Summary

Earth's climate varies over very long and comparatively short timescales. Before the Industrial Revolution, these variations were a consequence of natural variation in the physical processes that determine climate, including periodic variation in Earth's orbit, carbon dioxide in the atmosphere, snow and ice cover, and ocean currents. The warming since the late 1800s, however, is a consequence of human activities, such as burning fossil fuels and deforestation, that increase the concentration of heat-trapping gases in the troposphere. This global warming is causing significant changes in many ecosystems and threatens human well-being.

New technologies and changes in human behavior can reduce greenhouse gas emissions and slow global warming. Even so, historical emissions have committed Earth's atmosphere to some additional warming. In many places, adaptations will be required to minimize the effects of drought, flooding, and temperature extremes. Strategies to reduce greenhouse gas emissions and to adapt to inevitable change have been encouraged by changes in national and international policies, but additional actions are needed to prevent unpredictable and unsustainable outcomes.

8.1 Long-Term Climate Patterns

- Over the last several million years, Earth has experienced multiple ice ages. During the last 10,000 years, Earth's climate has generally been warming.

KEY TERMS

Pleistocene epoch, Holocene epoch, Medieval Warm Period, Little Ice Age

QUESTIONS

1. What caused concentrations of CO_2 to decline during cold glacial periods in the Pleistocene?

2. Scientists say that changes in Earth's snow and ice cover are both a cause and consequence of changes in climate. Explain what they mean.

3. Over the past 100,000 years, human migrations have been significantly influenced by changes in climate. Describe two important examples of such influences.

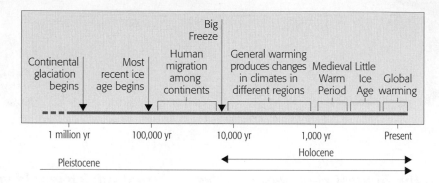

8.2 Measuring Global Warming

- Regional weather varies from year to year as a result of changes in ocean currents and events such as volcanic eruptions. In general, though, the average annual global temperature has increased significantly over the past 130 years.

KEY TERMS

temperature anomaly, global warming, El Niño/Southern Oscillation (ENSO)

QUESTIONS

1. Why is it difficult to measure directly the global average annual temperature?

2. Describe two factors that influence variations in annual temperature anomalies from one geographic location to another.

3. Why do El Niño conditions generally produce higher rainfall along the western slopes of the Andes Mountains?

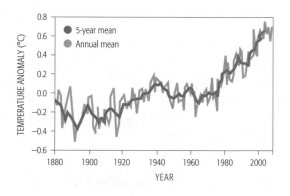

8.3 Causes of Global Warming

- Greenhouse gases such as water (H_2O) and carbon dioxide (CO_2) efficiently absorb infrared photons radiated from Earth's surface and warm the atmosphere.

- Human activities have increased the concentration of greenhouse gases in the atmosphere.

KEY TERMS

greenhouse effect, greenhouse gases, black carbon, global warming potential (GWP), carbon dioxide equivalents (CO_2e)

QUESTIONS

1. Water molecules absorb infrared light, but water vapor is not believed to contribute to global warming. Why not?

2. In what ways does deforestation contribute to the increase in greenhouse gases in the atmosphere?

3. What is responsible for the seasonal changes in CO_2 concentrations observed in data from the Mauna Loa Observatory?

4. CO_2e emissions differ widely among countries. Discuss four factors contributing to these differences.

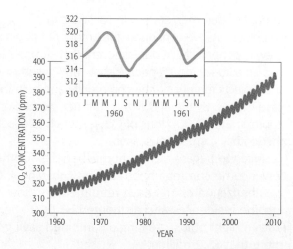

8.4 Consequences of Global Warming

- The effects of global warming vary among geographic locations; they include the retreat of glaciers and ocean ice packs, rise in sea level, and changes in ecosystems and populations of many species.

QUESTIONS

1. Global warming will likely increase rainfall in some places. Why might warming cause this change?

2. In some places, even if rainfall increases, it may actually become drier. Explain.

3. The melting of continental glaciers causes sea level to rise, but the melting of Arctic sea ice does not. Why is this so?

4. Global warming is having a significant effect on populations of many plant and animal species. Describe three ways in which this is occurring.

8.5 Forecasting Global Warming

- Three CO_2e emissions scenarios are frequently used to forecast future climate. They are (1) *Today's World*—immediate cessation of CO_2e emissions; (2) *Business as Usual*—no action taken, with CO_2 concentrations increasing beyond 900 ppm; (3) *Sustainable World*—actions are taken to slow CO_2e emissions so that atmospheric CO_2 stabilizes at about 540 ppm.

KEY TERMS

global circulation model (GCM), committed warming

QUESTIONS

1. General circulation models forecast that global warming will be most intense in polar regions and over continents. Why is this so?

2. The uncertainties for *Business as Usual* forecasts are much higher than those for *Today's World* or *Sustainable World* forecasts. Explain why.

3. Describe the potential differences between the consequences of *Business as Usual* forecasts and *Sustainable World* forecasts with regard to human health.

8.6 Mitigating Global Warming

- Mitigation refers to actions taken in *Sustainable World* scenarios that directly reduce the threat posed by CO_2e and associated global warming. However, no single mitigation strategy is sufficient to achieve *Sustainable World* conditions; a portfolio of actions will be required.

KEY TERMS

carbon capture and storage, photovoltaic (PV) cells, biofuels, biostorage

QUESTIONS

1. The Pacala and Socolow stabilization wedge model indicates that a portfolio of strategies will be required to mitigate the increasing rate of CO_2e emissions. Explain what this means and why it is the case.

2. It will eventually be necessary to lower CO_2e emission rates below their current level. Why?

8.7 Adapting to Global Warming

- The effects of committed warming and inevitable climate change will require humans to adapt in a variety of their endeavors.

QUESTIONS

1. Describe four strategies for adaptation to global warming that might be implemented in the area where you live.

2. Explain why many world leaders view adaptation to global warming as an environmental justice issue.

8.8 Mitigation and Adaptation Policies

- Actions to adapt to and mitigate global warming have significant costs; not acting has greater costs, but they are less immediate.

- Policy alternatives include regulatory actions, such as mandated fuel economy standards and use of biofuels, and market-based strategies, such as cap and trade on CO_2e emissions.

QUESTIONS

1. Some policymakers prefer cap and trade, whereas others prefer taxes as a means of lowering CO_2e emissions. What are the pros and cons of these two approaches?

2. Because the costs of global warming are uncertain and many of the consequences are decades away, some economists argue that we should delay action at this time. What is your response to such an argument?

Mastering**ENVIRONMENTALSCIENCE**

Go to **www.masteringenvironmentalscience.com** for homework assignments, practice quizzes, Pearson eText, and more.

9

Air Quality

The Killer Smog

What causes air pollution to become deadly?

The chill in the air early on Tuesday, October 26, 1948, left no doubt among the people of Donora, Pennsylvania, that summer was over. During the night, dense cold air had drained into the Monongahela River valley. This air was held in place by a stationary layer of warm air a few hundred feet above the river. As a result, the yellowish morning smog lasted through the noon hour. This cold air remained trapped, and each day the smog became more persistent and dense. By Friday, four days later, the smog did not clear at all (Figure 9.1).

Smoky, eye-burning fog was certainly familiar to Donora's 14,000 residents. Smog began to appear on cold mornings soon after completion of the steel mill in 1900. With the establishment of Donora Zinc Works, one of the world's largest zinc-smelting facilities, the smog events became more frequent and dense (Figure 9.2). For most people, smog seemed like a small price to pay for Donora's thriving economy and jobs. But this event in 1948 was different; this town was about to experience an environmental tragedy.

On Friday evening, October 29, hundreds of local residents began to crowd area hospitals, coughing and displaying other signs of respiratory distress. Through Saturday, conditions remained unchanged, and smog and traffic congestion prevented evacuation of the town. A local physician led an ambulance by foot through the darkened streets of Donora, trying to move seriously ill patients to hospitals and the dead to a temporary morgue.

By Sunday morning, 20 people had died, and thousands were ill. It was also on Sunday morning that the president of U.S. Steel ordered the shutdown of the smelter furnaces. Later that afternoon, a gentle rain began to clear the air. Even so, over the next month an additional 50 Donora citizens died of respiratory complications.

The "Donora killer smog" quickly became international news. Walter Winchell sensationalized it on his weekly radio show, and it was reported in newspapers around the world. Finger pointing began immediately. Local papers accused the zinc works of "murder" by releasing "an airborne poison" and called for the renovation or removal of the facility. The chemist for the Pennsylvania Smoke Control Bureau insisted that the smog was derived from multiple sources, including domestic and commercial furnaces, automobile emissions, and several mill operations. U.S. Steel argued that this event was caused by the unique weather conditions that trapped stagnant air in the valley and that it was not a direct consequence of their operations.

▶ **Figure 9.1 Midday Darkness**
A deadly smog had enveloped Donora, Pennsylvania, when this photograph was taken at noon on October 29, 1948.

▲ Figure 9.2 **The Price of Progress**
Donora's enormous smelting facility provided economic prosperity, but it was also the source of considerable air pollution.

Under political pressure from industry, state and federal agencies at first rejected requests by local officials for an investigation of this episode. When the Borough of Donora and the United Steelworkers Union initiated their own study, however, the U.S. Public Health Service reversed its earlier position. That agency's study of the Donora killer smog was the first comprehensive study of the health effects of air pollution. Over the months and years that followed, a clear picture of the causes of this tragedy emerged.

A variety of studies revealed that the toxic effects of the Donora smog were due to a mixture of chemicals, and nearly all of them originated from the smelter.

Initial roasting of zinc ores released sulfur compounds and heavy metal dusts. As molten zinc was produced, nitrogen oxides, carbon monoxide, and fluorine were also emitted from the zinc works. However, many of the most toxic components of the smog were formed by reactions among this mix of chemicals after they dissolved in fog droplets.

Yes, the meteorological conditions that trapped air in the Monongahela Valley prevented the dilution of these toxins and greatly exacerbated their impacts on human health. But such conditions are common in river valleys and topographic basins. Investigators concluded that emissions from facilities such as Donora Zinc Works must be managed with these weather conditions in mind. The Donora killer smog aroused public concern about the health effects of air pollution and led directly to the passage of the Air Pollution Control Act of 1955 and the Clean Air Act of 1967.

Much has changed in the Monongahela Valley since 1948. Unable to compete in a growing global market and to meet state and federal air-quality regulations, the Donora Zinc Works was closed in 1957. Donora's steel mill was shuttered in 1967. Today on cold October mornings, fog still forms in the Monongahela Valley, but this fog is laced with far fewer chemicals (**Figure 9.3**). The conditions in Donora in October 1948 were extreme, but they were by no means unique. Smog remains a common feature in cities across America and around the world.

Today, national laws and international treaties regulate many air pollutants; technologies such as scrubbers and catalytic converters reduce emissions of these pollutants. Nevertheless, air pollution remains a significant threat to the well-being of humans, plants, and animals. What are the sources of pollutants? What processes influence their dispersal in the atmosphere? By what means can air pollution be reduced or eliminated? These and other questions are addressed in this chapter.

▼ Figure 9.3 **Modern Donora**
Today, Donora's mills are gone, along with the smog that they produced. Its downtown features quaint storefronts dating from its mill town days.

9.1 Air Quality and Air Pollution

Air quality refers to the amounts of gases and small particles in the atmosphere that influence ecosystems or human well-being. The concentrations of some atmospheric gases are relatively constant, while those of many other gases vary from place to place and through time. Variable gases and particles have the potential to be air pollutants.

Air pollution refers to gases or particles that are present in high enough concentrations to harm humans, other organisms, or structures such as buildings or pieces of art. Some pollutants are released into the atmosphere directly from specific sources. Other pollutants are produced by reactions among chemicals that have been released into the atmosphere from more than one source. The dispersal of pollutants depends on the movement of gases in the atmosphere, the height at which pollutants are emitted into the atmosphere, and factors influencing the length of time that pollutants remain in the atmosphere.

Gases and Particles

The atmosphere contains a large array of gases and particles that vary in concentration over time and from place to place.

Three gases—nitrogen (N_2), oxygen (O_2), and argon (Ar)—account for over 99% of the gas molecules in Earth's atmosphere. The atmosphere also contains hundreds of **trace gases**, most of which are present in concentrations expressed in parts per million (1 ppm = 0.0001%) or parts per billion (1 ppb = 0.0000001%). The concentrations of some trace gases, such as helium (He), are quite stable from place to place. The concentrations of other gases vary widely from place to place, from day to day, from season to season, or over decades and centuries. It is these variable gases that have the potential to become pollutants.

Water and carbon dioxide are variable gases. Concentrations of water vary in association with proximity to bodies of water, air temperature, and other factors influencing the rate of evaporation. Concentrations of carbon dioxide are influenced by daily and seasonal patterns of photosynthesis and respiration, as well as emissions associated with the combustion of fossil fuels (8.2).

Nitrogen oxides (NO, NO_2, and N_2O) are also variable gases. In pre-industrial times, concentrations of nitrogen oxides varied because the balance of chemical reactions differed from place to place. Since the beginning of the Industrial Revolution, the burning of fossil fuels has caused concentrations of nitrogen oxides to gradually increase. Today, their concentrations are much higher near urban and industrial centers than in rural locations (Figure 9.4).

▶ Figure 9.4 **Air Pollution**
Riding through Lanzhou, China, this cyclist wears a protective mask to keep from breathing in the heavily polluted air. With its many factories, Lanzhou is one of China's most polluted cities.

Volatile organic compounds (VOCs) are organic chemicals that can vaporize into the air. VOCs are a diverse array of chemicals with a wide variety of properties. Some VOCs are produced naturally. Methane (CH_4) is the most abundant naturally occurring VOC. Other naturally-occurring VOCs give flowers and decaying organic matter their distinctive odors. Anthropogenic sources of VOCs include solvents, paints, gasoline, and the exhaust from automobiles.

In addition to gases, the atmosphere contains particulate matter—very small solid and liquid particles suspended in the air. These suspended particles are called **aerosols**. Aerosols originate from both natural and human sources. The most familiar aerosols are the tiny water droplets that make up clouds and fogs. Aerosols are considered to be pollutants if they pose a health hazard when inhaled or if they affect the regional or global climate.

The size of aerosol particles is significant because it determines how long they remain in the atmosphere, the way they scatter light, and how they are deposited in the lungs of humans or other animals. Aerosol particles vary in size from 0.001 to 100 micrometers (µm = 1 millionth of a meter). Particles smaller than 100 µm are called PM_{100}; those smaller than 10 µm are called PM_{10}; and so on. Because aerosols are denser than the air that surrounds them, gravity eventually causes them to settle

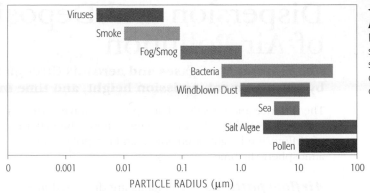

◀ **Figure 9.5 Common Aerosols**
Particles that are suspended in air vary in size. Larger particles fall out of the air faster than do smaller particles.

from the air onto the ground or other surfaces. The rate of such settling, or fallout, depends on the size of the particles; large particles settle much faster than smaller particles. PM_{10} and smaller aerosols are of particular concern to atmospheric scientists because they fall out of the atmosphere very slowly (**Figure 9.5**).

The variety of potential air pollutants is staggering. The U.S. Environmental Protection Agency sets standards, monitors, and controls nearly 190 individual air pollutants, but this is only a fraction of the variety of chemicals added to our air each day. For example, over 400 different kinds of hydrocarbons have been identified in the exhaust of automobiles alone.

Sources of Air Pollution

Some pollutants are released directly into the atmosphere, but many others are generated in the atmosphere by reactions among gases and aerosols.

We usually associate air pollution with human actions, but it can also arise from natural processes. For example, volcanic eruptions emit a variety of sulfur compounds. Radon, a radioactive gas that can have serious health effects in some indoor environments, is released naturally from rocks that are rich in uranium. Aerosols such as sea salt and pollen are produced naturally, whereas dust and smoke come from a combination of natural processes and human actions.

Primary air pollutants are chemicals or particles that are directly emitted from identifiable sources. The elemental mercury released by combustion and volcanic eruptions is an example of a primary air pollutant. Other primary pollutants include the carbon monoxide and sulfur dioxide that are released when fossil fuels burn and the chlorofluorocarbons (CFCs) that are released from various industrial processes.

Secondary air pollutants are chemicals or particles that are produced in the atmosphere as a result of reactions among chemicals or aerosols. The ozone (O_3) that accumulates in city air during the daytime is a secondary pollutant. It is produced by reactions involving

NO and O_2 in the presence of sunlight (**Figure 9.6**). Although its specific characteristics vary from place to place, urban smog is a complex aerosol that forms from the combination of smoke, water droplets, and a mixture of chemicals. Its formation is also facilitated by sunlight. Secondary pollutants whose formation is facilitated by sunlight are called **photochemicals** or photochemical aerosols.

A few chemicals can be either primary or secondary pollutants. For example, NO is released directly into the air by automobiles. It is also produced when NO_2 and O_2 react to form O_3.

Atmospheric scientists distinguish between point and non-point sources of air pollution. Point sources are stationary, localized sources, such as the smokestack of a factory. Each point source has the potential to produce large amounts of pollutants. Air pollution may also be generated by numerous non-point sources, each of which produces relatively small amounts of chemicals or aerosols. Non-point sources of air pollution may be mobile, such as automobiles, or stationary, such as residential fireplaces and agriculture.

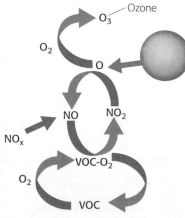

▲ **Figure 9.6 Forming a Secondary Air Pollutant**
During the daytime, primary air pollutants such as NO_2 and volatile organic compounds (VOCs) undergo a series of chemical reactions to produce ozone (O_3), a secondary pollutant that is toxic to humans.

9.2 Pollution in the Troposphere

The troposphere is the 15-km (9-mi) thick layer of air nearest to Earth's surface that is constantly mixed by convection (3.6). Here, three kinds of air pollution have the greatest effect on humans and ecosystems. (1) **Acid deposition** refers to the dispersion of acid-containing gases, aerosols, and rain onto soils, plants, buildings, and bodies of water. It forms from sulfur and nitrogen oxides that are emitted by the burning of fossil fuels. (2) Heavy metal pollution comes primarily from anthropogenic sources, including the combustion of coal, metal smelting, and waste incineration. Mercury and lead are particularly important heavy metal air pollutants. (3) Smog is a complex mixture of both primary and secondary pollutants.

Acid Deposition

Sulfur and nitrogen oxides dissolve in water to produce sulfuric and nitric acids.

In 1872, the Scottish chemist Robert Angus Smith published a book entitled *Air and Rain: The Beginnings of a Chemical Climatology*. By taking systematic measurements across Great Britain, Smith discovered that rain falling near industrial centers was considerably more acidic than rain falling in more remote locations (3.1). Smith was the first scientist to use the phrase "acid rain." He and others asserted that acid rain was responsible for the deterioration of buildings and statues in cities and the diminished production of nearby cropland (**Figure 9.9**).

In the 1950s, scientists began to suspect that acid rain was killing trees in the forests of northern Europe and causing the biological diversity of lakes and streams to decline. By 1970, nearly 25,000 lakes near Sweden's cities supported only those species of plants and animals that could withstand acidic conditions. In the 1960s and 1970s, aquatic and terrestrial ecosystems in New England and southeastern Canada began to display

similar symptoms (**Figure 9.10**). Acidic air pollution was the central focus of the United Nations Conference on Human Environment held in Stockholm in 1972. That conference catalyzed an international program to study and manage the issue.

What is the source of this acidity? Sulfur dioxide (SO_2) and nitrogen oxide (NO_x) gases are the most abundant acidifying pollutants. SO_2 reacts with oxygen and water to produce sulfuric acid; NO_x reacts with oxygen and water to produce nitric acid. These acids can adhere to dust particles or dissolve in cloud droplets. Eventually the acids are deposited on plants, soils, and waters. Because both dry deposition and precipitation deposit acids, scientists prefer to call the process acid *deposition* rather than acid *rain*.

Acid deposition has significant effects on both forest and freshwater ecosystems. Acidic rainwater leaches nutrients and minerals from vegetation and soils, diminishing the fertility of forests. Acidic rainfall and fog also damage the leaves of trees, causing defoliation. Increased acidity (lower pH) in lakes and streams increases the solubility of calcium carbonate, the major constituent of the shells of many invertebrates. Aluminum and other potentially toxic metals are usually bound to sediments but go into solution as waters become more acidic. High levels of aluminum clog the gills of fish and inhibit the development of their skeletons.

The primary sources of SO_2 are coal-fired power plants and metal smelters. NO_x is emitted from power plants and in vehicle exhaust.

▶ **Figure 9.9 Acidic Weathering**
These statues, called caryatids, have supported part of the Acropolis in Athens for over 2,000 years. But in the past century, sulfur and nitrogen emissions from nearby factories have greatly accelerated the weathering of these pillars.

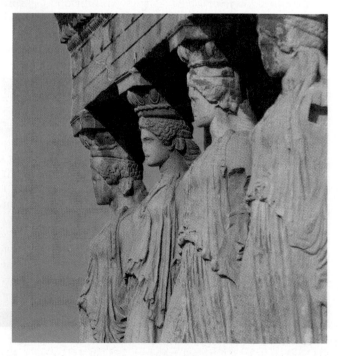

▶ **Figure 9.10 Acid Rain Affects Ecosystems**
Acid deposition was a major factor in the death of these fir trees.

Power plants typically have very tall smokestacks that are designed to disperse emissions and prevent high concentrations of pollutants from accumulating nearby. These tall smokestacks release pollutants into high prevailing winds that can carry them over long distances.

The pattern of acid deposition in North America is greatly affected by the location of the sources and prevailing winds. SO_2 emissions are especially high in the industrial cities in northern Illinois, Indiana, and Ohio. Emissions from Ohio alone are twice the combined emissions from all of New England, New York, and New Jersey. Prevailing winds carry acidic aerosols northeast, where they rain down on New England and southeast Canada (**Figure 9.11**).

Between 1980 and 1990, the United States, Canada, and the countries of the European Union enacted laws and regulations to limit sulfur emissions from coal-burning facilities. With these strong incentives, utilities and industries installed **scrubbers** on smokestacks that removed over 90% of SO_2. Scrubbers are devices that contain chemicals that react with pollutants such as SO_2 and filter them from industrial exhausts.

Today, sulfur emissions in North America and Europe are less than half of the 1980 levels. Emissions of NO_x from coal-burning facilities have also declined, but at the same time emissions from vehicles have increased. As a consequence, total annual NO_x emissions in developed countries have declined only a little over the past 30 years.

In most developed countries, overall acid deposition is declining, and the pH of many lakes and streams is

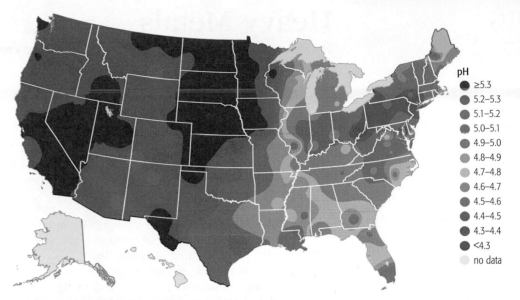

▲ **Figure 9.11 Acid Deposition in the United States**
Rainfall has a lower pH (is more acidic) near cities and industrial centers. The large amounts of SO_2 and NO_x emitted from industrial facilities and utilities in Illinois, Indiana, and Ohio are deposited as acid rain in regions to the east.
Source: http://nadp.sws.uiuc.edu/maps/Default.aspx

pH
- ≥5.3
- 5.2–5.3
- 5.1–5.2
- 5.0–5.1
- 4.9–5.0
- 4.8–4.9
- 4.7–4.8
- 4.6–4.7
- 4.5–4.6
- 4.4–4.5
- 4.3–4.4
- <4.3
- no data

beginning to rise. In developing countries such as China and India, however, coal consumption and automobile use are rapidly increasing: so are SO_2 and NO_x emissions and acid deposition.

Heavy Metals

Human activities have greatly increased the emission of heavy metals.

Heavy metals such as mercury and lead are quite toxic to humans and other animals, but normally they are present in Earth's atmosphere in only small amounts. These elements are emitted to the atmosphere from natural sources such as volcanoes, sea salt, and dust. Analyses of ice cores from Greenland and Antarctica reveal that atmospheric concentrations of heavy metals have increased severalfold over the past 200 years. This increase is the result of human activities, including the combustion of fossil fuels, waste incineration, and industrial processes such as metal smelting (**Figure 9.12**). The concentrations of heavy metal pollutants diminish with increasing distance from their sources.

Because it has become so widespread and its toxicity is so well understood, mercury may be the most notorious heavy metal air pollutant. Emissions of mercury are highest around the world's industrial centers. Most emissions to the air are in the form of mercury gas. Gaseous mercury is subsequently incorporated into droplets of water and dust, which eventually fall onto Earth's lands and waters (**Figure 9.13**).

In terrestrial and aquatic ecosystems, mercury is transformed into soluble methyl mercury. Methyl mercury is then absorbed and stored in the tissues of organisms. Mercury may accumulate to toxic concentrations in animals that feed near the top of food webs, such as carnivorous fishes (6.4). Humans take in mercury when they eat these fishes. Certain occupations, such as working near mines or coal-burning utilities, also expose humans to mercury. Prolonged exposure to even small amounts of mercury can cause damage to the nervous system, paralysis, and death.

Global emissions of mercury are estimated to be 5,000–7,000 metric tons per year. About one-third of this mercury comes from natural sources. Another third comes from human activities such as burning fossil fuels and smelting metals. Mercury is also reemitted to the environment from wildfires and ocean aerosols. This reemitted mercury is thought to be the source of the other third of annual emissions.

Over the past 30 years, anthropogenic emissions of mercury have declined. The major causes of this reduction have been the installation of smokestack scrubbers and the diminished use of mercury in industrial, commercial, and residential products and processes. In the United States in 1990, industries and medical facilities emitted more than 220 tons of mercury; today, annual emissions are below 100 tons. Nevertheless, mercury that was emitted in the past continues to be recycled through ecosystems.

Lead, like mercury, has numerous effects on the health of organisms. In humans, it can disrupt the synthesis of blood hemoglobin and cause brain and kidney damage. Even small amounts can cause neurological damage in young children.

To determine the history of lead emissions, scientists study sediments in bogs, areas of land saturated with moisture and covered with water. The levels of lead in these sediments provide a reliable record of the emissions of this pollutant over thousands of years. The pattern of change in these emissions is a telling case study of how human technologies affect air quality. Ten thousand years ago, annual global lead emissions were 100–200 metric tons. Most of this is thought to have come from volcanism and wind erosion of desert soils.

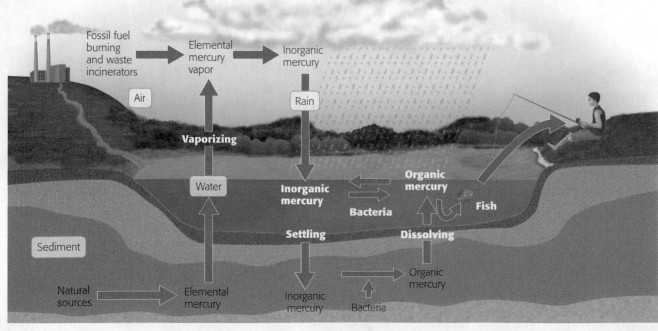

◀ **Figure 9.12 The Mercury Cycle**
Nowadays, human activities are the most important source of mercury to the environment. Because mercury is taken up by plants and animals and stored in their tissues, they play an important role in the cycling of mercury.

About 8,000 years ago, the concentration of lead in bog sediments gradually began to increase. This increase was a consequence of expanding agriculture, which put more lead-laden dust into the air.

Around 5,500 years ago, humans discovered how to smelt silver from lead-containing ores. Smelting caused total emissions to increase to about 1,000 metric tons per year. Silver coins and jewelry became widespread about 3,000 years ago in the Mediterranean region and elsewhere. This led to rapid expansion of silver mining and smelting and a tenfold increase in lead emissions. But as the Roman Empire declined, lead emissions diminished. When the industrial revolution began in the mid-18th century, levels of lead in sediments began to increase again.

Sediments laid down after 1920 show a dramatic increase in the amount of lead, suggesting that emissions had suddenly increased to well above 1 million metric tons per year. They had indeed. In 1921, industrial chemist Thomas Midgley found that adding tetraethyl lead (lead bonded to four ethylene molecules) to gasoline increased the power and efficiency of automobile engines and reduced engine knock. Gasoline containing lead was widely marketed as ethyl gasoline beginning in 1923. (Ironically, in that same year Midgely and three of his coworkers came down with serious cases of lead poisoning.) By the 1930s, 90% of all gasoline in the United States and Europe was leaded.

Some experts began to worry about the poisonous effects of lead emissions almost immediately. In 1925, a study commissioned by the U.S. Surgeon General concluded that, although there were no grounds for prohibiting the use of ethyl gasoline, further studies were needed. None were done for more than 30 years.

When the U.S. Public Health Service finally reexamined the health effects of leaded gasoline in 1959, they concluded that there was significant reason for concern. Despite mounting evidence to the contrary, car manufacturers and gasoline producers insisted that leaded gasoline was safe.

It was not until 1976 that the U.S. Environmental Protection Agency (EPA) began regulating tetraethyl lead as a pollutant. This change in policy came not so much from the many concerns about the health effects of lead pollution but rather from the fact that lead deactivates the catalysts in automobile **catalytic converters**. These devices were required on all cars manufactured after 1975 in order to reduce emissions of carbon monoxide, hydrocarbons, and nitrogen oxides.

Since 1980, the use of leaded gasoline has been banned in most countries, and global emissions have declined by nearly 90%. In the United States, for example, lead emissions were over 200,000 metric tons in 1970; today, they are below 2,000 metric tons, most of which comes from ore smelting and battery manufacture (**Figure 9.14**). However, a number of developing countries continue to use leaded fuel, and emissions in those countries remain quite high.

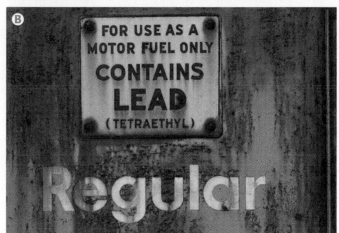

▲ **Figure 9.13 Lead Emission Sources**
Prior to 1920, smelters such as this Ⓐ were the primary source of atmospheric lead pollution. After 1930, the widespread addition of tetraethyl lead to gasoline Ⓑ increased global lead emissions several hundredfold. Today, use of leaded gasoline in fuels is forbidden in most countries.

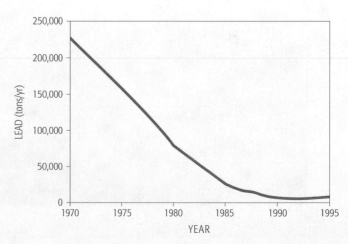

◄ **Figure 9.14 Getting the Lead Out**
The 1970 Clean Air Act required that leaded gasoline be phased out by 1980. This resulted in a 99% reduction in lead emissions in just 15 years.

Smog

The gases and aerosols in smog are a complex mixture of primary and secondary pollutants.

Do modern cars really add significantly to air pollution?

Sierra Seaman, University of Florida

Norm: *Catalytic converters on modern cars have reduced the amounts of some pollutants, but in most cities cars remain the most important source of the primary pollutants that form photochemical smog.*

QUESTIONS

1. What are the primary sources of acid deposition?
2. What are the primary sources of mercury emissions to the atmosphere?
3. Differentiate between industrial smog and photochemical smog.

The word *smog* was first used in the early 1900s to describe the dark, smoky fogs of many cities in the United Kingdom. Large cities such as London and Lancaster had been known for their "pea-soup" fogs since medieval times. With the advent of coal-fired furnaces and steam engines, episodes of foul air became particularly prolonged and intense. Smog was a major reason why people who lived in cities had significantly lower life expectancies than people who lived in rural areas.

Today, smog is a common feature of cities around the world. Atmospheric scientists differentiate between two kinds of smog, based on the mode of formation. **Industrial smog** is composed primarily of pollutants released in coal burning. These include carbon monoxide (CO), sulfur dioxide (SO_2), and particles of carbon soot. The illness and deaths in Donora, Pennsylvania, were a consequence of industrial smog.

Photochemical smog is formed when certain primary pollutants interact with sunlight. Many of these primary pollutants come from the exhaust of vehicles. Nitrogen oxide (NO) and volatile organic compounds (VOCs) are among the most important of these primary pollutants. These chemicals react with oxygen and water vapor to form a variety of secondary pollutants, including ozone (O_3) and nitrogen dioxide (NO_2). It is the NO_2 that gives photochemical smog its brownish color.

The formation of photochemical smog follows a daily cycle that is familiar to many city dwellers (**Figure 9.15**). Exhaust from morning traffic releases large quantities of NO and VOCs. On sunny, windless days these chemicals accumulate in the air and react to form O_3, NO_2, and a variety of other gases and aerosols.

Ozone formation is further enhanced by the presence of carbon monoxide and VOCs. Concentrations of these pollutants generally peak in the late afternoon.

The lifetimes of many of the constituents of photochemical smog are only an hour or two, so their high concentrations depend on their being constantly replenished. Thus, as the sun goes down and light-dependent reactions cease, concentrations of O_3 and NO_2 and other constituents of photochemical smog decline. But if the air is still or trapped by a temperature inversion, the chemical building blocks remain. With the morning sun and the addition of more automobile emissions, the photochemical smog that forms the next day may be even more intense.

Emissions of NO, CO, and VOCs have been greatly reduced in countries such as the United States where automobile exhaust systems are required to have catalytic converters. Metals such as platinum, palladium, and rhodium embedded in the interior surface of these devices catalyze the conversion of these primary pollutants to less toxic substances. Catalytic converters on new cars remove over 90% of the primary pollutants that contribute to photochemical smog.

In the world's developed countries, improved technologies such as smokestack scrubbers have diminished the extent of industrial smog. Although catalytic converters have reduced emissions of NO, CO, and VOCs from automobiles, the total number of vehicles on urban highways is growing, and photochemical smog continues to be a challenge. In developing countries with rapidly developing industrial and transportation systems and many fewer air-quality regulations, both kinds of smog are increasing.

▶ **Figure 9.15 Photochemical Smog**

In large cities such as Hong Kong, China, photochemical smog is common. In sunlight, nitrogen oxide, volatile organic compounds, oxygen, and water vapor interact through the day to form this complex secondary pollutant.

9.3 Pollution in the Stratosphere

The stratosphere is the layer of Earth's atmosphere extending 11.2–48.2 km (7–30 mi) above the troposphere. Although changes in the chemistry of the stratosphere do not directly affect the health of humans or ecosystems, they can influence climate and the amount of sunlight striking Earth's surface. Pollutants with particularly long atmospheric lifetimes can be dispersed through the troposphere and into the stratosphere. Some of these pollutants scatter or reflect sunlight back into space, reducing the amount of radiation reaching the troposphere and Earth's surface. Others chemically degrade stratospheric ozone. Reduced ozone results in increased amounts of ultraviolet light striking Earth's surface, with potentially serious consequences for ecosystems and human health. The Montreal Protocol on Substances That Deplete the Ozone Layer aims to halt the destruction of stratospheric ozone by these pollutants.

Aerosols and Climate

Sulfur compounds create aerosols in the stratosphere that reduce temperatures in the troposphere.

Many aerosols absorb or reflect light, thereby influencing climate. Some, such as dust blown from deserts and aerosols from volcanoes, are produced by natural processes. Others, such as smoke particles from burning wood and fossil fuels, are derived from human activities. In the troposphere, the lifetime of most aerosols is only a few days. In the stratosphere, aerosols may survive for several years.

Aerosols such as clouds, smoke, and dust block incoming sunlight, cooling Earth's surface. Because these particles have short atmospheric lifetimes, their effects on climate are generally brief. During the 1991 Gulf War in Kuwait, hundreds of oil wells were set afire, producing a plume of smoke that extended over much of the region (**Figure 9.16**). This dark haze produced cooler air temperatures during the months that the fires burned, but temperatures returned to normal soon after the fires were extinguished. Gaseous sulfur compounds released by volcanoes and human activities can have longer, more widespread effects on climate. Sulfur gases such as hydrogen sulfide (H_2S) and sulfur dioxide (SO_2) can be transported through the troposphere and into the stratosphere. There the gases are converted to an aerosol of tiny droplets of highly concentrated sulfuric acid (H_2SO_4). Because these droplets reflect sunlight back into space, variations in their abundance affect temperatures in the troposphere and at Earth's surface. For example, the massive eruption of Mount Tambura in Indonesia in 1815 caused 2 years of abnormally cold temperatures in places as distant as Europe and New England (7.7).

The combustion of fossil fuels, particularly coal, releases significant amounts of sulfur oxides; these chemicals have steadily increased the level of sulfur aerosols in the stratosphere. Atmospheric scientists

▲ **Figure 9.16 Aerosols Can Cool Earth's Surface**
In 1991, Kuwaiti oil wells burned for several months. The particulate emissions from these fires dispersed and produced cooler temperatures over large parts of the Middle East.

believe that these aerosols have moderated the global warming caused by increasing concentrations of greenhouse gases in the troposphere. This moderating effect has led some to suggest that we might be able to reduce or even halt global warming by putting more sulfur compounds into the troposphere. However, most scientists warn that this would have devastating effects on many of Earth's ecosystems and could pose significant health risks to humans.

Stratospheric Ozone Destruction

Chemicals produced by humans in the troposphere can have destructive effects in the stratosphere.

In the troposphere, ozone (O_3) is a pollutant that has serious health effects on both humans and ecosystems. But in the stratosphere, there is a layer of ozone that is essential to the health of both humans and ecosystems. This layer of ozone filters out most of the ultraviolet radiation coming from the sun. Without the ozone layer, ultraviolet light would destroy nearly all terrestrial life (3.6).

Chemicals that contain the elements chlorine and bromine can destroy the ozone in the stratosphere. Among the most important of these chemicals are chlorofluorocarbons (CFCs), which were once used as refrigerants and propellants in spray cans. In the troposphere where they are created, CFCs are quite stable. In the stratosphere, however, ultraviolet (UV) light causes them to release chlorine, which then catalyzes the destruction of ozone (**Figure 9.17**). Chlorine is constantly regenerated in this reaction. Thus, a single CFC molecule causes the destruction of thousands of ozone molecules, depleting the ozone in the stratosphere.

For reasons related to the circulation of gases in the upper atmosphere, ozone depletion is especially marked in the stratosphere over Antarctica and the Antarctic Ocean. This region is often called the Antarctic ozone hole, although it is not actually a hole but rather a zone of diminished ozone concentration (**Figure 9.18**).

Satellite measurements have verified the gradual global depletion of stratospheric ozone at the rate of about 4% per decade. As a consequence of depleted stratospheric ozone, levels of UV light have increased in Southern Hemisphere countries such as Australia and New Zealand. This increase is correlated with an increased frequency of skin cancers among people in these countries.

The issues related to CFCs highlight the challenges associated with many air pollutants. Chemicals that are relatively harmless in one setting can be quite destructive in another. This is particularly true of pollutants with long atmospheric lifetimes because they survive long enough to be dispersed widely. The atmospheric lifetime of a pollutant is also a major factor in the amount of time required to mitigate its effects.

▲ **Figure 9.17 Ozone Destruction**
In the stratosphere, ultraviolet light from the sun causes chlorine atoms to separate from CFCs. These atoms catalyze the conversion of O_3 to O_2. In the process, chlorine atoms are continually regenerated.

▲ **Figure 9.18 Ozone Depletion over the Antarctic**
The area of depleted stratospheric ozone known as the Antarctic ozone hole (blue-purple) is monitored each year by satellite. Although CFC emissions have declined significantly, the long residence time of these chemicals means that this hole has not diminished in size.

FOCUS ON SCIENCE: Laboratory Science Predicts Global Effects

How can throwing an old refrigerator into a landfill affect the stratosphere?

Why were CFCs developed? How were their destructive effects discovered? The answers to these questions show how seemingly benign chemicals can have unintended effects and the importance of basic scientific studies in identifying and remedying such effects.

In 1928, Thomas Midgley, the chemist of tetraethyl lead fame, synthesized the first chlorofluorocarbon, called CFC-12. Within days, the makers of Frigidaire refrigerators asked Midgley to find an inexpensive, nontoxic, insoluble, and nonexplosive gas to substitute for ammonia, which was then the most widely used refrigerant. CFC-12 fit these criteria exactly.

Within 3 years the chemical manufacturer DuPont had begun mass production of CFC-12 under the trade name Freon. Several similar CFC compounds were synthesized and manufactured soon after. By 1940, CFCs were in widespread use in refrigerators and air-conditioning systems around the world.

It is noteworthy that these refrigerants were only released to the atmosphere when coolants leaked or were drained. This changed in the 1940s when two other important uses for CFCs were discovered. Scientists found that these chemicals made ideal spray can propellants. Under pressure, the CFCs flowed out of spray can nozzles with a mist of other ingredients, such as paint, insecticide, or hairspray. CFCs were also found to be ideal "blowing agents" for various kinds of foam products, such as insulation and disposable cups. Here again, it was the nontoxic and stable characteristics of CFCs that made them ideal for these applications.

By 1970, chemical companies were selling over 1.5 million metric tons of CFCs each year. Sooner or later, all of it was making its way into the atmosphere. At that time, scientists assumed there would be no adverse environmental consequences because CFCs were nontoxic and did not react with other gases in the troposphere.

Then in 1973, chemists Frank Sherwood Rowland and Mario Molina began intensive studies of the fate of CFCs in Earth's atmosphere (**Figure 9.19**). They pointed out that because these chemicals are so stable, they have very long atmospheric lifetimes. This causes them to accumulate in the troposphere and eventually disperse into the stratosphere. Rowland and Molina demonstrated that ultraviolet light in the stratosphere would decompose CFC molecules, releasing chlorine atoms. Based on research by fellow chemist Paul Crutzen, they then hypothesized that this chlorine would catalyze the destruction of stratospheric ozone.

Rowland and Molina's hypothesis was hotly disputed by the chemical industry. The chairman of DuPont called it "a science fiction tale" and "utter rubbish." However, evaluations by the U.S. National Academy of Sciences supported the ozone depletion hypothesis. In fact, studies indicated that other chemicals containing chorine, bromine, and fluorine might also contribute to the destruction of stratospheric ozone. Such chemicals included carbon tetrachloride, used for dry cleaning, and methyl bromide, a widely used weed killer.

In 1985, a team of British scientists made measurements of ozone in the stratosphere over Antarctica. Their data left no doubt about the connection between CFCs and ozone depletion. Their studies were the basis for an international treaty banning the use of CFCs. (That treaty is discussed in the final module of this chapter.) For their important research, Rowland, Molina, and Crutzen were awarded the 1995 Nobel Prize in Chemistry.

F. SHERWOOD ROWLAND   MARIO MOLINA   PAUL CRUTZEN

▲ Figure 9.19 **Ozone Layer Heroes**
Based on their laboratory research, F. Sherwood Rowland, Mario Molina, and Paul Crutzen predicted that CFCs were destroying ozone in the stratosphere. Their prediction was verified with the discovery of the Antarctic ozone hole in 1985.

9.4 Indoor Air Pollution

During severe episodes of smog, people are often advised to stay indoors. Although staying indoors does protect us from some forms of air pollution, indoor air generally contains a greater quantity and variety of pollutants than does outdoor air. Indoor air pollution comes from many sources, including tobacco smoke, building materials, furniture, chlorine-treated water, ovens, and even the soil and rock beneath buildings. The long-term effects of these pollutants are compounded by the fact that we spend 80–90% of our time indoors. Efforts to make buildings more energy efficient by reducing ventilation can increase the concentrations of indoor pollutants.

Combustion By-Products

Burning organic matter releases a large number of inorganic and organic chemicals into the air.

The complete combustion of organic carbon in the presence of oxygen produces carbon dioxide and water. In most situations, however, the process of combustion is incomplete, and many additional gases and particles are also released into the air. These **combustion by-products** include carbon monoxide, nitrogen and sulfur oxides, hundreds of volatile organic compounds (VOCs), and particulate matter.

In developing countries, most people cook and heat their homes with open fires, particularly those fueled with wood or animal dung. These fires release large amounts of carbon monoxide and particulate soot. The World Health Organization estimates that the combustion by-products from open fires are responsible for respiratory and circulatory ailments that kill over 4 million people in developing countries each year (**Figure 9.20**). In many places, people are simply unaware of the health risks associated with open indoor fires. But even where the health effects are understood, people often cannot afford cleaner alternatives.

In developed countries, fireplaces and woodstoves are generally vented to the outside. However, improper installation, downdrafts, and leaks in flue pipes can cause indoor pollution. Pollutants can also be introduced by activities such as starting and stoking fires and removing ashes.

Gas stoves, furnaces, and space heaters used in developed countries also produce combustion by-products. Carbon monoxide and nitrogen oxide levels are generally higher in homes using gas stoves and furnaces than in those using electric stoves and heaters. However, studies of the relationship between gas cooking and respiratory ailments have been inconclusive. Unvented kerosene space heaters, which were popular in the United States in the 1970s and 1980s, also produce combustion by-products.

The U.S. EPA has developed emissions standards for all new gas- and wood-burning appliances. These standards have reduced emissions of combustion by-products to both outdoor and indoor environments.

▶ **Figure 9.20 Deaths Caused by Indoor Smoke** Death rates (number of deaths/million/year) from indoor smoke are especially high in countries where wood is commonly used as fuel. This includes India and most African countries. *Source:* World Health Organization. http://www.who.int/heli/risks/indoorair/en/webiapmap.jpg

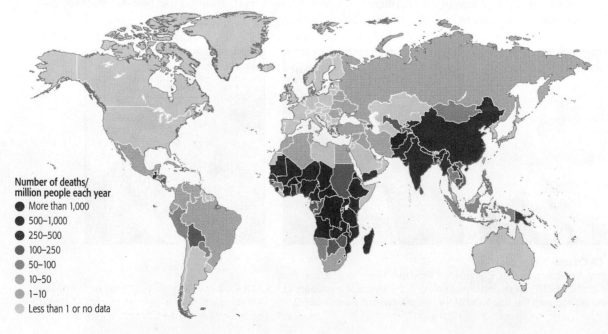

Number of deaths/
million people each year
- More than 1,000
- 500–1,000
- 250–500
- 100–250
- 50–100
- 10–50
- 1–10
- Less than 1 or no data

▲ **Figure 9.21 The Most Deadly Indoor Pollutant**
Tobacco smoke contains hundreds of toxic gases and aerosols.

Relatively inexpensive carbon monoxide detectors are available for those who may be concerned about levels of this deadly pollutant.

Tobacco smoke is one of the most common indoor air pollutants. It is also one of the most hazardous (**Figure 9.21**). The serious health risks associated with smoking have been known for decades. Smoking is the primary cause of lung cancer and pulmonary emphysema and significantly increases risks of heart disease. Environmental tobacco smoke, or secondhand smoke, poses these risks to smokers and nonsmokers alike. It is also linked to chronic bronchitis, asthma, eye and ear infections, and sudden infant death syndrome. The U.S. EPA estimates that each year environmental tobacco smoke is responsible for over 40,000 premature deaths from lung cancer and heart disease among nonsmokers.

Over 250 different toxic or cancer-causing gases and particles have been identified in tobacco smoke. These include particulate matter, nicotine, carbon monoxide, nitrogen oxides, formaldehyde, hydrogen cyanide, and a host of VOCs. For example, concentrations of respirable particulate matter are three to four times higher in homes with smokers than in homes where no one smokes. Such particulate matter is implicated in many of the negative health effects of tobacco smoke.

Smoking bans are the primary means for reducing pollution from environmental tobacco smoke. Many studies have demonstrated rapid improvement in both the air quality and the health of individuals working in areas where smoking is banned. Today, many cities and states have implemented laws requiring workplaces, restaurants, and bars to be smoke free. Similar laws are pending in many other locations (**Figure 9.22**). Efforts to set national standards and regulations for environmental tobacco smoke have, to date, been unsuccessful.

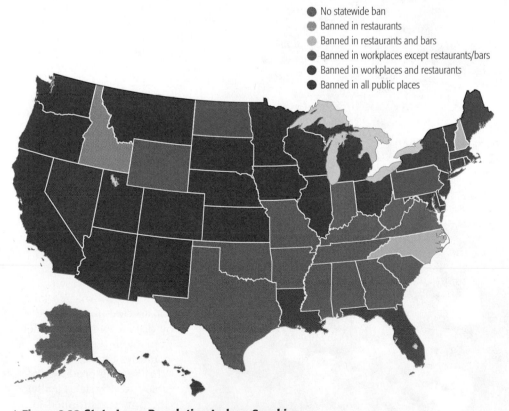

- No statewide ban
- Banned in restaurants
- Banned in restaurants and bars
- Banned in workplaces except restaurants/bars
- Banned in workplaces and restaurants
- Banned in all public places

▲ **Figure 9.22 State Laws Regulating Indoor Smoking**
As of 2010, thirty-six states have implemented either limited or complete bans on smoking in public places.

251

SEEING SOLUTIONS: Breathing Easier in the Kitchen

How can poor countries diminish indoor pollution while encouraging economic development?

▼ **Figure 9.23 Kitchen Pollution**
Each day, the wood fire used by this Masai family to heat water and food also exposes them to considerable indoor air pollution.

In developing countries worldwide, smoke from indoor wood-fueled stoves causes many health problems and more deaths than are caused from malaria (**Figure 9.23**). Wood smoke contains hundreds of different chemicals and aerosols, many of which cause cancer or encourage respiratory infections, tuberculosis, and other pulmonary diseases. Women and young children suffer the greatest exposure to wood smoke and so are at greatest risk. In addition to causing indoor pollution, burning wood for fuel increases the rate of deforestation in developing countries.

Using smokeless cooking fuels would greatly reduce the risks of indoor pollution. Unfortunately, kerosene and natural gas, the most reasonable alternatives, are too expensive for most families in the world's poorest countries. Charcoal is a more affordable option. Burning charcoal produces far less smoke than wood; it also releases more heat per unit weight. Charcoal can be stored indefinitely, whereas wood rots quickly in tropical regions. But charcoal is more costly than wood. And because charcoal is normally produced from wood, its use still encourages deforestation.

Enter the MIT students doing class projects in professor Amy Smith's Introduction to Development course. Amy and her students wondered whether charcoal might be produced from agricultural wastes such as banana leaves, corn cobs, and the sugar cane waste called bagasse. In many poor countries such wastes are abundant, but are simply discarded. This group understood that charcoal production needed to be cheap and easily implemented with minimum technology. They came up with a simple and economical three-step process. First, the wastes are dried in the sun for several days. Next, the wastes are heated in low-oxygen kilns constructed from discarded oil drums, which are widely available in most poor countries (**Figure 9.24**). This heating produces charcoal powder. Finally, the powder is formed into charcoal briquettes using a

▲ **Figure 9.24 From Agricultural Waste to Charcoal**
In the first step in charcoal manufacture, Amy Banzaert helps residents in an El Salvador village load sugar cane bagasse into an oil drum kiln. Banzaert did her master's degree research on technologies to optimize charcoal manufacture from this material.

paste made from cassava, a widely grown tropical crop. These briquettes are similar to those you might use at a backyard barbecue. They can be formed by hand, or they can be mass produced using relatively inexpensive briquetting presses.

One of the students who worked with Amy Smith on this project was business student Jules Walter, a native of Haiti. Walter teamed up with Amy Banzaert, a mechanical engineering graduate student researching this alternative charcoal and they, along with others, have created a company called Bagazo (Spanish for bagasse) that is dedicated to the development and application of this process. With ingenuity and lots of hard work, a class project has produced a solution to three challenging problems. Most immediately, it provides families in poor countries with the means to cook their meals, reduce indoor pollution, and preserve their respiratory health. Second, it provides an alternative to the use of wood as fuel and thereby promises to diminish deforestation. Third, it does all this while significantly diminishing health costs. This is truly sustainable development!

Building Materials

Construction materials emit many volatile organic compounds.

Many materials used in the construction of residential and commercial buildings emit pollutants that pose significant risks to human health. Adhesives, synthetic fabrics, paints, wood preservatives, and insulating materials are all sources of indoor pollutants. Formaldehyde (HCHO), volatile organic compounds, and asbestos are among the most important indoor pollutants derived from building materials. Sometimes the pollutants in a building will affect the health of so many people that the building is said to have "sick building syndrome."

Formaldehyde is a potent irritant of mucous membranes, and exposure can irritate the eyes and upper respiratory tract. Prolonged exposure can cause skin rashes, headaches, fatigue, and depression. Formaldehyde pollution began to receive public attention in the early 1980s, when homebuilders began to use more urea-formaldehyde foam insulation and pressed wood products, such as strand board. Since then, improved manufacturing processes have significantly reduced formaldehyde emissions from adhesives and pressed wood products. Furthermore, emissions from these products decrease significantly with time. Thus, risks from this pollutant can be diminished by allowing building materials to age.

Significant concentrations of more than 300 different volatile organic compounds (VOCs) have been measured in indoor air. Their sources include adhesives, solvents, paints, room deodorizers, carpets, draperies, and furniture. Carpets alone emit nearly 100 different VOCs. Because of the diversity of these compounds, specific cause-and-effect relationships with human health are difficult to establish. However, some VOCs, such as polycyclic aromatic hydrocarbons, are known to cause cancer.

Asbestos is a fibrous mineral that is resistant to heat and fire. Until the early 1970s, it was widely used for a variety of building applications, including spray-on fireproofing and insulation, fiberboard, roofing, and flooring (**Figure 9.25**). In 1973, the U.S. EPA determined that asbestos fibers were a direct cause of debilitating lung disease, lung cancer, and cancer of the lining of the chest and abdominal cavity. It has since been regulated as a hazardous air pollutant.

Today, the use of asbestos is banned in all new buildings. Nevertheless, asbestos-containing materials remain in place in hundreds of thousands of older buildings. The mere presence of asbestos fibers in building materials does not pose a health risk. However, if these materials are disturbed—as during building repair, renovation, or demolition—microscopic asbestos fibers are wafted into the air. Therefore, the EPA regulates the renovation and demolition of all older buildings made with asbestos-containing materials.

▲ **Figure 9.25 Asbestos**
The tiny fibers in asbestos-containing materials cause cancer in breathing passages and lungs.

Radon

Naturally emitted radon accumulates in closed indoor spaces and can cause lung cancer.

Radon (^{222}Rn) is a nontoxic gas produced from the radioactive decay of the element radium-226 (^{226}Ra). Radon itself undergoes radioactive decay to produce radioactive isotopes of the elements polonium (^{218}Po) and lead (^{214}Pb). Because they are electrically charged, polonium and lead bind to particulate matter, producing a radioactive aerosol. When inhaled, this aerosol can cause lung cancer.

Radium occurs in various amounts in most rocks, including granite, limestone, and basalt. As a result, it is present almost everywhere on Earth in or near soil. Outdoor concentrations of radon and its decay products are generally low and do not pose a significant health risk. However, concentrations of radon may be a thousandfold higher inside buildings with poor ventilation that are built on rock or soil that is high in ^{226}Ra (**Figure 9.26**). The EPA estimates that each year in the United States exposure to radon causes 15,000–16,000 lung cancer deaths.

Because of these health risks, the EPA and U.S. Surgeon General issued a public health warning in 1988 recommending that all homes be tested for radon and that steps to reduce radon levels be taken where warranted. Radon levels can be reduced by sealing basement walls and floors, installing check valves in drains, and ventilating crawl spaces.

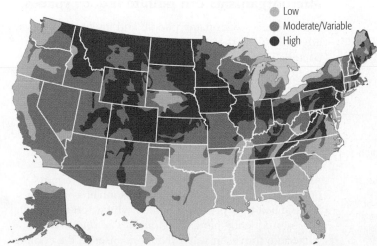

- Low
- Moderate/Variable
- High

◄ **Figure 9.26 The Geography of Radon Emissions**
Radon levels are largely determined by the geological strata upon which homes are built. For example, dark brown indicates areas with large amounts of the rock basalt, which is rich in the element radon.
Source: http://energy.cr.usgs.gov/radon/usrnpot.gif

Pesticides

Chemicals used to control insects and mold also affect human health.

We use a wide range of toxic chemicals in our homes and other buildings to control unwanted insects, rodents, molds, and bacteria. If misapplied, these chemicals may affect our health. Pesticides are especially hazardous to vulnerable populations, such as small children and the elderly (**Figure 9.27**).

Pesticides used to kill termites and preserve wood are of particular concern. Between 1948 and 1988, for example, the pesticide chlordane was injected into soils and cement slabs beneath houses to control termites. Significant amounts of this chemical accumulated in the air of houses that were treated. Chlordane affects the nervous and digestive systems of people and animals,

producing symptoms such as headaches, confusion, stomach cramps, and diarrhea. Since 1988, the EPA has banned all uses of this pesticide.

Prior to 1980, pentachlorophenol (PCP) was widely used to treat foundation timbers and the wood in log houses. Like DDT, PCP is a chlorinated hydrocarbon that accumulates in fatty tissues. People living in PCP-treated log houses were found to have from five to seven times as much of this pollutant in their tissues as people living in untreated homes. Such levels of exposure can produce symptoms similar to those produced by chlordane. Today, PCP-treated wood is no longer used in buildings, although it is still used to treat utility poles.

Biological Contaminants

Microorganisms can pollute indoor spaces.

Indoor spaces with many people and limited ventilation are more likely than outdoor air to contain viruses, bacteria, and fungi that cause illness in humans. Indoor environments may also favor the growth of bacteria and molds that cause allergies.

The 1976 outbreak of a pneumonia-like disease among some of the people attending an American Legion convention in Philadelphia is a dramatic example of building-related disease. Over several days, 221 individuals became ill and 34 died from this disease. After intensive study, the Centers for Disease Control determined that this outbreak of legionnaires' disease was caused by a previously unknown bacterium, which was subsequently named *Legionella pneumophila*. This bacterium thrives in water that condenses in the cooling

towers of air-conditioning systems. It was transmitted in aerosols from the air-conditioning system that served the hotel lobby.

Since then, hundreds of outbreaks of legionnaires' and similar diseases have been reported in office buildings, college dorms, cruise ships, hospitals, and industrial facilities. Regulations requiring large air-conditioning systems to be regularly cleaned and disinfected have reduced the incidence of these diseases.

In summary, many different kinds of air pollutants are more highly concentrated in the air inside buildings than in outdoor air. Activities such as cooking and heating are often the source of indoor pollutants. Building materials themselves may emit pollutants. Limited ventilation allows pollutants to accumulate indoors.

QUESTIONS

1. What are combustion by-products?

2. Describe four types of pollution generated inside your home or dormi room that might influence your health.

9.5 Air Pollution Policy and Law

Prior to 1963, the regulation of air pollution in the United States was delegated to individual states. State regulations were generally weak and highly variable. Laws passed between 1963 and 1970 moved the responsibility for managing air quality to the federal government. These laws required federal agencies to set standards for acceptable emission levels and to prescribe the best technologies to meet those standards. The 1990 amendment to the Clean Air Act implemented a market-based approach to reducing emissions of SO_2, although such an approach is not appropriate for managing all pollutants. International treaties have focused on the movement of pollutants across borders and on pollutants that pose global threats, such as CFCs. Such agreements require that countries perceive that pollutants threaten ecosystems and human health. They must also ensure that the costs and benefits of pollution reduction are shared among countries fairly.

U.S. Air Pollution Policy

Policies to control air pollution in the United States have evolved significantly over the past 60 years.

The public outcry over the 1948 tragedy in Donora, Pennsylvania, put pressure on politicians, but legislative action to reduce air pollution was slow in coming. As postwar industries grew and the population boomed, air quality in major cities declined rapidly. In 1952, a smog event in London resulted in the deaths of over 4,000 people. In 1953, a similar event in New York killed at least 200 people, and severe smog in Los Angeles resulted in dozens of deaths in 1954 (**Figure 9.28**).

Finally in 1955, the U.S. Congress passed the Air Pollution Control Act, which President Eisenhower signed. As it states, this act was intended "to provide research and technical assistance relating to air pollution control." While acknowledging that air pollution was a real danger, the act asserted that the responsibility for setting standards and regulating air quality should remain with the states. The federal government's role should be purely informational. The U.S. Surgeon General was given a small budget to conduct research and distribute information "relating to air pollution and the prevention and abatement thereof." Meanwhile, the nation's air quality continued to deteriorate.

Between 1963 and 1990, a series of new federal laws shifted the responsibility for regulating air pollution to the national government. The 1963 Clean Air Act focused on emissions of pollutants from stationary sources, such as power plants and steel mills. Subsequent legislation included regulations for emissions from mobile sources, such as trucks and automobiles.

In 1970, the Clean Air Act was amended. In fact, this "amendment" was nearly a complete rewrite of the original law. The amendment required that new and more stringent National Ambient Air Quality Standards (NAAQS) be set to protect public health. The amendment also set standards for emissions from motor vehicles. And perhaps most important, it allowed citizens to take legal action against any person,

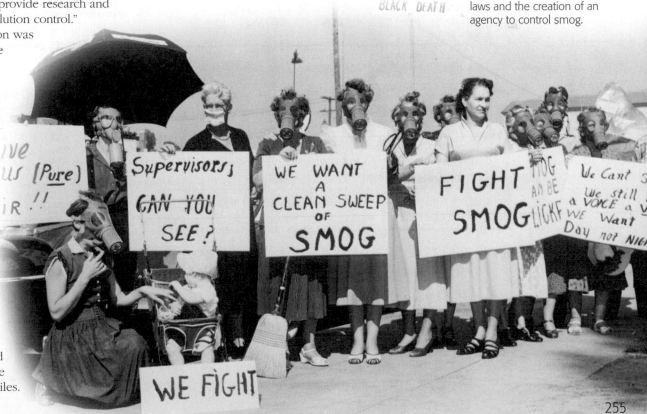

▼ **Figure 9.28 Protest for Change**
Following a deadly smog event in 1954, these Los Angeles women protested for air-quality laws and the creation of an agency to control smog.

organization, or government that violated emission standards (**Figure 9.29**).

What moved the federal government to act so swiftly and forcefully during this period? Certainly the rising tide of environmentalism played some role. More important, however, was the fact that a few leading cities and states—Los Angeles, California, New York City, and Pennsylvania—had begun to enact tough air pollution control regulations. Key industries such as utilities and automobile manufacturers found that it was in their interest to push for moderate and uniform federal standards that would preempt stringent or inconsistent state and local regulations.

In 1990, the U.S. Congress once again drastically amended the Clean Air Act. This amendment empowered the U.S. EPA to determine the best technologies for controlling specific pollutants and to require the installation of those technologies where appropriate. The act also specified actions to reduce emissions of chlorofluorocarbons in order to prevent stratospheric ozone depletion.

One of the most notable provisions of the 1990 Clean Air Act was the establishment of a cap-and-trade program for SO_2 emissions (2.5). These provisions authorized the EPA to set caps for SO_2 emissions from facilities such as power plants. Each facility can either take action to reduce its emissions to meet its cap or buy publicly traded emission credits. Emission credits are created and sold by facilities that have decreased their emissions below their caps. Ideally, emission caps would gradually be lowered and credits for emission reductions would eventually expire.

The sulfur emissions trading program was quite controversial because it allowed individual facilities

to continue to release high levels of SO_2. But advocates pointed out that the primary concern about SO_2 is its contribution to acid deposition. Individual facilities contribute only a small amount to this large-scale problem; rather, it is a consequence of the cumulative emissions from many facilities.

Since the SO_2 cap-and-trade program began, sulfur emissions have been reduced by more than a third (**Figure 9.30**). This cap-and-trade program has also served as a model for the reduction of emissions of other pollutants. A cap-and-trade program is now in operation for emissions of CO_2 under the Kyoto Treaty (8.8).

Despite the success of the SO_2 cap-and-trade program, environmental groups fiercely opposed George W. Bush's 2002 Clear Skies Initiative, which proposed extending the program to include emissions of NO_x and mercury. Why? Because cap-and-trade programs for NO_x and mercury would be likely to cause the effects of these pollutants to be felt more severely in some communities than others. Facilities that emit NO_x and mercury have a great impact on the environment in their immediate vicinity. NO_x is one of the most important primary pollutants contributing to photochemical smog and high concentrations of ozone in cities. And wet and dry fallout of mercury is much higher near coal-burning facilities than farther away. Environmental groups also opposed the Clear Skies Initiative because it extended the timetables for meeting emission standards for these pollutants. In the end, the 2002 Clear Skies Initiative was not approved.

Unlike standards for outdoor air quality, standards for managing indoor air quality are set by state and local governments. The Occupational Safety and Health Administration (OSHA) and the EPA provide guidelines for standards for important indoor pollutants, such as carbon monoxide, particulate matter, and radon. About half of the states have adopted legal standards and enforcement policies based on these guidelines.

▼ **Figure 9.29 Legislating Clean Air**
In just 27 years, a wide range of federal laws mandated cleaner air in the United States.

1963	Clean Air Act	Required that the secretary of Health, Education and Welfare (HEW) set firm standards for pollutant emissions from stationary sources such as power plants and steel mills.
1965	Motor Vehicle Air Pollution Control Act	Extended authority to mobile sources of pollutants such as trucks and automobiles.
1967	Air Quality Act	Expanded federal regulation of both stationary and mobile pollution sources, and it authorized the HEW secretary to seek immediate court-ordered abatement of air pollution that presented an "imminent and substantial danger."
1970	Clean Air Act Amendment	Required that National Ambient Air Quality Standards (NAAQS) be set to protect public health and welfare, and established New Source Performance Standards that strictly regulated the emissions from any new source in an area.
1990	Clean Air Act Amendment	Set new standards to reduce automobile emissions and established a cap-and-trade program for the reduction of SO_2 emissions.

▼ **Figure 9.30 U.S. SO_2 Emissions**
Emissions of SO_2 in the United States have been declining since 1970, but the implementation of the sulfur cap-and-trade program set up by the 1990 Clean Air Act has caused the decline to accelerate.
Source: US EPA. URL http://www.epa.gov/air/airtrends/sulfur.html. Last accessed 8-9-2011

International Air Pollution Policy

Agreements among countries to limit air pollution are based on a shared sense of need and an equitable distribution of costs and benefits.

Most international boundaries have little relationship to the movement of air masses or the pollutants they carry. Emissions from one country often affect the air quality of other countries downwind. In the case of pollutants such as CFCs and greenhouse gases, the combined emissions from many countries have significant global consequences.

Geneva Convention on Long-range Transboundary Air Pollution.

In the 1960s, scientists found that emissions from countries across Europe were contributing to the acidification of Scandinavian lakes. The need for action on this issue was highlighted in the 1972 United Nations Conference on the Human Environment held in Stockholm. This led the member states of the United Nations to craft the Geneva Convention on Long-range Transboundary Air Pollution, which was ratified in 1983.

The convention laid down general principles for international cooperation in reducing air pollution. It also created an institutional framework for research and policy collaboration. This framework has been the basis for the formulation and ratification of protocols to reduce SO_2, NO_x, VOCs, and heavy metal pollutants.

The 1999 Protocol.

The 1999 Protocol to Abate Acidification, Eutrophication, and Ground-level Ozone set 2010 targets for emissions of four pollutants—SO_2, NO_x, VOCs, and ammonia. The biggest cuts were required from countries whose emissions had severe environmental or health impacts and who could most easily afford abatement measures. The signatories to this protocol have reduced SO_2 emissions by 66%, NO_x emissions by 31%, VOC emissions by 40%, and ammonia emissions by 20% compared to 1990 levels.

The Montreal Protocol on Ozone.

One of the most successful international agreements to manage an air pollutant was the 1987 Montreal Protocol on Substances That Deplete the Ozone Layer. In it, countries agreed to a timetable for phasing out the production and use of chlorofluorocarbons and other ozone-destroying chemicals. Even the largest developing countries—China and India—ratified this protocol, although they were granted a longer phase-out period.

The Montreal Protocol was unusual in several respects. It incorporated procedures for continuing review and refinement as scientific knowledge changes. As a consequence, the timetable for reducing CFC emissions has been tightened several times. It also required that a strong majority of producer nations sign before it could be ratified.

The impact of the Montreal Protocol on global CFC production was immediate and significant (**Figure 9.31**). What factors made international action on the issue of ozone depletion comparatively easy? First, there was widespread agreement about the scientific evidence for the relationship between CFCs and ozone depletion.

▲ **Figure 9.31 Declining CFC Emissions**
Up to the 1987 Montreal Protocol, global emissions of CFCs were increasing. Countries and industries responded quickly to the protocol's mandates, and emissions declined by 80% in just 10 years.
Source: United Nations Environmental Program. URL http://www.unep.org/ozone/pdf/Press-Backgrounder.pdf. Last accessed 8-9-2011

Second, ozone depletion threatens everyone—there are not winners and losers. Third, it was easy to monitor production because there were relatively few CFC-producing companies, and they were concentrated in a small number of industrialized countries. Fourth, a number of substitute chemicals that did not interact with ozone were readily available. Many of the companies that produced CFCs played a very positive role in moving the protocol forward as they positioned themselves to produce substitute chemicals.

The Montreal Protocol has served as a model for the development of other agreements, including the Kyoto Protocol to limit greenhouse gas emissions. Unfortunately, the factors that led to rapid agreement on CFCs are not in place for greenhouse gases. Compared to the scientific consensus regarding CFCs, there has been much less certainty regarding the causes and consequences of greenhouse gas emissions. Unlike the risks of CFCs, the risks of global warming are not equally distributed among nations. Unlike the production of CFCs, the production of greenhouse gases occurs worldwide, and emissions from developing countries are accelerating. And finally, although there are energy alternatives to fossil fuels, there is social and economic resistance to their adoption (see Chapter 16).

In general, the development of policies and laws to limit air pollution requires agreement that the threats to the environment or human health are significant. Policies are more likely to succeed if they ensure that the costs and benefits of reducing emissions are shared equitably among stakeholders.

Is the risk from poor air quality greater in developed or developing countries?

Macy, Louisiana State University

Norm: *Generally, risks from poor air quality are higher in developing countries because of their lax regulations and standards. Developing nations also have large numbers of woodstoves and high rates of tobacco use.*

QUESTIONS

1. In general terms, how has U.S. policy toward air pollution changed since the tragedy in Donora, Pennsylvania?

2. How have the cap-and-trade provisions of the 1990 Clean Air Act Amendment operated to lower sulfur emissions?

AGENTS OF CHANGE: Mapping Pollutants in Little Village and Around the World

Marisol Becerra is studying public policy at DePaul University in Chicago, Illinois. As a volunteer with Chicago's Little Village Environmental Justice Organization (LVEJO), Marisol led a youth group in the creation of OurMap of Environmental Justice. This interactive online map helps educate the people of Little Village about sources of pollution and contaminants in their community. For her leadership and innovative use of social media to produce social change, Marisol was awarded a 2008 Brower Youth Award, a national award that recognizes the achievements of young environmental leaders. Marisol is also a Bill and Melinda Gates Millennium Scholar.

How did you first get the idea for OurMap of Environmental Justice?

I grew up in Little Village, a community in the southwest side of Chicago. I first became aware of the Crawford power plant at age three. Back then, I thought it was a "cloud factory" because of the large clouds of smoke emitted from its smokestacks. I would always ask my parents what the building was for. They couldn't tell me because they weren't sure themselves.

During my freshman year of high school, I accompanied my mom on a "community asset and toxic tour" led by the Little Village Environmental Justice Organization (LVEJO). As part of that tour we stopped by the industrial corridor in our community and

learned what each industry does and how it affects the environment.

It was then I learned the "cloud factory" was, in fact, a coal-fired power plant. I soon discovered that pollutants emitted from the plant increase the risk of several health problems, including asthma and other respiratory illnesses and premature death. When I realized the plant's emissions have an impact on the environment and public health, I was angry. I saw an injustice happening in my community, and I had to do something about it.

What steps did you take to get started?

I realized that if I wanted to make an impact, I had to get my peers involved. I decided to start Young Activists Organizing as Today's Leaders, a youth branch of LVEJO. Through this organization, I helped recruit and train local youths to give community asset and toxic tours and to lead strategy sessions for organizing the community. Members also helped to inventory and map environmental assets and toxins in Little Village.

Most of my peers interact and find information through the Internet, so I knew I could reach a larger audience that way. I decided to create a paper newsletter, which would also appear online in a blog format.

We then began to use Google Maps, YouTube, Flickr, and websites like Myspace and Facebook to disseminate information. Using those social media tools, we made an interactive, online tour of the community asset and toxic tours we had been conducting in person. This is how OurMap of Environmental Justice was created.

What are the biggest challenges you've faced with the project, and how did you solve them?

One of the biggest challenges was bridging the gap between youths and adults. We wanted adults to take us seriously. There was a misconception in our community that youths were not interested in being involved in community issues.

We bridged the generational gap by interviewing adults and sharing their stories on our map. We broke the stereotype of "lazy youths" who just play video games or cause trouble. The more work we did, the more the community saw that we were dedicated and hardworking and wanted to collaborate with adults to extend our project's reach.

How has your work affected other people and the environment?

OurMap of Environmental Justice is used nationally and internationally. It has connected people around the world. We now work with youth networks in the United States and other countries to educate communities about pollutants, toxins, and potentially harmful community issues, such as gang violence. The more that people know about what's going on in their community and how it's affecting others, the more likely they will be to organize and help stop the things that are adversely affecting people and the environment.

What advice would you give to students who want to replicate your model in their own communities?

Explore your community. You can do research through interviews and surveys. Identify the assets of your community, including the strengths of its members. Identify the toxic places causing harm to your community.

Then get to work by using the strengths and talents of community members. Everyone can contribute something, no matter how small. Work together to create your own Google map to shed light on community issues, to strengthen your case, and to change policy.

Summary

Air quality refers to the amounts of gases and aerosols—air pollutants—in the atmosphere that may degrade ecosystem processes and human well-being. Pollutants generated from various sources are dispersed in Earth's troposphere. Humans also generate pollutants that make their way into the stratosphere. Chlorofluorocarbons are particularly important because they cause the breakdown of stratospheric ozone, which in turn allows more UV radiation to reach Earth's surface. Indoor spaces confine air movement and provide important sources of pollution, including combustion by-products, building materials, radon, pesticides, and biological contaminants. During the past 60 years, U.S. policies regarding air pollution have evolved from programs of research and technical assistance to programs that set and enforce stringent standards for stationary and mobile sources of pollution. International treaties to manage air pollution have been most successful when the evidence for threats to human well-being is clearly established, when pollution sources are easy to identify and manage, and when cost-effective alternatives to pollutants are available.

9.1 Air Quality and Air Pollution

- Earth's atmosphere contains a large number of trace gases that vary in concentration over time and from place to place.

KEY TERMS

air quality, air pollution, trace gases, volatile organic compounds (VOCs), aerosols, primary air pollutants, secondary air pollutants, photochemicals, temperature inversion, dry deposition, wet deposition

QUESTIONS

1. Why are aerosols that are PM_{10} and smaller of particular concern to public health officials?

2. What is the difference between point- and non-point-source air pollutants? How might this difference influence their dispersion in the environment?

Pollutant types	Inorganic trace gases Volatile organic compounds (VOCs) Particulate aerosols
Pollutant sources	Primary, directly from specific sources Secondary, from reactions among other sources
Influences on pollutant dispersal	Air-flow patterns Emissions altitude Lifetime in the atmosphere
Pollution in the troposphere	Acid deposition Heavy metals Smog
Pollution in the stratosphere	Sulfur compounds CFCs
Indoor pollution	Combustion by-products VOCs and asbestos from building materials Radon Pesticides Biological contaminants

3. What is a temperature inversion, and why does it increase the threats posed by many kinds of air pollution?

4. Differentiate between wet and dry deposition.

9.2 Pollution in the Troposphere

- Acid deposition, heavy metals, and smog are the most important forms of pollution in the troposphere.

KEY TERMS

Acid deposition, scrubbers, catalytic converters, industrial smog, photochemical smog

QUESTIONS

1. Describe three significant impacts of acid deposition on freshwater and forest ecosystems.

2. Nowadays, acid rain contains more nitric acid than it did 25 years ago. Why is this so?

3. The concentration of lead in airborne dust increased dramatically in the 1930s, reached a peak in the early 1980s, and has been decreasing since then. Why?

4. Explain what causes the daily cycle in the amounts of photochemical smog in a large city on a windless day.

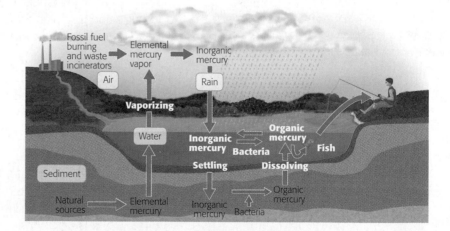

9.3 Pollution in the Stratosphere

- Sulfur-containing aerosols reflect solar radiation back to space, producing lower temperatures in the troposphere.
- Chlorofluorocarbons are stable in the troposphere but cause the destruction of ozone in the stratosphere.

QUESTIONS

1. Explain how sulfur-rich aerosols in the stratosphere influence Earth's energy budget.

2. Describe the chain of events that connects the release of CFCs from discarded refrigerators to increases in UV radiation at Earth's surface.

October 1980   October 1990

September 2000   September 2010

9.4 Indoor Air Pollution

- Indoor spaces constrain air movement and are also the sources of pollution from combustion by-products, building materials, radon, pesticides, and biological contaminants.

KEY TERMS

Combustion by-products, radon

QUESTIONS

1. Describe factors that increase the likelihood of adverse health effects from indoor pollutants.

2. What is the source of radon gas, and what is the basis for its effect on human health?

9.5 Air Pollution Policy and Law

- Policies to manage air pollution in the United States have evolved from research and technical assistance to the establishment and enforcement of air-quality standards.
- Agreements among countries to limit air pollution depend on a shared sense of need and an equitable distribution of costs and benefits.

QUESTIONS

1. Environmental groups have been united in opposition to the use of cap-and-trade programs to manage pollutants such as mercury and lead. Why?

2. What factors have contributed to the success in implementing the 1987 Montreal Protocol on ozone?

Mastering ENVIRONMENTALSCIENCE®

Go to **www.masteringenvironmentalscience.com** for homework assignments, practice quizzes, Pearson eText, and more.

261

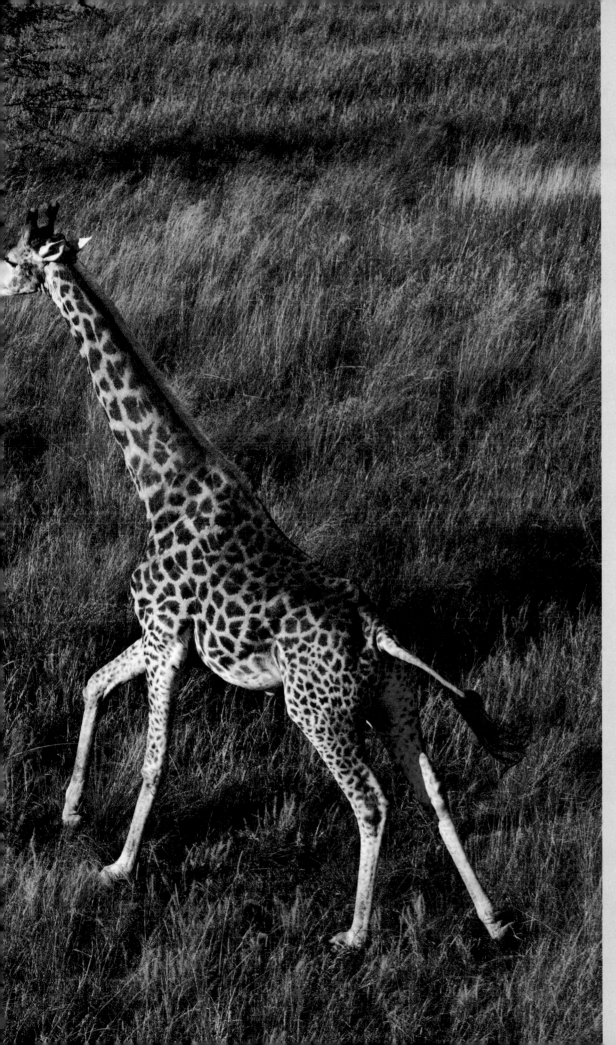

10

Geography of Terrestrial Life

Interstate Biogeography

How are human activities changing the kinds of plants and animals found in grasslands?

Have you ever driven cross country on Interstate 70? If so, you know that between Kansas City, Missouri, and Denver, Colorado, lie 600 miles of broad flat countryside (**Figure 10.1**). Although the highway climbs nearly 4,500 feet over this distance, it does so almost imperceptibly. As you travel west, perfectly rectangular fields of soybeans and corn gradually give way to circular plots of irrigated cotton and barley and finally to large expanses of heavily grazed grassland. There is little natural vegetation to be seen.

One hundred and seventy-five years ago, pioneer families traveled this same route for weeks, even months. What they saw on their journey was vastly different from the scenery on your day-long trek.

Covered wagons moved across eastern Kansas through a lush growth of waist-high grasses. The roots of these tall grasses formed a dense mat, or sod. Native sunflowers and other herbs grew among the grasses, and vast herds of bison grazed on the grasses and herbs. Traveling westward through central Kansas, the grasses and herbs became shorter in response to warmer and drier conditions. Bunchgrasses, species that grow in distinct tufts or tussocks, became increasingly common among the sod-forming species. Onto the high plains of western Kansas and eastern Colorado, wagons maneuvered around ancient clumps of bunchgrass and scattered bits of sagebrush. As the settlers moved west, the herds of bison grew smaller

▼ **Figure 10.1 Changing Vistas**
Today, traveling along Interstate 70 in Kansas and Colorado you will see far different vegetation than did the pioneers.

Short-grass Prairie

Mixed-grass Prairie

Tall-grass Prairie

and were joined by pronghorn antelope and elk. It is almost impossible to believe this route is the same one you travel today.

In the early 1800s, there were places on other continents where you could have traveled a similar distance through seemingly identical vegetation. In eastern Europe, for example, you would have seen a transition from tall grasses to short bunchgrasses if you had traveled from the Ukraine to the lower slopes of the Ural Mountains. In South America, you would have seen a similar transition by traveling westward from Argentina's coastal plain to the eastern slopes of the Andes. On close examination, however, you would have found that the species of grasses, herbs, and grazing animals were completely different from one continent to another (**Figure 10.2**).

Today, if you pull off Interstate 70 or travel back roads in eastern Europe or Argentina, you can still find remnants of natural grasslands; often they have been protected by government agencies or private land conservancies. Some of these grasslands look pretty much as they did to long-ago travelers. But in most places you will discover that human activities have brought significant changes.

At the University of Kansas Field Station, just 20 miles west of Kansas City, tall-grass prairie has been replaced by dense woodlands of oak trees. Suppression of wildfire and the loss of grazing animals such as bison have allowed these oak trees to invade the grassland. Much of the mixed-grass prairie of the Kirwin National Wildlife Refuge in central Kansas has been invaded by a non-native grass called smooth brome. Where smooth brome has become dominant, it has reduced the diversity of native grasses and herbs. This, in turn, has diminished the diversity of the animals that feed on the native plants. The U.S. Bureau of Land Management oversees numerous large tracts of bunchgrass-dominated prairie in eastern Colorado. Here, cattle grazing and climate change are encouraging the invasion of desert species, such as sagebrush (**Figure 10.3**).

As you travel along a stretch of road anywhere on Earth, you can recognize vegetation types, such as grassland or evergreen forest, although you may not be able to name the species. How has evolution among unrelated species of plants and animals produced such similar communities of organisms on different continents? Where do the different types of communities occur, and what determines their distribution? How have humans altered these patterns, and how are they likely to change in the future? This chapter provides the answers to these questions.

▲ **Figure 10.2 Similar in Appearance**
Rheas feed in the dry grasslands of Patagonia in southern Chile. These grasslands look remarkably similar to the grasslands of the Colorado high plans; however, the species are very different in these two places.

▲ **Figure 10.3 More Change**
Although protected from the plow, natural areas along I-70 have changed considerably since pioneer times. (**A**) In the east, former prairie is now woodland. (**B**) In the mixed-grass region, non-native species such as smooth brome have invaded mixed-grass prairie. (**C**) Overgrazing and climate change are converting parts of the short-grass prairie to desert.

10.1 Terrestrial Biomes and Climate

The process of natural selection has resulted in remarkable similarities among unrelated organisms—particularly plants—that grow in similar climates, even on widely separated continents. The communities of similar organisms found in a particular climate are called **biomes**. Earth's biomes occur in one of three general climate zones—the tropical zone, temperate zone, and polar zone. Specific climates within each biome are determined by patterns of temperature, precipitation, and evaporation. Climates within a region are also influenced by the presence of mountains and large bodies of water.

Characteristics of Biomes

Atmospheric circulation and climate determine the distribution of biomes.

If you visit places in North America, Europe, and Asia with warm, wet summers and cold winters, you will find an abundance of broad-leaved deciduous trees. These trees look alike, but they are not the same species. If you visit a desert in the American Southwest or South Africa, you are likely to find plants with thick, waxy leaves and stems that are armed with spines. But on closer examination, you'll discover that although the plants are similar in structure, they belong to entirely different groups of plants (4.7, **Figure 10.4**). Why do plants that grow so far apart look so much alike?

Evolutionary biologists have repeatedly observed that similar selective pressures tend to produce remarkably similar traits among species that have no close genetic relationship to one another. The process by which natural selection favors similar features among otherwise unrelated species is called **convergent evolution**. Because of convergent evolution, plants that grow in similar climates often have very similar adaptations. These include similarities in anatomy, in range of tolerance for variations in temperature and moisture, and in responses to seasonal change.

Biomes are defined by the key features of the plants that grow there. The most significant of these features are growth form (trees, shrubs, or grasses),

leaf type (broadleaf or needle-leaf), and adaptations to changing seasons (deciduous or evergreen leaves).

Biomes can be grouped into three climatic zones, which are defined by latitude and patterns of atmospheric circulation. The **tropical zone** straddles the equator and extends to about 25° latitude north and south, corresponding to the Hadley wind cells in the troposphere (3.7). Tropical rain forest, tropical seasonal forest, and tropical savanna biomes are found in the tropical zone. The **temperate zone** falls between 25° and 60° latitude north and south, which corresponds to the area of prevailing westerly winds in the Ferrel wind cells. Temperate deciduous forest, temperate evergreen forest, chaparral, and grassland biomes occur here. The **polar zone** extends above 60° latitude north and south; its climate is determined by the polar wind cells. Boreal forest and tundra biomes are located in the polar zone. Desert biomes are found in very dry areas of all three climatic zones.

Within these broad climatic zones, the distribution of biomes is influenced by the location of mountain ranges and by the proximity to large bodies of water, such as the ocean. Mountains alter the circulation of air and influence temperature and precipitation in the regions around them.

▶ **Figure 10.4 Look Alikes**
The *Euphorbias* growing in the African Namib Desert Ⓐ look very similar to the organ pipe cacti (*Stenocereus*) Ⓑ, which are abundant in North American deserts. However, the two plants are genetically unrelated.

- ● Temperate evergreen forest
- ● Temperate grassland
- ● Temperate deciduous forest
- ● Tropical rain forest
- ● Tropical seasonal forest
- ● Savanna
- ● Desert
- ● Tundra
- ● Taiga
- ● Chaparral
- ● Mountainous regions
- ○ Ice cap

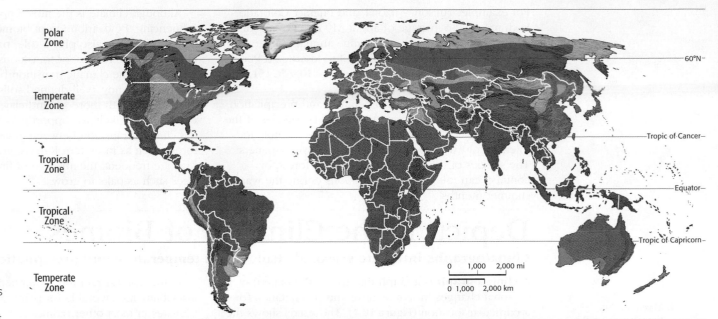

► **Figure 10.5 Earth's Biomes**
Ten major biomes plus mountainous regions are distributed across Earth's land masses. At this scale irregularities in boundaries and local variation in biome distribution are not represented.

Proximity to water affects climates because land surfaces heat up and lose their heat more quickly than do bodies of water. As a result, daily and seasonal temperatures vary widely at locations within the interior of continents, away from large bodies of water. Such areas are said to have **continental climates**. Within the temperate zone, most grassland and desert biomes have continental climates. Lands near oceans or large lakes experience much less variation in temperature; such regions have **maritime climates**. The climates of the temperate deciduous forest and temperate evergreen forest biomes range from maritime near coasts to continental inland.

At first glance, the distribution of biomes within a climatic zone and from continent to continent may seem to follow no discernible pattern (**Figure 10.5**). The tropical rain forest biome, for example, extends throughout northern South America, but is found primarily along the western coast and central regions of equatorial Africa. The temperate deciduous forest biome dominates western Europe, but in North America this biome is confined to the east.

A much clearer pattern emerges if you plot the distribution of biomes according to average annual temperatures and precipitation (**Figure 10.6**). All of Earth's biomes fall into the triangular area between cold and dry, warm and dry, and warm and wet. Note that because cold air cannot hold much moisture, there are no regions with low temperatures and high precipitation (3.6).

Total annual precipitation, in the form of rain or snow, is a major factor in the distribution of biomes. But the amount of water available to organisms in a biome is also influenced by the rate of evaporation (7.3). Evaporation rates generally increase with increasing temperature. This means that if two regions receive the same amount of rainfall, the region that is warmer will have less water available.

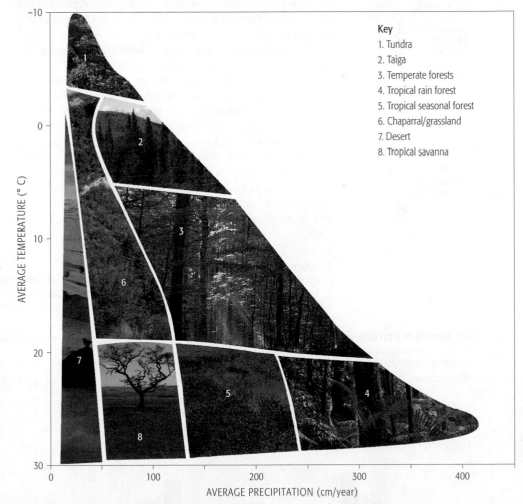

Key
1. Tundra
2. Taiga
3. Temperate forests
4. Tropical rain forest
5. Tropical seasonal forest
6. Chaparral/grassland
7. Desert
8. Tropical savanna

▲ **Figure 10.6 Temperature and Precipitation**
The numbered tiles in this triangular mosaic depict the range of climatic conditions for the corresponding biome.
Source: Whitaker, R.H., 1975. *Communities and Ecosystems.* New York: Macmillan.

For example, Tucson, Arizona, and Pueblo, Colorado, both receive an average of about 310 mm (12 in.) of rain each year. Tucson, with an average annual temperature of 20.3 °C (68.5 °F), is in the midst of a desert, and Pueblo, with an average annual temperature of 10.9 °C (51.7 °F), is surrounded by grassland.

Seasonal variations in temperature and precipitation also influence the distribution of biomes. Species of the temperate deciduous forest, for example, are able to tolerate cold winters but require warm, wet summers. The species of the temperate evergreen forest, by contrast, can tolerate drier conditions during the warm summer months.

Although climate is the most important factor influencing the distribution of biomes, factors such as soil and fire may also play a role, particularly where one climate region merges, or transitions, into another. For example, in the climatic transition between grassland and desert, sandy, well-drained soils are likely to support desert. Soils with more silt and clay, which retain more moisture, are likely to support grassland. Likewise, in the climatic transition between grassland and deciduous forest, such as in eastern Kansas, grassland occurs where fires are frequent; the absence of fire allows deciduous trees such as oaks to grow.

Depicting the Climates of Biomes

Climatographs integrate seasonal change with temperature and precipitation.

A **climatograph** is a graph that shows the pattern of seasonal changes in temperature and precipitation for a particular location (**Figure 10.7**). The x axis shows the months of the year, and the y axis indicates two factors—the average temperature and the average precipitation for each month. Temperature is shown by a line graph and precipitation by a bar graph.

The availability of moisture is not directly shown on a climatograph, but it can be inferred by comparing the temperature and precipitation curves. Evaporation is directly tied to temperature; higher temperatures produce higher evaporation rates. When evaporation is greater than precipitation, there is a moisture deficit, and the lack of water is likely to limit life functions. When precipitation is greater than evaporation, there is a moisture surplus, and plant and animal growth and reproduction are limited less by water and more by other factors, such as temperature or soil nutrients.

In the tropical rain forest, temperature and water availability are favorable for plant growth year-round. Climates in most other biomes are characterized by fairly distinct growing seasons, active growing seasons, when temperature and moisture conditions are favorable for growth, and dormant seasons, when climate limits the activities of plants and animals. In the seasonal tropical forest, the dormant season occurs when moisture is in short supply. In the temperate deciduous forest, the dormant season occurs in winter, when cold temperatures can damage growing tissues.

In summary, in climatically similar regions on different continents, the process of convergent evolution has produced plants and animals with similar adaptations. Biomes are communities of organisms with similar adaptations. Biomes are distributed among three general climatic zones; their specific locations are generally determined by seasonal variations in temperature and water availability.

▼ Figure 10.7 **Climatographs**
On climatographs, the *x* axis marks the months of the year and the *y* axis indicates mean monthly temperature and precipitation.
Ⓐ Warm conditions prevail year-round in tropical zone locations such as Nairobi, Kenya. However, rainfall is seasonal, and moisture deficits occur from June through October. Ⓑ In Madadan, Russia, average temperatures above freezing extend only from April through December, and moisture surpluses occur year-round.

Ⓐ **NAIROBI, KENYA (LATITUDE = 1° S)**

Ⓑ **MAGADAN, RUSSIA (LATITUDE = 59° N)**

FOCUS ON SCIENCE: Shifting Biomes

Wolfgang Lucht and his coworkers at the Humboldt University in Berlin, Germany, are interested in determining how climate change will affect the worldwide distribution of biomes (**Figure 10.8**). To do this, they are using images taken by remote sensors on Earth-orbiting satellites.

◄ **Figure 10.8 Biome Change Forecaster**
Wolfgang Lucht and his colleagues use data from satellite images and climate simulation models to predict the effects of global warming on the distribution of biomes.

Remote sensors are able to measure a variety of characteristics of Earth's surface. For example, the relative amounts of green and red light provide a direct measure of leaf cover. A satellite image is a composite of thousands or even millions of individual sample points, or pixels. Each pixel may represent anywhere from a few square meters to tens of square kilometers of land. By comparing satellite images obtained at different times, scientists can detect changes in the distribution of biomes on continents. This kind of analysis is widely used by scientists to measure changes in the past.

But Lucht is most interested in the future. How will the distribution of Earth's biomes change in the different scenarios of predicted greenhouse emissions and associated global warming? He and his associates address this question by using computer simulation models that are based on correlations between past changes in climate and the distribution of biomes and on knowledge of how plants respond to changes in temperature and moisture (8.5). These simulation models allow Lucht and his team to perform "what if" experiments about the likely redistribution of biomes given different scenarios for reducing greenhouse gas emissions.

Results from Lucht's Berlin research group form the basis for the Intergovernmental Panel on Climate Change's predictions of how global warming will cause biomes to shift (**Figure 10.9**). In North America and Asia, they predict that boreal forests will spread north into areas that are now tundra, while temperate grasslands will shift into areas where boreal forests now grow. The distribution of deserts and tropical biomes are also likely to undergo significant changes. The exact nature of these changes is difficult to predict because it is uncertain how warmer temperatures will affect rainfall.

To resolve these uncertainties, Lucht and his team are now working to incorporate knowledge from recent research into their computer models. They are also investigating how wildfire and human land use may affect future patterns of change.

Is global warming changing the distribution of plants and animals on Earth?

- Forest cover gain
- Shrub/woodland cover gain
- Herbaceous cover gain
- Desert amelioration
- Grass/tree cover loss
- Forest/woodland decline
- Forest type change

0 1,000 2,000 mi

0 1,000 2,000 km

◄ **Figure 10.9 Future Biome Shifts**
This map depicts likely changes over the next century in the distribution of general biome types if current trends in global warming prevail. Note particularly the increase in forest and woodland cover (green) in the Arctic and forest and woodland decline (orange) in central North America and Asia.
Source: Fischlin, A., G.F. Midgley, J.T. Price, R. Leemans, B. Gopal, C. Turley, M.D.A. Rounsevell, O.P. Dube, J. Tarazona, A.A. Velichko et al., 2007. Ecosystems, their properties, goods, and services. Climate Change 2007: Impacts, Adaptation and Vulnerability. Contribution of Working Group II to the *Fourth Assessment Report of the Intergovernmental Panel on Climate Change*, M.L. Parry, O.F. Canziani, J.P. Palutikof, P.J. van der Linden, and C.E. Hanson, eds. Cambridge: Cambridge University Press, 211–272.

10.2 Tropical Biomes

The tropical zone, which extends approximately 25° north and south of the equator, supports some of Earth's most productive and diverse biomes. Throughout the tropics, average monthly temperatures are warm (generally above 20 °C or 68 °F) and vary little throughout the year. Except in high mountains, temperatures do not drop below freezing. The total amount of precipitation and the intensity and duration of the dry season determine whether a tropical location supports a rain forest, a seasonal forest, or a savanna.

Tropical Rain Forest

Tropical rain forest is home to more plant and animal species than any other biome.

Tropical rain forest occurs where annual rainfall is greater than 2,000 mm (80 in.) and falls throughout the year (**Figure 10.10**). The plentiful rainfall and warm climate support an enormous diversity of plants and animals.

If you could stand in the middle of a tropical rain forest, trees would literally tower overhead. This forest has several distinct layers of plants (**Figure 10.11**). The leafy "ceiling" of the forest, called the canopy, consists of hundreds of species of broad-leaved, evergreen trees that grow 30–40 m (98–131 ft) tall. Occasional emergent trees poke above the rest of the canopy, reaching heights of 55 m (180 ft). Little light filters through the thick canopy to the layers of vegetation beneath. However, fallen trees create openings or gaps in the canopy; light that shines through these openings allows saplings, shrubs, and herbs to flourish.

Because of the consistently favorable climate, net primary production in tropical rain forests is much greater than in any other terrestrial biome (7.4). This high rate of productivity is a bit surprising, given that rain forest soils contain little organic matter and few nutrients. The high productivity depends more on rapid recycling of nutrients than on nutrients stored in soil. Insects, fungi, and bacteria rapidly decompose leaf litter in the warm, moist soil. The nutrients released by this decomposition are quickly taken up by the roots of the fast-growing plants.

The complex layers of vegetation in the rain forest provide favorable habitat for an enormous array of animals. In fact, the tropical rain forest hosts more animal species than any other biome. There are, of course, the species you expect,

▶ **Figure 10.10 Tropical Rain Forest Climate**
Although there is variation from month to month, rainfall is persistent and temperatures are continually warm throughout the year in tropical rain forests. These conditions support remarkably diverse forests of broad-leaved evergreen trees such as this one in the Amazon Basin near Iquitos, Peru.

IQUITOS, PERU (LATITUDE = 4° S)

▲ Figure 10.11 Complex Structure
The vertical arrangement of foliage of different trees in the tropical rain forest creates a diverse array of habitats for animals. For example, many monkey species spend their entire lives in the forest canopy, whereas other species are confined to the forest floor.

What can we do to prevent further loss of tropical forest biomes?

Christopher Shad, Cypress College

Norm: *Poverty in many tropical zone countries encourages deforestation. Many governments and environmental organizations are pursuing strategies to diminish poverty in ways that encourage tropical forest conservation.*

such as monkeys, snakes, tree frogs, and a great variety of birds. But the vast majority of the animals here are insects, many of which have yet to be described and named. A single hectare (2.47 acres) of tropical rain forest may support over 2,000 species of beetles.

Much of the diversity in the tropical rain forest is a consequence of evolution that has produced very specialized ecological niches (6.1). Tropical rain forests have a high degree of **endemism**, meaning that many species thrive within very restricted geographic ranges.

The value of species diversity in tropical rain forests cannot be overstated. The integrated activities of the many species sustain this biome's rapid nutrient cycling and high productivity. This diversity also provides a variety of benefits to humans. Many rain forest plants and animals have been used in the development of medicines and pest-resistant food crops (11.2). Given that well over half of all tropical species have yet to be discovered, we can conclude that many of these benefits have yet to be realized.

Deforestation, or the clearing of forests, is without question the greatest threat to tropical rain forests (14.3). Where endemism is high, destruction of forests eliminates a large fraction of habitat for many species, putting them at risk for extinction.

When tropical rain forests are cut and cleared for farming, nutrient cycles are disrupted and the topsoil is easily eroded. Without topsoil, crops fail. But where farming has compacted the soil and caused the loss of nutrients, it is very difficult to reforest the area by planting seeds or young trees. Deforestation also reduces the ability of the rain forest to take up and store carbon, thereby contributing to the increase in greenhouse gases in Earth's atmosphere.

Tropical Seasonal Forest

Seasonal drought determines the distribution of seasonal tropical forest.

Away from the equator and toward the Tropics of Cancer and Capricorn (23° N and S latitude), annual rainfall diminishes to 1,500–2,500 mm (60–98 in.), and months with little or no rain become more common. Here, **tropical seasonal forest** replaces tropical rain forest. The wettest regions of this biome support forests with canopy trees that are 40 m tall (131 ft); the driest regions support only scrubby woodlands.

In tropical seasonal forests, wet and dry seasons each span about half of the year (**Figure 10.12**). Dry seasons are longer and more intense in areas that are shielded from the wet easterly trade winds. Such areas include the west side of the Andes and the west sides of many tropical islands, such as Hawaii.

During the rainy season, plants thrive and animals find abundant food, water, and cover. But the organisms that live in this biome must also have adaptations that allow them to survive drought. When the dry season begins, many trees lose their leaves, a response that reduces their need for water. Such trees are said to be are drought deciduous.

During the wet season, large herbivores such as tapirs in South America and the black rhinoceros in Africa roam throughout the forests (**Figure 10.13**). In the dry season, large animals such as these must congregate within a few kilometers of permanent water holes.

Like tropical rain forests, large expanses of the world's tropical seasonal forests have been cut and burned and the land converted for agriculture. Disturbance and fragmentation of landscapes create challenging barriers to the seasonal migration of the animals of the tropical seasonal forest.

▼ **Figure 10.12 Tropical Seasonal Forest Climate**
A distinct dry season from December through March typifies the climate for the tropical seasonal forest in northern Venezuela. Many of the broad-leaved trees lose their leaves during this dry season.

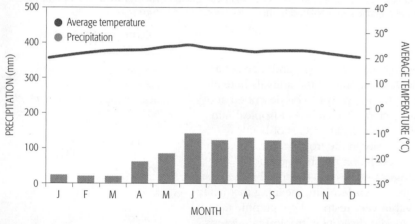

CARACAS, VENEZUELA (LATITUDE = 10° N)

- Average temperature
- Precipitation

▲ **Figure 10.13 The Brazilian Tapir**
Because of the deforestation of tropical seasonal forest, this once-common horse relative has disappeared over much of its range.

Tropical Savanna

Tropical savannas are habitat for large herds of animals that graze on the sea of grass.

Tropical savanna occurs where rainfall is highly seasonal and drought conditions generally persist for more than half of the year (**Figure 10.14**). This is an ecosystem of open grasslands and a few scattered trees. Like the trees of the tropical seasonal forest, many of the trees that live in the tropical savanna are drought deciduous.

In addition to seasonal moisture deficits, several environmental factors influence the growth and diversity of plants in this biome. Large herds of grazing animals feed on the plants, and frequent fires set the tinder-dry grasslands ablaze. The combination of animal grazing, drought, and fire limits the establishment of trees. As a result, there are sufficient light and soil resources to support the characteristic carpet of grass.

Tropical savanna is a biome of seasonal feast and famine. With the rain, grasslands flourish, supporting large populations of grazing animals. Kangaroos and emu thrive in the tropical savanna of Australia, and wildebeests, zebras, and giraffes roam the tropical savanna of Africa. These herbivores feed a diverse array of predators, including lions and cheetahs. When tropical savanna is dry, large wild herds of grazers and their predators migrate to where water and food are more abundant (**Figure 10.15**).

The climates where tropical savannas are found overlap considerably with those of tropical seasonal forests. This has led to debate among scientists about the role that humans have played in creating and maintaining tropical savannas. There seems to be growing agreement that, by grazing domestic animals and setting fires,

▲ **Figure 10.15 A Seasonal Resource**
The timing of rain varies across the large expanses of tropical savanna in Africa. The large mammals of this biome, such as these zebras, must therefore migrate throughout the year to obtain this precious resource.

humans have increased the extent of tropical savannas in parts of Africa, South America, and Australia.

The nomadic herders who live in tropical savannas have traditionally moved their cattle in imitation of the migratory patterns of the wild herds. Today, the boundaries of countries and private property limit such migration. This, in turn, promotes overgrazing, the loss of grasses, erosion, and eventually the conversion of savanna to desert. Global warming is accelerating this transition in many regions, especially in sub-Saharan Africa. The growing deserts threaten the well-being of millions of people.

QUESTIONS

1. Describe two factors that contribute to the very high productivity of the tropical rain forest biome.
2. How does being deciduous help trees survive the dry season?
3. The dry season is longer in tropical savannas near the Tropic of Cancer than in tropical savannas nearer the equator. Why is this so?

▼ **Figure 10.14 Tropical Savanna Climate**
An understory of grasses with scattered trees typifies the tropical savanna of northern Botswana. In this biome, the dry season is longer, and less rain falls during the rainy season than falls in the tropical seasonal forest.

MAUN, BOTSWANA (LATITUDE = 20° S)

10.3 Temperate Biomes

Over 60% of Earth's landmass is in the temperate zone, which extends from 25° to 60° latitude north and south of the equator. Most of this land is in the Northern Hemisphere on the continents of Asia, Europe, and North America. In the temperate zone, average annual temperature ranges from 5 °C to 20 °C (41 °–68 °F), and total annual precipitation ranges from 200 mm (8 in.) to over 2,000 mm (80 in.). The biomes in the temperate zone are temperate deciduous forest, temperate evergreen forest, chaparral, and temperate grassland. Seasonal changes in temperature and rainfall are especially important in determining the distribution of these biomes. Most primary production occurs during the warm growing season, which may be from 4 to 10 months long. Plant and animal activity is typically limited during the winter when temperatures may drop below freezing. Much of Earth's human population resides in the temperate zone, and human impacts on the biomes in this zone have been very significant.

Temperate Deciduous Forest

Little of the temperate deciduous forest remains undisturbed.

As the name of this biome suggests, the **temperate deciduous forest** is dominated by broad-leaved trees that lose their leaves in the autumn and grow a new set in the spring. Temperate deciduous forest occurs in regions with moderate summers and cold winters. Precipitation is relatively even throughout the year (**Figure 10.16**). Here, the growing season is generally defined as the period between the last hard frost (nighttime temperatures of less than 0 °C) in spring and the first hard frost in autumn. In the northern portion of this biome, the growing season extends from May through September; in Florida and along the Gulf of Mexico, many plants retain their leaves and grow from early March through November.

The spectacular display of autumn leaf colors is one of the most notable features of the temperate deciduous forest. The annual loss and regrowth of leaves has important consequences for the forest ecosystem. The decomposition of autumn leaves returns essential nutrients and adds organic matter to forest soils. The flush of new leaves in the spring provides abundant food for large numbers of plant-eating insects. These insects are, in turn, eaten by animals such as birds and bats. The deep layer of leaf litter provides cover for a wide variety of organisms inhabiting the forest floor; it also supplies the energy that supports very complex food webs of microbes, fungi, invertebrates, birds, and amphibians (6.4).

► **Figure 10.16 Temperate Deciduous Forest Climate** There is reliable rainfall year-round in the temperate deciduous forest biome, represented here by a forest in central Massachusetts. Most of the trees in this biome survive the cold winter by losing their leaves.

WORCESTER, MASSACHUSETTS (LATITUDE = 42° N)

Many bird species, such as the wood thrush, scarlet tanager, and ruby-throated hummingbird, migrate between tropical and temperate forests. They build nests and breed in the temperate deciduous forests of eastern North America during the spring and summer, when there is ample food. In the fall, they migrate to the tropical forests of Central and South America to feed over the winter months.

Human populations and urban development are particularly dense where temperate deciduous forests occur; as a result, very little of this biome has not been cut or otherwise disturbed at some time. In the eastern United States, for example, from 50 to 70% of the deciduous forest was cleared for agriculture between 1700 and 1900 (**Figure 10.17**). Most of the remaining forest was significantly altered by livestock grazing and logging for fuel wood. After 1900, farmers abandoned their farms, and much of the cleared land lay fallow. In the ensuing years, trees gradually began to reinvade, creating large expanses of immature forest.

Today, urban development and sprawl are causing the fragmentation and loss of temperate deciduous forests in much of the eastern United States, Europe, and China (14.4). Habitat loss and fragmentation in temperate deciduous forests, combined with deforestation in tropical forests, have caused the populations of many migratory birds to decline. Human disturbance has left the temperate deciduous forest particularly susceptible to invasion by non-native plant and animal species.

▼ **Figure 10.17 Changing Forests**
Temperate deciduous and evergreen forest cover in the United States has changed considerably in the last 400 years.
Source of 1620 and 1920 maps: Greeley, W. b. 1925. *The Relation of Geography to Timber Supply,* Economic Geography 1:1–11
Source of 2010 map: http://fia. fs.fed.us/slides/current-data.pdf

1620
Deciduous forests covered most of the eastern United States and evergreen forest dominated western landscapes, particularly in mountainous areas.

1920
In the 300 years that followed, logging and agriculture diminished the extent of forests in the West and left only scattered patches of forest in the East.

2010
Abandonment of croplands has resulted in a marked increase of forest cover in the East. Suppression of wildfire has contributed to the spread of evergreen forest in the West.

Temperate Evergreen Forest

The temperate evergreen biome includes some of the oldest and largest organisms on Earth.

The dominant trees of the **temperate evergreen forest** keep their leaves throughout the year. This biome generally experiences less precipitation and warmer temperatures than the temperate deciduous forest, and summer months are notably drier than winter months (**Figure 10.18**).

Where heavy winter rains and summer fogs provide ample year-round moisture, evergreen forests are especially lush and diverse, forming **temperate rain forests**. Temperate rain forests grow in coastal regions where temperatures are mild throughout the year. The coastal redwood forests of northern California and the cedar and fir forests of Oregon and Washington are examples of temperate rain forests.

In both the temperate evergreen forest and temperate rain forest, the dominant trees are towering evergreens; these evergreens include some of the largest and oldest organisms that have ever lived. The redwood and giant sequoia of California, the Douglas fir of the Pacific Northwest region of North America, and eucalyptus of Australia can grow to heights of over 90 m (295 ft) and live thousands of years.

In the temperate evergreen forests of the Northern Hemisphere, needle-leaved **conifers**—cone-bearing trees such as pine, spruce, and fir—dominate the canopy. Broad-leaved species, such as dogwood and maple, grow in the understory. In the Southern Hemisphere, the temperate evergreen forest is dominated by broad-leaved trees such as the kauri of New Zealand and eucalyptus trees of Australia.

Mature temperate evergreen forests and rain forests have a diverse array of tree sizes and ages, as well as an accumulation of dead trees, or snags, and fallen logs. Cavities in old trees, snags, and logs provide nesting sites for a variety of animals, including owls and flying squirrels. The woody litter of the understory provides favorable habitat for animals such as banana slugs and voles.

Logging and the suppression of wildfire are the two most important human impacts on these forests. Logging has diminished the extent of old-growth forests in many areas, threatening the survival of the animals that depend on the habitats found in mature and old-growth forests.

Prior to human intervention, naturally occurring wildfires played an important role in the ecology of the evergreen forests. But for the last century, humans—fearing for life and property and believing that fires were ecologically destructive—have worked to prevent or suppress wildfires. In many locations, the result has been the accumulation of litter and invasion of shade-tolerant trees. This has diminished the quality of habitat for many species, and has actually made these areas more prone to fire.

▼ Figure 10.18 Temperate Evergreen Forest Climate
This biome occurs in regions with cool, moist winters and comparatively dry summers, as are found in northwestern Oregon. The old-growth evergreen forest has a complex structure with a variety of trees sizes and an understory with abundant shrubs, herbs, and ferns.

PORTLAND, OREGON (LATITUDE = 45° N)

SEEING SOLUTIONS: Using Fire to Conserve a Biome

The U.S. National Park System is, without doubt, one of the world's most ambitious efforts to conserve the natural diversity of entire biomes. But since its establishment, the management of wildfires within the national parks has been a dilemma. A century ago, the idea of allowing even small natural fires to burn seemed contrary to the goals of park conservation.

By the late 1960s, it had become obvious to many ecologists and firefighters that the suppression of natural fires was having very negative consequences, especially in temperate evergreen forests. Nowhere was this more apparent than in the majestic groves of giant sequoias, a type of temperate evergreen forest restricted to a few areas in the California Sierra Nevada. Prior to protection, low-intensity fires set by lightning or Native Americans burned across the floors of these forests every 5–15 years, consuming woody debris that had accumulated since previous fires. Without harming large trees, these fires created the soil conditions necessary for the seeds of the giant sequoia to germinate (**Figure 10.19**).

Ironically, instead of protecting the giant sequoias, the suppression of fire actually threatened their survival. Because there had been few fires during the past 80 years, large quantities of woody debris had accumulated. Shade-tolerant trees, such as fir and incense cedar, had invaded most sequoia groves. Because their seeds had not germinated, there were virtually no young sequoias. It was especially disconcerting that the accumulated undergrowth increased the likelihood of unnaturally intense fires, which can actually kill the giant sequoias (**Figure 10.20**).

In 1968, the National Park Service initiated a controversial plan to reintroduce fire to these ecosystems. Low-intensity prescribed fires were introduced. To prevent intense fires and minimize the risk to life and property, smaller trees and brush were first cut down. In setting prescribed fires, managers take special precautions to keep the fires from spreading to nontarget areas and minimize the impacts of smoke on air quality. Following prescribed fires, managers must monitor the spread of invasive non-native plants; sometimes they weed out the invasive species.

Today, the giant sequoia groves have regained their open, parklike structure, and young giant sequoias are once again abundant. Fire management in the giant sequoia groves serves as a model for the design and execution of prescribed fires in other temperate evergreen forests.

How can fire conserve populations of Earth's largest trees?

▶ Figure 10.19 **Beneficial Fire**
Natural fires in giant sequoia groves primarily burned litter and woody debris on the soil surface. The heat from these fires did not damage the forest canopy.

▲ Figure 10.20 **Effects of Fire Suppression**
When fires are suppressed, litter accumulates and shade-tolerant trees invade. Ⓐ Light surface fire keeps the sequoia forest understory open and provides a suitable habitat for sequoia seedling establishment. Ⓑ The triangular fire scars that are found on nearly all mature sequoia trees are evidence of repeated fires over hundreds of years. They do not damage the tree.

Chaparral

Summer drought is an especially prominent feature of this biome.

Chaparral is a biome of evergreen shrublands and low woodlands. It occurs where winters are mild and moist and summers are hot and dry (**Figure 10.21**). These climatic conditions are often said to be Mediterranean because they are characteristic of the lands surrounding the Mediterranean Sea. Similar climatic conditions occur on the western edges of continents between 30° and 40° latitude in southern California, Chile, South Africa, and southwestern Australia.

Chaparral vegetation includes a diverse array of evergreen shrubs, such as the chamise and manzanita of southern California. Most of these shrubs are **sclerophyllous**, which means "stony leafed." The thick, hard leaves of sclerophyllous shrubs are resistant to water loss and wilting, an adaptation that protects the plants from intense summer drought (**Figure 10.22**). Sclerophyllous shrubs such as the rosemary and thyme of southern Europe contain fragrant resins and oils that make them valuable as spices.

Fires occur naturally during the hot summer drought. The dense shrubby growth of the chaparral allows fires to burn intensely and spread great distances. Oils and resins in the leaves of many shrubs contribute to this flammability. But the shrubs of this biome are also highly adapted to such fires. Some can sprout from fire-resistant roots within weeks of burning. Others produce seeds that remain inactive in the soil until the heat of a fire stimulates their growth.

Few large mammals are native to the chaparral, probably because the shrubs provide little forage. However, many rodents and birds feed on the seeds and young seedlings beneath the shrub canopy. In chaparral woodlands, grass and acorns provide food for deer and herds of domestic pigs, sheep, and goats.

Humans have been altering the chaparral around the Mediterranean Sea for thousands of years. In Portugal and Spain, for example, much of the chaparral has been transformed into a low-intensity agricultural landscape dominated by cork oak trees. Farmers sustainably harvest the bark of these trees to make wine corks and insulation materials. In many parts of the world, chaparral has been transformed into agricultural uses such as vineyards and olive orchards.

Urban and suburban development has increased the frequency of fires in this biome. Development has also fragmented and diminished the extent of the native chaparral and made way for the invasion of alien species of plants (19.2).

▼ **Figure 10.21 Chaparral Climate**
This biome, found in places such as southern California, is dominated by shrubs and low trees that are tolerant of severe summer drought.

SANTA BARBARA, CALIFORNIA (LATITUDE = 34° N)

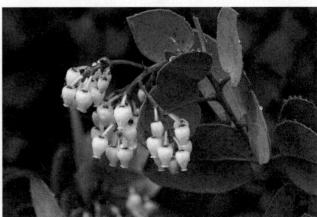

▲ **Figure 10.22 Stonelike Leaves**
The sclerophyllous leaves of manzanita (*Arcostaphylos*) are resistant to the hot, dry summers of California's chaparral.

Temperate Grassland

Agriculture and cattle grazing have altered over 90% of Earth's temperate grasslands.

Grasses interspersed with a diverse array of herbs are the dominant vegetation in the **temperate grassland** biome. The climate of this biome is too dry to support forest, woodland, or shrubland, but wet enough to prevent the land from becoming desert. In most temperate grasslands, the climate is decidedly continental—winters are often long and cold, but summers can be quite hot (**Figure 10.23**).

The productivity and diversity of temperate grasslands increases with increasing amounts of rainfall. In the wettest areas of the biome, such as the Pampas region of Uruguay and Argentina and the tall-grass prairies of Illinois and eastern Kansas, grasses may grow as tall as 2 m (7 ft). The roots of these grasses grow together in a dense mat, forming a sod that is very rich in organic matter.

▼ Figure 10.23 **Temperate Grassland Climate**
This biome has cold winters and hot summers and receives only modest rainfall. It is dominated by grasses and herbs whose roots form a dense sod. The Konza Prairie of eastern Kansas is one of the very few temperate grasslands that remains undisturbed by human activities.

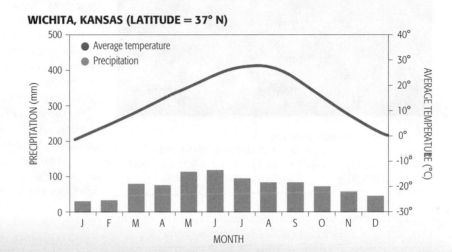

WICHITA, KANSAS (LATITUDE = 37° N)

1. The temperate deciduous forest biome growing season varies from four months in Maine to over nine months in northern Florida. What factor is most responsible for this variation?

2. Compare the climate of the temperate deciduous forest biome to that of the temperate evergreen forest biome. What is the main distinguishing factor?

▲ **Figure 10.24 Bunch Grasses**
These tussocks of fescue in the arid grasslands of Patagonia in southern Chile are adapted to drought and are able to grow back after being grazed. Individual clumps can survive for centuries.

In drier areas, such as the short-grass prairie of eastern Colorado, the tussock grassland of Patagonia, and the Russian steppe, grasses are comparatively short, growing to about 0.5 m (1.5 ft). Here, bunchgrasses that grow in distinct clumps or tussocks are common (**Figure 10.24**). The roots of these grasses typically do not form sod.

▼ **Figure 10.25 A Grassland Intruder**
Overgrazing in grasslands of the western United States has facilitated the invasion of non-native plants such as Russian knapweed. Because livestock cannot eat this plant, in places where it has become abundant it has diminished the value of grasslands for grazing.

In grasslands where summer rainfall is common, most of the grasses and herbs are perennials—plants that survive for several years. Where summers are dry, as in Mediterranean temperate grasslands, the dominant plants are annuals that grow from seeds each year, reproduce, and then die at the end of the growing season.

Much of the biomass of temperate grasslands is located underground in extensive root systems. These roots efficiently extract water and nutrients from the soil and protect the very fertile soil from erosion. As the harsh winter begins, grasses and herbs die back to the ground. In the spring, plants revive quickly, fed by the nutrients stored in their roots. Because of their ability to resprout from roots, grassland plants are resilient to fire and animal grazing.

Prior to the modern expansion of agriculture, the grasses and herbs of this biome supported a diverse array of herbivores, as well as the omnivores and carnivores that preyed upon them. Only remnants of those original ecosystems can be found today.

No biome on Earth has been so extensively transformed by human actions as the temperate grassland. More than half of the world's supply of plant and animal food comes from regions that once supported temperate grasslands. Virtually all (99.9%) of the North American tall-grass prairie has been converted to agriculture. Unfortunately, once the sod has been plowed, native grasslands can be restored only with great effort.

Short-grass prairie and steppe regions have been altered by cattle grazing or, where irrigation is available, by farming. In some regions, grazing and farming have proven unsustainable. These disturbances have also made the grasslands much more susceptible to invasion by non-native species (**Figure 10.25**). The combination of overgrazing and global warming is causing the driest temperate grasslands to change into desert.

10.4 Polar Biomes

Polar environments are generally found above 60° north or south latitude, where very cold annual temperatures (less than 5 °C, or 41 °F) and short growing seasons limit the abundance and diversity of life. Two biomes occur here—boreal forest and tundra. Most of these biomes are found in the Northern Hemisphere. Except for ice-covered Antarctica, the Southern Hemisphere has little land in the polar zone.

Boreal Forest

Boreal forest trees thrive in cold and wet conditions.

Where the growing season is less than four months and winters are long, dry, and bitterly cold, temperate forests and grasslands give way to the open coniferous forests of the **boreal forest** biome (**Figure 10.26**). Russians call the vast boreal forest of Siberia **taiga**. Similar forests are found in northern Europe, Canada, and Alaska. Annual precipitation in this biome varies between 500 and 1,500 mm (20 to 59 in./yr), most falling as snow. The cold weather here slows the rate of evaporation, and the relatively flat terrain impedes drainage. As a result, soils usually remain wet or boggy throughout the growing season.

Evergreen conifers such as spruce and fir dominate the boreal forest. Their small needlelike leaves and drooping branches shed heavy snowfall that otherwise would break their branches. The understory of the forest is composed of low shrubs, herbs, and lichens. Boreal forest trees and understory plants are able to photosynthesize at temperatures very near freezing, completing pollination and seed production within the very short growing season. The harsh climate actually "shelters" the biome from invasive alien species.

Animals such as moose, bear, lynx, snowshoe hare, and spruce grouse live in boreal forests year-round. These animals have well-insulated bodies. Many animals hibernate through the coldest winter months.

In summer, birds such as sparrows and warblers breed in the forest, migrating north from temperate and tropical biomes.

Near the transition from the temperate zone, boreal forest trees form a dense forest, reaching heights of over 20 m (66 ft). Farther north, **permafrost**, permanently frozen soil 30–100 cm (11–39 in.) beneath the soil surface, becomes more prevalent. The permafrost, harsher winters, and shorter growing season limit both productivity and species diversity. Here, the dominant black spruce trees grow only 3–5 m (10–17 ft) tall, and the understory is dominated by mosses and lichens.

Lightning-set fires naturally burn over millions of hectares of boreal forest each year. Therefore, the expansive boreal forest landscapes of Alaska, Canada, and Siberia comprise a mosaic of large patches of forest in different stages of succession (6.5). Global warming appears to be producing drier conditions in many parts of this biome. Consequently, fires have become more frequent and extensive (8.4).

▼ **Figure 10.26 Boreal Forest Climate**
The Russian word for this biome is *taiga*, which translates loosely as "little sticks." This is an apt description of the northern portions of this biome, as in Canada's Northwest Territory, where short growing seasons and boggy soils limit tree size.

FORT SMITH, CANADA (LATITUDE = 60° N)

- Average temperature
- Precipitation

In recent years, bark beetles have infested large areas of boreal forest in North America and Asia. These infestations are thought to be related to the longer growing seasons caused by global warming. Bark beetles normally complete only a single growth cycle in a year. Longer growing seasons allow them to complete two growth cycles each year, boosting their population.

In the southern boreal forest, logging for timber and paper pulp is the most important human disturbance. In Siberia, for example, over 12 million hectares (30 million acres) of boreal forest are cut each year, and the rate of cutting is increasing. Because of the cold, wet conditions, reforestation in taiga is very slow.

Tundra

Tundra exhibits a treeless landscape of subtle beauty and stark contrasts.

In the northernmost polar regions, where growing seasons are shorter than three months and winters are most harsh, boreal forest gives way to **tundra**, treeless landscapes dominated by grasses, herbs, and low shrubs (**Figure 10.27**). Tundra receives less precipitation than many deserts, only 100–500 mm/yr (4–20 in./yr). Even so, the limited evaporation associated with cold weather causes most tundra soils to remain wet through the summer.

The surface or active layer of tundra soil melts in spring, but permafrost remains 20–50 cm (8–20 in.) below the surface. As the active layer thaws, tundra plants rely on nutrients stored in their roots to begin to grow and flower while still surrounded by snow. Freezing temperatures may occur at any time during the summer. Tundra plants remain resistant to cold throughout the summer and can resume growth immediately when conditions are favorable.

The summer growth of tundra plants fuels a complex food web. Herbivores include rodents, such as lemmings, and grazers, such as caribou (called reindeer in Eurasia). Dominant carnivores include the arctic fox, wolves, and grizzly bears. The very moist summertime conditions provide extensive breeding habitat for mosquitoes and black flies, both of which prey on mammals.

Several human cultures have developed sustainable lifestyles in the seemingly inhospitable tundra. The North American Inuit sustain themselves as hunter-gatherers, and the Lapps of northern Scandinavia make a living by herding reindeer. More recently, humans have exploited some tundra areas for the mineral resources of oil and gas (**Figure 10.28**).

Because of global climate warming, the active soil layer receives more heat each year. As a result, the depth to the permafrost is increasing in many tundra regions (8.4).

▼ Figure 10.27 **Tundra Climate**
Although the growing season is less than three months and winters are harsh, tundra such as this in the Alaska Range supports a diverse array of shrubs and herbs.

POINT BARROW, ALASKA (LATITUDE = 71° N)

▲ Figure 10.28 **Human Intrusion into Tundra**
The Alaska Pipeline, which moves crude oil from the northern tundra near Prudhoe Bay to the state's southern seaport of Valdez, is emblematic of recent human disturbance in tundra.

10.5 Deserts

The foremost characteristic of the **desert** biome is a sustained and significant moisture deficit. Compared to other biomes, average annual temperatures in most deserts are high, but polar deserts can be quite cold. Deserts experience some of the widest daily and seasonal temperature fluctuations of any of Earth's biomes. Deserts occur in all three climatic zones, but are most extensive in the transition region between the tropics and the temperate zone. Desert plants and animals have a variety of adaptations that minimize water loss. Climate change, livestock grazing, and unsustainable agriculture are causing deserts to expand into other biomes, particularly grasslands and tropical savanna.

Defining Deserts

Deserts occur worldwide in very arid environments.

All deserts are dry. Annual rainfall is typically well below 250 mm (10 in.), and annual evapotranspiration, the loss of water vapor from soil and leaf surfaces, is much greater than annual precipitation (**Figure 10.29**).

You probably think of deserts as being very hot, and most are. But the range of annual average temperatures among the deserts of the world is quite large. For example, the central portion of the Sahara Desert in Africa has an annual average temperature of 27 °C (80 °F), whereas that of the Wright Valley in Antarctica is –20 °C (–4 °F), well below freezing. Yet both of these deserts are extremely dry, receiving far less moisture as precipitation than they lose by evaporation. Many temperate deserts have marked seasonal changes in temperature, with midday conditions varying from far below freezing in winter to over 40 °C (104 °F) in summer.

Deserts are noted for having large fluctuations in daily temperature. In California's Death Valley, for example, it may be below freezing at dawn and hotter than 38 °C (100 °F) by midafternoon. Within hours after sunset, desert temperatures may drop by 15–20 °C (27–36 °F). This is more than twice the usual temperature variation in a tropical rain forest over an entire year.

▼ **Figure 10.29 Desert Climate**
Desert is the driest biome on Earth and occurs in tropical, temperate, and polar zones. Deserts such as the dry valleys of Antarctica Ⓐ, the Asian Gobi Ⓑ, and the North American Sonoran Ⓒ support plant and animal species that are adapted to dry conditions and highly variable temperatures.

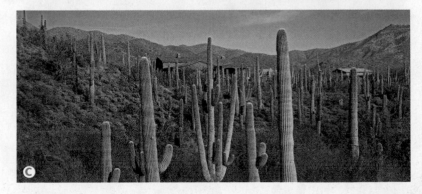

TUCSON, ARIZONA (LATITUDE = 32° N)

Are the desert biomes getting larger?

Breeann Sharma, Arizona State University

Norm: *Yes, climate change, overgrazing, and unsustainable agriculture are converting drier parts of the tropical savanna and grassland biomes to desert.*

QUESTIONS

1. In general, plant and animal diversity is highest in deserts where the warmest months correspond to the months of highest rainfall. How would you explain this relationship?

2. Why are some of the world's most extensive deserts found in the transition between the tropical and temperate climatic zones?

Deserts occupy more than one-fourth of Earth's land surface. Extensive deserts such as the Atacama of South America, the Sahara of northern Africa, and the Great Sandy of Australia extend from the tropical zones into the temperate zones. Their distributions correspond to the zone of dry air descending with the equatorial Hadley convection cells (3.7).

The Sonoran, Mojave, and Great Basin Deserts of North America and the vast Gobi Desert of central Asia are found in temperate zones, within the interior of continents. The aridity of these deserts is accentuated by rain shadows caused by air masses moving over large mountain ranges. (This process is described in Module 10.6.) The Gobi Desert, for example, extends to the north of the Himalaya and Hindu Kush Mountains. The southerly monsoon winds dump most of their moisture on the southern slopes of these mountains, so they bring only scant moisture to the Gobi.

The strong cold winds of the polar zone carry very little moisture, therefore creating deserts in some regions. Polar deserts are found on the Queen Elizabeth Islands of northeastern Canada and in the Wright Valley of Antarctica.

The composition of desert vegetation varies greatly, from just a few scattered herbs and lichens growing on barren soil to diverse shrublands with cacti the size of trees. Desert plants display a wide range of adaptations to dry conditions. Many desert plants are **succulent**, meaning they take up large quantities of water and store it in thick, fleshy stems or leaves. Succulents such as cacti possess abundant spines that protect them from thirsty herbivores.

Other desert plants become dormant during dry periods, growing and reproducing only after rain (**Figure 10.30**). In the driest deserts, herbs, mosses, and lichens form dense, drought-resistant "cushions," which are interspersed between bare ground and rock.

Despite the limited water and severe weather, a great many animals have adaptations that allow them to survive in deserts. The majority of desert animals are nocturnal, active primarily during the cool nighttime hours when water loss to evaporation is low. Some desert animals, such as the Mojave ground squirrel, hibernate in deep burrows during the hottest and driest months; their burrows are much cooler than the surface, and their bodies require far less water in this dormant state. Many desert rodents use water so efficiently that they are able to survive on water they generate from the respiration of their food. Such animals can live their entire lives without drinking a drop of water (**Figure 10.31**).

Although human activities have affected many desert locations, their impact is generally less extensive than in other biomes. Nomadic herders have lived in many of the world's deserts for thousands of years, moving their animals to find water and food. Until recently, farming in deserts was limited to areas close to springs, wells, or rivers. Today, humans use canals to divert large quantities of water to irrigate agricultural fields and to support rapidly growing cities. Although harsh conditions have limited their invasion, non-native species are increasingly common. Over 110 non-native annual herbs now grow in the California Mojave Desert.

Human actions are causing deserts to spread into transitional areas between other biomes. Overgrazing and poor farming practices in marginal grasslands and savannas allow desert plants to become established. Global warming is encouraging the expansion of deserts in Asia and sub-Saharan Africa.

▼ **Figure 10.30 Enduring Drought**
Some desert plants, such as the ocotillo of the Sonoran Desert, are leafless and dormant during dry periods. When rains come, however, they produce leaves and flowers within days and seeds within just a few weeks.

▼ **Figure 10.31 A True Nondrinker**
From birth to death, the kangaroo rat of North American deserts never drinks a drop of water. It is able to extract all the water it needs from its food.

10.6 Mountains and Coastlines

Mountains and coastlines are not biomes; instead, they are geographic features that are found in all three climatic zones. Both mountains and coastlines are characterized by gradients of environmental change that support distinctive communities of organisms. On mountains, changes in elevation produce a variety of climates. On coastlines, the most important gradients are those of salinity and water availability. Mountain and coastal communities are particularly vulnerable to human activities and global warming.

Mountains

One mountainside may experience as much climatic variation as does all of North America.

Hiking up the San Francisco Peaks in Arizona, you see some remarkable changes. In just a few miles, you travel from the Sonoran desert to a climate like the tundra. The change in plant and animal communities as you travel up the mountain is similar to the changes you might observe traveling from lower to higher latitudes (**Figure 10.32**). Biomes typical of warm, dry climates occur at lower elevations. At higher elevations, they are replaced with biomes associated with cooler, moister environments.

What causes these differences in climate and vegetation? The key factors are differences in air temperature and prevailing winds. The atmosphere becomes cooler with increasing elevation. Winds move air up a mountainside. (Prevailing winds blow toward the *windward* side of the mountain.) As air moves upward, it becomes cooler and its capacity to hold water vapor decreases (3.6). Clouds form, releasing rain or snow on the mountainside. On the other side of the mountain (the *leeward* side), the air moves downward, and the process is reversed. As the descending air warms, its capacity to hold moisture increases and it delivers less rainfall (**Figure 10.33**).

This movement of air results in two very predictable patterns of moisture distribution on mountain slopes. First, precipitation increases with increasing elevation. Second, where there are persistent prevailing winds, the windward sides of mountains are much wetter than the leeward sides. The dry region on the leeward side is called a **rain shadow**. Many of Earth's deserts occur in the rain shadows of high mountains.

The region of transition from boreal forest to tundralike vegetation on a mountain is called the **tree line**. In general, the tree line occurs at progressively

▲ **Figure 10.32 Mountain Biomes**
The distribution of biomes along the slopes of the San Francisco Peaks of Arizona is similar to what you might observe driving from Mexico to northern Canada.

▲ **Figure 10.33 Mountain Climates**
As air rises up a mountainside, it is cooled and produces rainfall. As it descends on the other side, the air is heated and dried.

285

▲ **Figure 10.35 Paramo**
Plants such as these *Espeletias* in the tundra-like paramo of the Colombian Andes are adapted to intense sunlight and very cold nights.

▲ **Figure 10.34 In the Clouds**
This cloud forest in the central mountains of Costa Rica receives ample rain and is often enshrouded in fog. Tree species similar to those found in some temperate forests grow here.

lower elevations in mountains at higher latitudes. On the mountains in the southwestern United States, trees grow as high as 3,650 m (12,000 ft); in the Olympic Mountains of Washington State, tree line is at 2,400 m (8,000 ft); in the Wrangell Mountains of southern Alaska, tree line is at about 1,000 m (3,280 ft).

The direction of the sun's radiation is also a factor in the distribution of vegetation on mountains. In the Northern Hemisphere, the sun's path tracks in the southern part of the sky. South-facing slopes, therefore, receive more direct sunlight and are considerably warmer than north-facing slopes.

These differences in temperature due to slope orientation increase with increasing latitude.

The pattern of biomes on mountains in tropical regions is distinctly different from the pattern in temperate regions, partly because tropical mountains experience comparatively little variation in temperature from season to season. The warm lower elevations of tropical mountains are generally covered with rain forest. At cooler and wetter elevations, rain forest is replaced by cloud forest, a region resembling the temperate biome (**Figure 10.34**). Even in the tropics, elevations above 4,000 m (13,200 ft) are too cold to permit tree growth; instead, the land is dominated by a tundra-like collection of grasses and herbs called **paramo**. Because the paramo is often above the clouds, plants that grow there must be adapted to frosty nighttime temperatures and sweltering daytime temperatures (**Figure 10.35**).

Ecosystems on mountain slopes are especially vulnerable to human activities. Logging and farming expose the already thin soils to high rates of erosion. Once trees are removed, the reforestation of steep slopes is especially difficult.

Global warming is having a direct effect on the distribution of mountain ecosystems. Long-term studies in the Andes of Colombia and Peru reveal that the ranges of many bird species have shifted to higher elevations over the past 50 years. Similar shifts in plant and animal species have been observed in the mountains of western North America.

For some species, these climatically driven shifts could result in the complete loss of habitat. In the Appalachian Mountains of Virginia, North Carolina, and Tennessee, spruce and fir trees grow at the very top of some peaks. Warmer conditions are allowing broad-leaved deciduous trees to grow at higher and higher elevations, encroaching on the range of the conifers. As a result, spruce, fir, and many of the species associated with them could disappear from these peaks (**Figure 10.36**).

▼ **Figure 10.36 Mountains and Global Warming**
As global warming causes climates to become warmer, ecosystems such as the deciduous forest on this mountain in Virginia are able to thrive at higher and higher elevations. Eventually, the deciduous forest may crowd out the coniferous forests growing on the peak.

Coastlines

Saltwater and wave action make the conditions in coastal environments especially harsh.

Coastal environments occur in the narrow transition between terrestrial biomes and the ocean. Salt-laden water droplets and the physical action of waves, currents, and storms make coastal environments especially challenging habitats. Because of the environmental stresses they share, the communities of organisms that live on coastlines around the world are remarkably similar. And because ocean currents can carry their seeds and fruits long distances, certain species of plants, such as coconuts and mangroves, grow on coastlines around the world (**Figure 10.37**).

The sea strand—the place where the ocean meets the land—is a high-salinity environment that is continually being reshaped by the rise and fall of tides and the action of waves and currents. Salt in sea spray is concentrated by evaporation and transported to land by wind off the ocean. This salt is then deposited on plants and soils near the coast, causing water to diffuse from plant tissues and limiting plant growth. Most plants are unable to grow in such a salty environment. But just a few minutes' walk away from the strand, and the salinity and availability of water change greatly. These gradients determine the distribution of grasses, shrubs, and trees.

Only a few annual plants can endure the harsh conditions near the water's edge (**Figure 10.38**). Once these annuals are established, they stabilize the sand and provide just enough protection from salt spray to allow perennial grasses, such as sea oats, to become established. When perennials gain a foothold, sand accumulates around their bases and sand dunes begin to form. As more and more perennials become established, the sand dunes grow larger and become more stable. Once stabilized, sand dunes protect plants from wind-borne salt. In fact, a diverse array of plants can grow on the leeward sides of sand dunes. Shrubs and even trees can grow in the most stable and protected coastal enviroments.

The animals that live in coastal environments are also affected by salinity and the availability of water. Terrestrial ghost crabs forage for small insects and small bits of vegetation amid coastal dunes. To survive in this salty environment, vertebrates such as rabbits and deer depend on the presence of small freshwater streams and ponds.

On some coastal islands there are large populations of domestic horses, pigs, and sheep; these animals were abandoned by sailing vessels over the centuries. Where these species are abundant, they have significantly altered the native vegetation.

Human developments in coastal areas, such as vacation homes and condominiums, have resulted in considerable loss of coastal environments. Human developments also destabilize sand dunes, rendering coastlines more vulnerable to major storms and hurricanes. The human impact on coastal environments is being accentuated by the rising sea levels associated with global warming (8.4).

► Figure 10.38 **Dune Pioneer and Stabilizer**
Sea rocket Ⓐ is an annual plant that can grow just above the high-tide mark along beaches. Its seeds are dispersed by ocean currents. Sea oat Ⓑ is a common grass on coastal dunes in eastern North America. Its fibrous roots stabilize the sand in the dunes against the forces of wind and moving water.

◄ Figure 10.37 **Coastline Vegetation**
Plants growing in coastal environments must be able to withstand the effects of salt carried in ocean spray as well as the physical effects of ocean waves.

Summary

Because of convergent evolution, places on Earth that have similar climates often support plants and animals with similar physical and physiological traits. These similar communities of organisms are called biomes. Most of Earth's biomes occur in one of three general climatic zones—the tropical zone, temperate zone, or polar zone. Climatogaphs show the yearly patterns of temperature and rainfall for specific biomes. Tropical biomes include tropical rain forest, tropical seasonal forest, and tropical savanna; these biomes experience warm temperatures throughout the year. Temperate zone biomes include temperate deciduous forest, temperate evergreen forest, chaparral, and temperate grassland; they are warm in the summer and generally cold during the winter. Boreal forest and tundra are polar biomes; they have short growing seasons with very cold winters. The desert biome occurs in all three climatic zones where there are significant, sustained water deficits. Mountains and coastlines are geographic features are not usually considered to be biomes. Nevertheless, they support unique communities of plants and animals that differ along gradients of elevation and salinity.

10.1 Terrestrial Biomes and Climate

- As a consequence of convergent evolution, organisms that live in similar environments tend to share similar features.
- Biomes are communities of similar terrestrial plants and animals that are adapted to a particular climate.

KEY TERMS

biome, convergent evolution, tropical zone, temperate zone, polar zone, continental climate, maritime climate, climatograph

QUESTIONS

1. Many biomes extend to higher latitudes in near-coastal locations compared to locations in the interior of continents. Why is this the case?

2. The climate–biome model reveals that every biome can exist at higher temperatures when the rainfall is also higher. Explain.

Key
1. Tundra
2. Taiga
3. Temperate forests
4. Tropical rain forest
5. Tropical seasonal forest
6. Chaparral/grassland
7. Desert
8. Tropical savanna

AVERAGE TEMPERATURE (° C)

AVERAGE PRECIPITATION (cm/year)

10.2 Tropical Biomes

- Tropical environments are distributed 25° north and south of the equator; tropical rain forest, tropical seasonal forest, and savanna biomes are found in this climate zone.

- Warm temperatures prevail year-round in tropical environments. Rainfall occurs year-round in tropical rain forest, while both tropical seasonal forests and tropical savannas experience seasonal drought.

KEY TERMS

tropical rain forest, endemism, tropical seasonal forest, tropical savanna

QUESTIONS

1. The diversity of tropical rain forest insect species is directly correlated with the diversity of trees and shrubs. Explain why.

2. The eastern slopes of mountains in Central America support tropical rain forest whereas the western slopes sustain tropical seasonal forest and savanna. What is the likely cause of this difference?

3. Without fire and grazing, tropical savanna would become woodland. Too much grazing by domestic animals has transformed some tropical savanna into desert. Explain both of these observations.

Emergents

Canopy

Midstory

Understory

10.3 Temperate Biomes

- Temperate environments extend north and south of the tropics between about 25° and 60° latitude.
- The distribution of temperate deciduous forest, temperate evergreen forest, chaparral, and temperate grassland biomes is determined by differences in winter temperatures and seasonal rainfall patterns.

KEY TERMS

temperate deciduous forest, temperate evergreen forest, temperate rain forest, conifers, chaparral, sclerophyllous, temperate grassland

QUESTIONS

1. Explain why soil nutrients and insect diversity change considerably throughout the year in temperate deciduous forest.

2. Why did the suppression of wildfires in temperate evergreen forests actually increase the risk of large, intense wildfires?

3. Where summers are hot, such as in Texas, grassland occurs where average annual rainfall exceeds 80 cm (32 in.). In cooler areas, such as Wisconsin, deciduous forest grows with this amount of rain. Why does one climate support grassland and the other support forest?

SANTA BARBARA, CALIFORNIA (LATITUDE = 34° N)

10.4 Polar Biomes

- Polar biomes of boreal forest and tundra are found above 60° in the Northern Hemisphere.
- In both the boreal forest and tundra, extreme cold and short growing seasons limit the abundance and diversity of life.

KEY TERMS

boreal forest, taiga, permafrost, tundra

QUESTIONS

1. Why is the total precipitation (rain and snow) lower in polar biomes than in most temperate and tropical biomes?

2. Even though precipitation is low, boreal forest and tundra are often wet and boglike. Why?

3. Describe three important effects of global warming on the boreal forest and tundra biomes.

10.5 Deserts

- The desert biome is characterized by arid conditions and widely varying temperatures.

KEY TERMS

desert biome, succulent

QUESTIONS

1. Why do deserts often occur adjacent to major mountain ranges?

2. Explain why deserts are expanding in many places.

10.6 Mountains and Coastlines

- Mountains and coastlines are not actual biomes; they are geographic features with unique communities of plants and animals that occur in all three climatic zones.

KEY TERMS

rain shadow, tree line, paramo

QUESTIONS

1. The eastern slopes of mountains in Central America support tropical rain forest whereas the western slopes sustain tropical seasonal forests and savanna. Explain why this is the case.

2. Describe three ways in which human activities affect coastal ecosystems.

Mastering**ENVIRONMENTALSCIENCE**

Go to **www.masteringenvironmentalscience.com** for homework assignments, practice quizzes, Pearson eText, and more.

289

11

Biodiversity Conservation

A Requiem for the Po'ouli

Why are Hawaii's unique bird species disappearing?

In 1973, students from the University of Hawaii discovered a new species of bird on the island of Maui. Biologists estimate that we have identified and named fewer than 10% of Earth's plants and animals. But because birds are often brightly colored and quite vocal, and because they are heavily studied by thousands of amateur bird-watchers and professional ornithologists, biologists think that the vast majority of Earth's birds have already been described. Discovering a new bird is a big deal!

The newly discovered bird was given the common name po'ouli. The po'ouli is a member of a subgroup of the finch family called honeycreepers (**Figure 11.1**). Honeycreepers are found on the Hawaiian Islands and nowhere else. Other honeycreepers have long bills and sip nectar from flowers. But the po'ouli has a short bill and eats only small snails that live in the moist tropical forests on the windward side of Maui.

Although this region of Maui had been heavily studied prior to 1973, the po'ouli had escaped detection for two reasons. First, it is cryptic—its coloring blends into its surroundings, and it has a very quiet call. Second, its population was very small. A census in the early 1980s located about 140 birds. Since then, the number of birds has diminished sharply. In 1994, only 6 birds could be located, and 2 years later only 3 were found.

In September 2004, a single po'ouli was captured in an effort to breed more individuals in captivity. Sadly, that bird died of malaria two months later. Despite intensive searches by dozens of professional and amateur birders, no po'ouli have since been found. It is likely that the bird is now extinct.

Unfortunately, the po'ouli's extinction is only one of a great many extinctions on these islands. Prior to human colonization, at least 140 unique bird species lived on the Hawaiian archipelago. When the Polynesians arrived about 800 years ago, many species, such as the flightless Hawaiian goose, were simply too easy prey for human hunters. Other species were displaced by animals that humans brought with them. Rats—those ever-present animals that accompany humans wherever they go—fed on eggs and nestlings, probably causing the disappearance of many species.

By the time the first European, Captain James Cook, set foot on the Hawaiian Islands, at least 43 bird species had already disappeared. Since Cook's landfall in 1778, an additional 18 species have disappeared. Habitat loss associated with urban and agricultural development caused some extinctions. In other cases, the introduction of yet more non-native predators, such as the mongoose, was responsible (**Figure 11.2**).

▼ **Figure 11.1 Alas, the Poor Po'ouli**
Discovered less than 40 years ago, the last po'ouli was seen in the wild in 2004.

Ornithologists believe the total number of bird species in Hawaii nowadays is approximately the same as it was 300 years ago, but many of these are species that have been introduced from elsewhere. Aggressive non-native birds, such as the common myna and the house finch, compete with native species, causing additional stress to struggling populations.

Today, only 71 of Hawaii's 140-plus unique bird species remain. Of those, 30 are listed as endangered under the United States Endangered Species Act. Fifteen of these are critically endangered, meaning that fewer than 500 individuals survive in the wild (**Figure 11.3**).

Conservation biologists estimate that in the absence of people, one bird species in the world would go extinct every 400–500 years. Yet over the past 800 years in Hawaii alone, bird species have gone extinct at the rate of one every 10–12 years. Worldwide, bird species are being lost at the rate of about one each year; many fear that rate could increase ten-fold by 2100.

Over the past several decades, efforts to reverse these trends in Hawaii have produced some success. The Hawaiian Endangered Bird Conservation Program is a unique partnership composed of the U.S. Fish and Wildlife Service, the State of Hawaii Division of Forestry and Wildlife, and the Zoological Society of San Diego. Since 1994, managers in this program have used a variety of strategies to save the native birds, including habitat protection, control of non-native species, and captive breeding. As a result, the populations of many endangered birds are beginning to increase (**Figure 11.4**).

The diversity of life is important for ecosystems and humans. How does this diversity vary around the globe and within specific regions? What factors are contributing to the extinction of species? Why are species that live on islands and in other isolated habitats especially vulnerable to such losses? What management and policy strategies can we employ to preserve Earth's diversity? These and other questions are addressed in this chapter.

▲ Figure 11.2 An Alien Threat
The Indian mongoose was introduced to Hawaii in 1872 in order to control rats. Because it also feeds on eggs and nestlings, it has had a devastating effect on populations of Hawaii's native birds.

◄ Figure 11.3 On the Brink
The akiapola'au is a honeycreeper found only on the big island of Hawaii. As a consequence of habitat loss and non-native predators, fewer than 500 individuals persist in the wild.

◄ Figure 11.4 Captive Breeding Success
In 1995, only about 20 puaiohi, a small thrushlike bird, survived on the island of Kauai. As part of the Hawaiian Endangered Bird Conservation Program, the San Diego Zoo has successfully hatched over 200 puaiohi, and nearly 150 have been reintroduced into the wild. The wild population is now growing.

11.1 What Is Biodiversity?

Biodiversity is the variety of life in all its forms and combinations and at all levels of organization. Biodiversity is important within ecological communities, such as forests and lakes, and at larger scales, such as entire regions and biomes (10.1). In communities, biodiversity is measured by the number of different species, their relative abundance, and their spatial distribution. Within populations of a single species, genetic variation among individuals is a measure of biodiversity. Environmental factors such as temperature, availability of water, and patterns of disturbance cause the numbers and kinds of species to differ from place to place.

Landscape Biodiversity

Environmental gradients and disturbances produce landscape biodiversity.

Landscape diversity refers to the differences in the variety and abundance of species from place to place. This diversity can be measured by collecting, or sampling, the species in particular locations and comparing those samples to other locations.

If you hike up a mountain, sampling the numbers and kinds of plants and animals as you go, you'll find that no two places are exactly alike. As you climb, new species appear and others disappear. If you compare the species living in places that are relatively close to each other—perhaps separated by an elevation of 100 m (328 ft)—you will find that they are quite similar. Over longer distances that similarity diminishes. Two thousand meters (6,560 ft) up the mountain, for example, the species in your sample will be totally different from those at the base.

If you hike up different sides of the same mountain, you will find similar patterns in the composition of species as the elevation increases. In the Appalachian Mountains of the southeastern United States, for example, the species of trees growing at low elevations include a diverse array of oaks and hickories. At higher elevations, beech, sugar maple, and buckeye are more prevalent; higher yet, spruce and fir trees predominate. Nevertheless, on each hike you would find significant differences in the composition of trees; on one hike you might find more pines and on another, more hemlocks.

After many such hikes, you could analyze your samples and classify them into communities composed of similar species. Ecologists often do this based on the prevalence of common trees and shrubs. You could then produce a map of the types of communities on your mountain. Such a map is a visual depiction of the mountain's landscape diversity (**Figure 11.5**).

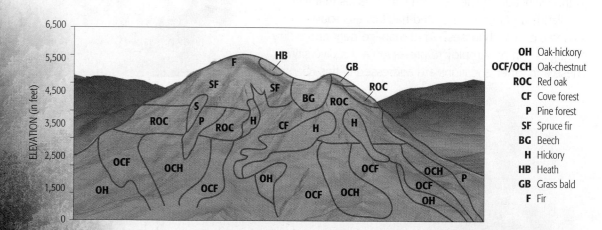

OH	Oak-hickory
OCF/OCH	Oak-chestnut
ROC	Red oak
CF	Cove forest
P	Pine forest
SF	Spruce fir
BG	Beech
H	Hickory
HB	Heath
GB	Grass bald
F	Fir

▲ **Figure 11.5 Landscape Diversity**
Ecologist Robert H. Whittaker sampled vegetation in the Great Smoky Mountains and then used his data to produce this generalized map of the forest communities on a typical mountain.
Source: Whittaker, R.H., 1956. *The Vegetation of the Great Smoky Mountains.* Ecological Monographs 26:1-80.

The pattern of landscape diversity is a consequence of two important factors—gradual changes in environmental factors and the history of local disturbances. Many environmental factors change from the base of a mountain to its peak. In general, as elevation increases, temperature and length of growing season decrease and precipitation increases (10.6). Temperature and length of growing season are also affected by the orientation of the mountain toward the sun, which causes differences in the amount of energy reaching north- and south-facing sides of the mountain. Other environmental factors include proximity to streams and variations in soil. For example, soils on high mountain ridges are often more acidic than those in low valleys. Each species has a specific range of tolerance for these environmental gradients; these tolerances determine its distribution on the mountain.

Unusual climatic conditions and types of soils often support species that cannot live in more typical environments. In the southern Appalachians, for example, some species of mosses and salamanders are restricted to the very moist conditions along streams; others live only on outcrops of particular rocks, such as granite or serpentine (**Figure 11.6**). Species that are found only in specific environments or restricted localities are said to be **endemic**. Landscapes with many unique species are said to have a high amount of endemism.

Disturbances such as fire and severe winds also affect landscape diversity. Disturbances create a mosaic

▲ **Figure 11.6 Endemic Salamander**
The moisture-rich forests of the southern Appalachian Mountains are an ideal home for Jordan's salamander, one of 20 salamander species found only in this location.

of patches undergoing different stages of succession (6.5). Plant and animal species vary from patch to patch, depending on the kind of disturbance and the length of time since the disturbance occurred (**Figure 11.7**).

Severe, crown-killing fire

Less severe, crown-scorching fire

Light fire, burned only in the forest understory

◀ **Figure 11.7 Disturbance Mosaic**
In 1988, fires burned over one million acres in Yellowstone National Park, creating a complex mosaic of patches. After the fire, the composition and abundance of plant species in these patches varied, depending on the severity of fire.

Community Biodiversity

The number of species, their relative abundance, and their arrangement in space are measures of community biodiversity.

At each location along your many mountain hikes, your samples of the flora and fauna provide measures of biodiversity. The simplest measure of biodiversity is **species richness**, the total number of species in each sample. Species richness varies greatly from location to location. In general, it is high in favorable environments, such as warm, moist climates and fertile soils along streams, and lower in less favorable locations, such as high-elevation ridgetops where growing seasons are short and soils are shallow. Fewer species are able to grow and reproduce in these less hospitable habitats.

Diversity can also be measured by **species evenness**, the relative abundance of the different species in a community. Suppose you collect butterflies from two different fields. Your samples might look like those in **Figure 11.8**. Since both samples have 12 butterflies

and four species, their species richness is the same. But because most of the butterflies in field A belong to the same species, the butterflies in field A seem less diverse than those in field B. In this example, field B has far greater species evenness than field A.

Among organisms such as trees, songbirds, and butterflies, species evenness is generally low. A few species are common, while the majority of species are represented by far fewer individuals. Within a deciduous forest, birds such as titmice and chickadees may be abundant, while most other species, such as tanagers and owls, are represented by many fewer individuals. At any given time, most of the species in a community are represented by relatively small numbers. Yet over the seasons and from year to year, there may be considerable variation as to which species are more abundant and which are less abundant.

▲ **Figure 11.8 Species Evenness**
The total number of butterflies and butterfly species is the same in these two samples, but species evenness is much lower in field Ⓐ than in field Ⓑ. Which sample appears more diverse to you?

Conservation biologists are interested in both species richness and species evenness. Loss of species richness indicates that some species are disappearing from an area and may indicate a threat to the survival of those species. Species evenness reflects a variety of interactions among species, so changes in this measure may signal important changes in the environment. For example, there is typically high species evenness among the algae in healthy streams. Nitrogen pollution often causes one or two species to grow more rapidly, crowding out the other species. In such cases, a decrease in species evenness indicates stream eutrophication (12.2).

Community biodiversity is also affected by **structural complexity**, the three-dimensional distribution of species and biological features. Structural complexity is relatively easy to visualize but difficult to quantify. The structure of an old-growth forest with its towering trees, shrub layer, snags, and logs is more complex than the single layer of vegetation in a grassland. The forest provides a much greater abundance of localized habitats such as tree cavities and rotting logs. As a consequence, it has more species than the single layer of grassland plants (**Figure 11.9**). Similarly, the structure of a coral reef, with its multiple layers extending upward for many meters, is far more complex than that of a flat, sandy ocean bottom (**Figure 11.10**). So it is not surprising that the reef is home to a great many more species than the sandy ocean bottom.

▲ **Figure 11.9 Complexity Causes Diversity**
A tropical forest Ⓐ has far more structural complexity than a grassland Ⓑ. It is easy to see how the complexity of the forest creates numerous habitats that support a great many species.

▲ **Figure 11.10 Marine Structural Complexity**
The structural complexity of coral reefs Ⓐ supports many more species than nearby sandy ocean bottoms Ⓑ.

Population Biodiversity

The diversity of populations is measured by genetic variation among individuals.

Within populations of individual species, biodiversity is measured by **genetic diversity**, the genetic variation among individuals. Genetic diversity within our own species is especially obvious in the variation of traits such as height, hair and skin color, and blood types. Genetic diversity in populations provides the variation upon which natural selection acts, allowing the evolution of new adaptations as environments change. Without genetic diversity, species could not compete in the different environmental conditions across their geographic ranges and through time (4.5).

In most populations of plants and animals, genetic variation is maintained by sexual reproduction. Sexual reproduction ensures that offspring are genetically distinct from one another and from their parents (4.2). Genetic diversity of populations is maximized by **outbreeding**, or mating between individuals that are not closely related. If the pool of potential mates is small, **inbreeding**, or mating between closely related individuals, is more likely.

Inbreeding reduces genetic diversity and increases the chance of genetic diseases and reproductive failures. In humans, for example, children whose parents are first cousins are more likely to suffer from genetic diseases such as hemophilia and cystic fibrosis. Maintaining genetic diversity is especially important in the conservation of threatened and endangered species. As species lose habitat, their populations decrease. The small remaining populations often lack the genetic variation needed for long-term evolutionary change. Smaller populations also increase the likelihood of inbreeding, and the adverse effects of inbreeding put these species at even greater risk. Where populations have become very small, conservation biologists may implement captive breeding programs to restore genetic diversity (**Figure 11.11**).

In popular publications, biodiversity is often equated to species richness. However, conservation of biodiversity must take into account the diversity of life at all levels of organization, including genetic diversity within species and the diversity of habitats and ecosystems across landscapes.

QUESTIONS

1. Describe two factors that influence landscape diversity.

2. What is the difference between species richness and species evenness?

3. Why does the genetic diversity of a population often decline when the populations become smaller?

▼ Figure 11.11 **Genetic Diversity Improves Survival**
Genetic variation within a species reduces the risk of the genetic disorders associated with inbreeding. These panther cubs were produced by crossbreeding endangered Florida panthers with mountain lions from Texas; as a result, these cubs have do not display the kinds of problems that commonly arise from inbreeding.

11.2 Why Biodiversity Matters

Biodiversity is valuable in many ways. A diverse array of species is required to maintain ecosystem functions and the services they provide to humans. Biodiversity helps ecosystems respond quickly to disturbances and environmental change. Biodiversity has great economic value to humans, who depend on many different species of plants and animals for food, fuel, shelter, clothing, medicines, and other purposes. Many people believe that Earth's species are valuable in their own right, independent of their value to humans. Therefore, the other creatures sharing our planet deserve our care and concern.

Ecosystem Functions and Services

Ecosystem functions and services depend on the actions of individual species and interactions among species.

How do different levels of species diversity affect ecosystem functions? For over 20 years, David Tilman and his colleagues at the University of Minnesota have been carrying out experiments to answer this question. In a prairie at the Cedar Creek Natural History Area, Tilman's team planted a number of experimental plots with different numbers of grass and herb species. In each plot, the team measured ecosystem processes such as net primary production, nutrient uptake, and water use (**Figure 11.12**). Their results showed that all of these processes were enhanced as species richness increased.

Similar studies carried out in a variety of ecosystems around the world have reached the same conclusion: Species richness increases ecosystem functions. Scientists do not fully understand *why* biodiversity enhances these functions, although the effects of sampling and complementarity appear to be important.

Species within a community vary with regard to their importance to particular ecosystem processes, such as net primary production. As biodiversity increases, there is a higher likelihood that an ecosystem will include a species that has a big effect on a process. This increased likelihood is known as the **sampling effect**.

Species within a community usually use resources in a complementary fashion. For example, the roots of various prairie herbs and grasses reach different parts of the soil, resulting in a fuller use of soil water and nutrients (6.1). Because individual species exploit different parts of the environment, groups of species are able to use resources more efficiently than would any single species. This is termed the **complementarity effect**.

Biodiversity affects many ecosystem services upon which humans depend. For example, water flowing through wetland ecosystems is purified of chemical wastes. Wetlands also moderate the flow of water, thereby minimizing floods. Over half of the wetlands in the United States have been destroyed or degraded, resulting in the loss of these services. The destruction of wetlands along the Gulf Coast is thought to have contributed to the devastating floods associated with Hurricanes Katrina and Rita in 2005. Where wetland ecosystems are being restored, these services are enhanced by increasing the amount of biodiversity.

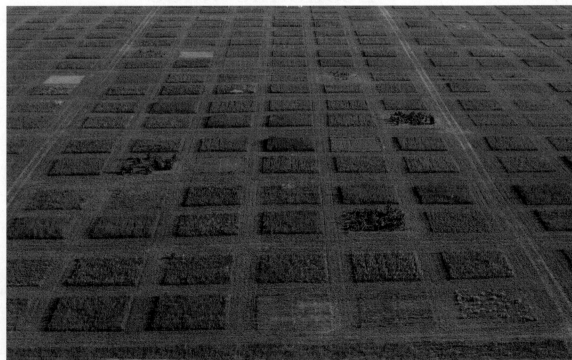

▲ Figure 11.12 **Diversity Increases Productivity**
David Tilman and his colleagues set up these experimental plots at the Cedar Creek Natural History Area in Minnesota. Note the differences in the growth and greenness of the plants in the different plots. The greenest and most productive plots also have the highest species richness.

What is the worst case scenario from losing one species of herbivore, for example?

Breeann Sharma, Arizona State University

Norm: *The loss of a single herbivore species can result in the loss of carnivore species that depend on it. If it was once common, its loss can also influence the diversity and abundance of various plant species.*

Ecosystem Stability

Biodiversity enables ecosystems to resist or recover from environmental changes and disturbances.

Biodiversity in a community buffers the effects of environmental change. During their long-term studies, Tilman's group noted considerable year-to-year variation in the annual rainfall in Minnesota. In general, lower rainfall produced lower net primary production, but greater species diversity lessened this effect (**Figure 11.13**).

Why were the more diverse communities more resistant to drought? Diverse ecosystems include species with different tolerances to the availability of water. Some plants grow better in wet years, and others grow better in dry years. From year to year, the abundance of the various species changes, but primary production remains relatively constant. However, the loss of just one or a few species from an ecosystem can diminish this buffering capacity.

Biodiversity also plays a significant role in the ability of ecological communities to recover from disturbances (6.5). Because they are widely dispersed and able to tolerate harsh open habitats, pioneer species such as mosses and lichens are able to colonize disturbed areas. In so

▲ **Figure 11.13 Diversity Increases Resistance to Drought** In Minnesota prairies, drought diminished net primary production (NPP) less in experimental plots where the species richness of plants was high. Drought resistance was measured here as the relative decline in NPP in dry versus wet years.
Data from: Tilman, D. and J.A. Downing, 1994. Biodiversity and stability in grasslands. *Nature* 367: 363–365.

doing, they stabilize and enrich soils, providing habitat for the establishment of other species. The total process of succession from bare rock to a forest involves many more species than are present at any single stage in that process.

Economic Value

Biodiversity provides goods and services that are essential to humans.

▼ **Figure 11.14 Ancient Hunters Depended on Biodiversity** These cave paintings in Lascaux, France, date from 15,000 to 20,000 years ago. They depict the diversity of animals upon which the painters depended for food and clothing.

Throughout human history, a great many species of plants and animals have been traded in economic markets (2.3). In traditional hunting and gathering cultures, wild plants and animals are the source of food and clothing (**Figure 11.14**).

In modern societies, biodiversity is harvested in activities such as lumbering, fishing, hunting and trapping for meat and furs, and collecting exotic animals and

tropical fish for sale as pets. Rare plants are dug for sale to gardeners or collected to make medicines. These direct uses of biodiversity are economically significant. For example, over 100 million metric tons of wild fish are captured each year, providing roughly 20% of the animal protein consumed by humans around the world. Forest trees provide wood for construction and fiber for paper. The many species within forests produce wild nuts and fruits, mushrooms, and understory plants that provide a continuous source of income for people living in forested regions.

Recreational hunting and fishing contribute significantly to the economies of rural communities. Money spent on equipment, fishing lodges, and hunting preserves multiplies the economic value of fish and game as much as 10-fold or even 100-fold. Every year, for example, the Platte River in Nebraska attracts 176,000 people who come to hunt game birds and 247,000 people who come to fish. An EPA-commissioned study estimates that the combined input of these outdoor enthusiasts contributes about $50 million to Nebraska's annual economy. The economic value of the various game species produces social pressure to save habitats in the Platte River valley, which otherwise might be lost to agriculture or development.

Biodiversity is an important attraction for many tourists, generating hundreds of billions of dollars of economic value annually. For example, Costa Rica derives more than 7% of its total GDP from **ecotourism**, in which visitors come to see the natural beauty and biodiversity of its extensive tropical forests (**Figure 11.15**). After the establishment of Costa Rica's national park system, visits by international tourists rose from 60,000 in 1983 to nearly 2 million in 2010.

Biodiversity-related tourism enhances the economies of industrialized nations, too. A recent study estimated that there are 46 million bird-watchers in the United States. These birders spend a total of $85 billion per year on equipment and trips, creating over 800,000 U.S. jobs.

Biodiversity is also of great value to biotechnology and medicine. About 20% of the pharmaceuticals distributed in the United States come from plants (**Figure 11.16**). Over half of all new drugs developed in the past 20 years come from chemical compounds derived from plants, animals, or microorganisms collected from ecosystems such as tropical forests, deserts, and coral reefs. In the year 2000, global sales of medicines developed from marine animals surpassed the $100 billion mark. A single product made from chemicals found in a sea sponge and used in treating viral infections is generating millions of dollars in profits each year. These examples suggest that the vast array of species that have not been studied represent an enormous pool of potential economic value.

Without the wealth of species found in the world's many natural ecosystems, biotechnology would have fewer chemical tools to use and limited sources of genetic information for research. Because biodiversity is the fundamental source of biological information, preserving it will further biotechnology and provide new treatments for human diseases.

Existence Value

Most people consider human-caused extinction of species to be unethical.

Many people believe that plants and animals have an intrinsic value, regardless of their importance to human beings. Therefore other living things deserve respect and protection. Most of the world's ethical and religious traditions recognize a moral obligation to protect other species. The idea of biodiversity's intrinsic value is also well established in the history and philosophy of science. This ethical belief is captured in the United Nations Charter for Nature, which states, "every form of life is unique, warranting respect regardless of its worth to man, and, to accord other organisms such recognition, man must be guided by a moral code of action."

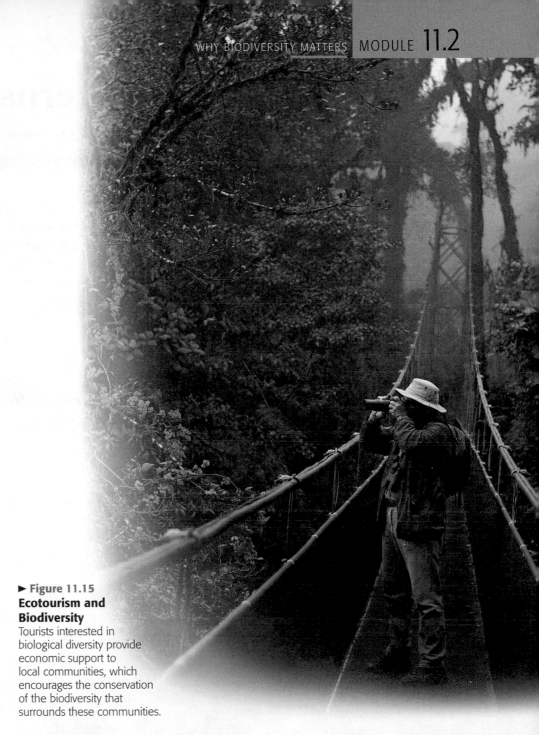

▶ **Figure 11.15**
Ecotourism and Biodiversity
Tourists interested in biological diversity provide economic support to local communities, which encourages the conservation of the biodiversity that surrounds these communities.

▶ **Figure 11.16**
Medicinal Plants
These are just a few examples of the many plants that provide drugs that improve human well-being.

Plant Name	Medicine	Treatment Use
Foxglove (*Digitalis purpurea*)	Digitalin and digitoxin	Heart failure
Goat's rue (*Galega officinalis*)	Metformin	Diabetes
May apple (*Podophyllum peltatum*)	Teniposide	Childhood leukemia
Octea (*Ocotea glaziovii*)	Glasiovine	Depression
Pacific yew (*Taxus brevifolia*)	Taxol	Cancer
Quinine tree (*Cinchona pubescens*)	Quinine	Malaria prevention
Rosy periwinkle (*Catharanthus roseus*)	Vincristine	Childhood leukemia
Sweet wormwood (*Artemisia annua*)	Artemisin	Malaria treatment
Velvet bean (*Mucuna pruriens*)	L-Dopa	Parkinson's disease
Willow (*Salix* spp.)	Aspirin (salicylic acid)	Pain, inflammation

QUESTIONS

1. In what ways does increased biodiversity increase ecosystem functions?

2. Diverse ecosystems are generally more stable than less diverse ecosystems in the context of year-to-year variations in climate. Why is this so?

11.3 Global Patterns of Biodiversity

Biologists estimate that Earth supports between 7 and 15 million species of eukaryotic organisms. This excludes species of bacteria, which may number in the millions. Our knowledge of this biodiversity is sadly incomplete. Only about 1.5 million species of plants and animals—10–20% of the total—have been described and assigned a scientific name. The vast majority of species live on land. Of these terrestrial species, over 80% are animals, most of which are insects. In the oceans, almost half of the species are algae. In general, species diversity is greatest near the equator and diminishes toward the poles. On land, tropical rain forest is the most species-rich biome; in the ocean, coral reefs are the ecosystem with the greatest diversity. Across the globe, conservation biologists have identified about 25 areas that possess high biodiversity, have many endemic species, and are particularly threatened by human activities.

Species Richness

The number of species in biomes decreases from the equator toward the poles.

Biological explorers who sailed around the world seeking new and exotic forms of life-forms noticed that species richness was greatest in the tropics and declined as they traveled north or south toward the poles (**Figure 11.17**). Tropical forests are the most species-rich communities on land, and tropical coral reefs are the most species-rich communities in the ocean. Forests, prairies, and ocean ecosystems in the temperate zone have intermediate species diversity. Communities in polar regions have low numbers of species. Within a particular region, species diversity is also affected by rainfall and elevation, but the underlying global pattern is clear.

Biologists have proposed that the global pattern in species richness can be attributed to differences in net

▼ Figure 11.17 **Global Species Diversity**
The richness of plant species in ecosystems (number per 10,000 km² or 3,860 mi²) varies widely from region to region. The patterns of variation in species richness are similar for other organisms, such as vertebrates and insects. How does diversity vary among places you have visited?
Source: Bartblott, W. et al. 1999. Terminological and methodological aspects of the mapping and analysis of global biodiversity. *Acta Botanica Fennica* 162: 103–110.

primary production, past disturbances, habitat gradients, and ecosystem complexity. There is spirited debate among biologists regarding the relative importance of these various factors. No factor alone provides a sufficient explanation. It is likely that all of these factors contribute to the variation in diversity from the tropics to temperate and polar regions.

Net primary production.
The warm weather and long growing seasons in equatorial ecosystems promote higher net primary production rates than those in temperate or polar ecosystems. Higher net primary production means that more energy is captured in photosynthesis and consumed by the entire ecosystem. This greater availability of energy has the potential of supporting more species at each successive trophic level of the food web (6.4).

Pleistocene climate change.
Ecologists once believed that tropical regions had high biodiversity because their climates had been relatively stable for hundreds of thousands or even millions of years. Thus, tropical biomes were spared the glaciers and shifting climates that probably wiped out many species in Arctic and temperate regions. However, recent evidence suggests that during the ice age, climate change in tropical regions, such as the Amazon, was much more significant than previously thought. Today, scientists are much less certain about how changes in climate have affected differences in species diversity among biomes.

Habitat gradients.
In temperate regions, the weather changes significantly through the course of the year, favoring the evolution of organisms that can tolerate a broad range of temperatures. As a result, temperate species are often widely distributed. For example, many species are able to live at a wide range of elevations on mountain slopes. In the tropics, the lack of seasonal temperature changes causes sharper distinctions among habitats, especially in mountainous areas. The tops of a tropical mountain are always cool, mid-elevations are always moderate, and lower elevations are always warm. Some biologists contend that the uniform conditions within elevation zones favor the evolution of species with narrow ranges of tolerance. As a consequence, tropical species have comparatively specialized ecological niches with respect to elevation. This allows species that are potential competitors to coexist along gradients of elevation.

Ecosystem complexity.
Tropical regions favor the evolution of organisms with structures that increase habitat diversity. The multiple layers of a tropical forest provide unique habitats for a diverse array of vines and epiphytes (plants that grow on other plants). These plants, in turn, provide habitats for a diverse array of animals. In the oceans, the complex structures of coral reefs provide distinctive habitats for many different species. Some biologists note that the diverse array of herbivores and predators in tropical ecosystems promotes diversity among the species upon which they feed (6.4).

Species per 10,000 square km²
- Fewer than 100
- 100–500
- 500–2,000
- 2,000–4,000
- More than 4,000

QUESTIONS

1. What factors likely contribute to the differences in species richness between ecosystems at the equator and those in polar regions?

2. What two features define a biodiversity hotspot?

Biodiversity Hotspots

The high biodiversity of certain places on Earth is especially threatened by human activities.

Certain places have an unusually large number of endemic species. Conservation biologists define such regions as **biodiversity hotspots** if the human threats to their habitats are especially high. Conservation International, an organization that uses biodiversity hotspots to prioritize its activities, classifies a region as a hotspot if it contains at least 1,500 species of endemic plants and at least 70% of its original area has been severely altered by human activities.

Using Conservation International's criteria, 25 biodiversity hotspots have been identified (**Figure 11.18**). These hotspots represent less than 12% of Earth's total area, but are estimated to contain at least 44% of all plant species and 35% of all vertebrate animal species. As of 2009, over 90% of the habitat in these regions had been destroyed or severely altered.

What makes a hotspot hot? In most hotspots, three factors have allowed the evolution of many endemic species—unique habitats, topographic diversity, and isolation. Two regions with Mediterranean-type climates, California and South Africa, provide excellent examples of the role of these factors (10.3). Both regions have complicated geologic histories that have produced a diverse array of unique soils. Natural selection has favored the evolution of a great many endemic species that are able to grow in these soils. Both regions also have several mountain ranges. The various elevations and differences between moist and dry slopes produce a

great diversity of mountain habitats. These unique habitats and the topography of the mountains isolate populations from one another. This geographic isolation limits interbreeding among populations within the hotspot, allowing new species to evolve (**Figure 11.19;** 4.6).

▲ **Figure 11.19 Hotspots Indeed**
In South Africa, shrubby ecosystems called fynbos include a great many endemic species, such as the red aloe in the center of this image. This picture was taken from atop Table Mountain near Cape Town.

▼ **Figure 11.18 Biodiversity Hotspots**
These 25 areas are designated as hotspots because of the high number of endemic species they contain and the large amount of habitat that has been lost from them.

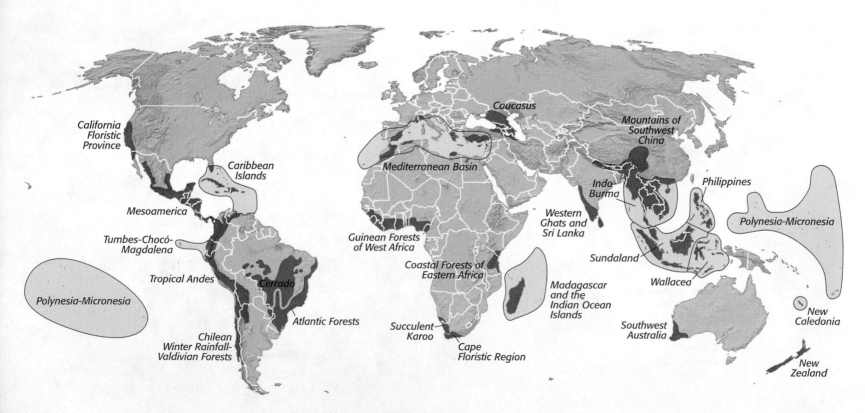

11.4 Differences in Biodiversity Among Communities

Biodiversity varies significantly among ecological communities. It is enhanced by differences in topography, geology, and ecosystem structure that create a variety of unique habitats. Interactions among species within communities, such as competition and predation, also influence biodiversity. Both natural and human disturbances can affect species diversity. The number of species in a community in an area is determined by the equilibrium between the rate at which new species migrate into the area and the rate at which species disappear from that area. Migration rates are influenced most by the degree to which areas are isolated from sources of new migrants. Extinction rates are influenced most by the size of islands or habitat patches.

Habitat Diversity

Topographic and geologic variation fosters diversity.

Within any region, environments with greater diversity of habitats support more species than environments with more uniform habitats. Landscape features such as mountains, valleys, lakes, and rivers create many kinds of habitats. This complexity of habitats increases the number of potential ecological niches, which in turn support a greater diversity of species. Coastal areas that are broken up into chains of islands tend to have more species than coastlines without islands. In marine ecosystems, areas of the ocean with rocks or reefs on the bottom support more species than areas with flat, featureless seabeds. Building an artificial reef on a flat seabed provides places to which organisms such as corals and algae can attach themselves, thereby increasing the number of species in the ecosystem (**Figure 11.20**).

Habitat complexity also affects ecosystems on a smaller scale. A natural stream with its riffles and deep pools, boulders, sunken logs, and sandbars can support a species-rich community of aquatic life. If dredging equipment is used to straighten out the stream and give it uniform depth, the resulting canal has fewer habitats and species richness declines significantly (12.2).

▼ Figure 11.20 Old Shipwrecks Support Marine Ocean Diversity
This old ship was sunk off the coast of Florida in order to create a complex habitat that will support many marine species.

Species Interactions

Competition and predation influence biodiversity.

In most ecological communities, many species compete for common resources, such as food, water, or nesting sites. Competitors are able to coexist because each species occupies a different ecological niche; that is, each species has a different range of tolerance for environmental factors or acquires its resources in a slightly different way (6.2). Changes in the environment, such as an increase in an essential nutrient, can alter the competitive relationships among species. This, in turn, can cause one or a few species to dominate the others, resulting in a loss of biodiversity. For example, nitrogen pollution in streams often results in the rapid growth of one or two species of algae. Other species of algae, which are no longer able to compete for light or other resources, decrease in number or disappear entirely.

Predators usually increase the biodiversity of an ecosystem. This fact may seem counterintuitive, but makes sense in the context of competition among the prey species. Predators usually feed most heavily on the species of prey that are most common and require the least energy to capture. As the numbers of a particular prey species decrease, resources become available for other species, which increase in number. Predators then focus their attention on the species that have become more common. Consequently, no individual prey species is able to become dominant (**Figure 11.21**). That is why the removal of top predators, such as wolves or large predatory fish, can have devastating effects on the diversity of the prey species upon which they feed (6.2).

Herbivorous animals appear to play an especially important role in the diversity of tropical rain forest trees. Ecologists have long been astounded at the diversity of tree species in these ecosystems, even within very small areas. Even an area as small as a single hectare (2.5 acres) may contain over 100 different tree species.

▲ **Figure 11.21 Predation Increases Diversity**
The diversity of savanna ungulates, such as these zebras and wildebeests, is partly the result of predation by carnivores, such as lions and leopards. No single prey species is able to become dominant enough to exclude its competitors.

Many different animals feed on the seeds and seedlings of these trees, and many of these herbivores feed only on a specific species of tree.

Experimental studies have demonstrated that the number of herbivores for any one particular tree species is highest in the vicinity of that species' largest trees. As a consequence, the probability that seeds from that species will successfully germinate and grow is lowest near adult trees of that species (**Figure 11.22**). Other tree species are more likely to germinate and grow there. When extended throughout a forest, this interaction between trees and their herbivores results in far greater diversity than would be found if the seedlings of each tree species grew best near their parent trees (Figure 11.22).

◄ **Figure 11.22 Herbivory Increases Tropical Tree Diversity**
Although the seeds of any particular tree species are more likely to fall near their parent tree, predators that feed on those seeds are also likely to be more abundant there. As a consequence, seedling survival increases with distance from the parent tree, and trees of the same species are much less likely to grow near one another.

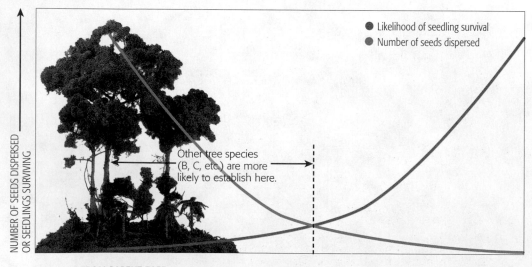

- Likelihood of seedling survival
- Number of seeds dispersed

Other tree species (B, C, etc.) are more likely to establish here.

NUMBER OF SEEDS DISPERSED OR SEEDLINGS SURVIVING

DISTANCE FROM PARENT TREE

Disturbance

Biodiversity is highest with intermediate levels of disturbance.

Natural disturbances such as fires or hurricanes occur in most ecosystems. Such disturbances initiate a succession of change in the abundance of different species (6.5). Certain kinds of disturbances also increase the diversity within individual communities. For example, frequent fires have been shown to maintain the high diversity of herbs and grasses in prairies and savannas (**Figure 11.23**). In these ecosystems, fire acts like an herbivore, preventing any particular herb or grass species from dominating competing species. Ecologists have shown that biodiversity in grasslands, savannas, and shrublands is highest at intermediate levels of disturbance. If fires are too infrequent, few species flourish. However, many species cannot tolerate fires that are very hot or too frequent, and such conditions also diminish biodiversity.

▲ **Figure 11.23**
Disturbance and Diversity
These summer wildflowers are growing in a recently burned Wisconsin prairie. Like an herbivore, fire limits competitive exclusion and helps increase diversity.

Local Immigration and Extinction Rates

The species richness of communities on islands is determined by the rates at which species arrive and disappear.

It stands to reason that the number of species in a particular place—say birds in a patch of forest or fish around a shipwreck—would be influenced by the rates at which new species appear and others disappear from that place. The rate of appearance is determined by the rate at which species immigrate, or move into an area. The rate of disappearance is determined by the rates at which populations of species in the area go extinct.

The importance of immigration and extinction was first clearly articulated by biologists R.H. MacArthur and E.O. Wilson in a theory to explain the variation in the number of species found on different ocean islands. Their **equilibrium theory of island biogeography** states that the diversity of species on an island is largely determined by two factors: the rate at which new species migrate to the island and the rate at which species disappear from the island, or become extinct.

MacArthur and Wilson hypothesized that distance from the mainland is the single most important factor influencing the rate at which species migrate to an island. Most plants and animals come to an island from the mainland, and it is easier to reach nearby islands than those that are far away. Therefore, islands close to the mainland have higher rates of immigration than more distant islands.

They also hypothesized that the rate of extinction on an island is most influenced by island size. Small islands support smaller average population sizes than large islands. As populations become smaller, they are more prone to extinction.

MacArthur and Wilson argued that the diversity of species on an island reaches an equilibrium number when the rate of new species immigration is equal to the rate of local extinction. At the equilibrium number,

individual species may come and go, but the total number of species remains relatively constant. Islands closer to the mainland have higher immigration rates. The smaller an island, the smaller the size of its plant and animal populations, and the higher their rate of extinction. Thus, the theory predicts that large islands close to the mainland will have more species than small, distant islands (**Figure 11.24**).

The equilibrium theory of island biogeography has become a fundamental principle in conservation biology.

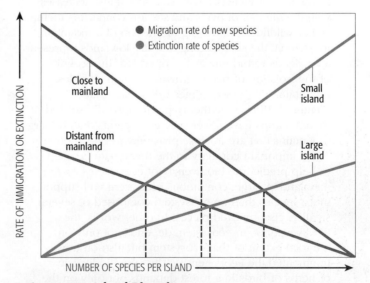

▲ **Figure 11.24 Island Diversity**
The number of species on an island is determined by the balance between its rates of immigration and extinction. The equilibrium number for each island occurs where the two lines cross. How does the equilibrium number for a large island close to the mainland compare with that of a small island that is distant from the mainland?

▲ Figure 11.25 Sky Islands
The peaks of the Santa Catalina Mountains near Tucson, Arizona, provide islands of unique habitat separated from one another by a sea of desert.

The importance of island size and distance to mainland to plant and animal biodiversity has been verified in clusters of islands around the world. On the hundreds of islands of Micronesia in the South Pacific, for example, the diversity of birds, mammals, and plants is highest on the largest islands and on islands closest to mainland sources of species.

The principles of this theory can be applied to other kinds of isolated habitat patches. For fish species, a pond is like an island of livable habitat surrounded by inhospitable dry land. As the theory predicts, the number of fish species living in lakes increases with lake size. It also increases with the number of rivers and streams connecting them.

For wildflowers growing on the top of a mountain in Arizona, the surrounding hot, dry lowlands represent a barrier as forbidding as the open sea. In recognition of the isolation of plants growing at high altitudes, mountaintops are sometimes called "sky islands" (**Figure 11.25**). As the theory predicts, small, isolated mountaintops have fewer species of plants than do larger mountains that are in close proximity to one another.

An important corollary to the theory can be used to help predict the biodiversity of ecosystems on the mainland. A large, continuous ecosystem will support more species than an ecosystem composed of several smaller, disconnected pieces. In other words, the biodiversity of an ecosystem depends not only on the total extent of that ecosystem but also on how fragmented the ecosystem is. Thus the biodiversity of herbs or birds in a forest depends not only on the total area of forest but also on how well the individual forest parcels are connected to one another. Large, interconnected forests have greater biodiversity than forests that are cut into disconnected patches by farms, roads, or housing developments (**Figure 11.26**).

▲ Figure 11.26 Fragmented Forests
Forest once covered this entire landscape. Agriculture, roads, and development now divide it into isolated patches and limit the migration of species between patches. Populations in small patches are more vulnerable to local extinction.

11.5 Threats to Biodiversity

Human activities are causing the global rate of extinction to increase at an alarming rate. Among those activities, the destruction of habitat is most important. Even where the amount of habitat remains high, development and land use are dividing it into smaller and more isolated fragments. As human populations have grown, they have exploited the populations of many plants and animals to extinction or near extinction. Humans have relocated species around the globe, introducing an array of invasive non-native competitors, predators, and diseases to every biome. Air and water pollution, altered patterns of disturbance, and global climate change present other threats to Earth's biodiversity.

Habitat Loss

Worldwide, the single greatest threat to biodiversity is habitat loss.

The foremost need of every species is abundant habitat in which to grow and reproduce. However, human activities have greatly reduced the extent of many ecosystems and the unique habitats associated with them. In 2011, the International Union for the Conservation of Nature (IUCN) "red listed" 562 of the world's bird species as critically endangered or endangered. Of these species, 82% were listed because of habitat loss.

The extent of habitat loss is particularly alarming in a number of ecosystems. By 1900, over 95% of North America's deciduous forests had been heavily logged or cleared for agriculture. Agriculture has taken an even higher percentage of North American grasslands. Only 3% of North America's mixed and tall-grass prairies remain, and most of these grow in small, isolated patches. Two hundred years ago, longleaf pine savannas covered millions of hectares in the southeastern coastal plain of the United States. Only 1% of this ecosystem is left today. Longleaf pine savannas provided habitat for very diverse endemic flora and fauna. Many of these species are now listed as endangered under the Endangered Species Act (**Figure 11.27**).

The dwindling tropical forests are of special concern because they contain a disproportionate number of the world's species. These forests were originally vast, covering more than 5 million square miles of Earth's tropical regions. Today about half of these forests are already gone, and the clearing of tropical forests continues at a rate of one-half million square miles every decade. The effects on biodiversity are sadly predictable. Biologists estimate that the current global extinction rate in tropical forests is 1,000 species per year.

RED-COCKADED WOODPECKER   GOPHER TORTOISE

▶ **Figure 11.27 A Vanishing Ecosystem**
Longleaf pine forests provide important habitat for many species. The loss of this habitat threatens the survival of the endangered red-cockaded woodpecker and gopher tortoise.

And this rate is likely to increase as survivors in small patches of forest suffer the inevitable fate of isolated populations whose numbers become too small to sustain themselves.

Marine ecosystems have also suffered significant habitat loss. Coastal development and boating have destroyed many of the extensive sea grass beds once found in shallow coastal waters. These beds provided habitat for a variety of shellfish, and their grasses were an important source of food for endangered sea turtles, manatees, and dugongs (**Figure 11.28**). Twenty percent of Earth's coral reefs, the most diverse of marine habitats, have been destroyed, and another 20% have been severely damaged. In some places the picture is much bleaker. In the Philippines, for example, only 10% of the coral reefs remain intact.

◀ **Figure 11.28 Sea Grass Beds**
Sea grass beds in shallow tropical waters are a critical resource for a great many animal species, including the endangered dugong.

Habitat Fragmentation

Human development and land use is dividing landscapes into ever smaller and more disconnected habitat islands.

I understand that there are more and more species going extinct. Is this mostly due to humans or is it just natural selection?

Jaymee Castillo, University of California, Los Angeles

Norm: *It is mostly due to humans. The current rate of species loss is more than 100 times greater than in pre-human times.*

As human activities expand across the globe, large areas of natural habitat are carved up into smaller fragments, which are increasingly separated from one another. Vast landscapes of adjoining habitat have been transformed into disconnected patches, or "islands," of habitat. In many places, urban development and road construction further isolate these patches from one another. Populations within habitat patches are smaller and more vulnerable to local extinction, and recolonization of patches is increasingly difficult. Habitat fragmentation is a major factor in the decline in the diversity of forest songbirds in the eastern United States (**Figure 11.29**).

When suitable habitat is divided into small patches, the life histories of many organisms are disrupted. Tree seeds, for example, tend to blow into adjacent land that is unsuited for seedling survival. The food and shelter needed by animals may be separated. Small animals tend to wander out of their preferred habitat and die before they find their way back. Larger animals may need breeding territories that are larger than the fragment of habitat they occupy. Migration pathways may be blocked by roads or highways, and the sound of traffic often disturbs or confuses wildlife.

Habitat fragmentation has also played a significant role in the loss of species in some aquatic ecosystems. For example, thousands of dams divide populations of fish that once lived in free-flowing streams. Dams also create insurmountable barriers for shad, salmon, and other fishes that migrate up and down streams as part of their reproductive cycles.

▼ **Figure 11.29 Forest Songbird Decline**
Based on censuses carried out for over a century, the number of songbird species, such as this scarlet tanager, is declining in forests across the eastern United States. Not only is there less forest habitat, but that habitat is also becoming much more fragmented.

Overharvesting

Species with large bodies, slow population growth rates, and flocking behavior are particularly vulnerable to human exploitation.

The human activities of hunting, fishing, trapping, and collecting endanger many plants and animals. The species in greatest danger are those that humans value as sources of food or medicine and those that are prized by collectors, such as exotic tropical birds and trophy animals.

There are many historical examples of overexploitation of natural populations. Whales once dominated the seas of the world, but several species have been driven near to extinction by fleets of whaling ships seeking their oil and meat. In the 1800s, passenger pigeons were so common that their migrating flocks darkened the sky for days as they passed over American frontier towns. These birds were hunted to extinction because they were easy to shoot, pack in salt, and ship to Europe where they were prized as an exotic meal. Ginseng plants have become rare in the Appalachians because poachers dig and sell the knobby ginseng root, which is especially valued as an herbal medicine in Asia.

The physical and behavioral characteristics of some species make them particularly vulnerable to overharvesting. Large animals are generally more vulnerable to human hunting. The populations of large animals typically have slow growth rates, so they are slow to recover if they are diminished. Animals that feed at higher trophic levels tend to have smaller populations and are also more vulnerable to harvesting pressure. Animals that live in herds, flocks, or schools are easier to capture and kill in large numbers than are solitary animals. Certainly, any trait that makes an animal economically desirable puts it at greater risk of overharvest.

The Nassau grouper, which is a highly valued item on seafood menus, unfortunately exhibits all of these traits. As a result, it has been fished nearly to extinction throughout most of its range in the Atlantic and Caribbean oceans (**Figure 11.30**). The Nassau grouper is a large predator at the top of the food chain of coral reef ecosystems. It takes up to 7 years for this fish to reach sexual maturity and decades for it to reach its maximum size.

Nassau groupers are solitary hunters, a feature that might make them difficult to exploit, except for one point in their reproductive cycle. Every winter month on the night of the full moon, groupers from hundreds of miles around gather at a common spawning site. These assemblies occur in the same locations year after year. From long experience, people fishing for grouper know exactly where and when to drop nets and lines to bring in boatloads of fish. The tragic impact of this fishing method is magnified because many of the spawning groupers are caught before they are able to release their eggs. As a result, the spawning ground can be fished out completely, and when it is, no groupers ever return to that site. Only strictly enforced fishing laws limiting the grouper season to nonbreeding months or forbidding all grouper fishing in some marine preserves will be sufficient to save this valuable fish from complete extinction.

▲ **Figure 11.30 Safety in Numbers?**
The Nassau grouper is a solitary reef fish for most if its life (A), but it is easily captured when hundreds of individuals gather in schools to spawn in the waters near some Caribbean islands (B).

Non-Native Invasive Species

Newly introduced species may be invasive because of their own biological traits or because of the absence of predators or competitors.

Through commerce and travel, humans have redistributed tens of thousands of Earth's species. Some of this redistribution has been intentional. For example, the plants on our farms and in our gardens come from locations around the world. Many other introductions have been unintentional, as when wood-boring beetles arrive in shipments of wood from other continents. But however they have come, non-native species now live in nearly every ecosystem on Earth. In some places they are more abundant than their native counterparts. In all too many ecosystems, the introduction of non-native species has had disastrous consequences for individual species and for biodiversity overall.

When introduced to a new place, most species have difficulty competing with native plants and animals, which are already well adapted to their environment. However, some introduced species have biological traits or behaviors that allow them to overcome such competition and flourish. These traits make them more likely to become invasive. These invasive traits give an introduced species a lopsided advantage in survival and reproduction, allowing it to outcompete native species. As the invader consumes most of the available resources, native species are driven to extinction.

Garlic mustard, for example, is a small biennial plant introduced from Europe that is displacing native wildflowers in woodland communities throughout the eastern United States. The secret to this invader's success is natural herbicides secreted by the plant's roots that change the microbial community in the soil and poison neighboring plants, allowing garlic mustard to take over the forest floor.

In other cases, the lack of competitors allows an introduced species to become invasive. For example, the brown tree snake has had a devastating effect on biodiversity, especially on tropical islands of the Pacific (**Figure 11.31**). Native to Indonesia, this active snake grows up to 8 feet long. It hunts at night in the branches of trees and is a deadly predator on birds and small animals. Brown tree snakes were accidentally introduced to Guam, the northernmost island in the Hawaiian archipelago, during World War II. Guam has no native snakes that

might compete with the brown snake. Furthermore, birds on the island had never developed behavioral defenses against this kind of predator. As a result, brown tree snakes devastated bird populations, driving 9 of Guam's 11 forest-dwelling bird species to extinction in the wild. Having eaten all the birds, the snakes then eliminated the small mammals and even the lizards on Guam. Thanks to extraordinary efforts by the U.S. Department of Agriculture and the state Division of Forestry and Wildlife, the brown snake has not reached other Hawaiian islands.

Introduced species can also become invasive because they encounter relatively few predators or diseases in the new community. Without natural enemies, the alien population can grow quickly, driving native species to extinction. This certainly seems to have been a factor with the introduction of European rabbits to Australia and their subsequent population explosion. In such cases, the introduction of a predator or disease from the invader's native ecosystem may help control its populations. For example, rabbit populations in Australia have been controlled to some extent by the introduction of a virus from European rabbits.

Wild animals and plants can usually tolerate diseases or parasites that their ancestors have encountered in generations past. Over time, natural selection builds genetically based resistance to common diseases. Just so, European hemlock trees are able to tolerate a European parasitic insect, the hemlock adelgid, which sucks nutrients from their leaves. However, the recent introduction of the hemlock adelgid to eastern North America threatens to eliminate the Canadian and Carolina hemlocks (**Figure 11.32**).

▼ **Figure 11.31 An Alien Predator**
The native birds of Guam have few natural defenses against the introduced brown tree snake, which crawls among branches and hunts for prey at night.

▲ **Figure 11.32 Hemlock Decline**
Across the eastern United States, populations of hemlock trees are in severe decline Ⓐ because of the introduction of the European hemlock adelgid. These aphidlike insects, which form white tufts on the undersides of needles and branches Ⓑ, rob trees of nutrients and ultimately kill them.

Pollution

Air and water pollution are directly responsible for the disappearance of some species.

Air and water pollution have directly affected biodiversity in many locations. Lichens, for example, are especially sensitive to air pollution, which has diminished their abundance in many areas, especially near cities. Lichens are symbioses between a fungus and an alga. They are able to survive in some of Earth's most extreme environments, such as on rocks in Antarctica and the driest deserts. Nevertheless, lichens are especially sensitive to chemical and particulate pollutants in the air—so much so that the U.S. Environmental Protection Agency (EPA) uses their diversity as an index of urban air pollution (**Figure 11.33**).

In many parts of the world, the biodiversity of streams has been diminished by sedimentation and chemical pollution. Many species of crayfish, mussels, and aquatic insects make their living by filtering pieces of dead organic matter from the flowing water. However, even small amounts of sediment can clog these animals' gills, and chemical pollutants quickly concentrate to toxic levels in their bodies. Of the 305 species of freshwater mussels in North America, 157 are now either extinct or in danger of extinction. Of the 330 species of North American crayfish, 111 are either extinct or threatened. The EPA has found that just as the diversity of lichens

▲ **Figure 11.33 Lichen Diversity and Air Pollution**
The many different species of lichen growing on this tree branch indicate that the tree is growing where air quality is high.

indicates air quality, the diversity of aquatic insects indicates water quality—their diversity diminishes directly in relation to increased water pollution.

Altered Patterns of Disturbance

Changes in the frequency and severity of fires are diminishing biodiversity in some ecosystems.

Some ecosystems have natural cycles of disturbance, such as periodic fires or windfalls from hurricanes. A great many species depend upon these regular disturbances to reproduce and grow. If humans alter these disturbance cycles, the biodiversity of the ecosystems may be diminished.

The jack pine ecosystem is one such example. The jack pine forests of Minnesota, Wisconsin, and Michigan once had frequent fires, which were set by lightning. These fires consumed both overstory trees and understory vegetation and converted leaf litter to mineral ash. Jack pines are well adapted to these ground fires. In fact, they cannot reproduce without fire because their cones do not release their seeds until after they have been subjected to its heat. A few days after a fire, the cones open and their seeds drop onto the burned ground. There the seeds sprout and seedlings thrive, establishing a new jack pine forest.

Other species that live in the jack pine ecosystem are also adapted to the cycles of fire. Most notable of these is a tiny songbird called Kirtland's warbler (**Figure 11.34**). These birds nest only in stands of jack pines that are between 5 and 20 years old; they do not nest in older stands. Suppression of fire and other timber management procedures have greatly reduced

the number of young stands. As a result, Kirtland's warbler is critically endangered; fewer than 500 birds persist in the wild.

▼ **Figure 11.34 Jack Pine and Kirtland's Warblers**
Young stands of jack pine are critical habitat for the endangered Kirtland's warbler. Because humans have limited the number of recent fires, few such stands of jack pines remain.

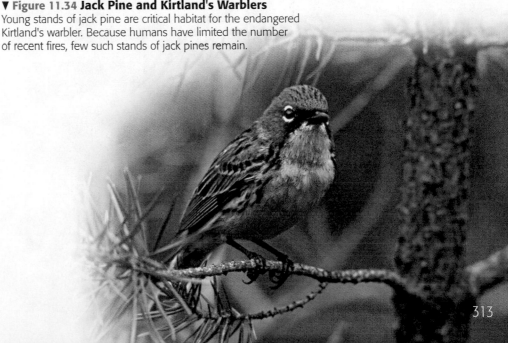

QUESTIONS

1. What is the single most important factor contributing to the loss of species worldwide?

2. Why does the fragmentation of landscapes diminish the diversity of species within habitat patches?

3. What factors influence the vulnerability of animal species to overharvesting by humans?

4. What are the factors that influence whether non-native species become invasive or not?

Climate Change

Disappearing sea ice endangers polar bears, and warmer seas threaten coral reefs.

There was considerable fanfare in late 2007 when Secretary of the Interior, Dirk Kempthorne, initiated the legal process of listing the polar bear as threatened under the U.S. Endangered Species Act. Kempthorne's decision was based on the persistent year-to-year decrease in sea ice off the Arctic shores of Canada and Alaska. Polar bears depend on sea ice as a platform from which to hunt for their most important prey, the Arctic seal (**Figure 11.35**). There had been much debate about whether the polar bear should be given legal status, and controversy over the decision continues. For example, critics point out that most polar bear populations elsewhere in the world are either stable or increasing. What is clear, however, is that Arctic climates are warming at a rapid pace, and this warming is having a significant impact on the tundra and the extent of near-shore sea ice (8.4).

Climate change also affects tropical oceans. A great many coral reefs have experienced a phenomenon called coral bleaching, which many scientists believe is tied to the warming of tropical oceans. Corals have a symbiotic relationship with algae that live inside their bodies. The algae are a critical source of nutrition for the corals; the algae also give the coral animals their color. If the water becomes too warm, the stressed coral animals expel the algae, causing the corals to lose their color. Bleached corals may regain algae, but often they simply die (**Figure 11.36**).

Climate changes due to carbon dioxide and other atmospheric pollutants will challenge living organisms for many years to come. Although humans can respond to climate change by planting different crops or moving to more hospitable areas, other species have fewer options. Alpine plants growing on a mountaintop cannot move to a higher mountain if climate change melts the snow and warms their habitat beyond their limits of tolerance. Stressed corals in a warming ocean cannot easily establish new reefs in cooler waters. Regardless of the eventual distribution of biomes such as deserts, savannas, temperate forests, and tundra, whenever their boundaries are shifted by climate change, biodiversity will suffer transitional losses.

▼ **Figure 11.35 Threatened by Climate Change**
The U.S. Fish and Wildlife Service has concluded that most polar bear populations in North America are threatened by the reduction in sea ice caused by global warming. Polar bears hunt seals—their primary prey—from ice floes.

▼ **Figure 11.36 Coral Bleaching and Climate Change**
Warming of tropical oceans appears to be one of the factors contributing to a significant increase in the worldwide bleaching of corals. Bleached corals expel their symbiotic algae. Some corals recover, others do not.

11.6 Strategies for Conserving Biodiversity

Effective conservation of diversity requires attention not only to the health of populations of individual species but also to the design and management of the places where they live. **Preserves** are areas that are protected by various means to maintain biodiversity. Based on the equilibrium theory of island biogeography, preserves should be as large as possible in order to minimize the rate of local extinctions within them. To maximize the migration of species into these areas, they should be connected to or close to other preserves. Lands that surround preserves should be managed to provide additional habitat and to allow species to migrate freely. Patterns of natural disturbances should be maintained. Human activities in preserves must be carefully managed. Healthy populations of wild species depend on the maintenance or restoration of habitat, the management of competitors and predators, and the maintenance of genetic diversity.

Preserves and Protected Areas

Conservation of biodiversity in preserves depends on principles of island biogeography.

The preservation of natural areas is a central strategy in the conservation of species and the habitats upon which they depend. Nearly every country has protected some of its lands and coastal waters as preserves, parks, or national forests.

Twelve percent of Earth's land surface—more than 17 million square kilometers (6.53 million square miles)—falls within legally designated preserves. This may seem like a considerable amount of land, but three factors limit the effectiveness of these preserves in conserving Earth's biodiversity. First, countries have tended to preserve areas with the least economic value, such as very arid, rocky, or mountainous lands. Land that is valued for other uses, such as agriculture or urban development, is far less likely to be preserved. As a result, many of the ecosystems that are most threatened by human development lack protected status. Second, even though an area has been designated as a park or preserve, it may not be protected from human activities. Poor countries may lack the economic or human resources to protect such areas. In other countries, lax administration and corruption allow or even encourage destructive activities within preserves (**Figure 11.37**). Third, preserves are often designed or managed in ways that do not offer optimal protection to their biodiversity.

► Figure 11.37 **Paper Parks**
Destructive activities in parks are more common than you might think. Here, a government helicopter flies over an illegal logging operation in the middle of a Brazilian national park.

Preserves are islands of sorts. Once their boundaries are delineated, pressure for human activities and development in the areas around them often increases (**Figure 11.38**). As a consequence, they become surrounded by a "sea" of less habitable, or even hostile, land. As islands, preserves are subject to the patterns of species diversity predicted by the equilibrium theory of island biogeography. Based on this theory, conservation biologists have developed six guidelines for designing and maintaining preserves in order to maximize biodiversity. They are as follows: (1) big is better than small, (2) connected is better than unconnected, (3) near is better than far, (4) buffers matter, (5) accommodate landscape change, and (6) manage people.

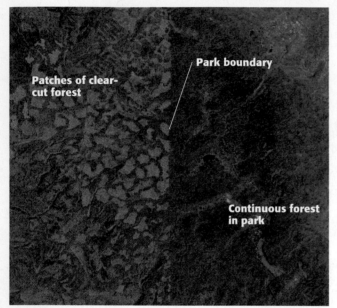

◄ Figure 11.38 Preserves Become Islands
This satellite photo shows part of the boundary separating the Targhee National Forest (left) from Yellowstone National Park (right). Clear-cutting in the national forest has transformed its landscape and isolated the forests within the park.

*Labels on image:*
Patches of clear-cut forest
Park boundary
Continuous forest in park

Big is better than small. Just as with oceanic islands, larger preserves have a more extensive representation of habitats and support larger populations of species than smaller preserves. Therefore, individual species are less likely to disappear from large preserves. For example, the country of Madagascar is identified as a world biodiversity hotspot because of its many endemic species and the rapid rate at which its interior forests are being cleared. An especially diverse community of lizards, including two-thirds of the world's chameleons, is found here and nowhere else. Exhaustive field studies show that each of these chameleon species has a unique set of habitat preferences. Large preserves that extend over more than one part of Madagascar are needed to conserve habitat for as many species as possible.

Connected is better than unconnected. Over time, the populations of species within a preserve fluctuate. The persistence of many species in an area depends on their ability to recolonize that area. This is especially important in relatively small preserves, where fluctuations are likely to result in the disappearance of a species.

A clever way to overcome the disadvantage of small preserves is to create **migration corridors** that connect several habitat fragments into one large preserve (**Figure 11.39**). Examples of migration corridors include strips of forest left along streams, greenways planted between city parks, and railroad rights-of-way containing remnants of native prairie vegetation in the American Midwest. Animals on the move tend to stay within their preferred habitat, so these strips of suitable vegetation encourage migration between habitat fragments. Animals moving through the corridors often help plants migrate through the corridors by dispersing their seeds. Populations of plants may also become established within the corridors, providing sources of seeds for the recolonization of the habitat fragments. If organisms are able to move among habitat fragments, outbreeding among unrelated individuals is facilitated, and the unfavorable effects of inbreeding are reduced.

▼ Figure 11.39 Connecting the Landscape
Hedgerows between these English pastures provide corridors that connect patches of forest. Many species of birds and mammals travel through these corridors.

Near is better than far. When local extinction occurs within a small preserve, reestablishment of the species depends on its migration from other sites. The closer a source population is to the affected preserve, the easier it is for immigrants to fill the void. Just as islands close to the mainland support a higher equilibrium species number, preserves located close to other preserves maintain greater biodiversity. The ability of species to migrate among preserves may play an important role in the survival of many species as climates change in the future.

The United States has formally designated over 109 million acres in over 750 locations in the National Wilderness Preserve System. In the western United States, wilderness areas are often located adjacent to or within tens of miles of one another, facilitating plant and animal migration among them. East of the Mississippi, there are many fewer wilderness areas, and they are often separated from one another by hundreds of miles. The large distances between preserves impede migration and increase the challenges for conserving biodiversity in the East.

Buffers matter. For biological preserves to be effective, their managers must pay attention to activities on the lands that surround them. Land use around the edges of a preserve has a significant effect on the biodiversity within the preserve. A city park bounded by busy streets is truly an island; it has no buffer around its edges to minimize the effects of disturbance from human activities (**Figure 11.40**). Pollution and noise from the city penetrate the park. Animals wandering across the park boundaries are unlikely to survive long enough to return. The result is a constant drain on species numbers. By contrast, a preserve surrounded by grazing land or forest managed for timber is buffered from many human activities and will support greater biodiversity than one would expect based on its size alone.

Accommodate landscape change. Disturbances such as fires, ice storms, strong winds, insect infestations, and fungal diseases are a natural part of nearly all ecosystems. These disturbances create a mosaic of landscape patches undergoing various stages of succession. These patches provide diverse habitats that allow many different species to survive (6.5). Conservation managers should accommodate disturbances that are similar to those that occur naturally and to which native plants and animals are

▲ **Figure 11.40 No Buffer to Be Found**
Central Park in New York City provides important habitat to many species, but there is no buffer around the park to protect its biodiversity from the human activities going on immediately adjacent to it.

adapted. However, disturbances that are significantly more frequent or intense than in the past are likely to diminish biodiversity.

Manage people. Managing the activities of people is one of the most vexing challenges in many parks and preserves. In the United States, for example, national parks are set aside to provide recreation for people as well as to conserve biodiversity. Reconciling the needs of wildlife with the demands of human recreation is a continuing issue for park managers.

In addition to conserving biodiversity, preserves in many countries are set aside to meet the needs of indigenous people, such as hunting, gathering, and subsistence agriculture. In some cases, the potential conflicts between the needs of indigenous people and those of the plants and animals that surround them appear to be well managed. For example, over half of the preserves in the Brazilian Amazon are officially designated for use and management by indigenous people. Illegal logging and poaching are less common in these indigenous preserves than in other types of preserves.

In other parts of the world, conflicts between humans and endangered species are a significant challenge. In India, for example, many preserves have been established to protect tigers and Asian elephants. Unfortunately, these animals often present a threat to people and their homes and farms. When elephants destroy crops or tigers attack livestock or children, people have little sympathy for the conservation of these threatened species (**Figure 11.41**).

▼ **Figure 11.41 Biodiversity and People in Conflict**
People live in and near many of India's preserves that are set aside for animals such as Asian elephants and tigers. Elephants can be very destructive to farms and villages and are often subject to poaching by people.

FOCUS ON SCIENCE: Conservation Corridors

Do connections between habitat patches actually increase their diversity?

The theory of island biogeography predicts that species diversity will be higher in preserves and patches of habitat that are connected to one another. For this reason, protected habitat corridors have become a central feature of management plans worldwide. But are these corridors really effective? Establishing and maintaining corridors is costly, so managers would prefer to devote their limited resources elsewhere if corridors are ineffective.

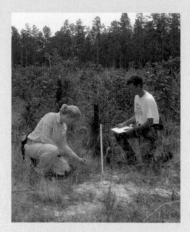

◄ **Figure 11.42**
Connection Science
Nick Haddad and Ellen Damschen of North Carolina State University were among the first scientists to carry out experiments confirming that corridor connections between habitat patches significantly increase biodiversity within those patches.

Until 2006, there was no experimental evidence addressing this question for whole communities of organisms because it was too difficult to design experiments on such a large scale. Then Nick Haddad and his student Ellen Damschen devised an innovative experimental design to address this question (**Figure 11.42**). With the cooperation of the U.S. Forest Service and the Savannah River Experimental Research

Park in southeastern South Carolina, they were able to implement this experiment.

Haddad and Damschen studied six 50-ha (120 acre) "landscapes" established in an extensive area of pine plantations. Within each landscape, trees were cleared from a central 1-ha (2.5 acre) patch. Four treatment patches were located about 150 m (492 ft) from the edge of the central patch. One of these treatment patches was connected to the central patch by a 25-m-wide corridor (82 ft). The other treatment patches were not connected to the central patch. Each of the other patches had the same total area as the connected patch and its corridor, but varied in shape so that the effect of the corridor without its connection could be accounted for (**Figure 11.43**).

All cleared patches were subsequently colonized by a variety of herbaceous plant species that were not found in the adjacent pine plantation. Over the next 5 years, Haddad and Damschen monitored the diversity of herbs in each treatment plot. By the end of the study, herb species richness was 20% higher in the connected patches than in the unconnected patches. There was no difference in species richness among the other treatments. These results clearly demonstrated that corridors connecting habitat patches do increase species diversity.

More recently, Haddad and Damschen have documented similar trends for species of butterflies. They continue to explore the factors that influence the effectiveness of corridors. Ellen Damschen is now on the faculty at University of Wisconsin.

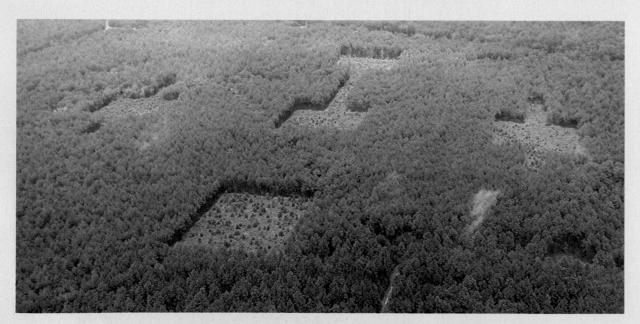

◄ **Figure 11.43 Experimental Design**
This is an aerial view of one of the six landscapes studied in Haddad and Damschen's experiments. The "winged" treatments are intended to simulate the effects of the corridor itself, without its connection.
Source: Damschen, E.L., N.M. Haddad, J.L. Orrock, et al. 2006. Corridors increase species richness at large scales. Science 313:1284–1286.

Managing Populations of Individual Species

Conserving individual species requires restoration and maintenance of healthy populations.

In many cases, conservation is focused on a single species or group of species. This is particularly true for species that have been designated as threatened or endangered. Conservation biologists determine whether a species is threatened or endangered by assessing its **population viability**, the probability that it will go extinct in a given number of years. The exact length of time used in a population viability assessment varies, although 100 years is widely used.

Several factors influence population viability; these factors are the focus of actions to conserve species. The foremost threat to species viability is usually diminished habitat; this includes both the availability and quality of habitat. Most recovery plans for endangered species focus on increasing the area and suitability of habitat. Sometimes introduced predators or diseases threaten population viability. In such cases, managers may work to remove or control the predator or disease.

When populations become very small, inbreeding and the associated loss of genetic diversity may be the greatest threat to restoring viability. In such cases, captive breeding programs are important. In these programs, care is taken to mate genetically different males and females, with the goal of enriching the genetic diversity of the captive population. Eventually, the captive animals

are reintroduced to their wild habitats. In 1987, the total population of the majestic California condor numbered only 27 individuals. A cooperative captive breeding program involving zoos and other facilities in Arizona, California, and Oregon has succeeded in building the total condor population to nearly 200 birds and the wild population to 50 birds (**Figure 11.44**).

Given the great number of organisms in ecological communities, it would be virtually impossible to monitor and manage the populations of every species. Instead, conservation biologists often focus their attention on **umbrella species**, species whose ecological requirements are highly correlated with the needs of many other species. Top carnivores are often good umbrella species because they require large territories and help ensure the diversity of lower trophic levels (**Figure 11.45**). Species such as freshwater mussels that are particularly sensitive to pollution or other environmental changes are also effective umbrella species.

▲ **Figure 11.44 Back from the Brink**
Captive breeding and reintroduction programs have doubled the wild populations of critically endangered California condors and increased their genetic diversity.

◀ **Figure 11.45 An Umbrella Species**
Much conservation in the tropical forests in Mexico and Central America has focused on the well-being of the jaguar. By doing this, conservation biologists believe they will meet the habitat needs of many other species, too.

QUESTIONS

1. Describe six general guidelines for the design and management of areas dedicated to preserving biodiversity.

2. How do conservation biologists determine if a species is threatened or endangered?

3. Why do conservation biologists consider large carnivores to be good umbrella species?

11.7 U.S. Policies for Conserving Biodiversity

The inauguration of the U.S. National Park System caused national interest in the conservation of biodiversity. As a result of the Wilderness Act of 1964, over 400,000 square km (100 million acres) of public land have been designated as wilderness areas. These lands are managed for biodiversity, and human intrusion is minimal. The earliest federal laws to protect biodiversity governed the buying and selling of species that had been legally protected by individual states. The 1973 Endangered Species Act provided much broader protection to threatened and endangered species on both public and private land. Management on private lands is playing an ever-larger role in the conservation of the nation's biodiversity.

National Parks and Wilderness Areas

The U.S. National Parks and Wilderness System set a high standard for managing biodiversity.

In 1864, President Abraham Lincoln signed a bill granting Yosemite and Hetch Hetchy valleys, as well as the nearby groves of giant sequoias, to the state of California to be conserved as a public park. In the words of the bill, the new park was to be "an inalienable public trust." Eight years later, the U.S. Congress established Yellowstone as the world's first national park. Although the legislation establishing Yellowstone called attention to the wildlife and beauty of the region, it emphasized that, above all else, this park was to be a "pleasuring ground for the benefit and enjoyment of the people" (**Figure 11.46**).

Over the next 40 years, Congress established more parks, including Sequoia, Mt. Rainier, Crater Lake, Glacier, and Grand Canyon. Unhappy with California's management practices, the federal government made Yosemite into a national park in 1896. The creation of new parks was heavily promoted by the railroads, which hoped to expand routes and recruit passengers. The railroads also managed many park concessions, such as hotels and restaurants. Park management was seen as "people management" and was carried out by the U.S. Army.

As early as 1900, John Muir and other preservationists began expressing concern that the parks' focus on people was having negative effects on the natural resources of the parks, including plant and animal diversity. The prolonged battle over the damming of the Tuolumne River in Hetch Hetchy Valley heightened these concerns.

After much political wrangling, President Woodrow Wilson signed the National Park Service Organic Act in 1916. This legislation brought all of the national parks under the management of a single agency in the Interior Department, the National Park Service. The Organic Act made it clear that the parks' natural resources and biodiversity were central to the National Park Service mission. Unlike the inscription on the entrance to Yellowstone, the new emphasis was on biodiversity, not people.

Today, the U.S. National Park System includes 58 national parks and nearly 340 other units, such as national monuments and national seashores, distributed among all 50 states. Conservation of biodiversity plays a central role in most of these preserves. In many regions, national parks provide the core for conservation planning. Parks are, of course, heavily visited by people, and accommodating human visitors remains one of the most important challenges to conserving the parks' biodiversity (**Figure 11.47**).

◄ **Figure 11.46 Focus on People**
Yellowstone was the nation's first national park. In the early days, park management emphasized human recreation, as indicated by the words engraved in its historic northern entrance: "For the benefit and enjoyment of the people."

In the 1950s, conservation professionals began to recognize that public lands outside the national parks, such as national forests, also supported a significant amount of biodiversity. They also noted that development for lodging and recreation diminished the biodiversity of some lands within national parks.

In 1964, with prodding from environmental groups such as the Wilderness Society, the U.S. Congress passed the Wilderness Act. This created a category of federal lands designated as wilderness, in which "the earth and its community of life are untrammeled by man." This act prohibits activities such as logging and mining in areas designated as wilderness; it also limits road construction. Initially, 36,000 square km (9 million acres) of national forest and national park land were designated as wilderness. Land can be added to this system only by congressional action.

As of 2010, the U.S. Wilderness Preservation System included more than 441,000 square kilometers (109 million acres). These lands are administered by four different agencies—the National Park Service (177,600 km² or 68,500 mi²), the National Forest Service (146,500 km² or 56,550 mi²), the U.S. Fish and Wildlife Service (83,760 km² or 32,310 mi²), and the Bureau of Land Management (35,200 km² or 13,590 mi²). The Wilderness System is extremely important to the conservation of biodiversity. In fact, wilderness lands are the core habitat of over 40% of the species that are listed as threatened or endangered in the United States (**Figure 11.48**).

▲ **Figure 11.47 Loving Them to Death**
National parks are greatly appreciated by visitors, but high visitation also produces many stresses, including the diminished diversity of wildflowers and other herbs, the introduction of non-native species, and threats to wildlife along trails.

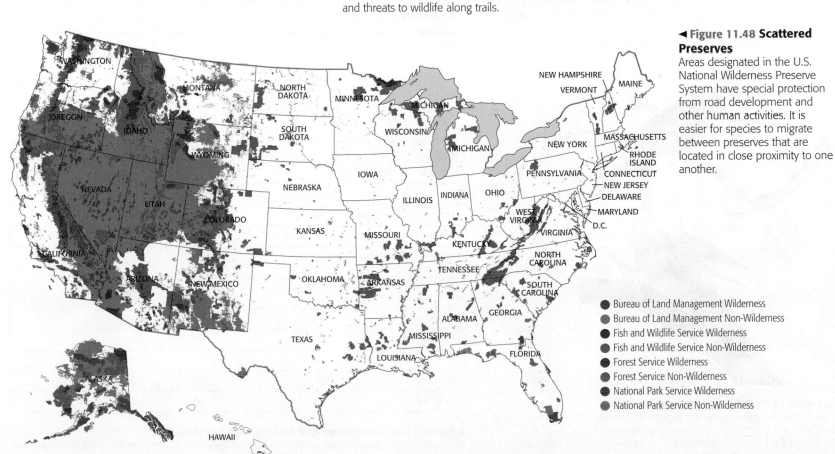

◄ **Figure 11.48 Scattered Preserves**
Areas designated in the U.S. National Wilderness Preserve System have special protection from road development and other human activities. It is easier for species to migrate between preserves that are located in close proximity to one another.

● Bureau of Land Management Wilderness
● Bureau of Land Management Non-Wilderness
● Fish and Wildlife Service Wilderness
● Fish and Wildlife Service Non-Wilderness
● Forest Service Wilderness
● Forest Service Non-Wilderness
● National Park Service Wilderness
● National Park Service Non-Wilderness

Legislation to Protect Species

U.S. laws limit commerce in and injury to threatened or endangered species.

In the early 19th century, the passenger pigeon was one of America's most common and most hunted birds. These birds flew in large flocks, were easily captured or shot, and fetched a high price in urban markets like New York, Boston, and Europe. As a result of market hunting and widespread deforestation, these once-abundant birds were extinct in the wild by 1900 (**Figure 11.49**). By that time, it was clear that other bird species were also threatened by overhunting and loss of habitat.

The declining numbers of birds and other wildlife led a number of states to pass laws prohibiting the hunting of threatened species. However, these laws varied enormously from state to state and were difficult to enforce. In response to the situation, the U.S. Congress passed the 1900 Lacey Act. This law, which has been reauthorized and amended several times since its passage, prohibits trade in wildlife, fish, and plants that have been taken illegally under either state or federal statutes. The U.S. Fish and Wildlife Service (USFWS) was created to enforce this law. The Lacey Act abruptly put an end to the market hunting that was responsible for the decline of so many wildlife species.

For the most part, the Lacey Act protects only those species that have commercial value and are threatened by overcollection or hunting. By the 1970s, it was clear that some sort of protection was needed for the vast majority of species that do not fall into this category. In 1972, the Marine Mammals Protection Act was passed. This law prohibits the taking of marine mammals such as seals, dolphins, and whales from U.S. waters; it also prohibits U.S. citizens from taking marine mammals from or in any other part of the world.

The following year, President Richard M. Nixon signed the Endangered Species Act (ESA) into law. As stated in its defining legislation, the ESA was intended "to provide a means whereby the ecosystems upon which endangered species and threatened species depend may be conserved and to provide a program for the conservation of such endangered and threatened species." The ESA prohibits the taking of any threatened or endangered species, regardless of land ownership. With the ESA, the term *take* has been interpreted very broadly to include not only capture, hunting, or collection, but any conduct that might damage a species' habitat or diminish its ability to reproduce.

Under the ESA, a species is considered endangered if it is in danger of extinction throughout all or a significant part of its range. A species can be listed as threatened if it is at risk of becoming endangered (**Figure 11.50**). The criteria for listing (and delisting) species as either endangered or threatened are set by the U.S. Fish and Wildlife Service. The USFWS must also develop a recovery plan for each species that it lists.

There is widespread agreement that the ESA is one of the most important pieces of biodiversity legislation ever passed by any country. Certainly it has played a significant role in slowing or halting declines in populations of many threatened and endangered species.

▲ **Figure 11.49 A Catalyst for Action**
The extinction of the passenger pigeon as a result of overhunting solidified public opinion about the need to protect species and catalyzed the U.S. Congress to pass the Lacey Act, which prohibits commerce in protected species.

Group	Endangered	Threatened	Total
Mammals (*polar bear, wolverine*)	68	13	81
Birds (*northern spotted owl, red-cockaded woodpecker*)	77	15	92
Reptiles (*pygmy rattlesnake, bog turtle*)	14	23	37
Amphibians (*tiger salamander, red-legged frog*)	13	10	23
Fishes (*delta smelt, chinook salmon*)	76	61	137
Invertebrates (*species of snails, insects, spiders, and shellfish*)	153	32	185
Animal Subtotal	401	154	555
Higher Plants	573	144	717
Ferns and Allies	24	2	26
Lichens	2	0	2
Plant Subtotal	599	146	745
Grand Total	1,000	300	1,300

▲ **Figure 11.50 Threatened and Endangered Species**
Many groups of organisms are represented by the species that are currently listed as threatened or endangered under the Endangered Species Act.

It has been responsible for the recovery and eventual delisting of other species, such as the bald eagle (**Figure 11.51**).

As it is currently written and administered, the ESA has many critics. This criticism comes from many quarters. Private property owners complain that it unfairly limits the use and economic value of some properties. Conservationists worry that such constraints may encourage landowners to take habitat-destroying actions, such as logging, before an ESA-listed species can be found on their property. Nearly everyone agrees that the processes by which species are eventually listed or delisted under the ESA can be bureaucratic and time-consuming.

Probably the greatest concern of conservationists is that the ESA provides no means to prevent species populations from declining to threatened or endangered status. No action to conserve or restore habitat is mandated until a species is listed as threatened or endangered. In some cases, this may be too late.

The ESA was initially scheduled for congressional evaluation and reauthorization in 1993, 20 years after its passage. Critics from all sides hoped to use the reauthorization process to amend the bill to their liking. However, each time an interest group has crafted reauthorization legislation, it has been blocked by other groups that fear their interests will be jeopardized. Without reauthorization, the 1973 ESA will remain in effect as written.

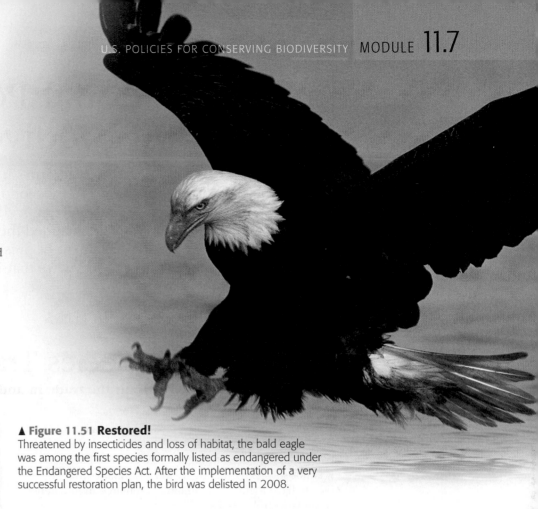

▲ **Figure 11.51 Restored!**
Threatened by insecticides and loss of habitat, the bald eagle was among the first species formally listed as endangered under the Endangered Species Act. After the implementation of a very successful restoration plan, the bird was delisted in 2008.

Conservation on Private Land

Private landowners can play a significant role in species conservation.

Public lands owned by state or federal governments account for about 20% of all land in the United States, although the proportion varies widely among states (76% in Nevada; less than 5% in most eastern states). The remaining land, which is owned by individuals and corporations, also contains important ecosystems and habitats. Thus, there is much interest in strategies to conserve biodiversity on privately owned lands.

Many cities, towns, and counties have zoning laws that encourage the conservation of biodiversity by limiting development or mandating greenspaces within developments. Greenspace zoning has helped many urban counties protect both habitat and water supplies (17.4). To date, Oregon is the only state to implement statewide zoning. The 1973 Oregon Land Use Act requires cities and counties to enact zoning plans using standards set by the state. Those standards include the needs of biodiversity in coastal, aquatic, and forest ecosystems.

As an alternative to zoning or mandated controls on land use, many environmental groups and government agencies rely on incentive-based programs. In some states, landowners receive tax exemptions if they protect threatened habitats or species. In other cases, private organizations or government agencies use conservation easements to purchase development rights from landowners. Under conservation easements, landowners give up certain rights to develop their land. For example,

to protect the biodiversity of a stream, a county might purchase a conservation easement on the forested land along the stream. That easement prevents the landowner from ever subdividing the land to build homes. Easements may also limit activities such as hunting and logging on the property. In any case, the landowner is directly compensated for the economic value of these activities.

In the United States there are over 1,400 separate **land trusts**, nonprofit organizations that are dedicated to conserving land for the protection of biodiversity and ecosystem services. Most land trusts focus on a few counties or a specific region within a state. A few, like the Trust for Public Lands, the Nature Conservancy, and the Conservation Fund, facilitate land conservation across the country. Land trusts use a variety of tools, including public education, conservation easements, and active land management, to conserve biodiversity on public and private lands.

In summary, the U.S. system of national parks, national forests, and wilderness areas has been effective in conserving biodiversity. Federal laws that regulate the trade in wild species and protect threatened and endangered species and their habitat have also been important. Biodiversity on privately owned lands is receiving more and more protection through conservation easements, greenspace zoning, and the activities of land trusts.

QUESTIONS

1. How do national parks, national forests, and formally designated wilderness areas differ from one another?

2. By what means does the Lacey Act preserve the diversity of species?

3. By what means does the Endangered Species Act preserve the diversity of species?

4. What is a conservation easement? How does a conservation easement benefit private landowners and help conserve biodiversity?

11.8 International Policies for Conserving Biodiversity

International cooperation to conserve Earth's biodiversity has been facilitated by legally binding treaties and conventions and by economic incentives and partnerships. Treaties designed to limit illegal trade and prevent overharvesting have focused on endangered species. Economic incentives have included the promotion of ecotourism and the exchange of debt for the establishment of preserves. In developing countries, biodiversity conservation must be undertaken in ways that help alleviate poverty and support economic development.

Endangered Species Trade and Harvest

International treaties that limit the trade in and harvesting of endangered species have had mixed results.

The global market for wild plants and animals and their products is very large. For example, the global trade in exotic birds alone is thought to exceed $50 billion. So it is not surprising that many international conservation efforts have focused on limiting commerce in threatened or endangered species.

The Convention on International Trade in Endangered Species of Wild Fauna and Flora (CITES), which was ratified by UN member nations in 1975, is the most notable example of an effort to conserve species by limiting trade. This treaty aims to ensure that international trade in wild animals and plants does not threaten their survival, whether they are traded as live specimens, fur coats, or dried herbs. Today, CITES accords varying degrees of protection to more than 30,000 species (**Figure 11.52**).

CITES has slowed the overhunting and collection of many plant and animal species, but its overall effectiveness is hotly debated among conservation biologists. They point out that in some developing countries enforcement of CITES is limited by corruption and the lack of resources. In some cases, international restrictions on commerce have driven up prices in black markets. These high prices encourage even more illegal hunting.

The illegal trade in rhinoceros horn appears to be such a case. In traditional Asian medicine, rhinoceros horn is highly valued as a treatment for many ailments, including fever and impotence. Daggers with rhino horn handles are prized in many areas of the Middle East. Trade in rhinoceros horn was banned under CITES in 1975. Over the next 15 years, however, the combined populations of the world's five species of rhinoceros declined by 85%, largely due to poaching. Today the black market value of rhinoceros horn is nearly $60,000 per kilogram, so

▲ **Figure 11.52 CITES Protection**
The international demand for bird feathers and plumage has resulted in overhunting of many species worldwide. CITES limits international trade in such items.

incentives for continued poaching are very high (**Figure 11.53**).

Other international treaties have successfully limited the harvest of endangered species. For example, the International Convention for the Regulation of Whaling was signed by 42 nations in 1946. This convention established the International Whaling Commission (IWC) to oversee the whaling industry and ensure the conservation of whale populations. With strong pressure from environmental groups opposed to whaling and like-minded governments, the IWC implemented a worldwide 5-year moratorium on all whaling. Despite objections from some member nations, this moratorium has been extended to the present. The actions of the IWC have helped to stabilize and even increase the populations of some whale species, although other species remain endangered due to factors such as coastal development and pollution.

Why has the ban on whaling been effective, whereas treaties to limit hunting of other species, such as rhinos, elephants, and tigers, have been less so? The most important factor is probably the capacity for treaty enforcement. Whaling is a complicated and expensive endeavor that requires large vessels and crews, so it is relatively easy to monitor.

▲ **Figure 11.53 CITES Criticism**
Limits on trade in rhinoceros horn and ivory from elephant tusks has driven up their price in black markets and led to continued poaching for these items.

Economic Incentives for Conservation

Conservationists strive for economic development that is compatible with the protection of biodiversity.

Much of the world's biodiversity is found in poor and developing countries. That biodiversity is often threatened by activities associated with economic development, such as road construction, mining, and logging. Conservation organizations and international development agencies are seeking ways to encourage economic development in these countries while maintaining or even enhancing biodiversity.

The Global Environmental Facility is an international agency that provides grants and loans for projects that promote both biodiversity and economic development. Nearly half of those projects encourage some aspect of ecotourism. Ecotourism has been a strong incentive for the creation of preserves and parks in a great many countries. It has the potential to bring economic resources to underdeveloped places and to create incentives for local people to conserve their natural resources. In some places, however, tourism has had a destructive effect on environments. In addition, the financial benefits of some ecotourism projects have not gone to the local communities or the countries that have hosted them.

The International Ecotourism Society has established criteria for sustainable ecotourism. First and foremost, ecotourism should encourage biological and cultural diversity by protecting ecosystems. It should also promote the sustainable use of biodiversity by local people and share the economic benefits with those people. Finally, ecotourism should minimize its own environmental impacts.

Many developing countries carry large burdens of debt to wealthier nations. **Debt-for-nature swaps** are international arrangements in which poor nations are forgiven a portion of their debts in exchange for the establishment of permanent parks and preserves. The debtor nation is then able to use more of its limited economic resources to meet local needs. Since their inception, debt-for-nature swaps in countries in Africa, South America, and Asia have converted nearly $500 million of debt into preserved habitat. All nations benefit from this conservation of global biodiversity.

The increasing number of preserves in developing nations is a hopeful trend. However, no wildlife preserve can protect biodiversity if the people living on its boundaries are desperately poor. When local people view the plants and animals in a preserve as the fuel and food they need to survive, the cause of biodiversity protection is lost. The World Commission on Protected Areas therefore promotes the idea that plans to protect biodiversity must incorporate the whole landscape, including its human inhabitants.

QUESTIONS

1. Describe the strengths and weaknesses of the Convention on International Trade in Endangered Species.

2. Describe three international policy initiatives aimed at encouraging both economic development and conservation of biodiversity.

Integrated conservation and development projects (ICDPs) promote both the protection of wildlife and the improvement of living standards for people who live within and around areas that are managed for biodiversity. ICDP projects begin with a comprehensive understanding of the local culture and economy. Investments and incentives are then structured to help the local people recognize that the protection of biodiversity is a source of prosperity. Poachers are hired as gamekeepers. Sustainable harvest of forest products is promoted. Slash-and-burn agriculture is replaced by steady jobs in ecotourism. Destructive overgrazing is replaced by buffer zones of sustainable farming. When local people begin to view biodiversity preserves as the source of their livelihood, conservation becomes a priority (**Figure 11.54**).

Because every region is unique, each ICDP begins with discovery. As landscape plans are designed, scientific knowledge must be supplemented with the knowledge of the indigenous people. As a project goes forward, continuous monitoring of biodiversity and ecosystem health provides new data. Stakeholders and participants continue giving feedback and receiving information. Economic development and activities to protect biodiversity are continuously adjusted as participants learn what does and doesn't work. Lessons learned are shared with other agencies and other nations in an effort to extend the management model to new sites.

In 1992, representatives from 179 countries participated in the United Nations Earth Summit in Rio de Janeiro, Brazil. The participants adopted Agenda 21, a set of principles and strategies for sustainable development in the 21st century. Agenda 21 emphasizes conservation of natural resources and biodiversity as well as the social and economic needs of specific human groups, including women, children, and indigenous peoples and their communities.

Much in keeping with Agenda 21's emphasis on both conservation and economic development, Earth Summit participants crafted the Convention on Biodiversity. This convention prescribes actions to conserve biological diversity and ensure the sustainable use of its components. It also provides guidelines to ensure that a fair portion of the economic benefits obtained from species in developing countries are returned to those countries. The Convention on Biodiversity was ratified by nearly all participants and is now international law. However, the United States has declined to ratify it, arguing that its economic guidelines are not fair.

The 2002 UN World Summit on Sustainable Development, held in Johannesburg, South Africa, marked the 10th anniversary of the Earth Summit. The agenda and products of this summit reflected developing countries' concerns that conservation efforts were not adequately considering the needs of poor people. Environmental justice remains a persistent theme in discussions of international environmental policy.

▼ **Figure 11.54 Changing Views of Conservation** In Kenya's Samburu National Reserve, endangered animals such as this baby rhino are tended by local residents, some of whom were former poachers.

SEEING SOLUTIONS: Butterfly Ranching

The Bird's Head Peninsula of western New Guinea in Indonesia contains one of the largest and most diverse contiguous expanses of tropical forest remaining in the world. The cloud forests of the peninsula's Arfak Mountains provide critical habitat for a remarkable array of plant and animal species (**Figure 11.55**). Yet, these habitats and the species they support are threatened by logging, mining, and hunting activities that are expanding into the region. To protect these habitats, the World Wildlife Fund (WWF) and the local Hatam people have formed a partnership to generate income from the forests while preserving a remarkable group of butterflies that are found there.

Birdwing butterflies are among the most striking insects in the Arfak cloud forests. These huge, showy butterflies are prized by collectors around the world (**Figure 11.56**). Birdwings spend most of their adult lives in the shadowy world of the cloud forest. After mating, they come to the edge of the forest to seek plants on which they can lay their eggs; these plants will be the food source for their young.

International trade of birdwings is limited by the United Nations' CITES. Before the partnership with the WWF, the Hatam poached and sold birdwings through the black market, with no regard for the species' survival. Now the Hatam tend gardens of the butterflies' breeding plants around their homes. They carefully protect and manage the birdwing caterpillars as they munch leaves and grow. When they reach full size, the caterpillars enter a resting stage called the *pupa*. The Hatam allow many pupae to metamorphose into butterflies, which fly back into the surrounding forest. Some of the pupae are sold to butterfly dealers with CITES permits that are arranged by WWF. These collectors pay a high price for birdwing pupae, which they allow to emerge as perfect butterfly specimens.

Because the adult butterflies range free in the mountains, this sustainable practice is called butterfly ranching rather than butterfly farming. By cultivating their food plants, the Hatam ensure that the butterfly populations remain healthy. The Hatam now value and protect the forests surrounding their communities because that is where the birdwings live. With an alternative income from butterfly ranching, the people of the Arfak Mountains live a better life while at the same time protecting the biodiversity of the cloud forest.

Can human exploitation of an endangered species serve as a conservation tool?

▲ **Figure 11.56 The Goliath Birdwing**
The Goliath birdwing butterfly has a wingspan of up to 26 cm (10 in). It is one of several birdwing species endemic to the Arfak Mountains. A single Goliath specimen may sell for more than a thousand dollars.

▲ **Figure 11.55 Bird's Head Peninsula**
The Bird's Head Peninsula at the western end of New Guinea supports as many species as all of Australia, although it has only one-tenth the land area. The cloud forests of its Arfak Mountains are especially diverse.

AGENTS OF CHANGE: Strengthening the Exchange to Protect a Unique Biodiversity Hotspot

Varsha Vijay is a recent graduate of Duke University with a degree in environmental sciences. While at Duke and after graduation, she worked in Ecuador with a remote Amazonian tribe called the Waorani. Varsha began an organization called Fortificando el Intercambio *(Strengthening the Exchange) in order to share academic findings about the rain forest with the Waorani and to give them technological tools to help them preserve their territories and their culture.*

How did you first get the idea for Strengthening the Exchange?

When I was a sophomore at Duke, I began working with Dr. Stuart Pimm's group of conservation biologists. My first assignment was at Ankarafantsika National Park in Madagascar. While conducting basic ecological monitoring at the field station, I became aware of the important role of the native villagers as stakeholders in the park's conservation efforts.

When an opportunity came up to work in the Amazon, I jumped at it. There I focused on conducting conservation projects to protect the Yasuní Man and Biosphere Reserve, a biodiversity hotspot that encompasses the ancestral territory of the Waorani tribe.

As I spent more time with the Waorani, I could not ignore the very real threats to their well-being. I saw the incursions on their territory by colonists following oil roads, the health effects of pollution from petroleum contaminants, and the threats to their food supply from overhunting and habitat loss.

I realized that if a conservation project was to be successful, it would need to show the Waorani how conservation and the sustainable use of resources are compatible with the tribe's own desire to preserve their culture and territory.

What are the biggest challenges you've faced with the project, and how did you solve them?

Initially, I felt extremely isolated due to the dramatic cultural differences between me and the tribe. I was raised as a vegetarian, but with them I found myself eating the traditional foods of monkey meat and saliva-fermented manioc beer.

At first I had to struggle to communicate, since I spoke no Spanish and had no familiarity with the Waorani tribal language, Wao terero. However, as I continued working with the tribe, I adapted and became fluent in Spanish. I still struggle with Wao terero, but the tribe is always quick to give me a hand.

By being open to these difficulties and sticking with it, I became closer with the Waorani than I ever thought possible.

How has your work affected other people and the environment?

I've worked to create a conservation ethic in Waorani communities by engaging them in projects that address the tribe's concerns for the sovereignty of their territory and the health and safety of their people, as well as their desire to profit from ecotourism.

Today my projects get younger members of the tribe back into the forest. They are learning and cataloguing the knowledge of medicinal plants possessed by the older members of their tribe. We also teach them the basic science of Amazonian ecosystems.

We have equipped them with GPS and other mapping technologies so they can map important places in their territories. These tools help them to conserve not only plants and animals but also important cultural landmarks.

The Waorani who have worked with me continue to talk about ideas they have for new conservation projects. I've influenced the dialogue and made conservation something that is interesting and fun!

What advice would you give to students who want to replicate your model?

You don't need an exotic location. Sometimes the best place to start is in your own community. Figure out what

environmental issue you're passionate about. Then find a mentor doing the kind of work you'd like to do. If you intend to do research, the best mentor is probably a professor. If you intend to be involved in environmental activism, there are many inspiring people doing great work in the NGO [nongovernmental organization] community.

Once you've found a problem that moves you, consult with your mentor and the people affected by that issue to find a way you can use your current skills to contribute, and then build from there. Don't expect to solve everything on your own. Once you get involved, you'll realize that just being part of a solution is rewarding enough.

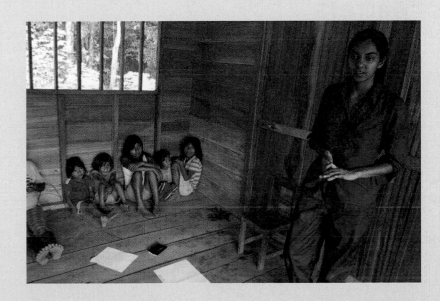

Summary

Biodiversity is the variety of life in all its forms, levels, and combinations. Biodiversity in landscapes is determined by variations in climate, soil, and patterns of disturbance. Diversity within communities can be measured by the numbers of species and the relative abundance of those species; the three-dimensional distribution of species and biological features is another measure of community biodiversity. Within populations, biodiversity is measured by genetic diversity. Biodiversity is important because it ensures the provision of ecosystem functions and services, imparts stability to ecosystem processes, and has economic value for humans. Most people believe that the protection of biodiversity is an ethical responsibility. In general, biodiversity is highest in tropical biomes and lowest in polar biomes. Biodiversity in communities is influenced by habitat diversity, competition, predation, and patterns of disturbance. The equilibrium number of species in a particular place is ultimately determined by the rates at which species immigrate and become extinct. The many threats to biodiversity include habitat loss and fragmentation, overharvesting of vulnerable species, competition and predation from non-native species, pollution, altered disturbance patterns, and global warming. Strategies for conserving biodiversity include the establishment of preserves, creating buffers and corridors to protect those preserves, and restoring healthy populations of species. A variety of domestic and international policies and laws has been established to reverse current trends that are accelerating the loss of biodiversity and to sustain the biodiversity that remains.

11.1 What Is Biodiversity?

- Biodiversity is the variety of species and their relative abundance across landscapes and within ecological communities, as well as the genetic diversity within populations.

KEY TERMS

biodiversity, landscape diversity, endemic, species richness, species evenness, structural complexity, genetic diversity, outbreeding, inbreeding

QUESTIONS

1. Describe the factors influencing landscape diversity for a park, national forest, or rangeland near your home.

2. When large top predators, such as wolves or mountain lions, disappear from an ecosystem, species richness and evenness among their prey often decline. Why?

3. Structural complexity of ecological communities is often correlated with species richness. Explain why this is so, using a specific community as an example.

4. Give two reasons why inbreeding within a species is a matter of concern to population biologists.

11.2 Why Biodiversity Matters

- Biodiversity is essential for ecosystem functions and services and for the stability of ecosystems.

- Biodiversity has great economic value, and most people agree that we have an ethical responsibility to conserve it.

KEY TERMS

sampling effect, complementarity effect, ecotourism

QUESTIONS

1. Describe two reasons why ecosystem functions, such as productivity, may increase as species richness increases.

2. Describe three ways in which biodiversity contributes to the economic well-being of communities.

11.3 Global Patterns of Biodiversity

- The species richness of Earth's biomes is highest in the tropical zone and lowest in the polar zone.

- Biodiversity hotspots are places with high numbers of endemic species that are especially threatened by human activities.

KEY TERM

biodiversity hotspots

QUESTIONS

1. Explain how the different rates of net primary production in tropical and temperate forests might be related to differences in the species richness of the forests.

2. What factors contribute to the endemism that is characteristic of biodiversity hotspots?

11.4 Differences in Biodiversity Among Communities

- Habitat diversity, interactions among species, disturbances, and rates of species immigration and extinction are responsible for differences in biodiversity among different ecological communities.

KEY TERM

equilibrium theory of island biogeography

QUESTIONS

1. Why do rugged mountainous areas generally support more species than relatively flat plateaus?

2. Explain how herbivorous animals are important to the diversity of the tropical trees that they feed on.

3. When temperate grasslands are not burned for several years, the diversity of grasses and herbs declines. Explain why this occurs.

4. In Minnesota, the Land of 10,000 Lakes, what factors might contribute to differences in plant and animal diversity among different lakes? Explain why.

11.5 Threats to Biodiversity

- Habitat loss and fragmentation, overharvesting, non-native species, pollution, altered disturbance patterns, and global warming are the most important causes of species extinctions.

QUESTIONS

1. What habitats are disappearing near your home? What species might be threatened by the loss of these habitats?

2. Tundra swans were hunted nearly to extinction 100 years ago. In the summer, breeding pairs nest in isolation from one another in the Arctic; in the winter, birds congregate in large flocks on a few lakes in the southeastern United States. How did this behavior make them vulnerable to overhunting?

3. Give an example of a non-native invasive species that is influencing biodiversity in ecosystems near your home. What factors contribute to its invasiveness?

4. Explain why global warming is likely to increase the threats to biodiversity posed by habitat loss and fragmentation.

11.6 Strategies for Conserving Biodiversity

- Biodiversity conservation depends on the establishment of preserves and protected areas, and the management of the ecosystems, disturbances, and people that surround them.

- Conserving endangered species requires restoration of healthy, genetically diverse populations.

KEY TERMS

preserves, migration corridors, population viability, umbrella species

QUESTIONS

1. Conservation biologists argue that to preserve biodiversity inside national parks, managers must pay as much attention to lands outside the park as they do to the land within the park. Explain why this is important.

2. Name three umbrella species for ecological communities that occur near your home. What features qualify them as umbrella species?

11.7 U.S. Policies for Conserving Biodiversity

- In the United States, biodiversity is protected by the system of national parks and wilderness areas, the Endangered Species Act, and actions by private landowners

KEY TERM

land trust

QUESTIONS

1. The U.S. Endangered Species Act prohibits the "taking" of a threatened or endangered species. What does this mean?

2. Describe three criticisms of the Endangered Species Act.

3. What laws or incentives exist to encourage private landowners to protect biodiversity on their lands?

11.8 International Policies for Conserving Biodiversity

- Most international policies to conserve biodiversity are based on treaties that prohibit trade in endangered species or economic incentives for conservation.

KEY TERMS

debt-for-nature swaps, integrated conservation and development projects

QUESTIONS

1. Although nearly all UN nations have signed agreements to protect their biodiversity, the actual level of protection varies considerably. What factors contribute to this variation?

2. How does the UN's Agenda 21 propose to deal with the fact that many of the world's most diverse ecosystems occur in very poor countries?

Mastering**ENVIRONMENTALSCIENCE**

Go to **www.masteringenvironmentalscience.com** for homework assignments, practice quizzes, Pearson eText, and more.

Water 12

A Disappearing Resource

How are human activities and climate change depleting an essential water resource?

The African Sahel is a 500- to 1,000-km (310–620 mi) wide ribbon of grassland, woodland, and savanna separating the very dry Sahara Desert to the north from the wet equatorial forests to the south. Near the dry northern edge of this ribbon, halfway between the Atlantic Ocean and the Red Sea, lies Lake Chad (**Figure 12.1**).

The Chari River, flowing from the much wetter savannas to the south, is the primary source of water for this shallow lake. Lake Chad has no outlet. Water is lost from it by evaporation and seepage into the deep sediments beneath its basin. Ninety-five percent of rainfall in the lake's basin comes between May and September.

During these months, the influx of water from the Chari exceeds water loss, and the volume of water in the lake increases. By late April, the end of the dry season, water flow in the Chari River is meager, and the lake may shrink to half its wet season size.

Five thousand years ago, this region was much wetter than it is today. The lands surrounding the lake were densely forested. At that time, Lake Chad was equal in area to Lakes Superior, Michigan, and Huron combined. This region has since grown drier, and for the past 2,500 years its climate has fluctuated around present-day conditions. Studies of lake sediments laid down over thousands of years reveal remarkable variations in the yearly rainfall and length of rainy seasons. Wet periods were punctuated by years of drought when streams barely flowed and the lake dried up.

Throughout this time, the nomadic people living nearby adapted their activities to changes in the lake ecosystem. These people subsisted by fishing, herding cattle and goats, and tending small, temporary farms. During the dry season, they took advantage of the new, fertile land exposed by the retreating lake. During the wet season, they were able to expand their activities farther from the lake's shore. Dry periods undoubtedly placed hardships on these people, but their nomadic lifestyles allowed them to migrate to more favorable locations during droughts.

In the late 1800s, the British and French colonized the Lake Chad region, establishing four protectorates—Cameroon, Chad, Niger, and Nigeria.

► Figure 12.1 **A Lake on the Edge**
Lake Chad and its drainage basin lie in the north central portion of the Sahel. The Sahel is a region of dry woodland and savanna that separates the Sahara in the north from moist tropical forests in the south.

► Figure 12.2 **People and Lake Chad**
Over the past 50 years, the human population surrounding Lake Chad has increased threefold, placing increased demands on its fisheries Ⓐ and the grazing lands that surround the lake Ⓑ. Irrigated agriculture in the former lake bed Ⓒ continues to draw water from the lake, as do growing cities such as N'Djamena Ⓓ.

The Europeans established land ownership and rigid national boundaries, notions that were foreign to the indigenous nomadic people. Between 1955 and 1965, colonial rule ended, and the four protectorates surrounding the lake became independent nations.

From 1968 to 1974, severe drought prevailed across the Sahel. National boundaries and patterns of land ownership prevented the migrations that had previously allowed the nomadic communities to adjust to drought. As a consequence, they overgrazed their lands and cut down their woodlands, converting vast areas of grassland and savanna to desert. Famine was widespread.

Today, over 5 million citizens of the four countries that border Lake Chad depend directly or indirectly on the resources it provides. Its waters quench their thirst and irrigate their crops, and its fish are a major source of their dietary protein (**Figure 12.2**). The wetlands and woodlands surrounding the lake support a diverse array of wildlife and waterfowl that are also important sources of food for this population.

Despite some recent increases in rainfall, the average size of Lake Chad has decreased over 90% since 1973 (**Figure 12.3**). Part of this decrease can be attributed to a gradual increase in temperature and the rate of evaporation in this region, perhaps associated with global warming. Deforestation of lands in the upper reaches of the Chari River has reduced rainfall in that region, so less water now flows in the Chari. However, most of the loss of water is due to increasing numbers of people diverting water from the river and lake to meet their growing needs.

Most of the people living around Lake Chad are poor. Sixty percent of them live on the equivalent of less than $1 per day, and more than 20% of their children perish before they turn 5 years old. Economic development is needed to improve the well-being of these people, but that development will depend on the restoration and sustainable management of Lake Chad's natural resources.

Recently, the four countries surrounding the lake created the Lake Chad Basin Commission to restore the lake and its ecosystems. To succeed, the commission will need to understand the natural and human processes that influence water flow. They will need to restore the features of the landscape that assure the reliable provision and use of water. They will need to secure collaboration among the countries that use the water and influence its flow. Finally, they will need to build resilience in the Lake Chad ecosystem and the human communities that depend upon it, so both can adapt to the inevitable changes that the future will bring. Indeed, the same needs are central to the sustainable use of water around the world and are important themes throughout this chapter.

▼ **Figure 12.3 Lake Chad 1973–2001**
The area of Lake Chad (the blue area in both images) has been reduced by more than 90% as a result of lower rainfall, diminished flows in the Chari River, and overutilization of lake water.

12.1 Water World

Our planet is called the "Blue Marble," and for good reason: Over 70% of Earth's surface is covered with water (**Figure 12.4**). Solar energy and gravity drive the hydrologic cycle, moving water between the oceans, the atmosphere, and the land (7.3). Precipitation that falls on land may evaporate back to the atmosphere, percolate down into aquifers in the ground, or flow back to the sea in one of the thousands of streams and rivers that course across every continent. Lakes and inland seas store water and slow its movement. Wetlands store and filter water. Human use of water affects every part of Earth's hydrologic cycle, including the distribution of water in streams, lakes, and groundwater and the kinds and amounts of chemicals and sediments that water carries.

The Global Water Budget

Water falling onto land may evaporate, percolate into groundwater, or flow in streams to the sea.

Hydrologists, the scientists who study Earth's waters, have compiled some remarkable statistics about the distribution and flow of water on our planet. For example, water is the third most abundant molecule in the atmosphere, yet only a tiny portion of Earth's water (about 0.001%) resides there. Every day about 11%, or 1,159 km^3 (278 mi^3) of that atmospheric water condenses into rainfall. And every day an equal amount returns to the atmosphere in evaporation. About 23% of the total global precipitation (271 km^3/day) falls on land (**Figure 12.5**). This is the total supply of fresh water for Earth's terrestrial ecosystems.

Water falling onto land has three fates: It may evaporate back to the atmosphere, flow over land in streams and rivers, or percolate through soil and rock into underground aquifers (7.3). A little less than two-thirds of the rain that falls on land evaporates back to the atmosphere from land surfaces and plant leaves. Hydrologists call this process **evapotranspiration**.

◄ **Figure 12.4 It's Mostly Water!**
The water that covers over 70% of Earth's surface is essential to all life.

The remainder (about 101 km^3 per day) represents the total renewable supply of liquid water on land. This water flows to the ocean in streams and groundwater. Two-thirds of it, about 68 km^3 each day, flows into rivers and streams. The rest makes its way into groundwater aquifers. Without human use, hydrologists assume that the volume of Earth's groundwater remains more or less constant. If so, the outflow from the world's aquifers into streams and the sea must equal the flow into them.

The majority of Earth's fresh water, 28.6 million cubic kilometers (6.86 million mi^3), exists in the form of ice. Most of that ice is in the massive glaciers over Greenland and Antarctica. Some fresh water is found in streams and rivers, which contain an average of about 1,200 km^3 (288 mi^3). Lakes store about 100 times that much. Earth's aquifers are thought to contain about 4 million cubic kilometers (0.96 million cubic miles) of groundwater.

These statistics give a sense of the magnitude of the global water budget. Yet the components of this budget vary greatly from region to region. Annual rainfall ranges from less than 10 cm (4 in.) in deserts such as the Sahara to more than 4 m in tropical regions such as the Amazon Basin and Southeast Asia. Annual potential evaporation, a measure of the ability of air to take up moisture, increases with increasing annual temperature. It ranges from 100 to 200 cm (40–80 in.) in Arctic regions to over 3 m in hot and dry locations.

The **water balance coefficient** is a general index of the availability of water in a particular region. It is calculated by subtracting potential evaporation from rainfall. In arid regions, water balance coefficients are negative. In the world's deserts, for example, they vary between –200 and –300 cm (–79 and –118 in.). In the humid rain forests of the world, water balance coefficients are positive and can exceed 400 cm (158 in.). In general, forests grow where water balance coefficients are positive; savannas, grasslands, and deserts are found in regions where coefficients are negative.

In many places, temperature and rainfall vary considerably from season to season. Water balance coefficients show those variations. In the southeastern United States, for example, rainfall is relatively constant throughout the year. During the autumn, winter, and spring, water balance coefficients are positive. But during the summer, warm temperatures produce high evaporation rates, resulting in negative water balance coefficients.

▼ **Figure 12.5 Water Budget** Only a minuscule fraction of Earth's water is in the atmosphere at any moment. Evaporation and precipitation move water through the atmosphere and bring fresh water to the land.

ATMOSPHERE
Contains 12,600 km^3 of water

Evaporation

271 km^3/day
About $1/4$ of total global precipitation falls on land

Precipitation

989 km^3/day

Most water flux occurs between the oceans and the atmosphere

170 km^3/day
Most of the rain that falls on land evaporates back to the atmosphere from land surfaces and plant leaves (evapotranspiration).

LAND
Streams, lakes, and groundwater contain 4,124,200 km^3 of water

888 km^3/day

101 km^3/day

Precipitation

Runoff and groundwater

OCEANS
Contain 1,230,000,000 km^3 of water

▶ **Figure 12.6 Watershed**
Because water inputs and outputs can be measured, watersheds are ideal units for the management and study of water flow.

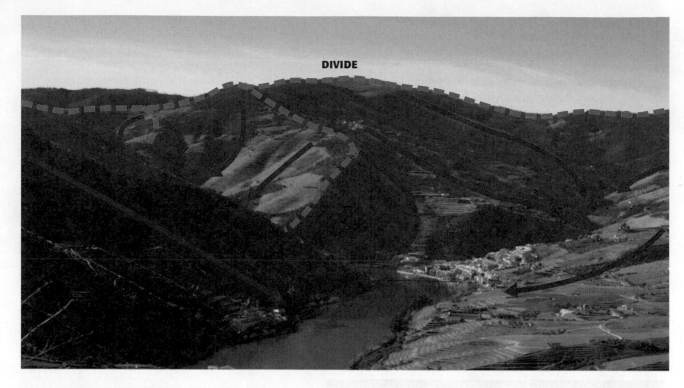

DIVIDE

Watersheds

Scientists study a river's drainage basin to measure water flow and evaporation.

A **watershed** is the area of land from which rainfall drains into a river or lake. Watersheds are also called drainage basins. Watersheds are separated from one another by mountain ridges and other topographic divides that direct the flow of water (**Figure 12.6**). The watershed is an ideal unit for understanding water flows and budgets at local and regional scales. Models of watersheds allow hydrologists to evaluate how human activities, such as irrigating crops and constructing reservoirs, affect the various components of water budgets.

How do hydrologists determine the flow of water in a watershed? To estimate the total amount of rain coming into a watershed, they use a number of widely distributed rain gauges. To estimate the amount of

liquid water leaving a watershed, they monitor the flow of streams and groundwater. Evapotranspiration can be measured directly or be estimated by subtracting stream and groundwater flow from rainfall.

Hydrologists study watersheds of many sizes. Watersheds of only a few hectares may be studied to learn how particular land uses or management practices affect the flow of water (7.8). Small watersheds drain into larger watersheds, which, in turn, drain into yet larger watersheds. Very large watersheds, such as that for the entire Mississippi River, can help us understand the cumulative effect of many activities over vast areas (**Figure 12.7**).

Open watersheds are drained by rivers that eventually make their way to the sea. These include the world's major river basins, such as the Amazon, Mississippi, Nile, and Congo. **Closed watersheds** are inland basins that do not drain to the sea. Rivers in closed watersheds often end in inland lakes or seas, such as the Great Salt Lake, Lake Chad, and the Dead Sea. Closed watersheds comprise about 18% of Earth's land surface.

Because the rainfall in its watershed is measured at hundreds of locations and its flows are closely monitored, the Mississippi River watershed provides an excellent case study of a water budget. The total volume of rain falling on its watershed (in cubic kilometers per year) is calculated by multiplying the average rainfall by the area of the watershed. The volume of water discharged from the mouth of the Mississippi River is measured directly. It includes flows from surface runoff and groundwater. Pumping from aquifers must also be included in the

▶ **Figure 12.7 Large Scale**
The watershed of the Mississippi River includes nearly one-third of the land in the lower 48 United States. This watershed can be subdivided into six smaller watersheds that correspond to the land drained by major rivers. Each of these is divisible into hundreds of yet smaller watersheds.

Missouri

Upper Mississippi

Ohio

Arkansas Red-White

Tennessee

Lower Mississippi

Gulf of Mexico

water budget; various measurements indicate that pumping from groundwater aquifers adds an average of 2 km³ to the Mississippi each year. Evapotranspiration over such a large area is impossible to measure directly. However, because the other components of the water budget are known, it can be calculated as evapotranspiration = rain volume + groundwater pumping – river discharge.

The flow of water within a watershed is influenced by a number of factors. Impermeable surfaces, such as pavement and rooftops, shed water and increase runoff. Plant cover slows the flow of water and increases evapotranspiration. Thus, in deciduous forests, a greater proportion of rainfall makes its way into streams in the winter than in the summer.

Human Water Use

Humans use over half of the water flowing in Earth's streams.

Since earliest history, humans have modified the flow of water in watersheds by irrigating crops, diverting and impounding streams, and draining wetlands. The agricultural production of many regions has been greatly increased by irrigation, the supplemental application of water to landscapes. Irrigation usually increases evapotranspiration and often alters water chemistry. Humans have also drained many wetlands and marshes to claim these lands for agriculture and urban development.

Diversions such as canals and aqueducts move water from areas where it is abundant to areas where it is scarce. In other words, diversions reduce the flow of water in one region to subsidize the need for water in other locations. Impoundments, such as human-made lakes and reservoirs, store water for a variety of uses, including irrigation, flood control, energy generation, and human consumption. Impoundments slow the movement of water and greatly alter the seasonal flow of the streams that drain them.

Activities that use water and then return it to streams or aquifers are said to be **nonconsumptive uses**. Such uses include the generation of hydroelectric power and the disposal of wastewater in septic systems. In other activities, such as irrigation and industrial cooling, a considerable amount of water evaporates into the atmosphere. These activities are said to be **consumptive uses** because much of the water is not returned to streams or aquifers.

Through our various activities, we humans have appropriated a large portion of Earth's water resources for our own use. Scientists estimate that about one-fourth of terrestrial evapotranspiration now comes from lands used to grow crops, graze livestock, or produce wood for human use. Dams and diversions control the flow of approximately half of the runoff (34 km³/day) in Earth's terrestrial ecosystems. A little more than half this flow is used for agriculture, industry, and municipalities. Approximately 82% of the water we use is returned to the streams from which it was taken, although the returning water often contains a variety of pollutants. The remaining 18% is lost to evapotranspiration.

The proportion of water used in agriculture, industry, and municipalities varies widely from country to country, depending on the climate and degree of economic development. In dry regions, agriculture usually accounts for a large proportion of water use. In regions with similar climates, economically developed countries use greater amounts of water for industrial purposes than do poorer countries (**Figure 12.8**).

In summary, Earth's water budget is determined by rates of evaporation and precipitation and the flow of water over the surface and underground. Watersheds provide an ideal unit for the study and management of these processes. Human activities have a significant effect on water budgets at both local and global scales.

QUESTIONS

1. What is the approximate residence time of a water molecule in Earth's atmosphere?
2. Describe the three fates of water that rains onto land surfaces.
3. Differentiate between an open and a closed watershed.
4. What is the difference between consumptive and nonconsumptive use of water by humans?

UNITED STATES (477.0) 13% 46% 41%

FRANCE (33.1) 10% 16% 74%

COLOMBIA (10.7) 4% 46% 50%

INDONESIA (82.8) 1% 8% 91%

◄ **Figure 12.8 Water Use** Patterns of water use vary among countries, depending on the importance of industry and agriculture in their economies. *Data from:* Pacific Institute

Water consumption by type of use (Total water use in billions of cubic meters)
- Industry
- Agriculture
- Domestic

12.2 Streams

Under the force of gravity, water flows across the surface of the land. Scientists use the word *stream* to describe all natural bodies of flowing water, regardless of their size. Depending on its size and location, a stream may be referred to more specifically as a rill, a brook, a creek, or a river. The total volume of water flowing in Earth's streams is 1,200 km^3 (288 mi^3), less than 0.0001% of Earth's water. Nevertheless, streams are a critical part of our planet's hydrologic circulatory system. Streams are shaped by erosion and sedimentation. These processes are, in turn, influenced by the volume and speed of the flowing water. The volume and speed of streams also determine the communities of plants and animals associated with them. Human activities have altered the flow and polluted the water of many streams and rivers.

Stream Flow

The volume and speed of flowing water determine a stream's structure.

Streams originate in the upper reaches, or headwaters, of watersheds. Most of their water comes from rain or snowmelt that runs off the surface. Streams may also be fed by groundwater that flows out of springs and seeps.

As streams flow down a watershed, they merge with other streams. Streams are classified by **stream order**, which refers to the number of streams feeding into them. A first-order stream flows from a single headwater; it does not have any other streams feeding into it. Two first-order streams converge to form a second-order stream. A third-order stream is formed

when two or more second-order streams converge, and so on.

The force of gravity pulls streams downhill. The speed of a stream is determined by the slope, or **gradient**, of the land. The gradient is calculated by dividing the change in elevation by the distance traveled. In mountains and hilly areas, gradients are high and streams flow quickly. Fast-moving water is able to carry large amounts of sediment; the energy of this water can also erode rocks and sediments in its channel. At lower gradients, water moves more slowly. Here water erodes its substrate more slowly, and its ability to suspend sediment is diminished. As a result, the sediments it had been carrying settle out of the water.

The amount of water flowing in a stream varies over time. **Perennial streams** flow all year round. Their flows vary from season to season and year to year, depending on the local patterns of precipitation and evapotranspiration. **Intermittent streams** flow only at certain times. Streams in headwaters are often intermittent, flowing only when it rains. Intermittent streams also occur where patterns of rainfall and evapotranspiration are distinctly seasonal (**Figure 12.9**).

A stream's **channel** is the waterway through which it normally flows. Next to the channel is the **floodplain**, the land that experiences periodic flooding. Hydrologists define the extent of a floodplain in terms of time. The annual floodplain is the area that is likely to be flooded each year. The 100-year floodplain is the area that is inundated during the extreme floods that are likely to occur only once every 100 years.

In headwaters where gradients are steep, streams have virtually no floodplains. Where the gradient is low, streams often have very broad floodplains. Stream channels in wide floodplains are often

▲ **Figure 12.9 An Intermittent Stream**
The Virgin River in Utah's Zion National Park has water only in the spring with melting snows and for a few days following summer storms.

bordered by embankments called **levees**. Natural levees form from sediments deposited during floods (**Figure 12.10**).

On broad, gently sloping floodplains, streams often flow in wide loops called **meanders**. As the stream flows, it erodes its banks and deposits sediment, eventually changing the shape of its channel (**Figure 12.11**). Thus, the floodplains and channels of large, naturally flowing streams and rivers are constantly being reshaped.

▼ **Figure 12.10 Floodplain Features**
A river's floodplain, shown here in cross-section, is shaped by the changing flow of water over thousands of years. Natural levees separate the stream channel from backwater areas, which receive water only during floods.

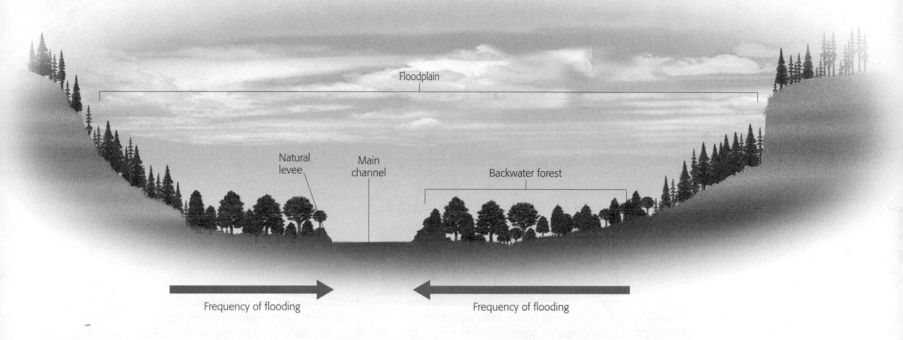

Floodplain

Natural levee

Main channel

Backwater forest

Frequency of flooding

Frequency of flooding

◄ **Figure 12.11 Meandering Stream**
Over many years, the channel of the Toad River in British Columbia, Canada, has meandered back and forth across its floodplain.

341

▶ **Figure 12.12 The Nile Delta**
Over millennia, the Nile has deposited sediments across its floodplain and delta (green). These regions are among the most fertile lands on Earth.

As streams enter lakes or the ocean, their gradient flattens out and they slow down. The slow-moving waters deposit large amounts of sediment. In some places, streams fan out into a complicated array of channels forming a **delta** (**Figure 12.12**). Stream and river deltas grow outward as sediment is continually deposited in stream channels. During floods, sheets of water may flow across large portions of the delta, adding even more sediment.

Stream Ecosystems
Streams influence ecosystem biodiversity and processes.

Scientists use the word **lotic** to describe ecosystems dominated by flowing fresh water. Lotic ecosystems include the communities of organisms that live in streams as well as the communities of organisms living on lands next to and beneath a stream. The flow of water greatly influences biodiversity and the movement of energy and nutrients through lotic ecosystems.

Food chains in streams. Much of the primary production that supports stream food chains comes from **detritus**, organic matter such as dead leaves and nutrient-rich soil that wash in from adjacent terrestrial ecosystems. Algae attached to rocks or other fixed structures in streams provide additional primary production (**Figure 12.13**).

In fast-flowing streams, planktonic algae are sparse because the flowing water quickly carries them downstream. Where the flow of water is slower or halted by natural or human-built barriers, planktonic algae are more abundant. Nitrogen and phosphorus in agricultural and municipal runoff also encourage the growth of algae. High quantities of these elements can lead to eutrophication (7.5).

The primary consumers in streams have three main ways of feeding (**Figure 12.14**). Scrapers are animals that use rasping mouthparts to feed on the algae attached to rocks. Snails and some species of insects and fish are scrapers. Shredders feed directly on leaves, small twigs, and other pieces of detritus that fall into streams. These consumers include the larvae of numerous insect species. Filter feeders use gills or netlike structures to

▶ **Figure 12.13 Energy Input**
Leaves and other debris from terrestrial ecosystems provide much of the organic energy for food chains in small streams.

▲ **Figure 12.14 Stream-feeding Strategies**
Freshwater snails Ⓐ use a rasplike tongue or radula to scrape algae from stream rocks. Blackfly larvae
Ⓑ shred leaf material with their grinding mouthparts. Freshwater mussels Ⓒ obtain their energy by
filtering organic matter from the water.

collect small bits of detritus carried in the flowing water. Freshwater mussels, worms, and the larvae of blackflies and caddisflies are common stream filter feeders. All of these animals have adaptations that allow them to grasp or attach themselves to rocks and avoid being carried downstream by the flowing water.

Stream herbivores and detritus feeders are very sensitive to changes in water chemistry and sediment load. Thus, their numbers and diversity are commonly used as indicators of stream pollution. When pollution is low, they are generally abundant. When pollution is high, they are scarce or absent.

The various herbivores and detritus feeders are the primary prey for a number of carnivorous animals, including crayfish, beetles, and the larvae of dragonflies and damselflies. They are also the primary source of food for many carnivorous fishes, such as trout. These carnivores depend on their mobility to capture their prey.

Riparian zone. The region of transition between a stream and the terrestrial ecosystems that surround it is called the **riparian zone**. The plants growing at the edge of the stream channel must be able to tolerate frequent flooding, saturated soils, and low levels of soil oxygen. Farther from the stream bank, flooding is less frequent and soils contain more oxygen. The farthest extent of the riparian zone is rarely flooded, but many of the plants growing there draw their water from the shallow aquifers associated with rivers.

Although riparian zones represent a relatively small part of the landscape, they include a diverse array of species and provide a variety of ecosystem services. Organic matter from riparian vegetation provides much of the energy that sustains stream food chains. Riparian vegetation and soils filter and cleanse the runoff and groundwater coming into rivers. This is particularly important in agricultural landscapes, where riparian zones absorb excess nutrients running off fields. During storms, the plants and soils in the riparian zone slow down the movement of water, thereby alleviating the effects of flooding.

The hyporheic zone. The region of saturated sediment next to a stream and immediately beneath it is called the **hyporheic zone**. A variety of microorganisms and invertebrate animals live in this zone.

Stream water is constantly flowing into and out of the hyporheic zone, filtering out organic matter and sediment as it does (**Figure 12.15**). This process facilitates nutrient cycles. Organisms in both the stream channel and the hyporheic zone cycle organic matter through the food web and return dissolved nutrients to the water.

◄ **Figure 12.15 Hyporheic Zone**
Stream water is constantly moving between the flowing channel (blue) and the saturated sediments beneath (brown). Nutrients and organic matter are filtered as water moves through the hyporheic zone.

Stream flow

Hyporheic zone

Hyporheic zone

Human Use and Impacts

Humans have modified streams in order to minimize the risks of both floods and water shortages.

Throughout history, much agricultural and urban development has depended on rivers. River floodplains provide some of the world's most fertile agricultural lands. The great majority of the world's cities have grown up beside rivers. Rivers provide energy, transportation for commerce, and water for industrial and residential use. Nevertheless, season-to-season and year-to-year variations in river flow have presented significant challenges to both urban and agricultural development. Throughout history, humans have taken a variety of actions to manage floods and periods of low flow. Many human activities have also polluted streams.

Dams and diversions. Humans construct dams to control floods, store water, and generate electricity. Dams restrict the flow of water on 180 of the 300 largest rivers in the world. In the United States alone, states regulate the activities of over 79,000 dams that are higher than 7 m (25 ft). There is no question that dams provide an important source of renewable energy. Their impoundments store water that can be diverted to irrigate dry places or used during dry seasons. But they also have significant negative impacts on stream ecosystems.

By design, dams create lakes with little or no water movement. When streams flow into these lakes, they slow down and drop the sediment they are carrying. As sediments accumulate, they reduce the capacity of the lake to store water. If streams carry high loads of sediment, lakes may fill in rather quickly. In contrast, streams flowing out of dams carry little sediment. Therefore, these streams cannot replace the sediments lost in the natural erosion occurring on stream banks below the dams (**Figure 12.16**).

Dams fragment stream ecosystems, often interfering with the movement of organisms such as salmon and other migratory fish. Today, most dams are required to have fish ladders or other means of moving fish over these barriers. But because most migratory fish use the flow of the water to navigate upstream, they may lose their sense of direction in the still waters behind dams (16.2).

Dams increase the total storage of water within watersheds, but they also reduce the flow downstream. Lower flows reduce the amount of water available across floodplains, thereby altering the composition of riparian forests and other floodplain ecosystems. Dams also alter seasonal patterns of water flow, which can have a significant effect on ecosystems downstream. Water diversions such as canals often move water to new locations, thereby diminishing the flow of water to the sea.

Dams and diversions have greatly altered water flows and ecosystems on the lower Colorado River. It is estimated that the Colorado River once delivered about 18.5 km^3 (4.4 mi^3) of water to Mexico's Gulf of California each year. Today, about 3.2 km^3 (0.7 mi^3) of this water is lost to seepage and evaporation in reservoirs such as Lakes Powell and Mead. Another 13.5 km^3 (3.2 mi^3) is diverted in canals to meet the needs of agriculture and

► **Figure 12.16**
Sedimentation
As rivers flow into lakes and reservoirs, they flow down and deposit the sediments they have been carrying. A delta of sediment is forming where the Colorado River meets Lake Powell Ⓐ. Downstream from Lake Powell's Glen Canyon Dam, the water contains very little sediment. As a consequence, sediments downstream are not replenished, and habitats such as sandbars that once supported willows and other riparian plants are now gone Ⓑ.

urban growth in California, Nevada, and Arizona. Only 1.8 km³ (0.43 mi³), about 10% of the historic flow, now makes its way to the Mexican border. This is exactly the amount guaranteed to Mexico by treaty with the United States.

Channelization and artificial levees.

Humans have built artificial levees and channelized rivers to control the flow of water, prevent flooding, and make floodplains available for agriculture and urban development. Artificial levees elevate stream banks in order to confine water to stream channels. When a river is channelized, it is dredged and straightened (**Figure 12.17**). Channelization usually involves the removal of most riparian vegetation.

While channelization and the construction of artificial levees may protect the lands adjacent to a particular part of a river, they also increase the speed of the water flowing through the modified channel. This greater speed increases the potential that flooding will occur downstream. In response to this threat, additional levees or channels may be constructed on lower parts of the river, until much of the length of the river is managed. For example, a series of 5- to 15-m (16–49 ft) tall levees extends along more than 1,600 km (995 mi) of the Mississippi River.

Pollution.

Chemicals from a variety of human activities pollute many rivers and streams. **Point-source** pollution comes from a specific location, such as an industrial facility or municipal sewage treatment plant. **Non-point-source** pollution comes from a variety of activities occurring at different places across landscapes. Urban stormwater and fertilizer-enriched runoff from croplands are examples of non-point-source pollution (**Figure 12.18**).

Pollution has devastated the biodiversity of many streams. Filter-feeding animals are particularly sensitive

▲ **Figure 12.17 Stream Channelization**
This segment of the River Brue near Somerset, England, has been straightened so that its floodplain can be used for agriculture.

to pollutants, especially sediments, which clog their gills. Today, over half of the 308 species of freshwater mussels native to streams in the United States are either endangered or have gone extinct.

Recognizing the intimate connection between water quality and stream life, the U.S. Environmental Protection Agency (EPA) has encouraged state water managers to develop bioassessment criteria for their streams. A **bioassessment** consists of surveys of aquatic biodiversity, including vegetation and animals such as fish, insects, and mussels, that are known to be sensitive to overall water quality. As of 2010, all 50 states had developed bioassessment programs, as well as criteria for ranking water quality based on those assessments.

QUESTIONS

1. How does water movement and sediment transport vary as streams move from areas of steep gradients to areas where the gradient is low?

2. Describe the three different groups of animals that feed on plant matter in streams.

3. Describe three important human impacts on stream ecosystems.

▲ **Figure 12.18 Sources of Water Pollution**
Stormwater runoff from streets and other impervious surfaces Ⓐ is an example of non-point-source pollution. Chemical-laden water released from industrial facilities Ⓑ is an example of point-source pollution.

12.3 Lakes and Ponds

Lakes and ponds are inland bodies of water that fill topographic basins. Although the water in them may slowly circulate, lakes and ponds have little or no directional flow or current. Scientists use the term **lentic** to describe the ecosystems associated with lakes and ponds. Lentic ecosystems account for 123,000 km^3 (29,000 mi^3) or 0.01% of Earth's water. About 30% of this water occurs in 120 lakes that have areas greater than 1,000 km^2 (390 mi^2). Humans have altered water storage in lentic ecosystems by creating artificial lakes and reservoirs and by diverting water from streams that feed natural lakes. In most lakes, primary production is limited by the availability of nutrients, particularly nitrogen and phosphorus. Runoff from agriculture, industry, and urban development has increased the concentration of nutrients in many lakes. Non-native species have also altered many lake ecosystems.

Characteristics of Lakes and Ponds

In geologic time, most lakes have a short life span.

There are over 300 million inland bodies of water on Earth, but most of them (90%) are smaller than 1 ha (2.5 acres). Bodies of water that are fed by a stream and have a surface area greater than 5 ha (12.5 acres) are generally classified as lakes; smaller bodies of water are called ponds. However, this distinction is somewhat arbitrary, and there is much overlap in the use of these terms, even by scientists.

Natural lakes are formed in several ways. In some places, the movement of tectonic plates along geologic faults produces basins that collect water. Lake Baikal in Russia and the lakes in the Rift Valley of Africa were formed this way. Other lakes form as a result of erosion and deposition by glaciers. Glaciers formed the Great Lakes, as well as many smaller lakes. Kettle lakes form in depressions left after large blocks of glacial ice melt.

▼ **Figure 12.19 Kinds of Lakes**
Africa's Lake Kivu Ⓐ formed in the rift valley created by the movement of Earth's tectonic plates; today it is 89 km (55 mi) long. Kettle lakes Ⓑ such as this one in the Yukon Territory, Canada, are formed by the melting of large pieces of ice left behind by the retreating continental glacier. Oxbow lakes Ⓒ, such as this one on the Blackfoot River, Montana, form in abandoned loops of meandering rivers.

Oxbow lakes are found in abandoned meander loops along the lower reaches of large streams and rivers. Oxbow lakes are typically shallow (**Figure 12.19**).

The life span of most lakes is relatively short, at least in terms of geologic time. Most lakes are considerably less than 1 million years old. Why are they so short-lived? Lakes are constantly receiving sediments from their banks and the rivers that feed them, causing them to fill in. Over time, stream deltas gradually invade their waters. Vegetation grows inward from the edges of lakes, increasing the rate at which mineral and organic sediments accumulate. Eventually sediments fill so much of the lake that it is transformed into a wetland, and other plant species become established.

Earth's oldest lakes are associated with tectonic rifts. Russia's Lake Baikal is thought to be over 30 million years old. Lake Baikal is actually growing larger. Geologic faulting is enlarging Lake Baikal's basin much faster than the rate at which sediments are being added to it.

The volume of a lake is controlled by the difference between inflow and outflow. Inflow is determined by the amount of rain that falls on the watershed that drains into the lake. Some lakes also receive inflow from groundwater. Outflow from a lake occurs when water flows into outlet streams, percolates into groundwater, and evaporates or is removed for human use. Climatic conditions influence these fluxes and cause seasonal and year-to-year changes in water level.

Most lakes have **open basins**, meaning they are drained by a well-defined stream or river. Lakes with **closed basins** have no outlet stream. They lose water by evaporation and percolation into underlying sediments.

▲ **Figure 12.20 Closed-basin Lake**
Water flows into Utah's Great Salt Lake from all directions, but it has no outflow streams. Over thousands of years, evaporation has concentrated the salts in the lake to the point of saturation.

As water evaporates, dissolved mineral salts such as sodium chloride are left behind. Thus, where evaporation is the prevailing mode of water loss, as in the Great Salt Lake, the Dead Sea, and the Aral Sea, lake water may be saturated with salts (**Figure 12.20**).

Lentic Ecosystems

Habitats vary across lakes and at different depths.

Scientists divide lake environments into the **benthic zone**, the waters and sediments on the lake bottom, and the **pelagic zone**, the waters that are not close to the bottom (**Figure 12.21**). The pelagic zone is further subdivided into the **photic zone**, which receives sufficient sunlight to support photosynthesis, and the **aphotic zone**, deep waters that receive little or no sunlight. The depth of the photic zone depends upon the clarity of the water, varying from 1 to 2 m (3.3–6.6 ft) in very turbid lakes to over 50 m (164 ft) where waters are very clear.

Energy and nutrients are cycled among these zones by sedimentation and the circulation of lake water. In the pelagic zone, the most important primary producers are **phytoplankton**, the various algae suspended in the water. Phytoplankton include cyanobacteria, green algae, and single-celled organisms called diatoms. Phytoplankton use the light energy in the photic zone to carry out photosynthesis—taking up dissolved carbon dioxide and manufacturing carbohydrates. This process also enriches the water in the photic zone with oxygen.

Zooplankton, the diverse array of tiny protists and animals suspended in pelagic waters, feed on phytoplankton as well as bits of organic matter (**Figure 12.22**). Phytoplankton and zooplankton are eaten by filter-feeding animals, such as insect larvae. Those animals are eaten by small fish, which are themselves prey for larger fish. Phytoplankton and zooplankton are constantly sinking into the aphotic and benthic zones.

There they may be eaten by small invertebrates and fish. These animals are preyed upon by larger fish.

The total pool of nutrients in a lake has a significant effect on its net primary production. The water and sediments of young lakes often contain low amounts of nitrogen and phosphorus. These conditions support only small populations of phytoplankton, so the waters of young lakes are generally clear. Over time, streams deliver additional nutrients to lake waters and sediment. As a result, primary production increases and the clarity of the water diminishes.

When organisms in the photic zone die, their remains drift to the bottom of the lake, causing nutrients such as nitrogen and phosphorous to accumulate in the sediments of the benthic zone. If lake waters remain unmixed, the pelagic zone gradually becomes depleted of nutrients. However, this does not occur in shallow lakes and ponds because winds continuously mix the water, moving it among the benthic, aphotic, and photic zones.

The waters of deeper lakes are mixed by seasonal changes in temperature and density. In the summer, sunlight heats the surface waters so they are warmer than the deeper waters, and therefore less dense. That difference causes the warm water to stay on the surface. In the fall, the surface waters cool more quickly than the deeper waters. As a result, the denser surface waters sink, forcing the nutrient-rich waters from the bottom upward. That change causes deep water to circulate to the surface. In such lakes, nutrient availability and primary production change from season to season (7.8).

▼ Figure 12.21 Lentic Zones
The open waters of a lake make up the pelagic zone. The bottom and near-bottom waters are the benthic zone. The littoral zone occurs near the shore where enough light passes through the water to support rooted vegetation.

Closterium, a desmid alga

Ⓐ

Ⓑ

Phyllopod water flea, *Polyphemus pediculus*

◄ Figure 12.22 Lake Plankton
Phytoplankton such as this green alga Ⓐ form the base of lake pelagic food webs. These primary producers are consumed by microscopic zooplankton such as water fleas Ⓑ.

Terrestrial plants

Emerged plants

Floating plants

Submerged plants

LITTORAL ZONE

PELAGIC ZONE (OPEN WATER)

BENTHIC ZONE

PHOTIC ZONE

APHOTIC ZONE

The **littoral zone** is the shallow, sloped area relatively close to shore. This zone is often defined by the distribution of rooted vegetation. Thus, its precise depth depends upon how murky the water is, which affects the penetration of sunlight. The aquatic plants in the littoral zone provide much of the primary production. Leaves and other detritus from nearby terrestrial plants are an additional source of food. This abundant organic matter supports a variety of invertebrates, including aquatic worms, snails, insect larvae, and crayfish. These animals are food for fish that live in the littoral zone, such as bass, and other animals, such as turtles, wading birds, and raccoons, that feed at the water's edge (**Figure 12.23**).

▲ **Figure 12.23 Shoreline Critters**
The primary production of the plants growing in the littoral zone supports complex food webs that include carnivores such as great blue herons and raccoons.

Human Use and Impacts

Human actions have altered the quantity and quality of water in lakes.

Worldwide, humans have dammed rivers to create hundreds of thousands of artificial lakes, or impoundments. These impoundments have added 10,800 km³ of water storage to land, nearly 9% of Earth's total lake water. If all this water were immediately returned to the ocean, sea level would rise by 3 cm (1.2 in.).

By diverting water from the streams that feed them, humans have diminished the amount of water stored in some natural lakes. For example, prior to 1960, the Aral Sea between Kazakhstan and Uzbekistan was Earth's fourth largest lake. The Amu-Dar'ja and Syr-Dar'ja Rivers provided water to the Aral Sea at approximately the same rate that water evaporated from its surface. Since then, water from both these rivers has been diverted to irrigate more than 8 million hectares (20 million acres) of former desert. As a result, the Aral Sea has shrunk to less than half its former size (**Figure 12.24**).

Streams carry nutrients from stormwater, factories, and farms to lakes, where they accumulate in the water and sediments. High levels of nutrients may cause eutrophication, in which populations of a few species of algae increase rapidly. In the photic zone of eutrophied lakes, concentrations of oxygen are high during the day when photosynthesis peaks, but very low at night due to high rates of respiration by the algae. In the aphotic and benthic zones of eutrophied lakes, the decomposition of large amounts of organic matter keeps the level of oxygen very low. These low levels of oxygen greatly reduce the biodiversity of eutrophied lakes.

Air pollution has altered the quality of lake waters in many locations. The combustion of coal releases significant quantities of sulfur and nitrogen oxides (SO_x and NO_x) into the atmosphere. These chemicals dissolve in rainwater to produce sulfuric and nitric acids, which may acidify lake waters.

Like many other ecosystems, some lakes have been significantly affected by the introduction of non-native species. For example, zebra mussels are native to the Black and Caspian Seas. In about 1987, ballast water from a ship accidentally released zebra mussels into Lake St. Clair, the lake that links Lake Huron and Lake Erie. Within just a few years, zebra mussels had spread throughout all of the Great Lakes. In addition to displacing native species of mussels, fast-growing colonies of zebra mussels have damaged harbors and boats and clogged the intake and discharge pipes of power plants.

In summary, lake ecosystems are influenced by the flow of water into and out of them, the nutrient content of their water, and the ecological communities that inhabit them. In many lakes, human activities have altered nearly all of these attributes.

QUESTIONS

1. Describe three ways in which lakes form.
2. Lakes have relatively short lives in geologic time. Why is this so?
3. What factors influence primary production in lakes?
4. Describe three important ways in which human activities affect lakes.

1973   1989   2000   2010

◄ **Figure 12.24 Aral Sea**
The Aral Sea is a closed basin lake that lies on the border between Kazakhstan and Uzbekistan. It has been rapidly shrinking since these countries began diverting water from the rivers that feed it in order to irrigate crops. In 1973, Aral Sea water covered 68,000 km² (26,300 mi²) but today it covers only about 10% of that area.

12.4 Groundwater

About 0.3%, or 10 million cubic kilometers, of Earth's water resides in rocks and soil beneath the ground. Most of this groundwater is in aquifers, layers of sediment and rock that are saturated with water (7.3). There is wide variation in the amount of water stored in aquifers and the rate at which that water moves through them. Aquifers in some regions formed long ago when the climate was wetter; little water enters these aquifers today. Humans use wells to extract groundwater for agricultural, industrial, and residential uses. The global rate at which humans use groundwater is about 38% of the rate at which this water is replenished. But in many locations, humans are withdrawing water far more rapidly than it is being replenished. Human activities have also polluted many aquifers.

Characteristics of Groundwater

Saturated sediments and rocks hold a significant portion of Earth's fresh water.

Rainwater that does not evaporate or run off percolates into the ground. There the force of gravity pulls it down through the pores and fractures in soil and rock. Eventually it arrives at the **water table**, the underground depth where rock and sediment are completely saturated with water. An aquifer is the saturated zone beneath the water table from which water can be extracted (**Figure 12.25**).

A **recharge zone** is an area where water flows directly between the soil surface and the water table. Aquifers located beneath recharge zones are said to be **unconfined**. In other aquifers, groundwater is trapped between layers of comparatively impermeable rock or sediment; such aquifers are said to be **confined**.

Once groundwater reaches an aquifer, it continues to flow downhill. Its rate of movement depends on the structure of the rocks and sediments in the aquifer. Generally, water moves most easily through coarse sediment or highly fractured rock. In these materials, water may move several meters per day. In clay sediments or unfractured crystalline rocks, groundwater may move less than a meter per year!

Groundwater flows to the surface in **discharge zones**. In places where a confined aquifer meets the soil surface, water may flow out freely in a seep or spring. Groundwater may also be discharged directly into rivers, lakes, or the ocean.

The size of aquifers varies considerably. On the major continents (excluding Antarctica), approximately 30% of the land is underlain by homogeneous aquifers that extend over thousands of square miles. However, most aquifers are smaller and are found where porous rocks and sediments are located.

There is also great variation in the recharge rate of aquifers. Recharge rates are high in aquifers near rivers and lakes and in regions with abundant rainfall. They are very low in aquifers in arid regions. In Germany, for example, average rainfall is about 800 mm (31.5 in.), of which about 15% or 120 mm (5 in.) recharges aquifers. In the desert country of Namibia, average rainfall is only 285 mm (11 in.), and only about 1%, or 3 mm (0.1 in.), percolates down to the water table.

► **Figure 12.25**
Groundwater
Aquifers are regions of saturated rock or sediment from which water can be pumped. Confined aquifers occur between relatively impermeable strata. The water in confined aquifers may be under pressure causing it to rise in wells above the level of the surface of the aquifer.

Recharge zone

Unsaturated zone

Stream

Unconfined aquifer

Confining layer

Confined aquifer

The residence time for groundwater varies from decades to thousands of years, depending on the rate at which the water flows and its depth in an aquifer. Where recharge and discharge rates are high, residence times are generally short. However, the natural recharge and discharge of many large aquifers is very slow. The water in such aquifers has accumulated over tens of thousands of years.

Some regions of the world have vast groundwater reserves that formed thousands of years ago when wetter climatic conditions prevailed. For example, the Nubian Sandstone Aquifer extends across 2 million square kilometers of the northeastern Sahara Desert, beneath the countries of Chad, Egypt, Libya, and Sudan. Water entered this aquifer during a wet period 5,000–10,000 years ago. Today, there is very little recharge to this aquifer (**Figure 12.26**).

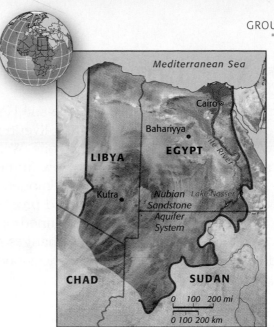

◀ **Figure 12.26 No Recharge**
Rain that fell over 5,000 years ago created the Nubian Sandstone Aquifer beneath the Sahara Desert. This aquifer is providing much needed water in this dry region, but accelerated use of this water could deplete this nonrenewable resource.

QUESTIONS

1. What is the difference between a confined and an unconfined aquifer?
2. What factors influence the residence time of water in aquifers?
3. Is the rate at which humans are currently using groundwater sustainable? Explain why or why not.

Human Use and Impacts

In many places, the extraction of groundwater for agricultural, industrial, and domestic uses far exceeds the recharge rate.

Humans access groundwater by digging or drilling wells. To understand how a well works, think of sipping water through a straw. A well is like a straw, and an aquifer is like the liquid in a glass. The analogy is even better if the straw has perforations that allow liquid to come in through the sides as well as the bottom. A pump provides the force to move water up the well casing. Water from the aquifer then moves into the well through perforations in the casing, replacing the water that has been withdrawn. If the rock or sediment in the aquifer is very porous, water moving through the aquifer quickly replaces the pumped water. In such situations, water can be pumped rapidly. Where groundwater flows slowly, the rate of pumping must be slower (**Figure 12.27**).

Sometimes a well is drilled into a confined aquifer where the groundwater is under great pressure from the weight of the water above it. The pressure causes the water to rise above the confining layer, without the need for pumping. Such a well is called an **artesian well**.

Today, humans pump approximately 3,800 km³ (912 mi³) of water from Earth's aquifers each year. Total annual recharge to Earth's aquifers is approximately 10,000 km³ (2,400 mi³). Thus, humans currently withdraw about 38% of the water that enters the world's aquifers each year. About 68% of this water is used to irrigate crops, 21% is used in industry, and 11% goes to domestic uses. This level of groundwater use appears sustainable, since the total natural recharge rate exceeds the total rate of human withdrawal. However, human dependence on groundwater is usually highest in arid regions where recharge rates are typically very low. As a result, aquifers in many arid regions are rapidly being depleted.

The Ogallala Aquifer of the High Plains region of the United States is an example of just such a situation. This vast groundwater reservoir extends from South Dakota to the Texas Panhandle. Most of the water in the Ogallala was deposited during the last ice age, over 12,000 years ago.

The present-day recharge to the Ogallala comes mostly from the Rocky Mountains to the west and is very slow. Significant pumping from the Ogallala began in the 1930s, following the droughts of the Dust Bowl era. Indeed, the water from this aquifer led to the restoration of agriculture in many parts of the Dust Bowl. The U.S. Geological Survey estimates that the total amount of water stored in the Ogallala has decreased by 10% since that pumping began. Where recharge rates are especially low, the loss has been much greater. In parts of Texas and southwestern Oklahoma, the volume of water in the aquifer is down by more than 40%.

Aquifers are the primary source of drinking water in many parts of the world. As water percolates through soil and rock, microorganisms and chemicals are filtered from it, making groundwater less polluted than surface water. Nevertheless, natural processes and a variety of human activities can alter the quality of groundwater. In some regions, groundwater contains high concentrations of natural chemicals, such as arsenic, nitrates, and sulfates, which limit its use by humans. Near seacoasts, overpumping can cause seawater to move shoreward, increasing the salinity of freshwater aquifers.

Despite the filtering ability of soils and sediments, human-generated pollutants can sometimes make their way into aquifers. If storage tanks for gasoline, oil, or other chemicals leak, their contents may seep into shallow aquifers. Household chemicals and bacteria and viruses from faulty septic systems sometimes make their way into groundwater, often through nearby wells.

▲ **Figure 12.27 Pumping Water**
As a pump pulls water from a well, water moves into the well casing from the surrounding aquifer. This pulls the water table down in the vicinity of the well. When pumping ceases, the water table recovers. The rate of recovery depends on the permeability of aquifer rocks and sediments.

351

12.5 Wetlands

Wetlands are ecosystems in which the soil is saturated with water all year long or for a significant part of the year. Plants that live in wetlands have adaptations that allow them to grow and reproduce in saturated soils. Wetlands are classified by the source of their water and the kinds of plants that grow in them. Wetlands provide a number of important ecosystem services: They store water, moderate the flow of water in floods and droughts, protect and improve water quality, and provide habitat for fish and wildlife. A large portion of the world's wetlands have been drained or altered to provide land for agriculture and urban development. In many areas, changes in drainage patterns and the supply of water have degraded the functions and services once provided by wetlands.

Types of Wetlands

Wetlands are a study in life on the edge, where land meets water.

Because oxygen diffuses through saturated soils very slowly, wetland soils contain very little oxygen. Most plants cannot survive in such low-oxygen conditions. Some species of plants have adaptations that allow them to grow in these anaerobic soils; these plants are called **hydrophytes**. The combination of saturated soils and hydrophytes separates wetlands from other kinds of ecosystems. Wetlands are classified into four categories based on the source of their water supply and the kinds of plants that grow in them.

Marshes are periodically or continuously flooded wetlands that are dominated by herbaceous plants, including grasses, rushes, reeds, and cattails. When marshes are flooded, water levels are generally shallow. Marshes may be tidal or nontidal.

Tidal marshes are found along seacoasts where the level of water is influenced by the daily and monthly ebb and flow of tides. The most extensive tidal marshes are salt marshes, which occur in salty water near the ocean. Although salt marshes are highly productive ecosystems, their vegetation is often dominated by a single species, cord grass (**Figure 12.28**). Upstream from salt marshes, but still within the influence of tides, are tidal freshwater marshes. Because their water is not salty, a great many more plant species can coexist in these marshes.

Nontidal marshes are also dominated by grasslike herbs. These freshwater marshes occur on floodplains and in poorly drained depressions. They also occur along the edges of streams and lakes, where they form part of the littoral zone. Nontidal marshes include temporary, or ephemeral, ponds, which contain water only during wet seasons or wet years. In regions with distinct wet and dry seasons, temporary ponds undergo predictable patterns of change as they fill up and dry out (**Figure 12.29**).

▼ **Figure 12.28 Tidal Marsh**
This salt marsh on the Edisto River near the coast of South Carolina is inundated by tidal waters twice each day. Tidal marshes may extend inland for 10–20 miles.

▼ **Figure 12.29 Fen**
The Everglades of south Florida is one of the world's largest fen wetlands. It is fed from waters that drain Lake Okeechobee.

Swamps are dominated by shrubs or trees and fed by flowing water. Forested swamps are found in the broad floodplains of many of the world's major rivers. They are especially common in the southeastern and south-central United States, where they are dominated by tall trees such as bald cypress and water tupelo (**Figure 12.30**). Mangrove swamps are found along seacoasts in many tropical and subtropical regions. The shrubs and small trees of mangrove swamps are able to tolerate high levels of salinity and regular tidal inundation (**Figure 12.31**).

The primary source of water in **bogs** is rainfall. Bogs are wetlands with peat deposits that support a variety of evergreen trees and shrubs. Sphagnum moss often forms a thick, spongy carpet under these woody plants (**Figure 12.32**). Rainfall contains few nutrients, and the soils of bogs typically have low level of nutrients. Bog plants have adaptations that allow them to grow and reproduce despite the limited supplies of nitrogen and phosphorus.

Fens are wetlands that are fed primarily by groundwater. They have peaty soils that typically support a variety of grasses and grasslike rushes, as well as occasional patches of woody vegetation. Fens are quite common in the glaciated regions of North America and Asia, but they also occur in subtropical settings. The Florida Everglades is one of the largest fens in the world.

Wetland Ecosystems

Wetland species are adapted to saturated, anaerobic soils.

Wetland plants possess a variety of adaptations that enable them to survive in anaerobic, water-logged soils. The roots of most hydrophytes are able to carry out anaerobic respiration, as do many bacteria and fungi. The stems of many wetland herbs contain specialized tissues that transport air down to their root systems.

The anaerobic conditions in wetlands slow the decomposition of dead plants and animals, limiting the availability of nitrogen and phosphorus. Wetland shrubs and herbs have root systems that are efficient at extracting limited nutrients from soil. In addition, most hydrophytes are quite frugal in their use of nutrients.

Many aquatic vertebrates and invertebrates thrive in wetlands, providing food for a host of other animals, including a notably wide range of water birds. Wading birds, such as herons and egrets, stalk their prey in the shallow waters, looking for frogs, small fish, and insect larvae. Swans and geese filter food from wetland sediments, and many ducks dive for fish in deeper water. Many reptiles, including freshwater turtles, crocodiles, alligators, and the anaconda (the world's largest snake), are at home in wetlands.

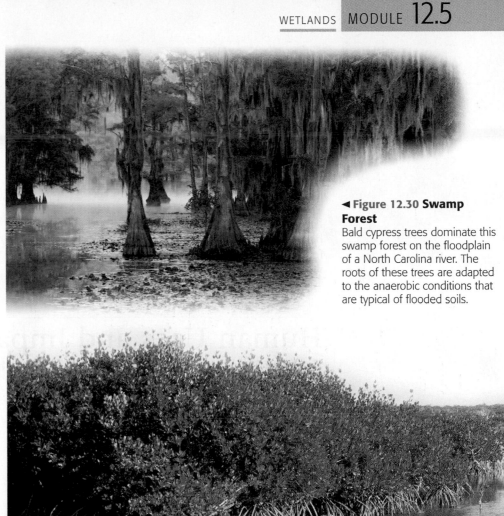

◄ **Figure 12.30 Swamp Forest**
Bald cypress trees dominate this swamp forest on the floodplain of a North Carolina river. The roots of these trees are adapted to the anaerobic conditions that are typical of flooded soils.

▲ **Figure 12.31 Mangroves**
Mangroves are able to tolerate anaerobic and saline conditions associated with regular tidal inundation by seawater.

▼ **Figure 12.32 Peat Bog**
The rainwater that feeds this shrubby bog in northern Minnesota contains very few nutrients.

Some mammals, such as beavers, muskrats, capybaras, water buffalo, and hippopotamuses, are wetland specialists, too (**Figure 12.33**).

Intact wetlands provide a host of ecosystem services. In addition to providing habitat for fish and wildlife, they store floodwater and protect water quality. Water flowing into wetlands is slowed down considerably by the dense vegetation, peat, and soils. Wetlands are like sponges; they soak up large amounts of water during storms and floods, and then release it slowly over time. Thus wetlands moderate the flow of water during both floods and droughts. Wetland soils and vegetation also have the capacity to absorb nutrients and sediment from the water they receive. These processes improve water quality.

▲ **Figure 12.33 Wetland Mammals**
Botswana's Okavango Delta wetlands support a diverse array of animals, including hippos that feed on submerged wetland vegetation.

Human Uses and Impacts

Worldwide loss and degradation of wetlands have greatly diminished the ecosystem services they provide.

QUESTIONS

1. Name and describe four types of wetlands.
2. What factors limit plant growth on wetland soils?
3. Describe three human actions that have degraded many wetlands.

More than half of the world's wetlands have been affected by human activities. Over the past 200 years, the total area of wetlands in the lower 48 United States has been reduced from 90 million to 40 million hectares (220 million to 100 million acres; **Figure 12.34**). Much of the remaining wetland has been degraded by drainage, changes in water supply, and vegetation damage. Wetlands in many areas have been drained to control mosquitoes or to facilitate nearby development and farming. Drainage can result in significant change in vegetation cover and accelerate the decomposition of peat. It can also make wetlands vulnerable to destructive wildfires.

There are limits to the capacity of wetlands to absorb nutrients and pollutants. Over long periods, sediment, fertilizer, sewage, pesticides, and heavy metals degrade the functions of these ecosystems. Overgrazing by animals, peat harvesting, logging, and the introduction of non-native plants can also degrade wetlands.

Many countries have set policies to halt the loss of wetlands. The United States, for example, has established a "no net loss" policy. Using powers provided under the Clean Water Act, the U.S. Army Corp of Engineers reviews actions by private landowners or public agencies that may impair or destroy a wetland. When those actions are approved and permits granted, the landowner or agency must compensate for the loss by restoring or protecting additional wetlands.

Wetland restoration projects are underway in many locations around the world (**Figure 12.35**). Scientists are developing techniques to create new wetlands and restore the ecosystem services that have been lost. Many of these projects and techniques show great promise. However, scientists agree that it is essential to protect Earth's wetlands that remain intact.

▶ **Figure 12.34 U.S. Wetland Loss**
Wetland loss has occurred during the past two centuries, but the rate of that loss was slowed by the passage of the 1972 Clean Water Act, which included provisions for wetland protection. In recent years, restoration efforts have been gradually increasing wetland area.
Source: Dahl, T.E. 2006. Status and trends of wetlands in the conterminous United States 1998 to 2004. US Fish and Wildlife Service, Washington, D.C. 112 pp. [From figure 26 on p. 46]

Chart — ACRES/YEAR vs YEARS
- 1950s–1970s: −458,000
- 1970s–1980s: −290,000
- 1980s–1990s: −89,550
- 1998–2004: +32,000

Legend: ● Gains ● Losses

▶ **Figure 12.35 Wetland Restoration**
Formerly drained and degraded, this 60-acre marsh near Seal Beach, California, has been restored and once again provides important ecosystem services.

FOCUS ON SCIENCE: Wetland Restoration Rates

To make up for the loss and degradation of wetlands, restoration projects have been undertaken in thousands of locations. In fact, in many states it is legal to change or even destroy wetlands to make room for developments or highway construction, as long as a previously degraded wetland is restored first. Such policies presume that newly restored wetlands have the same habitats and functions and provide the same services as the natural wetlands they replaced. However, few long-term data are available to determine how long it actually takes for a restored wetland to develop the characteristics of undisturbed wetlands.

Cornell University doctoral student Kate Ballantine and her mentor Rebecca Schneider investigated rates of change in wetlands in central New York State. They were particularly interested in the characteristics of soil in restored wetlands, since those characteristics influence net primary production and nutrient movement (**Figure 12.36**). In their investigations, they studied restoration projects in 30 marshes; these restoration projects ranged from 5 to 55 years in age. They sampled soils and vegetation biomass from each restored marsh. They also sampled soils and vegetation from five undisturbed natural marshes for comparison.

Ballantine and Schneider found that plant productivity increased rapidly in the first decade following restoration, but slowed thereafter. After 55 years, plant productivity in restored marshes was nearly equal to that in their natural counterparts. However, key soil characteristics were much slower to recover. Even after 55 years, the levels of organic matter in the soils of restored wetlands and the capacity of those soils to absorb nutrients were less than half of those in the natural wetlands (**Figure 12.37**). Their findings are significant because these soil characteristics are highly correlated with long-term fertility and important ecosystem services, such as the ability of a wetland to retain nutrients and pollutants.

Ballantine and Schneider's data represent one of the longest time intervals over which wetland restoration has been studied. They confirm what many scientists had suspected: Wetland restoration must be viewed as a long-term process because it takes decades, or even centuries, for many important wetland functions and services to return to their predisturbance condition.

How long does it take to restore a degraded wetland to its natural condition?

▲ Figure 12.36 **A Wetland Restoration Collaboration**
Research carried out by Kate Ballantine Ⓐ and Rebecca Schneider Ⓑ provides the best available data on the time it takes for the ecosystem functions and services in restored wetlands to recover to natural levels.

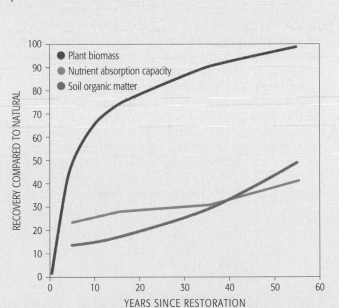

▲ Figure 12.37 **Soil and Vegetation Change**
Plant biomass, a direct measure of net primary production, recovers rapidly in restored wetlands. However, soil organic matter and nutrient absorption capacity recover much more slowly. At their current rates of recovery, these two characteristics of restored wetlands will not approach levels found in natural wetlands for a century or more.

Reference: Ballantine, K. and R. Schneider. 2009. Fifty-five years of soil development in restored freshwater depressional wetlands. Ecological Applications 19:1467–1480

12.6 Estuaries

Estuaries are partially enclosed bodies of water where fresh water from rivers and streams mixes with ocean water. The movement of water in estuaries is complex because it is influenced by tides, winds, and differences in the density of fresh and saltwater. Estuarine ecosystems are among the most productive ecosystems on Earth. Estuaries provide numerous direct benefits to humans, including fisheries, recreation, and scenic beauty. They also absorb sediment and nutrients and dissipate floods and storm surges. Human impacts on estuaries include shoreline development, dredging, pollution, and diminished inflow of fresh water.

Estuarine Water Flow

Estuaries are the transition from freshwater streams to the ocean.

Estuaries are known by a variety of names, including bays, lagoons, harbors, and sounds, although not all bodies of water called by those names are truly estuaries. True estuaries are regions of transition from freshwater to marine ecosystems. Their defining feature is the mixing of fresh water and saltwater. Familiar examples of estuaries include Chesapeake Bay, New York Harbor, San Francisco Bay, and Puget Sound.

The net movement of water in estuaries is from streams and rivers to the ocean. But because estuaries are usually broad and shallow, there is little difference in elevation from the river end to the ocean end. As a result, the flow of water is slow and often very complicated.

Fresh water is less dense than saltwater, so the fresh water from rivers generally flows out onto the surface of an estuary. As the surface water moves toward the ocean, it mixes with seawater and becomes saltier. This denser, saltier water then flows back toward the river along the bottom of the estuary (**Figure 12.38**). The time required for an estuary to completely cycle its contents from fresh water to saltwater is called **flushing time**.

A number of factors affect the circulation pattern and flushing time of an estuary. Tides produce a regular ebb and flow of ocean water. Winds and storms move surface water in various directions and contribute to the mixing of fresh water and saltwater. As a consequence, water levels and salinity at particular locations vary considerably from day to day, month to month, and season to season.

Nearly all estuaries are affected by seasonal variations in the rate of inflow from rivers. For example, the peak flow from the rivers that feed Chesapeake Bay occurs in the spring. By early autumn, their flow is much lower. This seasonal change in flow produces predictable changes in salinity and water movement (**Figure 12.39**).

Water movement and flushing time are also influenced by the size of the estuary, the complexity of its shoreline, and the number and size of the streams feeding into it. Chesapeake Bay and Puget Sound are fed by numerous rivers, each with its own pattern of flow. Where large rivers enter the sea, the mixing of fresh water and saltwater may extend well beyond the coast.

▼ **Figure 12.38 Estuarine Flow**
The complex flow of water in estuaries is influenced by water density. Fresh river water flows near the surface, mixing with saltier water as it moves toward the sea. Beneath it, denser saline water flows upstream.

Inflowing stream

Estuary

Ocean

Low-density fresh water moves seaward

mixing

Denser saline water is drawn up the estuary

► **Figure 12.39 Seasonal Salinity Change**

In the spring, heavy runoff in the rivers feeding Chesapeake Bay keeps salinity relatively low Ⓐ. Through the summer and into the fall, the rivers flow more slowly, allowing more seawater into the Bay and increasing its salinity Ⓑ.

Source: Chesapeake Bay Foundation. URL http://www.cbf.org/page.aspx?pid=943. Last accessed 8-9-11

Salt in parts per thousand
- 0–10
- 10–20
- 20–30

Estuarine Ecosystems

Estuarine production benefits adjacent terrestrial and marine ecosystems.

The complex topography and movement of water in estuaries usually results in a wide variety of habitats. Along the shore there may be sandy or stony beaches. At low tide, broad expanses of mudflats may be revealed. Most estuaries have salt marshes with tidal rivers flowing through them, and in warmer climates there are mangrove swamps. In the shallow water there may be beds of eel grass. Each of these habitats supports a distinctive community of plants and animals.

Abundant light and the constant input of nutrient-rich sediments make estuaries among the most productive ecosystems on Earth. The freshwater and saltwater wetlands that fringe most estuaries provide energy to their waters in the form of organic carbon. High rates of photosynthesis in seaweeds and phytoplankton also contribute to the high rate of net primary production. This productivity promotes an abundant and diverse array of other organisms.

The sheltered waters of estuaries provide ideal habitat for a variety of crustaceans, marine worms, and mollusks. Oysters, clams, and many kinds of worms filter organic matter from estuarine water and sediments (**Figure 12.40**). Crabs and other animals scavenge for larger bits of dead organic matter. These organisms are food for other animals, including a great variety of shorebirds and mammals such as raccoons and opossums (**Figure 12.41**). The abundance and variety of food make estuaries ideal stopping places for many migratory birds.

Many species of fish rely on the productive and sheltered waters of estuaries for spawning and early development. For this reason, estuaries are often called the nurseries of the sea. Fish such as flounder and striped bass may complete their entire life cycles within estuaries. Many large marine fish, such as mackerel, tuna, and swordfish, begin their lives in estuaries and then migrate to the open ocean.

▲ **Figure 12.40 Estuary Filters**
A single oyster filters about 200 L (about 53 gal) of estuary water each day. Where these animals have not been overharvested or depleted by pollution, their populations remain high. Healthy populations can filter the entire volume of an estuary in a week or two.

► **Figure 12.41 Estuary Critters**
The black oystercatcher Ⓐ is one of the many species of wading birds that feed on mollusks and crustaceans, such as the fiddler crab Ⓑ.

Human Use and Impacts

The many benefits of estuaries have encouraged human activities that threaten their sustainability.

Estuaries provide humans with a great variety of economic benefits and ecosystem services. Estuaries are important harbors for much of the world's commercial shipping. For example, New York, San Francisco, London, Tokyo, and Shanghai are built on estuaries. Coastal industries associated with estuaries, including recreation, tourism, and fishing, provide more than 30 million jobs. Estuaries provide habitat for over 75% of the world's commercial fish catch, valued at tens of billions of dollars each year.

Estuaries also provide important ecosystem services to coastal communities. Their diverse ecosystems filter nutrients and sediments from the water as it moves to the ocean. Their ecosystems also provide a buffer against floods and surges of ocean water, such as those associated with hurricanes.

Many human activities harm estuaries. Because estuaries occur at the base of large watersheds, pollutants of many kinds are carried into them. The excess nitrogen and phosphorus from farms eventually accumulates in estuaries, where it causes eutrophication. When the greatly increased populations of algae die, the high rates of respiration associated with their decomposition result in **dead zones**—regions of very low oxygen in which few marine animals can survive (7.5). Stormwater runoff from towns and cities adds additional pollutants that diminish ecosystem diversity and threaten human health.

Among the many human activities affecting estuaries, the alteration of the flow of the rivers that feed into them has been especially significant. Dams have changed the seasonal patterns of river flow. More importantly, the diversion of water has reduced the overall flow of fresh water into many estuaries.

San Francisco Bay provides a striking case study of such change. The primary sources of water for the northern portion of this very large estuary are the Sacramento and San Joaquin rivers. Prior to 1930, water flowing into San Francisco Bay from these two rivers averaged about 34 km^3 (8.2 mi^3) per year. Between 1930 and 1970, a number of dams and canals were installed to divert water to agricultural fields and urban centers. The California Aqueduct alone transports as much as 8 km^3 (1.9 mi^3) of water to the San Joaquin Valley and the city of Los Angeles. By 1980, the annual flow into the bay was less than 13 km^3 (3.1 mi^3), and today it is less than 10 km^3 (2.4 mi^3). Over that same time, springtime salinity in the northern portion of the bay has increased by three- to fivefold. These changes have altered the composition of species in the pelagic, benthic, and wetland communities in the northern portion of the bay and in the lower stretches of the rivers that feed it.

Many estuaries have been modified by dredging to facilitate boat traffic and by levee construction to allow the development of former wetlands. San Francisco Bay again provides a striking case study. Nearly 95% of the bay's tidal marshes have been lost to dredging and levee construction (**Figure 12.42**). The loss of these marshes has greatly diminished the ability of the estuary to filter out the municipal, industrial, and agricultural pollutants that rivers now carry to the bay.

Over the past few decades, communities have come to recognize the value of estuaries and the services they provide. As a result, they have begun to implement programs to restore and maintain these ecosystems. For example, the U.S. Environmental Protection Agency's (EPA's) National Estuary Program works with states and communities to restore shorelines and wetlands and to reduce the inputs of pollutants. The EPA also sets standards for salinity and regulates the minimum flow of rivers into many estuaries.

QUESTIONS

1. Explain how water in estuaries moves both toward and away from the sea.

2. The salinity of the water in estuaries is typically low in the spring and higher in late summer. Why?

3. Describe three important ecosystem services that estuaries provide to humans.

◄ **Figure 12.42 Estuarine Wetland Loss**
Destruction of most of the tidal wetlands in San Francisco Bay is a major factor contributing to the diminished water quality in this large estuary.
Source: Nichols, F. H., 2007. The San Francisco Bay and Delta–An Estuary Undergoing Change. U.S. Geological Survey. http://sfbay. wr.usgs.gov/general_factsheets/ change.html.

12.7 Oceans

Oceans cover 71% of Earth's surface and contain most of its water. Over millions of years, sodium chloride (NaCl) and other minerals have accumulated in ocean waters, so now their average salinity is about 3.5%. Earth's rotation, winds, and differences in water temperature produce currents that circulate water between the equator and the poles. Differences in water temperature and salinity produce currents that carry water between the ocean depths and surface. The vast majority of the oceans' biological diversity and primary production occur near land. Even so, a variety of ecosystems are found in deep ocean waters. Humans depend on marine ecosystems for an array of natural resources and ecosystem services. However, pollution, overexploitation, and climate change threaten these resources and services.

Ocean Currents

Ocean circulation redistributes heat around the Earth.

Oceans contain nearly 98% of the water on Earth. Oceans are the foundation of Earth's water cycle. They are the primary source of the water vapor in the atmosphere, and the ultimate destination for water flowing on and below the surface of all land. The circulation of ocean water has important consequences for all of Earth's ecosystems.

Surface currents. The surface waters of the ocean reach to a depth of about 400 m (1,320 ft). These waters circulate in currents that are driven by Earth's rotation, winds, and differences in water temperature. Near the equator, the high input of solar energy warms the surface waters of the Atlantic, Pacific, and Indian oceans. This warm water moves westward in currents propelled by the prevailing easterly winds.

As the equatorial currents approach the east coasts of South America and Asia, they are deflected toward the poles. From the poles, water flows back toward the equator along the west coasts of continents. This pattern of circulation produces the major ocean **gyres** (**Figure 12.43**). As a result of Earth's rotation, the circulation of these gyres is clockwise in the Northern Hemisphere and counterclockwise in the Southern Hemisphere.

Warm water loses heat as it moves toward the poles. Water that travels from the poles and back toward the equator is much colder. Thus, at any given latitude, the currents moving along the west coasts of continents are generally much colder than those moving along the east coasts. The temperature of these currents influences the climate of the nearby continents.

If you were to track a typical molecule of water, it would make a complete circuit in a surface gyre in about a year. The velocity of water in ocean surface currents

▶ **Figure 12.43 Major Ocean Currents**
The circulation of major ocean gyres brings warm water (red arrows) from equatorial regions along the east coasts of continents and toward the poles. They bring colder water (blue arrows) from the poles along west coasts toward the equator.

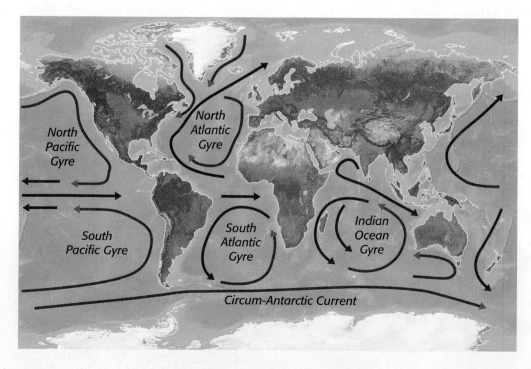

is generally less than 40–65 km (25–40 mi) per day. But some surface currents are much faster. The Gulf Stream, for example, flows out of the Gulf of Mexico between Florida and Cuba at about 160 km (100 mi) per day (**Figure 12.44**).

Vertical currents. Ocean waters also circulate vertically, exchanging surface water with water from the depths. This circulation, which is caused by differences in temperature and salinity, is called the **thermohaline circulation** (**Figure 12.45**). As water is transported to the North Atlantic, it loses heat and its salinity rises, causing it to become denser. Its greater density causes it to sink. This water then flows south, following a complex path along the ocean bottom. The water finally surfaces again in the Indian and Pacific oceans.

The thermohaline circulation moves very slowly—it takes hundreds of years for a water molecule to make a complete circuit. Yet because it moves vast quantities of water, the thermohaline circulation moves a great deal of heat. This transfer of heat has a significant effect on global climate.

On smaller geographic scales, the circulation of ocean waters is affected by coastal features, the inflow of rivers, and local winds. The interaction of these factors can result in areas of upwelling where cold, nutrient-rich water rises from the depths to the surface.

▶ **Figure 12.44 Gulf Stream**
Very warm water from the Gulf of Mexico, shown in red in this false-color satellite photograph, moves northward in the Gulf Stream along the U.S. east coast. This current moves at about 6 km/hr (4 mi/hr) along the Florida coast and slows to about 1.5 km/hr (1 mi/hr) farther north.

warm surface flow

cool subsurface flow

▲ **Figure 12.45 The Great Ocean Current**
Differences in water temperature and salinity affect water density, causing vast amounts of water to move in the thermohaline circulation. Cold water circulates in the deep ocean, rises in the North Pacific, then slowly travels back to the North Atlantic.

Ocean Ecosystems

Variations in ocean depth and nutrient availability produce a variety of habitats.

The ocean includes a great number of different habitats, including coastal regions and the open ocean. Like lakes, oceans are divided into pelagic and benthic zones.

Intertidal zone. The area along seacoasts that is submerged at high tides and exposed during low tides is the **intertidal zone**. Intertidal habitats include sandy beaches, mudflats, and rocky outcrops. Organisms in intertidal habitats must be able to tolerate regular cycles of wetting and drying as well as variations in water salinity. Rocky intertidal habitats support an especially diverse array of seaweeds and animals (**Figure 12.46**).

Pelagic zone. The pelagic zone comprises all of the oceans' open waters. It is subdivided into the photic and aphotic zones, based on the penetration of light sufficient to support photosynthesis (**Figure 12.47**). The depth of the photic zone varies considerably. In coastal waters that have large quantities of suspended plankton and sediment, light may only penetrate a few meters. Away from coasts and in deeper ocean waters, plankton are often sparse, and light may penetrate to depths greater than 200 m (660 ft). Even so, most of the ocean is in the aphotic zone—94% of ocean waters are deeper than 200 m (660 ft), and nearly half are more than 3,000 m (9,800 ft) deep.

Primary production by phytoplankton is highest near estuaries and in zones of upwelling where there is a steady influx of nutrients. These areas often support very complex food webs with as many as five trophic levels.

Organic matter (and the nutrients it contains) is constantly settling out of the photic zone into deeper waters. The constant "rain" of this organic matter, or detritus, and the daily migration of zooplankton between photic and aphotic zones support a diverse array of pelagic invertebrates and fishes. The rain of detritus diminishes with depth. At great depths, it is sparse indeed.

▼ **Figure 12.46 Between the Tides**
Rocks in the intertidal zone provide a secure substrate to which a rich variety of seaweeds and animals can attach themselves. However, these organisms must tolerate significant wet and dry periods as the tides move in and out.

▲ **Figure 12.48 Aphotic Zone Predator**
This angler fish has a luminescent "lure" projecting from its head that it uses to attract prey in dark deep ocean waters.

▲ **Figure 12.49 Kelp Forest**
In nutrient-rich coastal waters, giant kelp over 31 m (100 ft) long form a forest that sustains a diverse array of animals.

PHOTIC ZONE

APHOTIC ZONE

ABYSSAL PLAIN

Organisms here display a wide range of adaptations for acquiring energy (**Figure 12.48**).

Benthic zone. The ocean bottom, or benthic zone, includes a variety of habitats. In shallow waters where light is sufficient and the bottom substrate is firm, attached seaweeds are often abundant and a primary food source for animals (**Figure 12.49**). Few seaweeds can attach to shifting sandy substrates.

Coral reefs are by far the most diverse and productive benthic marine ecosystems. They occur primarily in shallow tropical waters between the equator and 30° north and south latitude. The predominant organisms are stony corals. The calcium carbonate skeletons of these invertebrate animals form the physical structure of the reef. Corals form symbiotic relationships with single-celled algae, which are the most important primary producers in coral reef ecosystems.

Coral reef ecosystems support extraordinary biodiversity, even though they are found in very nutrient-poor water (**Figure 12.50**). The close symbiotic relationship between the coral and its symbiotic algae ensures that scarce nutrients are efficiently recycled. Furthermore, many of these symbiotic algae are able to fix nitrogen (7.5).

In the deep ocean, the benthic substrate is a fine ooze composed of the skeletons of the planktonic organisms that have drifted down from the waters above. In these deep waters there is no light to support algae. Instead, invertebrates scavenge for bits of organic matter that rain in from the waters above.

The organisms that live on and near deep-sea thermal vents are part of the ocean's most peculiar ecosystem. Here, primary production is carried out by chemosynthetic bacteria, which use the chemical energy in the sulfur released from the vents to produce carbohydrates (4.1). These bacteria are consumed by filter-feeding clams and worms, which then become prey for other organisms (**Figure 12.51**).

◄ **Figure 12.51 Thermal Vents**
Chemosynthetic bacteria are the primary source of food for these peculiar segmented worms that live in the waters around deep-ocean thermal vents.

▼ **Figure 12.47 Ocean Zones**
As with lakes, Earth's oceans can be divided into several zones. Photosynthesis occurs in the photic zone. Primary production is highest in the coastal ocean zone because of nutrient inputs from estuaries.

▲ **Figure 12.50 Coral Reefs**
These ecosystems owe their remarkably high diversity to primary production by symbiotic algae associated with the coral animals.

OPEN OCEAN ZONE | COSTAL OCEAN ZONE | INTERTIDAL ZONE

SHELF

SLOPE

Human Use and Impacts

Pollution, overexploitation, and climate change threaten the sustainability of many ocean ecosystems.

Given their central importance to Earth's climate and water cycle, it is obvious that the oceans are essential to our survival. Humans have extracted food from the sea for millennia. Today, oceans provide more than 5% of the dietary protein consumed by the world's people; in many coastal locations, fish and shellfish make up more than half the food supply. Oceans also provide important natural resources, such as table salt, potassium, magnesium, and manganese. Over 30% of our crude oil and 50% of our natural gas come from ocean sediments. In addition, oceans provide transportation that is essential to global commerce.

Today, many resources and services provided by oceans are threatened by human activities. Pollution, overharvesting, and climate change are depleting ocean resources. And sadly, the impact of these activities on ocean ecosystems seems to be increasing.

Ocean pollution. The effects of pollution are particularly significant in the waters near land. Streams and estuaries transport nutrients from agricultural runoff and urban stormwater to coastal waters. There, those nutrients boost the primary production of algae, resulting in large quantities of organic matter. As bacteria decompose this organic matter, they deplete the oxygen dissolved in ocean water. This process creates dead zones, regions of ocean water that contain such low levels of oxygen that few animals can survive in them (**Figure 12.52**). The United Nations Environment Program estimates that there are over 200 such dead zones worldwide, nearly all of them associated with the mouths

of major rivers. Dead zones are a major contributor to the decline in the populations of many fish species.

Nutrient pollution has also had a devastating effect on coral reefs in many parts of the world. The addition of nutrients to these ecosystems encourages the growth of free-living algae, which may coat the corals and kill their symbiotic algae.

Even low concentrations of some pollutants, such as heavy metals, can be taken up by phytoplankton. When animals eat those phytoplankton, the pollutants are stored in the animals' tissues. At each successive trophic level, the concentration of pollutants in animal tissues increases. This process has significant consequences for top predators and humans (19.3). Mercury, for example, is released into ocean waters from coal-fired power plants and the dredging of some wetlands. When mercury is taken up by phytoplankton, its concentration increases at each higher trophic level of the food web. In some places, the concentration of mercury in the tissues of carnivorous fishes, such as shark, tuna, and swordfish, is sufficiently high to pose a threat to human health.

Oil is a significant pollutant in many of the world's oceans. Although this pollution is often the consequence of oil spills, such as that from BP's deepwater drilling rig in the Gulf of Mexico in 2010, the majority of the oil in the ocean comes from non-point sources, such as urban stormwater and leakage from marinas, small boats, and pipelines.

Each year, millions of tons of trash—anything from discarded fishing gear to plastic grocery bags—make

▶ **Figure 12.52 Dead Zones** This false color satellite photograph indicates phytoplankton activity in the Gulf of Mexico near the mouths of the Mississippi and Atchafalaya rivers. The red area corresponds to high amounts of algae and an extensive dead zone.

their way into the oceans. Winds and currents have concentrated nearly 4 million tons of this refuse in a 1.2 million square kilometers (0.5 million square miles) area about halfway between North America and Hawaii that has come to be known as the Great Pacific Garbage Dump. Chemicals that leach out of plastic are a significant source of pollution. More important, many marine mammals, birds, and fish die when they swallow or become entangled in this trash (**Figure 12.53**).

Overharvesting of fisheries.

The condition of the world's fisheries is a serious problem. The UN Food and Agriculture Organization (FAO) classifies the majority of the world's marine fish species as being fully exploited. For these species, any increase in fishing level will be unsustainable and will cause fish populations to decline. The FAO estimates that over 25% of fish species are being overfished; the stocks of these species are in rapid decline (**Figure 12.54**).

The cod fishery off Canada's east coast is representative of many declining fish stocks. Since the 1600s, cod caught there have been a very important source of food for people in both Europe and North America. During the century between 1850 and 1950, fishing fleets grew and the rate at which fish were taken gradually increased. Beginning in the late 1950s, new technologies allowed fishing boats to take cod from much deeper waters. By 1970, the harvest of cod from this fishery had increased by nearly fourfold; but soon afterward, the annual catch began a precipitous decline, reaching extremely low levels in the late 1980s. In 1992, Canada declared a moratorium on commercial cod fishing. Attempts to reopen the fishery have been unsuccessful. In 2003, it was closed indefinitely.

The cod fishery off the coast of New England has experienced a similar pattern of decline and closure.

Carbon dioxide and global warming.

Increasing atmospheric concentrations of carbon dioxide and global warming are also influencing ocean ecosystems. When carbon dioxide dissolves in seawater, it forms carbonic acid, which increases the acidity of the water. Ocean acidity has increased by a small but measurable amount since the beginning of the industrial revolution. Higher acidity increases the solubility of calcium carbonate in seawater. Because the shells and skeletons of so many marine organisms are composed of calcium carbonate, scientists fear that a continuing increase in acidity could severely disrupt marine ecosystems.

Global warming has increased the temperature of the ocean surface water and is contributing to a global rise in sea level due to the thermal expansion of water. Global warming is also a factor contributing to the widespread decline of coral reefs (11.5).

We humans may be a terrestrial species, but we depend on the oceans for a wealth of resources and ecosystem services. Pollution, overexploitation, and climate change are diminishing our ability to tap those resources and services in a sustainable fashion.

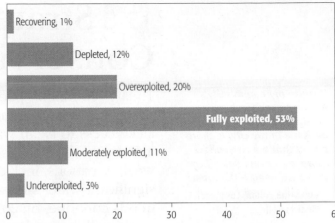

▲ **Figure 12.54 World Fisheries in Danger**
The world's fisheries have historically been treated as if they were inexhaustible. Today, 75% of the world's fish stocks are fully exploited, overexploited, or seriously depleted.
Data from: United Nations Food and Agriculture Organization, 2010. *State of the World's Fisheries and Aquaculture Report.* Rome, Italy: FAO.

QUESTIONS

1. Describe the general pattern of water circulation in the ocean's surface waters.

2. What is the thermohaline circulation, and how does it differ from ocean surface circulation?

3. In which part of the ocean is primary production likely to be highest? Why?

▼ **Figure 12.53 Ocean Garbage**
Each year, tens of thousands of tons of decomposition-resistant trash, such as fishing nets and plastic bags, end up in the ocean.

12.8 Strategies for Conserving Water

Despite my community's efforts to conserve water, droughts continue to be an issue in my city. Is there a way that we can collect more water and save it for when we need it the most?

Samantha Fullon, Los Angeles Mission College

Norm: *Many cities capture stormwater runoff in small reservoirs and allow the water to percolate down to recharge aquifers.*

Today, 700 million people live in countries experiencing water scarcity. More than three times this number lack access to clean water. The problem is most acute in some of the world's poorest countries, but many wealthy nations, including the United States, face significant water shortages (**Figure 12.55**). If current trends continue, by 2035 half of the people in the world will live under severe water shortage. The World Commission on Water estimates that human water use will increase by 50% by 2050. Increased water conservation is essential to the sustainable use of Earth's limited supply of fresh water. Strategies for reducing the use of water include improved techniques for irrigating crops, recycling of municipal and industrial water, and more efficient use of residential water. In some locations, it may be feasible to convert seawater to fresh water. Policies that determine the price of water also affect water conservation.

▲ **Figure 12.55 Water, Water Everywhere?**
Ten percent of Earth's people do not have enough water, and nearly 40% must drink and bathe in water that does not meet even the most minimal standards of sanitation.

Crop Irrigation

Improved technologies can reduce evaporation and runoff from agricultural fields.

About 70% of the water that humans use goes to irrigate crops and pastures. Much of that water is lost when it evaporates from the surfaces of reservoirs, canals, and flooded fields. The rate of that evaporation is greatest in arid regions, where irrigation is most widely used. Water that runs off into nearby streams is also lost. Yet there is much opportunity for improvement. In fact, engineers estimate that improved techniques could reduce the amount of water used to irrigate crops by more than 50%.

Irrigation efficiency is the percentage of the water applied to fields that is actually used by crop plants.

Agricultural scientists estimate that average irrigation efficiency of cropland is less than 40%. In other words, over 60% of the water applied to crops simply evaporates or runs off fields without being used by plants (**Figure 12.56**).

In most parts of the world, land is irrigated by flooding fields or channeling water down a field through shallow, parallel furrows. These methods distribute water so unevenly that farmers must apply excessive amounts of water to ensure that the entire field gets enough. Wasted water then runs off or evaporates. One method being used to reduce this waste is *surge irrigation*. In this method, an initial surge of water wets the furrows, partially sealing the surface of the soil. The next application of water flows across the field more evenly. Surge irrigation can decrease water use by 15% to 50%.

Aerial sprinkler systems are widely used in many arid regions, especially in the western United States. Much of the water distributed by these systems is lost to evaporation. New sprinkler designs can significantly diminish these losses. For example, low-energy precision application (LEPA)

◀ **Figure 12.56 Flood in a Dry Land**
Over 60% of the water applied by flood irrigation to this Arizona corn field will simply run off or evaporate without being used by the plants.

sprinklers have tubes extending down from the sprinkler arms that deliver water close to plants. In some situations, LEPA systems can reduce water use by more than 60%. And in addition to saving water, LEPA systems save the energy that would have been used to pump that water.

During the 1950s, scientists in Israel developed an entirely new method for delivering water to crops—drip irrigation. This process delivers water to the roots of individual plants through a network of perforated tubes or pipes installed at the soil surface. Since their conception, drip irrigation systems have become highly automated with monitors to ensure that the amount of

water delivered optimizes plant growth and minimizes losses to runoff and percolation. Irrigation efficiency for drip irrigation systems can exceed 95%.

The primary deterrent to the implementation of efficient irrigation systems is cost. LEPA systems cost approximately $500 per hectare, and drip irrigation systems can cost 5 to 10 times that amount. Such costs are prohibitive in the world's poorest countries. In developed countries, inappropriate pricing of water often means that there is little incentive for such investments.

Water Reuse

Reuse can conserve large amounts of water and provide other benefits as well.

Most industrial and residential uses of water are nonconsumptive. After the water is used, it is returned to a body of liquid water. For example, when you take a shower, the soapy water may eventually flow into a municipal sewer. From there it eventually makes its way into streams, estuaries, or groundwater. The primary challenge associated with the water coming from industries and residences is altered water quality. As Sandra Postel, Director of the Global Water Policy Project, points out, much water could be saved if we could "get the 'waste' out of wastewater." It may be difficult or impractical to restore the quality of wastewater to drinking water standards, but many uses of water do not require such high quality.

The idea of reusing municipal wastewater is not new. By 1800, wastewater from many English cities was being used to irrigate nearby farms. Such uses became limited in the early 19th century, as people came to understand how unclean water can spread disease and lead to other health problems. Modern methods of sewage treatment have put those health concerns to rest. Today, wastewater safely irrigates croplands near urban centers in many parts of the world (18.3).

Reusing wastewater provides several benefits in addition to conserving water. Because wastewater is high in nitrogen and other nutrients, using it to irrigate crops reduces the need for additional fertilizer. Furthermore, wastewater is purified as it percolates through soils. With additional treatment, wastewater can be recycled for a variety of other uses, including landscape and golf course irrigation, vehicle washing, toilet flushing, and groundwater recharge.

St. Petersburg, Florida, is one of a few cities that reuses all of its wastewater, releasing none into streams or the nearby ocean. The city has two water distribution systems. One delivers fresh water for drinking and other household uses. The other provides treated wastewater for irrigating public and residential lawns. Residents pay about 70% less for the treated wastewater (**Figure 12.57**).

There are important challenges regarding the reuse of municipal water. Wastewater is generated year-round, but

irrigation needs vary from season to season. Therefore, many municipalities must store wastewater in ponds, which adds to processing costs. In many developing countries, sewage treatment is inadequate and may be a significant risk to public health.

Reuse of water from commercial and industrial uses presents different challenges. Because wastewater from industrial facilities often contains heavy metals or organic pollutants, it cannot be used for irrigation. These chemicals can be removed from water, but the costs are often high. Nevertheless, in response to the increasing cost of water and tighter regulations regarding wastewater disposal, industries around the world have implemented a variety of techniques to reuse wastewater and increase the efficiency of water use. While industrial production in the United States has increased nearly fivefold since 1950, the amount of water used by industries has declined by more than 50%.

Can wastewater be recycled to drinking water standards? From a technical standpoint, the answer is yes. Pilot projects in water-stressed cities such as El Paso, Texas, and Denver, Colorado, have shown it to be possible. However, public acceptance of this idea is generally low, and the costs for such processing are very high. In most cases, it is far more economical to direct wastewater to uses that require far less treatment.

Is there actually a health benefit to bottled water compared to tap water?

Amber Apostal, Louisiana State University

Norm: *Bottled water is advisable in developing countries where water quality is poor. In the United States, however, tap water must meet the EPA's stringent quality standards, so there is no health benefit to drinking bottled water. In addition, plastic bottles generate waste and consume unnecessary resources.*

▼ **Figure 12.57 Reusing Water**
Nonsewage wastewater is treated and reused for yard sprinkling and similar purposes.

QUESTIONS

1. Describe two alternatives to aerial sprinkler systems that improve irrigation efficiency.

2. What are three benefits of water reuse?

3. Describe how a natural gas–generating facility can also function to desalinate water.

4. How do differences in household water-billing practices affect consumer water use?

Desalination

Converting seawater to fresh water is feasible in locations where water is scarce and energy is abundant.

There is, of course, a vast amount of water—albeit salty—in the ocean. The process of removing salts and other chemicals from seawater is called **desalination**. The two main methods of desalination are **distillation**, boiling water and condensing the steam, and **reverse osmosis**, filtering water through a selective membrane. Both methods require a great deal of energy and are, therefore, quite costly. Yet where other sources of water are scarce, desalination can be cost effective.

In many countries in the Middle East, water is particularly scarce and energy, especially oil and natural gas, is particularly cheap. Thus, it is not surprising that over 75% of the world's desalination capacity has been built in that region. For example, the world's largest desalination facility is in the United Arab Emirates near the port city of Dubai. This is a dual-purpose facility: It uses natural gas to generate electricity for Dubai and then uses the heat from the generators to distill seawater. Each year this facility produces 300 million cubic meters (79 billion gallons) of fresh water (**Figure 12.58**).

In other regions, there has been interest in desalinating brackish water, which contains less salt than seawater but is still too salty to drink. Removing the salt from brackish water requires considerably less energy and expense than desalinating seawater. Reverse osmosis systems are especially well-suited to this task. Hundreds of coastal communities use such facilities to produce fresh water from brackish groundwater.

Nearly 13,000 desalination plants have been built worldwide. Each year they produce 16.5 billion cubic meters (4.4 trillion gallons) of water. Although this may seem like a large amount, it represents only about 0.1% of total human freshwater use.

▼ **Figure 12.58 Desalination and Energy**
The Jabel Ali desalination plant in the United Arab Emirates generates electricity from natural gas. The heat from the natural gas turbines is used to desalinate seawater by distillation.

Getting the Price Right

Changes in water pricing are needed to ensure sustainable conservation.

Like the air we breathe, we generally consider access to water to be an entitlement. We rarely think about its value or its cost until we run short of it. Many of the world's water shortages are a consequence of our failure to value water appropriately.

Consider the example of water pricing in California. Agriculture uses more than 80% of the state's water resources, yet farmers pay only 5% to 20% of the costs associated with supplying that water. The remaining costs are subsidized by revenues from state and federal taxes. Some people argue that subsidies of this kind are appropriate because society at large benefits from a reliable and relatively inexpensive supply of food. But such subsidies externalize the real costs of water use. Because water is so inexpensive, farmers have little incentive to invest in technologies that conserve water, such as LEPA or drip irrigation.

Differences in pricing also influence the residential use of water. Studies show that water is used more efficiently in households with water meters than in households where water is priced at a flat monthly rate, regardless of the amount used. Many communities have increased conservation further by graduated pricing, in which the unit cost of water increases as more is used.

In summary, large amounts of water can be conserved by using water more efficiently, improving irrigation, and recycling municipal and industrial wastewater. In some locations, water supplies can be augmented by desalination. Yet many of these conservation strategies involve significant costs, so their implementation depends on how water is valued in economic markets.

SEEING SOLUTIONS: Improving Urban Water Use

Modern cities use a great deal of water, but much of it is wasted by inefficiency and negligence. For example, between 6% and 15% of the water used in U.S. cities is lost through leaks in distribution pipes and faulty plumbing. Depending on climate, between 30% and 60% of municipal water is used to irrigate lawns and landscaping.

In the past, cities usually implemented water conservation measures only in response to droughts or seasonal water shortages. But as water has become scarcer and costlier, cities around the world have begun to implement permanent measures to ensure that water is used more efficiently. Boston and nearby cities provide a case study of the kinds of strategies that can be used to conserve water.

The Massachusetts Water Resources Authority (MWRA) provides water to 2.5 million people and more than 5,500 businesses in the greater Boston metropolitan area. Nearly all of that water comes from the Quabbin and Wachusett reservoirs, which are located west of the city (**Figure 12.59**). In a 1986 study, MWRA determined that the rate at which water was being withdrawn from these reservoirs put the region at risk of a water shortage. In response to this situation, it immediately established an integrated water conservation program.

Distribution pipes belonging to the MWRA and local communities were checked for leaks and then repaired. Water meters were upgraded to help track and analyze community water use. Over 370,000 homes were retrofitted with low-flow plumbing devices, such as water-efficient showerheads. The MWRA even championed a change in the state plumbing code to require that water-saving, 1.6-gallon-per-flush toilets be installed in all new residential and commercial construction. Perhaps most important, the MWRA launched an extensive public information campaign and a variety of educational programs for the public schools. To pay for these programs, as well as the clean-up of Boston Harbor, the MWRA boosted the price of water. These higher rates have made Boston citizens much more aware of their water use.

By 1989, MWRA water use was below the safe reservoir yields. By 2010, water use had declined by more than 40% from 1986 levels (**Figure 12.60**). Furthermore, MWRA calculated that the conservation program cost only a fraction of what it would have had to spend to increase the capacity of the reservoirs or buy additional water from other watersheds in order to maintain the 1986 level of water use.

How can we reduce the amount of water that cities waste?

▼ **Figure 12.59 Boston's Water Supply**
The Quabbin Reservoir located in central Massachusetts provides two-thirds of the water used by the greater Boston metropolitan area. The volume of water in the reservoir varies from year to year, depending on the amount of rainfall in its watershed.

MWRA WATER SYSTEM DEMAND 1986–2009

SYSTEM SAFE YIELD

MILLIONS OF GALLONS PER DAY

YEAR

▲ **Figure 12.60 Sustainable Water Use**
The safe yield of water from the Quabbin and Wachusett reservoirs is determined by the average annual rainfall in their basins. In the mid-1980s, water use in Boston exceeded the safe yield. A combination of strategies implemented by the MWRA has lowered water consumption well below the safe yield.

12.9 Water Conservation Policy and Law

A variety of national and international policies and laws influence water use and water quality. In the United States, the laws governing access to water differ from region to region. Conflicts over water ownership and allocation are greatest in the West, where water is scarcest. Water quality in the United States is regulated by the Clean Water Act and the Clear Drinking Water Act. These laws provide a variety of measures for restoring and maintaining water standards. More than 60% of the water in Earth's streams and lakes is shared by two or more countries; many individual treaties regulate the use and quality of these waters. The member nations of the United Nations have passed a convention on the use of ocean waters, but have been unable to agree on laws to regulate the non-navigational uses of fresh water. Global warming is likely to diminish supplies of fresh water in many regions and increase international conflict over water use. These predictions reinforce the need for international agreements governing water use.

Water Use in the United States

Water allocation policies strongly affect water use.

In the United States, each state sets its own water rights policy. In the East, access to water is determined by **riparian water rights**, a system derived from English common law. Under this system, all landowners with property adjacent to a body of water have a right to make reasonable use of it. Reasonable use is defined by each state, but usually includes domestic water use, swimming, boating, and fishing. Rights can be sold or transferred only to adjacent landowners, and water cannot be transferred out of a watershed.

In most western states, where water is generally scarce, water use is determined by **prior appropriation water rights**. These rights are not connected to land ownership and can be bought and sold like other property. Under this system, the first person to use a quantity of water from a water source for a beneficial use has the right to continue to use that quantity of water for that purpose indefinitely. Rights to the remaining water are allocated to subsequent users, so long as they don't impinge on the rights of previous users. When water runs short, users with the earliest appropriation get their full allocation; later users may get no water. Conflicts regarding water use within and across state lines have been common, and they have increased in number as demand for water in the West has grown.

Prior appropriation water rights were created to accommodate the needs of farmers, miners, and frontier towns and cities. As a result, most states define beneficial use in terms of agricultural, industrial, or household purposes. When these laws were adopted, there was little awareness of environmental needs. Ecological purposes, such as maintaining floodplain ecosystems or conserving wildlife, are not classified as beneficial uses in most western states. For this reason, some argue that this approach to water allocation is outdated.

The 1973 Endangered Species Act (ESA) has often been used to protect the flow of water in rivers and streams that provide habitat for threatened or endangered species. For example, Friant Dam and associated diversions take water out of California's San Joaquin River for agricultural uses. The rights to that water have been assigned to farmers in the San Joaquin Valley. Since the completion of this project in 1984, long segments of the river have been completely dry (**Figure 12.61**). Recently, the U.S. Fish and Wildlife Service and the courts have required that water flows be restored to the San Joaquin River in order to recover stocks of endangered Chinook salmon. The courts also required the federal government to compensate farmers for water rights lost in association with this restoration.

▼ **Figure 12.61 Reviving a Dead River**
Diversions for agriculture have left this segment of California's San Joaquin River dry for over 60 years. Rights to this water have been allocated to area farms. To meet the conditions of the Endangered Species Act, the state must restore water to the river.

San Joaquin River

Water Quality in the United States

The Clean Water Act has evolved to emphasize integrated watershed strategies.

In the late 19th century, streams near cities were often so badly polluted that they were described as "open sewers." The 1899 Refuse Act was the first attempt to manage water quality at the national level. This law stipulated that wastes could not be dumped into navigable waters without a permit from the Army Corps of Engineers. The Oil Pollution Control Act of 1924 regulated the dumping of oil from boats but not from stationary sources, such as industrial facilities or oil wells. These laws may have diminished water pollution somewhat. But as the population continued to grow, pollution from rapidly growing cities and industries and the increasing use of agricultural fertilizers and pesticides continued to foul U.S. streams and lakes.

In 1948, the U.S. Congress passed the Water Pollution Control Act, the first comprehensive national law to manage water quality. This law provided grants to local governments to clean up water and authorized the federal government to sue polluters in court. It did not, however, set standards for water quality, nor did it assign the responsibility for enforcing the law to a specific agency.

The degradation of U.S. water supplies continued through the 1960s. A number of events, most notably the fire on Ohio's Cuyahoga River, garnered widespread public attention to this issue (**Figure 12.62**). This attention moved Congress to pass the Clean Water Act (CWA) of 1972 to ensure "the protection and propagation of fish, shellfish, and wildlife and recreation in and on the water." This act requires the EPA to regulate the discharge of pollutants into waterways. It also regulates activities that influence water quality, such as dredging streams, lakes, and wetlands. Indeed, the CWA now provides the basis for the protection of wetlands across the country. The Safe Drinking Water Act of 1974 required the EPA to set standards for allowable quantities of chemicals in water; it also authorized local governments to monitor and maintain these standards.

Since its original passage, the CWA has been modified to accommodate new knowledge and changing priorities. Programs have been implemented to engage landowners in efforts to minimize the runoff of pollution from their lands. Regulations have also been established to diminish pollution from urban storm-sewer systems and construction sites. Perhaps most important, instead of focusing on the regulation of individual sources of pollution, CWA programs now take an integrated approach to managing watersheds. These programs address a full array of issues and emphasize the involvement of communities and important stakeholders in the development and implementation of strategies to restore and maintain clean water.

International Water Law

A comprehensive international convention on freshwater resource use has yet to be ratified.

Rivers are among the most common boundaries between countries. Over 250 large watersheds are shared among two or more countries, and more than 400 international treaties regulate the allocation and use of those waters.

In 1997, the UN General Assembly adopted the Convention on the Non-navigational Uses of International Watercourses. This convention proposes three general principles for the use of international surface waters and groundwater. First, allocation among countries should be equitable and reasonable. Second, every country is obligated to prevent significant harm to those waters. Third, countries must notify and consult with one another regarding planned usage. However, this convention has yet to be ratified by enough countries to make it law.

The UN Convention on Law of the Sea was drafted in 1982. It requires countries bordering oceans to prevent the pollution of coastal waters and to cooperate in the management of living resources, such as migratory fish. This treaty was ratified by 60 countries and became international law in 1994. The United States signed the treaty in 1994 and recognizes it as general international law. But as of 2010, the U.S. Senate has not ratified the treaty because of concerns about how the convention would be governed.

Global warming is likely to cause even more challenges for allocating the use of water among nations. A 2007 report from the UN Intergovernmental Panel on Climate Change predicts that water availability will decline by 10% to 30% in many of the world's already water-stressed regions. Such shortages are likely to lead to international conflicts. These trends show the need for a comprehensive international water convention.

QUESTIONS

1. How do riparian water rights differ from prior appropriation water rights?

2. Describe three important changes to the original U.S. Clean Water Act passed in 1972.

3. What are the three primary principles for water management across international borders named in the 1997 UN Convention on the Non-navigational Uses of International Watercourses?

How can I determine the cleanliness of the water in my area?

Hayley Baum, University of Central Florida

Norm: *The U.S. EPA provides water quality data for your watershed and for drinking water in your community on its website.*

▼ **Figure 12.62 A Flaming River**
In 1969, flammable pollutants on the Cuyahoga River caught fire and burned for days. This event caught widespread public attention and catalyzed support for the passage of the Clean Water Act in 1972.

AGENTS OF CHANGE: The University of Oregon's Take Back the Tap

Chris Becker is majoring in architecture at the University of Oregon. Terra Smith is a recent graduate of the University of Oregon with a degree in environmental studies and a minor in nonprofit administration. Chris and Terra worked with the University of Oregon's Climate Justice League (CJL) to organize Take Back the Tap, a campaign that promotes drinking tap water from reusable containers instead of single-use plastic bottles. The purpose of the campaign is to reduce the environmental and economic waste caused by the use of bottled water. CJL Take Back the Tap is currently working to ban bottled water at the University of Oregon.

How did you first get the idea for Take Back the Tap?

We both had experiences in other countries where we saw firsthand the correlation between pollution, poverty, and low quality of life. Plastic bags and bottles littered the streets, the air and water were dirty, and people were unhealthy. We were frustrated by those conditions, but we also knew they weren't unique to developing countries. Similar issues show up right at home in this country, where processing plants and waste facilities are often located next to underprivileged neighborhoods.

The use of bottled water causes diverse problems, including pollution from the production, transportation, and disposal of plastic bottles and health risks from the chemicals that leak from thin plastic bottles. There are also issues related to water rights and the use of local water supplies. Plus, bottled water is an expensive duplicate of the great, clean tap water that flows from our faucets in Eugene, Oregon, and in most other parts of the country.

We got the idea for Take Back the Tap from other colleges that had done similar campaigns. Take Back the Tap is sponsored by a nonprofit organization called Food & Water Watch, which works with students across the country to address the social, economic, and environmental problems of bottled water on their campuses. At the time we began, all the other schools working on the project had been small liberal arts colleges—there were no large public universities like ours that had banned bottled water.

What steps did you take to get started?

We identified our objectives, created goals, and shared our vision with others to gain more support and team members. Our goal was to discontinue the sale of bottled water on campus within two years.

Later, we established a strategy for how we wanted to accomplish our goal. Our strategy included providing alternatives to bottled water, gaining the support of a diverse population of the campus's students and staff, and asking decision-makers to discontinue contracts and establish rules against the sale of bottled water on campus and at campus events. We also partnered with Food & Water Watch.

What are the biggest challenges you've faced with the project, and how did you solve them?

Our biggest challenge has been finding ways to keep volunteers involved while also moving the campaign forward. By identifying people's strengths and reasons for working on the campaign, we've been able to build positions where everyone can succeed and help us with the project.

How has your work affected other people and the environment?

To draw attention to our campaign, we built a huge sculpture out of empty plastic water bottles from

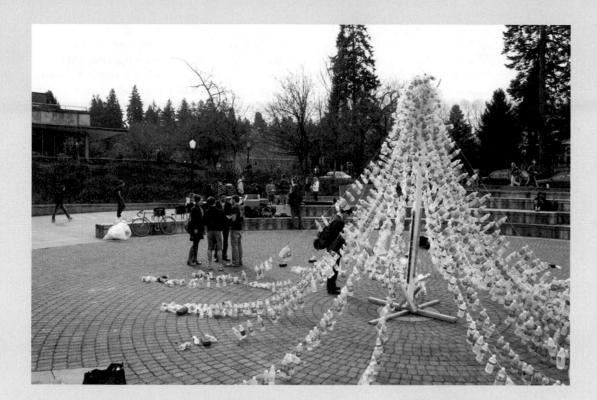

campus. The display informed students about the issues associated with the use of bottled water and told them what they could do to address the problem. Thanks to our display, we are now a well-known organization on campus. We've also attracted much media attention, including articles in the Associated Press. Most importantly, sales of bottled water on campus have decreased since the display of the sculpture.

What advice would you give to students who want to replicate your model on their campus?

Start talking with classmates about the issues they're most concerned about and what their vision for a successful campaign would be. From there you can begin gathering resources and building a leadership team. Food & Water Watch is a great resource if you want to start a bottled water campaign on your own campus.

The Sierra Student Coalition and the Energy Action Coalition and their partners have been very influential resources for environmental organizing and leadership growth. I would highly recommend checking them out for campaign ideas and models, as well as for training and workshops on organizing campaigns.

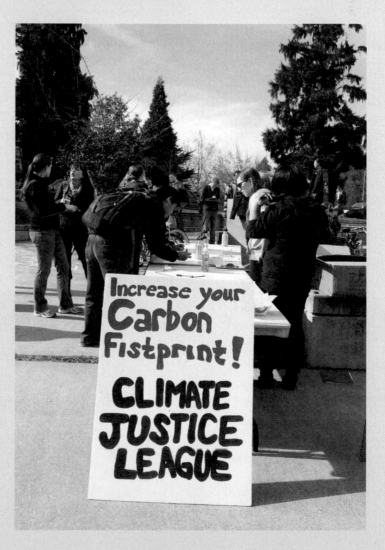

Summary

Energy from the sun and the force of gravity cause water to move between the oceans, atmosphere, and land. Watersheds define the area of land from which water drains into a stream, lake, or ocean. The speed of water in a stream is determined by its gradient. In addition to the organisms living in flowing waters, stream ecosystems include floodplains and the hyporheic zone beneath the streambed. Organic matter from surrounding ecosystems is an important source of energy for stream and river food webs. Humans have affected the majority of stream ecosystems by building dams,

diverting water, and releasing pollutants. Lakes fill topographic basins that may be either open or closed. Lake ecosystems are influenced by the depth to which sunlight penetrates the waters and the recycling of nutrients from lake sediments. Most of Earth's groundwater is found in aquifers. The flow of water through aquifers is very slow, as is the rate of recharge to aquifers. Many aquifers are threatened by overpumping and pollution. The soils of wetland ecosystems are saturated with water much of the year. These ecosystems provide important ecosystem services, including the control of water flows and removal of sediment and pollutants from water. Estuaries are partially enclosed bodies

of water where rivers and streams meet the sea. They support very diverse ecosystems that are an important source of seafood for humans. The vast majority of Earth's waters are contained in its oceans. Surface ocean circulation moves warm water toward the poles and colder water back toward the equator. The thermohaline circulation exchanges water from the ocean depths with surface water. The most productive and diverse ocean ecosystems are found near the coasts of continents and in upwelling zones where nutrients are most available. An increasing number of people are experiencing water shortages that require implementation of water conservation strategies and policies.

12.1 Water World

- Earth's water budget describes the movement of water among the oceans, the atmosphere, and the land.
- Humans use over half of the water flowing in Earth's streams.

KEY TERMS

hydrologist, evapotranspiration, water balance coefficient, watershed, open watershed, closed watershed, nonconsumptive use, consumptive use

QUESTIONS

1. How would you expect the water balance coefficient to change from season to season in a tropical seasonal forest? Deciduous forest? Tundra?
2. How do hydrologists typically estimate evapotranspiration for a watershed?
3. How would you expect human water use to differ between an industrial country, such as Japan, and a developing country, such as Bangladesh?

12.2 Streams

- The volume and speed of water flowing in streams determines the diversity of ecosystems in their channels and across their floodplains.

KEY TERMS

stream order, gradient, perennial stream, intermittent stream, channel, floodplain, levee, meander, delta, lotic, detritus, riparian zone, hyporheic zone, point source, non-point source, bioassessment

QUESTIONS

1. Explain this statement: Stream food chains are often sustained by terrestrial primary production.
2. Why is the hyporheic zone beneath a stream's channel important to the organisms in the waters above it?
3. Describe how a dam affects the amount of sediment carried by the streams flowing above and below the dam.

12.3 Lakes and Ponds

- Natural lakes form in a variety of ways, and they support a variety of ecological communities along their shores and at different depths.

KEY TERMS

lentic, open basin lake, closed basin lake, benthic zone, pelagic zone, photic zone, aphotic zone, phytoplankton, zooplankton, littoral zone

QUESTIONS

1. Why is the salt content generally high in closed basin lakes?
2. Why is primary production in ponds and lakes generally higher following periods of windy weather?

12.4 Groundwater

- The groundwater that resides in soil and rock beneath Earth's surface is replenished very slowly.

KEY TERMS

water table, recharge zone, unconfined aquifer, confined aquifer, discharge zone, artesian well

QUESTIONS

1. Explain why sometimes water will gush to the surface of a newly drilled well without pumping.
2. Why is groundwater considered to be a nonrenewable resource in many regions?
3. Why does groundwater pumping near coasts often result in the intrusion of seawater?

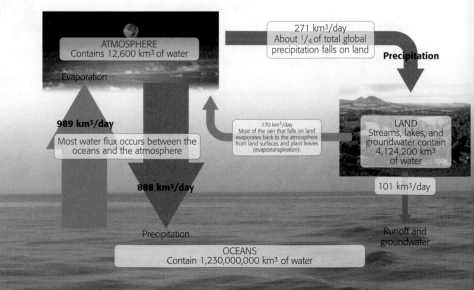

271 km³/day
About 1/4 of total global precipitation falls on land

Precipitation

ATMOSPHERE
Contains 12,600 km³ of water

Evaporation

989 km³/day

Most water flux occurs between the oceans and the atmosphere

170 km³/day
Most of the rain that falls on land evaporates back to the atmosphere from land surfaces and plant leaves (evapotranspiration).

LAND
Streams, lakes, and groundwater contain 4,124,200 km³ of water

101 km³/day

888 km³/day

Precipitation

Runoff and groundwater

OCEANS
Contain 1,230,000,000 km³ of water

12.5 Wetlands

- Soils in wetland ecosystems are regularly saturated with water, and they support plant species that are adapted to these conditions.

KEY TERMS

hydrophytes, marsh, swamp, bog, fen

QUESTIONS

1. Compared to swamps and fens, bog soils contain very low amounts of nutrients. Why?

2. Explain how wetlands moderate water flows and improve water quality.

3. What is the "no net loss" wetland policy, and what has been its effect?

12.6 Estuaries

- Estuaries are partly enclosed bodies of water where fresh stream water meets saline ocean water.

KEY TERMS

estuary, flushing time, dead zone

QUESTIONS

1. Why is the salinity of the water in estuaries typically lower in the spring and higher in the summer?

2. What factors influence the flushing times of estuaries?

3. Why are ecologists concerned that overharvest of oysters and other shellfish contributes to estuary pollution?

12.7 Oceans

- Ocean currents are powered by Earth's rotation, winds, and variations in water temperature and salinity; these currents redistribute heat around the world.

- The distribution of different marine ecosystems is influenced by the availability of nitrogen and phosphorus and the penetration of light into ocean waters.

KEY TERMS

gyre, thermohaline circulation, intertidal zone

QUESTIONS

1. Why are ocean waters along the east coasts of continents generally much warmer than those along the west coasts of continents?

2. Why does the photic zone in near coastal ocean waters generally extend to only 15–25 m (50–80 ft) compared to nearly 100 m (330 ft) farther from shore?

3. Explain why some of the ocean's most complex and diverse ecosystems occur in zones of upwelling.

4. Describe four ways in which humans are negatively affecting ocean ecosystems.

12.8 Strategies for Conserving Water

- At current rates of water use and population growth, half of Earth's people will experience water shortages by 2050.

- New crop irrigation technologies, water reuse, desalination, and changes in water pricing can reverse these trends.

KEY TERMS

irrigation efficiency, desalination, distillation, reverse osmosis

QUESTIONS

1. Why do drip irrigation systems increase irrigation efficiency by a factor of 10 compared to conventional systems?

2. Describe three challenges to water reuse.

3. Explain how current systems of pricing water externalize the costs of pollution and of inefficient water use.

12.9 Water Conservation Policy and Law

- Water use in the United States is influenced by laws determining water ownership and access.

- The Clean Water Act and its several amendments have done much to improve water quality in the United States.

- A comprehensive international agreement on water use has been drafted but has not yet been ratified by UN member nations.

KEY TERMS

riparian water rights, prior appropriation water rights

QUESTIONS

1. How are the rights to water in streams and rivers determined in the state where you live? How do these rights affect patterns of water use?

2. Explain how the Endangered Species Act has affected water use in many river systems.

San Joaquin River

Mastering**ENVIRONMENTALSCIENCE**

Go to **www.masteringenvironmentalscience.com** for homework assignments, practice quizzes, Pearson eText, and more.

Agricultural societies developed in three regions of North and South America between 5,000 and 6,000 years ago. People living in the Andes Mountains in parts of modern-day Peru, Bolivia, and Ecuador domesticated potatoes. People in central Mexico began the cultivation of corn. Native Americans in regions that are now part of the United States grew squash and sunflowers.

The general process of plant domestication was probably similar in these various regions. At first by trial and error and later by conscious effort, humans directed the evolution of domestic plants. They did this by selecting and planting seeds or tubers from individual plants that possessed traits that improved the growth of the plant or the quality of the food. For example, limited seed dispersal was a key step in the domestication of grains, such as wheat and corn. It was easier for early farmers to harvest seeds from stalks that failed to shatter; as a result, seeds from those plants were used to plant the next year's crops. Generation after generation, this artificial selection caused crops to have a greater percentage of plants with seeds that remained on the stalk until they could be harvested. Similarly, by planting and harvesting plants at uniform times, early farmers selected for uniform patterns of germination and growth in crop plants.

The origin of domestic corn, or maize, provides a case study for the evolution of domestic grain plants. The ancestor of modern cultivated corn is thought to be a grass called *teosinte*, which today grows wild in the highlands of central Mexico. Teosinte produces large grains that are arranged on a stalk that fragments and disperses the seeds when they ripen. Over several thousand years, humans selected the traits that produced modern corn (**Figure 13.7**). At the time of first European contact, over 300 varieties of corn were being cultivated by the indigenous people of the Americas.

About 12,000 years ago, humans began to domesticate animals. The first domestic animals were dogs. Dogs may have provided a source of food, but their primary role was to increase the effectiveness of human hunters.

Sheep, goats, and pigs were domesticated at about the same time that the first crops were cultivated in southern Europe and Asia. These animals undoubtedly provided

► Figure 13.7
Corn Evolution
Beginning over 5,000 years ago, indigenous people of Mexico transformed the grass teosinte into modern corn (maize) by selecting and sowing the seeds of plants with desirable traits.

① Teosinte
② Primitive corn
③ Modern corn

essential protein to the carbohydrate-rich diets of early farmers. Cattle were domesticated about 8,000 years ago; in addition to producing meat and milk, cattle provided much needed labor for the cultivation of crops.

Evolutionary biologists note that the wild animals that humans chose to domesticate generally shared five characteristics—a flexible diet, a fast rate of growth, an ability to breed in captivity, a calm disposition, and sociality. Most domestic animals are able to eat a wide variety of foods, mostly from plants. Compared to humans, domestic animals mature quickly. Domestic animals must be able to breed in captivity; animals that are reluctant to breed in captivity are unlikely candidates for domestication. Domestic animals must also be relatively docile. Large, aggressive animals, such as water buffalo, cannot be managed in captivity; it is also difficult to keep highly territorial animals in captivity. And finally, animals that live in herds with a social hierarchy, such as cattle and sheep, can also be herded by humans (**Figure 13.8**).

Who domesticated whom? By nurturing and protecting them, humans provided a significant competitive advantage to the plants and animals on the evolutionary road to domestication. But plant cultivation and animal care required significant changes in human behavior and even in human physiology. Thus, the origin and development of agriculture has been a coevolutionary process for us and the variety of plants and animals on which we now depend (6.2).

▼ Figure 13.8 **Herding Instinct**
The ancestors of domestic sheep were naturally social animals that moved in herds.

And Then What? Agricultural History

Technological and cultural innovations have increased agricultural production and reduced the need for human labor.

The earliest agriculture depended on stone tools. Later developments included ceramic, bronze, and iron tools, such as hoes, rakes, and sickles. Tending crops with these tools would have required a great deal of labor. The invention of the plow coincided with the domestication of cattle—cattle provided the power needed to pull plows, reducing the amount of human energy invested in producing crops.

Even with cattle, agriculture remained a labor-intensive enterprise that necessitated, but also supported, relatively large communities. For example, 5,000 years ago the town of Ur along the Euphrates River (in what is today Iraq) had a population of 6,000 people. A labor force of 2,500 men and many cattle cultivated 1,200 ha (3,000 acres) of land. In addition to these laborers, there were work foremen, harvest supervisors, and storehouse recorders. Land was cultivated with cattle-drawn plows and harvested with ceramic sickles. Wagons with solid wooden wheels were used to move the harvest from the field to storehouses. Fields were irrigated by a complex system of canals. Occasional flooding from the nearby river fertilized fields with nutrient-rich silt (**Figure 13.9**).

Aside from minor innovations, such as improved metal quality and the addition of wheels to plows, agricultural technology changed little for the next 4,000 years. In the Middle Ages (800–1300 AD), most of Europe was governed by feudal lords who controlled most of the land. Under the feudal system, laborers or serfs were virtual slaves and had few incentives to improve production. However, important changes did occur in regions of the Mediterranean that were under Muslim rule. Perhaps most important, Muslim leaders implemented a new program of land ownership that provided greater incentives to working farmers. At the same time, the development of sophisticated water-pumping devices, dams, and reservoirs allowed the exploitation of more land. Many new crops, including cotton, almonds, figs, and sugar cane, were also introduced during this period.

The period from A.D. 1400 to 1700 brought even greater changes. The discovery of the Americas resulted in an influx of new crops to Europe and Asia, such as corn, potatoes, tobacco, and tomatoes. The moldboard plow was imported from China. These plows turned soil more efficiently than conventional plows, greatly reducing the amount of labor required to grow crops. Farmers also began to use a system of crop rotation, alternating between grain and legume crops. Crop rotation sustained productivity because legumes fix nitrogen from the air, restoring nitrogen to soils that have been depleted by harvest.

Since the beginning of the 18th century and the Industrial Revolution, many innovations have combined to greatly increase agricultural production. Technologies such as improved plows and reapers facilitated cultivation and harvest, while diminishing the human labor required for these activities. Beginning in the late 1900s, tractors powered by fossil fuels greatly increased both the speed and scale of farming operations.

Throughout the industrial period, the rapidly increasing understanding of the science of plants, animals, and soils resulted in numerous agricultural innovations. **Agronomy** is the science that applies knowledge from fields such as genetics, physiology, chemistry, and ecology, to agriculture. For example, the growing understanding of genetics led to increasingly sophisticated plant and animal breeding. This resulted in varieties of crops and domestic animals that were better suited to particular environments, more resistant to pests and disease, and more productive.

Over the past 50–60 years, we have witnessed the **Green Revolution**, a period during which global agricultural productivity has increased many times over. Much of this increase in production is the result of the development of modern fertilizers, herbicides, and pesticides. Synthetic nitrogen fixation using the Haber-Bosch process significantly increased the availability of nitrogen fertilizers and also decreased their cost (7.5). Because production of most crops is limited by nitrogen availability, the widespread use of such fertilizers has been central to the Green Revolution. Herbicides and pesticides have allowed farmers to control weeds, insects, and various diseases that compete with humans for the productivity of farm fields.

Several trends are apparent in the history of agriculture. Domestic plants and animals have coevolved with human cultures and technologies. The increased food production provided by agriculture has allowed the growth of human populations and increasingly complex human societies. During the Industrial Revolution and the Green Revolution, the use of energy from fossil fuels brought even greater productivity to agricultural systems.

QUESTIONS

1. What factors are likely to have played a role in the origins of agriculture?

2. What kinds of adaptations would have made a plant an ideal target for domestication?

3. What features made certain animals likely candidates for domestication?

4. Why did the earliest agricultural communities begin in the floodplains of rivers?

▼ **Figure 13.9 Ancient Ur**
Crops irrigated by water from the nearby Euphrates River supported Ur's large and complex population. The city included a large temple (distant), storehouses, palaces, and housing for thousands of people. Crop fields were located outside the city walls.

383

13.2 Agroecosystems

Today, we humans make up less than half of 1% of the total biomass of Earth's animals but use more than 30% of Earth's terrestrial net primary production to obtain food and fiber. In fact, humans have converted about one-third of Earth's ice-free land surface to agriculture. Agriculture focuses on the production of a few species of plants and animals. Yet, like all species, the species that humans grow are parts of larger ecosystems. **Agroecosystems** include crops and domestic animals, the physical environments in which they grow, and the communities of other organisms associated with them. Agroecosystems depend on energy flow and nutrient cycling, following the same principles that govern natural ecosystems (7.1). Because they are prone to significant losses of water and nutrients, agroecosystems often require significant inputs of these resources. The development of agriculture and improvements in productivity have been major factors in the growth of human populations. Thus, the well-being of human communities throughout the world depends on the sustainable management of agroecosystems.

Energetics

Agroecosystems funnel energy into the production of plants and animals that are most useful to humans.

In agricultural systems, as in nearly all ecosystems, photosynthesis in green plants converts solar energy to the chemical energy of carbohydrates (3.3). In agroecosystems, this primary production is at the base of relatively simple food chains in which humans feed directly on plants or on herbivorous animals, such as cattle (**Figure 13.10**; see 6.4). In either case, the main objective in managing agroecosystems is to funnel a

▶ **Figure 13.10**
Agroecosystem Food Chain Because they eat plant materials directly, vegetarians function as primary consumers in the agricultural ecosystems. Meat-eaters function as secondary consumers. Eating meat requires more cropland per person because of the energy lost in the livestock step of the food chain.

TROPHIC LEVEL

Secondary consumers — Meat-eater

Primary consumers — Vegetarian

Energy is lost to nonedible plant and animal parts, waste, and heat.

Primary producers

major portion of the ecosystem's net primary production into food or fiber for human use. Simplification is a key strategy to meet this objective in agroecosystems. Farmers and ranchers work to optimize the growth of a single species and to minimize the growth of other species that may eat it or compete with it.

Humans cannot use all of the primary production of a crop or all of the biomass of a herd of domestic animals. The fraction of total production that can be used by people is the **harvest index**. For example, the net primary production of a cornfield can be as high as Earth's most productive natural biomes. But more than half of that primary production is located in corn stalks and roots, which have limited use to humans. The harvest index for the part of the corn plants that can be eaten—the kernels—is between 30% and 50%. The harvest indexes for other grain crops are similar.

When humans eat meat, such as beef or chicken, they are in effect consuming net primary production that has been transformed by another organism. As with any such transformation, energy is lost in the process. The amount of energy lost depends on the domestic animal's **trophic-level efficiency**, the fraction of energy the animal consumes that is actually stored as biomass.

For most wild animals, trophic-level efficiency is less than 10%. Thus, 1 unit of energy in the meat of a deer or wild bison required more than 10 units of primary production (6.4). Trophic-level efficiency for domestic animals is generally greater than 10%. This is partly due to selective breeding. In addition, domestic animals are often given food that is more easily digested than the food that is available to animals in the wild. Another reason for the higher efficiency of domestic animals is that they are usually confined, so they expend less energy searching for food than their wild counterparts. For enclosed chickens, trophic-level efficiency may exceed 40%. In other words, 40% of the energy in the corn eaten by a chicken is transformed into the tissues of the chicken.

Trophic-level efficiency for corn-fed cattle in enclosed feedlots may be as high as 20–25%. A corn-fed steer in an enclosed feedlot requires energy equal to the annual net primary production of about 0.56 ha (1.4 acre) of cornfield (**Figure 13.11**). Yet in the context of modern agriculture, this energy calculation is incomplete. For example, it does not include the energy required to fuel the machines used in irrigation, cultivation, harvest, production, and distribution. These energy investments vary widely with growing conditions and cropping practices, but can be as high as 30–70% of net primary production. The energy expended by the farmers who grow the corn or care for the cattle could also be added to these calculations of energy expenditures.

An average steer gains **320 kg** (700 pounds) per year in an enclosed feedlot.

To gain that weight, the steer consumes about **2,240 kg** (4,900 pounds) of corn feed.

At 40% harvest efficiency, total annual net primary corn production of **5,600 kg** (12,250 pounds) is required to produce this corn feed.

▲ **Figure 13.11 Beefy Energetics**
Net primary production in an average cornfield is about 10,000 kg (22,000 pounds) per ha. A single steer requires the net production from about 0.56 ha (1.4 acres) of cornfield.

Nutrient Cycling

Agroecosystems are prone to nutrient losses.

Like their natural counterparts, agroecosystems depend on the cycling of nutrients. The productivity of crop and pasture lands depends on the continuous supply of nitrogen, phosphorus, and other essential nutrients. These nutrients are equally important to the growth and reproduction of domestic animals and humans. Nevertheless, the rate at which nutrients are lost or flow out of agroecosystems is often considerably greater than the rate at which they are naturally replenished. To maintain high levels of primary production, farmers must compensate for this imbalance between nutrient inflow and outflow by applying fertilizers. Nutrient losses from agroecosystems are typically high for four reasons: harvest, continual disturbance, irrigation, and low biodiversity.

Harvest. Each year, nutrients that are stored in plant tissues are harvested and removed from the ecosystem. The rate at which these nutrients are harvested is generally far greater than the rate at which they are restored. Thus, with each successive year, nutrient stocks decline, resulting in diminished primary production.

Continual disturbances. Many systems of growing crops require that soils be plowed, sometimes more than once each year. This continual disturbance makes soils more vulnerable to erosion and nutrient loss. The movement of air into soil improves the growth of some plants, but it also causes some nutrients to escape into the atmosphere more rapidly.

Irrigation. In many regions, crops receive supplemental water in the form of irrigation. Water that runs off agricultural fields or seeps into groundwater

▲ **Figure 13.12 Water Erosion**
Irrigation water flowing through the exposed soil of this cornfield has cut into the ground and carried away sediments and nutrients.

often carries large amounts of water-soluble nutrients that have been leached from the soil (**Figure 13.12**). This depletes the soil of nutrients. It can also pollute streams and lakes that are fed by the runoff (7.5).

Low biodiversity. Agroecosystems are composed of only a few species, those that maximize the production of the things that humans need or want. This low biodiversity affects key ecosystem functions, such as nutrient retention and efficiency of water use (11.2). For example, the root systems of most crops are far simpler than those of the diverse plant communities that they have replaced (6.1). As a consequence, nutrients and water are not absorbed as efficiently and are more likely to be lost.

Dynamic Homeostasis

Agroecosystem simplicity causes vulnerability.

In natural ecosystems, biodiversity increases the capacity for dynamic homeostasis—the ability of an ecosystem to maintain stable values of key processes, such a net primary production and nutrient cycles, despite change or disturbance (1.3). Since the biodiversity of agroecosystems is very low, it is not surprising that their capacity for dynamic homeostasis is often significantly diminished. In other words, compared to most natural ecosystems, agroecosystems are quite fragile. This weakness is obvious when agroecosystems suffer from periods of low rainfall or harsh weather. It is also evident in their vulnerability to invasion by competing weeds or voracious herbivores.

There can be no question that the dominance of our species on this planet is largely due to our development and management of agroecosystems. That said, we must be mindful that their productivity is finite. The energy they can provide us depends on how well we manage them. Agroecosystems are vulnerable to the loss of essential nutrients and water, and their sustainable productivity inevitably requires subsidies of these materials. Finally, we must keep in mind that the key feature to their success, simplicity, also makes them fragile and vulnerable to collapse.

13.3 The Growth of Crop Plants

Like all green plants, agricultural crops depend on photosynthesis for growth and reproduction. The process of photosynthesis requires light, water, and carbon dioxide. Plants also require various nutrients that they obtain from soil. The ecological needs and tolerances of the various crop species determine when and where particular crops can be grown. The growth of crops is greatly influenced by the community of other organisms living in the agroecosystem. Interactions with some of these organisms are beneficial to crop production, while interactions with others are detrimental.

Plant Growth and Reproduction

Plants depend on light, water, and essential nutrients.

The energy for plant growth is provided by photosynthesis, which is carried out in the green tissues of leaves and some stems (4.1). In photosynthesis, plants use the energy from light to convert carbon dioxide and water into high-energy carbohydrates. Carbon dioxide from the atmosphere enters leaves through tiny pores in the leaf surface called stomata. As stomata open to take in carbon dioxide, they also allow water vapor to escape from the moist interior of the leaf. Continued photosynthesis requires replacement of this water.

Most plants obtain water from the soil. Water moves from the soil into the plant's roots, up through the stem, and into the leaves. The movement of water from soil to plant to air is called **transpiration** (**Figure 13.13**). Transpiration transports upward against the force of gravity. How this is accomplished was once a puzzle because a plant does not have a physical pump like a heart. It is now understood that the energy for transporting water upward is provided by a diffusion gradient; that gradient is between the water-saturated air in the spaces between cells in leaves and the drier air outside. As water evaporates from the leaf, it literally pulls water up from stems and roots. The rate at which a plant loses water to the air, its transpiration rate, depends on the plant's leaf area. It also depends on the amount of moisture in the soil and the relative humidity of the air: The drier the air in comparison to the leaf's moist interior, the faster the plant transpires.

Plant growth also depends on a steady supply of nutrients, most of which come from the soil. The plant's root system brings fine roots into direct contact with soil particles and soil water. The cells of these fine roots have extensions called root hairs that greatly increase the surface area through which water and nutrients can be absorbed. The energy that roots need to grow comes from carbohydrates that are made in the leaves and then carried downward through specialized tissues in the stems.

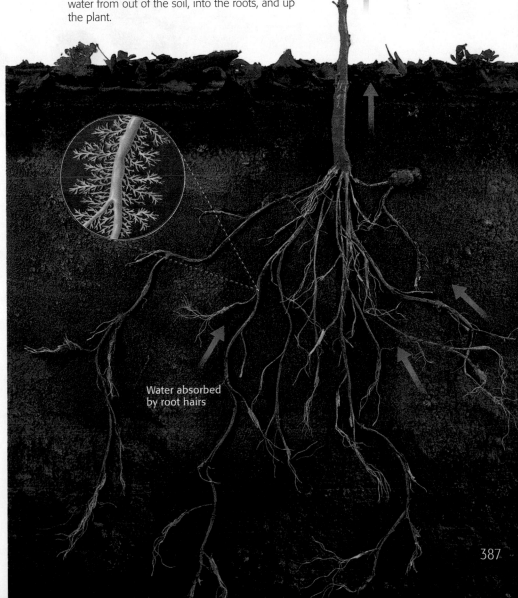

Water lost by transpiration from leaf surfaces

▼ **Figure 13.13 Supplying Soil Water to Leaves**
The force of transpiration from leaf surfaces pulls water from out of the soil, into the roots, and up the plant.

Water absorbed by root hairs

What Are You Eating?

Patterns of growth and reproduction determine the parts of plants that humans use.

Most crop plants have been selected for domestication because they store carbohydrates, proteins, and oils that can be harvested and used by humans. The particular part of a plant in which these nutrients are stored varies greatly from species to species. The following are examples of that variation (**Figure 13.14**).

Because photosynthesis takes place in leaves, it seems natural that leaves might be an important source of food. The leaves of many plants do indeed store large amounts of carbohydrates. Lettuce, spinach, cabbage, and celery are leaf crops. Onions are also leaves, although they may seem more like stems or roots. The thick, fleshy layers of an onion are modified leaves that store large quantities of carbohydrates; this adaptation made onions especially attractive for domestication.

Eat asparagus or a potato, and you are eating a plant stem. The young shoots of asparagus are harvested early in the growing season while they are still tender. Potato plants are native to high elevations in the Andes of South America. The underground stems of these plants—the potato—store carbohydrates that allow these plants to survive extended periods of cold.

Many domesticated plants store carbohydrates in their roots. Examples include carrots, radishes, beets, turnips, and sweet potatoes. Many crops grown in the tropics, such as taro and yams, are root crops.

In other cases, plants have been domesticated because of the nutrients stored in various reproductive parts. Broccoli and cauliflower are subspecies of cabbage, but we consume primarily their flowers. Grains such as wheat, barley, and oats are seeds, as are peas, peanuts, and soybeans. Fruits are the ripened flower ovary, the fleshy structure that surrounds and protects seeds. String beans, tomatoes, peppers, apples, and squash are fruits.

▼ **Figure 13.14 What Am I Eating?**
The part of a plant that serves as human food varies widely among crop species.

Leaves

Roots

Stems

A flower

Seeds

Fruits

What Grows Where and Why?

The climate and soil of a region determine which crops can be grown.

Most agriculture takes place in the tropical and temperate climate zones. The seasonal patterns of temperature and rainfall within a region determine when crops can be grown. In tropical countries such as Brazil, seasonal changes in temperature and rainfall are small, and crops can be grown year-round. The long, warm growing season creates the potential for high primary productivity in agricultural systems, just as it does in natural ecosystems in the tropics. In subtropical climates such as the Sahel of Africa, the growth of crops must be timed to coincide with wet seasons and avoid dry seasons. In temperate latitudes, the growing season is usually limited to the months when freezing temperatures are unlikely.

Like plants in natural biomes, each kind of crop plant has a specific set of tolerances for climatic factors, including temperature, rainfall, and length of day. For this reason, the climate of a region is a major factor in determining where each crop may be grown (**Figure 13.15**). Wheat, for example, grows best in cool, moist conditions. It can tolerate relatively short growing seasons, but the soil must remain moist for complete grain development. In the southeastern and midwestern United States, wheat is grown in early spring and harvested before the onset of hot weather. In the northern United States and southern Canada, wheat is grown in the summer. Wheat also thrives in much of northern Europe and Russia at this time.

Corn requires much warmer growing conditions than does wheat. Its photosynthetic chemistry works most efficiently at temperatures above 30 °C (86 °F). Thus, in North America, corn is grown farther south and planted later in the growing season than wheat.

Rainfall is a critical factor in determining the kinds of crops that are best suited to a region. Some kinds of plants use water more efficiently than others, allowing them to thrive in dry conditions. Desert-adapted food plants, such as the prickly pear cactus, can be grown without irrigation in regions that are far too dry for most crops.

Plant growth can be limited by too much water as well as by too little water. Plant roots need oxygen to grow, and soils saturated by water contain little oxygen. As a result, most crops cannot grow in water-logged soils. On the other hand, cranberries are cultivated in standing water that would suffocate the roots of most other crop species.

The characteristics of soil limit the kinds of crops that can be grown in a region. Soil moisture and chemistry also differ from place to place, depending on the prevailing weather, topography, past vegetation, and character of the underlying bedrock. For example, acid soils tend to form in regions where granite is the underlying material. Blueberries thrive in acidic soils and are a good choice in those sites. Limestone tends to produce more alkaline soils. Asparagus is a crop plant that requires alkaline soil.

Plants vary greatly in their need for soil nutrients. Crops such as corn and wheat require large amounts of nitrogen, which is often supplied by fertilizers. Because of their symbiotic relationship with nitrogen-fixing bacteria, legumes such as soybeans and peanuts can be cultivated on nitrogen-poor soils without the heavy application of nitrogen fertilizer.

Agricultural systems and the human cultures they support ultimately depend on the local climate and soils. The traditional foods of a region, such as the rye bread of northern Europe, the corn of Central America, and the rice of Southeast Asia, come from plants with adaptations that make them well suited for cultivation in that part of the world.

▼ **Figure 13.15 Climate Limits the Growth of Crops**
Each kind of crop has a particular range of tolerance for rainfall and temperature.

Key
● Potato
● Wheat
● Corn
● Rice
● Sugar cane
● Bananas

Key
1. Tundra
2. Taiga
3. Temperate forests
4. Tropical rain forest
5. Tropical seasonal forest
6. Chaparral/grassland
7. Desert
8. Tropical savanna

AVERAGE TEMPERATURE (°C)

AVERAGE PRECIPITATION (cm/year)

The Role of Other Organisms

The growth of crop plants is influenced by the surrounding ecological community.

QUESTIONS

1. Explain how transpiration from plants is related to the uptake of CO_2.

2. Name three crop plants that are typically grown in tropical climates and three that are typically grown in temperate climates.

In most agroecosystems, attention is focused on the production of a single crop species. However, the growth of those species also depends on a community of other organisms, including pollinating insects and bacteria and fungi that live in the soil.

Before seeds and fruits can form, the flowers of crop plants must be pollinated. Insects pollinate vegetable, fruit, and nut crops, which make up about a third of the human diet (6.3). Honeybees are among the most important insect pollinators, but the loss of most of their natural habitat has resulted in very sparse populations of wild honeybees. Instead, domesticated honeybees raised by beekeepers provide this essential ecosystem service for many crops (**Figure 13.16**). Today, honeybee populations are threatened by the inappropriate use of pesticides, bacterial diseases, and infestations of parasitic mites. Because honeybees are so important to our food supply, these threats must be taken very seriously.

A diverse community of soil organisms recycles the nutrients in agroecosystems. Earthworms are among the most visible of these soil organisms. Because they feed on bits of dead leaves and other organic materials in the soil, earthworms are very effective at converting agricultural crop residues back into inorganic soil nutrients. Their vertical burrows allow air to enter the soil and improve soil texture, vastly benefiting plant roots. Because earthworms constantly bring soil from deep in the root zone back to the surface of the ground, they play an important role in soil development.

Mites, springtails, beetles, and a myriad of beneficial bacteria also contribute to the breakdown of crop residues and the subsequent release of nutrients to the soil (**Figure 13.17**). Fungi are particularly effective in this recycling function. They release enzymes that break down the cellulose in plant cell walls, a biochemical task few other organisms can do. Without fungi in the soil, most of the carbon, nitrogen, and other elements essential for life would remain in dead leaves and stems, and new growth would come to a standstill.

Scientific understanding of the community of soil organisms and the beneficial work they accomplish is limited, and this is a challenge for agricultural science. We do know that the excessive application of herbicides and fungicides to agricultural fields can kill vital soil organisms and interfere with nutrient cycling. The result is a greater need for artificial cultivation and fertilizers and a lower net yield.

Organisms involved in mutualistic associations with plant roots are especially important to agricultural productivity. Remember from Chapter 6 that specialized fungi form associations with plant roots called mycorrhizae. The fine threads of these fungi grow through the soil and around the hairlike extensions of the roots of their plant partners. The mycorrhizal fungi transport inorganic nutrients such as phosphates and potassium into plant roots, while the plants provide the fungi with carbohydrates derived from photosynthesis. In nutrient-poor soils, many crop plants fail to thrive without symbiotic fungi.

Of course, not all of the organisms in agroecosystems are beneficial to the growth of crops. A variety of non-crop plants also grow in agricultural fields, where they compete with crops for nutrients, water, and light. Farmers call these competitors weeds. Many animals that grow in fields feed upon crops, thereby lowering production. Herbivorous insects may consume a significant portion of crop production. Roundworms called nematodes may parasitize roots. Crop plants are also susceptible to a variety of viral, bacterial, and fungal diseases. Much of agroecosystem management is aimed at minimizing the impacts of these competitors, herbivores, and parasites.

▼ Figure 13.16 Domestic Honeybees
Each spring, hives of honeybees are brought to orchards to ensure that the trees are pollinated and produce fruit. In addition to honey, domestic honeybees provide pollination services to farmers worth over $14 billion each year.

▲ Figure 13.17 Soil Fauna
Earthworms and a diverse array of smaller soil organisms speed up the recycling of nutrients as they feed on organic material in the soil.

13.4 Managing Soil Resources

The uppermost layer of Earth's crust is **soil**, a mixture of organic matter and mineral particles. Soils are formed by the constant weathering of the rocks below them and the continual addition of organic matter from the plants that grow on top of them. As a consequence, soils have abundant mineral-rich, rocklike particles at the bottom and nutrient-rich, organic materials toward the surface. The roots of plants draw nutrients and water from soil. The ability of a soil to support plant growth is determined by the total amount of nutrients it contains, as well as by factors such as acidity and aeration that determine how well plants can use those nutrients. In ecosystems such as forests and prairies, natural processes maintain soil structure and fertility. In agroecosystems these processes are compromised, so humans must take steps to sustain soil productivity.

Soil Origins and Structure

Organic matter and weathered rock form soil.

Soil is composed of mineral particles and organic matter, as well as water, air, and a community of living organisms. The mineral particles are classified according to their size (**Figure 13.18**). **Sand** particles are 0.05–2 mm in diameter. Although sand particles have been weathered, the chemical structure of these fragments still resembles that of their parent bedrock. **Silt** particles are smaller, with diameters ranging from 0.002 to 0.5 mm. These particles have experienced more weathering than sand. The smallest particles are **clay**; their diameters are less than 0.002 mm. Clay particles are highly weathered and have unique physical and chemical properties. Clays are composed of platelike particles of silicon-rich minerals. Electrical charges in these minerals allow clay particles to bind to one another and to form chemical bonds with many soil nutrients. Soil **texture** refers to the relative amounts of sand, silt, and clay in soil.

Soils form as a result of two processes: the weathering of the bedrock below them and the deposition and decomposition of leaf litter and other organic materials from the soil surface. As a consequence, natural soils have a distinctive vertical structure, or **soil profile**. At the surface, soil profiles are dominated by organic matter. Beneath that is a mixture of organic matter and weathered mineral particles. At the bottom, soil profiles contain mainly less-weathered, rocklike particles. Soil scientists divide soil profiles into layers called *horizons* (**Figure 13.19**).

The **O horizon** at the very top is not really soil, but consists of leaves or crop residue and the organic products of their decomposition called **humus**. Below this, the **A horizon** is made up of a mixture of organic matter and mineral particles. The upper, humus-rich portion of the A horizon is topsoil. Hundreds of species of organisms live in the O and A horizons, including bacteria, fungi, and animals such as worms and insects. The activities of these organisms contribute to the decomposition of humus, breaking down complex organic molecules and releasing inorganic nutrients, such as ammonia and phosphate. Plant roots are most dense in the A horizon because essential nutrients are most abundant here.

Below the A horizon is the **B horizon**, also called subsoil. This layer is often especially rich in clay particles that form from minerals dissolved from the weathering of sand and silt. Minerals such as iron and aluminum leach out of the A horizon and accumulate with the clays in the B horizon. These minerals often give the B horizon a red or yellow hue.

The lowest layer of soil is the **C horizon**, or weathering zone. Here bedrock is broken up by the action of soil water and sometimes by the growth of plant roots. The C horizon differs substantially from place to place, depending on the type of bedrock.

Horizons

O

A

B

C

▲ **Figure 13.19 Soil Horizons**
Weathering and the decomposition of organic matter create distinct soil layers. The O horizon is primarily organic matter and fine roots. The A horizon contains organic matter and highly weathered mineral soil. The B horizon contains abundant clay and far less organic matter. The C horizon contains particles of rock that are chemically similar to the parent rock.

Ⓐ   Ⓑ   Ⓒ

► **Figure 13.18 Soil Texture**
Depending on their size, the mineral particles in soil are classified as Ⓐ sand, Ⓑ silt, or Ⓒ clay.

Soil Fertility

The ability of soils to support plants depends upon the essential nutrients they contain, as well as their pH, degree of aeration, and overall structure.

Soil fertility refers to the variety of soil characteristics that support plant growth. These characteristics include the availability of essential nutrients, the pH (or acidity), the amount of aeration, and the overall structure of the soil.

The availability of nutrients in soil is influenced by several factors. The total amount of a nutrient in soil sets an upper limit on its availability. Because organic matter contains significant amounts of nutrients such as nitrogen and phosphorus, fertility is often correlated with the amount of organic matter in soils. Clay tends to increase soil fertility, because clay particles chemically bind to many nutrients and prevent them from being leached away by water flowing through the soil.

Fertility is also determined by factors such as moisture content, aeration, and pH. These factors influence the decomposition of organic matter and the chemical transformation of nutrients. Many of the nutrients in soil are not in a form that plants can use. For example, over 99% of soil nitrogen exists in complex organic molecules, such as proteins, that plant roots are not able to access. Before plants can take up this nitrogen, it must be converted to either ammonium (NH_4) or nitrate (NO_3).

The acidity, or pH, of soil is a measure of the concentration of hydrogen ions (H^+) relative to other cations (positively charged ions) in the soil (3.1). Soils with pH values less than 7 are acidic, and those with pH values greater than 7 are alkaline. The decomposition of organic matter is the largest source of H^+ to the soil. Continual decomposition of organic matter tends to increase soil acidity, or lower soil pH. Weathering of soil particles releases large amounts of other cations, such as potassium (K^+), calcium (Ca^{++}), magnesium (Mg^{++}), and iron (Fe^{+++}). These cations tend to raise soil pH, making it more alkaline. Other factors, such as the amount of

▼ Figure 13.20 Desert Soil
High rates of evaporation cause salts to accumulate in desert soils. As a result, desert soils have higher pHs than soils that develop in moister conditions.

aluminum and sulfur, also influence soil pH. In some wetlands where organic matter abounds and other ions are in short supply, soil pH may be as low as 3. In contrast, in many deserts where soils have only small amounts of organic matter and large amounts of other ions, pH may approach 9 (**Figure 13.20**).

Soil pH influences a variety of processes that determine soil fertility. It has a direct effect on the growth of plant roots. The roots of some species grow best in acidic soils, whereas others do best in alkaline soils. Soil pH also affects the growth of many bacteria and fungi, thereby influencing the decomposition of organic matter and the availability of essential nutrients. Finally, because it affects how clay particles bind to one another, soil pH can directly influence the physical structure of the soil itself.

Soil aeration refers to the ability of atmospheric gases, particularly oxygen and carbon dioxide, to diffuse into and out of the soil. As a general rule, as soil aeration increases, so does the growth of roots and soil microbes. Poor aeration can affect the availability of some essential nutrients. For example, when soils lack oxygen, certain bacteria convert the nutrient nitrate to nitrogen gas (N_2). This nitrogen diffuses to the atmosphere and cannot be used by plants.

Soil aeration is directly related to a structural property of soil that farmers call tilth. **Tilth** refers to the physical arrangement of soil particles that facilitates aeration, seedling emergence, and the growth of roots. Tilth is affected by soil texture, particularly by the extent to which clay particles are bound to one another. Tilth is generally improved by increased amounts of organic humus. This is one of the main reasons that gardeners often add organic amendments such as peat moss or cow manure to soil.

Soil Conservation

Agroecosystems lack many of the processes that sustain soil fertility.

Soil structure and fertility are sustained in natural ecosystems in a variety of ways. Prairies and forests build topsoil through the constant accumulation of organic matter and protection of the soil already in place. Organic matter is constantly being added to soils, replenishing the stock of essential nutrients such as nitrogen and phosphorus. Plant roots and chemicals from organic decomposition ensure steady weathering of soil minerals, thereby ensuring continual soil development. The leaves and stems of plants intercept rainfall and their root systems hold soil in place, minimizing erosion.

All of these processes are altered when natural ecosystems are replaced by agroecosystems. The sustained productivity of agroecosystems, therefore, depends on sustainable management of soils. Because it is the zone of highest fertility and root activity, topsoil is especially vital to agricultural production. However, most agricultural practices diminish the organic inputs that build topsoil and accelerate the loss of organic matter and erosion of topsoil. When the majority of crops are harvested, most of the aboveground organic matter is removed from fields. Crop residues are often burned. Plowing and tilling increase humus decomposition, further depleting stores of nutrient-rich organic matter. These activities also expose topsoil to rain and wind, accelerating erosion.

Water erosion is equally problematic in unprotected agricultural fields. When raindrops fall on exposed soil, they dislodge soil particles, which are carried along by the surface flow of rainwater. If running water cuts a small gully, more water moves through that channel, cutting the gully deeper and deeper. If agricultural practices fail to prevent erosion, topsoil built up over several centuries can be lost in a few heavy rains.

Not only does this diminish fertility of agricultural fields, but it also contributes large amounts of nutrients and sediment to the waters of streams and rivers.

Several conservation practices can prevent the loss of nearly all topsoil. The addition of humus from domestic animals or other organic waste can restore lost organic matter and the nutrients it contained. Plowing along the contour of the land, rather than up and down hills, can prevent the formation of gullies (**Figure 13.21**). Maintaining strips of grass-covered ground between crop rows is another conservation strategy. Since grass roots hold soil more effectively than many crop plants, these strips recapture soil particles before they wash out of the field. Cover crops such as winter wheat or rye, planted to hold soils that would otherwise lie exposed to the weather between seasons, reduce erosion for the same reason (**Figure 13.22**).

Many farmers now use planting and harvesting technologies that minimize or eliminate the need to plow soil. At the end of the growing season, crop residues are left on the ground, so the soils are never exposed to erosion. To plant new crops, implements called seed drills sow seeds through the protective layer of the previous year's crop residue.

▼ Figure 13.21 Preventing Soil Erosion
Planting crops in rows that follow the contours of hills causes water to flow across the rows and minimizes erosion. Here, strips of perennial alfalfa (green) alternate with strips of annual corn (gold) to further limit erosion.

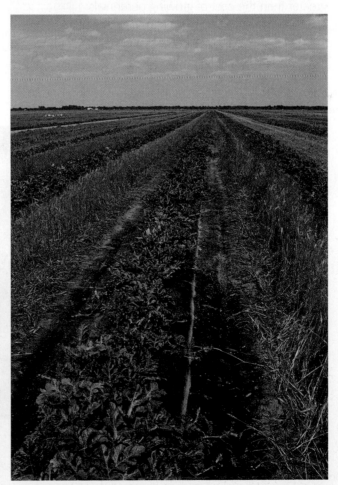

◄ Figure 13.22 Cover Crop
A wheat cover crop between the rows of immature watermelon plants (green) prevents erosion and protects the young plants from blowing sand.

1. What are the characteristics of the different layers in a soil profile?

2. What four factors influence soil fertility?

3. Why are the soils in agroecosystems prone to erosion and loss of fertility?

FOCUS ON SCIENCE: Developing Perennial Wheat

Can perennial wheat provide a more sustainable replacement for annual wheat?

In 2010, *National Geographic Magazine* recognized Jerry Glover as one of "five crop researchers who could change the world." The change Glover most wants to see is a decreased dependence on annual crops and an increased use of perennial crops. (Annual plants complete their life cycle in a year; perennials live for many years.) At present, foods from annual crops, such as wheat, corn, rice, and beans, make up about 70% of the calories in the human diet.

Annuals die when they are harvested, so each year they must be replanted from seed. Why is that a problem? First and foremost, before planting annual crops, fields are usually cleared and plowed. These activities expose the soil, increasing runoff and soil erosion; year by year, these processes diminish soil fertility. Second, because plant leaves, stems, and roots are constantly being removed, the permanent store of carbon on annual croplands is minimal. Third, adverse weather conditions, such as drought or late frost, pose a greater risk to annual plants than to perennial plants. Finally, the need to clear, plow, fertilize, and plant new crops each year makes the cost of growing annuals greater than the cost of growing perennials.

Unlike annual crops, perennial crops maintain important ecosystem services throughout the year. For example, they intercept, retain, and efficiently use rainwater, diminishing the need for irrigation. Perennials have longer growing seasons and deeper roots and thus maintain more carbon on the land.

Today, annual crops are grown on over 70% of Earth's croplands. As Glover sees it, soils on the world's best croplands are at low risk of being degraded by the cultivation of annual crops, but these lands make up only about 20% of farmable land. The remaining land is at high risk of degradation by the production of annual crops. Much of this lower-quality cropland is capable of producing perennial crops.

Glover and his colleagues at the Land Institute have focused much of their attention on a perennial called intermediate wheatgrass (*Thinopyrum intermedium*), which they hope can be used as a replacement for annual wheat (**Figure 13.23**). The roots of this grass live for many years; each year its roots send up new shoots that produce seeds. Intermediate wheatgrass has many of the characteristics required of crop plants, such as having stiff, erect stems, synchronous seed maturation, and large, easily harvested nutritional seeds.

Glover has developed several breeding populations of intermediate wheatgrass. The seeds of these plants have a nutritional value that is similar to or better than that of the seeds of annual wheat, although the seeds of wheatgrass cannot be used alone to make leavened bread. Total seed production from wheatgrass does not yet match that of annual wheat, but Glover is confident that continued selective breeding will improve yield and the dream of perennial wheat will be realized.

SEPTEMBER DECEMBER MARCH JUNE

- 0 ft
- -2
- -4
- -6
- -8

■ Annual winter wheat
■ Perennial wheatgrass

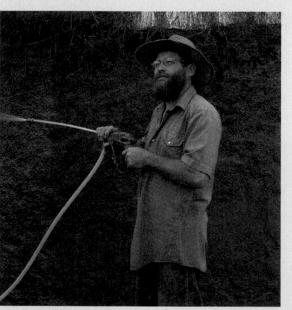

◀ **Figure 13.23 Agricultural Revolutionary**
Jerry Glover is an agroecologist at the Land Institute, an independent agricultural research center near Salina, Kansas. Seasonal development of annual winter wheat is compared with that of intermediate wheatgrass. Winter wheat must be harvested in late spring and replanted in the fall, so it is absent in the summer.

PHOTOS: REBECCA HALE, NGM STAFF
GRAPHIC: LAWSON PARKER, NGM STAFF. SOURCE: LAND INSTITUTE

13.5 Water and Agriculture

Water is essential to the growth of all plants. In fact, modern agroecosystems account for 70% of all human water use. Water availability in a region is determined by rainfall and factors that influence runoff and percolation into soil. The amount of water that soil can hold against the force of gravity depends on soil texture and organic matter content. Irrigation has greatly increased the productivity of many existing agroecosystems and allowed the expansion of agriculture in dry regions. At the same time, irrigation has increased demands on water supplies, diminished water quality, and increased the salinity of soils in many places. Increasingly, agronomists and farmers are growing crops and using technologies that make use of water in more sustainable ways.

Water in Soil

The surfaces of soil particles hold water against the force of gravity.

Water comes into an agricultural field as rainfall or irrigation. As in natural ecosystems, the water entering an agroecosystem has three fates. It may simply run off the soil surface into nearby streams; it may percolate through the soil to groundwater aquifers; or it may evaporate to the atmosphere from soil or plants (7.3). Interactions between soil and water play a key role in determining the ultimate path of this water.

How much of the water that falls on a field becomes runoff and how much percolates into the soil? The answer depends on several factors, including the topography of the land and the porosity of the soil. Runoff is generally greater on steep slopes than on gradual slopes. The more porous the soil, the easier it is for water to percolate through it. In general, sandy soils are more porous than clay soils. Organic matter usually increases porosity. Soils that have been compacted by heavy equipment or other activities are less porous and have increased runoff.

Water that percolates into the soil is drawn downward by the force of gravity. Water flowing through the soil is called **gravitational water.** However, the soil is able to hold some water against the force of gravity.

Soil holds water because of hydrogen bonding between soil particles and water molecules (3.3). Electrical charges on the surface of soil particles bind very tightly to water molecules, forming a microscopic film of water around the soil particles. The water that is bound to soil particles is called **hygroscopic water.** Water molecules that are bound to soil particles also form hydrogen bonds with other water molecules, and those molecules form bonds with other water molecules. The water that is held in the soil by water-to-water hydrogen bonds is called **capillary water** (**Figure 13.24**). A sponge holds water against

▼ **Figure 13.24 Soil Water**
Field capacity is the amount of water that a soil can retain against the pull of gravity. When soil water is below the wilting point, plants are no longer able to take up water fast enough to keep pace with transpiration.

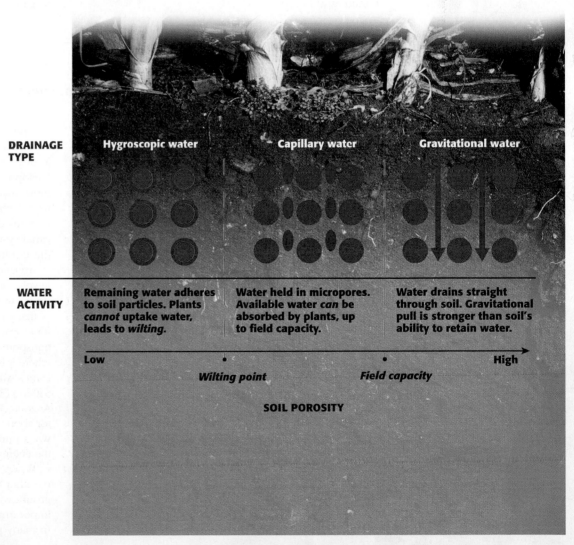

| DRAINAGE TYPE | Hygroscopic water | Capillary water | Gravitational water |

| WATER ACTIVITY | Remaining water adheres to soil particles. Plants *cannot* uptake water, leads to *wilting*. | Water held in micropores. Available water *can* be absorbed by plants, up to field capacity. | Water drains straight through soil. Gravitational pull is stronger than soil's ability to retain water. |

Low   *Wilting point*   *Field capacity*   High

SOIL POROSITY

the force of gravity by exactly the same principle. Water molecules form strong bonds with the surfaces of the sponge, and more water is retained due to water-water hydrogen bonding.

The amount of energy required to remove the water in soil depends on the proximity of water molecules to soil particles. Within the hygroscopic film, the molecules closest to the surface of soil particles are held most tightly. Those further away are held less tightly. At some distance from a soil particle, the electrostatic forces holding water are weaker than the force of gravity, and water percolates downward.

Field capacity is the amount of water that a given volume or weight of soil can hold against the force of gravity. The capacity of soil to hold water is determined primarily by the total surface area provided by soil particles. The tiny particles of clay have much greater surface area relative to their volume than the coarser particles of silt and sand. Thus, fine-textured soils have higher water-holding capacities than coarse soils. Particles of organic matter have a great deal of surface area, so water-holding capacity generally increases with increasing organic matter content. This is another reason why gardeners and farmers amend soil with organic manure and humus.

To obtain water from soil, plants must overcome the forces that hold water to soil particles. When soils are near their water-holding capacity, plant roots can easily take up water. However, as soil water is depleted, the soil holds the remaining water ever more tightly. At some point, plant roots cannot take up enough water to keep pace with the rate of transpiration from their leaves. When this occurs, the plant wilts. **Soil wilting point** is the point at which there is no longer enough water in the soil for plants to replace the water that is being lost to transpiration.

If you forget to water your houseplants, the soil surrounding their roots gets drier and drier. Eventually their roots cannot take up water fast enough to replace the water evaporating from their leaves, and they wilt. The soil in their pots is then at its wilting point. If you water your plants promptly, they can recover; if you do not, they will dry out completely and die.

As water moves through soil, it is altered chemically. Chemicals from soil particles or produced by the decomposition of organic matter may dissolve in water. Other chemicals dissolved in the water may adhere to soil surfaces. Thus, the chemical properties of the water that runs off agricultural fields or percolates into groundwater are different from those of the rainwater or irrigation water coming into the fields.

Irrigation

Water diversion and pumping has allowed humans to grow crops where natural supplies of water are limited.

The **potential evapotranspiration (PET)** in a region is an estimate of the average amount of water that would evaporate from a hypothetical agricultural field over the course of a year. Like rainfall, PET is measured in centimeters or inches. In many dry regions of the world, PET is much higher than total rainfall. In these regions, the availability of water is a major factor limiting the productivity of agroecosystems. Even where annual rainfall exceeds PET, water deficits may occur during seasons when there is little rainfall, high temperatures, and low relative humidity, which increases the rate of evaporation.

◄ **Figure 13.25 Irrigating the Desert**
The Central Arizona Project is an 80-ft-wide concrete-lined canal that supplies water to central Arizona farms and to the rapidly growing desert cities of Phoenix and Tucson.

Irrigation has increased crop production in many regions where water deficits are common. Irrigation has a long history, dating back over 5,000 years to farms in Mesopotamia and Egypt. As pipelines and pumps became more sophisticated, humans were able to transport water from sources that were farther and farther away. The Romans developed remarkable aqueducts and complex canal systems to transport agricultural water; many of these remain in use to this day.

Irrigation has played a significant part in the Green Revolution of the past six decades. It has improved the productivity of existing croplands and allowed the cultivation of many lands that were previously thought to be unsuitable for crops. Recently, humans have gone to great lengths to transport water from places where it is abundant to places where it is not. The Central Arizona Project, for example, carries water from the Colorado River up and down hills for 540 km (336 miles) to irrigate agricultural fields in central Arizona (**Figure 13.25**). Indeed, crop irrigation now accounts for about 70% of all human water use. In the United States, water pumps for crop irrigation account for nearly 20% of the energy used in agriculture.

Where surface water is unavailable or is consumed by other demands, crops may be irrigated by pumping groundwater. Such irrigation has been especially important in the High Plains region of the American West. In many places, the rate at which aquifers are recharged

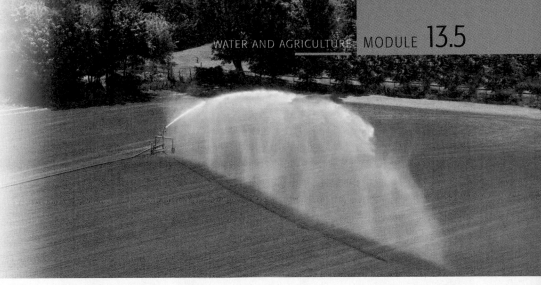

is considerably less than the rate at which water is being pumped for irrigation (12.4).

Throughout history, the most common form of irrigation has been surface irrigation, in which gravity moves water through canals, across fields, and between furrows. The energy costs for this form of irrigation are comparatively low. However, surface irrigation often distributes water across fields unevenly and is not effective on complicated topography.

Today, overhead spraying is widely used to put water onto fields (**Figure 13.26**). Overhead spraying distributes water across fields more evenly than surface irrigation but requires more energy to pump water. Water sprayed from overhead evaporates more rapidly than water flowing along the soil surface. As a result, overhead spraying often decreases the efficiency of water use.

In addition to its impact on the water supply, irrigation presents other environmental challenges. Most conventional methods of irrigation apply more water to fields than can be held by the soils or used by plants. The excess irrigation water drains into streams and rivers or percolates into groundwater, carrying with it the chemical residues from fertilizers and pesticides.

Where evaporation rates are high, the salts contained in irrigation water are concentrated and crystallize on the soil surface. This process is called **soil salinization**. Over many growing seasons, irrigated lands accumulate more and more salts. Eventually, salts brought in with irrigation water build up to the point that they interfere with plant growth. This is a longstanding problem that is difficult to solve. Arid farmlands rarely have access to sufficient irrigation water to wash these excess salts out of the root zone.

▲ **Figure 13.26 Aerial Irrigation**
Although aerial spraying distributes water evenly, it is inefficient because much of the water evaporates before reaching the plants.

Conserving Water in Agroecosystems
New irrigation technologies can provide significant water savings.

As energy costs and competing demands for limited supplies of water increase, the need to conserve water resources in agroecosystems has become clear. To increase water use efficiency, decrease runoff, and minimize salinization, many farmers are turning to drip irrigation. In **drip irrigation** systems, pipes with small openings at the base of each plant feed water to the root zone, avoiding the dry air above the field (**Figure 13.27**). This approach not only reduces the amount of water wasted by evaporation but also reduces the rate of salinization, because less evaporation means that less salt is left behind in the soil.

An even more efficient approach to farming in arid regions is to plant crops that demand less water. Plant breeding programs have developed new genetic strains of many common crop plants that are able to grow on reduced amounts of water.

Many agronomists suggest that entirely different plant species that are adapted to arid lands should be substituted for water-demanding species, such as corn and soybeans. For example, jojoba (pronounced ho-HO-ba) is a large shrub that thrives in the Sonoran desert of Arizona, southern California, and northwestern Mexico. Jojoba produces edible seeds from which fuel-quality oil can be extracted. Thus, jojoba produces products that are similar to corn and soy but requires virtually no irrigation.

◀ **Figure 13.27 Efficient Water Use**
Drip irrigation systems deliver water directly to the roots of crop plants. These systems use water 5 to 10 times more efficiently than conventional irrigation systems.

QUESTIONS

1. What soil factors determine the amount of water that is available to crop plants?

2. Why does irrigation cause soil salinization in arid regions?

13.6 Domestic Animals

Domestic animals represent a significant portion of agricultural operations, both in industrialized and developing nations. Humans use nearly 20% of Earth's land as pasture to support domestic animals. In addition, nearly 30% of cropland—another 5% of Earth's land—is used to produce hay and grain to feed livestock. Trophic-level efficiency, the ability of domestic animals to convert their food into food that humans can consume, varies among animal species. It also varies within species, depending on the source of food and level of activity. The environmental impacts of raising large numbers of domestic animals include air and water pollution, significant emissions of greenhouse gases, and the spread of disease.

Trophic-Level Efficiency

The amount of energy that livestock provide to humans depends on the species, the quality of their food, and the activity level of the animals.

With the exception of dogs and cats, most domestic animals are herbivores. Much of the diet of herbivores is composed of cellulose, a polymer of glucose that is not easy for most animals to digest (3.2). As a consequence, the trophic-level efficiency of herbivores is generally lower than that of carnivores.

The trophic-level efficiency of domestic species varies considerably, depending on how efficiently their digestive systems manage cellulose. Ungulates, cloven-hoofed herbivores such as cattle, sheep, goats, and pigs (mammal order Artiodactyla), have complex, multichambered stomachs that harbor symbiotic microbes that assist in the decomposition of cellulose. Horses and their kin (mammal order Perissodactyla) have much simpler digestive systems that are less efficient at using cellulose (**Figure 13.28**). For this reason, an area of pasture can support about twice as many cattle as horses. Because of the inefficiency of their digestive systems, horses are usually fed a grain-rich, low-cellulose diet.

The source of food also influences trophic-level efficiency. Cattle raised on wild rangeland typically have trophic-level efficiencies well below 10%. Cattle raised on pastures that have been planted with more digestible grasses and herbs have higher trophic-level efficiencies. The highest efficiencies are found in cattle that are fed

► **Figure 13.28 Comparing Digestive Systems**
Compared to an ungulate, a horse has a relatively simple digestive tract that is not very efficient at digesting high-cellulose plant material.

Rectum

Small colon

Large colon

Esophagus

Stomach

Teeth

Cecum

Intestine

◄ **Figure 13.29 Activity Burns Calories**
Cattle raised on the open range Ⓐ burn many more calories than cattle raised in the confined conditions of a feedlot Ⓑ.

grains, such as corn and oats. Chickens do not have especially efficient digestive systems, but they have high trophic-level efficiencies because they feed almost entirely on grain.

Finally, the trophic-level efficiency of domestic animals is strongly influenced by their degree of physical activity.

Like humans who exercise often, domestic animals that can move about freely burn much of the energy in their food in respiration (**Figure 13.29**). Many nutrition experts consider the leaner meat from such animals to be a healthier food choice than the fattier meat from animals that are raised in confined conditions.

Environmental Impacts

Managing animal waste is a significant challenge, but it can be an opportunity too.

The production of large numbers of domestic animals presents a number of challenges to the environment and human health. These include issues related to waste production, changed patterns of land use, and the transmission of diseases.

Waste production. The wastes produced by large livestock operations have significant environmental impacts. According to the Environmental Protection Agency, the amount of waste generated by 700 dairy cows is equivalent to the septic waste produced by a city of 10,000 people. Hog lots and poultry operations also produce large quantities of waste. In the United States, most states have regulations regarding the management of waste for large numbers of animals raised in confined conditions. Such management is, however, far less stringent than the management of human waste.

When exposed to the weather, animal manures release inorganic nutrients, such as nitrates, ammonia, and phosphates, that can pollute surface waters and nearby streams. The decay of livestock waste also

releases a variety of foul-smelling organic gases. Typically, wastes from cattle feedlots and the confined feeding facilities for swine and poultry are stored in open lagoons. Such lagoons may leak or fail completely, particularly during rainy periods. Wastes are allowed to decompose over a period of months, after which they are often distributed as fertilizer onto cropland or pasture. Once again, excess nutrients often find their way into nearby streams.

Methane gas is a by-product of the digestion of nearly all herbivores. Methane is a powerful greenhouse gas that is 23 times as effective as CO_2 at trapping heat in the atmosphere (8.3). The Intergovernmental Panel on Climate Change estimates that methane emissions from domestic livestock account for more than 5% of current global warming.

Methane, which can be burned like natural gas, is a potential source of energy (15.5). Realizing this, managers of many domestic animal facilities are beginning to install waste treatment systems that capture methane and use it to generate electricity (**Figure 13.30**). Such facilities

◄ **Figure 13.30 Energy from Waste**
In many livestock operations, animal waste is simply pumped into lagoons Ⓐ, where it decomposes. As an alternative to lagoons, waste can be processed in containers that trap methane gas, which can then be burned to generate electricity Ⓑ.

399

▲ Figure 13.31 Domestic Animals and Human Health

Humans share many diseases with their domestic animals. For example, some strains of flu originate in poultry that are raised and sold in unsanitary conditions.

QUESTIONS

1. How does a domestic animal's digestive system influence its trophic-level efficiency?

2. Describe three important environmental challenges associated with livestock production.

have the added benefits of providing more secure waste confinement and releasing fewer odors than conventional lagoons.

Land use. Livestock operations often cause major changes in land use. In many developing nations, forests are being cleared to create pasture for sheep, goats, and cattle. Such deforestation contributes to climate change, reduces watershed protection, and threatens biodiversity. As human populations that rely on herd animals grow, there is a need for larger herds. As a result, more and more animals are concentrated in a fixed area of land, resulting in overgrazing. Under crowded conditions, sheep, goats, and cattle degrade pastures by stripping vegetation from the ground and exposing soils to erosion.

Transmission of diseases. As humans and their domestic animals have coevolved, so have the disease organisms that infect them. Livestock such as sheep, cattle, and pigs are alternative hosts for parasites that can also afflict humans, such as flukes and tapeworms. Forms of the bacterium *Escherichia coli* that can cause deadly infections in people grow in the intestines of cattle, goats, and pigs with no ill effects. Viruses that are responsible for various forms of flu can originate from populations of swine or domestic birds (**Figure 13.31**).

Confined animals easily spread disease from one individual to another. To reduce outbreaks of these diseases and to maximize animal growth rates, antibiotics are often added to animal feed. Because low levels of these drugs are administered routinely, the only strains of bacteria that can thrive in the poultry house or feedlot are those that are resistant to antibiotics. In the absence of competition with drug-susceptible strains, the resistant bacteria spread throughout the animal population. Thus the animals become a source of antibiotic-resistant bacteria. These bacteria may spread to human food supplies when the animals are slaughtered; they may also spread to surface waters when animal manure washes into rivers and streams. Most of these bacteria do not cause human disease, but some, such as *E. coli*, can cause severe problems. Diseases caused by antibiotic-resistant bacteria are a special concern because traditional antibiotic therapies may not work against them.

The impact of human diets. Some argue that a vegetarian diet would eliminate environmental consequences of livestock. Others argue that meat and animal products such as milk, cheese, and eggs have long been a part of most human diets and are important sources of protein and other nutritional elements. In regions that are too arid, hilly, or cold to support row crops, domestic animals are vital to human subsistence. If managed properly, livestock can facilitate nutrient recycling from crop residues back to fields, and grazing animals can be an environmentally sensible means to control weeds. However, high-density animal facilities, where thousands of animals are raised in confined spaces, present significant challenges to sustainable management.

In general, as societies develop economically, the proportion of their population's total diet that comes from meat and animal products increases. Thus the economic development and human population growth expected over the next several decades suggest that the management of domestic livestock will continue to be important.

13.7 Managing Genetic Resources

With industrial development, humans have become dependent on fewer and fewer species of crops and domestic animals. While selective breeding, hybridization, and cloning have produced highly productive strains of domestic plants and animals, they have also reduced the genetic diversity within some species. This genetic homogeneity increases the risk that disease or environmental change could jeopardize human food supplies. Older varieties of crops and livestock have greater genetic diversity; that diversity may be important for future adaptation. Genetic engineering holds promise for improving the rate of production, nutritional value, and disease resistance of some crops. However, care must be taken to avoid negative environmental impacts that could arise from these genetic techniques.

Genetic Diversity and the Stability of Agroecosystems

Deliberate efforts to enhance crop diversity are creating a more stable food supply.

Plantings of a single crop species are called **monocultures**. The practice of planting multiple crop species in the same field is called **interplanting** or **polyculture**. Just as in natural ecosystems, increasing the diversity of crop plants in an agroecosystem can increase the ecosystem services that it provides. Differences in root systems of interplanted crops can result in more efficient uptake of water and nutrients and increased productivity (6.2). Furthermore, the spread of insect pests and fungal diseases is retarded in polycultures compared to monocultures.

The diets of people in industrialized societies include foods from only a few of the thousands of plants that have been used for food throughout human history. In the United States, most of the items on the shelves of grocery stores are from only six commonly cultivated plants: corn, wheat, rice, potatoes, soybeans, and barley.

In addition, modern plant breeding has reduced the genetic diversity within the few species that are grown. Today, agronomists use hybridization, asexual propagation, and cloning to produce crop plants that are genetically similar, or even identical, to one another. The goal of this standardization is to maximize production and ensure homogeneous growth. The high production that comes from this homogeneity has contributed to the greatly increased food production of the Green Revolution. However, this homogeneity also can lead to serious problems when diseases or other unanticipated threats arise.

The story of the Irish potato famine illustrates the danger of relying on a single source of food. In the early 1800s, potatoes became the primary food source for nearly all of the peasant farmers of Ireland. In 1845, a rapidly spreading fungal blight attacked the Irish potato crop, creating a terrible famine across the country. One million people starved to death. Two million people fled from Ireland, many of them immigrating to the Americas during the famine years between 1845 and 1855.

Reliance on a few crop species is unnecessary because farmers around the world continue to cultivate a rich variety of food plants and research into new crop species with great agricultural potential is ongoing. Grain amaranth (also called quinoa) and the winged bean of Southeast Asia are two examples of new crop plants with exceptional promise (**Figure 13.32**).

Diversity within varieties and genetic strains of crops is also important. In the 1970s, for example, corn plantings in the American Midwest were so genetically homogeneous that a fungus called the southern corn leaf blight raced through field after field of identically susceptible corn stalks. Millions of bushels of grain were lost. Although other corn hybrids were resistant to the blight, too many square miles of farmland had been planted with a single, susceptible strain. A more biologically diverse planting strategy that incorporated many crop varieties would have afforded greater protection against the blight.

Like any population of organisms, an agricultural crop depends on genetic diversity to adapt to changing environmental conditions. The strategy of planting crops that are genetically diverse helps slow the spread of routine diseases and also helps prevent total crop failure within a region. Diseases tend to attack particular genetic strains of a particular plant species. When susceptible strains are separated by other types of plants, diseases cannot spread as easily from field to field. In agriculture, as in the rest of nature, biological diversity promotes ecosystem stability.

Fungal disease in bananas illustrates the ongoing consequences of low genetic diversity in crops. Most of the bananas that are grown for export around the world belong to a single variety, the Cavendish. This variety

▶ **Figure 13.32 Alternative Crops**
Grain amaranth, a staple food of the ancient Aztecs, is high in carbohydrates and protein. Agronomists are interested in expanding the use of amaranth plants because they are resistant to the diseases and pests associated with other crops.

is very productive, and its bunches simultaneously ripen to uniform yellow color. Because Cavendish bananas were bred to be seedless, they cannot be sexually propagated. New plants are started from cuttings, so every plant is genetically identical to all others. Because Cavendish bananas have no genetic diversity and because so many acres are planted in this one variety, the banana crop is vulnerable to the same kind of disaster that befell the U.S. corn crop in 1970.

A deadly fungus known as race 4 Panama disease has now begun to spread through Cavendish banana plantations around the world. Affected plants wilt and die. Spores of the fungus spread in the wind and remain in the soil, where they can attack future plantings. This new strain of Panama disease threatens Cavendish banana plantations worldwide and could mean the loss of the banana most Europeans and Americans have come to love. In places where people depend on bananas as a staple food, such as Uganda, this loss would cause economic hardship and hunger.

Fortunately, there are a number of other varieties of bananas, many of which are resistant to the Panama fungus (**Figure 13.33**). Banana growers often cultivate these other varieties for local consumption but not for export. If the Cavendish can no longer be grown, other varieties are available to take its place. Rather than replacing the Cavendish with a single variety, it would be wise for growers and consumers around the world to rely on a greater diversity of bananas.

Until the last century, there was great genetic diversity among most cultivated plants. These "heirloom" varieties may not be ideally suited for today's agricultural economy, but they retain genetic variations that may be needed in the future. To preserve the option of incorporating genetic diversity in future crops, the rich variety of these older vegetables, grains, root crops, and orchard trees must be protected. To preserve heirloom crops, seed banks have been established in the United States and in other countries. Seed banks store the seeds from heirloom crops in cold, dry conditions that can keep them viable for many years (**Figure 13.34**).

Heirloom breeds of animals, such as Icelandic sheep and bourbon red turkeys, are also valuable to the preservation of agricultural diversity (**Figure 13.35**). Hobby breeders and small-scale farms provide an international service by maintaining breeds of domestic birds and animals that would otherwise be allowed to go extinct. Heirloom breeds may provide much needed genetic diversity if disease or environmental change threatens the limited variety of breeds used in large farming operations today.

▲ **Figure 13.34 Storing Genetic Variation for the Future**
A technician at the Kew Botanical Garden's Millennium Seed Bank in the United Kingdom checks seeds stored at −4 °F. Seeds stored in cold, dry conditions remain viable for many years. In the future, these seeds may provide the genetic resources needed to improve crop plants.

▼ **Figure 13.35 The Cost of Genetic Homogeneity**
Nearly all domestic turkeys are of the same white-feathered variety Ⓐ, which has been selected for its high production of breast meat and rapid growth in crowded conditions. These turkeys must be artificially inseminated, hatched in incubators, and raised in protected houses. Bourbon red Ⓑ is an heirloom breed that retains the ability to fly, to find its own food, to mate, and to rear offspring.

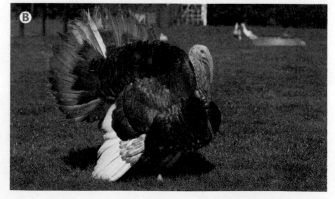

► **Figure 13.33 Banana Varieties**
Though only a few banana varieties are grown for international export, small farms in the tropics grow many different varieties. This genetic diversity ensures that we shall continue to enjoy bananas in the future.

13.8 Managing C[...]

Weeds, pests, and plant diseases have pl[...]
Scientific studies of eight of the world's [...]
of all agricultural productivity is consume[...]
all the pesticides and pest control practic[...]
wide variety of organisms and often led t[...]
Today, agronomists try to use pesticides [...]
of species. They also time the application[...]
effect. To avoid the undesirable effects o[...]
controls, such as organisms that parasitiz[...]

Chemical Pest Cont[...]

Pesticides have been both a salvation an[...]

A variety of organisms in the agroecosystem compe[...]
with humans for the energy in crop plants. Speciali[...]
insects attack every part of the growing plants. Wir[...]
worms attack roots, cutworms eat stems, caterpillars[...]
and locusts eat leaves, beetles munch on flowers, a[...]
fly larvae bore into fruits. As a result of these pests,[...]
total production is reduced and fruit and vegetables[...]
look unappetizing. In some cases, insects can cause[...]
entire crops to fail. Outbreaks of pests, such as the [...]
locust plagues described in the Hebrew Bible, have [...]

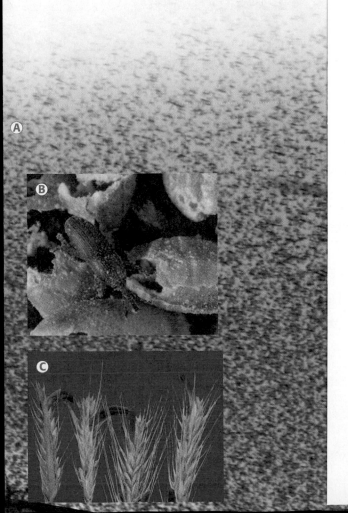

Genetically Modified Organisms

Genetic engineering of domestic plants and animals can improve yields but also is controversial.

Since the dawn of agriculture, humans have used artificial selection and crossbreeding to create new genetic combinations in plants and animals. Over time, this has created the great majority of domesticated plants and animals grown today. However, this form of breeding is limited to selecting among the traits that already exist within a species or within closely related species that can be crossbred.

With the development of modern genetic technology, it has become possible for geneticists to develop new varieties by taking a gene from the cells of one species and inserting it into the cells of another. The new varieties have a mixture of traits not achievable with more traditional methods of genetic improvement. An organism whose DNA has been altered in this way is a **genetically modified organism**, or **GMO**.

In principle, GMO technology is similar to traditional methods of crop breeding and animal husbandry. The difference is that molecular biology now allows DNA from any organism, even human DNA, to be inserted into the genetic code of a crop plant or farm animal. The speed and precision of genetic modification using the new molecular techniques gives agricultural scientists unprecedented power to change the species we use for food.

One of the most promising examples of GMO technology is a newly developed variety of rice. Rice is a staple crop that is especially important in Southeast Asia. In the poorest communities of that region, rice may provide up to 80% of dietary calories. Although rice is nutritious, its grains do not contain beta-carotene, a yellow pigment that the human body converts to vitamin A. Because vitamin A is essential for healthy skin and vision, insufficient dietary beta-carotene can lead to skin disorders, blindness, and other health problems.

Beta-carotene is made in the leaves of rice plants but is not normally stored in the edible part of the seed. In 1999, scientists inserted into the genetic material of an existing variety of rice two genes that control the production of beta-carotene, thereby creating a genetically modified plant called golden rice. Golden rice grows as any other variety of rice and provides as many calories, but it offers the additional nutritional value of beta-carotene. Beta-carotene lends its yellow color to the rice grain, giving the variety its name (**Figure 13.36**).

GMO technology also led to the development of corn plants with the capacity to produce an insect-killing compound that is normally found only in bacteria. The bacterial source of the gene is *Bacillus thuringiensis*, or Bt, so the plant is called Bt corn. *B. thuringiensis* causes disease in insects, attacking the lining of their gut with a protein toxin. This toxin is not soluble in the human stomach, so it does not harm people. Organic gardeners have long used spores of *B. thuringiensis* to prevent insect damage in their gardens. Geneticists were successful in

inserting the gene for this toxin into the DNA of corn plants. These corn plants produce the toxin in all of their tissues. Insects feeding on Bt corn are poisoned, so there is no need to apply additional pesticides.

Genetic tools have also been used in efforts to improve farm animals. Scientists at the Baylor College of Medicine in Houston found that genes for human growth hormone introduced into the chromosomes of pigs cause the animals to grow larger and mature more rapidly. As they mature, these genetically modified swine consume 25% less food and generate less manure than traditional breeds, thus there are potential environmental benefits to growing these genetically modified pigs.

Some people embrace GMO technology for its potential to produce more food at lower costs and to reduce environmental impacts. Others wonder about the nature of the foods they are eating and the long-term effects on the diversity of life. There are also persistent questions about the safety of meats modified by human hormones, of plant foods with modified biochemistry, and the environmental side effects of genes moved from one species to another. American consumer groups are demanding labeling of GMO foods, and the European Union has banned imports of American crops produced with GMO technology.

Some negative side effects of GMO have been demonstrated. Pollen from Bt corn can limit the growth of butterfly larvae, although direct effects of butterfly populations have not been demonstrated. Most of the objections to GMO foods, however, are based on more philosophical concerns. For example, some people question the right of humans to change the fundamental nature of other species. Others question the ethics of eating animals that have been artificially endowed with human genetic material.

Like any powerful technology, genetic modification of plants and animals conveys both great promise for improvements and serious concerns for health and safety. Although some fears may be overblown, public concerns in the United States and abroad have generated careful examination of the potential good and the potential problems that could arise from biotechnology in agriculture.

QUESTIONS
1. Why is genetic diversity within crop species important to sustainable agriculture?
2. How does GMO technology differ from traditional methods of plant and animal breeding?

▼ **Figure 13.36 A Golden Opportunity**
Unlike the grains of normal rice Ⓐ, the grains of golden rice Ⓑ contain beta-carotene, which colors them yellow. The plants of golden rice have been genetically modified to produce seeds that contain beta-carotene, an essential nutrient for humans.

SEEING S...

Can Brazil's dry lands be sustainably farmed by planting multiple species together?

Rainfall is quite variabl...
of Brazil and in the su...
during the rainy seaso...
by extended dry perio...
unsustainable farming...
making it susceptible...
to rapid drying thereaf...
species crops, such as...
More often than not, s...
of the economy in this...
model, economic dev...

Agronomist and tea...
region in the 1990s, t...
not being caused by la...
farmers needed strate...
noted that in single sp...
individual plants are w...
runoff during rains and...

Hanzi recommende...
monocultures with sy...
combines multiple sp...
short plants, annual c...
of the soil. Legumes...
to eroded soils.

In 1999, Hanzi's co...
Institute established t...
recommendation. The...
demonstration fields...
this region to establish...
Besides food crops s...
include castor beans,...
for petroleum, and sis...
Other species, includi...
peas, can also be add...

---

paper containers to get at stored foods. If moisture is not adequately controlled, molds can grow. Molds reduce the palatability of crops and may also produce toxins that threaten humans with disease.

Since ancient times, farmers have used various measures to control insect pests. In the mid-20th century, the developments in chemistry brought a variety of powerful pesticides to agriculture. Foremost among these developments was dichloro-diphenyl-trichloroethane, or DDT, an artificially synthesized, oil-soluble toxin. DDT belongs to a class of chemicals called chlorinated hydrocarbons, so named because chlorine is attached to a chemical framework of carbon and hydrogen atoms.

DDT is a **broad-spectrum pesticide**, meaning it kills a wide variety of insects. In the 1950s, it was discovered to be much more toxic to insects than to humans, so scientists concluded it could be used to control pests with little harm to human populations. It was, therefore, enthusiastically adopted for many uses. Often it was applied by aerial spraying, in which low-flying airplanes released the pesticide over wide areas. Public health agencies sprayed DDT on marshes and around swimming pools and parks to control the spread of mosquito-borne diseases, such as malaria. Farmers sprayed DDT on fields to control crop pests, and foresters sprayed woodlands to control destructive caterpillars. DDT was readily available to homeowners to kill cockroaches, flies, and other household pests.

However, DDT is a long-lived compound that is not easily broken down in soil or water. As a result, it continues to kill insects for a long time after its application. Since it easily attaches to body fats and does not break down in the body, DDT is stored within the body. All the DDT consumed over the lifetime of an animal remains in the animal's fatty tissues. Thus, as one organism eats another, it consumes tissues laden with DDT. The higher an organism is in the food chain, the greater its exposure to DDT. The process by which the concentration of a pesticide increases through a food chain is called **biomagnification** (**Figure 13.40**).

The negative consequences of the wholesale application of DDT were not noticed until wildlife began to succumb to its long-term effects. Birds, particularly birds such as eagles and hawks that are at the top of the food chain, were the first indicators that DDT was poisoning the entire food chain. In some cases, DDT killed the birds outright. In other cases, high levels of DDT in the bodies of adult birds made their eggshells so thin that they could not reproduce.

Author and naturalist Rachel Carson became alarmed by the loss of birds and began investigating other signs of ecological trouble. Her best-selling book, *Silent Spring*, alerted the public to the dangers of the indiscriminate use of toxic chemicals in agriculture. She warned that DDT might exterminate birds from the natural world, resulting in a "silent spring" devoid of their songs. As a result of Carson's work, the United States banned the use of DDT in 1972.

Carson also pointed out that broad-spectrum pesticides can be counterproductive in the long run. As a result of

DDT concentration: increase of 10 million times

DDT in fish-eating birds 25 ppm

DDT in large fish 2 ppm

DDT in small fish 0.5 ppm

DDT in zooplankton 0.04 ppm

DDT in water 0.000003 ppm

▲ **Figure 13.40 Biomagnification**
When DDT is sprayed on agricultural fields and marshes, it eventually flows into streams, lakes, and oceans. Because DDT is stored in animal tissues and not metabolized, it becomes increasingly concentrated in the tissues of animals at higher trophic levels.

natural selection, pests develop adaptations that allow them to live in their chemically altered environment. For example, when DDT is first sprayed on a wetland for mosquito control, most of the mosquitoes die. The only mosquitoes that survive are those that happen to carry genes conferring some tolerance to the pesticide. These resistant mosquitoes reproduce, producing more resistant mosquitoes. Over time, the mosquito population becomes increasingly resistant to DDT. Eventually, the pesticide becomes useless against its intended target.

Broad-spectrum pesticides disturb agroecosystems by killing helpful insects as well as harmful ones. Before DDT was widely used, farmers had little appreciation for the level of pest control exercised by naturally occurring insect predators. Helpful insects, such as robber flies, assassin bugs, and praying mantises, protect crops by consuming large numbers of insect pests (**Figure 13.41**). After DDT is sprayed on a field, pest species generally breed much faster than predatory insects, so their populations rebound more quickly. Released from the control of their natural enemies, pest populations commonly reach higher levels after spraying than before.

Today, agronomists recognize the long-term effects of DDT and similar pesticides. As a result, they have adopted three strategies to maximize the effects of pesticides while minimizing their environmental impacts. First, agronomists try to use short-lived pesticides that break down quickly in water or soil. Such pesticides do not remain in the environment to kill other organisms after their intended targets have been killed. Second, they try to use pesticides that target particular species or groups of species. Pesticides with greater specificity kill the pest but not other organisms, so they cause fewer damaging side effects. Third, they study pest life cycles carefully to determine the time and place that pests are most susceptible to control measures. If the pest populations are too low to cause significant damage, no chemical control is used.

▲ Figure 13.41 **A Little Help from Our Friends**
A praying mantis protects a crop by consuming insects such as this grasshopper.

Biological Pest Control

Biologically based methods can control pests without collateral damage to ecosystems.

A detailed understanding of agroecosystem food chains can often lead to biological controls that avoid chemical toxins altogether. Most pests have predators and parasites, and farmers can use these helpful species to consume pests. The use of predators and parasites to manage pests is called **biological pest control**.

For example, braconid wasps are widely used by farmers and gardeners to control populations of leaf-eating caterpillars (**Figure 13.42**). These tiny parasitic wasps lay their eggs beneath the skin of a suitable host caterpillar. When the wasp larvae hatch, they feed on the live tissues of their host. Caterpillars that are parasitized by braconid wasps fail to mature. Thus the tiny wasp prevents the population growth of these garden pests without harming other species in the environment.

Another mechanism of biological control involves manipulating the chemical communications of pest insects. Many insects use volatile chemical signals, called **pheromones**, to send a message from one individual to another. Pheromones may be used to communicate the location of mates or sources of food. A common garden pest called the Japanese beetle uses pheromones to find both mates and food sources. Traps baited with artificially synthesized Japanese beetle pheromones can lure hundreds of beetles. Beetles attracted to the scent bump into the top of the trap and fall through a funnel into a container.

One of the greatest success stories in the realm of pest management involved the release of millions of pest animals. The screwworm fly was once a virulent pest on livestock across much of the southern United States. Beginning in the 1960s, the U.S. Department of Agriculture (USDA) began a program of releasing massive numbers of male flies that had been sterilized by radiation. When wild female flies mated with these sterile males, their eggs were not fertilized and produced no offspring (**Figure 13.43**). Populations rapidly dropped, and by 1966 the USDA declared the country free of this pest. Since then, this approach has been successfully employed for a variety of other fly pests.

▼ Figure 13.42 **Biological Pest Control**
These braconid wasp juveniles (white) are parasitizing a hornworm caterpillar, which is an important crop pest.

◄ Figure 13.43 **Controlling Screwworm Flies**
Screwworms lay their eggs on livestock; when the eggs hatch, their larvae invade the tissues of the livestock. Populations of this pest have been controlled by releasing large numbers of sterile male flies.

Agroecosystem Management of Pests

Control of the crop plant's environment further reduces the need for pesticides.

Good agricultural practice can make the crop environment less suitable for pest species, reducing the need for chemical pest control. Such practices include careful crop inspection and quarantines, crop rotation, and integrated pest management.

Quarantines prevent the introduction of exotic pests in the first place. The laws of agricultural states such as Florida, Arizona, and California require fruits and live plants to be carefully inspected before they are brought across their borders. These laws are intended to prevent the importation of agricultural pests, such as the Mediterranean fruit fly, from sub-Saharan Africa, and the citrus canker, believed to have originated in Southeast Asia. If even one of these pests is detected inside the state boundaries, agricultural officials act quickly to isolate it and destroy all fruit or vegetation that might harbor it before the pest is able to spread.

Crop rotation—the practice of replacing one crop species with another on an annual planting cycle—helps to control pest populations because pests tend to specialize in particular crop plants. European corn borers, for example, are insects that feed only on corn plants. The adult corn borer is a moth that lays eggs on corn leaves. Young larvae feed on corn leaves and stems, and older larvae spend the winter in crop debris.

If corn is grown in the same field year after year, corn borer populations build up over time. Farmers can easily disrupt this buildup by planting soybeans every other year. When corn borer larvae emerge in the spring, they will encounter young soybeans, a crop they cannot use as food. In the following year, the pests that grew on bean plants will encounter nothing but corn. Thus crop rotation helps to control all types of specialized insect pests. For insects like the corn borer that require crop residues for shelter during the winter, burning crop wastes or plowing them under can remove residual pests before the crop is planted again.

Weeds can be controlled by mechanically disturbing the soil between crop rows or by mulching. Mulching involves covering the ground with plant material or plastic sheeting to prevent weeds from germinating near crop plants. Organic mulches, such as wheat straw and pine needles, have the added benefit of shading the soil and reducing the loss of moisture during hot weather. Organic mulches also degrade over time, releasing nutrients back into the soil.

Chemical, biological, and cultural means of pest control are not mutually exclusive strategies. **Integrated pest management (IPM)** is the effective combination of all three kinds of pest control, carefully designed to manage pest populations while also minimizing environmental damage (**Figure 13.44**). It is not the goal of IPM to completely eradicate pests. Rather, IPM aims to keep pest populations small enough to minimize crop loss while also controlling the cost of their management. This approach requires extensive knowledge of the pest species, including its life cycle, population levels, and interactions with other organisms in the agricultural ecosystem. IPM methodologies usually must be tailored to the natural enemies, plants, soils, topography, and climate of a particular region. Extensive local experimentation is essential. IPM is a growing movement for the environmentally sound management of row crops, orchards, and even lawns and golf courses.

QUESTIONS

1. Why is the concentration of DDT magnified in organisms at higher trophic levels?

2. Describe three examples of biological pest control.

Integrated Pest Management

Pest resistant crop species and genotypes

Cultivation practices that discourage pests

Careful monitoring of pest populations

Natural control of pest populations

Prudent use of pesticides

◄ Figure 13.44 **Agroecosystem-based Management**
Integrated pest management (IPM) combines several strategies to minimize impacts of weeds and pests on crop species. This not only reduces negative environmental effects of pesticides, it also saves farmers money.

13.9 The Ecology of Eating

As omnivores, we humans can choose where to eat on the food chain. From well-established principles of ecosystem ecology, we know that our ecological footprint increases as we shift from an herbivorous to a carnivorous lifestyle (5.4). Our ecological footprint is also influenced by the place where our food is grown and the amount of energy and materials used to transport, process, and market our food. The choices that individual consumers make about what to eat and how it is produced affect the environment and directly influence human well-being.

The Food Footprint

What we eat determines a significant portion of our ecological footprint.

Historically, humans chose diets that reflected the biological resources that were available where they lived. In southern Asia, where the land and climate are suitable for agriculture and farms support dense human populations, many cultures have a largely vegetarian diet. At the other extreme, in Arctic environments where cultivation of crops is impossible, indigenous cultures depend primarily on animal foods. In Lapland, for example, people obtain almost all their food from reindeer, fish, and other animals (**Figure 13.45**).

In general, as countries and regions develop economically and become more urban, meat and fish make up a larger proportion of their total diet. In most developed countries, food accounts for about 25% of the total ecological footprint. That proportion is greater in countries where people consume lots of meat.

Placing ourselves toward the top of the food chain by consuming more meat and less plant foods poses two potential problems. First, it takes more land to produce a unit of energy in meat than to produce an equivalent unit of energy in vegetables or fruits. The amount of

land required to support one person who gets half of his or her dietary calories by eating meat from domestic animals could instead be used to support two or three vegetarians. The second problem is that eating meat may expose humans to toxins, such as mercury, that undergo biological magnification in the food chain. Poisoning from mercury in tuna, for example, results from humans choosing to eat large carnivores from the sea, thereby placing themselves at the end of a long food chain.

The ecological impact of our diet is also influenced by the energy and materials required to transport, process, and market our food. In places where people consume food that comes mostly from local or regional sources, the contribution of food transportation to total ecological footprint is comparatively small, just 1–2%. However, in places where a large amount of food is acquired from distant regions, the cost of transporting food may exceed 10% of the footprint.

A century ago, even in industrialized countries, diets varied with the seasons. Peas, asparagus, strawberries, and many leafy vegetables were available in the spring and early summer. Tomatoes, corn, and squash were

◄ **Figure 13.45 Contrasting Diets**
Thai farmers grow a diverse array of fruits and vegetables. Along with rice, these plants make up the majority of food purchased in this floating market in Thailand Ⓐ. In Lapland, where there are few options for agriculture, reindeer herders rely on animals for most of their food and clothing Ⓑ.

▲ Figure 13.46 Tasty But Costly
Tomatoes purchased during cold winter months are likely to have been grown far away or in a hothouse. In either case, considerably more energy is required for their production than is needed to produce tomatoes that are purchased when they are in season locally.

Are organic foods and liquids really better for us?

Stanley Uzoka, Kennesaw State University

Norm: *The health benefits of organic foods are debated, but there is no question that their production has a much lower impact on the environment.*

summer and autumnal fare. Winter meals typically included lots of cabbage and potatoes. Nowadays, go to the fresh vegetable section of a supermarket in Seattle, Chicago, or Boston, and you are likely to find all of these items, regardless of season.

We enjoy this dietary diversity at a price. The fresh tomatoes we eat in midwinter are either grown in the Southern Hemisphere or in a heated greenhouse (**Figure 13.46**). In either case, considerable energy from fossil fuels is needed to produce and transport them. Refrigeration and chemicals that retard fruit and vegetable ripening add additional costs in energy and materials. If fresh items are unavailable, canned or frozen forms certainly are available. Such processing can add an additional 20–30% to the energy cost of processed foods compared to the alternative of consuming them fresh, in season, from local sources.

You can also alter your dietary ecological footprint by considering how what you eat is grown or raised. **Organic foods** are produced to meet strict standards that limit the use of fertilizers and pesticides. In the United States, these standards are set by the USDA, and plant and animal products must be independently certified against these standards to bear the "organic" label (**Figure 13.47**). Nutritional experts disagree as to whether or not organic foods are healthier than nonorganic foods. Yet there is no doubt that a diet of mostly organic vegetables and meat has a much smaller ecological footprint than a conventional diet. A recent study in Britain revealed that eating a diet of 87% organic vegetables and meats reduced the "food ecological footprint" by 23% compared to an average diet containing about 1% organic foods.

The methods by which foods are grown and processed have significant consequences for public health. In September 2006, hundreds of people in 26 states became seriously ill. Symptoms included vomiting, severe diarrhea, and, in the most severe cases, kidney failure.

Within a month, 4 people had died and over 100 were hospitalized in critical condition. The U.S. Food and Drug Administration (FDA) quickly determined that the culprit was a particularly virulent strain of *E. coli* known as O157:H7. They also found that the infected individuals had eaten fresh spinach that had been grown and bagged at a single farm in the Salinas Valley of California. The food preparation plant that sent out the contaminated spinach processed 26 million servings of salad vegetables every week, mixing the contaminated spinach from one farm with uncontaminated spinach from many others. As a result of this mixing, it was very difficult to trace the widespread incidence of food poisoning to its source.

The ultimate source of this virulent *E. coli* was cattle reared in feedlots near the fields where the spinach was grown. This strain of *E. coli* is common in cattle, where it causes no illness. However, the widespread use of antibiotics in cattle at feedlots has favored the development of strains of *E. coli* that are resistant to antibiotics. These bacteria are passed in the cattle's waste, so the water draining from large cattle feedlots often contains high concentrations of bacteria. Thus, the resistant strain of bacteria was especially common in the water used to irrigate the spinach on the farm. The problem was further exacerbated by placing spinach in closed plastic bags, which provided ideal conditions for bacterial growth.

International shipping of processed food and food commodities has raised concerns as well. In 2007, a substantial amount of processed wheat imported from China was discovered to be contaminated with a toxic fertilizer additive called melamine. In the United States, the melamine-contaminated vegetable protein caused sickness in animals that ate imported pet foods but did not affect people. However, the following year melamine was discovered in baby food and other processed foods that originated in China. Its source was contaminated milk. China exports more than $2 billion of food per year to the United States, and that number has been growing rapidly. Since the FDA has personnel sufficient to inspect only 1% of these imports, any lapse in the food safety systems of our international trading partners raises health concerns among U.S. consumers as well.

► Figure 13.47 It's Organic
American consumption of organic foods for production has increased tenfold over the past decade. To be labeled as organic food, products must meet strict standards.

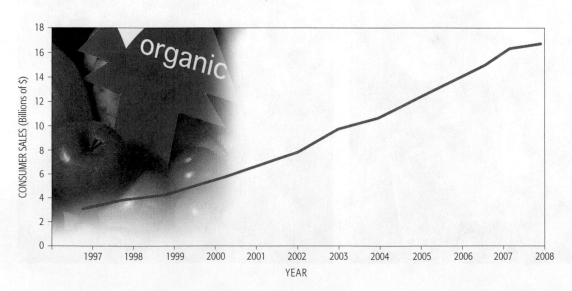

Ecological Eating

Consumer choices can diminish environmental impacts and encourage sustainable agriculture.

The challenges of sustainable agriculture can be met by improved education and more environmentally conscious public policies. But the most powerful force for change lies in the choices made by consumers. The agricultural economy follows preferences of grocery shoppers and restaurant patrons, so each person exercises choices in the way food is grown, processed, and delivered. People in wealthy countries have the luxury of a great variety of food choices, which in turn influences agricultural practices. With these choices come a greater responsibility to support a sustainable future. Here are some simple approaches for you to consider when making choices about foods.

Eat more plant-based foods. By placing yourself toward the bottom of the agricultural food pyramid, you reduce the size of your ecological footprint. Less agricultural land is required to support a vegetarian diet than one based largely on meat and dairy products. The fewer animal foods that are eaten, the fewer tons of grain must be harvested, dried, and transported to feedlots, and the fewer tons of manure threaten water quality. Fewer cattle mean that fewer tons of heat-trapping methane are released into the atmosphere. The choice between a plant-based diet and an animal-based diet does not need to be an either-or choice. Shifting even a little of your diet toward plant-based foods makes a difference.

Eat local foods. Find out where farmers are producing foods close to your home. Most cities have farmer's markets that specialize in local, in-season produce (**Figure 13.48**). You may be surprised at the difference between locally grown fruits and the supermarket varieties, which are sometimes bred more for their capacity for shipping and storage than for their taste.

Eat fresh foods. Fresh foods, particularly those grown locally, involve less processing and throw-away packaging. They also generally require less energy to move from the farm to the dinner table.

Enjoy foods in season. Learn what is in season in your area, and enjoy each fruit and vegetable in its own time. Out-of-season foods, such as fresh strawberries in the middle of winter, are a luxury purchased at the high cost of the fuels consumed to refrigerate and ship them long distances—sometimes halfway around the world.

Choose sustainably grown products. Choosing products of sustainable agriculture is not usually a significant sacrifice for the consumer but may require some research. Coffee, for example, can be grown in the shade of native trees that continue to provide wildlife habitat, or it can be grown in monoculture stands that replace local trees. Certified organic produce and pasture-raised meats are generally considered more environmentally sound choices as well. Find out which producers are making real efforts at environmental protection. Be wary of green-sounding labels that do not represent real differences in agricultural practice. Although there are strict standards for organic foods, there are no standards for foods labeled "natural."

Ask questions. When dining out, ask about the sources of food. Many restaurants support local growers and select sustainably grown foods. Some items on the menu may be more sustainably produced than others. Environmental publications and websites provide a wealth of information about foods that support or undermine sustainability in agriculture. At the meat counter or vegetable cooler of your grocery, ask where foods came from. The more you seek information, the more responsive the growers and sellers will be to your desire to support the safety and sustainable production of food.

◄ **Figure 13.48 Eat Local Foods**
Shopping for produce at a local farmers' market can help decrease your ecological footprint.

QUESTIONS

1. Describe how changes in our eating habits have increased our ecological footprints.

2. What are the advantages of eating locally grown food?

411

13.10 Food for the Future

There is no doubt that the increased agricultural production associated with the Industrial Revolution and Green Revolution has played a significant role in the growth of human populations over the past century. Nevertheless, a significant portion of Earth's people remains undernourished. Can we provide adequate nutrition to all of the world's people? Current trends suggest that our numbers will increase by another 2 to 3 billion over the next 40 years. Can we ensure adequate food to match this growth in human numbers? In order to answer these questions in the affirmative, we must sustainably manage our agricultural resources, minimize waste, and ensure that the worldwide distribution of food is equitable. Sustainable agriculture includes careful stewardship of soil and water resources, maintenance of the genetic diversity of crops and domestic animals, the fostering of stability in agroecosystems, and minimization of negative environmental impacts. Most of all, sustainable agriculture requires long-term vision. To guarantee adequate food for all people, governments must develop stable, honest institutions, manage economic markets for agricultural economies, and improve systems of food distribution and storage.

Sustainable Agriculture

Feeding a growing human population on a finite supply of arable land without causing irreparable damage to our environment is a major challenge.

Fossil fuels, synthetic pesticides, chemical fertilizers, and genetic modification of crops help boost crop yields. Each of these innovations has contributed to increases in the world food supply. Yet each also has the potential to cause problems that reduce the productivity of agricultural lands and damage the environment upon which they depend. Sustainable agriculture aims to provide future generations with high levels of food production with minimal environmental costs.

Agricultural production ultimately depends on the quality of soil and the availability of water resources. The sustainability of each of these resources is very dependent on the management of the other. Most erosion and loss of soil nutrients is associated with the movement of water across farmlands. This process is also the source of virtually all water pollution associated with agriculture. The demand for irrigation water can be greatly diminished by growing drought-resistant crops. Strategies such as drip irrigation can decrease water use by as much as 90%, while reducing erosion and nutrient loss by similar amounts. Cover crops and no-till cultivation and harvest techniques further protect soil resources. Although there are short-term costs involved in the implementation of some of these technologies and strategies, in the long term they are both economical and ecologically sustainable.

As discussed several times in this chapter, the simplicity of agroecosystems is a major factor in their productivity. Yet that simplicity makes them especially vulnerable to disturbance. To ensure resistance to and resilience from competitors, pests, and climate change, sustainable agriculture seeks ways to maintain the genetic diversity of domestic species. In many situations, the overall productivity and stability of agroecosystems can be increased by planting polycultures of multiple species. Greater diversity can also diminish the demand for water, fertilizer, and pesticides.

Sustainable management of agroecosystems pays attention to landscape-level factors that influence production and environmental impacts. Polyculture farming and crop rotation increase production and reduce the loss of soil nutrients. Adequate buffers of native prairie or forest along streams and rivers reduce the impacts of agriculture on water resources.

Agronomists have understood each of these elements of sustainable agriculture for several decades. One might, therefore, ask why they have not been universally applied. In some cases, as in some developing countries, this is a consequence of ignorance or misunderstanding among local farmers. In those cases, education alone may transform agricultural practice. More often, however, the problem is that management changes involve significant short-term costs. These may be capital costs, such as investment in drip irrigation or no-till cultivation technologies, or they may be costs in the production of specific crops, such as those associated with polyculture and intercropping. In either case, management that is focused solely on short-term economic gain or that ignores off-site environmental impacts will not recognize the benefits of sustainable agricultural practices.

Government policies and laws can provide incentives and regulations to ensure the sustainable management of agroecosystems. In the United States, the Farm Bill that is reauthorized by Congress about every 5 years

◄ **Figure 13.49 Conservation Reserve Land**
This diverse community of prairie plants (inset) has stabilized soils and diminished erosion on former cropland that is now protected under the Conservation Reserve Program. This map displays the amount of land enrolled in the program. Note that much of the conserved land is in the basins of the Mississippi River and its many tributaries.

2011 CRP
enrollment acres
Less than 10,000
10,000–25,000
25,000–75,000
75,000–150,000
More than 150,000

provides many incentives and regulations. Farm Bill programs provide economic incentives for soil and water conservation, as well as funding for the maintenance of genetic resources. The 1985 Farm Bill, for example, established the Conservation Reserve Program (CRP), which pays farmers to retire eroded croplands and lands near streams from production for a minimum of 10 years (**Figure 13.49**). Natural succession on these lands restores soil resources and protects water quality. As of 2008, about 140,000 km² (about 34.5 million acres) of land were enrolled in CRP.

Farm Bill critics point out that this legislation often creates incentives for actions that are counter to

sustainable agricultural management. For example, crop subsidies may encourage the cultivation of crops in marginal areas. Such subsidies often influence the supplies and prices of agricultural commodities in ways that create disincentives for sustainable cultivation both here and in other countries. Other critics see the 5-year time horizon as a significant challenge to the long-term vision required for sustainable management. Programs such as crop subsidies and the CRP are constantly revisited, and policies lack the permanence implied by sustainable management. This has caused some to argue that the major provisions of the Farm Bill should be formulated over a 50-year rather than 5-year timeline.

Feeding a Hungry World

Food distribution is a major obstacle to feeding all of Earth's people.

Are we going to be able to keep up with demand for food as the population increases, cities grow, and the amount of available farmland decreases?

Rebecca Sullivan, University of West Florida

Norm: *To meet future nutritional needs we must ensure equitable food distribution by fighting poverty, corruption, and the potential negative effects of global markets.*

According to the World Food Organization, chronic hunger affects more than 10% of the people in the world. In sub-Saharan Africa, undernutrition is a perennial concern. Uncertain rainfall, political unrest, and growing populations have seriously threatened food security in the region. Children are especially vulnerable to shortages of protein, which is needed for normal development (**Figure 13.50**). At the other end of the food availability spectrum, overnutrition is an increasing problem in the United States, where nearly 20% of children and adolescents are overweight. The Centers for Disease Control reports that obesity occurs in one out of three American adults. Since obesity increases the risk of developing coronary artery disease, stroke, type 2 diabetes, and some kinds of cancers, overnutrition is also a serious health issue.

Careful analysis of current agricultural trends indicates that there currently is not a global shortage of food per se: In total, global agriculture produces sufficient yields to feed the world's people. However, factors such as poverty, corrupt governments, economic markets, and problems in shipping and storing food prevent the food supply from reaching everyone.

Sustaining agricultural productivity in the world's poor countries remains a significant challenge. Ignorance of sustainable farming practice is only part of that challenge. Even where knowledge exists, people often lack the resources needed to improve local agriculture. The United States Agency for International Development (AID), Heifer International, and many other international relief organizations are working to help people in famine-prone countries increase their food security through locally controlled agricultural projects.

Where economic hurdles prevent impoverished people from starting a farm, small loans from international development agencies have proven especially effective. Microcredit programs supply start-up costs for orchards, farms, and livestock operations in the poorest nations (**Figure 13.51**). A loan as small as $100 can help an African farmer purchase seeds, oxen, and a plow to begin farming. As these small loans are paid off, the capital can be loaned out again to help others.

Corrupt and unstable governments can interfere with the equitable distribution of food. This was a problem as far back as Roman times and remains a serious problem in several countries today. Prior to 2000, Zimbabwe was a major producer of food, particularly corn, for much of southern Africa. In that year, the government of President Robert Mugabe implemented policies that were supposedly aimed at reallocating the ownership of farmland in a more equitable fashion. Instead, corruption in that process resulted in land ownership shifting to Mugabe's political cronies. Between 2000 and 2004, Zimbabwe's corn production declined by nearly 75%. As of 2011, attempts by relief agencies to distribute donated food to this country's many starving people have been thwarted by corrupt government officials.

▼ **Figure 13.50 Malnutrition in a World of Plenty?**
These children in Yama, Niger, are starving because of a regional shortage of food. Nevertheless, the global supply of food is sufficient to feed everyone adequately.

▼ **Figure 13.51 A Little Help Can Go a Long Way**
A microloan helped this Cambodian woman start her watermelon shop in Phnom Penh.

Food donation by wealthy nations is the most direct effort to address world hunger. The United States government spends over $19 billion per year in international aid. Though these gifts make up a small part of the gross domestic income of industrialized economies, and though they have been criticized for promoting various political agendas, there is no doubt that government-funded assistance has saved the lives of many hungry people.

People around the world have become directly or indirectly dependent on global markets for their food. For this reason, policy decisions in food-producing countries that affect the supply and price of food commodities can affect the availability of food worldwide.

For example, U.S. government policies have encouraged the increased use of corn to produce ethanol as a supplement to gasoline to fuel automobiles (16.5). By 2008, one-third of U.S. corn production was being used to make ethanol. In just two years, the price of corn had increased by 300%. Corn is a staple food that is also used to feed livestock and is a basic ingredient in many processed foods. Because of the wide use of corn and because the United States produces 40% of the world's corn, the U.S. policies regarding corn affected the price and availability of a wide variety of foods around the world.

Global markets and competition have a significant effect on agriculture within individual countries. In Mexico, for example, 2 million small-scale farming operations, most smaller than 5 ha (12 acres), have historically produced most of the corn needed for domestic consumption. Tortillas and other traditional foods made from corn comprise one-third of the national diet. This traditionally stable agricultural system is changing rapidly as a result of the international trade stimulated by recent trade agreements. Small, hand-cultivated corn crops cannot compete with grain imported from the subsidized and heavily mechanized farming operations in the United States. More and more of Mexico's farmers are abandoning their traditional way of life and seeking work in cities or across the border in the United States. The traditional Mexican farm economy and the connection between local crops and local foods have been destabilized by unprecedented global competition.

Similar losses of small farms are occurring in the United States. Economies of scale and Farm Bill subsidies have favored large farms instead of small family operations. As a result, the average size of an American farm has increased, while the numbers of farmers has declined (**Figure 13.52**). As more and more family farmers sell out to larger-scale operations, fewer and fewer farms control more and more of agricultural markets.

Finally, food distribution and storage are often major obstacles to feeding the world's hungry. Grain is a prime example. While grain supplies are scarce in the hungry nations of the world, grain-producing nations sometimes produce such large surpluses that the food spoils for lack of storage and shipping capacity (**Figure 13.53**).

The effects of global warming on world food supply are at best uncertain. Warmer temperatures and longer growing seasons may increase crop production in temperate climates. However, most climate models indicate that conditions for agriculture will worsen in tropical regions. It is these regions where food shortages are most acute.

In summary, sustainable management of Earth's agroecosystems is essential if we are to provide adequate nutrition to all people in the future. Some experts worry that increases in agricultural production may not be able to keep pace with projected growth in human populations. Climate change adds to this uncertainty. Sustainable management requires a long-term vision and a willingness to accept short-term costs that will reap long-term benefits. Providing food for all will also require economic and social policies that ensure that food is distributed equitably.

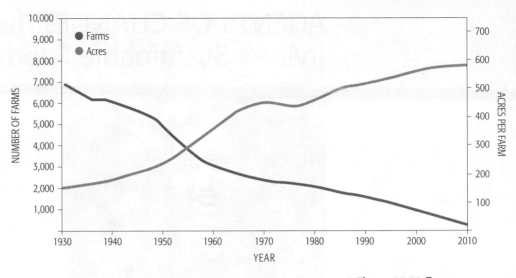

▲ Figure 13.52 **Fewer, Bigger Farms**
Today there are 60% fewer farms than in 1930, but the average farm is now three times larger.
Source: http://nationalatlas.gov/agriculture_files/agr_graph.gif

QUESTIONS

1. Discuss three strategies that farmers can use to make their activities more sustainable.

2. Describe four factors that may limit the supply of food to people in poor countries.

▶ Figure 13.53 **Wasted Food**
Grain produced in a bumper-crop year is piled on the ground due to lack of storage capacity.

423

AGENTS OF CHANGE: The University of California Irvine's Sustainable Food Initiative

Hai Vo incubated the University of California, Irvine, chapter of the Real Food Challenge (RFC) and has helped students throughout California and the nation participate in the creation of sustainable food systems. The RFC is a national campaign to increase the amount of ecologically sound food served on college campuses. RFC organizes students to advocate for food systems that support the health of consumers, producers, the environment, and local communities.

How did you first get the idea for the Real Food Challenge?

By the time I was 18, I weighed 250 pounds. I was unhealthy, depressed, and unmotivated. Two months before high school graduation, I put on a headband, short shorts, and mentally prepared myself for healing. I walked and ran as long as I could. I also began removing processed foods and drinks from my diet. Once I started eating healthy, I gained energy and confidence and began soaring academically. By the end of the summer, I had lost over 100 pounds.

That summer I moved to Davis, California, and spent time at a housing cooperative, cooking and preparing healthy foods. I also worked at the Davis Student Organic Farm and Garden, where I grew food for the first time. I made weekly deliveries to the UC Davis cafeteria, where I asked questions and learned about their food production and distribution model so I could bring their ideas back to UC Irvine.

At the same time, I worked with friends who were students throughout California to organize a leadership summit, bringing almost two dozen students from across the country to California to organize a campaign that we all knew needed to happen. As college students, we understood our privileges and responsibilities and realized that universities can be levers of community change.

What steps did you take to get started?

There were three main program components to incubating the Real Food Challenge at UC Irvine. The first component was research. We wanted to understand how much of the food procured for our campus dining halls was sustainable. We collaborated with dining staff and researched the top 25 food items purchased in eight food groups: red meat, poultry, dairy, fish, beverages, preserves, produce, and frozen foods. Using the Real Food Calculator, we determined that about 10% of those items were "real" foods—foods that had been produced in ways that are ecologically sound, community-based, humane, and fair.

Our second program component was education. During the academic year we hosted several dinners and roundtables with different campus and community groups to ask questions about our food system and introduce new ideas. We also organized an educational series that included panel discussions on the future of the UC Irvine and Orange County food systems. That series ended with a presentation of our research using the Real Food Calculator.

The third component of our program was advocacy. We used our research to make recommendations

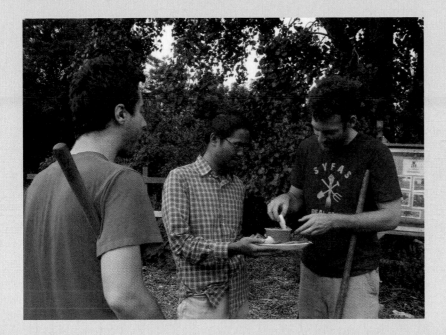

about procurement policies and practices for UC Irvine Dining. Students received a commitment from UC Irvine Dining to establish a Food Systems Working Group that would meet on a quarterly basis to determine policy recommendations and track metrics toward their goals for sustainable food.

Since my graduation, the RFC at UC Irvine has championed the first-ever student-led garden and partnered with freshman introductory writing courses to facilitate dialogue and research about food-system topics.

What are the biggest challenges you've faced with the project, and how did you solve them?

One of the most significant challenges I have faced is apathy and hopelessness. Initially, university students may not understand how their lunch at the dining hall relates to global climate change. At first, the managers of college food services may tremble at the idea of serving organic food, which often costs more than traditional purchases.

Through this project, I have overcome these challenges by listening and helping to organize the ideas of those who were originally apathetic and hopeless. This project has garnered spaces for people to come together, voice their interests and concerns, and educate themselves and others about food sustainability.

How has your work affected other people and the environment?

RFC at UC Irvine has engaged over 500 campus and local community members in leadership development, dinner and roundtable events, educational series events, and online networks. It has also facilitated discussions with the University of California Sustainable Food Systems Working Group to develop institutional policies on food purchasing. That working group has set a goal of 20% real and sustainable food by 2020.

Procuring healthy and sustainable food ensures the protection, conservation, and revitalization of ecosystem services, animal habitats, and land. It reconnects us with our ecological surroundings.

What advice would you give to students who want to replicate your model on their campus?

Learn about the food systems in your region, nationally, and internationally, and then begin a place-based project where you go to school. Connect with other students who are interested in food issues. Seek a teacher, administrator, or staff member who will help mentor and connect you. Reach out to people working in the cafeteria. Ask bold questions. Be true to your story, who you are, and why you're interested in changing our food system.

Summary

Agriculture originated independently in several regions of the world during the past 10,000 years as a consequence of changing climates, diminishing wild resources, and changing human cultures and technologies. People chose for domestication plants and animals that possessed useful features; over many generations, humans enhanced those features through the process of artificial selection. During the Industrial Revolution, new technologies increased the production of existing farmlands and allowed agriculture to expand onto new land. In the past century, fertilizers and pesticides have increased both the productivity of land and the amount of land in production. Today, humans use over 30% of Earth's terrestrial net primary production to obtain food and fiber. Agroecosystems support many fewer species than natural ecosystems and are prone to nutrient loss because of harvest removals, continual soil disturbance, irrigation, and low biodiversity. Crop plants depend on light, water, and nutrients to produce carbohydrates, proteins, and oils that are stored in various plant parts. The range of tolerance for climatic and soil conditions determines where and when particular crops are grown. Although agroecosystems are usually focused on the growth of a single species, those species depend on other species, including pollinating insects and soil insects, bacteria, and fungi. Characteristics of soil that influence plant growth include the depth and structure of soil horizons and soil fertility. Soil characteristics such as texture and organic matter also influence the availability of water. Irrigation has allowed farming in regions where evapotranspiration far exceeds rainfall. A large amount of agricultural production from crops and pasture land supports domestic animals that provide meat and other products to humans. Their environmental impacts include increased demand for land, disposal of waste products, and spread of disease. Artificial selection, cloning, and genetic engineering have diminished the genetic diversity of many crop plants and domestic animals, putting some agroecosystems at greater risk of failure due to disease and climate change. Pesticides and herbicides can improve crop production, but they also present significant environmental challenges. Integrated pest management is an alternative approach designed to control weeds and pests and minimize environmental impacts. What we chose to eat, where our food is grown, and how our food is processed have an important effect on our ecological footprint. Although the world's food supply appears to be adequate to feed our current population, chronic hunger and poor nutrition are problems for many people. Climate change adds more uncertainty. Sustainable agriculture and equitable food distribution are important to meet future needs.

13.1 Origins and History of Agriculture

- Agriculture originated independently in several places as a consequence of climate change, cultural progress, and human population growth.
- Humans, crop plants, and domestic animals have coevolved.

KEY TERMS

agriculture, agronomy, Green Revolution

QUESTIONS

1. It appears that early agriculture may not have improved human well-being compared to hunting and gathering. What was the likely driver for adoption of this style of life?

2. Describe how artificial selection likely led to the evolution of crop plants and domestic animals.

3. Explain how the Industrial Revolution and the Green Revolution significantly increased human supply.

13.2 Agroecosystems

- Agricultural ecosystems depend on energy flow and nutrient cycling following the same principles as natural ecosystems.

KEY TERMS

agroecosystem, harvest index, trophic-level efficiency

QUESTIONS

1. Why is trophic-level efficiency generally higher for domestic animals than for their wild counterparts?

2. Why are agroecosystems generally more fragile than natural ecosystems?

13.3 The Growth of Crop Plants

- Plant growth and reproduction depend on adequate light, water, and nutrients.
- Differences in range of tolerance for environmental factors determine when and where different crop plants grow best.

KEY TERM

transpiration

QUESTIONS

1. Explain how evaporation from a leaf surface pulls water up a plant.

2. Describe the various plant parts that you might find at a typical salad bar.

3. Name four ways that other organisms influence the growth of crop plants.

13.4 Managing Soil Resources

- Soil structure, organic matter, and nutrient content influence the growth of crop plants.

KEY TERMS

soil, sand, silt, clay, texture, soil profile, O horizon, humus, A horizon, B horizon, C horizon, soil fertility, tilth

QUESTIONS

1. Why is soil nutrient content highest in the O and upper A horizon of the soil profile?

2. Describe three ways that soil pH influences plant growth.

3. What actions can a farmer take to minimize soil erosion and fertility loss?

13.5 Water and Agriculture

- Soil texture and organic matter determine the amount of water that a soil can hold against the force of gravity.
- Irrigation has allowed expansion of agriculture in arid regions, but it creates important environmental challenges.

KEY TERMS

gravity water, hygroscopic water, capillary water, field capacity, soil wilting point, potential evapotranspiration (PET), soil salinization, drip irrigation

QUESTIONS

1. Why do sandy soils generally have a lower field capacity than soils rich in clays and organic matter?

2. Why does irrigation tend to diminish soil fertility?

3. Describe three ways in which modern farmers are using water more efficiently.

13.6 Domestic Animals

- Domesticated animals present several environmental challenges, including land for food and range, waste disposal, and disease.

QUESTIONS

1. Why is the trophic-level efficiency of free-range livestock considerably lower than similar animals reared in feedlots?

2. The use of antibiotics to control disease in domestic animals has increased significantly over the past 20 years. Why is this a matter of concern among many public health experts?

13.7 Managing Genetic Resources

- Selective breeding, cloning, and genetic engineering have produced very productive strains of domestic plants and animals, but they have also diminished genetic diversity.

KEY TERMS

monocultures, interplanting, polyculture, genetically modified organism (GMO)

QUESTIONS

1. What are the advantages of genetically uniform agricultural species?

2. Why is such uniformity risky in the long term?

3. What are the potential advantages and disadvantages of GMO technologies?

13.8 Managing Competitors and Pests

- Pesticides are used to control populations of weeds, pests, and disease organisms that compete for the food in agroecosystems.

KEY TERMS

broad-spectrum pesticide, biological magnification, biological pest control, pheromone, integrated pest management (IPM)

QUESTIONS

1. Describe three changes in pesticide development and use that are designed to minimize negative environmental impacts.

2. Rather than simply relying on pesticides, many agronomists recommend the use of integrated pest management (IPM). What does this involve?

13.9 The Ecology of Eating

- The ecological impacts of our eating habits are determined by whether our diet emphasizes plants or meat, how our food is grown, and where our food comes from.

KEY TERM

organic foods

QUESTIONS

1. What factors have increased the incidence of outbreaks of strains of bacteria such as *E. coli* and *Salmonella* due to consuming tainted meats and produce?

2. Describe six ways that you can diminish the ecological impacts associated with your food.

13.10 Food for the Future

- Chronic hunger affects more than 10% of Earth's people even though global food production is sufficient to feed everyone.

QUESTIONS

1. If you had the opportunity to craft a farm bill for congressional consideration, what four features would you include to improve the sustainability of farming practices?

2. Explain why increased use of biofuels in automobiles has resulted in food shortages in many countries.

3. Give two reasons why advocates of sustainable agriculture are concerned about the current trends in the number and size of farms.

MasteringENVIRONMENTALSCIENCE®

Go to **www.masteringenvironmentalscience.com** for homework assignments, practice quizzes, Pearson eText, and more.

Forests

14

The Tragedy of Forest Loss in Haiti

Why does a country need its forests?

On September 16, 2004, tropical storm Jeanne settled over the Dominican Republic and Haiti. In the next 48 hours, Jeanne dumped over 50 cm (20 in.) of rain on the northern portions of both countries. That's more than a third of the rain that typically falls during an entire year. In the Dominican Republic there was flooding and damage to crops, but the effects were short lived. In a matter of days, life was back to normal. However, just across the border in Haiti, Jeanne created a national disaster. There was extensive flooding, and massive landslides buried entire villages (**Figure 14.1**). Over 3,000 lives were lost in the city of Gonaïves alone.

Why did the same storm produce such different consequences for two countries on the same island? It was not because Haiti received more or heavier rain than the Dominican Republic—if anything, less rain fell on Haiti. The difference, in one word, was forests. The Dominican Republic has forest, and Haiti has virtually none (**Figure 14.2**).

Haiti was not always a barren land. When Christopher Columbus visited its shores in 1492, 75% of Haiti's land was covered by lush tropical rain forest, and the rest of the land supported seasonal dry tropical forest. Even a century ago, despite land clearing and agricultural development, 60% of the country was covered by forest. At that time, Haiti's coastal waters were known for their azure clarity, diverse coral reefs, and productive fisheries. Like Cuba and Jamaica, Haiti was a premier tourist destination in the Caribbean, and its economy was thriving.

In the first half of the 20th century, commercial logging and the expansion of rubber plantations increased the pace of deforestation. From 1957 to 1986, Haiti was under the brutal dictatorships of François (Papa Doc) and Jean-Claude (Baby Doc) Duvalier. Under their regimes, trees were cut for revenue and the process of deforestation accelerated greatly. Loss of tourism dollars

▶ Figure 14.2 **Jeanne Hammers Hispaniola**
Tropical storm Jeanne brought heavy rains to the island of Hispaniola, causing severe damage to the people of Haiti.

▼ Figure 14.1 **The Haiti-Dominican Republic Border**
The countries of Haiti and the Dominican Republic are located next to each other on the island of Hispaniola. The river channel in this photo forms the boundary between deforested Haiti on the left and the Dominican Republic on the right. The Dominican Republic still has over 60% of its forests.

and corruption during the Duvalier dictatorships inaugurated a spiral of poverty that further amplified deforestation.

By 1990, most of Haiti's forests had been cut. As the country's rapidly growing population has become increasingly dependent on subsistence farming, the forests have been almost completely cleared from the mountainous terrain (**Figure 14.3**). The meager forest that remains has been logged, mostly to produce charcoal, the primary source of fuel for 90% of the population. Today, less than 3% of Haiti's landscape supports forest, compared to 60% in the Dominican Republic, a country that did not suffer the same unfortunate political rule.

Haiti's deforestation has had catastrophic consequences. Without forests to retain and filter water, even moderate rains produce floods, and groundwater and stream waters are laden with sediment and pollution. In most places, Haiti's once clear coastal waters are now murky and filled with sediment. The sediment and pollution have degraded estuarine and coastal ecosystems; coral reefs are nearly gone, and the fish have nearly vanished.

It is no coincidence that Haiti is, by a considerable distance, the poorest country in the Western Hemisphere. Over 60% of its national income comes as aid from foreign countries such as the United States, and 65% of its people survive on less than $1 a day.

There have been a great many efforts to restore forests to Haiti, but the task has proved daunting. Agriculture and erosion have stripped soils of their capacity to hold water and nutrients, and as a result, many native trees have difficulty growing. Goats, pigs, and chickens, which are the primary source of protein for Haiti's rural poor, feed on the sprouts of most tree seedlings, making it even more difficult for new trees to become established (**Figure 14.4**). When trees do manage to survive a few years, they are usually harvested to provide fuelwood and charcoal. As if all this were not enough, Haiti's governmental institutions are corrupt and lack the capacity to mount and enforce countrywide programs of reforestation and conservation.

Extreme poverty is itself one of the root causes of deforestation, as well as one of the most powerful barriers to forest restoration and sustainable management. A forest manager working on a reforestation project in Haiti's southern mountains recalls meeting a group of men who were cutting the young trees from a demonstration plot. The forester pointed out that if the trees were left to grow for a few more years, they would provide much greater economic value as

structural timber and would also restore soil fertility. To this, the men replied, "But we must boil our water and cook our food today," describing the immediate services the young trees would supply.

A long-term perspective is essential to sustainable forest management. Yet obsession with immediate needs is one of the unfortunate consequences of poverty. Herein lies a great paradox: Much of Haiti's poverty and human suffering is the result of the loss of its forests, but in order to restore those forests, poverty must be alleviated.

Community-based programs that couple forest restoration with economic development and the well-being of families are beginning to show promise. The 2010 earthquake that killed over 225,000 people in Haiti attracted considerable international attention to the plight of the Haitian people. Today, there is widespread agreement that long-term plans for Haiti's recovery must include the restoration of its forests.

Haiti's history demonstrates the importance of forests to humans. In addition to providing economically valuable resources, forests provide ecosystem services that are essential to human well-being. Forest restoration and conservation require scientific understanding of climate, soils, and tree growth. As Haiti's story shows, these actions also depend on an understanding of human behavior, economic incentives, and institutional structures. These lessons are the central themes of this chapter.

▲ **Figure 14.3 An Impoverished Land** Deforestation and subsequent erosion have left Haitian soils with little organic matter or nutrients. As a result, the steep, poorly terraced slopes support only meager crops.

▼ **Figure 14.4 A Tree's Enemy** Rural Haitians depend on domestic animals, especially goats, for nutritional protein. However, these animals contribute to deforestation by eating most tree seedlings.

14.1 The Values of Forests

Biologists estimate that forests contain over 75% of the world's biodiversity. This biodiversity is essential to many of the ecosystem services that forests provide, such as flood prevention, maintenance of air and water quality, carbon storage, recreation, and aesthetic beauty. Forests provide humans with a diverse array of wood and paper products that add over $500 billion to the world economy each year. In addition, forests are an important source of food, clothing, and medicine.

Ecosystem Services

Forests prevent floods and provide clean water, recreation, and inspiration.

The world's forests provide a remarkable array of ecosystem services. They purify water and regulate its flow, they withdraw and store carbon from the atmosphere, they conserve biological diversity, and they are a source of beauty and inspiration to humans. Most of these services are not bought and sold, so they are not assigned a monetary value. However, when forests are disturbed or destroyed, the loss of these services can have significant economic consequences.

Forests play a key role in the hydrologic cycle. The roots of trees absorb large quantities of water, which eventually evaporates from their abundant leaves. In a tropical forest, transpiration from a single large tree can release over 11,000 L (2,905 gallons) of water each day. Forests are constantly recycling moisture back into the atmosphere, where it can then recondense as rain. Thus, when forests are lost over a large area, rainfall in that area often declines (7.3).

When rain falls on a forested landscape, the trees, shrubs, and complex mixture of plants and leaf litter on the forest floor slow the flow of water across the land. Thus, even on steep slopes, forests diminish the direct runoff of rainfall and reduce the risk of flooding (**Figure 14.5**). Before entering streams, most water flows through the litter of the forest floor and then through soils. The litter and soil remove sediment and dissolved chemicals from the water. A study of hundreds of streams in locations across the United States revealed that streams flowing through forests contained less than 15% of the nitrogen and phosphorus found in streams flowing through farmland or cities (7.8).

Forests store large amounts of carbon. In the process of photosynthesis, forest plants remove carbon dioxide from the atmosphere to make organic compounds, which are stored in the tissues of their roots, stems, and leaves. Worldwide, forests store approximately 400 petagrams ($1 \text{ Pg} = 10^{15} \text{ g}$) of carbon; each year they take up and release approximately 60 Pg of carbon (7.4).

▶ **Figure 14.5 Forests and Water**
As water flows through a forested watershed, trees moderate its flow and remove dissolved minerals from it.

Deforestation reduces total global forest carbon storage by about 1 Pg each year. That carbon is eventually added to the atmosphere as CO_2. This CO_2 accounts for about 15–20% of the anthropogenic carbon emissions that are contributing to global climate change (8.3).

Forests are the primary habitat for 60–80% of all terrestrial species—microbes, plants, and animals. (Scientists are uncertain of the exact proportion because so many species have yet to be identified.) In addition to its intrinsic value, this biodiversity provides humans with an array of foodstuffs, clothing, and building materials, as well as hundreds of medicinal chemicals (11.2).

Few would disagree that forests are among the most beautiful places on Earth, and many of us are quite willing to pay for this beauty. Ecotourism in forested regions adds over $200 billion to the global economy each year. In many developing countries, it is a major portion of the gross domestic product (GDP) (11.2). In fact, ecotourism is one of the few ways in which we humans pay directly for the ecosystem services that forests provide.

Wood Products

The wood extracted from Earth's forests is primarily used for fuel, paper, and building materials.

Each year, approximately 3.4 billion cubic meters of wood are harvested from the world's forests. To give a sense of scale, a log that is 2 feet in diameter and 16 feet long yields about 1 cubic meter of wood; an oak tree with a trunk that is 2 feet in diameter might yield 2 to 3 cubic meters of wood. The volume of the world's annual harvest of wood is equivalent to about 1.5 billion of those oak trees.

Nearly 53% of the world's wood consumption is used for fuel, about 31% is used to produce various building materials and solid wood products, and about 16% is used to make paper. However, these proportions vary widely among different regions of the world (**Figure 14.6**).

▼ Figure 14.6 **Wood Use**
In North America and Europe, most wood is harvested for lumber used in construction and for fiber used to make paper. In Asia and Africa, most wood is harvested to meet local needs for cooking and heating.
Source: Food and Agriculture Organization of the United Nations (FAO), 2010. *Global Forest Resources Assessment 2010: Main Report.* FAO Forestry Paper Rome: FAO, 378 pp.

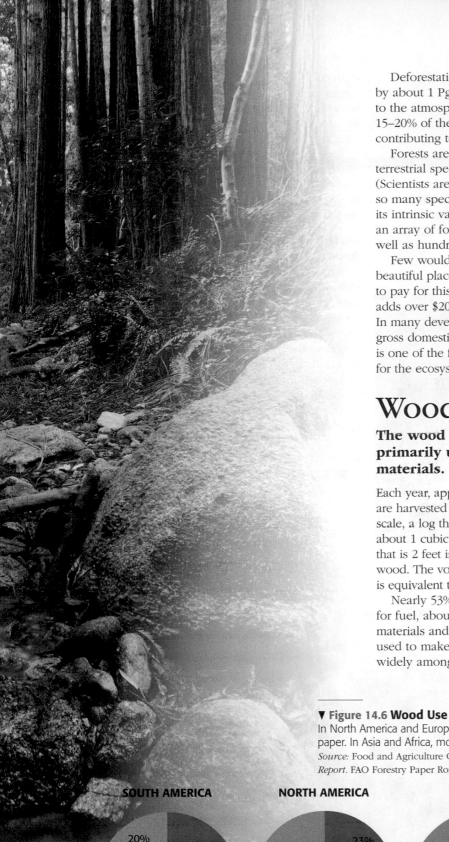

Fuel from wood. More than half of all the wood that is harvested is burned for fuel or made into charcoal. Yet wood provides less than 7% of the total energy used by humans. In Africa, nearly 90% of wood is consumed for fuel, representing over 40% of the continent's total annual energy use.

In many developing countries, wood is harvested, dried, and burned to heat homes, food, and water. More often, wood is converted to charcoal, which is then burned for fuel (**Figure 14.7**). Charcoal has many advantages over wood. Weight for weight, it produces nearly twice as much heat as wood. Once it is lit, charcoal burns much more evenly than wood. Because charcoal is impervious to termites, it can be stored for longer periods. Charcoal fire releases less smoke than wood fire, so it is far less irritating to use, particularly indoors (9.4). Unfortunately, burning charcoal does release pollutants, such as carbon monoxide.

Rapidly growing interest in biofuels is encouraging the development of technologies that convert wood cellulose into methanol or ethanol (16.5). When fully developed, these technologies are likely to create a much greater demand for wood from the world's forests.

Building materials and solid wood products. Each year just over 1 billion cubic meters of wood are harvested for use as building materials and solid wood products, such as furniture, containers, and musical instruments. About 68% of this wood comes from temperate evergreen and deciduous forests—43% from North America, 21% from Europe, and 4% from Asia. The majority of wood from these regions is used for lumber, plywood, and strand board used in construction.

Tropical forests account for only 32% of solid wood production, most of which comes from Southeast Asia and South and Central America. Yet tropical forests provide some of the most valuable woods, such as ebony, teak, mahogany, and jacaranda. Most often these woods are harvested from old-growth forests (**Figure 14.8**).

► **Figure 14.7 Charcoal Production**
Charcoal is produced by burning wood in special kilns that provide only limited oxygen.

▼ **Figure 14.8 High-value Wood**
Trucks loaded with teak logs that have been cut illegally from old-growth tropical forests are a common site on Burma's (Myanmar's) roads. Each log is worth hundreds of dollars.

| Timber | Chipping | Mechanical or chemical pulping | Product |

Chemical treatment

▲ **Figure 14.9 Producing Paper**
Wood fiber is the primary ingredient in paper manufacture. Each of the steps involved in producing paper has significant impacts on our environment.

Paper production. Each year, approximately 540 million cubic meters of wood are converted into pulp, which is then used to produce over 388 million tons of paper. That is equal to about 50 kg (110 lb) of paper for every person on Earth, although per capita use of paper varies widely among regions. Each year an average American uses over 320 kg (700 lb) of paper, whereas an average African uses less than 1 kg (2.2 lb).

In general, only smaller logs (less than 30 cm, or 12 in., diameter) are used to make paper. A significant amount of this wood comes from plantations of fast-growing trees, such as pine and eucalyptus, that are grown for this purpose.

To produce paper, wood is cut into small chips, which are then chemically digested into pulp (**Figure 14.9**). Wood is made up of cellulose fibers that are bound together by a chemical called lignin. Chemical digestion separates the cellulose fibers from one another. After digestion, the fibers are washed and concentrated into a slurry. The slurry is then dried and pressed to produce large roles of paper. To make white paper, the fibers must be chemically bleached before they are added to the slurry.

In addition to using great quantities of wood fiber, the production of paper has a significant impact on air and water quality. The production of a ton of paper requires 7,000–10,000 gallons of water, and the caustic chemicals used in the process are a source of both water and air pollution (**Figure 14.10**).

Because paper can be decomposed to its original fibers and then reconstituted, it is relatively easy to recycle. A ton of paper containing 100% recycled fiber saves about 20 plantation trees (8-in. diameter) and about 5,000 gallons of water. The production of recycled paper uses fewer caustic chemicals than that of regular paper and so causes far less air and water pollution (18.2).

Are we making a dent in the number of trees being cut down each year with paper recycling?
Rebecca Sullivan, University of West Florida

Norm: *Currently, each person in the United States recycles about 350 lb of paper per year. That saves three mature trees and over 1,000 gallons of water for each person.*

▶ **Figure 14.10 Paper Production and Water Pollution**
Waste water from a paper mill pours directly into Dongting Lake in China's Hunan Province. Such pollution has severely damaged the lake's ecosystems.

Non-wood Forest Products

Hundreds of non-wood commodities come from forests.

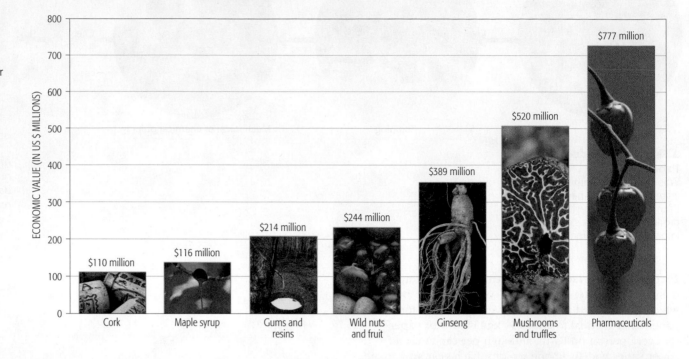

Economic value (in US $ millions)

Cork	$110 million
Maple syrup	$116 million
Gums and resins	$214 million
Wild nuts and fruit	$244 million
Ginseng	$389 million
Mushrooms and truffles	$520 million
Pharmaceuticals	$777 million

A remarkable variety of natural products can be harvested from forests without cutting trees. Wild nuts and fruits, mushrooms, and understory plants provide a continuous source of supply for people living in forested regions. Eighty percent of people in the world's poorest countries depend on non-wood forest products directly for health and nutrition (**Figure 14.11**).

Non-wood products are also important sources of income. The United Nations Food and Agriculture Organization (FAO) lists over 150 non-timber forest products that total over $17 billion in international trade each year. Such products include honey, gum Arabic, rattan and bamboo, cork, forest nuts, mushrooms, essential oils, and raw materials to make pharmaceuticals.

Natural rubber is an especially important forest product. Nearly 6.0 million metric tons (6.5 U.S. tons) of rubber, with an estimated value of $14.7 billion, is extracted from tropical forest trees each year (**Figure 14.12**). Today, most natural rubber comes from Asia and Africa, where extensive tracts of tropical forest have been replaced by plantations of rubber trees.

In summary, forests are critical to human well-being in a variety of ways. They provide many essential ecosystem services, including carbon storage and the provision of clean water. They provide wood for fuel, paper, and construction, as well as a wide array of non-wood products.

▶ **Figure 14.12 Natural Rubber Extraction**
Latex tapped from rubber trees (*Hevea brasiliensis*) provides nearly half of the world's rubber. Although some latex is harvested from trees in natural forest ecosystems, most of it comes from plantations.

Nutritive
tissue

Embryonic
leaf

Embryonic
stem

Seed
coat

14.2 Forest Growth

Much of a forest's value is derived from its patterns of growth. Forest growth can be considered at three levels—individual trees, communities of trees called forest stands, and large forested landscapes. Individual trees begin as seeds; as they grow taller, they are supported by growth in the diameter of their stems. Forest stands develop over time, undergoing predictable patterns of changes in species composition and structural complexity. Across landscapes, natural and human-caused disturbances produce a mosaic of forest patches undergoing different stages of development. Such mosaics influence forest biodiversity as well as the goods and services provided by the forest.

The Life History of a Tree

Wood supports the vertical growth of trees.

The life of an individual tree begins as a **seed**, which is an embryo embedded in nutritive tissue and surrounded by a protective coat (**Figure 14.13**). Trees are classified into two major groups, based primarily on the structures in which their seeds are produced. The seeds of **conifers** grow in cones. The seeds of **angiosperms** grow in flowers and are eventually surrounded by fruits that grow from the flower (**Figure 14.14**). Each species of tree has its own particular type of cone or flower.

When seeds are mature, they may be dispersed by the wind or by animals. If a seed lands in soil that has the appropriate temperature and moisture conditions, it will germinate, extending a shoot and leaves above ground and roots into the soil. This upward and downward growth is called **primary growth**. Primary grow occurs because of rapid cell division and growth at the very tips of the shoots and roots (4.2).

When a tree first germinates, the carbohydrates it needs to grow come from the nutritive tissues in the seed.

▲ **Figure 14.13 Trees Begin Here**
The seed of a tree includes the embryonic root, shoot, and leaves surrounded by nutritive tissue and a protective seed coat.

▼ **Figure 14.14 Cones or Flowers**
In conifers, such as ponderosa pines, seeds are borne in cones Ⓐ. In angiosperm trees, such as tulip poplar, seeds are produced in flowers Ⓑ.

Ⓐ

Ⓑ

As the tree continues to grow, its leaves begin to produce carbohydrates by photosynthesis. These carbohydrates are transported down the stem to the roots in a thin layer of specialized tissue called **phloem**. Phloem encircles the stem just beneath the protective bark. The tree's roots absorb water and mineral nutrients from soil and transport them up the stem in a specialized tissue called **xylem**. Most of the cells of xylem are hollow with very thick walls comprising the chemicals cellulose and lignin. The woody interior portion of the stem is made up of xylem (**Figure 14.15**).

Each year, tree stems produce additional layers of xylem, or wood, which causes them to increase in diameter. This increase in girth is called **secondary growth**. In temperate and boreal climates with distinct growing seasons, the xylem cell walls that grow in spring are thinner than those that grow in summer, so the wood produced in spring is less dense than that produced in summer. This difference in growth results in distinct annual rings. In tropical climates where trees grow year-round, xylem cell walls have uniform thickness throughout the year, so annual rings do not form.

▲ Figure 14.15 Tree Stem Cross Section
Most of the interior of a tree stem or bole is comprised of xylem. Phloem, the tissue that transports food and nutrients up and down the tree, forms a thin layer just beneath the bark.

▼ Figure 14.16 Forest Stand Development
Establishment Stage. After a disturbance, the seeds of numerous trees sprout and begin to grow. Fast growing trees are most competitive in this stage Ⓐ. **Thinning stage.** The loblolly pine trees in this stand have formed a tight canopy of leaves that casts dense shade on the understory. The fallen logs are the remains of smaller trees that were unable to compete for light and soil resources Ⓑ. **Transition stage.** As trees begin to die, gaps in the canopy allow light onto the forest floor, allowing new trees to become established. These new trees start to grow at different times Ⓒ. **Old-growth forest.** Old-growth is the most mature stage of forest development. Living trees and abundant logs and understory litter create high structural complexity and promote biodiversity Ⓓ.

The Life History of a Forest Stand

Communities of trees develop through four stages that differ in structure and species composition.

After a disturbance such as fire or logging, a forest goes through four distinct stages of succession (**Figure 14.16**). In the **establishment stage**, tree seedlings germinate on the recently cleared ground. Tens of thousands of tree seedlings may germinate on just one acre. With abundant light and soil resources, seedlings quickly grow into saplings. The species of trees that grow best under these conditions are likely to grow quickly and to compete effectively for water and nutrients in the soil.

Toward the end of the development stage, the leafy branches of the fastest growing trees begin to close in on one another, creating dense shade beneath the canopy. Competition for soil resources becomes more intense. Because of the intense competition for light and water, trees grow more slowly; trees that are not able to access enough light or water die. This inaugurates the next stage of forest development, the **thinning stage**.

During the thinning stage, the largest trees continue to grow and the total biomass in the forest continues to increase, but the total number of trees declines from several thousand per acre to a few hundred. Intense competition for light and soil resources severely limits the establishment of new trees. Because most of the trees became established at about the same time, thinning stage forests are said to be even-aged.

Toward the end of the thinning stage, trees are much larger and much more widely spaced. As these large trees die, they leave large openings, or gaps, in the forest canopy. At this point, total biomass may actually decline somewhat. This is the beginning of the **transition stage** of forest development; this stage is also called the gap phase.

In the transition stage, increased light and reduced competition for soil resources allow tree seedlings to become established in the gaps. A gap receives less light than an open field, so shade-tolerant trees typically grow best there. The total biomass of the forest remains relatively constant as new trees replace those that die, but the diversity of trees usually increases. Because trees die at different times, the ages of the new trees vary greatly. Therefore, transition forests are said to be uneven-aged.

If transition forests are not cut or visited by a major disturbance, such as a fire or hurricane, they accumulate ever-increasing amounts of decaying logs and standing snags. The **old-growth stage** of forest development is characterized by very large, old trees and abundant standing and fallen woody debris. The complex structure of the old-growth forest creates a variety of unique habitats. Many kinds of animals feed on and nest in the woody debris. The rich soils and favorable moisture and temperature conditions in the understory of such forests support a very diverse assemblage of herbs.

The time span over which these four stages occur varies considerably, depending on the conditions influencing tree growth and longevity. Where tree growth is rapid, as in tropical climates, the change from establishment to thinning stages may occur in less than 5 years, and transition stage forests may be less than 100 years old. In most temperate evergreen and deciduous forests, the succession takes longer. In these cooler climates, it usually takes several hundred years for old-growth forests to develop.

The Life History of a Forested Landscape

Landscapes are a mosaic of patches representing different stages of forest stand development.

Forested landscapes may extend over thousands of square kilometers. Such forests are made up of a mosaic of individual forest stands in various stages of development as a result of past disturbances. These forest patches vary in size because natural and human-caused disturbances occur at different spatial scales. An individual tree fall may produce a patch that is 20 m wide, whereas a large wildfire may produce a patch that extends over many kilometers.

Because some plant and animal species are typically found in establishment or thinning stage forests, and others occur primarily in transition or old-growth stands, you might predict that a mixture of patch types would promote biodiversity. However, transition and old-growth forest patches generally support many more species with very specific habitat requirements than do patches of less mature forest. The biodiversity of transition and old-growth forest patches tends to diminish if they are small and disconnected from one another (11.4).

Across large landscapes, net primary production and many other ecosystem processes are determined by the arrangement and relative abundance of patches in different stages. Patches in the establishment stage store relatively little carbon and have lower net ecosystem production than do later stages. Patches in this stage may be subject to erosion and nutrient loss. If many patches in the establishment stage are located near streams, large amounts of sediment and nutrients may be lost to stream waters.

In stands that are in the thinning stage, trees are rapidly taking up nutrients, and the loss of nutrients to streams is small. Net ecosystem production is high, and the amount of stored carbon is rapidly increasing. By the time patches reach the transition and old-growth stages, net ecosystem production slows. Even so, these stands store large amounts of carbon. They also take up large amounts of water, thereby limiting sedimentation and flooding in streams.

Over time, the forest landscape changes as disturbances create new patches and individual patches undergo succession from one stage to the next. Ecologists define the **minimum dynamic area** of a forested landscape as the area necessary to maintain all different patch types and populations of their associated species, given the typical patterns of disturbance. Minimum dynamic area is an important consideration in forest management and conservation.

If disturbances occur at irregular intervals and vary widely in size, the relative abundance of different patch stages will change considerably over time. The minimum dynamic area for such landscapes will be quite large. This appears to be the case in some regions where large fires naturally occur at intervals of hundreds of years. For example, researchers William Romme and Donald Despain have used tree ring data to determine how the lodgepole pine forest of the Yellowstone Plateau changed between 1700 and 1985. Their data show a steady progression of change in the dominance of different stages between major fire events (**Figure 14.17**). To conserve the biodiversity and ecosystem processes of such an area, forest parks and preserves must be quite large.

In contrast, where disturbances occur at regular intervals and create similar-sized patches, the relative abundance of patch stages is more constant through time. As a result, the minimum dynamic area is smaller. Where forests are managed for timber, harvests are usually carried out to maintain this kind of landscape. This ensures a steady supply of trees ready for harvest and constantly replenishes the landscape. However, trees are usually harvested during the middle or late thinning stage. Transition and old-growth patches are rare or missing, as are their unique biodiversity and the ecosystem services that they provide.

In summary, forests can be considered at scales ranging from the growth of individual trees, to stands initiated by a single disturbance, to landscapes with many stands in different stages of development following disturbances. Sustainable forest management involves decisions and actions at each of these scales.

QUESTIONS

1. What is the difference between primary and secondary tree growth?

2. Describe the four stages of forest stand development following a disturbance such as fire or logging.

3. Explain the minimum dynamic area concept.

▼ Figure 14.17 **Forest Landscape Change**
Between 1700 and 1740, fires burned over half of the forest landscape of the Yellowstone Plateau. In 1740, the landscape was predominately covered by establishment and thinning stage forest. By 1850, transition forest patches became the dominant forest patch type. Just prior to the extensive 1988 Yellowstone fires, old-growth forest patches had become the most abundant forest stage on this landscape.
Source: Romme, W.H. Romme and D.G. Despain, 1989. Historical perspective on the Yellowstone fires of 1988. *BioScience* 39: 695–699.

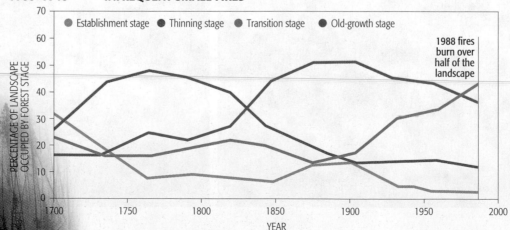

EXTENSIVE FIRES 1700–1740   INFREQUENT SMALL FIRES

● Establishment stage ● Thinning stage ● Transition stage ● Old-growth stage

1988 fires burn over half of the landscape

PERCENTAGE OF LANDSCAPE OCCUPIED BY FOREST STAGE

YEAR

FOCUS ON SCIENCE: CO$_2$ and the Growth of Forest Stands

At its current rate of increase, the concentration of CO$_2$ in our atmosphere will reach 550 parts per million (ppm) by the year 2050. That will be twice the concentration of CO$_2$ prior to the Industrial Revolution. Emissions of CO$_2$ from burning fossil fuels and clearing land are the most important factors influencing global warming, yet atmospheric CO$_2$ is also used in the process of photosynthesis. Is it possible that higher concentrations of CO$_2$ will increase the net primary production (NPP) of the world's forests?

In highly controlled greenhouse conditions, ecologists have shown that a variety of forest plants have significantly higher rates of photosynthesis at CO$_2$ concentrations of 550 ppm than at current concentrations. Would such high concentrations of CO$_2$ increase NPP in natural forest ecosystems where other factors, such as variations in water and soil nutrients, are also important? This is the question that Ram Oren and his colleagues at Duke University sought to answer.

Oren and his team carried out their research at Duke University's Forest-Atmosphere Carbon Transfer and Storage (FACTS) facility, which is a natural forest area in which it is possible to control the level of CO$_2$ in the atmosphere (**Figure 14.18**). FACTS consists of a set of eight experimental plots inside a forest dominated by loblolly pine. Around each plot is a ring of large pipes that can pump CO$_2$ into the atmosphere; each ring has a diameter of 30 m (98.4 ft). Inside each ring, sensors measure CO$_2$ concentrations at numerous locations, from the soil surface to the top of the forest canopy. A system of computers adjusts the pumping rates to maintain the desired concentrations of CO$_2$.

In four of these rings, Oren and his team set the concentration of CO$_2$ at 550 ppm. The four other rings received no additional CO$_2$ and tracked current concentrations (360–380 ppm). Then every year from 1994 to 2010 the team monitored the growth of trees, shrubs, and herbs in each of the eight rings.

Oren hypothesized that increased CO$_2$ would have its greatest effect on NPP when water and soil nutrients, such as nitrogen, were abundant. During dry periods and in places where nitrogen was in short supply, he predicted that elevated CO$_2$ concentrations would have little effect. To test this prediction, the levels of soil water and nitrogen were monitored in each ring. Two of the 550-ppm rings were divided in half; each year nitrogen fertilizer was applied to one half and the other half received none.

Over the course of the study, elevated CO$_2$ concentrations increased NPP by 22–30%. As predicted, this effect was highest in the places with the highest levels of soil nitrogen. The effect of elevated CO$_2$ on NPP also depended upon the availability of water: The effect was high in wet years and very low in dry years.

These results suggest that increasing concentrations of CO$_2$ will enhance the ability of forests to take up and store carbon. However, this effect will not be uniform. Rather, it will depend on the availability of other resources, such as water and soil nitrogen.

Source: McCarthy, H.R., R. Oren, K.H. Johnsen, et al., 2010. Re-assessment of plant carbon dynamics at the Duke free-air CO2 enrichment site: interactions of atmospheric CO$_2$ with nitrogen and water availability over stand development. *New Phytologist* 185: 514–528.

Can rising carbon dioxide increase net primary production in forest ecosystems?

▼ **Figure 14.18 Enriching the Air**
Ram Oren and his team used the FACTS facility at Duke Forest to determine how increased levels of CO$_2$ affect forest productivity. Ring-shaped arrays of vertical pipes deliver and maintain CO$_2$ at concentrations that are nearly twice the normal level. Control arrays disperse air with normal levels of CO$_2$.

14.3 Deforestation

The United Nations Food and Agriculture Organization defines **deforestation** as "the removal of forest cover to an extent that allows alternative land uses." More simply, deforestation is the process in which humans replace a forest with something else, such as agriculture, pasture, or pavement. Each year 0.2% of the world's forests disappear, but within particular regions and forest types the rate of deforestation is often much higher. Reasons for deforestation include the expansion of agriculture, unsustainable harvest, road construction, and urban development. These activities are driven by poverty, misguided government policies and subsidies, and corruption.

Historical Change

The pace of deforestation is quickening.

Scientists estimate that 10,000 years ago, forests covered 65.8 million square kilometers (25.4 million square miles), or nearly 50% of Earth's land. By 1750, the beginning of the Industrial Revolution, about 16 million square kilometers of forest had been permanently cleared for agriculture and urban centers (**Figure 14.19**). Today, the remaining forests cover less than 40 million square kilometers (15.5 million square miles), or 30% of Earth's land.

In southern and central Europe, deforestation began more than 2,000 years ago as human populations spread across the continent. Initially, most of the land was cleared for agriculture. Between A.D. 1100 and 1500, the rate of forest loss accelerated greatly in association with the warmer climate and rapid population growth. Cities were often built in the midst of forests, which were quickly cut for building materials and fuelwood.

Between 1500 and 1700, shipbuilding created an enormous demand for forest products (**Figure 14.20**). Much of the remaining forests were cut to provide wood, resins, and turpentine for the rapidly expanding navies and commercial shipping industry. Forest cutting expanded even more between 1750 and 1800, as industrial development and steam power increased the demand for wood and charcoal. By 1800, coal had

What percentage of the world's forests have we lost in the past 10 years?

Sierra Seaman, University of Florida

Norm: *Between 2000 and 2010, Earth lost about 2% of its remaining forests. The rate of loss for tropical and old-growth forests was considerably greater.*

▼ **Figure 14.19 The Loss of Ireland's Forest**
The Nephin Beg Mountains in Ireland were once covered by dense forest. Between1600 and 1800, this forest was systematically cut for fuel and shipbuilding.

▲ **Figure 14.20 Built of Oak**
The development of Europe's navies contributed to widespread deforestation. Six thousand mature oak trees were required to build a single fighting ship such as these.

replaced wood and charcoal as the industrial fuel, but the process of deforestation continued as agriculture expanded and the demand for wood for paper and construction increased.

Today, worldwide rates of deforestation remain high. In each year between 2000 and 2010, about 130,000 km^2 (50,100 mi^2) were deforested. Taking into account the amount of land that was naturally or artificially replanted in trees, this resulted in an annual net loss of 52,000 km^2 (20,100 mi^2), an area about twice the size of Massachusetts.

The rate of net forest loss between 2000 and 2010 was actually somewhat less than the average net loss between 1990 and 1999, which was 83,000 km^2 (32,000 mi^2) per year. This decrease was partly due to a nearly 20% decline in the rate of deforestation in countries like Brazil and Indonesia. It was also due to increased efforts to reforest lands in several countries. For example, between 2005 and 2010, China reported planting trees on 2,000 km^2 (772 mi^2) of land.

Causes of Deforestation

Conversion to agriculture and overharvest are accelerated by poverty, road and urban development, and political corruption.

Deforestation occurs as a direct consequence of agricultural expansion, wood harvest, road building, and urban development. There is much controversy regarding the relative importance of each of these causes. This is partly due to limited data. It is also due to the fact that these causes often operate simultaneously or sequentially on the same landscape.

Agricultural expansion. In tropical regions, the conversion of forested land to cropland and pasture accounts for more than half of all deforestation. Much of this conversion is attributed to the practice of **shifting agriculture**. In shifting agriculture, forests are cut and the slash—the downed logs and branches—is burned. Burning releases nutrients to the newly cleared soil, making it ideal for agriculture. Crop production is initially high in such clearings. But after several years, productivity declines and fields are abandoned. Given enough time, the bare fields undergo natural succession and forest and soil productivity are eventually restored (6.5; **Figure 14.21**).

Shifting agriculture has been practiced around the world for thousands of years. Where farmed plots are small and the periods between forest cutting are long enough for soil productivity to be restored, this form of agriculture may be sustainable. However, the ever-growing numbers of subsistence farmers and changes in their cutting and cultivation practices are having unsustainable consequences. Today 300–500 million people practice some form of shifting agriculture. This has encouraged cutting on marginal lands as well as ever-shorter periods between forest cutting. Under these conditions, fragile tropical soils become infertile and are easily eroded, and reforestation becomes very difficult.

Expansion of global markets for livestock and cash crops, such as sugar, soybeans, and corn, is also responsible for much deforestation. Approximately 60% of the deforestation in Brazil is associated with the establishment of cattle ranches that provide beef to meet growing global demand, driven in part by the fast-food industry. In recent years, much of this pasture has been converted to high-value crops, such as soybeans and sugar cane, which are used for ethanol. This conversion has, in turn, led to even more forest cutting to create more pasture.

Wood harvest. Much deforestation is a direct consequence of cutting trees for fuelwood. Over 90% of fuelwood is consumed in underdeveloped countries, where wood is the most important source of energy for heating and cooking. For example, most of the wood harvested in Africa is used locally for fuel. Population growth has increased this demand and has been responsible for the conversion of millions of acres of arid woodland to desert.

Global demand for wood and wood products has also accelerated deforestation in many places. Commercial logging for high-quality woods, such as teak, mahogany, and ebony, is a major factor in the loss of tropical forests in Indonesia and Central America. Global demand for paper has encouraged the clearing of large tracts of old-growth forest and their replacement with plantations of fast-growing trees, such as eucalyptus and pine.

Roads and urban development. Although it accounts for a comparatively small portion of total global deforestation, the development of urban infrastructure, including roads and housing, contributes to significant forest loss in many areas. This is particularly true in many developed countries. In the United States, approximately

▶ Figure 14.21 **Shifting Agriculture**
This mosaic of landscape patches in Indonesia is the result of shifting agriculture. The lighter color patches are being actively farmed. Other patches have been abandoned and are undergoing succession back to forest.

1,800 km² (695 mi²) is deforested for transportation and urban development each year.

To encourage economic development, many countries have undertaken road and railway expansion projects that have a significant effect on deforestation. Such projects are often done with financial support from developed countries. Such roads provide access to uncut forest and facilitate more forest cutting (**Figure 14.22**).

▲ Figure 14.22 **Roads Facilitate Deforestation**
In this satellite photograph, the primary forest is green. The photograph shows the clear relationship between road development (tan lines) and deforestation (large tan areas of open land) in the Amazon Basin of Brazil. Note the smoke plume from a fire where trees have been recently cut.

Economic and political causes. Poverty is often cited as the most important factor contributing to deforestation worldwide. This is, however, an oversimplification. It is true that deforestation rates are highest among the world's poorest countries and that poverty causes people to migrate to forest frontiers where they engage in subsistence farming. But many other political and economic factors also drive deforestation and forest degradation. For example, agricultural subsidies and timber concessions aimed at encouraging economic growth often encourage deforestation in developing countries.

Many developing countries have passed laws to regulate forest cutting and limit deforestation. However, deforestation continues because of governmental corruption and because governments have very limited capacities to enforce their laws. Illegal logging accounts for over half of the timber production in Brazil, Russia, Indonesia, and many African countries. This logging accelerates deforestation and deprives the citizens of these countries of much needed revenue. Illegal logging is estimated to cost the world's developing countries $15–$20 billion each year.

Industrial countries in North America and Europe lead the world in rates of reforestation. At the same time, their patterns of consumption are driving deforestation in many developing countries. Although this is partly due to the demand for wood and wood products, it is also driven by the demand for other commodities. The high consumption of beef associated with the expansion of the fast-food industry, for example, is directly correlated with deforestation to create pasture for cattle in the Amazon. Likewise, the demand for gasoline in developed countries creates incentives to replace forests with agricultural crops that can be used to produce gasoline substitutes, such as ethanol.

How Can Deforestation be Halted?

Slowing worldwide deforestation will require multiple strategies.

QUESTIONS

1. What factors have contributed to the global loss of forests over the past 10,000 years?

2. Describe the factors that can cause shifting agriculture to lead to extensive and permanent deforestation.

3. Why has road development accelerated deforestation in many countries?

Just as deforestation has multiple causes, controlling deforestation will require multiple strategies. These include improved agricultural practices, management of population growth in forested regions, reduced demand for virgin wood products, reforestation and sustainable forest management, and sustainable economic development.

Intensive agricultural methods, including the use of high-yield crops and the heavy application of fertilizers, herbicides, and pesticides, can produce more food on less land, thereby decreasing the need to convert forests to farmland. However, these methods depend on chemical and energy inputs that often have unsustainable consequences.

Many governments and development organizations are working to limit migration and population growth in frontier forest areas. To be successful, such efforts must be coupled with programs that alleviate poverty and provide economic opportunity. Ecotourism has provided important economic incentives for forest conservation (11.8).

In the past, governments and development banks, such as the World Bank, have had inconsistent policies, often funding projects to establish forest preserves through one department while funding road construction projects through other departments. Recently, there has been an increasing awareness of the need to limit road construction through regions where forest preservation is a high priority.

Perhaps most important, many organizations and governments in developed countries are implementing programs aimed at limiting the consumption that encourages deforestation. For example, McDonald's restaurants has committed to using recycled paper products and to procuring its beef only from suppliers that can certify that their activities have not caused deforestation.

SEEING SOLUTIONS: Restoring Forests and Community Well-Being in Haiti

How can very poor people in rural communities be engaged in the restoration of their forests?

In places such as the Mayan Forest of Central America and the Atlantic Forest of Brazil, halting deforestation and conserving the biodiversity and ecosystem services of the remaining forests is the central priority. In Haiti, opportunities for forest conservation are all but gone, and there are numerous obstacles to forest restoration. Perhaps most important, the extreme poverty of Haiti's people provides few incentives for forest stewardship.

Recent community-based initiatives show some promise of overcoming these obstacles. Near the watersheds of Cormier and Fond de Boudin in Haiti's southern mountains, the Comprehensive Development Project (CODEP) is working with nearly 650 farmers to reforest entire watersheds. The first steps in this reforestation are to stop further erosion and to reclaim soils (**Figure 14.23**). This is done by digging contour canals. Grasses are planted on the downhill side to prevent erosion, and fast-growing trees like eucalyptus are planted above. The canals retain water and collect falling leaves. In two or three years soils have been reclaimed from the natural compost in these canals so that gardens, fruit trees, and coffee can be grown successfully (**Figure 14.24**). CODEP works with individual communities to provide training in techniques to limit erosion, restore soil fertility, and plant trees.

Just as important, CODEP programs focus on economic development and the well-being of families. Participating communities volunteer a day of work each week. In return, they are guaranteed to receive much-needed pay for working an additional 2 days. As they work, farmers earn credits that can be used to purchase water cisterns, hurricane-proof homes, tools, and fertilizer. They are also eligible for micro-economic loans. CODEP helps support several schools that teach environmental reclamation and have plant nurseries on site.

CODEP also works with communities to build and maintain fish ponds. Tilapia grown in such ponds provide a source of dietary protein, thus diminishing the demand for grazing animals. When the fish are sold at market, they provide families with additional income.

As of 2011, CODEP partners have 8 million trees in the ground, of which approximately 20% are fruit trees. They have also constructed 30 working fish ponds, over 200 water cisterns, and 20 homes. In response to the 2010 earthquake, they have built an additional 39 homes in partnership with the Building Goodness Foundation of Charlottesville, VA. CODEP has been working in the area for 20 years and it is just now ready to say that the development effort is working—because the people are healthier, have improved nutrition, and young folks are staying in the area rather than moving to large cities like Port-au-Prince. The community-based approach, CODEP people believe, is the key to restoring both forest ecosystems and the economy.

▼ **Figure 14.23 First Steps**
On barren slopes, CODEP volunteers have dug canals along the contours of barren slopes. Grass has been planted on the downslope side of canals, and eucalyptus trees will be planted upslope. Canals will slow water flow and capture tree litter to restore soil organic matter.

▼ **Figure 14.24 New Forest**
This five-year-old patch of forest now protects part of Haiti's Cormier watershed. Farmers are able to grow coffee in the shade of these trees.

14.4 Forest Degradation

As the total area of Earth's forests is decreasing, the overall condition of the remaining forests is also deteriorating. **Forest degradation** occurs when changes in forest structure and composition diminish biodiversity or key ecosystem services. Logging and other human activities are simplifying once complex and diverse native forests. Agriculture, roads, and urban development are dividing forest tracts into ever-smaller pieces and preventing the migration of forest species. In many forests, natural disturbance processes have been altered, thereby reducing biodiversity. The introduction of non-native species threatens forest biodiversity and the ecosystem services it supports.

Forest Health in Peril

Cutting, fragmentation, altered disturbance patterns, and non-native species have degraded forests worldwide.

Worldwide, only about 19 million square kilometers (7.3 million square miles) of **primary forest**—forest that has been largely unaffected by humans—remains. Currently, over 60,000 square kilometers (23,000 square miles) of this forest are cut each year. Although some of this land is naturally or artificially reforested, the **secondary forests** that grow back have less biodiversity and provide fewer ecosystem services than primary forests.

Cutting and high grading. The temperate deciduous forests of eastern North America provide an informative example of the large-scale conversion of primary forest to secondary forest. Between 1650 and 1850, nearly 65% of the forests in this region were cut and converted to farmland. Forests that were not cleared were degraded by heavy livestock grazing and **high grading**, the removal of particular species of trees for fuel and construction.

Following the Civil War, many of the farms in this region were abandoned. Old fields were quickly invaded, first by herbs and shrubs, and then by forest trees (6.5). Today, forest cover in the eastern United States is about 70% of what it was in 1600. However, less than 0.1% of that cover is primary forest, and most of that primary forest is in small, widely scattered patches. Much of

the forest may have been restored, but it is not nearly as diverse as it once was. Recent studies show that the diversity of herbs and birds in primary deciduous forests is 20–50% higher than in secondary forests.

Fragmentation. Throughout the world, forest landscapes are becoming increasingly fragmented. Forest patches are smaller and more widely separated. Roads, agriculture, and urban development make it more challenging for forest organisms to travel between patches. As predicted by the equilibrium theory of island biogeography, this fragmentation has diminished the biodiversity of animal groups such as migratory songbirds (11.5).

Altered patterns of disturbance. The natural frequency and severity of fires vary widely among different forest types. Nearly everywhere, humans have modified fire regimes, often in ways that have degraded forests.

In the mid-19th century, the ponderosa pine forests extending from eastern Washington and Oregon to Arizona and New Mexico were typically quite open with a grassy understory. This pattern of growth allowed light fires to burn across the surface of the ground every

▶ **Figure 14.25 Fire Suppression Creates Risk** In the absence of fire over the past century, young trees and shrubs have invaded the understory of this stand of ponderosa pine in northern Arizona. Today, if a fire ignites in this forest, it will likely be very severe.

Ⓐ Prior to 1900, frequent fires in the understory kept this ponderosa pine stand open.

Ⓑ In the absence of fire, young trees and shrubs had invaded the understory of this stand of ponderosa pine by 1949.

Ⓒ Today, if a fire ignites in this forest, it will likely be very severe.

2–4 years. Between 1880 and 1900, heavy livestock grazing in much of this region reduced the grass cover so that there was insufficient fuel to carry fires. Without regular fires, tree seedlings and woody shrubs invaded the forest floor (**Figure 14.25**). In the decades that followed, fire suppression favored the continued growth of the woody understory. Because of this accumulation of potential fuel, when fires occur in these forests nowadays, they are often quite large and severe (**Figure 14.26**).

Recently, there has been a dramatic increase in the frequency of fires in tropical rain forests, such as those in the Amazon and Central Africa. Usually these fires are set by humans; many are set to burn slash left after logging, but the fires escape into nearby forests. These fires may not totally destroy the forest overstory, but they do kill many understory species and diminish overall forest diversity.

Non-native species. The introduction of non-native species has degraded forests in many regions. In 1748, a gardener introduced an Asian species called tree of heaven to Pennsylvania. Tree of heaven thrived in the climates of the eastern United States and spread rapidly. It is particularly well adapted to colonizing and outcompeting native trees in forest openings and disturbed locations.

Non-native fungal diseases and insect pests have had an especially significant impact on the health of many forests. The fungus that causes chestnut blight has all but eliminated the American chestnut from eastern forests,

and the aphidlike hemlock adelgid seems to have doomed the eastern hemlock to the same fate (11.5).

The gypsy moth is native to Europe and Asia, where its larvae feed on the spring leaves of deciduous trees, especially oaks. It was introduced to the United States in 1868 by a Massachusetts scientist interested in crossing it with silkworms. Lacking its natural predators, the moth has quickly spread through eastern forests. Gypsy moth larvae can consume nearly all of the leaves of canopy trees (**Figure 14.27**). This weakens trees and makes them more vulnerable to drought and fungal disease.

Halting deforestation—keeping forests as forests—is a central conservation goal. However, that objective alone is insufficient to preserve the biodiversity and ecosystem services provided by Earth's forests. Sustainable forest conservation must also include the protection of primary forests, connection of forest patches, restoration and maintenance of natural disturbance cycles, and management of non-native species.

▲ Figure 14.27 **Gypsy Moth Caterpillar**
Caterpillars of gypsy moths feed on the leaves of many trees, often completely defoliating them. Over several years, such defoliation can result in the death of trees.

QUESTIONS

1. How do primary and secondary forests differ in terms of biodiversity and the history of past human disturbances?

2. Describe two threats to the health of forests in the United States.

◄ **Figure 14.26 Megafire**
This is a satellite photograph of the 2004 Rodeo-Chediski fire, which burned nearly 200,000 ha (500,000 acres) in northern Arizona. The unprecedented size and severity of this fire was a consequence of the great quantities of brush and other flammable materials that had accumulated in the forest understory during many years of fire suppression.

14.5 Defining Sustainable Forest Management

Can harvesting trees for heating fuel be a sustainable process?

Glenda Sullenger, St. Charles Community College

Norm: *Yes it can, so long as harvest rates do not exceed the rate of forest recovery and care is taken to manage pollution from wood smoke.*

Humans manage forests for different purposes and with differing impacts. At one end of the spectrum, forest preserves are managed with minimal human intervention. At the other end of the spectrum, plantation forests are intensively managed to produce wood and fiber, somewhat like an agricultural crop. Between these two extremes, mixed-use forests are managed at varying levels of intensity and for a variety of objectives, including wood fiber, recreation, water protection, and biodiversity. Forest management across this spectrum entails four categories of action: allocation, harvest, rationing, and investment. Each of these categories includes decisions and actions that influence forest sustainability.

Allocation

Which forests should be dedicated to which management goals?

Allocation is the management of the location and relative amounts of different forest uses. Managers allocate forestlands according to their specific objectives. For example, landowners who are interested in using forests for commercial wood production, but are also interested in maintaining water quality, might locate forest preserves near streams and plantations away from streams. To sustain biodiversity, they might create corridors of mixed-use forests to allow wildlife to migrate between the preserves (**Figure 14.28**).

Environmental features such as topography, economic factors, and legal restrictions influence the allocation of forest use. Approximately 17% of the world's forests are designated as preserves, although the extent of protection of such forests varies widely from country to country. Currently, 8% of the world's forests are allocated to intensively managed plantations; such forests produce approximately 30% of commercial wood fiber.

Environmentalists disagree about the role that plantations should play in global conservation. Because plantations provide fewer ecosystem services than other forms of forest management, some argue that it is undesirable to allocate increased amounts of forests to plantations. Others note that plantation forests provide a disproportionate amount of the world's commercial wood fiber. If the allocation to plantations were increased to 12%, such forests could possibly meet over 50% of the global demand for wood fiber. Such an increase could relieve the pressure for legal and illegal logging in other forests.

▼ **Figure 14.28 Forest Corridors**
The corridors of forest across this landscape provide migration routes for a variety of forest species, thereby helping to maintain biodiversity.

Harvest

How should goods and services be taken from forests?

Harvest is the extraction of goods or services from a forest. Often there are several ways to harvest goods. When deciding what methods of extraction to use, managers must consider what is to be taken from the forest and also what is to be left behind.

Clear-cutting, the complete removal of forest canopy trees, is certainly the most controversial harvest practice. In the western United States, extensive clear-cutting of old-growth forests growing on steep slopes has produced significant erosion and flooding; it has also had devastating effects on wildlife habitat. In other types of forests, however, clear-cutting on relatively flat terrain may be a sustainable practice (**Figure 14.29**). In such situations, forest managers often clear-cut patches of 10–20 ha (25–50 acres) to simulate patches of natural disturbance. Because clear-cutting produces trees of uniform age, it is often called **even-aged management**.

To sustain biodiversity and ecosystem services, forest managers are increasingly turning to less intensive methods of harvest. In various kinds of **uneven-aged management**, trees, dead snags, and woody debris are retained to produce a more heterogeneous age structure. The simplest forms of uneven-aged management are seed cuts, in which select trees are left behind to provide seeds to accelerate the establishment stage of forest recovery. Other forms of uneven-aged management preserve habitat for particular species and maintain tree diversity while allowing the removal of valuable wood or non-wood forest products.

▶ **Figure 14.29 Unsustainable Harvest?**
Clear-cutting forests on steep terrain and fragile soils can produce severe erosion and very slow rates of reforestation Ⓐ. The loss of forests from such landscapes increases runoff and This form of harvest may be appropriate on gentle slopes to regenerate forests in which the dominant trees like these pines grow best in full sunlight Ⓑ.

Rationing

What amount of forest goods and services should be dedicated to human use?

Rationing is the process of determining the amounts of forest goods or services to be used by humans. Examples of rationing include decisions about how much wood to harvest each year and how many visitors to allow in a forest reserve.

In intensively managed forests, rationing of wood harvests is determined by **rotation time**, the average interval between successive cuts. Rotation time depends on the intended use of the wood. For example, paper is most efficiently produced from small-diameter trees; therefore, rotation times for paper production are quite short, varying from 7 to 25 years. In contrast, the production of high-quality wood for furniture requires much larger trees and may require rotation times exceeding a century (**Figure 14.30**).

Increasingly, forest managers are striving to select rotation times that are similar to the return times that forest communities require to recover from natural disturbances, such as fires. Research has shown that such rotation times are more likely to support overall forest biodiversity.

Sustainable forest management must address human use of all forest goods and services. Thus, managers set bag limits on wildlife and the collection of understory herbs. Restrictions on intensive recreational use of public forests and preserves are an important type of forest rationing.

► **Figure 14.30 Short and Long Rotation Times.**
Plantations of eucalyptus trees, which are grown for wood fiber, grow so rapidly that their rotation times may be as short as 7 years Ⓐ. In contrast, oak forests that are grown for high-quality timber may have harvest times that are longer than 100 years Ⓑ.

Investment

What actions should be taken to restore and maintain forest health and productivity?

Investment is the wide array of management strategies used to restore, maintain, or protect a forest's capacity to provide goods and ecosystem services. The application of fertilizer to improve plantation growth and the construction of roads for logging access are examples of investments directed toward commercial wood extraction (**Figure 14.31**). Understory thinning and managed burns to reduce the risk of catastrophic wildfires are examples of investments aimed at forest protection. The installation of nesting cavities in longleaf pine trees for endangered red-cockaded woodpeckers is an example of a management investment aimed at conserving biodiversity.

Sustainable forest management begins with a clear articulation of management goals and then ensures that those goals can be achieved for generation after generation. Such goals set priorities for the forest goods and services that managers wish to sustain. Most important, sustainable forest management is not defined by specific choices, such as clear-cutting versus selective harvest. Rather, it encompasses a broad array of choices about when, how much, and by what methods products are harvested, and what to invest to restore and maintain forests.

▼ **Figure 14.31 Investing in the Future**
Replanting trees following harvest is an example of investment management.

QUESTIONS

1. Describe four categories of forest management activities.

2. For each forest management category, describe a management action that would likely have unsustainable consequences.

442

14.6 Sustainable Forestry—A Global Vision

Global economic development and human population growth are likely to increase the demand for the goods and services that forests provide. Sustainable forest management is essential to meet these demands and to ensure that these resources are available for future generations. The United Nations has established seven criteria to guide nations in the sustainable management of their forests. Forest certification programs aim to create demand for forest products that are managed according to these criteria.

Criteria for Sustainable Forest Management

Sustainable forest management is concerned at least as much with what it leaves behind as with what it extracts.

The future of the world's forests was front and center in discussions at the 1992 United Nations Conference on Environment and Development in Rio de Janeiro. Attendees agreed that sustainable forest management maintains the biodiversity, productivity, capacity for regeneration, and health of forests and forest landscapes for current and future generations.

In 1994, delegates from United Nations member nations meeting in Montreal, Canada, used this definition to identify standards for sustainable forest management. They proposed seven criteria for setting sustainable management goals. For each criterion, they identified several indicators of success. These have come to be known as the Montreal Criteria and Indicators. The seven criteria are listed below, followed by a description of some of the important indicators.

1. Biological diversity

Sustainable forest management requires biological diversity from the level of individual trees to entire forested landscapes. Managers should maintain genetic diversity within individual tree species. Individual stands should be managed to maintain a diverse number of native species and to minimize the impact of invasive non-native species. At the landscape level, indicators include the abundance of different forest types and stages of succession, the extent of forest fragmentation, and the protection of rare, unique, or threatened forest types.

2. Productive capacity

Individual forests and forested landscapes should maintain their capacity to produce a full array of forest goods and services. Indicators of productive capacity include the total area of forested land, the relative amounts of lands managed for different purposes, and the rate at which timber and non-wood products are removed from forests compared to the rate at which they are regenerated.

3. Ecosystem health and vitality

Factors that present risks to individual species and forest landscapes must be managed. Indicators of ecosystem health include the amount of forest that is abnormally affected by factors such as insects, disease, fire, land clearance, and animal grazing, and the amount of forest that is affected by air pollutants, such as ozone or nitrogen oxides.

4. Soil and water resources

Soil and water resources, as well as forests' protective and productive functions, must be conserved. Key indicators of unsustainable management include the area of forestland with significant soil erosion, the area of forestland with significantly diminished soil fertility, and the number of polluted streams and lakes in forested regions. An indictor of sustainable management would be the area of forestland managed for protective functions, such as water management and avalanche control (**Figure 14.32**).

5. Global carbon cycles

Forests should be managed to maintain their ability to take up and store carbon. This criterion recognizes the important role of forests in determining CO_2 fluxes to and from the atmosphere. The most important criterion is the total amount and stability of biomass on forested landscapes.

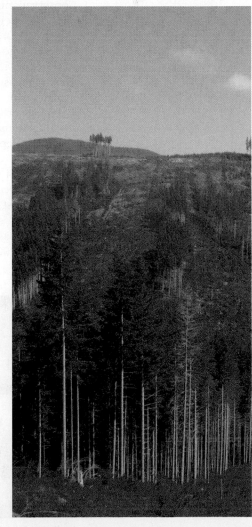

▶ **Figure 14.32 Sustainable Forestry**
This aerial photo shows forested buffers along streams in an area that is being logged. Such buffers prevent erosion and protect water quality.

Other criteria include the patterns of carbon uptake and loss among different forest types and successional stages and the storage of carbon in forest products, such as paper and wood.

6. Socioeconomic benefits

Forests should maintain and enhance long-term socioeconomic benefits. One set of indicators is the amounts of wood and non-wood forest products produced, consumed, and recycled. Another set of indicators is the amount of forestland conserved for recreation and tourism and the rate of their visitation and use. Other indicators include direct and indirect employment in forest management or the production of forest products, the amount of land used for subsistence purposes, and the amount of forestland managed to protect cultural, social, and spiritual needs. In this regard, an especially important indicator is how well forest-dependent communities, including indigenous peoples, adapt to changing economic conditions.

7. Legal, institutional, and economic framework

The overall policies and laws of countries should facilitate the conservation and sustainable management of forests. Legal indicators include laws that clarify property ownership and the rights of indigenous people; laws should also set standards and regulate management practices. The ability to enforce those laws and regulations is an important measure of success, as is the public's involvement in and awareness of forest management issues. Other important indicators include the capacity to measure and monitor changes in forest status and the capacity to conduct and apply research to improve forest management and the delivery of forest goods and services.

Forest Certification

The marketplace can be a powerful force for sustainable forest management.

Conservationists have long argued that deforestation and forest degradation would be slowed greatly if consumers of wood and paper products demanded that those commodities come from forests that are managed sustainably. The Montreal Criteria and Indicators provide general measures for assessing the sustainability of forest management.

Forest certification programs create standards for the sustainable management of specific forest types and provide independent assessments of forest management activities. Certification programs also increase public awareness of the issues and build demand for certified forest products.

The Forest Stewardship Council (FSC) is one of several independent organizations that certify forest owners and managers. Based on the Montreal Criteria and related principles, FSC has established standards for forest management that include allocation, harvest, rationing, and investment activities. Their standards also aim to ensure worker safety and the fair treatment of indigenous people. Under FSC supervision, independent auditors assess the operations of forest product companies and individual forest managers relative to these standards. If these stringent standards are met, products are allowed to carry the FSC label (**Figure 14.33**).

Consumer demand for certified wood products is critical to the success of certification programs. Although public awareness in the United States remains low, many retailers who sell or use wood or paper products require some form of sustainability certification. Many European countries now require that these commodities be certified. The European Union has not yet adopted mandatory certification requirements but strongly encourages member countries to do so.

◄ **Figure 14.33 Certified Sustainable** Retailers and consumers increasingly expect that construction wood and finished wood products produced in a sustainable manner. This chair, for example, carries the FSC label indicating that it was produced from sustainably managed and harvested trees.

Forests and Greenhouse Gas Emissions

Reforestation and forest restoration are likely to become important strategies to limit global warming.

International negotiations aimed at reducing greenhouse gas emissions and global warming have paid particular attention to forests and forest management. An integral part of all proposed climate change treaties is country-by-country accounting of net carbon emissions. In most of the proposed accounting systems, reducing emissions from deforestation and degradation (REDD) is a prominent feature. In REDD accounting systems, countries could receive credits for reducing carbon emissions by reducing their rate of deforestation, by reforesting lands that are now unforested, and by increasing carbon storage in existing forests.

There is general agreement among policy makers and forest scientists that actions within a REDD program must meet four standards if they are to be truly effective in reducing greenhouse gas emissions. First, they must be *additive*—they must involve actions to restore forests or reduce deforestation that would not have happened otherwise. Second, they must be *permanent*—there must be a commitment that increased stores of forest carbon will remain in place over long periods of time. Third, they must be *verifiable*—it must be possible to measure accurately the amount of carbon stored in forests and to audit these amounts over large areas. Fourth, they must avoid *"leakage"*—activities such as reforestation or reduced deforestation in one location cannot encourage

deforestation elsewhere. (With the ever-increasing demand for wood products, diminished logging in one place often creates incentives to simply relocate that logging to another place.)

Those who support the use of REDD argue that this approach is less expensive than other options for reducing carbon emissions, such as switching from fossil to non-fossil fuels or increasing the use of renewable energy sources. Proponents note that REDD may be an especially attractive way for developing countries to comply with stricter greenhouse-gas emission standards. Others acknowledge a potential role for REDD in a combination of actions to limit global warming, but warn that this approach presents a number of serious challenges and may not be particularly cost effective.

Strategies for sustaining the world's forests include management practices that support biodiversity and ecosystem services, such as soil fertility and productivity. National and international policies should ensure fairness in the distribution of the benefits that forests provide, as well as the legal and economic tools to implement and maintain these strategies. Forest certification provides a means for economic markets to encourage sustainability. Global initiatives to reduce emissions of greenhouse gases may include incentives for sustainable forest management (**Figure 14.34**).

QUESTIONS

1. What are the seven Montreal criteria for sustainable forestry?
2. Explain how sustainable forest management can reduce emissions of greenhouse gases.

▶ **Figure 14.34 Reforestation** Reforestation can help to slow the rate of global warming. China has made significant progress in reforestation.

AGENTS OF CHANGE: The University of Winnipeg's Campaign Against Logging in Parks

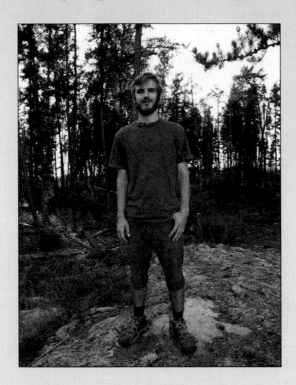

Robin Bryan lives close to the boreal forest of the East Shore Wilderness Area in Manitoba, Canada, the world's largest single land storehouse of carbon and most abundant source of fresh water. While a student at the University of Winnipeg, Robin led a campaign against logging in Manitoba's provincial parks. In 2008, he was rewarded for his efforts when Manitoba banned logging in four of the five parks that had logging operations.

How did you first get the idea for a campaign against logging in Manitoba's provincial parks?

I grew up in Manitoba, a province containing the largest tracts of pristine, unbroken forests on the planet. Here a network of clean rivers and lakes interweave among lush evergreen foliage and rock ridges. These vast forests are also home to a host of hoofed mammals, migratory birds, amphibians, and fish.

In my early teens, I learned that commercial logging and mining operations were degrading these amazing natural wonders. Our provincial government was not taking meaningful actions to protect the forests, and their lack of sustainable ecological planning was placing our ecological resources in danger. As I began learning about this situation, I knew I had to change it.

I became involved with the Wilderness Committee, where the mentors I met sparked my interest in becoming a conservation activist. My first major activity was to participate in the ongoing campaign to stop clear-cut logging in Manitoba's provincial parks.

What steps did you take to get started?

First, I carried out vigorous research to gain knowledge and perspective on the issues of sustainability, logging practices, and policies. This knowledge helped build my confidence as an activist. I continued by learning about volunteer management, fundraising, organization structure, persuasive writing, and public education.

Because the success of my work rested upon my ability to speak clearly, I began practicing speaking to people one-on-one and to small groups. Eventually I overcame my fear of public speaking and was able to feel confident when speaking with elected officials, organizing rallies, delivering classroom presentations, raising funds, and organizing volunteers to write and send letters to the government.

What are the biggest challenges you've faced with the project, and how did you solve them?

As a person who is passionate about environmental protection, it is hard to face the fact that many people in our communities have limited education or experience with environmental issues, causing them to misunderstand how important these issues are. Because of public apathy, things are often slow to change, especially when it comes to changing major legislation.

During long periods of hard work with little visible impact, burnout can become a problem for many activists. When facing burnout, it was important for me to set boundaries between my activism and my personal life. I needed to recognize the limits of what I could contribute to the cause while meeting my own needs and maintaining my health. It was also important to surround myself with peers who could share my inspirations and frustrations. This was crucial to unwinding and feeling connected to my community. In the end, I learned to push through the rough patches and stick to the campaign until the job was done.

How has your work affected other people and the environment?

Over the course of four years, I worked with the Wilderness Committee and the community to end the practice of logging within provincial park boundaries. In 2008, legislation banned logging in all but one park in Manitoba, resulting in the protection of one million acres of land. I truly feel that ending logging in Manitoba's parks was a victory for all Manitobans. It was a testament to the power of proactive citizen involvement in a democracy to build change for our ailing environment.

What advice would you give to students who want to replicate your model on their campus?

Start by gaining knowledge. Build confidence in your abilities as an activist by practicing your communication skills. You must overcome any fears you have of speaking out in public about contentious subjects such as environmental protection.

To see a major campaign succeed, be prepared to work hard for years until your vision is realized. Be prepared to face apathy and criticism, but try to have compassion for those who oppose what you stand for.

Getting involved with an organization as a volunteer or an entry-level employee is a great way to access resources and begin developing your own vision. Books like *The Young Activist's Guide to Building a Green Movement and Changing the World*, recently produced by Sharon Smith of the Brower Youth Awards and Earth Island Institute, can also provide insight.

The most important thing is to find your passion, aim high, and don't give up until you've done what you set out to do.

Summary

Forests contain a major share of Earth's biodiversity. They also provide essential ecosystem services on which we all depend. They are the source of wood products such as fuelwood, lumber, and fiber for paper manufacture; they are also the source of numerous non-wood products, including pharmaceuticals, mushrooms, wild nuts and fruits, and rubber. Forests grow as individual trees germinate from seeds and then increase in height and girth. Forest stands develop through a series of successional stages, beginning with the establishment of young trees and culminating in uneven-aged, old-growth forest. Forested landscapes comprise patches representing these different developmental stages. Over the past 10,000 years, the total area of Earth's forests has been reduced by 40%, with much of the deforestation occurring over the past 300 years. Most deforestation is due to clearing for agriculture and harvest for fuelwood and commercial wood products. In many countries, poverty, road construction, and corruption encourage deforestation. The health of the world's remaining forests is declining due to unsustainable management, fragmentation, altered patterns of disturbance, and invasion of non-native diseases and pests. Sustainable forest management includes decisions about which forests should be used in which ways, how much product should be harvested, how harvesting should be done, and what investments should be made to ensure future forest diversity and productivity. As human demands for forest products and ecosystem services increase, sustaining Earth's forests will require clear standards for the management of forests and the marketing of forest products. The world's forests can play a key role in mitigating and adapting to global warming.

14.1 The Value of Forests

- Forests provide essential ecosystem services, including flood protection, clean water, recreation, and inspiration.
- Humans depend on forests for both wood and non-wood products.

QUESTIONS

1. Explain how diminished biodiversity influences three important ecosystem services provided by forests.

2. How is demand for various wood products likely to change as economic development proceeds in poor regions, such as sub-Saharan Africa?

Fuelwood ● Fiber ● Lumber

14.2 Forest Growth

- Trees germinate from seeds and grow in both girth and height.
- Over time, populations of trees in forest stands undergo successional development, passing through the establishment, thinning, transition, and old-growth stages.
- Forested landscapes comprise patches representing different forest stand developmental stages.

KEY TERMS

seed, conifer, angiosperm, primary growth, phloem, xylem, secondary growth, establishment stage, thinning stage, transition stage, old-growth stage, minimum dynamic area

QUESTIONS

1. If the bark and phloem are stripped away from the trunk of a tree, it will quickly die. Why?

2. How does the likelihood of a tree seed germinating and becoming established change through the various stages of forest stand development?

3. What role might the minimum dynamic area concept play in determining ideal boundaries for conservation preserves?

14.3 Deforestation

- Each year, more than 130,000 km² are deforested, primarily as a consequence of agricultural expansion and overharvest.
- Deforestation is facilitated by poverty, road construction, and governmental corruption.

KEY TERMS

deforestation, shifting agriculture

QUESTIONS

1. The rate of net forest loss has diminished in the past decade compared to the previous decade. Why has this occurred?

2. In what ways does poverty contribute to deforestation?

14.4 Forest Degradation

- Cutting, fragmentation, altered disturbance patterns, and non-native diseases and pests have degraded forests worldwide.

KEY TERMS

forest degradation, primary forest, secondary forest, high-grading

QUESTIONS

1. How has the suppression of fire degraded the health of forests in parts of the western United States?

2. Why are non-native tree diseases and insects often so much more virulent than their native counterparts?

14.5 Defining Sustainable Forest Management

- Sustainable forest management requires the determination of which forests should be dedicated to which goals, how goods and services should be harvested, what amounts of goods and services should be harvested, and what investments should be made to ensure future forest health.

KEY TERMS

allocation, harvest, clear-cutting, even-aged management, uneven-aged management, rationing, rotation time, investment

QUESTIONS

1. Investment is a particularly important part of sustainable forest management. Why?

2. For forests near you, give an example of actions associated with each forest management category.

14.6 Sustainable Forestry—A Global Vision

- The Montreal Criteria for sustainable forest management include indicators for biological diversity, productive capacity, ecosystem health and vitality, soil and water resources, the global carbon cycle, socioeconomic benefits, and policy frameworks.

- Forest certification programs provide oversight of forest management by independent auditors.

- Reducing emissions of greenhouse gases from deforestation and forest degradation (REDD) is a significant part of proposed international treaties aimed at curbing global warming.

QUESTIONS

1. How do forest certification programs operate to ensure sustainable forest management practices?

2. To be successful, REDD programs must meet four standards. What are they?

Mastering**ENVIRONMENTALSCIENCE**®

Go to **www.masteringenvironmentalscience.com** for homework assignments, practice quizzes, Pearson eText, and more.

449

15

Nonrenewable Energy and Electricity

The History of an Oil Field

Just when is a nonrenewable resource exhausted?

How much oil is available on Earth? In theory, the answer to that question is determined by the absolute amount of oil in Earth's geologic strata. In reality, the answer is far more complicated. Oil deposits exist in different forms and in different places. Some of these deposits are easily tapped, and others can be exploited only with great difficulty. Thus, how much oil is actually available is determined by the technologies that are available to extract it and the cost of applying those technologies relative to the price of the resource. Many of those costs are directly related to the health of the environment. These principles are demonstrated in the history of the human use of a single oil field—the Kern River Oil Field near Bakersfield, California (**Figure 15.1**).

The Yokut Indians were the first humans to benefit from the Kern River Oil Field. As far back as 8,000 years ago, they mined the tar that oozed from the ground and used it as an adhesive and for waterproofing and decoration. The Yokut traded the oily substance far and

wide and thus might be considered to be one of the first oil-exporting nations. Of course, they had no idea of the vast amounts of oil that had accumulated hundreds and thousands of feet beneath their villages. Even if they had, they lacked the technologies to extract and use it.

When California was granted statehood in 1849, people of European descent were still collecting thick oil and tar from seeps and shallow pits in this area. But the demand and, therefore, the price for this goo were modest and brought little income to the local farming economy. No one saw uses for this resource beyond those devised by the Yokuts.

In the mid-1800s, lamps and small heaters were fueled with whale oil. With growing human populations, the demand for whale oil was increasing. This demand caused the price of whale oil to rise, just as whale populations around the world were rapidly declining.

The first oil well was drilled in Pennsylvania in 1859. Within a few years, technology was developed for distilling crude oil to make a fuel called *kerosene*, which could be used as an alternative to whale oil. The demand for kerosene grew quickly, as did its price. This demand spurred interest in locating and exploiting other sources of oil.

In 1899, this interest motivated James and Jonathan Elwood to dig by hand a 45-ft-deep well on the banks of the Kern River. They celebrated as thick crude oil squirted up to the surface. News spread quickly, and soon others began digging and drilling nearby with similar success. Within a couple years, the small farming community of Bakersfield was crowded with men seeking their fortune in the Kern River Oil Field, one of the largest oil reservoirs ever discovered in the United States.

By 1906, oil production in the Kern River Oil Field had risen to a peak of 16 million barrels a year. But the oil wells built at that time could access only the oil that flowed easily through rock pores and fractures. During the next 50 years, oil production declined, slowly at first and then more rapidly (**Figure 15.2**). By 1955, most of the 9,183 wells on the oil field were producing considerably less than 1 barrel of oil each day. Early on, little energy was required to harvest large amounts of energy-rich oil. After a half-century of constant drilling, however, energy and economic costs exceeded the benefits of production. Most of the wells were shut down.

Those decades of oil exploitation had a long-lasting impact on the environment. The large amounts of sediment- and chemical-laden water that were pumped to the surface with the oil were simply dumped into small streams, which eventually flowed into the Kern River. Spilled oil was seen as a wasted resource but not

▼ **Figure 15.1 Kern River Oil Field**
California's southern San Joaquin Valley sits atop oil-rich sediments that were laid down 2–5 million years ago.

In compliance with clean-water regulations, wastewater was managed in treatment facilities and no longer dumped into streams. Thus, the cost of pollution was captured in the cost of oil production. Indeed, clean water from these treatment facilities was sold to nearby farmers for irrigation. On the other hand, these new technologies spewed large amounts of pollutants into the air. That air pollution was viewed as a cost of doing business; its environmental costs were ignored.

Oil prices collapsed in the 1980s. At the same time, air-quality regulations were becoming stiffer. Operations at the Kern River Oil Field were again tenuous. Yet once again, technological innovation provided a fix. Oil companies built facilities to generate electricity that were fueled by natural gas, which burns cleaner than oil. This electricity was a source of revenue. The electric facilities also supplied steam that was used to increase production from the wells. In 2000, the Kern River Oil Field produced nearly 40 million barrels of oil. However, this level of production could not be sustained. Since then, production has fallen to less than 30 million barrels each year (**Figure 15.3**).

Since 1899, over 2 billion barrels of oil have been extracted from the Kern River Oil Field. Scientists estimate that this field could yield another 475 million barrels. But actually producing that much oil will depend on continuing improvements in technology and high oil prices.

Like many of the resources upon which we depend, oil is being consumed by humans at a rate that is thousands of times faster than the rate at which it is being produced. What are the factors that influence the total amounts of such resources? How do technology and economic factors affect the availability of those resources? What are the environmental consequences of their use? These questions are central to this chapter.

▲ **Figure 15.2 Kern River Oil Field 1910**
By 1910, over 9,000 wells had been drilled, but production had already peaked and was beginning to decline.

as a pollutant. Consequently, the costs associated with this pollution were never captured in the price of the oil from this field. To this day, large areas of land in the oil field remain contaminated with oil residues.

In the 1960s, oil companies began experimenting with new technologies to coax oil from unproductive wells. At the Kern River Oil Field, companies began using large boilers and complex pumping systems to inject steam into wells. Technology to the rescue! By 1968, the field was producing 68,000 barrels of oil each day, or nearly 25 million barrels each year. With these new technologies, the cost of producing each barrel of oil was several times higher than those produced by conventional operations. Oil-fired steam generators required 25–40% of the crude oil pumped from the field. But the rapid growth of the U.S. population and its affection for automobiles increased the demand for oil and kept the price high. The high price of oil more than compensated for the costs of the new technologies.

▼ **Figure 15.3 Not Dead Yet!**
New technologies and increasing demand and prices for oil have provided the incentives to maintain production in the Kern River Oil Field.

15.1 Energy Production and Consumption

Day in and day out, the average American family uses an amount of energy equivalent to the work that 3,000 laborers and 400 draft horses would have provided on a large farm 200 years ago. Most of that energy comes from resources that exist in limited amounts. The energy from those resources usually undergoes one to several transformations before it can be used. The kinds of energy resources that people use vary considerably from region to region and within different economic sectors. In the future, patterns of energy use will depend heavily on the relative supply of different energy sources. They will also depend on the environmental impacts associated with the extraction, conversion, delivery, and consumption of these energy resources.

Energy Sources

For more than 100 years, most of our energy needs have been met by nonrenewable fossil fuels.

Scientists classify energy by its source. **Primary energy** is the energy contained in natural resources, such as coal, oil, sunlight, wind, and uranium. Primary energy can be changed into other forms of energy through the process of **energy conversion**. These other forms of energy are classified as **secondary energy**. Thus, the primary energy of oil may be converted into secondary forms of energy, such as electricity or the kinetic energy of an automobile. Further energy conversions are also possible. For example, electricity may be used to charge a battery to power a cell phone. **End use** refers to the final application of energy, such as running an appliance or driving a car.

Remember that the second law of thermodynamics dictates that no energy conversion is 100% efficient (3.2). In each conversion step, some energy is lost as heat. **Energy conversion efficiency** is the percentage of primary source energy that is captured in a secondary form of energy. For example, when coal is burned to generate electricity, 70% of the primary energy in the coal may be transformed into heat. Therefore the efficiency of the conversion from chemical to electrical energy is only 30%.

Energy end-use efficiency is the product of the efficiencies of all the energy conversions from the primary source to the end use. For example, if 100 units of coal

▼ **Figure 15.4 End-Use Efficiency**
The electricity to power an incandescent lightbulb comes from the potential chemical energy of coal that is burned in a coal-fired power plant. In these energy transformations, 100 units of chemical energy produce 1.2 units of visible light energy. End-use efficiency for this process is, therefore, 1.2%.

70 units lost as heat as the chemical energy in coal is transformed to electrical current

6 units lost as heat due to resistance in transmission lines

Primary Energy Source

Chemical energy 100 units

Generation

Electric current 30 units

Transmission

energy are converted to electricity, which is then used to power an incandescent lightbulb that produces 1.2 units of light energy, the end-use efficiency is 1.2% (**Figure 15.4**).

Energy experts further classify forms of primary energy as nonrenewable or renewable. **Nonrenewable energy** is derived from sources that exist in limited quantities. When nonrenewable energy sources are consumed, they are unavailable for future use. Forms of nonrenewable energy include the three main types of fossil fuels—crude oil, coal, and natural gas—as well as nuclear energy. Fossil fuels were formed from the remains of organisms that lived hundreds of millions of years ago; the geological processes that formed those fuels took millions of years. Nuclear energy relies on uranium, an element that exists in limited amounts. In contrast, **renewable energy** is derived from sources that are not depleted when they are used, such as sunlight and wind, or that can be replenished in a short period of time, such as fuelwood.

Through history, the primary energy sources used by humans have changed considerably. In prehistoric times, human populations were sustained primarily by solar energy, which kept them warm, and the chemical energy in their food and fuelwood. As agriculture developed, people began to rely on the energy from domestic animals, flowing water, and wind. In regions where coal and oil were available, people began to use these fossil fuels for heating and refining metals, such as iron.

With the advent of the industrial revolution in about 1750, an ever-greater proportion of the energy that humans used came from fossil fuels. Indeed, by the beginning of the 20th century, fossil fuels dominated the human energy portfolio. The high-energy content of fossil fuels and the ease with which they could be used allowed the energy consumption of the average person to increase 20-fold. The widespread use of fossil fuels led to a revolution in personal mobility, comfort, and lifestyle (**Figure 15.5**).

Today, the human energy system continues to be dominated by fossil fuels, despite growing concerns over their supply and impact on the environment. Fossil fuels currently meet 79% of global energy needs. Biomass energy, primarily fuelwood used in poor countries, accounts for about 10% of primary energy use. Nuclear energy provides about 5% of primary energy use, and hydropower contributes 6%. Other forms of renewable energy, such as wind and solar, contribute less than 1% (**Figure 15.6**).

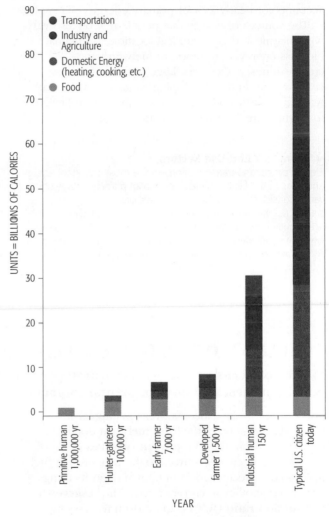

◄ **Figure 15.5 Energy Consumption Estimates**
The average amount of energy consumed per person has dramatically increased over the past several hundred thousand years. An ever-larger portion of that energy is consumed to grow the food we eat and to transport us from one place to another.
Source: World Nuclear Organization. URL http://www.world-nuclear.org/images/education/yu-graph.gif. Last accessed 8-9-11

End use

▲ **Figure 15.6 Primary Energy Sources**
Fossil fuels—coal, oil, and natural gas—account for nearly 80% of the primary energy resources consumed by humans.

End Use

Primary energy sources vary in importance, depending on the country and specific end use.

The rate of primary energy consumption per individual varies widely among countries, depending largely on the wealth of the country. The world's 50 richest countries have only 20% of the global population but account for more than 70% of all primary energy consumption. The United States, with less than 5% of the global population, accounts for nearly 20% of primary energy consumption.

The sources of energy that people rely on also vary with income and geographical location. Traditional biomass energy, in the form of fuelwood and crop residues, makes up a large portion of energy consumption in the developing nations of Africa and Asia. In wealthy countries, such as the United States, fossil fuels are far more important.

◀ **Figure 15.7 End-Use Sectors**
Electric power and transportation are the most important energy end uses in the United States. The mix of primary energy sources varies considerably among end-use sectors.
Source: U.S. Environmental Protection Agency. 2007. Energy trends in selected manufacturing sectors: Opportunities and challenges for environmentally preferable energy outcomes. URL http://www.epa.gov/sectors/pdf/energy/ch2.pdf. pages 2–4. Last accessed 9-9-11.

Consumption of primary energy sources is typically broken into four end-use sectors: transportation, industry, residential/commercial, and electricity. The residential/commercial sector comprises energy used for heating and appliances in homes, stores, and office buildings that use primary energy sources such as natural gas, coal, or petroleum. The industrial sector comprises the energy used in an array of activities, from mining to manufacturing. The transportation sector includes energy used by cars, trucks, trains, and planes. The electricity sector is composed of primary energy sources used to generate electricity. Each of these sectors relies on a different mix of energy sources (**Figure 15.7**). For example, the transportation sector relies almost exclusively on oil, whereas the industrial sector relies largely on coal.

The relative importance of different energy sectors varies widely from country to country. Because the world's 50 wealthiest nations have high rates of car ownership, their transportation sectors consume over 33% of the energy used in those countries. In contrast, transportation accounts for less than 20% of energy used in developing countries.

Future Energy Portfolios

Future patterns of energy use will depend on the size of reserves and rate of extraction, market prices, and environmental impacts.

QUESTIONS

1. Differentiate between energy conversion efficiency and energy end-use efficiency.

2. Contrast the primary energy sources used for transportation and those used for electricity generation in the United States.

3. Explain how energy experts estimate the number of years that a nonrenewable resource is likely to be available.

There are growing concerns over the long-term availability of some of the fossil fuels that dominate our energy economy. The **proven reserves** of a nonrenewable energy source are the quantities of fuel that could be recovered from known deposits using current technology at current prices. The **reserves-to-production ratio (R/P)** is the proven reserves figure for a given fuel divided by a particular year's level of production or use. The reserves-to-production ratio provides an estimate of how many years a fuel will last if the level of production remains constant and no additional reserves are discovered. For example, proven reserves of oil are currently estimated at 1.2 trillion barrels, and current rates of production are about 31 billion barrels a year. Therefore the reserves-to-production ratio for oil is about 40. In other words, at current rates of production, we will exhaust the current reserves of oil in 40 years. The current reserves-to-production ratio for natural gas is about 67 and for coal it is close to 120.

In reality, we do not really know when we will "run out" of oil, natural gas, or coal. Estimates of proven reserves are constantly changing as new deposits are discovered and new technologies are developed to extract more fuel from deposits already in production. In fact, it is unlikely that we will ever totally exhaust the supply of these primary energy

sources. Before the last barrel of oil is pumped or the last lump of coal dug out of the ground, it will cease to be economical to exploit that source of energy.

The **energy return on investment (EROI)** is a measure of the useful energy produced from an oil well, natural gas field, or coal mine divided by the amount of energy it took to obtain it. For example, reserves of oil that can easily be extracted from the ground have a high EROI. As more and more oil is pumped from the ground, the EROI declines, since the remaining reserves of oil are likely to be in more remote places and will be more difficult to extract.

Environmental constraints are providing strong incentives to reorder the relative proportions of renewable and nonrenewable energy sources that we use. Discovering, extracting, transporting, refining, and consuming nonrenewable energy sources create a range of environmental problems. The burning of fossil fuels is the single most significant contributor to air pollution and production of the greenhouse gases that cause global warming (8.3). The use of nuclear energy relies on mining activities that damage the land, and it requires the long-term storage of highly radioactive spent fuel. The environmental impacts associated with the use of nonrenewable energy resources present external costs to society, costs that are not easily captured in the economic markets for these resources.

15.2 Electric Power—Generation, Distribution, and Use

Of all the forms of energy we use in our lives, none is more ubiquitous than electricity. Electrical energy is the energy of flowing charged particles. Electricity is a secondary energy source that is derived from primary energy sources, such as coal, wind, or flowing water. Electricity may also be generated by chemical reactions in batteries. Most electric power is distributed to end users through complex transmission networks. The environmental impacts of electric power are largely associated with the primary energy sources used to generate it, its transmission, and its distribution. The chemicals in batteries are also important pollutants.

Generating Electricity

Most of the electricity we use is generated by the kinetic energy that comes from primary energy sources.

An electric current is the flow of electrically charged particles, usually electrons flowing through a wire. Current is measured in **amperes**, or amps. The current is the amount of charge that passes through the wire in a certain amount of time. The charged particles in an electric current flow between terminals that have opposite electrical charges. A flowing current has the capacity to do work. The amount of work depends on the difference in electrical potential energy between the two terminals. Electrical potential is measured in **volts**. The greater the difference in electrical potential between the terminals, the greater the voltage of the current.

The flow of electricity through a wire is similar to the flow of water in a river. The total amount of water flowing in the river is the river's current. The difference in elevation between the river's source and its outlet to the sea determines the potential energy of the river.

The rate at which energy is converted from one form to another is power. Electric power is the rate at which electricity is generated or the rate at which electrical energy is transformed into another form, such as heat or the kinetic energy of an appliance. Electric power is measured in **kilowatt hours** (kWh) and is calculated by multiplying electrical potential (volts) by current (amperes). Power (kWh) = voltage (volts) × current (amperes).

Most electricity is generated in power plants. First, primary energy sources, such as coal or natural gas, are used to heat water and produce pressurized steam. This steam is then used to turn a turbine that drives an **electric generator**, or dynamo. Alternatively, water or wind may move the turbine that drives a generator. Spinning magnets in the generator produce an **electromagnetic field,** or area of electromagnetic force. The electromagnetic field provides the voltage that causes an electric current to flow (**Figure 15.8**).

Most power plant generators operate at low voltage. Power plant **transformers** reduce the amperage and increase the voltage so that electricity can move efficiently over long distances through transmission lines. The voltage in major transmission lines is high, 230,000–760,000 volts. At distribution points, transformers convert

▼ Figure 15.8 **Steam Turbine**
Steam turbines are attached to electrical generators. Turbines are turned by pressurized steam that can be generated from a number of primary energy sources.

high-voltage electricity back to lower voltage electricity. The local overhead transmission lines that deliver electricity to your home typically carry 20,000–30,000 volts. Transformers near your home reduce that voltage for end use. In the United States, the electric circuits in a home typically operate at 110–120 volts.

Multiple power plants, perhaps drawing on a variety of primary energy sources, supply electricity to a network of transmission lines and transformers called an **electric power grid** (**Figure 15.9**). Electric power grids can be quite large. The ever-present power lines that surround us are part of an electric power grid that may extend over tens of thousands of square miles.

All of the components of an electric power grid create some resistance to the flow of electricity, much as friction with the walls of a pipe retards the flow of water through it. This resistance converts some of the electrical energy into heat, which is then lost to the system. The total energy loss associated with the transmission of energy from a power plant to its end use is about 7–10%.

In natural ecosystems, the flow of energy is regulated by feedbacks among the various ecosystem components, such as predators and their prey. The flow of energy in electrical power grids is also regulated by feedbacks, although these feedbacks are often incomplete and not nearly as efficient as their natural counterparts. For example, the flow of electricity through a grid is driven by consumer demand. The more electricity used by consumers, the more electricity the power plants on the

grid must generate. If total demand exceeds production, the system can fail, producing brownouts or blackouts.

Base load is the minimum amount of electricity that a utility must provide to meet consumer needs. The demand for electricity never drops below this level. **Peak load** is the maximum amount of electricity demanded by consumers. Peak load is often several times greater than base load, and the variety of power plants on the grid must be able to adjust quickly to this difference in demand. For example, in the middle of winter the demand for electricity is usually highest on cold nights, when the use of heating and lighting increases consumption. The problem of matching supply with demand is compounded by the fact that peak output from some renewable energy sources, such as wind, solar, and tides, may not coincide with periods of peak consumer use.

In general, the cost of generating electricity increases as demand increases from base load to peak load. Nevertheless, most residential and commercial consumers pay a single price for their electricity based on the total energy used over a month as measured by an electric meter. These consumers have neither the knowledge nor the incentive to adjust their energy use to these varying energy costs. In contrast, many industrial consumers negotiate variable rates with utilities. They are able to meter their electrical use throughout the day and maximize use during periods when rates are low.

Most existing electric power grids were designed to distribute electricity from large, centralized generating

▼ Figure 15.9 The Grid
Electrical grids are composed of the transformers and transmission lines that distribute electricity from power plants to homes and businesses. Grids generally extend over several states and include multiple generating facilities.

facilities that derive their power from nonrenewable sources of energy, such as coal, natural gas, and nuclear energy. Facilities that use renewable energy sources, such as wind, solar, or wave energy, usually generate less power and are often widely separated rather than being centralized in one location. Redesigning electrical power grids to make efficient use of these widely separated power sources is an important challenge to the transition from nonrenewable to renewable sources of energy.

In addition to the electric power grid, we obtain a considerable amount of electrical energy from batteries. Batteries use chemical reactions to produce electrical potential. The alkaline battery in your flashlight, for example, has zinc powder at the negative terminal and magnesium oxide at the positive terminal, all soaked in an alkaline solution (**Figure 15.10**). A car battery takes advantage of the electrical potential between plates of lead and lead oxide that are immersed in sulfuric acid.

As a battery is used, the strength of its electrical potential diminishes. In disposable batteries, this potential cannot be reestablished. In other kinds of batteries, such as those in a car or cell phone, the electric potential can be restored by using the electric current from a charger.

Positive

Zinc powder

Graphite rod

Magnesium oxide

Negative

◀ **Figure 15.10 Stored Electrical Energy** Alkaline batteries use zinc powder and magnesium oxide to generate electrical potential.

SEEING SOLUTIONS: SmartGridCity

Can we improve the feedback between the producers and consumers of electricity?

Do you know how much electricity you used today? You probably don't, and neither does the utility that provides your electricity!

There is widespread agreement that the existing system of demand-driven generation and distribution of electricity is wasteful and inefficient and that the root of this problem is poor communication. Consumers communicate to the utility that provides their electricity once a month, when their meter is read. Utilities communicate back to consumers once each month through electric bills. Moment to moment, individual consumers have little sense of how much energy they are using or at what cost. Similarly, utilities have no way of knowing the moment-to-moment use of energy by individual consumers. If consumers knew when they were using the most energy or what activities required the most energy, they could change their habits.

Boulder, Colorado's, new SmartGridCity project aims to remedy this communication problem. In late 2009, Xcel Energy, Inc., finished the infrastructure and software development needed to provide Boulder with the world's first full-scale smart grid. Xcel's goal is to adjust the price that consumers pay for electricity every 5–10 minutes, according to demand. Customers will pay more during periods that are equivalent to grid "rush hours."

For this real-time pricing to work, customers must be able to see the price changes. To meet that need, homes will be outfitted with a small device that is in constant wireless communication with the grid. The device signals green when power on the grid is abundant and cheap. It signals red during spikes in demand, when power is more expensive. Consumers will also have the option of receiving this information on their cell phones via text messaging.

For consumers who prefer more personalized control, Xcel is providing metering systems that allow certain appliances to be controlled remotely (**Figure 15.11**). Consumers will go to the utility's website and set goals for energy use. For example, they will be able to set conditions under which the utility might turn their heater up or down, or turn their dishwasher or clothes dryer on. They might decide to turn off the hot water heater during daytime hours when no one is home.

Xcel and the city of Boulder expect to see immediate benefits in energy savings. They also recognize that this technology opens the door to other important changes. SmartGridCity technology will allow individual homes to become power plants in their own right. For example, consumers who have home solar panels could sell their excess solar power back to Xcel. In the future, homeowners may be able to use batteries to store bargain electricity purchased during slack hours for use at peak times. Systems such as this will certainly provide much more efficient feedback between energy production and energy demand.

▶ **Figure 15.11 A Meter That Communicates** An electric meter such as this allows Xcel, Boulder's electricity provider, to adjust usage of the homeowner's heating and air-conditioning systems and appliances to lower costs and improve efficiency. The homeowner provides guidelines for these adjustments through Xcel's website.

QUESTIONS

1. What do volts and amperes measure with regard to the flow of electrical energy?

2. Describe two problems with the existing electric power grid that reduce the efficiency of electricity use.

3. Describe two ways that the current system of distributing electricity impacts the environment.

Environmental Impacts

Transmission of electricity has uncertain health effects.

The most significant environmental impacts of using electric power are associated with the primary energy sources used to generate electricity. The transmission and distribution of electricity also have important environmental consequences.

The electric currents flowing through transmission lines create electromagnetic fields around the lines. There has been considerable debate over the possible effects of those electromagnetic fields on human health. Some scientific evidence suggests that children who live near power lines may have a higher frequency of leukemia. Other alleged health effects have, however, been hard to substantiate. In any case, no cause-and-effect relationship has been established between electromagnetic fields and specific ailments.

One of the most important environmental impacts of electric power transmission is associated with the transformers used to modify line voltage. The electrical wiring within a transformer is immersed in liquid coolants. Until recently, these coolants often contained chemicals called polychlorinated biphenyls (PCBs). Although effective as coolants, PCBs have a wide range

of toxic effects on wildlife and humans. Today the use of PCBs is banned, but in the past, large quantities of these chemicals were released into the environment. Because they are remarkably resistant to decomposition, PCBs remain a threat to ecosystem health in many places (**Figure 15.12**).

The use of batteries also affects the environment. Most of the chemicals used in batteries are potentially toxic, and their disposal is a significant source of water and soil pollution. Although batteries make up less than 1% of an average city's solid waste, they contribute a much larger share of its chemical pollution. For example, the nickel-cadmium batteries commonly used in electronic games and cell phones account for 75% of the cadmium found in landfills. Discarded batteries are also a major source of mercury and lead pollution.

In summary, nearly half of all energy end uses take advantage of electricity. Current systems of generating and distributing electricity rely on large power plants and complicated electric power grids. This results in the inefficient use of electric power and constrains efforts to limit environmental impacts.

▼ **Figure 15.12 PCB Contamination**
Wastewater from a General Electric Corporation facility that manufactures electrical transformers has contaminated the sediments over large stretches of the Hudson River with PCBs. Pollutants are no longer being released, but contaminants will remain in the sediment for hundreds of years.

15.3 Coal

Humans have been using coal longer than any other fossil fuel. Archeological evidence shows that coal was being used in England as far back as the Bronze Age, 5,000 years ago. Energy from coal inaugurated the Industrial Revolution in the mid-1700s, and today it meets 28% of global energy needs. Coal is the most abundant of all fossil fuels, and supplies appear to be sufficient to meet global demands for at least a century. Today, most coal production is used to fire power plants that generate electricity. The use of coal is associated with a variety of environmental problems. Coal mining alters the landscape and pollutes water; the combustion of coal emits air pollution and greenhouse gases. Technological advances are already reducing some of these impacts and have the potential to diminish others.

Sources and Amounts

Coal was formed from plants that lived in swamps hundreds of millions of years ago.

Most of Earth's coal began to form between 300 and 400 million years ago, when large swampy forests covered much of the land. Over millions of years, large volumes of plant material fell into these swamps, where it was covered by sediment and water before it could decay. At first, this organic material was broken down anaerobically into a wet, partially decomposed mixture material called **peat**. Over time, thick layers of sediment accumulated on top of the peat. The higher temperature and pressure caused by the weight of these sediments forced much of the water out of the peat and packed the carbon compounds from the original plants closer together, eventually forming coal. Today, the layers of sediment that contain coal are called **coal seams**.

The carbon and energy content of coal deposits vary, depending on their age and the extent to which they were subjected to underground heat and pressure. **Lignite** is a type of coal that is generally associated with younger deposits that have not been subjected to as much heat and pressure as other types of coal. As a result, lignite is comparatively soft and contains considerable moisture. Lignite contains 25–35% carbon; when it is burned, it yields 3.0–4.5 kWh of energy per kilogram (3.0–4.5 kWh/kg). **Sub-bituminous coal** contains 35–45% carbon; its energy content is, therefore, somewhat higher than lignite.

Bituminous coal is much denser than either sub-bituminous coal or lignite because it was formed under higher heat and pressure. Bituminous coal has a carbon content of 45–86% and yields 5–8 kWh/kg when it is burned. **Anthracite coal**, which is dark and shiny, has the highest energy content of all types of coal. It has a carbon content of 86–97% and yields more than 8 kWh/kg when burned. Anthracite is much less common than bituminous coal and is usually found only at great depths (**Figure 15.13**).

Bituminous coal and anthracite coal are often called "hard coal." Because hard coal has higher energy content and contains fewer polluting chemicals, it is more widely used as a primary energy source than either lignite or sub-bituminous coal.

Worldwide, the proven reserves of coal equal nearly 850 billion metric tons. About 54% of this is anthracite

Lignite
3.0–4.5 kWh/kg

Bituminous
5–8 kWh/kg

Anthracite
8 or more kWh/kg

▲ **Figure 15.13 Types of Coal**
Energy content per unit weight is highest in anthracite coal, less in bituminous, and lowest in lignite.

and bituminous coal; the remainder is sub-bituminous coal and lignite. In 2010, approximately 7 billion metric tons of coal were mined and used, giving a reserve-to-production ratio of 120. Even with coal production increasing at a rate of 1.5–2.0% each year, most experts agree there is enough coal to last for at least another 120 years.

The geological conditions under which coal formed were geographically widespread, so coal deposits are widely distributed. Nevertheless, 67% of the world's recoverable coal reserves are located in just four countries—the United States, Russia, China, and India (**Figure 15.14**).

▶ **Figure 15.14 World Coal Reserves**
Coal is relatively abundant in North America, Europe, and Asia, although much of it is low-quality sub-bituminous and lignite.
Data from: British Petroleum, 2010. *Statistical review of world energy June 2010.* URL www.bp.com/statistical review, p. 32. Last accessed 8-9-11.

Coal Extraction and Processing

Shaft mines and strip mines provide access to coal seams.

In what year will the supply of coal be depleted?

Ciara Tyce, Georgia Southern University

Norm: *At current rates of use, coal supplies could last more than 120 years. But can Earth's ecosystems tolerate the effects of burning that much coal?*

Coal is removed from sediments by underground and surface mining. Underground mining is used to extract coal from sediment layers that are deeper than 200 ft. The deepest coal mines are well over 1,000 ft below the surface. Mine operators dig multiple shafts into the ground to reach the coal seam (**Figure 15.15**).

Surface mining, also called strip mining, is used to extract coal from deposits that are less than 200 ft underground. Giant earth-moving machines remove topsoil and layers of rock, known as overburden, to expose coal seams (**Figure 15.16**). Mountaintop removal is a particularly controversial form of surface mining. In this form of mining, the entire top portion of a mountain is removed to expose a coal seam, and the overburden is dumped into surrounding valleys. This mining technique has been widely used in the Appalachian Mountains of West Virginia and Kentucky.

Once coal is removed from the ground, it usually undergoes some form of cleaning to remove dirt, rock, sulfur, and other impurities. It may then be transported to coal-burning power plants and other end users by train, barge, or truck. Alternatively, it may be mixed with water and sent by pipeline.

Globally, 67% of coal production is used to generate electricity, whereas 30% goes to industrial end uses, mostly steel production. In the United States, more than 90% of coal production is used to generate electricity. This is a considerable change from 50 years ago, when more than half of U.S. coal production was used to heat homes and for industrial purposes.

The chemical energy in coal is transformed into electricity in coal-fired power plants. There coal is burned to produce heat, which is captured in a boiler and used to produce pressurized steam. This steam turns a turbine, which then powers an electrical generator (**Figure 15.17**). After passing through the turbine, the steam condenses into liquid water; its remaining heat is dissipated to the air in cooling towers or to the water in a river or lake.

Exhaust from burning coal is passed through a **scrubber**, a device designed to remove particulates and some pollutant gases. The exhaust then rises through a smoke stack and is released into the atmosphere. The solid material left behind is called **fly ash**.

In the oldest coal-fired power plants, energy conversion efficiency may be less than 25%. This means that only a quarter of the chemical energy in coal is converted into electrical energy. In contrast, most modern coal-fired facilities use more efficient combustion and heat-capturing technologies, which allow them to operate at efficiencies of more than 40%.

▼ **Figure 15.15 A Modern Coal Mine**
Underground coal mines employ an array of shafts and tunnels. Some of these are used to transport miners and equipment into and out of the mine, and others are used to ventilate the mine and haul the coal out.

▼ **Figure 15.16 Open Pit Mine**
This open pit coal mine is in the Kemerovo region of Russia.

Hot pressurized steam

Boiler

Generator turbines

Electric power

Cooling water

"Scrubbers" remove NO_x and SO_x

Coal supply

Firing system

Coal hopper

▲ Figure 15.17 **Coal-Fired Power Plant**
In a modern coal-fired facility, coal is burned at high temperature to produce pressurized steam, which turns turbines attached to a generator. Scrubbers remove pollutants such as SO_x and NO_x from the exhaust, which is then released from the smoke stack. The aerial photo shows a modern lignite-fired power plant in Grevenbroich, Westphalia, Germany.

Environmental Impacts

Coal mining and coal combustion present significant environmental challenges.

From its extraction to its various end uses, coal presents more environmental challenges than any other energy source. Underground coal mining is a very hazardous occupation due to the potential for cave-ins, flooding, dust, and gas explosions. In 2010, mining accidents in the United States killed 48 coal miners, and accidents in China killed more than 2,400 miners. Even without accidents, thousands of underground coal miners suffer from respiratory diseases caused by inhaling coal dust.

Mining activities also have a direct effect on the environment. **Mine tailings**, the rock and debris from mining operations, often contain high concentrations of sulfide. When exposed to oxygen, these sulfides are transformed to sulfuric acid, which runs off into nearby streams, where it harms fish and other aquatic organisms (**Figure 15.18**).

Surface mining destroys the terrestrial ecosystems above the coal seams. Once the coal is removed, however, the overburden may be returned to the pit. The overburden can then be covered by a layer of topsoil and the area revegetated to prevent erosion (**Figure 15.19**). The coal-mining industry in the United States has received especially harsh criticism for the environmental impact of mountaintop removal. In West

Virginia and Kentucky, overburden from mountaintop removal has permanently buried more than 700 miles of mountain streams, thereby affecting wildlife, flooding nearby communities, and degrading water quality far downstream (**Figure 15.20**, on page 464).

The exhaust and fly ash from coal fires contain a number of toxic chemicals that are harmful to many organisms, including humans. Fine particulate soot from coal fires can cause respiratory distress. In addition, coal usually contains mercury, which accumulated in the wetlands where it was formed. When coal is burned, this mercury is released into the atmosphere. Concentrations of mercury in streams and lakes near coal-fired power plants are often several times higher than normal. Eventually, this mercury accumulates in the tissues of animals, including

▲ Figure 15.19 **Mine Restoration**
This former open pit mine in Wyoming is in the early stages of surface reclamation.

▲ Figure 15.18 **Acid Mine Drainage**
Sulfides leached from coal mine tailings pollute a nearby stream.

QUESTIONS

1. Compare the four types of coal with regard to their carbon content and energy yield when burned.
2. What are the environmental impacts of the most commonly used methods of coal mining?
3. Describe three environmental challenges caused by the combustion of coal.

humans, where it can cause a number of serious health problems, including neurological disorders. Fly ash also contains high concentrations of mercury, which makes it complicated to dispose of.

Coal contains significant quantities of organic nitrogen and sulfur, which are converted to nitrogen oxides (NO_x) and sulfur oxides (SO_x) when coal is burned. In rainwater, nitrogen oxides are converted to nitric acid, and sulfur oxides are converted to sulfuric acid (9.2).

New technologies can reduce the amounts of pollutants produced by coal-powered plants. Improved methods of cleaning coal before it is burned reduce the emissions of many pollutants. In many modern plants, sophisticated scrubbers remove significant amounts of NO_x, SO_x, and mercury from the exhaust. However, the efficiency of such scrubbers varies from pollutant to pollutant. For example, scrubbers can reduce SO_x emissions by over 90% but typically reduce mercury emissions by only 30–40%.

The burning of coal is a significant contributor to global warming. For every unit of energy production, the combustion of coal releases 80% more carbon dioxide (CO_2) than is released by the combustion of natural gas and 30% more than is released by the combustion of oil (8.2). A typical 1,000-megawatt coal-fired power plant produces 6 million metric tons of CO_2 a year. This is equivalent to the annual CO_2 emissions from 2 million cars.

By 2030, the worldwide consumption of coal is expected to double, largely to meet growing demands for electricity in developing countries. China alone is currently commissioning two to three new large coal-fired power plants each month. The coal-fired power plants that are slated for construction between now and 2030 are expected to emit as much CO_2 as all the coal consumed in the world since the beginning of the Industrial Revolution to the present. Such projections have caused great

interest in the development of alternative energy sources to offset coal consumption.

Carbon capture and storage (CCS) technologies may provide a means to reduce CO_2 emissions from the combustion of coal and other fossil fuels. Instead of releasing the CO_2 from combustion into the atmosphere, CCS facilities capture and store CO_2 in underground geologic formations (8.6). Several CCS pilot projects are underway. If they are successful, it may be possible to exploit remaining coal reserves while diminishing emissions of greenhouse gases.

However, critics point out that CCS technologies add considerable cost to coal-produced energy. They also note that there is considerable uncertainty about the ability of geologic formations to hold CO_2 and prevent eventual leakage to the atmosphere.

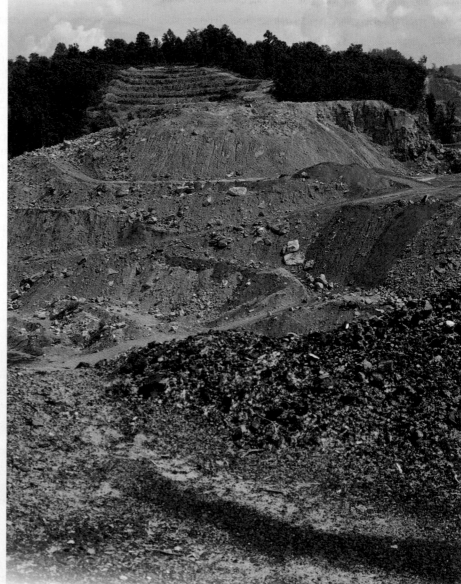

▶ **Figure 15.20 Moving Mountains** In mountaintop removal coal mining, the upper portion of a mountain is removed to expose the coal seam. The overburden is then deposited in surrounding stream valleys.

15.4 Oil

The first commercial oil well was drilled in Titusville, Pennsylvania, in 1859. Within 100 years, oil had surpassed coal as the leading source of energy in the world economy. Today, oil accounts for nearly 35% of global energy use. However, oil reserves were created under very specific conditions, so they are unevenly distributed around the world. Experts disagree about the exact amount of oil left in these reserves but are in complete agreement that supplies are dwindling as demand is growing. The majority of this growth is associated with economic development and increased demand from the transportation sector. There are growing concerns over the environmental impacts of oil use, as well as with the effects of its supply on international security and conflict. Despite these concerns, it appears that our reliance on oil will remain high for some time due to the lack of affordable and reliable alternatives.

Sources and Amounts

Oil formed under specific geologic conditions that were common in only a few places.

Like coal, oil is a fossil fuel. Crude oil, or petroleum, is formed from the remains of microorganisms that lived in shallow seas and coastal swamps hundreds of millions of years ago. These areas were characterized by abundant algae and zooplankton and very high rates of net ecosystem production (7.4). As these microorganisms died, they sank to the bottom, forming thick layers of organic matter. Over the course of tens of thousands of years, this organic matter was buried under layer after layer of sediment, where it was cut off from oxygen and subjected to increasing amounts of heat and pressure. These conditions transformed the organic matter into

kerogen, a waxy substance that is a precursor to oil and natural gas. Underground at depths of 2–3 miles and at temperatures above 200 °F, kerogen became crude oil.

The newly formed crude oil was less dense than the water in the ground, so it tended to migrate upward through porous sediments. In some places, oil reached the surface, where it seeped out into tar pits. In other places, oil became trapped beneath a layer of impermeable rock, forming **oil reservoirs (Figure 15.21)**. Natural gas, which is discussed in the next module, is often trapped in these reservoirs. Oil reservoirs are not underground lakes of oil. Rather, reservoirs are composed

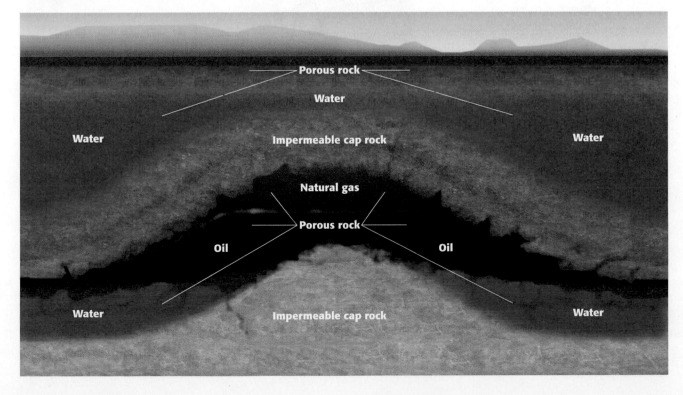

◄ **Figure 15.21 Oil Reservoirs**
Oil reservoirs formed in porous rocks where there was enough organic matter to produce oil and an impermeable rock cap prevented its upward movement. Natural gas, which is lighter than oil, was trapped above the oil.

Porous rock

Water

Water

Impermeable cap rock

Water

Natural gas

Porous rock

Oil

Oil

Water

Impermeable cap rock

Water

When are we going to run out of gas?

Alexcis Zare, University of Houston

Norm: *We cannot set an exact time when gasoline will no longer be available, but there is no doubt that the supply is rapidly diminishing while the demand is increasing.*

of permeable rocks with numerous pores and fractures that hold small drops of oil, similar to the way in which a sponge holds water.

For an oil reservoir to form, three geological conditions must be met. First, there must have been a shallow sea with large quantities of aquatic organisms that died and sank to the bottom, were covered in sediments, and were eventually converted to oil by subterranean heat and pressure. Second, the oil had to migrate upward into a porous rock reservoir. Third, the reservoir rock must have been covered by a layer of impermeable rock, or a cap rock, that prevented the oil from seeping to the surface.

The geological conditions necessary for oil formation are not evenly distributed on Earth. As a consequence, oil reservoirs are concentrated in certain geographical regions. Today, close to two-thirds of known oil reservoirs are located in just five Middle Eastern countries—Saudi Arabia, Iran, Iraq, Kuwait, and the United Arab Emirates. The United States possesses less than 2% of Earth's known oil reserves (**Figure 15.22**).

Like all nonrenewable resources, the world's supply of oil is finite. Over the past century, global oil consumption has totaled about 1.1 trillion barrels of oil. Proven oil reserves that can be recovered using available technologies at current world oil prices are estimated to equal about 1.2 trillion barrels. Worldwide, about 31 billion barrels of

oil are produced each year. At this rate of production, the reserve-to-production ratio is 38.7, which suggests that oil reserves will be depleted in less than 40 years. However, the global demand for oil is increasing so rapidly that global consumption could exceed 100 billion barrels per year in less than 20 years. At that rate of consumption, the lifetime of known reserves would be much shorter.

Peak oil is a term used to describe the point of maximum production in an oil field; beyond that point, production from the field will only decline. The peak oil concept can also be applied to oil production at the national scale. Oil production in the United States peaked around 1970.

Some energy experts argue that we are near the peak of global oil production (**Figure 15.23**). Whether or not global oil production is near its peak depends on how much oil remains in the ground, an area of much uncertainty. Some analysts predict that continued exploration and the development of new technologies will cause proven reserves to increase to as much as 2 or even 3 trillion barrels. For example, new technologies and higher prices may facilitate access to oil in geological deposits that are not currently being exploited on a large scale. Such deposits include the oil shale in the Rocky Mountains and the Dakotas, and the tar sands in Canada and Venezuela; both the oil shale and tar sands contain significant amounts of low-grade petroleum.

◄ Figure 15.22 World Oil Reserves
The world's proven reserves of oil are concentrated in Middle Eastern countries, including Saudi Arabia, United Arab Emirates, Kuwait, and Iraq.
Data from: British Petroleum, 2010. *Statistical review of world energy June 2010.* URL www.bp.com/ statistical review, p. 6. Last accessed 8-9-11.

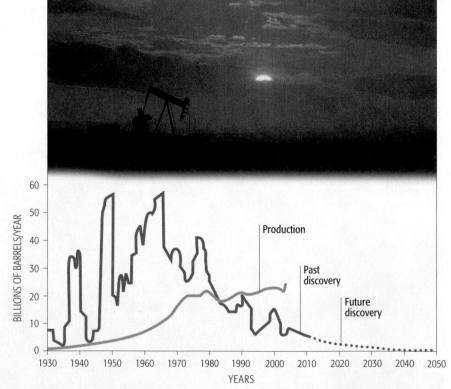

▲ Figure 15.23 Peak Oil?
Until about 1975, the rate of new oil discovery kept pace with the rate of new production. Since then, new oil discovery has declined while oil production and consumption have continued to increase.
Source: Longwell, H.J. 2002. The future of the oil and gas industry: Past approaches, new challenges. *World Energy* 5:100-104

Oil Extraction and Processing

Crude oil is pumped from underground reservoirs and sent to oil refineries to be converted into gasoline and other petroleum products.

When an oil company locates an oil reservoir, it begins drilling oil wells. At first, there is usually enough built-up pressure in the reservoir to push crude oil up to the surface. This is called **primary oil recovery**, and it can usually extract about 20% of the oil in a reservoir. After that, oil companies use secondary recovery methods to extract oil, such as injecting water into a reservoir to increase pressure. **Secondary oil recovery** methods can extract another 10–20% of the oil reservoir. Historically, oil wells were capped once primary and secondary methods had extracted as much oil as they could. Even though 60–70% of a reservoir's original oil remained, the cost of retrieving it was too high.

Nowadays, most of the world's easily accessible oil reservoirs have been tapped. As a result, oil companies are often resorting to tertiary recovery methods to extract another 10–20% of oil from depleted reservoirs. **Tertiary oil recovery** methods stimulate the flow of additional oil by injecting CO_2, steam, or hot water into reservoirs (**Figure 15.24**).

The cost and amount of energy required to extract oil depends upon the method of extraction. For primary oil recovery, energy return on investment (EROI) is about 25, meaning that 25 units of energy are obtained for each unit of energy used to extract it. For secondary oil recovery, EROI is 8–12, and for tertiary oil recovery, EROI may be as low as 2.

Some rocks, such as oil shale, are not porous enough to allow sufficient amounts of oil to flow into wells. In recent years, engineers have extracted oil from these deposits by using high-pressure liquids to fracture the rocks. This process, called **hydraulic fracturing**, or fracking, has a very low EROI. Furthermore, the chemicals in the fracking fluids can pollute aquifers.

After crude oil is brought to the surface, it is sent through pipelines or loaded onto oil tankers and shipped to refineries, where it is converted into a variety of useful petroleum products. Crude oil is made up of hundreds of different types of hydrocarbon molecules, each with a different boiling point. To separate crude oil into different products, oil refineries heat it in large distillation columns, also called *cracking columns* (**Figure 15.25**). Lighter hydrocarbon molecules, which have relatively low boiling points, evaporate and rise as vapor to the upper portions of the distillation column, where they cool and condense into a liquid. These lighter hydrocarbons are used for jet fuel and gasoline. Heavier hydrocarbon molecules, which have higher boiling points, condense near the bottom of the distillation column. These heavier hydrocarbons are used for fuel oil, lubricating oil, and asphalt.

About 15–16% of oil production is used for nonenergy products, such as solvents, fertilizers, pesticides, and plastics. The remainder is converted into products that are burned as sources of energy.

About 70% of every barrel of crude oil sent to a refinery is converted into gasoline or diesel fuel to power cars and trucks. The chemical energy in these fuels is converted to kinetic energy by **internal combustion engines**. In such engines, fuel is explosively ignited in a closed cylinder, driving a piston that turns a crankshaft. In gasoline engines, an electric spark ignites the fuel. In diesel engines, high pressure ignites the fuel.

Aviation fuel accounts for 7% of all oil production. Most of this fuel is used to power commercial and military jet airplanes. Jet engines are a form of internal combustion engine in which ignited fuel is used to create the high-pressure stream of air that provides thrust.

Forty years ago, residential and commercial heating consumed nearly 15% of total oil production. Since then, there has been a significant increase in the amount of electricity and natural gas used for heating. Today, the kerosene and oil used to heat homes and offices account for less than 7% of oil production.

Crude oil

20 °C — Petroleum gas
150 °C — Gasoline (petrol)
200 °C — Kerosene
300 °C — Diesel
370 °C — Industrial fuel oil
400 °C — Lubricating oil, parafin, and asphalt

FURNACE

▲ **Figure 15.25**
Refining Oil
In oil refinery cracking towers, the components of crude oil, such as diesel fuel and gasoline, are separated by their different boiling points.

Injection well   Storage   Production well

❶ CO_2 is injected along with steam

❷ Oil flow is stimulated

◄ **Figure 15.24 Tertiary Oil Recovery**
By injecting CO_2 into depleted oil reservoirs, additional oil can be extracted. This has the additional benefit of removing CO_2 from the atmosphere.

QUESTIONS

1. Describe the process by which oil reservoirs form.

2. Differentiate among primary, secondary, and tertiary methods of oil recovery.

3. Describe two important environmental consequences of the extraction and combustion of oil.

Environmental Impacts

Transportation of oil and combustion of the products derived from it have significant environmental consequences.

Locating and drilling for oil present risks to habitat and wildlife in sensitive environments. In 2008, a large area of Amazon rain forest bridging the countries of Colombia, Ecuador, Peru, and Brazil was approved for oil exploration. Conservation scientists fear that roads, pipelines, and associated activities will threaten biodiversity across this area.

Since 1977, there has been ongoing debate about drilling for oil in the Arctic National Wildlife Refuge (ANWR) on Alaska's northern coastal plain. A U.S. Department of Energy Report suggests that the sediments beneath ANWR's tundra might yield as much as 876,000 barrels a day by 2025 if drilling were to begin immediately. Environmentalists argue that this represents a small portion of U.S. oil demand and that the threats to the refuge's fragile ecosystems from a possible oil spill are too great relative to this potential benefit (**Figure 15.26**). This controversy is likely to continue in the years ahead.

Many oil reserves are located in coastal waters. Drilling for these reserves puts ocean ecosystems at risk. In 2001, the world's largest floating oil-drilling rig suffered an explosion and sank in waters about 150 km (93 mi) from the coast of Brazil. Although little crude oil escaped into the environment, more than 1 million liters (265,000 gallons) of diesel fuel used to power the rig were released into surrounding waters, killing thousands of fish and seabirds.

In April 2010, an explosion and fire on BP's Deepwater Horizon drilling platform killed 11 workers and began spewing oil into the waters of the Gulf of Mexico. In the months that followed, nearly 5 million barrels (270 million gallons) of crude oil were released, severely affecting the fisheries of the Gulf, as well as its beaches and coastal wetlands (**Figure 15.27**).

Major spills associated with oil transportation have done serious damage to wildlife and important marine fisheries. The 1988 grounding of the oil supertanker *Exxon Valdez* spilled over 40 million liters (10.6 million gallons) of crude oil into Prince William Sound, Alaska. The impacts on coastal and marine ecosystems were

▲ Figure 15.27 Infamous Oil Spill
The 2010 Deepwater Horizon disaster produced the largest marine oil spill in the history of the petroleum industry. Here, oil is washing onto a sea island off the coast of Louisiana.

immediate and tragic; plant and animal communities still have not recovered fully.

The emissions from gasoline and diesel-powered automobiles and trucks are a major cause of air pollution. The exhaust of automobiles contains toxic pollutants, such as carbon monoxide, hydrocarbons, and nitrogen oxides (NO_x). Most diesel engines also release hydrocarbon particulates into the atmosphere. Along heavily used transportation corridors, NO_x interacting with oxygen produces high levels of ozone (O_3) in the lower atmosphere.

The combustion of oil-derived fuels accounts for 31% of all human-caused greenhouse gas emissions. Currently, there are about 600 million automobiles and trucks in the world. With economic development and population growth, that number could double by 2030. However, higher gasoline prices and more stringent regulatory standards are creating incentives for the production of more fuel-efficient vehicles and vehicles powered by alternative energy sources. If these trends continue, some experts predict that greenhouse gas emissions from transportation could actually decrease.

More than 60% of world oil reserves are located in countries with political conditions that could constrain oil exploration and production, thereby threatening the global oil supply and international security. Such political conditions could also lead to actions that would have significant implications for the environment. For example, as Iraq retreated from Kuwait in 1991, it set afire hundreds of oil wells. These fires devastated the surrounding ecosystems and created enormous amounts of air pollution. Today, many drilling, transportation, and refining facilities are vulnerable to terrorist attacks. A single oil terminal in Saudi Arabia, for example, loads almost 10% of the world's oil production onto oil tankers every day. A successful attack on such a facility would have serious economic consequences and devastate many Persian Gulf ecosystems.

There is much discussion but little consensus about future trends in oil supply and demand. Nevertheless, the words of a former Saudi Arabian oil minister ring true: "The stone age did not end for lack of stone, and the oil age will likely end long before the world runs out of oil."

▼ Figure 15.26 The ANWR Controversy
Significant reserves of oil sit beneath this portion of the U.S. Fish and Wildlife Service Arctic National Wildlife Refuge on Alaska's northern coastal plain. Environmentalists fear that drilling for oil and the construction of an oil pipeline would have serious consequences for this fragile ecosystem.

FOCUS ON SCIENCE: Monitoring Pollution from Oil Spills

In November 2002, the oil tanker *Prestige* sank in heavy seas, releasing 60,000 metric tons of crude oil into the marine environment along Spain's northwest coast. There was no doubt whatsoever that this oil spill had serious immediate impacts on coastal ecosystems. Although much of the visible evidence of the spill was gone by the summer of 2004, scientists suspected that residual pollutants from the oil spill were continuing to affect these ecosystems.

Alberto Velando and his colleagues at the University of Vigo, in Vigo, Spain, hypothesized that chemical analysis of the red blood cells of seagulls might provide a means of monitoring the levels and distribution of oil spill pollutants in these ecosystems (**Figure 15.28**). They focused their attention on the concentrations of specific polycyclic aromatic hydrocarbons (PAHs) that are released as crude oil decomposes. These chemicals accumulate in animal tissue and have well-known toxic effects. Because gulls feed near the top of marine food webs, their tissues are likely to accumulate larger amounts of PAHs than animals at lower trophic levels (**Figure 15.29**).

Velando chose to examine PAH concentrations in red blood cells because these cells are produced constantly and have a life span of less than 4 weeks. Thus, PAH in a gull's red blood cells indicates that it has had relatively recent exposure to the pollutants. In addition, blood samples can be obtained without harming the gulls.

Gulls feed within 40 km (25 mi) of their nesting sites; therefore, any differences in their blood PAH levels reflect the environment in that geographic range. In 2004, Velando's team found that PAH concentrations in red blood cells of gulls nesting near oiled areas were more than 120% higher than those from gulls nesting outside the oil spill zone. Over the next two breeding seasons, PAH levels dropped by nearly a third.

To verify that differences in PAH concentrations were a direct consequence of oil pollution, separate field experiments were conducted outside the range of the oil spill. One set of gulls was given food that contained small amounts of oil, and a control group of gulls was given food that did not contain oil supplements.

Then blood samples were taken from each group of gulls. PAH levels were indeed higher in the treatment group than in the control group. As expected, PAHs generally decreased within 4–6 weeks of exposure, reflecting the death and replenishment of red blood cells.

The research by Velando's group provides evidence for the distribution and duration of oil pollution following this particular oil spill. More importantly, it suggests that nondestructive sampling of blood from seabirds such as gulls can provide an effective means of monitoring oil pollution.

Source: Perez, C., A. Velando, I. Munilla, M. Lopez-Alonso and D. Oro, 2008. Monitoring polycyclic aromatic hydrocarbon pollution in the marine environment after the Prestige oil spill by means of seabird blood analysis. *Environment Science & Technology* 42: 707–713.

How can tissues from seagulls be used to measure oil pollution?

▲ **Figure 15.28 Team Leader** Alberto Velando is a professor at Spain's Vigo University, where he studies the effects of pollution from oil spills on seabird populations.

▲ **Figure 15.29 Biomonitors** Levels of PAHs in seagull blood indicate whether birds have been feeding in areas polluted by spilled oil. This graph shows that blood PAH levels in gulls from colonies located near polluted sites were twice those found in gulls from colonies near unpolluted sites.

15.5 Natural Gas

For hundreds of years, people were aware of the potential of using natural gas as a source of energy, but challenges related to storing and transporting gas limited its use. Today, natural gas meets about 24% of the world's energy needs. Natural gas reserves were created under conditions very similar to those that created oil reserves. Also like oil, there are concerns about the long-term supply of natural gas. The greatest demand for natural gas is often in places with the least supply, so transportation is often a challenge. Power plants fueled by natural gas are providing a growing portion of the world's electricity because of their efficient energy conversion and lower emissions of pollutants and carbon dioxide. Even so, concerns remain regarding the plants' overall impact on greenhouse gas emissions and global warming.

Sources and Amounts

Natural gas was formed in much the same way as oil and is usually found in association with crude oil deposits.

Natural gas is a mixture of hydrocarbon gases composed mainly of methane (CH_4), with smaller amounts of other gases, such as propane. Natural gas is colorless and odorless. However, gas distributors add small amounts of sulfur-containing chemicals so that gas leaks can be easily detected by the smell.

Like oil, natural gas was formed from the remains of marine microorganisms buried in layers of sediments. Over time, their remains were converted into kerogen. In deep deposits, where kerogen was subjected to very high temperatures and pressures, it was converted into natural gas rather than oil.

Natural gas is usually associated with crude oil reservoirs (**Figure 15.30**). Because gas is less dense than crude oil, it tends to rise to the top of the reservoir and become trapped just below the cap rock that holds the oil in place. Natural gas may also be found in large quantities in coal seams and sedimentary rocks such as shale in the absence of oil.

In 2007, global proven reserves of natural gas were estimated to be 6,300 trillion cubic feet. Current world production of natural gas is about 97 trillion cubic feet per year. At this rate of consumption there is a 65-year supply (R/P = 6,300/97 = 65). However, if the rate of natural gas use continues to accelerate as it has over the past decade, by 2030 production will exceed 182 trillion cubic feet per year. As with oil production, many experts speculate that we may pass the peak of natural gas production within a decade or two.

Nearly 75% of known natural gas reserves are located in the Middle East, Russia, and southern Asia (**Figure 15.31**). The location of these reserves presents an important challenge because they are quite distant from the regions with the greatest demand for natural gas. Countries in North America currently consume close to 30% of the annual natural gas production but possess only about 4% of the global reserves. In contrast, the countries of the Middle East consume about 4% of total production but possess over 40% of the reserves.

◄ **Figure 15.30 Burning off Natural Gas**
Because it is comparatively light, natural gas accumulates in reservoirs above oil and beneath cap rocks. Often, the excess gas released from oil wells is simply burned.

PROVEN NATURAL GAS RESERVES (Trillions of m³)

Middle East · Russia · Africa · Asia and Australia · Europe · North America · South America

► **Figure 15.31 Natural Gas Reserves**
The world's largest natural gas reserves generally occur where oil is also abundant.
Data from: British Petroleum, 2010. *Statistical review of world energy June.* URL. 2010. www.bp.com/statistical review, p. 22. Last accessed 8-9-11.

Natural Gas Extraction and Processing

Natural gas is pumped from underground reservoirs, purified, and sent to consumers by pipeline.

In the past, it was not uncommon for oil companies to simply burn the natural gas that escaped in the process of drilling for oil. Today, oil companies recognize the value of this resource and capture it in pipelines. If a pipeline is not available, the gas is reinjected into the oil deposit. In Alaska, for example, oil well operators reinject the natural gas pumped from oil wells with the expectation that it will be reextracted when a natural gas pipeline connects the oil field to consumers. In addition, the reinjected gas helps to increase pressure in the oil reservoir, thereby improving the production of the oil field.

After natural gas is pumped from underground reservoirs, it is sent by pipeline to gas refineries, where impurities, such as water, sulfur, and other gases, are removed. The purified natural gas is then delivered to consumers through another set of pipelines.

Transporting natural gas across the ocean presents a significant challenge, since it is not possible to send it through pipelines. Instead, it is often converted into **liquefied natural gas (LNG)**. When natural gas is chilled to −160 °C (−260 °F), it changes into a liquid that takes up only 1/600th of the space that it occupies in its gaseous form. In this form, it becomes economical to use ocean tankers to carry natural gas to distant consumers (**Figure 15.32**). LNG production does, however, add to the EROI costs of using natural gas.

Globally, the largest consumers of natural gas are the industrial and electric power sectors. Industry consumes 44% of total natural gas production. Much of this is used in high-temperature boilers and furnaces, but some is used as the raw material, or feedstock, for the production of plastics, fertilizers, synthetic fiber, and other products. About 25% of natural gas production is used as a direct source of energy in households.

Currently, 31% of global natural gas production is being used to generate electricity, and that fraction is growing. Although their construction and operating costs are high compared to coal-fired power plants, natural gas–fired power plants release many fewer pollutants and no fly ash. Two main kinds of power plants are fueled by natural gas. **Gas-turbine power plants** burn natural gas to turn turbines that are directly connected to electrical generators. **Combined-cycle power plants** power their generators in two ways; they burn natural gas to power their turbines directly and also use steam boilers and steam turbines driven by the heat from the natural gas–powered turbines.

▼ Figure 15.32 **Shipping Liquefied Natural Gas** The tanks hold very cold liquefied natural gas (LNG), which is transported around the world in specially designed ships.

Environmental Impacts

The use of natural gas has fewer effects on ecosystems than the use of either coal or oil.

As with oil, hydraulic fracturing or fracking has allowed extraction of natural gas from rock formations where conventional technologies have been unproductive. High prices for natural gas have encouraged the use of these more costly technologies. But there are concerns among some scientists that the fluids used in fracking can contaminate aquifers. These fears are shared by many who live near drilling sites, along with concerns about noise and air pollution associated with drilling equipment. The natural gas industry counters that such risks are minimal and are more than balanced by the environmental benefits of natural gas. Research is underway to evaluate these risks more thoroughly.

Because it is a versatile and relatively clean fuel, and because it is abundant in North America, the demand for natural gas is likely to grow. Since natural gas is mostly methane and contains only traces of nitrogen and sulfur, power plants fired by natural gas release few toxic pollutants into the atmosphere. Of course, those power plants do release CO_2, but only about half as much as is produced by coal-fired plants that generate the same amount of electricity. Thus, nearly all plans to reduce greenhouse gas emissions include replacing coal-fired power plants with power plants fueled by natural gas.

Despite these desirable attributes, the leakage of methane in the production of natural gas is a persistent concern. Scientists estimate that the extraction and processing of natural gas account for about 15% of the methane that human activities add to the atmosphere each year. Because the global warming potential (GWP) of methane is 25 times higher than that of carbon dioxide, these methane emissions have the potential to cancel out the favorable effects of natural gas use on CO_2 emissions (8.3). Work is underway to reduce leakage at each stage of natural gas extraction and processing.

QUESTIONS

1. How does the formation of natural gas differ from that of oil?

2. Combined-cycle power plants have a higher energy conversion efficiency for producing electricity than simpler gas turbine power plants. Explain why.

3. What is hydraulic fracturing and why has it become an attractive technology for natural gas extraction?

15.6 Nuclear Power

The awesome power of nuclear energy was first demonstrated when the United States dropped atomic bombs on Japan near the end of World War II. Only a few years later, scientists were developing peaceful uses for nuclear energy, including electric power generation. The first commercial nuclear power plants were commissioned in the 1950s. In nuclear power plants, fission of an isotope of the element uranium is used to heat and pressurize steam, which then turns generator turbines. Although the activities associated with the use of nuclear power emit few greenhouse gases, mining activities and waste management have real and potential environmental impacts. Concerns about the safety of nuclear power plants have been heightened by accidents at reactors in Japan and the former Soviet Union. Nevertheless, the increasing costs of other forms of nonrenewable energy and concerns about global warming are increasing interest in the expanded use of nuclear power.

Sources and Amounts

Nuclear power takes advantage of heat generated by the fission of atoms.

Nuclear energy is energy contained in matter itself (3.3). In nuclear fission, the nucleus of an atom splits to form two or more smaller nuclei. In the process, a small amount of the nucleus's mass is converted to electromagnetic and kinetic energy. Nuclear fission occurs naturally when radioactive isotopes of particular elements decay. Indeed, it is this process of decay that heats Earth's interior.

Electrical power is generated by taking advantage of the energy released by the decay of an isotope of the element **uranium**, ^{235}U. When a ^{235}U nucleus is split, high-energy neutrons are emitted, as well as other forms of energy. If these neutrons are sufficiently abundant, they can split the nuclei of other ^{235}U atoms. The newly split atoms emit yet more neutrons, which then split more nuclei, creating a chain reaction. If this reaction is uncontrolled, the result is explosive, as in the atomic bomb. In nuclear power plants, the chain reaction is controlled, and the released energy is used to generate electricity.

Uranium ore is found in many locations. Each year, mining operations produce about 42,000 metric tons of uranium oxide (**Figure 15.33**). Most of this is produced in

Australia, Kazakhstan, Russia, South Africa, and Canada (**Figure 15.34**). Proven economically accessible reserves of uranium ore are thought to contain about 5 million metric tons of uranium oxide. This suggests that reserves are sufficient for another 120 years (R/P = 5,000,000/42,000 = 120). However, experts note that at higher energy prices or with improved technologies, uranium may be economically extracted from lower grade deposits or even from seawater.

Today nuclear reactors supply about 16% of global electricity. This electricity comes from 443 nuclear reactors, located in 31 countries. The United States has the highest number of nuclear reactors: 104. These reactors supply about 18% of its electricity. In contrast, France has 59 nuclear reactors, but they provide over 80% of that nation's electricity.

The cost of operating a nuclear power plant is generally much lower than the cost of operating a power plant fueled by coal or natural gas. However, the start-up costs for a nuclear power plant are much higher—a typical nuclear power plant costs $3–5 billion. In most countries, including the United States, these costs are met with the help of various government subsidies.

▼ **Figure 15.33 Proven Uranium Reserves**
Like other primary energy sources, uranium ore is found in large quantities in a relatively few locations around the world.
Data from: UN International Atomic Energy Agency. 2009. World Distribution of Uranium Deposits (UDEPO) with uranium deposit classification. Vienna, Austria – IAEA. Page 11.

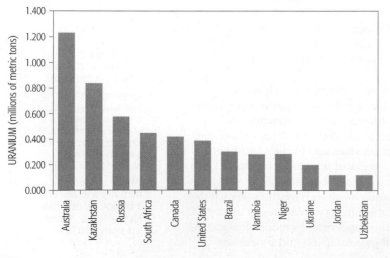

▼ **Figure 15.34 Uranium Excavation**
Uranium open pit mine near Arlit, Niger, has environmental impacts on nearby ecosystems and human communities.

Uranium Extraction and Processing

Uranium processing aims to increase the relative amount of ^{235}U.

Uranium is obtained from open pit mines. Uranium ore is rarely concentrated in large amounts, so mining operations must excavate large amounts of material to recover comparatively small amounts of ore. Once mined, uranium ore is crushed into a fine powder and washed with a strong acid or alkali. Next, the resulting liquid is concentrated into a solid material called *yellowcake*, which contains about 75% uranium oxide. Yellowcake is then transported to plants where it will be processed to produce uranium fuel.

At the processing plant, the uranium in the yellowcake is converted to gaseous uranium hexafluoride. Only 0.7% of the uranium in yellowcake is composed of ^{235}U atoms. The remainder is ^{238}U atoms, which do not undergo nuclear fission. Atoms of ^{235}U weigh slightly less than atoms of ^{238}U, and this weight difference is exploited to increase the amount of ^{235}U relative to ^{238}U. When the uranium hexafluoride is spun in cylindrical centrifuges, the heavier isotope moves to the cylinder walls, and ^{235}U is concentrated in the center. In this way, the relative concentration of ^{235}U is increased to about 4%.

This material is then formed into fuel pellets that are about the size of jelly beans. Fuel pellets are packed into corrosion-resistant metal tubes to make fuel rods. Next, 100–300 fuel rods are joined together to make a fuel assembly (**Figure 15.35**). Inside a nuclear power plant, up to 3,000 fuel assemblies are placed in a **reactor core,** where the fission reaction will take place.

In the reactor core, the fuel assemblies produce high energy neutrons that set the nuclear fission chain reaction in motion. The energy released through the fission process heats pressurized water or liquid sodium surrounding the assemblies. Control rods in the fuel assemblies regulate the rate of the chain reaction and the heat generated from the fuel assemblies. Control rods contain elements such as silver or cadmium that intercept fast neutrons. That hot liquid is then used to convert water into steam. In the final step, this steam is used to turn turbines to generate electricity (**Figure 15.36**).

Note that the steam that drives the turbines does not come into direct contact with the radioactive fuel assemblies. Nuclear power plants operate much like power plants that use fossil fuels, except the water is boiled with heat from a nuclear reaction instead of heat from burning fossil fuels.

▼ **Figure 15.36 Electricity from Atoms**
Heat generated from the controlled fission of ^{235}U is used to heat either water or sodium to very high temperature and pressure. This superheated liquid is then used to generate steam that powers turbines, which in turn drive electric generators.

Reactor core   Steam generator   Reactor

Steam

Turbine

Generator

35 °C

−25 °C

Primary water   Secondary water   Condenser

Pump

▼ **Figure 15.35 Producing Nuclear Fuel**
Uranium ore is formed into yellowcake Ⓐ, which is then spun in a centrifuge. The uranium is then formed into fuel pellets Ⓑ that are inserted into fuel rods Ⓒ. Fuel rods are joined together into a fuel assembly Ⓓ.

Ⓐ

Yellow cake

Ⓑ

Fuel pellets

Ⓒ

Fuel rod

Ⓓ

Fuel assembly

473

Environmental Impacts

The environmental benefits and costs of nuclear power are hotly disputed.

Given the damage to the plants in Japan, is nuclear energy worth the risks?

Patty Adams, Metro State College of Denver

Norm: *Nuclear energy is appealing because it generates few greenhouse gas emissions. However, public resistance, the high cost of building new plants, and the difficulties associated with disposing of radioactive waste are serious impediments to its use. The disaster at Japan's Fukushima nuclear facility raised additional concerns that safety risks have been underestimated.*

Even in the best mining sites, uranium is not very plentiful. To extract the uranium, large quantities of rock must be removed and processed. This process disturbs large areas of land and leaves behind extensive piles of rock and soil, or mine tailings. As with other kinds of mines, water flowing through these tailings is often acidified and has severe negative impacts on nearby streams and lakes.

Excess heat is one of the important by-products of nuclear power plants. This heat is most often dispersed in the form of steam that exits through large cooling towers. In fact, those cooling towers have come to symbolize nuclear power generation itself. In some cases, heat is dissipated in nearby river, lake, or ocean water (**Figure 15.37**). Elevated temperatures in waters near nuclear plants can alter the quality of the aquatic or marine habitats and diminish biodiversity.

Even without controls, the rate of fission in a nuclear reactor is not sufficient to trigger a nuclear explosion. However, if reactor temperatures are not carefully monitored and managed, the core can overheat and literally melt. That is what happened in 1979 when equipment malfunctions and operator errors resulted in the meltdown of a reactor at the Three-Mile Island nuclear power plant in Pennsylvania. In this incident, the reactor dome—the heavily reinforced container that holds the reactor—remained intact and prevented any significant leakage of radioactive materials. Had the dome failed, a large area of southeastern Pennsylvania and parts of New Jersey could have been contaminated.

In April 1986, a far more serious failure occurred in a reactor at the Chernobyl nuclear power plant in the former Soviet Union (now Ukraine). Operator negligence resulted in a reactor meltdown, accompanied by a gas explosion that destroyed the reactor dome (**Figure 15.38**). Additional explosions sent a plume of highly radioactive material into the atmosphere and over parts of western Russia, Belarus, eastern Europe, and Scandinavia. The amount of radioactive fallout released in this disaster was more than 30 times greater than that released by the atomic bombs in Hiroshima and Nagasaki, Japan. A large area of Ukraine, Belarus, and Russia was contaminated, causing the evacuation and relocation of over 330,000 people. This event was the direct cause of at least 50 deaths. The World Health Organization estimates that radioactive fallout from Chernobyl has caused at least 9,000 extra cases of cancer among the 7 million people exposed to the highest levels of radiation.

Three Mile Island and Chernobyl alerted government agencies and people around the world to the potential risks associated with the use of nuclear power. In the United States, public opinion regarding the use of nuclear power turned strongly negative. Electric utilities shelved all plans to build additional nuclear power plants. Some European countries, such as Italy and Spain, decided to decommission existing nuclear facilities and forego construction of any new facilities.

A recent catastrophe in Japan has raised new doubts about reactor safety. On March 11, 2011, a major earthquake struck the northeastern island of Honshu. Although the nearby Fukushima Daiichi nuclear power plant had been designed to withstand such events, it was flooded by the subsequent tsunami. This caused explosions and fires, damaged at least two reactor containment vessels, and resulted in the release of large amounts of radioactivity. As many as 50 workers were exposed to high levels of radiation; communities within 20 km (12 mi) were evacuated.

Despite these events, soaring energy prices and concerns about greenhouse gas emissions and global warming have caused governments, power utilities, and even some environmental groups to reconsider their opposition to nuclear power. While acknowledging the significant consequences of accidents at nuclear facilities, these groups note that the specific circumstances that led to the accidents at Three Mile Island and Chernobyl have been remedied by improved technologies and management protocols. They also argue that the Fukushima Daiichi catastrophe occurred in a unique geologic setting; in most locations, nuclear plants are not at risk for the same level of earthquakes and tsunamis.

Others argue that the benefits of nuclear power do not outweigh its potential and real costs. They note that there are still many uncertainties about the safety of

▼ Figure 15.37 Keeping Things Cool
The Diablo Canyon Nuclear Power Plant on the California Coast uses water from the Pacific Ocean to dissipate heat.

► Figure 15.38 1986 Meltdown
As a consequence of operator error, Reactor number 3 at the Chernobyl nuclear power plant overheated and exploded, spreading radioactive fallout across much of northern Europe.

Chernobyl

nuclear power plants, including the threat of terrorist attacks.

Nuclear power does not rely on the combustion of fossil fuels and, therefore, does not directly emit CO_2 or other atmospheric pollutants. Small amounts of emissions are associated with uranium mining, transportation, and processing. Scientists calculate that annual greenhouse gas emissions could be reduced by 1 billion metric tons if there were a threefold increase in the number of nuclear power plants built to replace coal-fired plants. By itself, this is not sufficient to avoid significant future global warming, but it might be included in a portfolio of actions to achieve this goal.

One of the most daunting challenges associated with nuclear energy is the safe disposal of **spent nuclear fuel.** Through time, the ^{235}U in fuel rods is depleted and the rods' heating capacity diminishes. Although they can no longer be used in reactors, these fuel rods accumulate an array of other radioactive elements and will remain radioactive for over a million years. This spent fuel must be carefully stored and monitored. Currently, it is stored in storage pools or stainless steel casks at the nuclear reactor sites (**Figure 15.39**).

The total volume of high-level radioactive waste can be reduced through **fuel recycling**, in which the fuel is reprocessed and the elements that can undergo fission chain reactions are reconcentrated. This process has been implemented by some countries, such as France. However, as waste is recycled, concentrations of another element, **plutonium**, increase. Like uranium, plutonium can undergo fission and generate heat. Unlike ^{235}U, however, it is relatively easy to use plutonium to make nuclear bombs. Because of the threat that plutonium production poses to global security, the United States and most other European countries have agreed not to recycle spent fuel.

Many experts believe that geologic repositories are the best long-term solution to the spent fuel challenge. A **geologic nuclear repository** is a layer of rock in which nuclear waste and all of its toxic and radioactive elements are permanently sealed from the biosphere. By international agreement, a repository must protect humans and the ecosystems they depend on from exposure to radiation for hundreds of thousands of years.

The United States has been working to develop a geologic nuclear repository at Yucca Mountain, Nevada, a site about 90 miles north of Las Vegas (**Figure 15.40**). Here, spent fuel in durable stainless steel casks would be placed in long tunnels dug through dry rocks about 300 m (1,000 ft) below the surface. Sweden and Finland have established experimental geologic nuclear repositories in granite, 300 m beneath their coastal waters.

At present, no geologic nuclear repositories are in operation. This is partly due to political and social opposition to their location. As of 2011, for example, funding for the Yucca Mountain repository has been frozen by Congress, in large part because of opposition to the project by the state of Nevada. In addition, significant uncertainties remain regarding the ability of repositories to meet minimum performance standards over long spans of time.

▲ **Figure 15.39 Storing Spent Fuel**
In wet storage, spent fuel assemblies are immersed in large pools of water. In dry storage, fuel assemblies are encased in thick-walled metal casks.

It is fair to say that no primary energy source generates more controversy and debate than nuclear power, especially in the United States. Because of high construction costs and widespread public resistance, no new nuclear facilities have been built in the United States since 1978. In recent years, however, many electric utilities have begun planning for new reactor facilities in order to reconcile rising demands for energy with the need to mitigate global warming. As a result, the debate over nuclear power has intensified. Meanwhile, 30 new nuclear reactors are under construction around the world, mostly in Asia, and additional growth appears very likely.

QUESTIONS

1. Describe the chain reaction that provides energy for nuclear power plants.
2. How is the heat from the fission of ^{235}U captured to produce electricity in a nuclear power plant?
3. Describe two ways of storing high-level radioactive waste.

▼ **Figure 15.40 Geologic Nuclear Repository**
For the past 25 years, the U.S. Department of Energy has been working on technology and protocols to place over 70,000 metric tons of spent fuel in the geologic strata of Yucca Mountain in southern Nevada.

Summary

Primary energy sources are natural resources such as sunlight, oil, and coal that have not been transformed by humans. These sources may be converted to secondary energy sources such as electricity or the kinetic energy of machines. Energy conversion efficiency is the percentage of energy in a primary source that is captured in a secondary energy form. Energy end-use efficiency is the amount of energy available to the final application compared to the amount of energy from the primary source that was required to produce it. Nonrenewable primary energy resources are used more rapidly than their natural rates of production. Proven reserves of such resources are the amounts that can be recovered in the future from known sources using existing technologies at current prices. The ratio of proven reserves to extraction or production rate provides

an estimate of the length of time that reserves of a nonrenewable resource will last, assuming current technologies and prices. Electricity is the flow of electrons through a material, such as a wire. Electricity can be generated by spinning magnets in an electric generator or by chemical reactions in a battery. Electric power is distributed through a system of transmission lines and transformers called the electric power grid. Environmental impacts of electrical transmission include energy lost in the power grid and chemical pollutants from transformers and batteries. Coal, the most abundant of all fossil fuels, occurs in underground reserves that vary in energy yield per unit weight. Environmental impacts of coal use include the direct effects of mining, as well as air and water pollution from coal-fired power plants. Crude oil is formed from the remains of marine microorganisms that were slowly transformed by heat and pressure in geologic

sediments. Oil is pumped from underground reservoirs and refined to produce fuels such as gasoline, kerosene, and diesel oil. Environmental impacts include spills associated with oil drilling and transportation as well as pollution associated with the combustion of oil-derived fuels. Natural gas is primarily composed of methane, and it is usually found in association with crude oil reserves. The combustion of natural gas produces fewer pollutants than that of coal or oil, but the methane released when natural gas is extracted and transported contributes to global warming. Nuclear power uses the heat released by the fission of uranium to produce steam that is used to drive turbines and generate electricity. Environmental impacts of using nuclear power include the direct effects of uranium mining and the potential release of highly radioactive elements and toxic chemicals into the air and water.

15.1 Energy Production and Consumption

- Over the past century, our demand for energy has increased dramatically. That demand has largely been met through the use of nonrenewable fossil fuels, such as coal, oil, and natural gas.

- Primary energy is the energy contained in natural resources. Secondary energy sources, such as electricity, are produced by transforming primary energy sources.

- The future availability of nonrenewable energy sources depends on the size of proven reserves, the rate of extraction, and the prices consumers are willing to pay.

KEY TERMS

primary energy, energy conversion, secondary energy, end use, energy conversion efficiency, energy end-use efficiency, nonrenewable energy, renewable energy, proven reserves, reserves-to-production ratio (R/P), energy return on investment (EROI)

QUESTIONS

1. The end-use efficiency of most cars is less than 10%. What does this mean, and what factors are likely to be contributing to this low number?

2. What factors influence the relative sizes of the end-use energy sectors in different countries?

3. The current reserves-to-production ratio (R/P) for crude oil is about 40 years. What factors are likely to influence the actual future availability of oil?

4. How does energy return on investment (EROI) generally change as reserves of a nonrenewable energy source are depleted? Why?

15.2 Electric Power—Generation, Distribution, and Use

- Primary energy sources are used to power electric generators and produce electric currents.

- Electricity is distributed on power grids, and its flow is regulated by feedbacks between consumers and power generators.

- The direct effects of the electromagnetic fields generated by the power grid on human health are uncertain.

KEY TERMS

amperes, volts, electric generator, electromagnetic field, kilowatt hours, transformers, electric power grid, base load, peak load

QUESTIONS

1. What is the role of the transformers in power grids?

2. What do base load and peak load refer to, and how are they related to the cost of electricity?

3. Design experts complain that electricity is wasted because of poor feedback between consumers and producers of electricity. Why?

15.3 Coal

- Coal was produced by the incomplete decomposition of plants that grew in swamps 300–400 million years ago.
- Underground mining and surface coal mining have direct impacts on the environment. The combustion of coal produces a variety of air and water pollutants.
- Coal burning is a major source of greenhouse gas emissions.

KEY TERMS

peat, coal seams, lignite, sub-bituminous coal, bituminous coal, anthracite coal, scrubber, fly ash, mine tailings

QUESTIONS

1. What is mountaintop-removal mining and why is it a major concern to environmentalists?
2. What is carbon capture and storage (CCS), and how might it be used to reduce greenhouse emissions from coal burning?
3. Why are some energy experts concerned about the use of CCS?

15.4 Oil

- Oil is produced from the decomposition of microbes in ocean sediments; oil is trapped in reservoirs beneath impermeable cap rocks.
- Oil is pumped from reservoirs and refined to produce a variety of fuels and other products.
- Major environmental impacts caused by the use of oil include spills associated with oil pumping and transportation and air and water pollution from fuel combustion.

KEY TERMS

kerogen, oil reservoir, peak oil, primary oil recovery, secondary oil recovery, tertiary oil recovery, hydraulic fracking, internal combustion engine

QUESTIONS

1. What evidence supports the conclusion that we are very close to peak global oil production?
2. What factors might extend the availability of oil beyond the time spans estimated by the ratio of production to reserves?

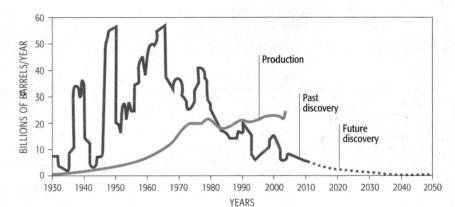

3. Describe three reasons why continuing to rely heavily on oil-derived fuels presents serious challenges for future sustainability.

15.5 Natural Gas

- Natural gas is composed mainly of methane.
- Natural gas is formed in nearly the same way as oil and is usually associated with crude oil reservoirs.
- Natural gas combustion releases relatively few pollutants, but methane is a powerful greenhouse gas.

KEY TERMS

liquefied natural gas (LNG), gas-turbine power plants, combined-cycle power plants

QUESTIONS

1. Why is the conversion of natural gas to electricity considerably more efficient than the conversion of coal to electricity?
2. Describe two ways in which the use of natural gas adds to global warming.

15.6 Nuclear Power

- The controlled fission of ^{235}U in the core of a nuclear reactor heats water or sodium, which is then used to produce steam to drive turbines and generate electricity.
- Many decision makers are concerned about the potential for accidents and the release of highly radioactive materials from nuclear power plants.
- The normal operation of nuclear power plants produces significant amounts of spent nuclear fuel that remains radioactive for hundreds of thousands of years.

KEY TERMS

uranium, reactor core, spent nuclear fuel, fuel recycling, plutonium, geologic nuclear repository

QUESTIONS

1. What are the steps required to transform uranium ore to the fuel rods that power a nuclear reactor?
2. Describe two kinds of high-level radioactive waste storage. How does such storage differ from a nuclear waste repository?
3. How does a high-level nuclear waste storage facility differ from a nuclear waste repository?
4. What are the primary environmental advantages of nuclear power compared to fossil fuels?
5. What are the primary environmental concerns associated with nuclear power?

Mastering ENVIRONMENTALSCIENCE

Go to **www.masteringenvironmentalscience.com** for homework assignments, practice quizzes, Pearson eText, and more.

More Efficient Lighting and Appliances

Opportunities for energy efficiency exist in virtually all sectors.

About 25% of the energy consumed by the residential sector goes to lighting. Most of that lighting comes from **incandescent lights**, in which an electric current causes a thin metal filament to glow. Only about 10% of the energy consumed by these fixtures is used to produce visible light. The remaining 90% is radiated as heat, as is painfully familiar to anyone who has accidentally touched a lit incandescent bulb.

Compact fluorescent lights (CFLs) emit three to five times more light per kilowatt-hour than do incandescent fixtures. Although CFLs are more expensive than incandescent lights, they last 8 to 15 times longer. **Light-emitting diodes (LEDs)**, which emit light as electric current passes through silicone-rich materials, are even more efficient and long-lasting than CFLs. At present, the cost of LEDs is beyond the reach of most households but is likely to be competitive in the future.

Electric motors, which are found in a wide variety of machines and appliances, consume more than 60% of all generated electrical energy. In conventional electric motors, about 80% of that electricity is converted to kinetic energy; the remaining 20% is lost as heat. Electric generators, in which the kinetic energy of a turbine is converted to electricity, also have an efficiency of about 80%. A new generation of motors and generators has a very different wiring configuration that allows them to operate at efficiencies of greater than 90%. This technology is already widely used to maximize the efficiency of generators in wind turbines.

Some appliances waste energy even when they are not being used. When many modern appliances are turned "off," they are not actually off but instead are on "standby." They continue to use energy to run timers, clocks, computer memory, and remote on/off switches. This hidden use of energy is often called "vampire power." The typical American home has about 20 appliances that, turned off, add $2–$300 to the electric bill each year. Efficient clocks and switches can diminish vampire power use; however, the best strategy is to unplug devices such as computers and coffeemakers when they are not in use.

Automobile Efficiency

Automobiles with traditional internal combustion engines are extremely inefficient in their use of energy.

No machine has caused greater awareness of energy efficiency than the automobile. "What's the mileage?" is one of the first questions asked by most car buyers these days. Yet today's cars are highly inefficient. In an average automobile with an internal combustion engine, only about 10% of the chemical energy in a tank of gasoline is used to drive the wheels. The rest of the energy is lost to idling, dissipated as heat, or used for air conditioning and other accessories.

There is general agreement that the automobiles we produce today can be made far more efficient. In the years after the 1973–1974 oil embargo, there were dramatic improvements in automobile fuel efficiency. By the 1990s, however, fuel efficiency gains had leveled off. In the years between 2000 and 2003, truck and SUV fuel efficiency even declined (**Figure 16.40**). While automobile design and engineering have improved since the 1980s, most of the focus has been on increasing power and comfort, not on fuel efficiency. Recently, however, the high cost of oil and concerns about its environmental impact have renewed public interest in automobile efficiency.

One of the most direct ways to improve fuel efficiency is to make vehicles lighter.

◀ Figure 16.40 **Fuel Efficiency**
Gas mileage for U.S. cars (red) and trucks (blue) improved dramatically between 1970 and 1985. Since then, it has remained flat, or even decreased.
Source: U.S. Bureau of Transportation Statistics 2010 Transportation Statistics Annual Report. URL: http://www.bts.gov/publications/transportation_statistics_annual_report/2010/ Last accessed 8/25/2011.

Physicist Amory Lovins advocates the use of ultralight carbon composite materials that are also ultrastrong. These composites can reduce the weight of an automobile by over 50% without sacrificing safety. Automobiles with lighter bodies could use smaller engines to achieve the same performance, greatly increasing fuel efficiency. Lovins responds to concerns about the safety of such vehicles by pointing out that increased weight does not always translate into increased safety—otherwise bicycle helmets would be made of steel rather than lightweight foam and carbon fiber material.

Much greater fuel efficiency is being realized by significant changes in the way cars are powered. Over the past decade, **gasoline-electric hybrid automobiles**, such as the Toyota Prius, have become widely available. Hybrid vehicles are equipped with both an internal combustion engine and an electric motor powered by batteries. On the highway, a hybrid's primary source of power is its gasoline-powered engine. When only low power is needed, as in slow city driving, the hybrid relies on the batteries to power its electric motor. The batteries are constantly being recharged by the kinetic energy created when the car brakes, a process known as regenerative braking.

Recently, **plug-in hybrid automobiles** have become available commercially, including a new model of the Toyota Prius. These vehicles operate much like gasoline-electric hybrids, except their batteries can be recharged by plugging the car into an electric outlet. This gives the electric motor of plug-in hybrids greater range and power than that of other hybrid vehicles, thereby increasing gas mileage. Fully electric automobiles such as the Chevrolet Volt have just recently become widely available.

The development of economical and efficient batteries has been the limiting factor in the development of both plug-in hybrid and all-electric cars. New battery technologies promise to increase the use of these vehicles. Whether these vehicles actually reduce pollution or greenhouse gas emissions depends on the source of electricity. If it comes from coal-fired plants, then pollution and emissions are only being shifted from one place to another.

Although they do not really represent a new automobile design, high-efficiency diesel vehicles operate with far greater fuel efficiency than did earlier diesel engines. Earlier diesel engines were already more fuel-efficient than conventional internal combustion engines. High-efficiency diesel vehicles combine that advantage with improvements that make them cleaner, quieter, and more powerful than earlier models.

Can we power automobiles without using fossil fuels? Advocates of vehicles that rely on hydrogen-powered fuel cells believe that we can. Rather than burning gasoline in an internal combustion engine, hydrogen vehicles use onboard fuel cells to convert the chemical energy in hydrogen fuel to electric power. Prototype vehicles are already in operation, and manufacturers expect to begin commercial production as soon as 2012 (**Figure 16.41**). The greatest challenges to the widespread development of this technology are obtaining and distributing hydrogen. The production of hydrogen from organic materials or water requires energy that must be generated from some other primary source. Furthermore, the infrastructure for distributing and selling hydrogen remains limited and may be very costly to develop on a large scale.

▼ **Figure 16.41 Coming Soon to a Showroom Near You**
Korean car manufacturer Hyundai has announced that it will begin retail sales of hydrogen fuel cell cars in late 2012.

1. Does energy conservation always require personal sacrifices? Explain.

2. What is economic energy intensity and why does it vary among countries?

3. Why are compact fluorescent light fixtures more energy-efficient than conventional incandescent lightbulbs?

4. How does a gasoline-electric hybrid car differ from a conventional car? From an electric car?

16.9 Sustainable Energy—Economics and Policy

Even with the utilization of renewable energy sources, is it really possible to maintain our current lifestyle without further degrading the environment?

Darrik Carlson, Citrus Community College

Norm: *Sustainability will require a greater use of renewable energy sources, as well as more efficient use of all energy resources.*

Renewable energy resources and energy conservation strategies would appear to offer many environmental, economic, and social benefits. So why haven't we pursued these resources and strategies to a greater extent? Part of the challenge to the wider use of renewable energy and energy conservation is simply being able to assess their advantages. Additional challenges include the lack of familiarity with new technologies, economies of scale, externalized costs, and limited consumer knowledge and understanding. Meeting these challenges will depend in large part on political decisions and the economic circumstances surrounding different energy technologies. Government policies can encourage conservation and boost the demand for and production of renewable energy resources. Such policies can help to build a diversified and decentralized energy economy that addresses the environmental and economic issues of our current energy system.

Measuring Economic Benefits

Conservation metrics compare the costs and benefits of energy conservation and renewable energy use.

Energy conservation is often described as a "win-win" approach: It saves money and also has many environmental advantages. There are many ways to improve energy efficiency, such as replacing inefficient lightbulbs, installing sensors for lights and heating, upgrading to more efficient appliances, and installing more insulation or energy-efficient windows. Yet all of these actions cost money. How can we assess the benefits of each action relative to its cost?

Energy economists use a measure known as the **cost of saved energy (CSE)** to compare the costs and benefits of energy alternatives. CSE is calculated by dividing the cost of an energy investment by the energy savings resulting from that investment. For example, a company is considering spending $2,000 to install CFL lighting in its office. That upgrade would save 10,000

kilowatt-hours of electricity per year. In this case, the CSE would be $2,000 divided by 10,000, which equals $0.02, or $0.02 per kilowatt-hour. If the company is already paying $0.08 per kilowatt-hour for its electricity, then it could save $800 a year by going ahead with the lighting upgrade. This investment would pay for itself in less than 3 years (3 × $800 = $2,400).

By themselves, actions such as upgrading lighting, adding insulation, or installing solar panels may not be cost-effective. But taken together, a combination, or portfolio, of actions might be cost-effective. Energy experts use conservation supply curves to evaluate such portfolios (**Figure 16.42**). These graphs compare the amount of energy saved by each action to the cost of that energy. Whether or not to include an action in a portfolio can be judged in relation to the price of energy.

▶ **Figure 16.42 Conservation Savings**
This is a hypothetical conservation supply curve for energy-saving features that might be used in building a new home. Each step represents savings from a conservation feature. Measures 1–4 are cost-effective because the cost of conserved energy is less than the price of energy (dashed line). If energy prices were to rise, then measures 6 and 7 might become cost-effective.

1. Add insulation to walls
2. Add insulation to ceiling
3. Compact fluorescent lighting
4. Triple paned windows
5. Solar hot water heating
6. Ground source heat pump
7. Gas stove

COST OF CONSERVED ENERGY (%/kWh)

Current energy price line ($/kWh)

ENERGY SAVED (kWh/year)

Policy Options

A variety of policy options can be used to speed the transition to greater energy conservation and use of renewable energy.

Advocates of renewable energy sources and increased energy conservation frequently argue for the elimination of existing government subsidies to oil, coal, and natural gas as a critical first step in the energy transition. If even a small portion of the hundreds of billions of dollars in annual government subsidies were redirected to the development of renewable energy resources, then some of the investment and production barriers limiting the expanded use of those resources could be overcome.

Some energy economists favor imposing taxes on fossil fuels so that their price reflects more of the external costs the use of those fuels imposes on society. Currently, federal and state gasoline taxes are used to fund highway construction and maintenance, thereby capturing those costs in the price of automobile fuel. The same could be done to cover costs associated with the mitigation of pollution or global warming.

One way that governments can encourage the development and commercialization of renewable energy resources is through **renewable portfolio standards (RPS)**. An RPS is a government mandate that a certain percentage of energy use come from renewable energy resources. RPS mandates help build market demand for renewable technologies, increasing their scale of production and driving down costs. As of 2010, there were 35 states in the United States with an RPS mandate in place. Governments can also mandate energy-efficiency standards for automobiles, appliances, and even buildings. For example, both the United States and the European Union set gasoline-mileage standards that must be met by automobile manufacturers.

To help consumers make informed decisions, utilities could be required to disclose the energy sources they use to generate electricity and the amount of emissions released in the process. Consumers could then have the option of selecting an electric power provider that is using a greater amount of renewable energy to generate electricity.

To improve consumer awareness of energy efficiency, greater effort could be put into consumer education and labeling. The U.S. Environmental Protection Agency's Energy Star program is an example of a very successful education program. The Energy Star program certifies the energy efficiency of virtually all home appliances, from hot water heaters, to refrigerators, to air-conditioning systems (**Figure 16.43**). Manufacturers are encouraged to build appliances that incorporate innovations such as improved insulation, solid-state technologies, and more efficient motors and compressors. The Energy Star program provides consumers with easily understood measures of energy efficiency to guide their purchases. In 2010 alone, energy savings from this program exceeded $18 billion and avoided greenhouse gas emissions equivalent to 33 million cars.

The combination of policy, economic, technical, and social changes described above can accelerate the use of renewable energy resources and energy efficiency. However, because this shift will take decades, it is important to begin the process now.

By itself, no single renewable energy resource is going to replace our dependence on fossil fuels; there is no single technical "silver bullet" to meet our energy needs. Instead, we need what one energy expert has called a "web of innovations"—improved energy efficiency and advances in many energy technologies, including the development of new ways to store and use the energy obtained from renewable resources. It is this web of innovations that holds the promise of ushering in the transition needed to address the great environmental challenges of our current energy system.

QUESTIONS

1. Explain how cost of saved energy (CSE) is calculated.

2. Describe three policies that might be employed to encourage greater energy conservation.

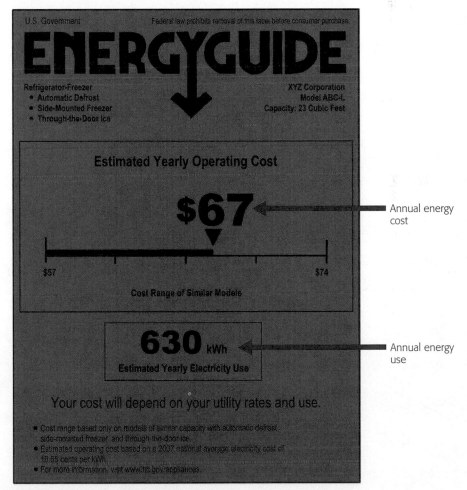

▲ **Figure 16.43 Incentives for Efficiency**
Appliances that meet EPA Energy Star standards clearly display average annual energy use and likely annual cost to the consumer.

AGENTS OF CHANGE: Morehouse College's Let's Raise a Million

Tony Anderson is a recent graduate of Morehouse College in Atlanta, Georgia, where he served as vice president of the student body and initiated a campus-level dialogue about sustainability and climate change. Marcus Penny holds a B.S. in psychology from Morehouse College. Together, Tony and Marcus founded Let's Raise a Million (LRAM), a student-led, nonprofit organization committed to introducing energy-saving solutions to low-income households. LRAM conducts free energy audits and energy-efficient lightbulb retrofits for residents of modest means. The pilot phase of the project is currently underway in West Atlanta and at Grambling State University in Grambling, Louisiana. Once the pilot phase is complete, the goal of LRAM is to retrofit homes with 1 million "clean" bulbs within 4 years.

How did you first get the idea for Let's Raise a Million?

On Mother's Day 2007, Tony exchanged his mother's incandescent lightbulbs with compact fluorescent light bulbs (CFLs), after being inspired by a speech given by Nia Robinson, Director of the Environmental Justice and Climate Change Initiative. That week Tony overheard a conversation between his mother and grandmother, then another conversation she had with a colleague. The exact words his mother shared with her mother and colleague were "Girl, I'm going green!"

Later that month, another intergenerational act of household greening occurred: Tony's mother gave more than 30 CFLs to her mother. This provided confirmation that one small act can lead to other small acts. Taken together, many small acts can result in great energy savings. Tony then thought, "What if I provided CFLs to other households of modest means? What would happen then?" That's when Let's Raise a Million was born.

What steps did you take to get started?

The initial funding for Let's Raise a Million came from a fellowship that Tony received from the Compton Foundation, in conjunction with a grant from the National Wildlife Federation. While the initial funding was enough to pay for many of the supplies, the project had to rely on the efforts of the students and volunteers to serve the community. It all began with campus education and an awareness campaign that focused on the importance of environmental sustainability and environmental justice. Once students became aware of the environmental injustices occurring throughout communities of modest means, their passion for the work grew and we were able to conduct a successful Retrofit Day of Action.

What are the biggest challenges you've faced with the project, and how did you solve them?

The most difficult part of starting was having a group of volunteers stay actively engaged and continue to be dedicated to the work. Like many other grassroots efforts, we have had to rely on volunteers because our initial budgets do not have room for full-time staff.

It has been important to highlight the impact that each volunteer is having on the community, because ultimately this is what keeps individuals engaged. Compensation goes beyond money. It goes back to the roots of human interaction and an understanding of the greater good.

How has your work affected other people and the environment?

In 2008, LRAM received the inaugural Clinton Global Initiative University Award for Environmental Projects. Tony shared the story of his mother "going green" with President Clinton and his chuckle and words of advice were the highlight of attending the award ceremony.

LRAM has engaged more than 3,200 volunteers in delivering CFLs, high-efficiency water kits, and low-flow toilets to more than 4,300 apartments and homes, free of charge. By providing these "direct-to-door" services to households of modest means, LRAM has realized a water and energy savings of $7.3M that will be returned to engaged households. Currently, we are working on building the infrastructure of the program to expand in other cities. We hope to hit our goal of delivering 1 million CFLs by 2013.

What advice would you give to students who want to replicate your model on their campus?

It is important to directly connect with the communities you would like to serve. Get to know local organizations that have been supporting the community for years, such as churches, schools, and other community-based organizations. In addition, local politicians can serve as champions of the work and help provide access to local resources and media outlets.

Make sure you set up a structure that doesn't "parachute" into the community and then "evacuate" after the installation. Develop relationships, and you will have community support for years to come.

Let's Raise a Million has developed a local affiliate network to help train and support students who are interested in bringing this model to communities near them. If you are interested, please contact us at info@letsraiseamillion.org.

Summary

Renewable energy refers to energy sources that are replenished continuously as they are used. These sources include the energy contained in falling water, wind, sunlight, biomass, Earth's heat, and ocean tides. Renewable energy sources are abundant on a global scale, but they are often diffuse and intermittent at particular locations. The environmental impacts associated with the use of renewable energy sources are generally less than those associated with the use of nonrenewable energy sources. Nonetheless, they do affect the environment. Use of renewable energy sources is growing rapidly with the development and scaling up of new technologies. The growing costs of nonrenewable energy sources, such as oil and coal, as well as concerns about their environmental impacts, are contributing to the increased use of renewable energy. These factors are also encouraging the development of strategies and technologies to conserve energy and diminish per capita energy demand. Increased energy conservation will include the development of technologies for more efficient use of electrical energy and more fuel-efficient transportation, as well as policies to encourage the adoption of these technologies.

16.1 Renewable Energy Overview

- There is great potential for growth in the use of renewable energy if we can meet the challenges of technology development, economies of scale, externalized costs, and limited consumer knowledge.

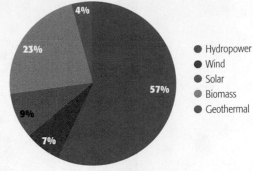

QUESTIONS

1. Many renewable energy sources are intermittent. What does this mean, and why does it pose a challenge to the use of these sources for electrical power generation?

2. The economic costs of environmental impacts of energy use are often externalized. Why does this favor the use of nonrenewable energy sources compared to renewable energy sources?

3. How might improved consumer knowledge result in greater use of renewable energy sources?

16.2 Hydropower

- The use of hydropower produces few greenhouse gas emissions and air pollution, but it has significant impacts on river and floodplain ecosystems.

KEY TERMS

hydropower, hydroelectric power, impoundment hydroelectric plant, run-of-river hydroelectric plant

QUESTIONS

1. What region of the United States produces the greatest amount of hydroelectricity? Why?

2. Hydroelectric dams increase sediment accumulation upstream and diminish it downstream. Explain why this is true and the problems that it presents.

3. What factors are likely to limit the growth in the number of hydroelectric facilities in the future?

16.3 Wind Power

- Wind power use is rapidly growing, but wind energy is intermittent and concerns remain regarding its effects on wildlife.

KEY TERMS

wind power, wind turbines, wind farms

QUESTIONS

1. In what regions of the United States has wind energy been exploited to the greatest extent? Why?

2. What are two challenges to the distribution of electricity that are presented by wind power?

Rotor blade
Gear box
Rotor break
Blade pitch mechanism
Electrical switch boxes and controls
Directional system
Tower
Grid connection
Foundation

16.4 Solar Energy

- Solar energy has relatively few environmental impacts, and its use will likely grow as its costs decrease.

KEY TERMS

passive solar technology, active solar technology, solar power, photovoltaic (PV) technology, concentrating solar power (CSP) systems, solar photovoltaic farms

QUESTIONS

1. Describe how a building might be designed to passively collect solar energy.

2. PV cells and CSP systems are both used to generate electricity. How do they differ?

Silicon layer with phosphorus

Silicon layer with boron

16.5 Biomass Energy

- Fuelwood, charcoal, and ethanol are the most widely used forms of biomass energy.

KEY TERMS

biomass energy, bioenergy, biofuels, biodiesel, cellulosic ethanol

QUESTIONS

1. In theory, biofuel combustion releases CO_2 but does not contribute to global warming. Why is this so?

2. How might the production of biodiesel and ethanol contribute to global warming?

3. What are the challenges and advantages of producing ethanol from a perennial plant, such as switchgrass, compared to corn?

16.6 Geothermal Energy

- The energy of Earth's heat can be used directly to heat and cool buildings and to generate electricity.

KEY TERMS

geothermal energy, ground-source heat pumps (GSHPs)

QUESTIONS

1. How do flash-steam geothermal power plants differ from dry-steam power plants?

2. What factors limit the widespread use of geothermal energy?

16.7 Ocean Energy

- The kinetic energy of tides and ocean waves, as well as differences in ocean temperature, can be tapped to produce electricity.

KEY TERMS

tidal power, tidal barrage, tidal fence, tidal turbine, wave power, ocean thermal energy conversion (OTEC) systems

QUESTIONS

1. Explain how tidal power systems capture the energy of horizontal water movement whereas wave power systems capture the energy of vertical water movement.

2. Differentiate between open-cycle and closed-cycle OTEC systems.

16.8 Energy Conservation

- Energy conservation requires an understanding of the trade-offs between perceived and real benefits and the costs of consumption.

KEY TERMS

economic energy intensity, incandescent lights, compact fluorescent lights (CFLs), light-emitting diodes (LEDs), gasoline-electric hybrid automobile, plug-in hybrid automobile

QUESTIONS

1. Economic energy intensity is much higher in some states than in others. Describe two factors that might cause such differences.

2. What is vampire power? Give examples of this kind of energy use in your home and classroom.

3. Is it true that electric automobiles produce no greenhouse emissions? Explain.

16.9 Sustainable Energy—Economics and Policy

- Conservation metrics and a portfolio of policies can speed the transition to increased conservation and renewable energy use.

KEY TERMS

cost of saved energy (CSE), conservation supply curve, renewable portfolio standard (RPS)

QUESTIONS

1. Explain how to use cost of saved energy (CSE) to determine the time required to repay the cost of investments in conservation or renewable energy.

2. How would you use a conservation supply curve to determine the best combination of conservation strategies relative to cost?

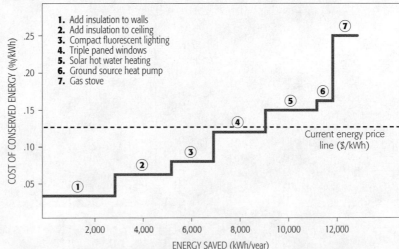

Mastering ENVIRONMENTALSCIENCE

Go to **www.masteringenvironmentalscience.com** for homework assignments, practice quizzes, Pearson eText, and more.

17

Urban Ecosystems

Malmö—Sweden's Green City

How can a deteriorating city become a model for sustainability?

In Malmö, Sweden, the Turning Torso building towers above all other buildings (**Figure 17.1**). This new apartment building is intended to mimic the human body in motion. In many ways, this building also symbolizes the remarkable history of change in this, Sweden's third largest city.

In A.D. 1275, Malmö was established as a fortress city to monitor and control the movement of ships through the narrow Øresund strait that connects the Atlantic Ocean with the Baltic Sea (**Figure 17.2**). By 1450, Malmö's strategic location had allowed it to become a major center of fishing and commerce in northern Europe. During the 19th century, Malmö became Europe's largest shipbuilding center. This industry anchored Malmö's economy and supported the development of other industries, including food processing and the production of leather goods and textiles. Throughout these seven centuries, Malmö enjoyed continuous growth in its population and economy.

In the late 20th century Malmö fell on harder times. An economic recession in the 1980s forced the closure of Malmö's shipyards. The economic impact of their closure cascaded through the city's economy. By 1990, the city's population had declined by 15%. Buildings and public spaces were beginning to deteriorate. Indicators of community decline, such as unemployment, crime, and suicide rates, were the highest in all of Sweden.

Despite these challenges, Malmö began to renew itself as a sustainable city, spurred on by strong leadership from its City Council and a variety of citizen groups. In 1997, the City Council adopted its Local Agenda 21, modeled after the Global Agenda 21, which was adopted by the United Nations at its 1992 Conference on Environment and Development. Global Agenda 21 is a set of strategies that countries can use to diminish their negative impacts on the environment while increasing the well-being of their people. Similarly, Malmö's goal was to improve its local environment and thereby the health of its population. The city was committed to becoming a model for sustainable development in both its day-to-day work and its long-term planning.

Public education and engagement are key components of Local Agenda 21. Immediately after its adoption, the city implemented an outreach and public participation program to assist in the development of its environmental strategy. Exhibitions showcasing elements of urban

◄ **Figure 17.1 A Symbol of Sustainability**
The Turning Torso Building designed by architect Santiago Calatrava dominates the skyline of Malmö's Western Harbor District. This building's environmentally sustainable design is emblematic of planning and renewal throughout the city.

▼ **Figure 17.2 A Strategic Location**
Malmö's location on the Øresund Strait contributed to its development as an important European center for fishing and commerce. The Øresund Tunnel-Bridge that connects Copenhagen to Malmö is visible in this satellite photograph.

sustainable development were displayed at local libraries, and a monthly Agenda 21 newsletter highlighting new projects was distributed throughout the city. The city published a children's book focusing on environmental issues and human health and distributed it to every Malmö child between the ages of six and eight. With widespread public support, Malmö has inaugurated programs to improve energy use, reduce waste, and develop open space.

The spirit of Turning Torso is especially evident in Malmö's partnership with the private utility eON Sweden, which has the goal of obtaining 100% of the city's electrical energy from renewable sources. Today, solar collectors provide hot water to most of the city's buildings. Innovative heat pump systems extract heat from seawater and groundwater. A 2-megawatt wind turbine and photovoltaic panels on a single large building provide virtually all of the electricity to over 5,000 homes and businesses in the city's Western Harbor area. All of the apartments in the Turning Torso have units to convert organic waste into biogas, which is used for cooking and to fuel vehicles. With these sources of renewable energy and the expansion of its biofuel-powered public transportation system, Malmö aims to reduce its emissions of carbon dioxide to 25% less than 1990 levels by 2012.

Outside, Malmö is using its existing system of city parks to anchor a larger system of green spaces. Sixteen green corridors provide streams, ponds, and other wildlife habitats, as well as a system of bike trails (**Figure 17.3**). These green spaces are designed to improve stormwater infiltration, increase biological diversity, and provide opportunities for recreation and human-powered transportation.

All of these efforts have been coupled with programs to improve the health and well-being of Malmö's citizens. These include educational programs dealing with nutrition, smoking, and alcohol use. With support from the Swedish government, the city has begun a program to offer affordable housing to singles, small and large families, and the elderly in compact developments with easy access to public spaces and services.

Malmö's Local Agenda 21 is also a commitment to influence sustainable development in other communities. In 1998, the city founded Malmö University, which has a focus on environment,

conservation of natural resources, ethnicity, and gender. In addition, Malmö has joined other cities in the Sustainable Energy Communities and Urban Areas in Europe (SECURE) project, which provides opportunities for sharing ideas for energy conservation.

Yet challenges remain. In parts of the city, crime rates are still high. A rapid influx of immigrants presents challenges on many fronts. The city is struggling to meet its Agenda 21 goals in the suburban communities that are rapidly growing outward from the city center. Nevertheless, Malmö's health indicators are improving, and its unemployment rate has dropped from 16% in 1996 to less than 8% in 2011. Since then it has remained below 8% despite a worldwide economic recession.

The problems of decaying urban infrastructure and rapid suburban growth are shared by cities around the world. What are the causes of these changes? Can cities around the world become sustainable from the standpoint of the environment, economy, and community? These are some of the questions that are central to this chapter.

▼ **Figure 17.3 The City of Parks**
Malmö's Green Plan envisions 16 green corridors connecting the city's many parks.

17.1 Urbanization

The earliest permanent human settlements were located in areas with abundant supplies of food and other resources. Some settlements eventually became cities—centers of trade and industry. **Urbanization**, the concentration of human populations in densely populated cities, depended on the growth of the food supply, as well as the development of technologies to deliver water and remove wastes. Thus the process of urbanization hastened the development of technology, market economies, and social and governmental structures. Cities facilitated the growth of human populations, yet the poor sanitation and diseases associated with them often diminished the well-being of city dwellers. The Industrial Revolution has hastened the growth of cities, and improved technologies for food production have allowed a dwindling number of people to produce enough food to support the rapidly growing urban populations. The growth of cities has, however, significantly altered Earth's ecosystems and led to the unsustainable use of many resources.

Ancient Cities

Permanent settlements require abundant and reliable resources.

When did humans start to live in cities? Archeological evidence of the earliest permanent human settlements dates from nearly 15,000 years ago. Such settlements have been found in several locations in Europe and Asia. These simple communities were located where food was both ample and reliable, such as along seashores and near the mouths of rivers.

Although these early settlements had some of the characteristics of cities, archeologists prefer to call them villages or hamlets. Like cities, these villages supported relatively high population densities. There was division of labor among their residents, and not all of them gathered or grew their own food. These settlements functioned as centers for the import and trade of raw materials, although this trade was carried out by barter rather than by the exchange of money. These villages did not have organized governments or public buildings.

The first true cities—settlements that supported societies with governments, public buildings, laws, and taxes—arose in several different regions about 6,000–8,000 years ago. The development of cities coincided

with the domestication of plants and animals and the development of agriculture. Residents of these early cities practiced specialized trades and bought and sold goods in monetary economies. That trade encouraged the development of writing and systems for keeping records.

About 4,500 years ago, large cities with populations greater than 40,000 sprang up in the valley of the Indus River in what is today Pakistan. These cities were laid out on regular grids. They also had sophisticated sanitation and sewage systems, including well-planned drainage and flush toilets.

Between 4,000 and 2,000 years ago, the number and size of cities grew, and their organization became increasingly complex. These ancient cities depended upon the flow of energy and materials from surrounding regions. Urban populations were ultimately limited by their supplies of resources, such as water and the amount of food produced in surrounding agricultural lands. In the first century, for example, the million plus residents of ancient Rome obtained their food from farms that were located across southern Europe and northern Africa. Water was brought into the city by a system of 11 large aqueducts, and waste was carried out by a complicated system of sewers known as the Cloaca Maxima (**Figure 17.4**).

People were probably drawn to cities by economic or social opportunities, but most city dwellers did not enjoy a higher quality of life than their rural kin. Cities were infested with insects, rats, and other vermin. Despite sewers and other amenities, water supplies were badly polluted. These conditions facilitated the spread of disease. Studies of birth and death dates recorded on thousands of tombstones from across the Roman Empire reveal that people in cities had significantly reduced life spans. During the period from A.D. 0 to 300, for example, the life expectancy for people living in Rome was about 27 years, and that of people living in rural areas was more than 40 years.

▼ Figure 17.4 **Water for Ancient Rome**
Two thousand years ago, a complex system of canals and aqueducts carried water from distances of more than 50 km (31 mi) to supply the millions of people living in Rome and other urban centers.

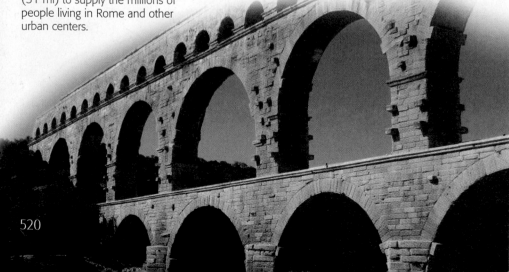

Evolution of Modern Cities

Urban development was both necessary for and dependent on the growth of industry and technology.

As empires waxed and waned over the succeeding centuries, cities continued to be centers of commerce and governance. During medieval times, many cities and towns were politically independent from the surrounding rural lands. Because cities provided freedom and opportunity from the oppressive rule of feudal lords, they attracted many residents. Even so, life for the citizens of most cities was difficult, even brutish. Death rates from the multiple plagues that swept Europe between the 13th and 17th centuries were 5 to 10 times higher in cities and towns than in the rural countryside.

Beginning around 1750, the Industrial Revolution increased the rate of urbanization. An ever-increasing proportion of the population was employed in urban industry, either directly or indirectly. At the same time, technological advances in agriculture made it possible for fewer people to produce increasing amounts of food. This increased supply of food fed the growing numbers of city dwellers. In 1740, nearly 70% of British laborers were farmers. By 1840, only 25% of British workers were employed on farms; the rest worked in cities.

The rapid urbanization of the early Industrial Revolution did not necessarily bring improved well-being to city dwellers. Up to 100 years ago, large industrial cities in Europe and North America were characterized by polluted air and water, extensive poverty and slums, and high rates of a great many diseases.

Around the year 1900, the health penalty for urban life began to diminish. In many regions, cities built water and sewage treatment facilities, established effective methods of waste management, and improved health care facilities. About the same time, the use of fossil fuels, fertilizers, herbicides, and pesticides produced the Green Revolution, allowing fewer and fewer people to produce the food needed by the rapidly increasing number of urban dwellers (12.1).

Current Trends

Urbanization has been the cause for substantial change in Earth's ecosystems.

Since World War II, urbanization has proceeded at a rapid rate. In 1950, 29% of the world population lived in urban areas. Today, 51% are urbanites, and that number is projected to reach nearly 70% by 2050. This change has been directly coupled to the global demographic transition in which birth, death, and population growth rates have declined.

The growth of urban populations has been especially notable among the world's wealthiest nations. Currently, 75% of the people in these countries live in cities. In the United States, 82% of the people live in urban areas. The population of U.S. cities is growing by 0.55% each year, while the population of rural areas is shrinking by 0.77% each year.

Over the next 30 years, the vast majority of Earth's human population growth will reside in its poorest countries. Urban populations in the world's 50 poorest countries are growing by more than 4% each year; at this rate, city populations in these countries will double in 17 years. Because they have limited economic resources, many cities in these countries are unable to provide clean water, manage wastes, or meet the basic health needs of their citizens (**Figure 17.5**). As in ancient times, urban citizens in most poor countries

▼ Figure 17.5 **The Urban Poor**
Port Au Prince, Haiti, has nearly 2 million residents, many of whom suffer from infectious diseases because they lack sanitary facilities and have no access to clean water.

suffer diminished well-being, as indicated by high rates of infectious diseases and infant mortality and diminished life expectancy.

The redistribution of Earth's people into urban centers has affected nearly every aspect of human life. It has redefined the nature of human interactions and communities, powered economic growth, and altered modes of recreation and education. It has required increased growth and complexity of local governance, a change that many sociologists argue has been a major factor in global trends toward increased democracy.

The effects of urbanization are especially evident in **megacities**, urban areas with more than 10 million inhabitants (**Figure 17.6**). Although less than 10% of Earth's urban population dwells in such cities, megacities have a disproportionate importance in terms of global commerce, governance, and environmental impacts.

Urbanization has caused significant restructuring of Earth's ecosystems. In order to support the needs of large concentrations of people, humans have redirected the flow of water all over the world. For example, over 90% of the water consumed by the people in the Los Angeles metropolitan area comes from more than 200 miles away. Most of the energy that powers cities comes from nonrenewable fossil fuels. Most of the food for the growing number of city dwellers is supplied by a dwindling number of farmers but requires increasing investments of energy from fossil fuels. Some urban wastes are altering the chemistry of our water and air. Other urban wastes accumulate in huge garbage heaps and landfills that rival the extent of natural geological deposits.

Clearly, many of these urbanization trends are not sustainable. Nevertheless, it is impossible to imagine Earth supporting our current numbers, much less our projected population growth, without cities. To resolve this paradox we must pursue strategies that lower the demand for water, energy, and other natural resources and reduce the streams of waste. With appropriate investments, urbanization in the world's poorest countries can actually improve human well-being.

QUESTIONS

1. Name three features that early villages and hamlets share with modern cities.

2. What is the evidence that the well-being of people living in ancient Rome was less than that of people living in rural areas?

3. How do urban population growth rates differ between wealthy and poor countries?

▼ **Figure 17.6 Megacities**
Megacities, metropolitan areas with over 10 million people, are widely distributed and growing in number.

17.2 Urban Ecosystems

An **urban ecosystem** is a region of Earth that supports the commerce, industry, and community interactions of a large number and high density of humans. Urban ecosystems include the physical environment and all the organisms living in cities and the less-densely populated areas that surround them. It may seem paradoxical to refer to a city as an ecosystem. Yet, like natural ecosystems, cities are dependent on the interactions among living and nonliving components and on the sustained flows of matter and energy (1.3). Human business and communities require a constant inflow of energy and materials and produce a steady stream of wastes. Urban ecosystems are sustained by the inflow and outflow of ideas and money and by the movement of people along transportation corridors. The physical environment and biodiversity of urban ecosystems change significantly from city centers to surrounding rural areas. The complex structure and artificial surfaces of city centers produce climates that are very different from those in rural locations. Urban development also alters the flow of water and diminishes water quality.

Defining Urban Ecosystems

Urban ecosystems change gradually from city centers to rural landscapes.

On many maps of the United States, Los Angeles appears as a large "smudge" on the southwestern coast of California. In satellite photographs, it is easily identified by its sprawling developments and freeways. Yet if you drive across this urban landscape, it is hard to define the actual boundaries of Los Angeles.

Pulling out of your parking place in front of the Los Angeles City Hall, you know you are in a city. Skyscrapers tower above you as you maneuver through the heavy traffic on South Main Street toward Interstate 10, known locally as the San Bernardino Freeway. Heading east on Interstate 10, tall buildings give way

to a mixture of residential and commercial neighborhoods (**Figure 17.7**). Less than a mile from downtown, the freeway crosses a wide, dry, concrete-lined canal—the Los Angeles River.

About 4 miles farther on, a small sign at the edge of the freeway reads "Los Angeles City Limits." Aside from that sign, nothing in the landscape indicates a boundary. You journey east for miles through a mix of housing subdivisions, shopping malls, and commercial districts, such as West Covina, Pomona, Ontario, and Fontana, although no clear lines delineate these communities. As you proceed east, buildings and homes seem to have

◀ Figure 17.7 **Downtown L.A.**
Los Angeles city center is dominated by high-rise buildings and traffic-clogged streets with few natural spaces.

been built more recently, and patches of shrubs and desert become more frequent (**Figure 17.8**).

After an hour of driving, the traffic thins and you pass more open spaces. Finally, about 15 miles east of the city of Palm Springs, the strip malls, new housing developments, and golf courses give way to the vast, wild expanses of the Mojave Desert (**Figure 17.9**).

Scientists who study urban ecosystems would use more specific terminology to describe your drive. The U.S. Census Bureau defines an area as **urban** if its average population density is at least 500 people per square mile. Certainly, the entire corridor from Los Angeles city center to Palm Springs qualifies as urban. A city is an area that is defined by legal and political boundaries and has a formal government and taxes (**Figure 17.10**). Thus, the city of Los Angeles is bounded by very precise city limits. Within those limits, the decisions and policies of the Los Angeles

◄ **Figure 17.9 Exurban Los Angeles**
The community of Palm Desert currently marks the outer fringe of the Los Angeles metropolitan area. Here development is concentrated along major transportation corridors and is surrounded by desert.

METROPOLITAN AREA

► **Figure 17.10 Urban Terminology**
Geographers define metropolitan areas, cities, suburbs, and exurbs in specific ways. Note that a suburb might also be defined as a city if it has legally defined boundaries and a government.

SUBURB

CITY

EXURB

SUBURB

-- Metropolitan Area: The city and the adjacent communities that have a high degree of social and economic interaction

● City: Defined by legal boundaries, government, and taxes

● Suburb: Densely populated centers outside the central city, but in the metropolitan area

● Exurb

mayor and city council hold sway over its 3.8 million residents.

Extending out in all directions from the city center is the Los Angeles metropolitan area. A **metropolitan area** is a large central city and the adjacent communities that have a high degree of social and economic integration with that central city. The transportation system and movement of commuters between the suburbs and the city is a very visible example of such integration. **Suburbs** are the parts of the metropolitan area located outside of the central city. Beyond the suburbs are **exurbs**, areas where patches of commercial and residential development are interspersed in a matrix of rural land.

All of these areas are components of an urban ecosystem. For those who associate ecosystems with the natural world, the phrase "urban ecosystem" may seem contradictory. Built by humans and supporting only the most meager biodiversity, cities seem to exemplify the unnatural. Indeed, cities exist for totally anthropocentric purposes—to support the commerce, industry, education, and social interactions necessary to sustain the well-being of their human residents. Nevertheless, thinking of a city as an ecosystem can help to clarify its inner workings (**Figure 17.11**).

One characteristic of natural ecosystems is the biodiversity represented in food webs—the complex array of interactions that affect the flow of matter and energy through ecosystems. Interactions among different organisms also play important roles in urban ecosystems, although urban ecosystems typically have far less biodiversity than natural ecosystems. Urban ecosystems also include the complex array of interactions represented in the structure of economies, governments, and society—the human community.

Compared to natural ecosystems, urban ecosystems are especially open to inflows and outflows of matter and energy. Indeed, urban ecosystems depend on inputs of resources from local, regional, and global locations, whereas urban outputs have consequences ranging from local to global scales.

The city limits established by urban governments define the boundaries of certain political and legal jurisdictions, but often have little to do with patterns of transportation, energy use, or citizens' sense of community. Rather than trying to draw firm boundaries, urban ecologists prefer to describe gradients of change from the city proper into its rural surroundings. Thus urban centers grade into less densely populated suburbs and then into sparsely populated exurbs. In the centers of large cities such as Hong Kong, human population density may exceed 10,000 people per square kilometer (25,900 per square mile), while in rural areas there may be 1–10 people per square kilometer.

The processes that support urban ecosystems often extend beyond the boundaries defined by population density. According to the U.S. Census Bureau, the metropolitan area of Los Angeles ends near Palm Springs, but more than 90% of the water and electricity consumed within that area come from places that are more than 100 miles from that boundary. This urban ecosystem is clearly larger than its metropolitan area.

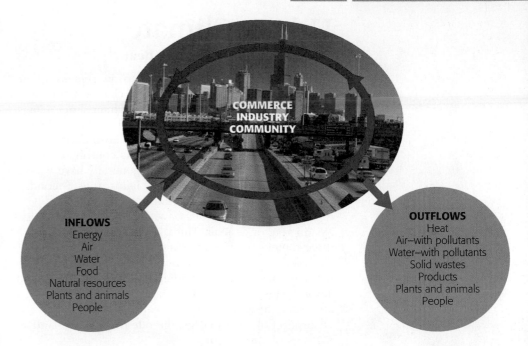

INFLOWS
Energy
Air
Water
Food
Natural resources
Plants and animals
People

OUTFLOWS
Heat
Air–with pollutants
Water–with pollutants
Solid wastes
Products
Plants and animals
People

▲ Figure 17.11 **The Urban Ecosystem**
Ecosystem components and processes are as applicable to cities as they are to natural areas. The ecosystem services that sustain commerce, industry, and community well-being in urban ecosystems depend on inflows and outflows of matter and energy, as well as the movements of plants, animals, and people into and out of the city.

Material Flows

Urban ecosystems require large inputs of material resources and produce considerable outputs of waste.

Every day, significant amounts of materials are moved into and out of cities. Raw materials are brought into cities and used to manufacture finished products. Finished products are then exported for sale. Many different kinds of finished products are imported to support the needs of city residents. Some amount—often a large amount—of this material eventually ends up as waste.

The United States leads the world in the amount of solid waste generated by the average city dweller—720 kg (1,580 lb) per year. Across all U.S. cities, more than 90% of this waste ends up in sanitary landfills; less than 9% is recycled. In comparison, Sweden's municipal waste is an average of 360 kg (790 lb) per person. Only 19% of this waste ends up in landfills; about 58% is recycled, and 23% is incinerated to generate electricity (18.2).

Urban Climate

Cities capture and retain heat.

The transition from rural landscape to city center is often correlated with significant changes in climate. Cities are generally warmer and less humid than their surroundings. The tendency for cities to be warmer than surrounding rural areas is called the **urban heat island** effect.

Four characteristics of urban areas contribute to the heat island effect. First, cities generally have less vegetation than rural and natural landscapes. Second, in many parts of cities, naturally porous soils are replaced with impervious or sealed surfaces, such as concrete or asphalt. Third, cities have complex three-dimensional structures of buildings and "street canyons" that tend to trap heat. And fourth, a great many urban sources, including buildings, factories, and automobiles, release gaseous and aerosol pollutants and waste heat. These urban features affect the absorption of solar energy, the rate of heat loss, and patterns of air movement.

The urban heat island effect is generally larger at night than during the day and larger in the winter than in the summer. During the day, the surfaces of buildings and pavement absorb solar radiation. These surfaces become hotter than soils and vegetation, which are cooled by evapotranspiration. During the night, these artificial surfaces lose heat by radiation. However, the loss of heat is slowed because buildings block the way to the cold night sky (**Figure 17.12**). The radiated heat simply cannot escape the buildings in a large city. Tall buildings also block winds, further inhibiting cooling. In addition, waste heat from air conditioning, automobiles, and industrial operations contribute to the warmth in cities.

Heat island effects vary across urban ecosystems, depending on building density, vegetation cover, and other urban features. City parks and urban greenways decrease urban temperatures (**Figure 17.13**). Even with this variation, urban heat island effects extend for some distance from urban centers. A recent study of 70 urban areas in the eastern United States and Canada revealed that the effects of urban heating extended up to 10 km (6 mi) from cities. Within this range, the growing season for vegetation—the interval between the last spring and first autumn frost—was about 15 days longer than in more rural areas.

The heat island effect has significant consequences for human communities. Buildings and homes in cities require more air conditioning and refrigeration than in rural locations. For example, the heat island effect is estimated to add more than $100 million to energy costs in the city of Los Angeles. Furthermore, death rates due to heat-related illness are much higher in large urban areas than in rural settings.

Cities also affect the amount of rain falling on the landscapes that surround them. In areas that are located downwind from large cities (within 60 km, or 40 mi), the total average rainfall is generally 10–30% higher than

Temperature (°C)

0   50

▲ **Figure 17.12 Urban Heat**
In this thermal satellite image of Atlanta, Georgia, buildings and streets appear yellow to red, indicating that they are warmer than the open lawns, parks, and ponds, which appear green to blue Ⓐ. Heat absorption by buildings and streets and shading by tall buildings produce complex variations in climate within city centers Ⓑ.

URBAN HEAT ISLAND PROFILE

▲ Figure 17.13 **Urban-Rural Temperature Gradient**
On average, temperatures are highest in city centers and cooler in suburban and rural areas.
Adapted from: Figure 2, p 157 of Lemmen, D.S. and F.J. Warren (eds.), 2004: Climate Change Impacts and Adaptation: A Canadian Perspective. Climate Change Impacts and Adaptation Program, Natural Resources Canada, Ottawa, ON, 174 pp. URL:http://adaptation.nrcan.gc.ca/perspective/pdf/report_e.pdf. Last accessed 10-1-2011.

in areas that are upwind (**Figure 17.14**). This is partly due to the heat island effect, which causes air to rise as it passes over a metropolitan area. It is also due to urban aerosols that serve as nuclei for the formation of raindrops.

Air movement in urban centers is limited. This facilitates the formation of stagnant air masses and temperature inversions that hold polluted air over cities. The negative health effects of this air pollution are exacerbated by the higher temperatures of cities (9.1).

◄ Figure 17.14 **Urban Rain Shadow**
In this image of eastern Texas, blue indicates areas of higher rainfall. Note the blue patches downwind (to the east) of the major cities along Interstate Highway 35. As moisture-laden air moves from west to east, the warm air in urban centers causes it to rise, become cooler, and release more moisture to the east of the cities.
Source: NASA URL:http://www.nasa.gov/centers/goddard/images/content/94790main_texas.jpg. Last accessed 10-1-2011

Urban Hydrology

Impervious surfaces limit stormwater infiltration and increase runoff.

A dominant trend along the rural-to-urban gradient is **impervious surface cover (ISC)**, the percentage of land covered with sealed surfaces, such as pavement and rooftops. These impervious surfaces affect the hydrologic cycle in several ways.

High ISC diminishes the infiltration of stormwater, thereby increasing surface runoff (**Figure 17.15**; 7.3). Overall, the higher the ISC, the more precipitation that turns into surface runoff. This greater surface runoff increases the likelihood of flash floods. During non-storm periods, high ISC reduces the recharge of aquifers and the flow in streams. This reduces groundwater storage and the amount of water available in streams and lakes between storms. In addition, high ISC diminishes water quality. Between rainstorms, impervious surfaces collect a wide variety of chemicals and particles, which then pollute stormwater.

Impervious surfaces alter the pattern of water flows in urban streams. Increased ISC shortens flood lag time, the time delay between the peak of rainfall and the peak of flow in streams. Because infiltration is limited, a greater proportion of rainfall runs off, so peak flows are much higher. Thus, urban streams are more prone to flooding.

In addition to ISC, construction and other forms of urban development alter the flow of urban streams.

Over periods of years, the height and width of stream channels change in response to changes in the amount of sediment and patterns of flooding. During the early stages of urban development, streams carry heavy loads of sediment from construction sites; these sediments tend to fill stream channels. Once a city is built up, the source of new sediment diminishes and the frequency and intensity of flooding increases. This increases erosion, causing stream channels to become deeper and wider.

To facilitate development and manage peak water flows, many urban streams have been straightened or channelized. In many cases, their channels have been sealed with concrete to halt erosion. These changes further slow the infiltration of water, diminish flood lag times, and increase peak flows.

Urban development generally decreases water quality. Urban runoff and streams usually carry increased concentrations of chemicals. However, the specific effects on water quality depend on whether the development is residential, commercial, or industrial, on the presence and management of wastewater treatment plants, and on the extent of stormwater drainage. Illegal waste discharges, leaking sewer systems, and failing septic systems are also important sources of pollution (12.2).

QUESTIONS

1. What is the difference between a metropolitan area and a city?
2. What factors cause the climate of urban areas to differ from that of rural areas?
3. How does the relative amount of rainwater runoff in cities differ from that in nearby rural areas? What is responsible for this difference?

▼ Figure 17.15 **Urban-Rural Water Budget Gradient**
The relative amounts of impervious surface and forest cover are significantly different in rural areas and city centers. These differences affect the fate of rainwater falling on the landscape.
Source: http://www.cleanwatermn.org/resources/Hard-Surfaces.aspx. Last accessed 10/10/2011.

NATURAL GROUND COVER | 10–50% IMPERVIOUS SURFACE | 75–100% IMPERVIOUS SURFACE

40% evapotranspiration
35% evapotranspiration
30% evapotranspiration
10% runoff
30% runoff
55% runoff
25% shallow infiltration
20% shallow infiltration
10% shallow infiltration
5% deep infiltration
15% deep infiltration
25% deep infiltration

FOCUS ON SCIENCE: Long-Term Urban Ecosystem Science

Can ecosystem studies be used to characterize variations in human well-being in a large city?

Do ecosystem scientists really study cities? If you had asked that question 20 years ago, the answer would have been yes, but only a very few. Then in 1996, the National Science Foundation (NSF) announced that it would consider proposals to fund integrated studies of urban ecosystems as part of its Long-Term Ecological Research (LTER) Network. The LTER Network promotes long-term synthesis and comparative studies at sites that represent a unique approach to ecosystem research.

Nancy Grimm and her colleagues at Arizona State University responded to the NSF by proposing to create the Central Arizona-Phoenix (CAP) LTER site (**Figure 17.16**). The central goal of the CAP LTER program is to understand the interactions among urban climate, ecosystems, and social systems in the Phoenix metropolitan area. Since its inception, the themes of climate, biogeochemical fluxes, and people have inspired dozens of linked CAP LTER research projects.

▲ **Figure 17.16 Urban Ecosystem Science**
Nancy Grimm, the principle investigator of the Central Arizona-Phoenix Long-Term Ecological Research Site, works with nearly 30 colleagues and students on comprehensive studies of ecosystem processes in the Phoenix metropolitan area.

CAP LTER has placed particular emphasis on understanding how urban sprawl in the Phoenix metropolitan area has altered water flows and water quality. This work has provided the foundation for the development of a computer model called WaterSim, which allows decision-makers to simulate the consequences of their policies on water flows and the overall water supply in the Phoenix area.

Biogeochemical studies have included the flow of materials into and out of Tempe Town Lake. This artificial lake provides flood control and recreational opportunities for a large part of the Phoenix metropolitan area. CAPLTER studies reveal that the amounts and kinds of carbon compounds dissolved in the lake vary through the year as a result of changes in stream flow and the amount of runoff from rooftops and road surfaces (**Figure 17.17**).

Studies of the interactions between people and biodiversity in the Phoenix metropolitan area are designed to determine how human activities and values modify the distribution and abundance of plants, fungi, birds, and arthropods. The diversity of native plants and birds is highest in affluent neighborhoods and considerably lower in poorer parts of Phoenix. Surveys of residents reveal that these differences influence aesthetic perceptions of different neighborhoods.

For more than 12 years, CAP LTER has shown that cities can serve as excellent laboratories for the study of basic ecosystem processes. More important, the insights garnered from these studies are directly relevant to policy decisions regarding the well-being of people in urban settings.

Source: Grimm, N. B., S. H. Faeth, N. E. Golubiewski, et al., 2008. Global change and the ecology of cities. *Science* 319: 756–760; Grimm, N. B., D. Foster, P. Groffman, et al., 2008. The changing landscape: Ecosystem responses to urbanization and pollution across climatic and societal gradients. *Frontiers in Ecology and the Environment* 6: 264–272.

▲ **Figure 17.17 Water in a Desert City**
As this aerial photo shows, Tempe Town Lake is surrounded by an array of impervious surfaces that influence the chemistry of its waters. This artificial lake provides a number of important ecosystem services to Phoenix residents, including water storage, recreation, and aesthetics.

17.3 Urban Land Use

Around the world, urban populations have grown more rapidly than rural populations. However, the amount of land occupied by urban regions has grown far more rapidly than urban populations, particularly in developed countries such as the United States. This means that each urban citizen is using an ever-increasing amount of land. This widespread growth in the amount of land included in metropolitan areas is called **urban sprawl**. Sprawl is a consequence of people moving out of city centers and is influenced by the nature of transportation networks and commercial and residential development in suburbs. Government policies, economic factors, cultural issues, and poor planning often encourage urban sprawl. Sprawl wastes energy, increases air and water pollution, and results in the loss of farmland and natural ecosystems. It also has negative economic effects on human communities and is correlated with diminished human health and well-being.

Urban Population Distribution

Is the population in cities just getting too dense?

Jaymee Castillo, University of California, Los Angeles

Norm: *Actually, in many parts of the United States, it is the diminishing density of cities and growing suburban sprawl that present the greatest environmental challenges.*

Suburban land use is increasing because of population growth and changing patterns of development.

In the 1970 U.S. census, the population of Las Vegas, Nevada, was 358,000 people. This number did not include the 100,000 tourists who were then visiting the city each day. At that time, the urban area of Las Vegas encompassed about 38 square kilometers (15 square miles). By 2010, Las Vegas's population had grown fourfold, to over 1.9 million people, with an additional population of 500,000 tourists each day. Yet during those 40 years, the area of the city had grown nearly twice as fast as its population, ballooning to 290 square kilometers (112 square miles), nearly 7.5 times its 1970 size (**Figure 17.18**).

Over the past 40 years, the population in Las Vegas has grown more rapidly than that of any other large American city. However, the disproportionate growth in its land area is, in fact, typical of most metropolitan areas in the United States and around the world.

Between 1950 and 2000, the amount of land in the United States that the Census Bureau classified as urban grew from 60,700 square kilometers (23,430 square miles) to 242,000 square kilometers (93,400 square miles). That is an increase of about 300%. Some of this change was the result of the growing population in urban areas, which increased by 168%. But if urban population growth had been the sole factor determining this change, total urban land area would have increased to only 162,000 square kilometers (62,600 square miles). The remaining urban sprawl was caused by a significant increase in per capita land consumption.

▼ **Figure 17.18 Sprawling Las Vegas**
This sequence of satellite photos shows the urban development (green) of Las Vegas in 1973, 1991, and 2006. Over this time, the population of the city grew from 358,000 to more than 1,700,000.

1972

1991

2006

Per capita land consumption is the average amount of land used by each urban citizen. It includes the land occupied by homes, as well as the land used for other urban purposes, including roadways, public buildings, factories, commercial offices, and public parks. In 1950, per capita urban land consumption was 0.071 ha (hectares; 0.18 acres) per person. By 2000, it had increased to 0.11 ha (0.27 acres) per person, an incremental change of 0.04 ha per person.

For census purposes, metropolitan areas are divided into a central city, the largest incorporated unit in the area, and the suburbs outside of the central city. Since 1950, central cities have increased in size, but the number of people living within them has shrunk by almost 8% (**Figure 17.19**). The decline in the population density of central cities has been even more striking. In 1950, population density in U.S. central cities averaged about 2,900 people per square kilometer (7,500 per square mile). In 2000, the population density of central cities was only 1,160 people per square kilometer (3,000 per square mile). During this same period, the number of people living in suburbs increased by 43%, and population density increased slightly, from 66 people per square kilometer (175 per square mile) to 77 people per square kilometer (200 per square mile).

These data reflect two persistent trends. First, people have been moving out of U.S. urban centers. In many cases, this decline in population has been coupled with **urban decay**, a decline in economic and community activity and deterioration of infrastructure (**Figure 17.20**). Second, urban populations have shifted to the suburbs, pushing commercial and residential development ever farther from central cities.

PERCENT OF U.S. POPULATION LIVING IN METRO-POLITAN AREAS AND IN THEIR CENTRAL CITIES AND SUBURBS: 1910 TO 2000

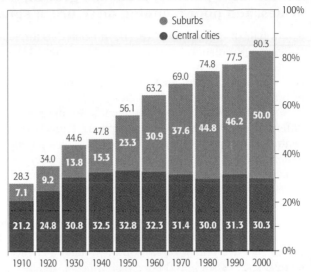

Legend: ● Suburbs ● Central cities

Year	Total	Suburbs	Central cities
1910	28.3	7.1	21.2
1920	34.0	9.2	24.8
1930	44.6	13.8	30.8
1940	47.8	15.3	32.5
1950	56.1	23.3	32.8
1960	63.2	30.9	32.3
1970	69.0	37.6	31.4
1980	74.8	44.8	30.0
1990	77.5	46.2	31.3
2000	80.3	50.0	30.3

◄ **Figure 17.19 Urban-Suburban Shift**
The percentage of the total U.S. population living in metropolitan areas has grown steadily over the past century. Within these areas, growth has been greatest in the suburbs.
Source: Hobbs, Frank and Nicole Stoops, U.S. Census Bureau, Census 2000 Special Reports, Series CENSR-4, Demographic Trends in the 20th Century, U.S. Government Printing Office, Washington, DC, 2002 URL:http://www.census.gov/prod/2002pubs/censr-4.pdf Last accessed 10-1-2011

► **Figure 17.20 Urban Decay**
This abandoned apartment building in Detroit is emblematic of the diminished economic activity and the deterioration of buildings and infrastructure in many city centers.

Causes of Urban Sprawl

Government policies, economic growth, transportation, cultural issues, and poor planning drive urban sprawl.

About half of urban sprawl can be attributed to the increased population in urban areas. The remainder is the result of an overall increase in per capita land consumption in urban areas combined with decreased population density in central cities.

Most suburban land development is significantly different from the high-rise development typical of central cities. In the suburbs, residential development generally focuses on single-family dwellings, often situated on relatively large lots. Suburban commercial development is usually spread out in office parks and strip malls with large parking lots. All of these suburban developments are served by transportation networks that take up even more land. This shift in land use has been driven by government policies, transportation, economic factors, social issues, and poor planning.

Many government policies have encouraged dependence on automobiles and population decentralization. Federal and state funding have encouraged the development of a complex system of highways and secondary roads, allowing sprawl to leapfrog farther and farther away from central cities (**Figure 17.21**).

Commitment to the automobile as the primary mode of transportation is a major factor contributing to U.S. urban sprawl. Limited public transportation alternatives reinforce this commitment, which in turn discourages investment in public transportation. It has been further reinforced by comparatively low gas prices; owing to low gas taxes, U.S. gas prices are about half those in most European countries.

Government policies have facilitated the "American Dream" of single-family home ownership. For example, various government programs have facilitated home financing, and the U.S. tax code allows homeowners to deduct mortgage interest and property taxes. Much of this growth has been spurred by the rapid growth in per capita wealth. In the United States, per capita GDP has more than doubled over the past 30 years. Much of that economic development has been driven by the expansion of the automobile industry and home and commercial construction.

Many of the economic costs associated with sprawl are externalized from the direct costs of commercial and residential development. These external costs often include those associated with road construction and maintenance and the mitigation of environmental impacts, such as increased air and water pollution (**Figure 17.22**). Because property taxes are a large source of their revenue, and many of the costs of sprawl are

▶ Figure 17.21 **Transportation Encourages Sprawl**
The construction of Interstate 15 at the top of this photo has encouraged the development of many suburban communities, such as Corona, California. From Corona, people may commute for more than an hour each way to work near the center of the Los Angeles metropolitan area.

externalized, urban governments often have strong incentives to encourage sprawl.

A variety of social issues have also encouraged flight from central cities and suburban sprawl. These include push factors, such as high crime rates and environmental deterioration associated with urban decay, as well as pull factors, such as the dream of owning a home, better school systems, and the beauty of natural landscapes in some suburban areas. It must also be said that racism has been a factor in the flight of many white Americans from inner cities.

Finally, much sprawl is a direct consequence of poor planning. Political and legal concerns about the rights of private property owners and developers are often powerful disincentives for effective planning. To improve the planning process, decision-makers should envision a city as an integrated ecosystem in which decisions about specific locations have consequences for the functioning and well-being of the entire system and its human communities. Such a vision is discussed in the final module of this chapter.

▼ Figure 17.22 **Sprawling Consequences**
Sprawling developments, such as in this community north of Denver, Colorado, encroach on rural and natural landscapes Ⓐ. In addition to the direct effects of increased air and water pollution, sprawl and its associated dependency on the automobile discourage physical activity Ⓑ.

Consequences of Sprawl

Sprawl degrades human environments, reduces biodiversity, and diminishes human health and well-being.

Urban sprawl affects the human and natural environment in numerous ways. Most important among these impacts are the inefficient use of energy and increased pollution. Increased automobile traffic increases urban air pollution, including photochemical smog and high concentrations of ozone (9.3). It is also an important source of the greenhouse gases that cause global warming. The great expanses of impervious surfaces associated with many sprawling developments increase stormwater runoff and diminish water quality.

Roadways and suburban developments result in the loss and fragmentation of rural and natural landscapes, leading to the loss of biodiversity. In many regions of the United States, the loss of forest cover is directly attributable to urban sprawl. Often the open areas and green spaces within suburban landscapes are too small and disconnected to maintain diverse communities. The loss and fragmentation of open lands has been an important factor in the decline of songbird diversity in many suburban and exurban areas.

Many studies have shown that sprawl has a negative impact on human health. In addition to the direct effects of increased air and water pollution, sprawl and its associated dependency on the automobile discourage physical activity. Obesity and high blood pressure are more prevalent in sprawling communities than in communities where walking and bicycling are encouraged. Several studies suggest that sprawl also takes a psychological toll on human communities. Crowded highways increase commuter "road rage." For those who cannot drive, access to nature and to public services and facilities is often very limited.

Economic studies show that the per capita costs of building and maintaining neighborhood facilities, such as schools, public buildings, and emergency services, increase as per capita land use increases and population density decreases. These costs are reflected in the rapid growth in house sales and property taxes in many sprawling metropolitan areas.

As human populations grow, some sprawl is inevitable. However, continued growth in per capita land use is not inevitable. Today there is widespread agreement that the growth of urban areas can and should be more sustainable, from the standpoint of both the environment and human communities. Careful planning aimed at reducing the many negative consequences of urban sprawl is a critical step in this direction. Changes in urban transportation can help diminish negative environmental and community impacts. These issues are discussed in the following modules.

QUESTIONS

1. What is the connection between per capita land consumption and urban sprawl?

2. Describe five factors that have contributed to increased rates of urban sprawl.

3. What are four negative consequences of urban sprawl for ecosystem and human well-being?

17.4 Urban Planning

Until the last century, the growth of most towns and cities involved only minimal planning, and that planning was done by government officials. Today, much urban growth is guided by professional urban planners. The profession of **urban planning** unites disciplines such as engineering, architecture, ecology, and sociology to design built environments and transportation systems for urban communities. In most cities, urban planners work with city governments, developers, and citizen groups. Urban planners use zoning and growth boundaries to plan for growth and limit sprawl. To be environmentally, economically, and socially sustainable, urban planning must pursue a portfolio of strategies.

Urban Plans and Planning

Dedicated planning programs are relatively new to many cities.

Most towns and cities, even ancient ones, reflect some amount of planning. Most cities are organized around one of three basic plans—a grid pattern, a radial pattern, or a combination of the two. Cities that developed near rivers are often organized in a grid pattern, in which streets and alleys were laid out at right angles and buildings were grouped into rectangular blocks. Cities built on hilltops or around a central market are typically arranged in a radial pattern. In cities with radial patterns, major boulevards radiate outward from the city center like spokes of a wheel; these boulevards are connected by a complicated array of cross streets. Many large cities built during the Industrial Revolution have a combination of radial and grid patterns (**Figure 17.23**).

Historically, defense and aesthetics have also been important themes in the organization of cities. Fortress cities were common throughout medieval Europe, and

▼ **Figure 17.23 Urban Organization**
San Diego, California Ⓐ, provides an example of an urban grid, whereas Paris, France Ⓑ, has a radial organization. Barcelona, Spain Ⓒ, combines grid and radial patterns of organization.

their protective walls imposed constraints on both city design and growth. During Renaissance and Enlightenment times, parks, town squares, and other public areas became more common features of many cities. This led to the development of park or garden cities, in which aesthetics was a fundamental organizing principle (**Figure 17.24**).

During the last century, urban planning became an organized profession and area of academic scholarship. Today, nearly all cities with more than a few thousand residents have urban-planning departments. These agencies are staffed by professionals with interdisciplinary training. Much of the work of these planners is focused on the linked challenges of urban decay and sprawl.

In the beginning, formal urban planning in most cities was a top-down process. Urban planners created plans that were then approved and enforced by city governments. In recent decades, urban planning in most communities has become more democratic. In addition to providing technical input, urban planners engage citizens groups, community organizers, and developers in many important decisions.

In the past, most urban planning focused on transportation and the design of public spaces. Today, urban planning seeks to influence development on private lands by setting rules for the location of residential and commercial development and by using strategies to encourage more sustainable development.

◄ **Figure 17.24 A Garden City**
Savannah, Georgia, is an example of a U.S. city that was planned around garden parks.

Bounding Growth

Zoning and urban growth boundaries provide ways to manage planned growth and development.

One of the most important tools for urban planning is **zoning**, the designation of specific areas for different categories of land use. Urban zones are usually based on categories of land use, such as residential, commercial, industrial, or open space. Zones may also regulate other characteristics, such as building height, the size of residential lots, and residential density. In the United States, zoning is usually regulated by counties or municipalities, although sometimes states set rules or limits on urban zoning. In France and Germany, zoning is regulated by strict national codes.

Zoning serves several purposes. Planners use it to segregate types of land use that are thought to be in-compatible. For example, zoning typically separates residential neighborhoods from high-density commercial developments. Zoning may also be used to provide open space or to conserve the historical character of communities.

Some landowners believe that zoning restricts their right to use their land as they see fit. However, most home and business owners feel that such constraints are more than outweighed by the assurance of knowing what sort of development can occur near their property.

An **urban growth boundary** (**UGB**) is a form of regional zoning that delineates areas suitable for high-density urban development from those suitable for lower-density development. Large residential and commercial developments are permitted only within the boundaries of the UGB. Essentially, UGBs determine where there will be cities and where there will be rural landscapes. In nearly all cases, UGBs are administered by a regional planning agency established by a county or state government with the goal of limiting urban sprawl. Often, the

area affected by a UGB involves metropolitan areas with multiple jurisdictions.

Many urbanized areas across the United States have designated urban growth boundaries, but Oregon is the only state that has mandated UGBs for all of its metropolitan areas. The UGB for the 25 cities and three counties composing the Portland, Oregon, metropolitan area, for example, is administered by the Metropolitan Service District, or Metro. Many credit the Metro and UGB with the revitalization of downtown Portland, including increased urban employment and investment in urban infrastructure and mass transit (**Figure 17.25**).

Few doubt that UGBs can contain sprawl and many of the costs associated with it. However, critics point out that UGBs can cause a shortage of housing units within city limits, which often results in home prices escalating beyond the reach of poorer families. Faced with rapid population growth, regional planning agencies are frequently forced to relocate UGBs. Indeed, since 1979 the population of the Portland metropolitan area has grown by nearly 50%, causing the Metro to enlarge the Portland UGB more than 40 times.

▼ **Figure 17.25 An Urban Renaissance**
Portland, Oregon, has revitalized its city center by limiting suburban sprawl through the use of urban growth boundaries and by investing in transportation and development.

Sustainable Urban Growth

Sustainable planning depends on a combination of different strategies.

In many metropolitan areas, population growth is inevitable. With it, some growth in the amount of land occupied by the city is probably desirable. Yet urban planners recognize that such growth can only be economically, socially, and environmentally sustainable if a combination of 10 "smart growth" strategies is employed. These strategies are the basic tenets of a social movement called **new urbanism.**

How can we encourage people to move to urban centers without resulting in sprawl?

David Oborn, University of Tennessee, Chattanooga

Norm: *Urban growth can occur without sprawl by planning compact communities that offer a variety of transportation choices and focus on human well-being.*

Promote compact communities.

Urban planning should encourage human settlement patterns that lower per capita land consumption. The goal is not wall-to-wall high-rise development. Rather, planners should encourage compact developments embedded in natural landscapes with abundant green spaces. In some ways, this is a return to the modes of city development embodied in many European cities and older U.S. communities (**Figure 17.26**).

Mix land uses.

In the 1960s and 1970s, much planning and zoning focused on razing old neighborhoods and separating residential areas from commercial functions. Among other things, this reinforced dependence on automobiles. New urbanists argue that diversity is important in urban neighborhoods and urban environments. While incompatible land uses should not be combined, an intermix of residential, commercial, and public land uses promotes a sense of community and minimizes transportation needs (**Figure 17.27**).

Create a range of housing opportunities and choices.

Traditional zoning and urban growth boundaries often have the effect of increasing property values so that housing is out of reach for low-income and even middle-income families. Today, many cities promote the urban village strategy, in which relatively dense, multifamily housing and single-family housing are located near shops, cafes, and public transit stops. Tax incentives and subsidized mortgage financing are often used to ensure that housing remains accessible to lower-income families.

Foster communities that provide a sense of place.

Much criticism of suburban sprawl is directed at its environmental, cultural, and social sameness. Strip malls and suburban housing projects all seem to look alike. Sustainable cities thrive on community involvement and promote the appreciation of unique environments and the social context of cities.

Conserve open space, farmland, natural beauty, and critical environmental areas.

The intensive development of rural lands results in the loss of biodiversity and many ecosystem services. Yet all too often, zoning regulations and even UGBs leave the impression that most rural land will eventually be open to development. One strategy for halting rural land loss is for communities to purchase open spaces. A less costly alternative is to buy the development rights to open spaces but allow farming and forest management activities to continue on the land.

Strengthen existing communities first.

Sustainable development gives priority to the efficient use of land that is already committed to urban development. This is the impetus for the renewal of inner-city spaces. Urban infill is a planning strategy that focuses on the development and restoration of undeveloped or abandoned building lots. By strengthening existing communities, cities can grow and evolve through many incremental changes.

Provide a variety of transportation choices.

Urban sprawl encourages dependence on automobiles. Yet dependence on automobiles increases the need for more roads, thereby encouraging more sprawl. To break this cycle, transit-oriented development has become a guiding theme for many urban planners. Strategies for balanced transportation systems are discussed later in this chapter.

▼ **Figure 17.26 Compact Communities**
This urban development in Toronto, Canada, has used architecture and land planning to create appealing urban spaces.

QUESTIONS

1. Describe the three basic plans of urban organization.
2. What are urban growth boundaries, and how are they generally administered?
3. What are the basic tenets of the new urbanism movement?

Create walkable neighborhoods. Pedestrian-oriented neighborhoods promote human well-being and a sense of community. They are also necessary to the success of public transportation systems. Cities such as New York have converted abandoned rail lines to pedestrian corridors.

Make development decisions predictable, fair, and cost-effective. Many cities are notorious for decision making that occurs in secret in backrooms and involves only few powerful individuals. Such processes are likely to be driven by special interests and usually do not serve the broader interests of the environment and the community. In sustainable development, decision making should be transparent, or open to the public, and should take account of environmental, social, and cultural needs.

Encourage community collaboration in development decisions. To be sustainable, decisions regarding urban planning and development require public support and participation (**Figure 17.28**). Involving and empowering the community not only draws on local knowledge and resources but is also critical to resolving the inevitable conflicts over appropriate goals and the means of achieving them.

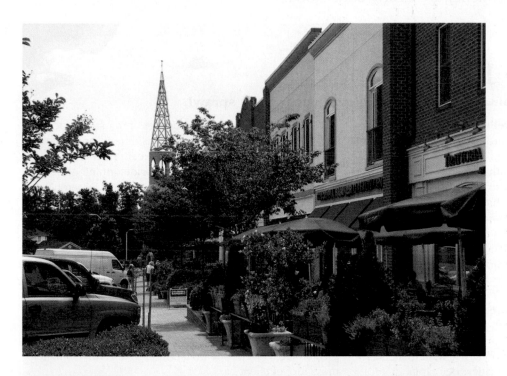

◄ Figure 17.27 **A New Urban Community**
Southern Village, near Chapel Hill, North Carolina, was developed based on new urbanism principles. Shops, churches, and other meeting places populate the town square and are in easy walking distance from surrounding homes and apartments.

◄ Figure 17.28 **Community Involvement**
Open hearings such as this before the Boise, Idaho, City Council provide an opportunity for public input into development decisions. That involvement builds a true sense of community among citizens.

17.5 Urban Transportation

Over the past 50 years, suburban sprawl has been a major factor in the exponential growth in automobile use. In a vicious cycle, sprawl creates automobile dependency and road construction, which then encourage additional sprawl. Automobile dependency has high economic, environmental, and social costs. Public transportation, including bus and train systems, is more cost-effective and has many fewer environmental impacts than travel by automobile. Successful public transportation systems depend on strategies to encourage nonmotorized travel, such as walking and bicycling. Economic incentives for balanced transportation systems are greatly enhanced when the environmental, economic, and social costs of automobile dependency are clearly understood.

Automobile Dependency

Sprawl encourages automobile use, which in turn encourages sprawl.

Around the world, urban dwellers have become increasingly dependent on automobiles. Urban sprawl has been a driving factor in this trend. But this trend is also the result of the unwillingness of city and national governments to invest in alternative forms of transportation. Most cities have responded to the growing dependence on automobiles by expanding road and highway systems. Investment in roads and highways has, in turn, displaced other modes of transportation, including public transit, bicycling, and walking. Furthermore, the proliferation of roads facilitates additional sprawl. Travel times increase, and people become even more dependent on automobiles.

As might be expected, cities in developing countries have considerably lower automobile use than those in developed countries. In such cities, per capita consumption of gasoline in private autos is less than 100 L (26 gal) each year. In contrast, in most American cities per capita consumption of gasoline in private autos is 1,000–3,000 L (260–800 gal) per year. Yet a country's level of wealth is not the only factor determining the degree of automobile dependency in cities. Differences in suburbanization and the availability of transportation alternatives are also important.

Among large cities in the world's wealthiest countries, automobile use generally increases as per capita land consumption increases (**Figure 17.29**). However, there are significant differences among cities with similar per capita land consumption. For example, individuals in U.S. cities consume nearly twice as much gas in their cars as the citizens of Australian cities with similar levels of per capita land consumption. This is because Australians have greater access to public transportation and drive smaller, more fuel-efficient cars.

Increased dependence on automobiles diminishes the environmental, economic, and social sustainability of urban ecosystems. Among the many negative consequences of automobile dependency are increased congestion and accident rates and diminished air quality. Automobile exhaust is the primary cause of photochemical smog and other air pollutants in cities. Increased automobile use also drives up greenhouse gas emissions. The spiraling relationship between suburban sprawl and automobile use contributes to the loss of forest and farmland. Expanding roadways increase the proportion of impermeable surfaces and the management challenges of stormwater runoff.

The economic consequences of automobile dependency include high infrastructure costs and lower worker productivity due to increased commute times. Social impacts include diminished neighborhood and community interactions, diminished public safety, and increased road rage and other health problems.

▼ **Figure 17.29 Suburban Automobile Dependency** The amount of gasoline used by each person generally increases with per capita land consumption. However, at any level of per capita land consumption, gasoline consumption in U.S. cities (green dots) is higher than that in Australian cities (blue dots). This is a consequence of easier access to public transportation in Australia.
Data source: Newman, P.W.G., and J.R. Kenworthy, 1989. Gasoline consumption and cities. *Journal of the American Planning Association* 55: 24–37.

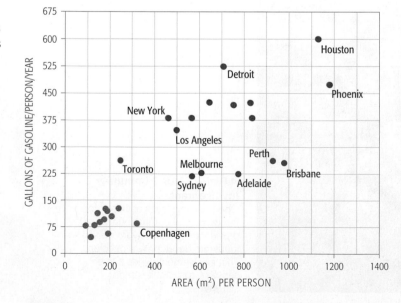

- U.S. Cities
- European Cities
- Australian Cities
- Canadian City

Balancing Transportation Options
Pedestrian-oriented cities encourage public transportation.

In U.S. cities, fewer than 5% of trips are made by walking or biking. Only 12% of trips take advantage of public transportation. In European cities, 20% of trips rely on walking and biking, and 35% rely on public transit.

To diminish the vicious cycle of automobile dependency and sprawl, many U.S. cities are taking actions to provide a more balanced array of transportation options. Different modes of public transit vary in their energy use and operating cost. However, on a per passenger basis, they are all more efficient than automobiles.

Public buses have long been a transit option in many cities, although poorly maintained vehicles and limited routes have often kept ridership low. Where bus fleets have been modernized and routes configured to maximize efficiency, bus use has increased dramatically, with a corresponding decrease in automobile use. In large cities, for example, planners estimate that each bus replaces 250–300 cars each day.

Many larger urban centers depend on rail transportation systems. Extensive subway and aboveground rail systems, such as Washington D.C.'s Metro, New York's subways, Atlanta's MARTA, and San Francisco's BART, carry 25–35% of commuters in these cities. Light rail systems using smaller, electrically powered trains, are being developed in many metropolitan areas (**Figure 17.30**).

Bus rapid transit (BRT) combines many of the advantages of bus and rail transportation. In BRT systems, buses move on dedicated roadways that allow them to bypass road traffic. BRT roadways feature sophisticated boarding stations, prepaid fares, and frequent service (**Figure 17.31**). BRT systems can be developed at much lower cost than light- or heavy-rail systems.

One of the biggest economic impacts of automobiles is the space they require for parking and roads. Automobile-dependent cities require three to five parking spaces for each car. Automobiles on a multilane freeway can transport about 2,500 people per hour. In contrast, a bus lane can transport 5,000–8,000 people, light rail can carry 10,000–20,000 people, and a heavy-rail subway can transport up to 50,000 people each hour.

Urban planners have found that for public transit to be successful, nonmotorized transportation, such as walking and bicycling, must be facilitated. Heavy traffic and multilane streets present major obstacles to pedestrian and cycle activity. This discourages movement within communities and also limits access to centers of public transportation.

Transit-oriented development emphasizes the role of streets for more than automobile traffic. Many cities have transformed former roadways into pedestrian malls. This has

▲ **Figure 17.30 Light Rail**
The new light-rail system in Houston, Texas, is one factor accounting for diminished automobile traffic congestion in its city center.

▼ **Figure 17.31 Bus Rapid Transit**
As with BRT systems in other cities, buses in Bogotá, Colombia, have their own lane and boarding stations, which are physically separated from the rest of the highway.

encouraged walking and cycling and revitalized downtown business districts (**Figure 17.32**). In an increasing number of American cities, dedicated walking trails and bike lanes encourage nonmotorized transportation.

To transform cities into pedestrian-oriented places, attention must be given to aesthetics and safety. Many cities have undertaken downtown beautification programs that include trees, gardens, and public art. Some cities require that a small portion of the budget of public buildings be dedicated to landscaping and public art. To improve public safety, cities can provide public call boxes, efficient evening lighting, and pedestrian police.

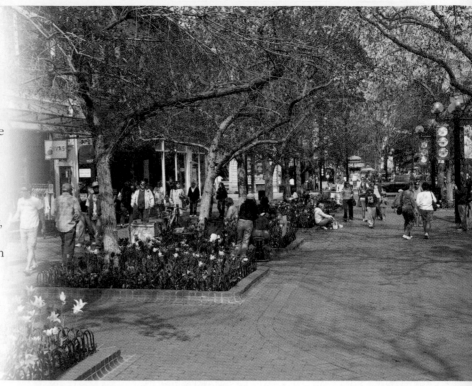

► Figure 17.32 **Pedestrian Friendly**
This pedestrian mall in downtown Boulder, Colorado, has become an attraction for residents and visitors and has moved automobile traffic out of the city center.

Urban Transportation Economics

Sustainable balanced transportation requires a full assessment of the costs of automobile dependency.

One of the reasons for our continued dependence on automobiles is that neither car owners nor city officials have to pay many of the costs associated with auto use. Low U.S. gas prices and free parking make automobiles appear to be the cheapest form of transportation. Drivers do not factor the cost of roadways into their travel expenses. When cities construct new roads and freeways, the costs are often subsidized by national governments. The costs of pollution on the health of humans and ecosystems are also externalized.

Development of balanced transportation systems can be greatly facilitated by strategies to internalize the real costs of auto dependency. Gasoline taxes and road tolls are mechanisms to do this. In its Central City Plan, Portland, Oregon, placed a cap on the number of downtown parking spaces and increased parking fees. In addition, several large parking lots were torn out and replaced by pedestrian parks.

Congestion pricing can be useful in limiting automobile traffic and encouraging the use of public transportation. Oslo, Norway, for example, raises about $75 million each year by charging drivers for the privilege of driving in the city. London has developed the world's largest market-based system to limit automobile use. For the privilege of driving in certain downtown areas between 7 A.M. and 6:30 P.M., private vehicles must pay $12 each day. Closed-circuit television monitors record license plate numbers, and fines for nonpayment can exceed $200 (**Figure 17.33**).

The success of green transportation systems ultimately depends on changes in planning goals. As former New Jersey Governor Christine Todd Whitman commented, "If you design communities for automobiles, you get more automobiles. If you design them for people, you get walkable, livable communities."

▼ Figure 17.33 **Congestion Market**
During peak traffic hours, motorists driving in London's Central Zone are charged a daily fee.

QUESTIONS

1. Describe two factors that influence automobile dependency worldwide.

2. Compare the proportion of automobile use relative to public transportation in U.S. and European cities.

3. How do different modes of transportation in urban areas compare with regard to the number of people transported past a place per hour?

17.6 Urban Biodiversity

Urban ecosystems provide habitats for a diverse array of native and non-native plants and animals. Most of these species are able to tolerate a wide range of environmental conditions and to disperse easily from one location to another. Food webs in urban ecosystems may be complex and are often subsidized by garbage and other inputs from humans. Because they moderate urban climates and provide habitat for other species, trees play a key role in cities. Urban gardens and green roofs provide food for humans, diminish stormwater runoff, and improve water quality. City parks, greenways, and waterways are especially important in the maintenance of urban biodiversity and ecosystem services. Such habitats are most effective when they are designed and managed as integrated networks.

▲ **Figure 17.34 An Urban Tree**
Tree of heaven produces tiny windblown seeds that can germinate and thrive in a variety of urban habitats.

Urban Wildlife

Urban species are widely dispersed and tolerant of a wide range of environments.

In Betty Smith's acclaimed 1943 novel, *A Tree Grows in Brooklyn*, a single tree—a specimen of tree of heaven (*Ailanthus altissima*)—is a central metaphor for a young city woman's coming of age. Tree of heaven is also representative of the vast majority of plant and animal species that thrive in urban environments (**Figure 17.34**). The tree is well known for its ability to tolerate a wide variety of environments, including sidewalk cracks, abandoned industrial sites, and disturbed roadsides. It is an invasive, non-native tree that produces large numbers of small, wind-dispersed seeds. It is also a source of food and habitat for squirrels, a variety of birds, and a host of insects that share its broad range of tolerance and ability to disperse.

Ecologists classify most of the species that thrive in urban centers as habitat generalists. These organisms are able to take advantage of habitats that have no counterpart in nature, such as mowed lawns, vacant fields, garbage dumps, and even cracks in sidewalks. Tree of heaven, dandelion, and other "weedy" species are habitat generalists, as are robins, rats, and pigeons. Certain falcons and hawks are interesting exceptions. Although these birds of prey are highly specialized, they are able to feed on the plentiful birds and rodents that live in cities; sometimes these birds nest in skyscrapers, finding the canyon-like structure of many downtown areas similar to rocky cliffs in their native habitat (**Figure 17.35**).

The diversity of creatures in urban ecosystems decreases along the gradient from rural areas to the city center. In exurban and suburban communities, flora and fauna from surrounding farms and wild lands may be common and have a significant impact on ecosystems. Some of these impacts, such the grazing of ornamental plants by deer or the raiding of garbage containers by raccoons and opossums, might be classified as benign nuisances. Others, such as attacks on humans and pets by mountain lions and bears or the spread of infectious diseases such as plague and rabies, are much more serious.

Moving from highly urbanized to suburban and rural settings, the number of plant and animal species generally increases (**Figure 17.36**).

▼ **Figure 17.35 An Urban Predator**
Rodents and pigeons provide abundant food for urban peregrine falcons such as this one.

◀ **Figure 17.36 Urban-Rural Gradient Biodiversity**
In a study of 15 London gardens, the diversity of insects was found to increase with distance from city center.
Source: Davis, B.N.K. 1979. The ground arthropods of London Gardens. London Naturalist 58:15–24. London Naturalist is published by the London Natural History Society http://lnhs.org.uk/

Even so, urban food webs can be quite complex. Plants, including ornamental flowers and shrubs, street trees, and weeds, provide some of the energy for such food webs. Human inputs, in the form of pet food, birdseed, and all sorts of garbage, are often a major source of energy for urban food webs.

The mowed lawn is a distinctive urban habitat that is useful for recreation and urban aesthetics. Because lawns are permeable surfaces, they facilitate water infiltration. However, in their fervor to have the greenest lawn in the neighborhood, homeowners often use large quantities of fertilizer and pesticides that contribute to stormwater pollution. Furthermore, lawn maintenance can require large quantities of water in dry climates (**Figure 17.37**).

▲ **Figure 17.37 Xeriscaping**
Many communities in arid regions are encouraging homeowners to replace water-demanding lawns with drought-adapted landscaping.

Green Infrastructure

Trees, gardens, and green roofs provide a great many services in urban ecosystems.

In 1792, the citizens of Philadelphia took the revolutionary step of petitioning their city government to plant trees along city streets. Prior to that time, planting trees near buildings was prohibited for fear that they would increase the risk of catastrophic fires. The Philadelphians argued that the risk of fire was more than offset by the "healthful effects" of trees—summer shade and beauty. Street trees were not widely used in either European or American cities until the 1800s.

Today, the benefits of urban trees are widely recognized. In summer, deciduous trees provide shade that mitigates urban heat island effects. In winter, they allow solar warming. Evergreen trees are used as windbreaks in many cities. Both deciduous and evergreen trees provide nesting sites and habitat for a diverse array of urban birds. Perhaps most important, trees are greatly valued by city dwellers for their beauty (**Figure 17.38**).

Most cities have programs for planting and managing trees that are overseen by professionals trained in urban forestry. Urban foresters must understand which tree species are most likely to thrive at a particular location within a city. Some tree species require open, sunny environments, and others can thrive in shady urban canyons.

Urban environments present special challenges for tree growth. Urban soils are often compacted and poorly drained and are likely to be surrounded by impermeable surfaces. Without special soil preparation and drainage, tree roots may experience extremes of wet and dry conditions. Pollution and disease also affect the growth of urban trees. For example, moderate to high levels of ozone and photochemical smog can

▶ **Figure 17.38 Urban Forests**
Trees such as these in Philadelphia beautify cities; they also provide shade and diminish the urban heat island effect.

suppress the growth of many trees (**Figure 17.39**). Because urban trees are often grown in single species groups, they are particularly vulnerable to insect and fungal pathogens.

Urban gardens have become an increasingly prominent feature in many urban landscapes. In many inner cities, trash-strewn vacant lots have been transformed into productive vegetable gardens. Urban flower gardens can beautify abandoned industrial sites, and urban vegetable gardens can provide fresh, healthful food for local neighborhoods. Such gardens have been the focus for recreation and educational activities in many communities (**Figure 17.40**).

Green walls represent an interesting variation on the urban garden theme. A nonprofit group, the Urban Farming Food Chain, uses recycled materials to manufacture modular wall panels that provide a growth medium and automatic irrigation for vegetables. "Edible walls" can provide fresh produce to communities that lack open land.

In the 1960s, German architects began to experiment with green roofs. These are roofs constructed of multiple layers to support the growth of plants while protecting the interior of the building from the elements (**Figure 17.41**). Green roofs greatly diminish stormwater runoff. Properly designed, they can retain up to 75% of rainwater, gradually releasing it back to the atmosphere through evaporation. Green roofs also decrease the urban heat island effect.

▲ **Figure 17.39 Air Pollution and Urban Trees**
Ozone pollution has damaged the green needles of this urban white pine, causing the tips to turn brown.

▲ **Figure 17.40 A Community Resource**
This garden in downtown Philadelphia is one of its nearly 500 community gardens. Across the city, 600 families produce about $1.5 million worth of food each year.

▲ **Figure 17.41 Green Roof**
The plant cover on top of Chicago City Hall keeps summer roof temperatures 14–44 °C (25–80 °F) cooler than the roofs of surrounding buildings. Engineers calculate that on hot days Chicago temperatures would be as much as 7 °C (13 °F) cooler if all downtown buildings had green roofs.

City Parks, Greenways, and Waterways

In integrated networks of urban green spaces, parks and preserves are connected by greenways and waterways.

The 1859 dedication of Central Park in New York City was a landmark event that has since affected the planning of hundreds of other cities. Central Park was designed by Frederick Law Olmstead, a pioneer in city planning who was committed to blending the natural environment of a place with the aesthetic qualities of the city. To give city dwellers access to nature, Olmstead included wooded landscapes, open "pastures," and ponds in the design of Central Park. His design also integrated transportation corridors, public spaces, and formal gardens (**Figure 17.42**). Olmstead was well aware of other important benefits of city parks, including water management and escape from city heat.

Although Olmstead is best known for his design of city parks, he also played a central role in the development of the profession of city planning. He recognized the importance of streams and wetlands in managing stormwater flows and maintaining water quality. In the late 19th century, he planned the Back Bay Fens project for the city of Boston. This project converted a sewage-filled swamp into a network of artificial streams and restored marshes that were able to hold floodwater. The project also reduced the flow of pollution into the Charles River. The Fens was carefully planted so that it would serve as a recreational park.

It is ironic that many of the streams and rivers that once nurtured the growth of cities have since been nearly destroyed by the urban and industrial development associated with those cities. In many places, urban streams have been channelized, and riparian vegetation has been cut back or eliminated entirely. In recent years, a number of cities have undertaken major projects to restore streams and rivers. These projects aim to reduce pollution and flooding and are also central components of comprehensive urban renewal (**Figure 17.43**).

Biodiversity, ecosystem services, and community benefits can be greatly enhanced when urban green spaces are viewed as hubs and links in an integrated network. Hubs include parks and preserves that are large enough to harbor sustainable populations of flora and fauna and to sustain key ecosystem processes. Hubs are also destinations for human recreation and educational activities, such as gardens and museums. Greenways and stream corridors are the links that provide pathways for wildlife and people to travel between the hubs. Greenways also provide people with opportunities for recreational activities, such as biking and hiking. Historic sites might also be included in such a network.

▼ Figure 17.42 **Central Park, New York**
Frederick Law Olmstead's design for New York City's signature park set a standard for the development of city parks around the world.

▼ Figure 17.43 **A Restored Urban River**
In the early 1920s, the City of San Antonio, Texas, considered paving over its namesake river. Instead, it invested resources to restore the San Antonio River and develop the River Walk that is now lined with shops and restaurants. Flows are maintained in this dry city using treated municipal wastewater.

SEEING SOLUTIONS: Restoring the Los Angeles River

In about 1781, Spanish settlers in southern California established a small village, El Pueblo de Nuestra Señora la Reina de los Ángeles del Río de Porciúncula—the town of Our Lady the Queen of the Angels on the Porciuncula River. By 1850, this growing village and its nearby river had been renamed Los Angeles (**Figure 17.44**).

Through much of its early history, the Los Angeles River was the only source of water for the town. A complex series of canals diverted its water to the town and nearby agricultural fields. During the dry season, however, the river carried little water, and by 1900 the growing city needed more water than the river could supply. In 1913, the city began receiving large quantities of water from the Owens Valley far to the north. Then in the 1930s, two major floods on the Los Angeles River caused significant property damage and more than 50 deaths. In response, city and federal agencies straightened the river and lined it with concrete to more quickly carry its floodwaters to the sea (**Figure 17.45**).

By the year 2000, Los Angeles and its suburbs had grown up to the river's edge, and most vestiges of riparian vegetation had disappeared. Although the concrete river still served to divert rainwater, it had become an eyesore and collection place for trash. Most Angelinos were unaware that a once vital river flowed through their community.

But some Angelinos did value the river, and these citizens formed groups such as the Friends of the Los Angeles River to encourage its restoration. In 2002, the Los Angeles City Council responded to the advocacy of these groups by establishing the Committee on the Los Angeles River. In 2005, Mayor Antonio Villaraigosa signed the Los Angeles River Revitalization Master Plan, a 20-year plan for restoring and revitalizing the river. The plan relies on an integrated network of over 200 individual projects. Where possible, concrete will be removed and the meandering river channel restored. Over a 32-mile stretch, the river will serve as a corridor linking small lakes, recreational areas, and wildlife parks. In addition, the plan identifies housing, transportation, and development opportunities that will enhance nearby communities. In May 2007, the Los Angeles City Council unanimously approved $2 billion for the plan.

The environmental benefits of restoring the river will certainly be significant. The economic and social benefits are likely to be significant, too. As one member of the council commented, "revitalization of the Los Angeles River enables us to redevelop and utilize existing open space to create recreational opportunities for the diverse communities throughout Los Angeles connected by the river."

▼ **Figure 17.44 The Los Angeles River**
This river flows 82 km (51 mi) from the foothills of the San Gabriel Mountains to the sea.

▼ **Figure 17.45 A River No More?**
To avoid floods, the Los Angeles River was straightened and lined with concrete. Today, it is little more than a canal.

17.7 The City as a Sustainability Strategy

Over the next 30 years, an additional 2 billion people will be added to our population. Most of that growth will occur in cities. Cities concentrate human impacts in ways that are not sustainable. Cities rely on inputs of energy, natural resources, and water, and the per capita demands for these inputs are increasing. Cities are the source of nearly 80% of human greenhouse gas emissions and 90% of Earth's garbage. They are also home to a disproportionate share of human poverty and misery. Yet it is impossible to imagine the current or future human population living on Earth without cities. It is within cities that human societies have developed much of their social and economic complexity and cultural diversity. By mimicking important features of natural ecosystems, cities can play a central role in the development of a sustainable future. These features include the efficient capture and use of energy and matter, biodiversity, self-regulation, adaptation, resiliency, and beauty.

Efficient Capture and Use of Energy and Matter

Cities' metabolism should be circular rather than linear.

Writer and urban philosopher Herbert Girardet notes that unlike natural ecosystems, which have circular metabolism, most urban ecosystems have linear metabolism. By this, Girardet means that cities require very large inputs of energy and materials; cities then use or recycle these resources inefficiently, resulting in large outputs of waste.

The circular metabolism of a forest ecosystem is illustrated by the cycling of nitrogen (7.5). A nitrogen molecule entering the forest in a raindrop has a likely residence time of a century or more. When the raindrop reaches the ground, the molecule is rapidly absorbed by a plant root, transported up a stem, and used to make a protein in a leaf. In autumn, this leaf falls from the plant and decays on the forest floor, returning the nitrogen molecule to the soil. In spring, the nitrogen will again be taken up by a plant

root and serve another term as a leafy protein. This circular process can go on indefinitely. If the leaf is eaten by an insect, the nitrogen molecule might make a journey through the forest food web. But ultimately it returns to the soil to be recycled. The annual influx of nitrogen is meager in comparison to the amount of nitrogen needed to sustain the forest's annual productivity. Most of the nitrogen the forest needs comes from recycling.

In contrast, the residence times for most materials coming into urban ecosystems are measured in weeks or months. Natural resources are often manufactured into finished goods that are used only once before they are discarded (**Figure 17.46**). Thus, the annual per capita demand for resources is particularly high among urbanites. This rate of resource use is unsustainable environmentally. This inefficiency and waste also inflict economic and social costs.

REUSE

INPUTS   OUTPUTS   RECYCLE   WASTE

ENERGY CONVERSION

▶ **Figure 17.46 Linear Metabolism**
Compared to natural ecosystems, the flow of materials and energy in urban ecosystems is linear.

Stockholm, Sweden, is one of several cities that have implemented strategies to circularize their metabolism. In its model ecological district, Hammarby Sjöstad, residents put their solid wastes into an underground collection system that separates recyclable materials from organic and other wastes. The recyclables are sold to factories, which use them to manufacture new products. Combustible wastes are burned, and their heat is used to generate electricity and hot water. The hot water is used for residential heating. Stormwater from streets is collected in a special purification system, and green roofs purify water flowing from buildings. Wastewater, including human waste, is kept separate from stormwater and treated so that methane can be captured and used as a source of energy. All of Hammarby Sjöstad's buses are powered by this biogas (**Figure 17.47**).

▶ **Figure 17.47 Circular Metabolism**
The waterfront at Hammarby Sjöstad Ⓐ resembles that of other European cities. But all waste is collected in special bins Ⓑ and subsequently sorted Ⓒ to maximize recycling. Along streets, stormwater is diverted through gardens to moderate flows and remove pollutants Ⓓ.

Diversity

Biodiversity enhances ecosystem services and provides economic and social benefits.

Ecosystem functions and services, such as the cycling of nutrients and the purification of water, depend on a diverse community of microbes, plants, and animals (1.3). Yet historically, urban planning has given little attention to biodiversity. The resulting lack of biodiversity in urban centers has economic costs, since humans must pay to restore lost ecosystem services. It also has social costs. In his book *Last Child in the Woods*, Richard Louv argues convincingly that separation from nature harms the physical and emotional health of both children and adults.

Parks, greenways, and streamside buffers can help restore biological diversity to urban ecosystems. They also diminish urban heat island effects, mitigate flooding, and purify stormwater. In addition, they provide opportunities for recreation and education.

Many urban planners envision even greater integration of natural ecosystems in urban settings. Urban Biosphere Reserves that have been proposed for several European cities have the goal of connecting rural landscapes with networks of urban green spaces. The Green-Blue Meander in the heavily populated Ranstad region of the Netherlands is one such project. Here, waterways connect fragmented areas of forest and pastureland to urban greenways and parks (**Figure 17.48**). Legacies of past industrialization, such as abandoned lands, are being restored. Planners are optimistic that this development will enhance the quality of life in urban centers and serve as a boundary for sprawl.

▲ **Figure 17.48 The Green-Blue Meander**
This system of parks, greenways, rivers, and lakes will provide important ecosystem services and beauty to several of the largest cities in the Netherlands, including Leiden, The Hague, and Rotterdam.

Self-regulation and Adaptation

Knowledge and learning are critical prerequisites to adaptation.

Natural ecosystems have a remarkable capacity for self-regulation and adaptation. Forests moderate weather extremes and ensure regular flows in the streams that drain them. While individual species may appear and disappear in particular places, the total number of species and their relative abundance remain remarkably constant. This all happens largely as a consequence of feedback between an ecosystem's diverse species and their abiotic environment (1.3).

In urban ecosystems, responses to environmental change and patterns of growth are determined largely by human interactions—economic, political, and social. These human interactions lack many of the homeostatic mechanisms of natural ecosystems. For example, many of the everyday costs of sprawl are not included in the economics and commerce associated with development. Therefore, incentives for sustainable

urban planning and development are often overlooked. Political boundaries, such as city limits, often result in processes of decision making that are fragmented across urban landscapes. Urban governments that are not fully engaged with all elements of their community are prone to make decisions that are socially inequitable and unsustainable.

Sustainable self-regulation of urban ecosystems requires a connection between planning and process. The scales over which decisions are made must be in agreement with the scales over which processes occur. The flows of water and waste materials, for example, often extend across city and county boundaries. In response, cities and counties in many urbanized areas have organized metropolitan or regional councils that have authority to coordinate planning and regulate the actions of their members.

When environmental costs are included, urban planning and development decisions are far more likely to be sustainable. For example, congestion pricing is a very successful strategy for decreasing automobile use and encouraging use of public transit.

Knowledge and learning are critical prerequisites for adaptation. Cities and agencies within urban ecosystems must systematically monitor key processes and measures of sustainability. They must also foster the means to learn from this information and adapt. Citizens, too, must be well informed. Thus, educational programs such as those implemented by Malmö are critical elements in sustainable urban development.

Resiliency

Change is inevitable, and renewal must be seen as a continuous process.

Through the process of succession, natural ecosystems have a remarkable capacity for renewal and restoration (6.5). Disturbances such as fires and hurricanes have occurred in Earth's ecosystems for millions of years. Species in natural ecosystems possess myriad adaptations that make individual populations, communities, and ecosystem processes resistant to disturbance and allow ecosystems to recover rapidly when disturbances occur. The inevitability of disturbance and the capacity for recovery are, quite literally, embedded in the DNA of the ecosystem.

Disturbances are inevitable in urban ecosystems, too, a fact that must be acknowledged in urban planning. Many communities have developed plans for responding to the effects of earthquakes, fires, and floods. Yet as demonstrated by the hurricanes of 2005 in New Orleans and the massive earthquake and tsunami of 2011 in Japan, such planning is often inadequate. Cities often fail to plan for economic and social disturbances, such as the collapse of major industries or major shifts in the demand for residential housing.

Diversity—the variety of species and complexity of processes—has been found to be an important factor contributing to the ability of ecosystems to resist and recover from disturbance. Urban diversity—variety in the economy, the culture, and political view-points—has a similar effect on the resilience of cities.

▶ **Figure 17.49 Walkable Communities**
New York's High Line Park was completed in 2009. This 1.5-mile-long park was created from the rundown and decaying remnants of a railway system in the lower west side of Manhattan.

Beauty

Just as in their natural counterparts, beauty is a key element in the sustainability of urban ecosystems.

We are drawn to nature by its beauty. Indeed, that beauty is often a central motive for our preservation of natural places. Beautiful urban environments also motivate preservation and sustainable behavior.

For example, the recent transformation of an abandoned railway line into a park on the west side of Manhattan has greatly increased the beauty of that community. The new High Line Park is a garden walkway that offers a view of the Hudson River and a retreat from noisy city streets. Foot traffic is heavy, and residents and visitors often compare the new High Line Park to Central Park (**Figure 17.49**).

In an essay called "The Land Ethic," Aldo Leopold wrote, "A thing is right when it tends to preserve the integrity, stability, and beauty of the biotic community." Cities are human-centered ecosystems. They exist to support the complex environmental, economic, and social needs of human communities. Nevertheless, if we were to craft an urban ethic, its central creed might well be that *a city is sustainable if it tends to imitate the integrity, stability, and beauty of a natural community.*

QUESTIONS

1. What is meant by the assertion that a city's "metabolism" should be circular rather than linear?

2. Describe three ways in which the governance of urban ecosystems can become more sustainable.

AGENTS OF CHANGE: Stockton College's Greening Urban Spaces

Jessica Franzini is a graduate of Stockton College in Galloway, New Jersey. While an undergraduate, she led New Jersey Water Watch's first vacant lot transformation project, working with municipal officials and volunteers to transform a vacant lot in Galloway City, New Jersey, into a green community space. After that project, Jessica went on to serve as codirector of Stockton College's Stockton Save. Since graduating, she has served as program director of the New Jersey Tree Foundation's Urban Airshed Reforestation Project (UARP). In that role, she has planted hundreds of trees in Camden, New Jersey, while uniting communities, improving the environment, and teaching community members about urban forestry.

How did you first get the idea to green an urban space?

Early in college, I took a class called Global Issues. When we learned how small the U.S. population is in relation to how much we contribute to environmental degradation, I realized that it's not just the size of a country's population that's important. What really matters are the lifestyle choices that the people in a country make—individually and collectively. In that course we learned about sustainable cities and areas in Europe where public transit is used more frequently, people bike instead of drive, and food is grown locally. I realized that

was the kind of place I wanted to live, and that there was a lot I could do to make a difference, especially in an urban space.

What steps did you take to get started?

I began by getting involved with a student club on campus. Through that club, I volunteered to coordinate a project. I decided to look in Galloway, New Jersey, near Stockton College, where I found some abandoned land covered in trash. My goal was to transform that desolate area into a green community space. I started meeting with city officials and other municipal workers, who went out to show me which areas were owned by the city. The muncipality was able to help with all the things that seemed so overwhelming to me as a student.

Having to organize volunteers really intimidated me. But as I started attending weekly meetings, I saw that everyone loved my idea, so I became more confident. We got a lot of support by setting up tables at one of the busiest areas on campus. We then rolled out a large canvas and provided paint for people to illustrate something they wanted to change about the environment. Everyone who came over to the table got information on all the projects we were doing, including my vacant lot project in Galloway. Surprisingly, people really responded to this!

I then organized volunteers from Stockton and Galloway to recycle and dispose of the trash, build park benches, and plant a dogwood tree and the other plants that had been donated to us. If you have a positive attitude and invite others to share part of themselves right away, you will find people who want to support you.

What are the biggest challenges you've faced with the project, and how did you solve them?

I have to admit, I didn't know what I was doing when I started this project. My project was the first of its kind, and I didn't know how to take care of trees, and neither did anybody else. We weren't quite prepared.

We got lots of volunteers, beautiful plants, and had an awesome event. However, we soon realized you can't just plant a tree and expect it to grow. You have to take care of it. We hadn't thought of a maintenance plan at all. Because of this experience, we learned and were able to do further greening projects more effectively.

How has your work affected other people and the environment?

With the New Jersey Tree Foundation, I've worked with community members to plant hundreds of trees in some of the most dangerous communities in the country. I've seen the trees and the people come alive. Being able to help people who want to improve their community and take pride in it makes me feel like I make a difference every day.

What advice would you give to students who want to replicate your model on their campus?

First, see what's already out there that could be a good resource. Connect to a group that is already active, or create your own group with specific goals in mind.

When you know what materials or support you will need, you can seek donations or small grants to make the work happen! I recommend reaching out to local businesses, organizations, and corporations as partners. And remember, volunteers go a long way! A small group of people can work together to motivate more students as volunteers to move projects forward in a fun and exciting way.

Regular meetings are helpful. Decide a project point-person, and develop a committee with those who want to help. Part of being a good leader is learning to delegate and share tasks with others. Students are empowered when they can truly contribute to the planning and implementation of a worthwhile project.

Most importantly, continue to build experience. Without my college experience in organizing the community to transform a vacant lot, I would never be able to do the work I do now. Learning about plants in class just isn't enough. You have to get out in the world and do something!

Summary

Over the past 10,000 years, the development of industry and technology has encouraged the growth of human populations in urban centers. The growth of cities has, in turn, stimulated the development of new industries and technologies. Urban ecosystems support the commerce, industry, and community interactions of a large number and high density of humans. By 2050, 70% of the world's population is expected to live in cities. Cities have precise political boundaries, but the climate, hydrology, and flow of materials in metropolitan areas change along a gradient from densely populated city centers to less densely populated suburbs, exurbs, and rural areas.

Over the past 50 years, government policies, economic growth, dependence on automobiles, and poor planning have produced urban sprawl in many metropolitan areas. Increased urban land consumption has resulted in the loss of ecosystem services and diminished human well-being. Urban planners can use zoning, urban growth boundaries, and the development of parks and greenways to reduce urban sprawl and many of its negative impacts. Sustainable urban planning must provide alternatives to the use of automobiles, such as pedestrian corridors, subways, light rail, and bus rapid transit. Economic incentives such as higher parking fees and congestion pricing can diminish the use of automobiles in city centers.

Urban ecosystems provide habitats for a diverse array of organisms; however, urban food webs are generally simple. Cities that support biodiversity and use energy and matter more efficiently are more resilient, adaptable, and beautiful. Such cities will be essential to the well-being of future human populations.

17.1 Urbanization

- The first cities with formal governance and division of human labor appeared in association with the domestication of plants and animals and the development of agriculture.

- Urbanization is an especially prominent trend in the world's poorest countries; in these countries the well-being of people living in cities is less than that of people living in rural areas.

KEY TERMS

urbanization, megacities

QUESTIONS

1. The people who lived in ancient cities were often less healthy than people who lived in the rural countryside. What factors contributed to this so-called urban health penalty?

2. In what ways has urbanization required the restructuring or reorganization of many of Earth's ecosystems?

17.2 Urban Ecosystems

- Urban ecosystems are dependent on interactions between organisms and their nonliving environment and on the flows of matter and energy; they are also sustained by the flows of ideas and money and by the movement of people.

KEY TERMS

urban ecosystem, urban, metropolitan area, suburb, exurb, urban heat island, impervious surface cover (ISC)

QUESTIONS

1. In the United States, rainfall is often higher to the east of large metropolitan areas. What causes this pattern of rainfall?

2. Why must urban ecosystems be especially open to flows of energy and materials?

3. How would you expect the water quality of urban streams to differ from that of streams in nearby forested areas?

17.3 Urban Land Use

- Many U.S. metropolitan areas have experienced urban sprawl; human populations have grown most in suburban and exurban areas, resulting in diminished population density in urban centers and increased per capita land consumption.

- Government policies, economic growth, highway development, and poor planning have encouraged urban sprawl.

- Urban sprawl has diminished biodiversity and a variety of ecosystem services that influence human well-being.

KEY TERMS

urban sprawl, per capita land consumption, urban decay

QUESTIONS

1. Urban sprawl has produced very different trends in urban centers than in suburbs. Explain.

2. Many experts feel that urban sprawl has been encouraged by a variety of economic externalities. Explain the basis for this viewpoint.

3. Describe the impacts of urban sprawl on three ecosystem services that influence human well-being.

PERCENT OF U.S. POPULATION LIVING IN METRO-POLITAN AREAS AND IN THEIR CENTRAL CITIES AND SUBURBS: 1910 TO 2000

17.4 Urban Planning

- Urban planning involves strategies to manage development in order to maintain key ecosystem services and human well-being while facilitating economic growth.
- The smart growth, or new urbanism movement, has 10 main strategies for strengthening human communities and sustaining ecosystem services.

KEY TERMS

urban planning, zoning, urban growth boundary (UGB), new urbanism

QUESTIONS

1. How would you describe the organization of your hometown or the town nearest to your home? What factors might have contributed to that organization?

2. Some criticize urban zoning as an infringement on the rights of private land owners. What public benefits does zoning provide that outweigh such criticism?

3. In what ways do urban planners in your hometown engage the public?

17.5 Urban Transportation

- Many urban areas are turning to public transportation to diminish automobile dependency, revitalize central cities, and slow sprawl.

QUESTIONS

1. What factors contribute to the higher dependence on automobiles in most U.S. cities compared to most European cities?

2. Why is the use of public transportation often dependent on facilitation of nonmotorized transportation, such as walking or bicycling?

3. What role do externalized costs play in urban automobile dependency?

17.6 Urban Biodiversity

- Biodiversity generally increases along the gradient from urban centers through suburbs to rural landscapes.
- Green infrastructure—trees, gardens, and green roofs—provides important ecosystem services, including improved water quality and flows, climate moderation, and aesthetics.

QUESTIONS

1. Describe three strategies to increase the biological diversity of cities.

2. In what ways do green roofs improve urban environments?

3. Describe the benefits that are provided by a park or greenway in your hometown.

17.7 The City as a Sustainability Strategy

- Urbanization has created many challenges to ecosystem and human well-being, but thoughtfully planned cities will be essential to accommodate future human populations in a sustainable fashion.

QUESTIONS

1. Describe three strategies to increase the residence time for materials coming into urban ecosystems.

2. Why must beauty be an important part of planning for urban sustainability?

REUSE

INPUTS · OUTPUTS · RECYCLE · WASTE

ENERGY CONVERSION

Mastering**ENVIRONMENTALSCIENCE**

Go to **www.masteringenvironmentalscience.com** for homework assignments, practice quizzes, Pearson eText, and more.

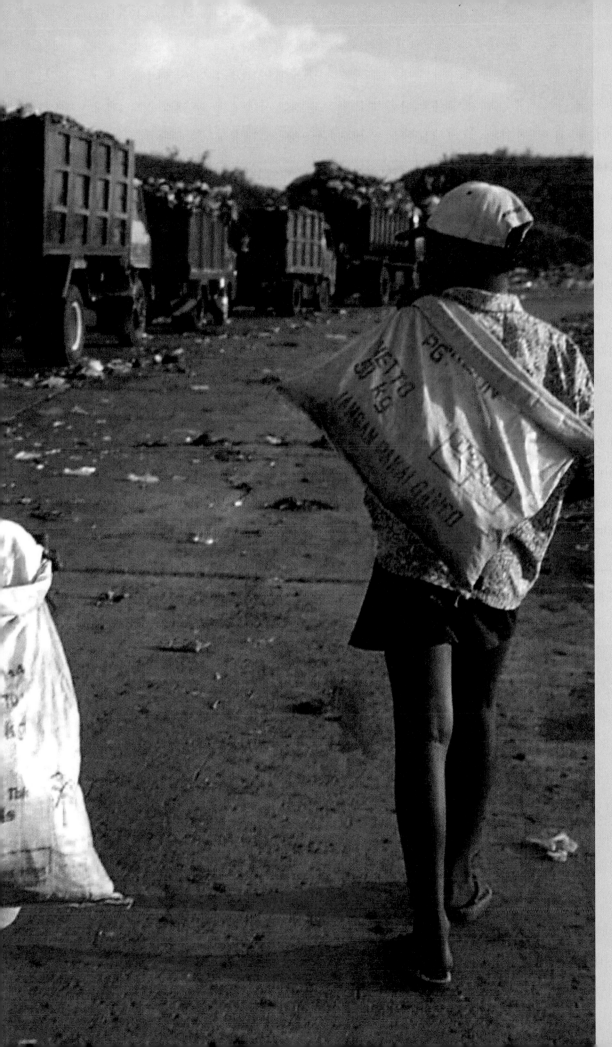

Waste Management

18

What a Waste!

What are the environmental costs of throwing a soda can in a landfill?

Aluminum cans were first used by soda manufacturers and brewers in 1965. Compared to tin and steel cans, these new cans had a number of advantages: Aluminum resists corrosion and has an extended shelf life. The sides and tops of aluminum cans are made of a single piece of metal, which simplifies their manufacture and packaging. Their structure allows businesses to imprint labels, logos, and other information on designs that surround the entire can. No wonder the aluminum can has become one of the most ubiquitous and durable products of our age!

Despite their advantages, aluminum cans have also become a prominent emblem of one of our most widespread environmental challenges: the disposal of waste. Each year, the typical American uses and disposes of an average of 340 aluminum cans. This is equal to about 5 kg (11 lb) of aluminum per person. In 2010, that added up to 101 billion cans, or 1.54 million metric tons of aluminum.

The production of aluminum cans has significant impacts on Earth's environment, as can be seen by analyzing their life cycle (**Figure 18.1**). The first step in making a can involves the mining of aluminum-rich bauxite ore. Bauxite forms from highly weathered soils in moist tropical areas. Because it is found near the surface, bauxite is generally strip-mined in a process that permanently removes forests from large tracts of land: for every ton of bauxite, more than 5 tons of earth must be moved. In addition, bauxite mining often results in water pollution and acid mine drainage.

In the second step, bauxite ore is treated with steam and caustic soda to yield aluminum oxide, or alumina. This refining process requires thousands of gallons of water for each ton of bauxite. It also produces considerable waste in the form of a toxic slurry called "red mud." One metric ton (2,200 lb) of bauxite yields about 500 kg (1,100 lb) of alumina.

In the third step, alumina is smelted to remove metallic aluminum, which is then formed into bars, or ingots. The smelting process requires very strong electrical currents and uses large quantities of electrical

---

▼ **Figure 18.1 Life Cycle of an Aluminum Can**
Aluminum is concentrated in tropical soils as bauxite. Bauxite mining requires the removal of forests and causes a range of other environmental impacts Ⓐ. Bauxite is then refined to alumina in facilities such as this one in Queensland, Australia Ⓑ. In smelting "pots" the oxygen in alumina reacts with carbon electrodes that carry a very high electrical current, producing carbon dioxide and pure aluminum metal. Each ton of aluminum requires a great deal of electricity and nearly a half ton of carbon electrodes Ⓒ. The ingots of aluminum are rolled into sheets, then formed into cans. At the end of its life cycle, an aluminum can is either discarded as waste or recycled Ⓓ.

energy—by itself, aluminum production uses about 1% of the electrical energy generated in the United States. For this reason, smelting facilities are usually located close to large power plants.

In the fourth step, ingots of aluminum are heated and rolled into aluminum sheets. These sheets are then formed and cut to make cans. Once the cans are complete, they are labeled, filled with beverages, shipped to stores, and sold to the public.

In the final step of an aluminum can's life, it is either recycled or discarded. Although aluminum cans are 100% recyclable, every year more than 640,000 metric tons of these cans are discarded in landfills, incinerators, and along roadsides. The high volume of these discarded cans obviously increases the demand for waste management facilities and the cost of maintaining those facilities.

Each step in the production of aluminum cans uses energy and produces waste. Yet when aluminum cans are produced from recycled material, steps one through three—mining, refining, and smelting—are skipped. If all aluminum cans were recycled, bauxite mining and its associated environmental impacts would be greatly reduced. Skipping the refining and smelting stages would

save 10 billion kilowatt hours of electricity each year, an amount equivalent to the electricity used by 600,000 U.S. homes. Recycling would also avoid the emission of between 25 and 50 metric tons of greenhouse gases, or 0.3–0.6% of the total human-caused emissions per year. However, it must be noted that the process of recycling aluminum does require energy and has some environmental impacts.

If throwing away something as simple as a soda can creates these environmental impacts, imagine the effects of discarding complex consumer items, such as automobiles and kitchen appliances. In addition to generating solid waste, Americans generate large amounts of sewage and wastewater, electronic wastes, and industrial wastes. Each of these forms of waste requires a different approach to management. Furthermore, some wastes contain toxic chemicals that are a potential threat to human health.

How can waste streams be managed to minimize environmental impacts? What strategies can we pursue to minimize the amount of waste we generate in the first place? What policies and incentives can we employ to ensure those strategies are pursued? These questions are central themes of this chapter.

18.1 Wastewater

Wastewater includes sewage, water from sinks and other household uses, stormwater runoff, and water used by manufacturing facilities and other industries. In the past, wastewater was simply dumped into nearby waterways. Today, in most developed countries, wastewater is treated to protect the environment and prevent the spread of disease. Unfortunately, in some poor countries, wastewater still goes untreated. The most common methods of managing wastewater are municipal sewage treatment plants and septic systems. Recently, some municipalities have become interested in treating wastewater with methods that mimic the biogeochemical processes of natural ecosystems.

Municipal Wastewater Treatment

Wastewater treatment varies widely among developing and developed countries.

Until recently, sewage and other forms of wastewater were simply dumped into nearby waterways. Over time, dilution and natural processes would eventually decompose the sewage and purify the water. In areas with concentrated human populations, however, this approach was problematic because natural processes cannot break down large amounts of waste in a short period of time. Untreated sewage and wastewater harm natural ecosystems, threaten human health, and contaminate surface water, the source of drinking water for the vast majority of Earth's people.

In most developed countries, wastewater is now treated before it is returned to the environment. In less developed countries, the treatment of wastewater is more variable. Many poor countries have virtually no wastewater treatment. In Latin America, only about 15% of the wastewater that is collected is treated in some manner. The lack of adequate water treatment is a major cause of the high rates of waterborne illnesses, such as cholera and typhoid fever, in developing countries.

In the United States, most cities and towns manage their wastewater and sewage in **municipal sewage**

▼ Figure 18.2 Wastewater Treatment
When wastewater enters a modern treatment plant, it is pretreated to remove solids. Primary treatment removes additional solids and produces a homogeneous liquid that is high in organic compounds. In secondary treatment, microorganisms decompose those organic compounds.

treatment plants (MSTPs). These plants use a stepwise process to remove wastes and chemicals from the water (**Figure 18.2**). When water first enters a sewage treatment plant, it is pretreated to remove large solids, such as rags, feminine hygiene products, sand, and gravel. Insoluble chemicals, such as grease and oils, may also be removed.

Pretreated wastewater then flows into large settling tanks, where it undergoes **primary treatment**. Particles in the wastewater settle to the bottom of the tank, forming a sediment called sludge. The main purpose of primary treatment is to produce a relatively homogeneous liquid that can be treated biologically and a sludge that can be processed separately. In many developing countries, municipalities return wastes to the environment after primary treatment.

In **secondary treatment**, bacteria and other microorganisms are used to break down the organic material dissolved in the wastewater. In most treatment plants, secondary treatment takes place in aerated tanks and basins. Some facilities use membranes or gravel filters to further separate solid and liquid wastes.

PRETREATMENT — Air — Scum — Trash

PRIMARY — Sludge

SECONDARY — Aerated tank

Some treatment plants also use **tertiary treatment** to remove inorganic nutrients from wastewater. In this stage of treatment, wastewater is passed through sand and charcoal filters to remove residual solids and toxins. Next, the water is stored in human-made ponds or lagoons where microorganisms remove significant amounts of dissolved nitrogen and phosphorus. Finally, the water is disinfected with chlorine, ozone, or ultraviolet radiation to reduce the number of microorganisms.

Solid wastes, or sludge, accumulate at each step in this treatment process. Most often, sludge is subjected to digestion by microorganisms, which reduce the volume of organic matter and the number of disease-causing microbes. Sludge is then dried so that it can be transported and disposed of off-site. Usually, it is dumped into landfills or spread onto open land. However, a growing number of treatment plants convert sludge into pellets that can be used as fertilizers; these pellets are often sold to local gardeners and farmers.

In place of traditional MSTPs, some communities are beginning to use natural or constructed wetlands to purify wastewater that has had primary treatment. Wetlands are very effective at purifying water (12.5). As water slowly percolates through wetland soils, solid materials are filtered out and microbes decompose the organic matter. Nutrients, such as nitrogen and phosphorus, and contaminants, such as heavy metals, are adsorbed by soil particles or taken up by plants and stored in their tissues.

Household wastewater
Distribution box (optional)   Septic tank
Nonperforated pipe
Drain field
Perforated pipe   Vent pipe   Gravel or crushed stone

▲ **Figure 18.3 Household Septic System**
Rural wastewater is often treated in on-site septic systems. Wastewater flows into the septic tank, where solid materials are decomposed. Liquid wastes flow out of the tank and into a system of perforated pipes in the leach field, where dissolved organic chemicals are broken down by microbes.

On-Site Wastewater Treatment

If properly maintained, septic systems can isolate waste and protect water supplies.

In less densely populated areas, households often rely on **septic systems** to treat their wastewater and sewage. In these systems, sewage and household wastewater flow to an underground septic tank outside the home. Solids settle to the bottom of the tank, and wastewater flows to a series of perforated, underground pipes. Water released from these pipes flows into a **leach field**, where microorganisms in the soil break down the waste materials (**Figure 18.3**). Periodically, the solids that settle in the septic tank need to be pumped out and disposed of in a landfill.

Nearly 25% of the households in the United States rely on septic systems to treat their wastewater. When properly maintained, septic systems are an effective means of isolating wastes and protecting water supplies.

Maintenance includes monitoring leach fields and occasional pumping of septic tanks. However, about 10% of these systems are not functioning properly. In communities where soil conditions prevent effective leaching, failure rates may exceed 70%. The U.S. Environmental Protection Agency (EPA) reports that failed septic systems are the third most common cause of groundwater contamination.

An increasing number of households are using plumbing systems that separate **graywater**, the water from washers, sinks, and baths, from blackwater, the water from toilets and garbage disposals. Graywater can be used to flush toilets or water gardens. Reusing graywater reduces the demand for water and extends the life of septic systems.

SLUDGE DIGESTION   METHANE

Filter

Dry sludge finishing

SEEING SOLUTIONS: Eco-Machines

Can simulated wetland ecosystems purify wastewater?

John Todd has been long impressed with the capacity of complex ecosystems to adapt to constant change over the eons (**Figure 18.4**). That capacity has been the basis for Todd's development of Eco-Machines, systems that use living organisms from diverse ecosystems to treat wastewater. His systems are solar powered by photosynthesis; they are also self-organized and adaptable to change by virtue of their diversity.

Todd's first Eco-Machine was installed in the Cape Cod town of Harwich, Massachusetts. Prior to 1990, nearly all of the town's wastewater was dumped into Flax Pond, a small lagoon in a 25-foot-deep layer of sand just above the aquifer that supplies the town's drinking water. The pond was filled with an organic "soup" that was at least 35 times more concentrated than municipal wastewater. The pond water also contained high quantities of nearly every one of the priority pollutants listed by the U.S. EPA.

To restore Flax Pond, Todd set up a series of 21 large tanks that were connected by pipes. Into these tanks he put thousands of species of plants and animals collected from a great variety of natural habitats, including pig wallows, swamps, salt marshes, and forests. Water from the pond was circulated through the tank. Each aquarium received a different strength and composition of waste, with those nearest the pond receiving the highest concentrations. Soon, a different ecological community organized itself in each tank. This process of community development was encouraged by the installation of racks for plants and special surfaces for microbes (**Figure 18.5**).

Within 13 days, the water flowing out of the system was free of all EPA-listed pollutants, and concentrations of heavy metals in the water were well within the standards for safe drinking water. After inspection, the state of Massachusetts approved Todd's installation for septic waste treatment.

Since 1990, Todd has installed over 80 Eco-Machines in schools, offices, and factories in the United States and elsewhere. An Eco-Machine may be housed in a small greenhouse or deployed over several hectares, depending on the amount of effluent it must process. Each machine is divided into a number of individual tanks that correspond to various kinds of wetland habitats you might encounter in nature.

Wastewater treatment begins in anaerobic tanks that support blue-green algae, bacteria, fungi, and protozoa. Successive tanks support higher plants, which are often suspended in reinforced fiberglass racks. The final processing tanks support shrubs and trees, as well as snails, clams, and fishes. As water moves through the system of tanks, the concentrations of wastes diminish and the concentration of oxygen increases.

Todd has received numerous awards for his innovation, including the Environmental Merit Award from the EPA and the Buckminster Fuller Challenge Award for visionary thinking, which seeks "to solve humanity's most pressing problems in the shortest possible time while enhancing Earth's ecological integrity." That visionary thinking is captured in Todd's own words: "The world is a vast repository of unappreciated or unknown biological strategies that have immense importance for humans if we can develop a science of integrating the stories embedded in nature in the basic systems that sustain us."

▲ **Figure 18.4 Wastewater Visionary**
John Todd has applied his understanding of the ecological processes in a variety of natural wetlands to the design of Eco-Machines, which use communities of living organisms to purify waste-laden water.

▲ **Figure 18.5 An Eco-Machine in Action**
In this greenhouse-size system, waste is removed from water as it moves from anaerobic ecosystems in the tanks through tanks such as these that support ecosystems that are successively more diverse and increasingly aerobic.

18.2 Solid Waste

Municipal solid waste (MSW) consists of materials that are discarded from homes, small businesses, and institutions such as hospitals and universities. This is the stuff that most of us call trash or garbage. In the past, MSW was dumped into open pits, rivers, or the ocean, or burned in open piles. In developed countries such as the United States, these methods of disposal have been replaced by modern landfills and incineration facilities, which have fewer negative environmental impacts. In less developed countries, however, open dumping and burning of MSW still persist. In some regions, recycling programs have slowed the rate of growth of MSW, and composting is being used to turn waste into material that is useful for gardening and landscaping. **Industrial solid waste (ISW)** is produced in industrial processes before a finished product reaches the consumer. The amounts of ISW that are generated are heavily influenced by patterns of consumption and management.

Municipal Solid Waste

The quantity and composition of solid waste differ from region to region and have changed over time.

Each year, an average American generates 720 kg (1,590 lb) of municipal solid waste. About 33% of this waste is paper products, including newspapers, office papers, unwanted mail, and books (**Figure 18.6**). Other important components of this waste stream are yard wastes, food scraps, and containers, packaging, and other objects made of plastic, metal, rubber, and cloth.

The United States ranks high among the world's developed countries in per capita waste production. For example, although the composition of their MSW is similar, the average French person throws away only 520 kg (1,145 lb) of trash per year, about 28% less than the average American.

In less developed countries, the consumption of packaged goods is not as pervasive and reuse is more common. Consequently, waste streams are smaller and are dominated by food wastes. For example, the average person in Bangalore, India, generates about 146 kg (320 lb) of MSW each year. This is less than a quarter of the amount generated by an average American. Approximately 75% of the Indian MSW is food scraps and other organic wastes that decompose rapidly. Paper accounts for less than 2% of their waste. In Mexico City, the average person produces 248 kg (547 lb) of MSW each year; about 60% of this is food waste and 11% is paper.

The composition and amount of the waste we generate have changed over time. One hundred years ago, trash collectors in New York City picked up about 530 kg (1,170 lb) of waste per resident each year. Nearly three-fourths of that waste was coal ash from home furnaces. The rest was mainly food waste and paper. By 1960, most homes were heated with fuel oil, natural gas, or electricity, and the per capita production of MSW in America had fallen to about 435 kg (960 lb) per year.

Over the next 30 years, the increased use of packaging and disposable containers in the United States contributed to a 40% increase in the per capita waste stream. Since 1990, local, state, and federal agencies, as well as many environmental groups, have mounted campaigns to reduce the annual flow of waste. As a consequence, per capita MSW production peaked in 2000. Since then, it has declined (**Figure 18.7** on page 562).

A century ago, MSW was simply dumped into open pits or rivers; often it was burned in open piles. The decaying garbage attracted animals such as rats and produced foul odors that could be smelled for miles around. Furthermore, these dumps were a source of hundreds of polluting chemicals. Large cities, such as New York City, loaded their waste onto barges and dumped it into the ocean a few miles from shore.

▼ **Figure 18.6 MSW Stream**
Paper accounts for nearly one-third of all municipal waste in the United States. Nonorganic wastes such as metals and glass make up less than 15% of the waste stream.
Source: U.S. EPA. 2009. Municipal solid waste (MSW) in the United States: Facts and figures. URL:http://www.epa.gov/osw/nonhaz/municipal/. Last accessed 10-6-2011

● Paper
● Yard trimmings
● Food waste
● Plastics
● Metals
● Rubber, leather, and textiles
● Wood
● Glass
● Other

► Figure 18.7 **Growth of Waste**

Although the total MSW stream has continued to increase, per capita MSW in the United States has remained nearly constant since 1990. This indicates that campaigns to encourage reuse and recycling are having a positive effect.

Source: U.S. EPA. 2009. Municipal solid waste (MSW) in the United States: Facts and figures. URL: http://www.epa.gov/osw/nonhaz/municipal/. Last accessed 10-6-2011

As urban populations grew, so did the amount of waste, odors, and associated pollution. In developed countries, public opposition to open dumps and ocean dumping has resulted in better methods of managing MSW. These include sanitary landfills, waste-to-energy incineration, composting, and programs to reduce waste at its source. In many developing countries, however, dumping into pits and rivers and open burning persist to this day (**Figure 18.8**).

GROWN TOGETHER

▼ Figure 18.8 **Open Dumps**

In many parts of the world, municipal solid waste still ends up in open dumps such as this one in Manila, the Philippines. The very poor scour such dumps for any reusable or recyclable items.

Trench method

Final earth cover   Compacted solid waste   Daily earth cover   Original ground

◄ **Figure 18.9 Sanitary Landfill**
Municipal solid waste is dumped and compacted into open trenches. These trenches are lined and capped with a dense layer of clay.

Sanitary Landfills

Modern landfills have many advantages over traditional open dumps.

In 1935, the city of Fresno, California, opened the first modern landfill in the United States. Managers at that landfill pioneered the use of trenches to deposit waste, compacting to compress waste into smaller chunks, and daily burial to keep waste underground and combat rodent and odor problems. Because it set the standard for all subsequent landfill construction, the Fresno Municipal Sanitary Landfill was declared a National Historic Landmark in 2001.

Modern **sanitary landfills** are designed to prevent the contamination of groundwater and to reduce the odors and aesthetic problems associated with open dumps. To protect groundwater, landfills are lined with impervious clay. Collection systems in landfills prevent liquids from leaching, or leaking, into nearby water supplies. (Liquids that leak from landfills are called *leachates*.) To reduce odors, layers of waste are covered with soil, coal fly ash, or similar materials. The layers of waste are then allowed to settle and decompose (**Figure 18.9**).

When sanitary landfills are full, they are capped by a thick layer of clay. Over time, the waste in the landfills decomposes, releasing large amounts of methane gas. Because this gas is flammable and explosive, it can be dangerous if it is not properly vented, as is often the case in older landfills. Today, as landfills are covered, they are fitted with pipes to vent the gases. At most retired landfills, gas wells capture this methane, which can then be used for energy. When retired landfills are deemed safe, they can be converted into public parks, open spaces, or even building sites (**Figure 18.10**).

Sanitary landfills are a definite improvement over open dumping, but they do pose environmental risks and challenges. If landfill liners are punctured or the systems for collecting leachate fail, groundwater may be contaminated. The methane leaking from landfills is a greenhouse gas with a warming potential that is more than 1,300 times greater than that of carbon dioxide. Even where there is abundant open space for landfills, public opposition to their location is often fierce. Thus, some cities must transport their garbage hundreds of miles for disposal. Finally, the loss of useful materials (such as paper, plastics, and metals) in landfills is not sustainable over the long run.

▼ **Figure 18.10 Landfill Restoration**
This public park near Virginia Beach, Virginia, was once a sanitary landfill. Appropriately enough, the park is called Mt. Trashmore.

Waste Incineration

Modern incineration facilities are an improvement over the open burning of waste that was practiced in the past.

To reduce its volume and weight, MSW can be incinerated at high temperatures. This process consumes organic matter, producing heat, solid ash, flue gases, and aerosols. The volume of the ash produced by incineration is only 5–6% of the original volume of waste. Modern incinerators are designed with complex filtration and scrubbing devices to reduce emissions of particulate aerosols and other gaseous pollutants.

Incinerators known as **waste-to-energy facilities** use the heat produced by burning waste to generate electricity. In these facilities, MSW is fed into furnaces that operate at temperatures near 850 °C (1,560 °F). At these temperatures, toxic organic gases are broken down entirely to carbon dioxide and water. The super-hot flue gas from these furnaces is used to heat pressurized steam to over 400 °C (752 °F). The steam is then transported to turbine generators to produce electricity (**Figure 18.11**).

Waste incineration and waste-to-energy facilities are particularly common in Japan and some European countries where land is scarce. Denmark and Sweden have been leaders in the development of combined heat and power facilities. Those facilities pipe the excess heat from electric power plants to district heating systems, which use it to heat homes and businesses. Currently, Denmark obtains about 5% of its electricity and nearly 14% of domestic heat from the incineration of waste.

Despite improvements in the design of incineration facilities, emissions of hazardous chemicals and the highly toxic nature of incinerator ash make this approach to waste management less than optimal. Public opposition to incinerator sites is often higher than for landfills.

Composting

Organic waste can be changed to garden mulch through natural decomposition.

Yard trimmings and food waste constitute about 25% of MSW in the United States. With relatively simple treatment, most of this material can be converted to **compost**, decayed organic matter that can be used to improve the fertility and water-holding capacity of soil, thereby improving plant growth.

Small-scale or household composting can significantly reduce the volume of yard waste and food scraps entering the waste stream. Such composting can be carried out even in small yards. Organic wastes such as yard trimmings and most kitchen wastes are collected in piles or composting bins, where they are allowed to "cure," or decompose into stable humus. This compost can then be tilled into garden soils to improve aeration and fertility or used directly as a medium for plant growth.

Municipal governments and institutions often carry out large-scale composting, sending food and yard waste to facilities designed to speed up the decomposition process (**Figure 18.12**). This procedure diverts foul-smelling, methane-producing organic matter from landfills and cuts the cost of waste disposal. It also produces compost that can be used for landscaping or sold for additional revenue.

◄ **Figure 18.12**
Composting
Many municipal waste facilities are now able to convert mixed paper, food, and yard waste into compost, which can be sold commercially.

▼ **Figure 18.11 Waste to Energy**
Modern waste incinerators capture heat to generate electricity while reducing the volume of waste by more than 95%. Filters and scrubbers remove pollutants from the exhaust.

Turbine generator

Electric power

Boiler

Steam

Hot gases

Exhaust filters

Crane

Receiving area

Ash

Refuse storage

Reducing the Waste Stream

Reduce, reuse, and recycle to diminish the original source of municipal solid waste.

Over the past 50 years, the total amount of waste generated by each person in the United States has nearly doubled. Much of this increase is a direct consequence of our growing affluence. We acquire more things, and those things are eventually discarded. Many of the products we buy, such as disposable bottles, diapers, cameras, and razors, are not designed to last beyond a single use. Pop songs, editorials, and blogs refer to ours as the "disposable society."

A significant portion of the recent increase in MSW is related to changes in how the things we acquire are packaged. Each year, the United States produces nearly 25 billion kilograms (56 billion pounds) of plastic for containers and packaging. That is enough plastic to make 2 trillion plastic water bottles; if arranged end-to-end, those bottles would reach from Earth to Mars and back. Certainly, some of this plastic serves a useful, perhaps even essential, function. But much of it is simply cosmetic. Currently, less than 10% of plastic containers are recycled; the rest ends up in landfills or incinerators.

Reducing waste. Given the immense volume of packaging waste, even small changes in individual containers can have a significant effect on the waste stream. Over the past two decades, the net weight of a 2-liter plastic soda bottle has been reduced from 68 to 51 grams (2.4–1.8 ounces) by using less plastic. This has eliminated 110 million kilograms (243 million pounds) of plastic from the waste stream each year. Consumers can diminish their individual waste streams by buying in bulk with less packaging or by purchasing items in reusable containers. These practices save money, too.

In response to requests from environmental groups and consumers, major retailers are beginning to put pressure on manufacturers and suppliers to reduce packaging waste. In 2006, the world's largest retail chain, Wal-Mart, announced policies to reduce its solid waste by 25%. It pledged to work with suppliers to create less packaging overall and to increase package-recycling rates. Although complete data are not available, Wal-Mart has made significant progress on its promise. In 2010, these stores recycled 25 billion pounds of cardboard that previously would have been sent to landfills.

Reusing waste. The U.S. EPA estimates that 2–5% of the material sent to landfills is actually reusable. This includes items as varied as shopping bags, clothing, furniture, and building materials. For hundreds of years, poor people in both developing and developed countries have made a meager living harvesting items from trash heaps (**Figure 18.13**). Even some rather well-to-do people save money and diminish their demands on the environment by retrieving usable items and material from dumps. In the United States, more than 6,000 reuse centers run by organizations such as Goodwill, the Salvation Army, and Habitat for Humanity provide an outlet for the resale of these materials. This reuse reduces the waste stream, or at least delays adding materials to it.

Recycling waste. Recycling programs have become common in many developed countries. In 2009, the rate of recycling in the United States reached 33.8%, more than double the rate in 1990 (**Figure 18.14**). In Japan and many European countries, recycling rates exceed 50%. Recycling diverts material from landfills and incinerators, allowing it to be remanufactured into other items. This reduces the demand for resources and saves energy. For example, making a can from recycled aluminum requires only 5% of the energy needed to make a can from bauxite ore. Recycling items such as batteries and

How much recycling gets done in the United States every year compared to what is consumed?

Shayna Jefferson, University of Florida

Norm: *The United States recycles about one-third of its municipal waste. Japan and many countries in Europe recycle more than half of their waste.*

Are landfills obsolete?

Macy, Louisiana State University

Norm: *Unfortunately not. But municipal waste streams are diminishing, and the demand for new landfills is declining.*

▼ **Figure 18.13 Dumpster Diving**
Considerable amounts of economically valuable material that enter the MSW stream are retrieved for reuse or recycling. Such trash picking is estimated to add nearly $1 billion to the U.S. economy.

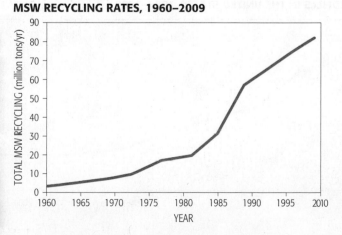

MSW RECYCLING RATES, 1960–2009

◄ **Figure 18.14 Recycling Rates**
Since 1985, recycling rates have increased significantly. This increase is a result of a growth in the technology and infrastructure to support recycling. In addition, many communities have mounted programs to encourage recycling.
Source: U.S. EPA. 2009. Municipal solid waste (MSW) in the United States: Facts and figures. URL: http://www.epa.gov/osw/nonhaz/municipal/. Last accessed 10-6-2011

▲ Figure 18.15 **Closed-Loop Recycling**
The aluminum in cans can be recycled to produce new cans.

▲ Figure 18.16 **Open-Loop Recycling**
The wood-like boards in this park bench are a composite material made from recycled plastics.

electronics also diminishes the flow of toxic chemicals into the environment.

In **closed-loop recycling**, the material in a product is reused to produce a similar product (**Figure 18.15**). Metals such as aluminum can be recycled in this manner. In **open-loop recycling**, a material is recycled into a different product that is discarded after it is used. Recycled glass and plastics, for example, are often not suitable for their original use but may be manufactured into a variety of other products (**Figure 18.16**). Other factors may also limit recycling. For example, most paper can only be recycled a few times before its fibers are degraded beyond further use.

Critics of recycling often argue that the costs associated with collecting and transporting recyclable materials offset the savings in production. However, such calculations usually do not account for most of the environmental and social costs associated with landfills and incinerators.

The economic feasibility of municipal recycling programs depends on several factors, including markets for recycled material, the cost of waste disposal, and government policies toward resource extractive industries, such as mining and forestry. Consumer demand for recycled products is important: Recycling programs cannot succeed if consumers do not purchase recycled products.

Where they have been implemented, deposits on recyclable products such as bottles and cans have significantly increased recycling rates. Convenience also influences the rate of recycling. Communities with curbside collection and disposal facilities that require a minimum of sorting have higher recycling rates than those that do not.

Over the past two decades, increased public awareness and aggressive community programs have resulted in significant increases in the amount of MSW being recycled. Recycling rates in some communities exceed 60%. As a consequence, the amount of material being discarded in landfills has remained flat, and per capita waste generation is actually decreasing (**Figure 18.17**).

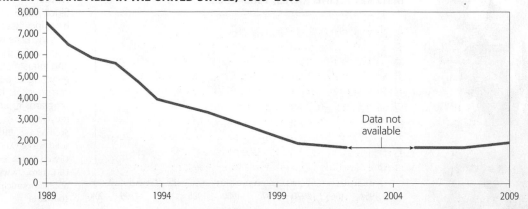

NUMBER OF LANDFILLS IN THE UNITED STATES, 1989–2009

Data not available

◄ Figure 18.17 **Landfills on the Decline**
Increased reuse, recycling, and composting have diminished the amount of MSW going to landfills, allowing more than 6,000 facilities to close. Since 2000, however, the rate of recycling has begun to level off, and the number of landfills has remained nearly constant.
Source: U.S. EPA. 2009. Municipal solid waste (MSW) in the United States: Facts and figures. URL: http://www.epa.gov/osw/nonhaz/municipal/. Last accessed 10-6-2011

Municipal Solid Waste Management Policy

MSW policies and management decisions are mostly set by city and county governments.

The U.S. federal government first became involved in MSW policy with the passage of the 1899 Rivers and Harbors Act. Using the federal government's constitutional authority over navigable streams, this act prohibited the dumping of waste materials into rivers without a permit. Nevertheless, until 1965 policy related to the disposal of waste was mostly a local affair. Landowners and city and county governments typically disposed of wastes in the most expeditious way, usually in open dumps.

Federal oversight of MSW management increased considerably with the passage of the Solid Waste Disposal Act in 1965. This legislation provided funding for the development of state and local waste management programs. More important, it required states to set standards for the regulation of MSW facilities. The 1970 Resource Recovery Act transferred the oversight of MSW to the U.S. EPA. The following year, EPA Administrator William Ruckelshaus initiated his Mission 5,000 Program, aimed at closing 5,000 open dumps within a year and converting all open dumps to sanitary landfills by 1980. Mission accomplished!

Federal and state regulations remain an important element in waste management. In 2009, a bill to mandate glass and aluminum recycling was introduced to the House of Representatives, but was not brought to a vote. State-level programs to encourage recycling range from simple education programs, to prohibitions on placing recyclable aluminum or plastic in landfills, to mandatory deposits on bottles and cans.

Nevertheless, most policies and decisions affecting the reduction, reuse, and recycling of MSW remain with local governments. Limited space for landfills and public objections to their expansion have created powerful incentives for reducing waste and recycling.

The use of strategies such as placing collection centers in convenient locations and providing curbside collection of glass, aluminum, and paper have helped many cities and counties increase their recycling rates to well above 50% (**Figure 18.18**).

Changing attitudes about the economics of these programs have been a major factor in their success. In the past, local governments often considered composting and recycling programs to be an economic drain if they were not able to generate revenues to cover their costs. However, when governments calculate all of the costs involved in building and maintaining landfills, they discover that programs to reduce, reuse, and recycle municipal solid waste are a true bargain.

▼ **Figure 18.18 Encouraging Recycling**
Convenient and prominent disposal systems are the foundation of successful recycling programs.

Industrial Solid Waste

Waste streams from manufacturing facilities vary widely.

There is a wide variety in the kind and amount of waste generated in making the many thousands of products used today. ISW includes all of the waste generated at each stage in the manufacture of products. For example, the ISW generated in making an aluminum can includes the wastes associated with mining bauxite, refining bauxite into alumina, smelting alumina into aluminum ingots, shaping aluminum ingots into rolls of aluminum sheeting, and forming and cutting sheeting into cans.

In contrast to municipal waste, industrial waste is usually collected and managed by the private sector. This makes data on the specific nature and fate of ISW difficult to acquire. However, the U.S. EPA estimates that

U.S. industrial facilities generate and dispose of about 7.6 billion tons of waste each year. Large industrial facilities typically discard their waste in specially designed ISW facilities. Yet some industrial waste, particularly that generated by smaller enterprises, makes its way to municipal landfills. Most municipal facilities have a special area dedicated to wastes from building construction and demolition.

Many industries have recognized that simply treating ISW as something to be discarded has adverse consequences for the environment as well as for their economic bottom line. Thus they are seeking opportunities to reduce, reuse, and recycle at every stage of production (18.4).

QUESTIONS

1. How do municipal solid waste streams in developed countries differ from those in less-developed countries? What causes these differences?

2. How does a sanitary landfill differ from a simple open garbage dump?

3. What is the difference between closed-loop and open-loop recycling?

4. How does industrial solid waste differ from municipal solid waste?

FOCUS ON SCIENCE: Garbology

▲ Figure 18.19 An Archeological Treasure
Dr. William Rathje began his career studying archeological evidence from ancient waste sites to learn about the early Mayan civilization in Mexico. He and his Garbage Project colleagues used similar methods to study current landfills and learn about the connections between modern cultures and the waste they produce.

According to the *Oxford English Dictionary*, the word *garbology* was first used in the 1960s by garbage collectors as a polite way to describe their vocation. Today, this word is more commonly used to describe the kind of research carried out by anthropologist William Rathje. Anthropology may seem very distant from studies of municipal solid waste, but in fact, anthropologists garner some of their most significant information from ancient sewers and trash heaps.

Early in his career, it occurred to Rathje that modern landfills might provide important information about the societies they serve, as well as about the characteristics and decomposition of the waste itself. Rathje began the Garbage Project (le Project du Garbage) at the University of Arizona, Tucson, in 1974 and continued this area of research until his recent retirement (**Figure 18.19**).

In one major study, Rathje and his Garbage Project colleagues sampled the waste in eight different landfills in Arizona, California, Florida, and Illinois. The locations of the landfills were selected to represent a range of climates, cultures, and economic conditions.

The team took advantage of the fact that the structure of landfills is very well-defined. Waste is deposited in layers. At the end of each workday, compacted waste is covered with a thin layer of soil. After about a year, when the accumulated refuse is around 7.5 m (25 ft) thick, the layers are capped with 1.5 m (5 ft) of soil; this yearly deposit is called a lift. Once a lift is capped, a new lift is begun above it or in another location. Lifts and their daily layers can be readily excavated and studied using standard archaeological methods.

Rathje and his colleagues verified the age of individual layers using the dates on newspapers and bottles. Waste was then sorted into different categories of materials, such as paper, plastics, and fast-food packaging. Materials in each category were measured by weight, volume, and moisture content. The research team found that the various landfills differed significantly in the relative amounts of trash in each category; they attributed these differences to cultural and economic differences among the communities served by the landfills.

Rathje's team found that the rate of decomposition varied considerably among landfill sites. The rate was highly correlated with moisture—the wetter the site, the faster the rate of decomposition. They also found that some organic materials, such as newspapers and disposable baby diapers, decompose very slowly. Even after 20 or 30 years of burial, newspapers could be easily read (**Figure 18.20**).

The data showed a gradual increase over time in the amounts of certain items, such as fast-food packaging and disposable baby diapers. These waste items have received considerable attention in the news media, yet the amounts of these items were actually quite small compared to other materials, such as paper and construction waste. Rathje has suggested that instead of focusing on particular kinds of waste, it would be more beneficial for policymakers to promote large-scale recycling and composting.

The study also showed that the volume of the waste in landfills is not closely correlated with its weight. Why does this matter? Nearly all measures of MSW deposited in landfills are based on weight. In addition, tipping fees—the amount that trash collectors pay to dump refuse into landfills—are based on weight. However, the lifetime of a landfill is determined by the volume of MSW coming into it, not the weight of MSW. Given the diminishing capacity of many landfills and the challenges of locating and maintaining these facilities, it would make more sense if the auditing and pricing of waste disposal were based on volume rather than weight.

Rathje and his collaborators concluded their landmark paper by noting that archaeologists have been studying the past by scouring the garbage dumps of antiquity for more than a century. The archaeology of contemporary waste facilities has just begun.

Source: Rathje, W.L., W.W. Hughes, D.C. Wilson, et al., 1992. The archaeology of contemporary landfills. *American Antiquity* 57: 437–447.

▶ Figure 18.20 Durable Trash
Because of the arid climate, even newspaper is slow to decompose in Tucson, Arizona's landfill. This very readable 1957 copy of the Arizona Daily Star was unearthed 50 years later in 2007.

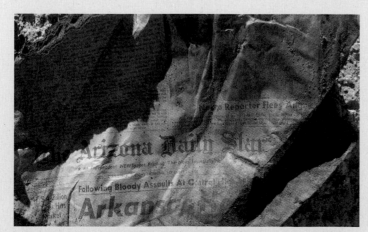

18.3 Hazardous, Electronic, and Radioactive Waste

Although they make up only a small portion of the waste stream, hazardous and radioactive wastes pose grave threats to the environment and public health. Until recently, the regulation and management of hazardous waste sometimes allowed for significant public exposure to these materials, resulting in concerns about public health and calls for tighter regulation. Because of the risks associated with hazardous materials, they require specialized handling and disposal and are costly to manage.

Hazardous Waste

The management of dangerous waste has evolved from unregulated dumping to permanent storage, reduction, and remediation.

Wastes that are flammable, corrosive, explosive, or toxic are classified as **hazardous wastes**. The U.S. EPA identifies three categories of hazardous wastes. *Source-specific wastes* are wastes that are unique to particular industries. Sludge and wastewater from petroleum refiners and pesticide manufacturers are examples of source-specific wastes. Over 70% of the hazardous wastes produced in the United States come from the chemical and petroleum industries. *Nonspecific-source wastes* are produced by common manufacturing and industrial processes. Solvents used for cleaning or degreasing, including those used by commercial drycleaners, are examples of nonspecific-source wastes. *Discarded commercial chemical wastes* include commercial chemical products, such as unused paints, pesticides, and pharmaceuticals.

Although less hazardous than the waste produced by industry, household hazardous wastes also represent a significant threat to the environment and public health (**Figure 18.21**). Household hazardous wastes include unused paint, old batteries, solvents, cleaners, and other chemicals. On average, each person in the United States generates about 5 kg (11 lb) of household hazardous waste each year. Compared to the stream of municipal solid waste, this may seem like a small amount. However, household wastes tend to accumulate in basements and garages. The EPA estimates that the average home has accumulated more than 45 kg (100 lb) of hazardous waste. As much as 90% of this hazardous material is eventually poured down drains or into storm sewers or discarded in municipal dumps or landfills.

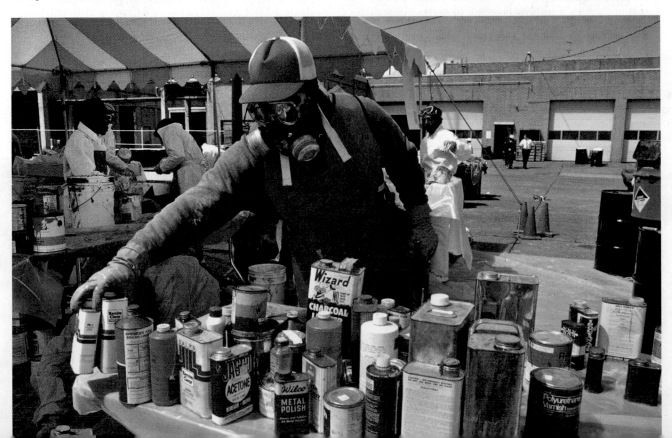

◄ **Figure 18.21 Household Hazardous Waste**
Each year, U.S. households dispose of more than 1.5 million tons of hazardous waste. This waste comes from a diverse array of products.

Approaches to managing hazardous waste have changed as our understanding of the dangers associated with them has grown. In the past, most hazardous wastes were simply dumped into the ocean, lakes, and rivers or open pits and allowed to degrade. This misguided approach to waste management is neatly described by this bit of doggerel: "The solution to pollution is dilution." In the 1970s, several highly publicized cases brought the health risks associated with such dumping to public attention. For example, in 1976 the Love Canal housing development in Niagara Falls, New York, was discovered to have been sited on a former dump site with soils contaminated by a variety of toxic chemicals. Public outcry over these cases led to stricter regulations for the management of hazardous wastes, as well as efforts to clean up previously contaminated sites. Nevertheless, thousands of previous dumping sites continue to contaminate soils and groundwater.

Current methods for disposing of hazardous wastes have greatly reduced the risks to the environment and public health. Many hazardous wastes are stored in specially designed landfills called **permanent retrieval storage sites**. Wastes in these sites can be monitored to ensure they do not leak into the surrounding environment; they can also be recovered for reuse or for more permanent disposal. Chemical processing and high-temperature incineration are used to convert some hazardous waste to less toxic ash or gases.

In **bioremediation**, living organisms—microorganisms, plants, or fungi—break down hazardous wastes into less harmful components. The most common use of bioremediation is to clean up spills of hazardous chemicals. For example, many kinds of microorganisms can break down the various hydrocarbons in petroleum products. Mixtures of such organisms are now routinely used to help clean up oil spills (**Figure 18.22**). In other situations, microorganisms growing in reactor vessels are used to break down particular hazardous chemicals.

Another way to dispose of liquid hazardous wastes is to inject them into deep wells in porous rocks, such as sandstone or limestone. More than 375,000 waste-injection wells have been formally listed with the EPA. These wells are tightly regulated and monitored to prevent the contamination of groundwater. Nevertheless, there is concern that considerable amounts of hazardous liquids are being injected into unregulated wells.

In the United States, the management and disposal of hazardous wastes are governed by two federal laws. The Resource Conservation and Recovery Act (RCRA) of 1976 set national goals for protecting human health and the environment from hazardous wastes; it also sets goals for conserving natural resources by reducing and recycling wastes. This law required the EPA to set standards for the disposal of hazardous wastes.

The Comprehensive Environmental Response, Compensation, and Liability Act (CERCLA) of 1980 established the "polluter pays" principle. Under this principle, companies are responsible for the cost of cleaning up sites that have been polluted by their activities. CERCLA also created a tax on industries that produce hazardous wastes; revenues from this tax go into the Superfund. The Superfund pays for cleaning up hazardous wastes in abandoned sites when the company that was responsible for the pollution cannot be identified (**Figure 18.23**). There are estimated to be hundreds of thousands of such sites in the United States. Between 1980 and 1985, the Superfund grew to $1.9 billion. When CERCLA was reauthorized in 1986, Congress increased the Superfund to $9.3 billion, to be funded by the federal government.

As many businesses, governments, and environmental leaders now recognize, the best approach to managing hazardous waste is to not generate it in the first place. In late 2006, the European Union adopted the Registration, Evaluation, and Authorization of Chemicals (REACH) policy. REACH requires chemical manufacturers and other industries to test for toxic and hazardous substances and to avoid using them.

▼ **Figure 18.22 Bioremediation**
These workers are spraying nutrients to encourage the growth of oil-devouring microbes on the shores of Prince William Sound. This area was polluted with oil spilled from the *Exxon Valdez* tanker in 1989.

▼ **Figure 18.23 Superfund Site**
Vermont's Elizabeth Copper Mine is now closed, but the pollution from it remains. The company that opened the mine no longer exists, so the cleanup of this site is being funded by the EPA Superfund.

Electronic Waste

E-waste regulation has only recently begun in some places.

One of the most prominent trends of the past 50 years is our ever-increasing dependence on electronic devices, such as computers and cell phones. When these devices eventually fail, they are discarded. Changes in technology, such as the transition from analog to digital television, have made millions of electronic devices obsolete; these devices must also be discarded or put into storage. Old televisions, computers, cell phones, and other electronic devices disposed of by businesses and households make up electronic waste, or **e-waste** (**Figure 18.24**).

E-waste is classified as hazardous because it contains large amounts of heavy metals, including lead, cadmium, and mercury. Although e-waste represents only 2% of the trash in landfills, the EPA estimates that it is the source of more than 70% of the heavy metals in them. Some devices that were manufactured before 1977 also contain significant amounts of polychlorinated biphenols (PCBs), which were used to insulate electronic components such as capacitors and transformers. PCBs have been shown to cause cancer in many animals and in humans.

In 2009, the stream of e-waste in the United States exceeded 2.1 million metric tons (about 2.4 million U.S. tons). More than 80% of discarded e-waste ends up in landfills. A significant amount of e-waste is simply "stored," although most stored devices are eventually discarded. About 75% of all of the personal computers ever sold are now gathering dust in storage.

Despite the fact that discarded electronic devices contain many valuable components and chemicals, recycling rates for e-waste are less than 20%. Nevertheless, electronic-waste processing is a rapidly growing commercial sector, and the rate of recycling is increasing. Shiploads of e-waste are sent to Asia and Africa, where products are disassembled and useful materials are salvaged (**Figure 18.25**). While shipping these hazardous wastes overseas benefits the U.S. environment, it often presents significant health risks to workers in underregulated facilities elsewhere.

At present, the United States does not have national regulations regarding e-waste. However, several states

have taken action to diminish its stream. For example, in 2003 California passed its Electronic Waste Recycling Act. This act requires manufacturers to reduce the amounts of certain hazardous substances in the electronic devices they produce. It also requires consumers to pay recycling fees when they purchase new devices. The recycling fees go to a special fund to compensate for the costs of e-waste recovery and recycling. Finally, the act mandates that state agencies purchase their electronic devices from manufacturers that have reduced the amount of hazardous chemicals in their products.

The European Union and some states in the United States have recently adopted take-back legislation for some electronic devices. **Take-back regulations** require electronic manufacturers to take back their products at the end of their useful life, at no charge to the consumer. The manufacturer is then responsible for the proper disposal or reuse of the materials in those products.

▲ **Figure 18.24 E-Waste**
The electronic age has generated new forms of hazardous waste. Circuit boards and other components contain heavy metals.

◄ **Figure 18.25 Recycling e-Waste**
Workers in Asia reuse and recycle parts from electronic devices that have been discarded in developed nations. These activities expose the workers to many toxic substances.

QUESTIONS

1. What is a permanent retrieval storage site? How does it differ from an open dump?

2. Describe California's approach to managing e-waste.

3. What is the difference between low-level and high-level radioactive waste? How does this difference affect the methods of managing these wastes?

Radioactive Waste

Radioactive wastes require long-term monitoring to avoid human health risks.

Radioactive waste is material that is contaminated by isotopes of elements that emit destructive forms of radiation. Because most radioactive waste decays very slowly, the management of this waste poses unique challenges.

Low-level radioactive waste contains comparatively low amounts of radioisotopes and represents the greatest volume of radioactive waste. Much of this waste comes from the use of radiation in hospitals and research laboratories; typically, it consists of contaminated clothing, filters, wiping rags, medicinal swabs, tubes, syringes, and the carcasses and tissues of laboratory animals. The residues from nuclear reactor wastewater are also an important form of low-level waste. Radioactivity in such wastes ranges from just above natural background levels to very high levels, as in the case of parts from inside nuclear reactors. Much low-level waste is stored at the site where it is generated. When enough material accumulates, it is shipped to low-level radioactive disposal sites (**Figure 18.26**). In the United States, such sites are licensed and regulated by the Nuclear Regulatory Commission.

Nuclear power plants and weapon labs generate **high-level radioactive waste**, which is far more dangerous than low-level waste and requires long-term storage and management. The majority of high-level waste is in the form of used fuel rods from nuclear power plants. At present, all high-level radioactive waste is stored on-site at the world's nuclear power facilities. In the United States, such waste is stored at over 100 locations. When spent fuel is first removed from reactor cores, it is placed in wet storage pools. At most facilities it is subsequently transferred to stainless steel casks for dry storage (15.6, **Figure 18.27**).

Wet and dry storage facilities must be licensed and are constantly monitored. Nevertheless, there are concerns that such storage may be hazardous to the health of people working at the sites. Such facilities may also be vulnerable to terrorist actions. Although spent fuel is not explosive, it does contain large amounts of radioactive and toxic heavy metals. It is feared that a terrorist attack could disperse these metals widely in the environment.

The safe management of high-level nuclear waste is one of the most daunting challenges associated with the use of nuclear power. Because spent fuel rods still contain large amounts of radioactive material, some experts argue that they should be recycled into new fuel rods. However, reconstituted fuel rods are enriched with plutonium, an element that can be used to manufacture nuclear bombs. For security reasons, therefore, most nations have agreed that such recycling should not be pursued.

Several nations are pursuing the development of permanent geologic repositories for nuclear waste. In these repositories, used fuel rods encased in special metal casks would be placed in deep caves in geologic strata that are isolated from the biosphere. It is hoped that the metal casks and the surrounding rock will prevent the radioactive materials from leaking into the environment. In the United States, one such repository has been proposed for Yucca Mountain, Nevada. However, licensing of this repository and of those in other locations has been hampered by uncertainties about the eventual fate of the nuclear waste and by opposition from nearby communities and other interest groups.

In summary, industry, electrical utilities, and electronic devices of all kinds generate wastes that pose a direct threat to human health. While current management focuses on the containment and storage of these hazardous wastes, increasing emphasis is being placed on reducing the production of such wastes and reducing their toxicity in the environment.

▼ **Figure 18.26 Storing Low-Level Nuclear Waste**
Articles that are contaminated with low levels of radioactive chemicals are encased in concrete casks or containers, which are then buried in clay-lined excavations.

▼ **Figure 18.27 Storage Sites for High-Level Nuclear Waste**
High-level nuclear waste is stored at nuclear power facilities across the United States. At first, spent nuclear fuel is stored in pools of water. After the fuel has cooled for a couple of years, it may be transferred to dry storage casks.

Yucca Mountain

• Sites storing spent nuclear fuel, high-level nuclear waste, and/or surplus plutonium destined for geologic deposition

18.4 Managing Product Life Cycles

Much of waste management focuses on how to properly dispose of waste once it is produced, but a more compelling approach is to avoid generating waste in the first place. Generating less waste reduces the pressure on landfills and other waste management facilities and saves the energy and resources that are lost when items are discarded. It also minimizes the environmental impacts that occur throughout the life of products. A variety of approaches can be used to reduce the stream of waste and to produce products that can be recycled.

Life-Cycle Assessment

Product manufacture, use, and disposal systems should be less linear and more circular.

Energy and matter move through natural ecosystems in a circular fashion. Energy and matter stored at one trophic level are consumed and stored by organisms at the next level. When organisms decompose, the energy and nutrients stored in their tissues are recycled back into the ecosystem's food web. In most ecosystems, the ultimate wastes are simply carbon dioxide, water, and heat.

In contrast, most human systems of manufacturing products and managing waste are largely linear. Raw materials, often from nonrenewable sources, are used to create products. These products are then packaged and marketed in ways that demand additional raw materials. Products are used until they are either obsolete or worn out and are then discarded in landfills or burned in incinerators. This approach to manufacturing is energy intensive and has significant environmental impacts.

Life-cycle assessment (LCA) is a method used to evaluate the environmental impact of a product throughout all the stages of its life, from resource extraction to manufacturing, distribution, use, and final disposal. This allows the environmental and resource impacts of a product to be examined in the context of the entire system.

A complete LCA includes three interrelated components: an inventory analysis, an impact analysis, and an improvement analysis. An *inventory analysis* quantifies the inputs of energy and raw materials and the outputs of solid waste, air and water pollution, and other environmental releases associated with each stage of a product's life cycle. An *impact analysis* evaluates the various impacts—environmental, economic, health, social, and cultural—associated with the inputs and outputs identified in the inventory analysis of a product. An *improvement analysis* identifies opportunities to reduce the environmental impact of a product by changing aspects of its life cycle, such as its design, manufacturing methods, consumer recycling, and waste management.

Life-cycle analysis begins with the creation of a flow diagram representing the steps in the system used to manufacture, use, and dispose of a particular product. Each step is then analyzed in finer detail as a subsystem with its own set of energy and material inputs and outputs (**Figure 18.28**).

An analysis of the life cycle of a cathode-ray tube (CRT) computer monitor illustrates the current linear approach to product manufacture, use, and disposal.

▼ Figure 18.28 Life-Cycle Analysis
The full consequences of the manufacture, use, and disposal of a product can be understood by a close examination of the inputs and outputs associated with each stage of its life, from "cradle to grave."

Inputs:
Water
Raw material
Energy

PRODUCT STAGES
Manufacturing, processing, and function
Distribution and transportation
Use, reuse, maintenance
Recycle
Waste management
Raw materials acquisition

Outputs:
Usable products
Water effluents
Airborne emissions
Solid wastes
Other
Environmental releases

The initial manufacture of the monitor requires a wide variety of raw materials, including large amounts of water. Wastes produced in manufacturing and distributing the monitor include packing materials and a variety of by-products associated with its manufacture. The average lifetime of a computer monitor is about 5 years; during that time, the monitor consumes electrical energy. The production of that electricity requires more than 500 kg (1,100 lb) of raw materials and 1,100 L (300 gal) of water. When a monitor fails or, like many CRT devices, becomes obsolete, it is usually discarded; fewer than 10% of computer monitors are recycled. Disposal consumes more energy and materials (**Figure 18.29**).

The environmental costs associated with a computer monitor also include outputs of air and water pollution. From "cradle to grave," a single monitor is responsible for the production of 664 kg (1,465 lb) of air pollutants, such as ozone, NO_x, and SO_2, and 19.8 kg (43.6 lb) of water pollutants. Greenhouse gas emissions associated with the monitor's manufacture, use, and disposal are equal to about 700 kg (1,540 lb) of CO_2.

The overall consequences of these life-cycle impacts are magnified by the sheer number of computer monitors in various stages of life. For example, today more than half of all U.S. households have at least one computer and CRT monitor stored in a closet or garage, awaiting disposal.

▼ **Figure 18.29 A Complex Life**
The environmental impacts associated with manufacturing, using, and disposing of a computer monitor extend well beyond the materials in the monitor.

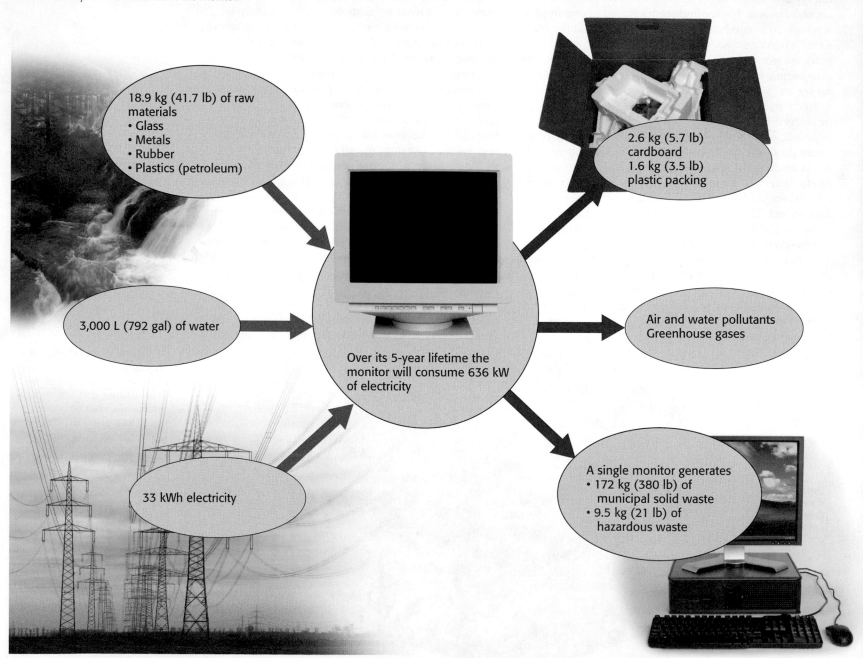

18.9 kg (41.7 lb) of raw materials
• Glass
• Metals
• Rubber
• Plastics (petroleum)

3,000 L (792 gal) of water

33 kWh electricity

Over its 5-year lifetime the monitor will consume 636 kW of electricity

2.6 kg (5.7 lb) cardboard
1.6 kg (3.5 lb) plastic packing

Air and water pollutants
Greenhouse gases

A single monitor generates
• 172 kg (380 lb) of municipal solid waste
• 9.5 kg (21 lb) of hazardous waste

Reimagining Product Life Cycles

Waste is just that!

In recent years, a variety of alternative approaches to product manufacture and disposal have gained recognition. Even small gains in efficiency at each life-cycle stage can have significant effects on environmental impacts and total energy and resource use. Greater efficiency may come from reduced resource use, reuse, or recycling.

Many manufacturers have embraced a concept called **industrial ecology**, in which industrial systems are redesigned to mimic the cycling of materials in natural ecosystems. Industrial ecology combines the disciplines of engineering, chemistry, biology, and economics to find ways to make the generation and disposal of industrial wastes more efficient and environmentally sustainable. Wherever possible, the industrial wastes generated by one process are used in another process or by another industry.

One of the most famous applications of industrial ecology occurs in an industrial zone in Kalundborg, Denmark. Dozens of companies in Kalundborg transfer some of their industrial wastes to other companies, which use those wastes in other processes. For example, sulfur removed from natural gas at an oil refinery is shipped to a nearby producer of sulfuric acid. Ash from a power plant is used to make cement. Waste heat from a power plant is used to heat greenhouses that grow vegetables year round.

Interest in industrial ecology is growing because it makes economic as well as environmental sense. However, industrial ecology requires cross-functional thinking and collaboration; these activities are often restricted when companies are geographically separated.

Thinking of a product as a service can also transform patterns of manufacturing, marketing, and consumption. For example, Interface Carpet, a flooring company based in Atlanta, Georgia, leases "flooring services" instead of selling carpeting. Companies that purchase Interface's flooring services never actually own any carpets, so they do not have to worry about how to dispose of them after their useful life. When it is time to replace carpeting, Interface takes back their old carpets and remanufactures them into new carpets. Over 1.5 million tons of conventional carpeting is dumped into U.S. landfills each year. The Interface model for carpet life-cycle management could significantly reduce this waste stream.

Many companies are using **design for disassembly (DfD)** to circularize the life cycle of their products. DfD involves the development of materials, components, construction techniques, and management systems that permit the easy recovery of parts and materials from discarded products. DfD allows most of the material in a discarded product to be reused in a new product, thereby reducing the need for raw materials. DfD is most effective when manufacturers institute programs that encourage consumers to return products at the end of their useful life.

Electronics companies such as Sony and Dell are now producing televisions, telephones, and computers that are designed for easy disassembly and recycling (**Figure 18.30**). DfD principles are also being incorporated into the design and architecture of many new buildings.

Germany requires that automobiles be manufactured so that all parts can be easily recycled. This has encouraged German manufacturers, such as BMW, Mercedes-Benz, and Volkswagen, to use DfD principles in the design of their automobiles. BMW, for example, has designed its cars so that potentially polluting fluids, such as oil, brake fluid, and coolants, can be easily removed; the design of their electronics allows the easy recovery of recyclable parts. At specialized take-back facilities, over 95% of the car's mass is recycled.

Extended product responsibility (EPR) is based on the belief that those who make a product should assume responsibility for that product throughout its life cycle. EPR regulations sometimes take the form of electronic take-back laws, which require manufacturers to take back electronic devices at the end of their useful life, at no cost to the consumer. The state of Maine recently enacted take-back legislation requiring manufacturers to pay for the return and reuse or disposal of computers and televisions.

At present, several economic and policy barriers are limiting the shift to a more cyclical approach to manufacturing. For example, many of the costs associated with mining and generating energy are not captured in the price of raw materials. This can make the use of raw materials appear to be cheaper than reuse or recycling. Laws that regulate the transport of hazardous materials complicate the development of industry-wide take-back programs. In the United States, these economic and policy barriers are being addressed on a state-by-state basis, but this can lead to confusion for manufacturers operating at the national and international level.

All of these approaches promote new modes of thinking about product manufacture, use, and disposal. Rather than designing consumer products with a once-through, linear model of resource use, products are designed with reuse in mind from the very start. A more cyclical approach to product life cycles has multiple environmental and economic advantages over current approaches. Fewer raw materials need to be extracted to manufacture products, reducing the environmental impacts of mining and land disturbance. Energy-intensive refining and processing are reduced or even eliminated, thereby saving energy and money and reducing pollution. Life-cycle approaches also tend to emphasize durability in consumer products, which reduces the total production of all forms of waste.

▲ **Figure 18.30 Designed for Disassembly**
This cell phone is designed so that its parts can be easily separated into reusable or recyclable units.

1. Describe the three components of a complete life-cycle assessment for a product.

2. Wastes can be reduced by marketing a product as a service. What is meant by this statement?

3. Describe two barriers to the adoption of cyclical approaches to product manufacture.

AGENTS OF CHANGE: Plymouth State University's Initiative to Reduce Waste on Campus

Kelly Donovan studied environmental science and policy at Plymouth State University in New Hampshire. In 2009, Kelly was chosen as PSU's coordinator for RecycleMania, a nationwide competition that promotes recycling and the reduction of waste on college campuses. With Kelly's leadership, PSU placed first in New Hampshire in the Waste Minimization category, which recognizes the institution that produces the least amount of waste per student.

How did you first get the idea for RecycleMania at PSU?

I wanted to get involved in RecycleMania because I recognized that nearly 50% of the material entering the landfill at Plymouth State University could be reduced. Also, I realized that many members of the PSU community were unaware of what is and is not recyclable. I knew that education from peers was the best way to reach out to students and help them understand the importance of recycling.

What steps did you take to get started?

I worked with PSU's Office of Environmental Sustainability to educate and involve faculty, staff, and students in RecycleMania. We used both top-down and bottom-up approaches.

To target faculty and staff, we met with the university president's cabinet, as well as the vice-president's committee, which includes representatives from various departments across campus. I also attended faculty and operating staff meetings.

In the residence halls, a team of students called sustainability fellows designed educational programs, created bulletin boards, and collected recyclable materials. During their interactions with residents, they would begin conversations about the importance of recycling and explain how students could "go green" on a college budget.

Our team and the Office of Environmental Sustainability worked very closely with three other departments—Physical Plant, Residential Life, and Information Technology. These departments helped us spread awareness across campus. For example, Physical Plant allowed us to use the shuttle service to pass out flyers about recycling, and Residential Life helped us to hang banners in the student union. Information Technology helped us change the MyPlymouth login to a RecycleMania theme and send e-mail reminders to students.

What are the biggest challenges you've faced with the project, and how did you solve them?

The biggest challenge was creating behavioral change. There were already a variety of proposals and pilot programs for encouraging recycling on campus, such as centralizing waste containers and decentralizing recycling bins. This made it less convenient for people to throw things away and more convenient for them to recycle.

Strategies for creating behavioral change were largely associated with the University Environmental Committee and the President's Commission on Environmental Sustainability. The two groups worked with departments and students, encouraging them to make small changes in their daily lives, such as printing double-sided, bringing reusable coffee mugs to school and work, thinking of waste prevention when purchasing materials, and switching from paper to online. We also encouraged people to discuss how behavioral changes would benefit both the environment and their wallets!

How has your work affected other people and the environment?

By promoting behavioral change, we saw a significant decrease in consumption across campus, as well as an increase in the amount of materials being recycled. We were certain of our success when PSU placed first in New Hampshire in the Waste Minimization category of RecycleMania for the 2009–2010 academic year. RecycleMania helped reduce the amount of waste being put into the university landfill and proved that environmental awareness and the motivation to live in an eco-friendly way are on the rise at Plymouth State University.

What advice would you give to students who want to replicate your model on their campus?

If you're interested in registering your school for the RecycleMania competition, contact helpline@ recyclemaniacs.org. At www.recyclemaniacs.org you can find tools for planning your participation in the competition and developing a strategy for winning. This site also tells how to measure waste and recycling and how to promote recycling on your campus.

I would suggest approaching students and departments as an assistant, rather than as an enforcer. Create positive energy and encourage everyone you talk with to live as sustainably as possible!

Summary

Human activities generate large amounts of waste including wastewater, municipal and industrial solid waste, and hazardous waste. Wastewater includes sewage, stormwater runoff, and water used in industry and manufacturing. In the United States, most wastewater is treated in a stepwise process that first removes sediment and then removes dissolved organic matter and inorganic chemicals. In many rural areas, wastewater is treated in on-site septic systems. Municipal solid waste (MSW) consists of wastepaper, food wastes, yard trimmings, plastics, and the like. The composition of the MSW waste streams varies widely among countries, reflecting different patterns of resource consumption and waste management practices. In many developing countries, MSW ends up in open dumps or in rivers. In the United States, the majority of MSW is now placed in sanitary landfills that are lined with clay to minimize leakage of pollutants to aquifers. In some places, MSW may be incinerated in waste-to-energy facilities. In many cities and counties, organic wastes such as food and yard trimmings are separated and converted to compost for gardens and agricultural uses. Over the past four decades, the U.S. EPA and environmental groups have worked to reduce the MSW stream by the reduction of packaging; the reuse of items such as shopping bags, clothing, furniture, and building materials; and the recycling of materials such as paper, plastic, glass, and metals. As of 2010, the total MSW stream had begun to decline and recycling rates were nearing 35%.

Wastes that are flammable, corrosive, explosive, or toxic are designated as hazardous. These wastes include a variety of household materials such as paints, insecticides, and solvents, as well as toxic wastes from industry. In recent years, discarded electronic devices (e-waste) have added significantly to the hazardous waste stream. In the past, hazardous wastes were simply dumped into open pits or rivers. Today, laws in the United States and other countries regulate the generation and disposal of hazardous waste, as well as the cleanup of polluted sites. In the United States, the disposal of such waste is monitored and regulated by the EPA. Hazardous wastes may be placed in permanent retrieval storage sites or disposed of in licensed incineration facilities. Radioactive wastes are generated by hospitals, laboratories, and nuclear reactors; these wastes require special treatment and long-term monitoring.

Much waste management remains focused on the disposal of waste and the isolation of hazardous materials from the environment. In contrast, life-cycle assessment examines the full life cycle of products from materials acquisition to disposal. This approach also identifies ways to minimize waste at each step in the process.

18.1 Wastewater

- In many of the world's poorest countries, wastewater is simply dumped into waterways or the ocean. In the cities of most developed countries, wastewater is treated in municipal sewage treatment plants to remove sediments and chemicals.

- In rural areas, properly designed and maintained septic systems can isolate waste and protect water supplies.

KEY TERMS

municipal sewage treatment plant, primary treatment, secondary treatment, tertiary treatment, septic system, leach field, graywater

QUESTIONS

1. You can find information on wastewater treatment in your county on its government website. What is the ultimate fate of treated wastewater and sludge from your county's sewage treatment plant?

2. What is graywater and how can it be used?

PRETREATMENT   PRIMARY   SECONDARY   SLUDGE DIGESTION   METHANE

Air
Scum
Trash
Aerated tank
Sludge
Filter
Dry sludge finishing

18.2 Solid Waste

- The United States ranks high among the world's countries in per capita generation of MSW, but the total and per capita waste stream appears to be decreasing.

- Most MSW is now treated in sanitary landfills that are designed to prevent pollutants from leaking into groundwater.

- Programs to reduce packaging, reuse products, and recycle the materials in products are further reducing the amount of MSW.

Trench method

Final earth cover
Compacted solid waste
Daily earth cover
Original ground

18.3 Hazardous, Electronic, and Radioactive Waste

- Although management of hazardous waste has improved in the United States, thousands of historically polluted sites remain to be cleaned up.

- E-waste—discarded electronic devices—is a growing part of the global waste stream. Take-back regulations and recycling laws are encouraging the reuse and recycling of e-waste.

- Radioactive wastes are contaminated with isotopes of elements that emit dangerous radiation. These wastes are generated by hospitals, research and weapons laboratories, and nuclear power plants.

KEY TERMS

hazardous waste, permanent retrieval storage site, bioremediation, e-waste, take-back regulation, low-level radioactive waste, high-level radioactive waste

QUESTIONS

1. Describe three approaches for the disposal or cleanup of hazardous waste.

2. How has U.S. legislation regulating the generation, disposal, and cleanup of hazardous waste changed over the past 50 years?

3. What is the difference between a high-level nuclear waste storage facility and a permanent waste repository?

18.4 Managing Product Life Cycles

- New approaches to waste management focus on ways to avoid generating waste by assessing the full life cycles of products.

- The generation of waste can be diminished by treating products as services, by designing products for disassembly and recycling, and by extended product responsibility.

KEY TERMS

life-cycle assessment (LCA), industrial ecology, design for disassembly (DfD), extended product responsibility (EPR)

QUESTIONS

1. Industrial ecology strives to make manufacturing processes circular rather than linear. What is meant by this statement?

2. Describe three approaches that industrial ecologists are using to diminish manufacturing waste.

19

The Environment and Human Health

DDT vs. Malaria

How can we diminish the risk from one health hazard without increasing the risk of another?

Malaria has afflicted the people living near the headwaters of Africa's Kafue River for more than 100,000 years (**Figure 19.1**). Today, this region is in the nation of Zambia. It is here that the deadliest form of the disease, falciparum malaria, is thought to have evolved; historically, it was more prevalent here than anywhere else on Earth. This disease, noted for intense fevers that recur at 2-day intervals, can infect anyone, but is especially deadly in small children. Although actual numbers are unknown, it appears that prior to 1900 malaria killed about 20% of the children in the region who were younger than five years old. Since that time, changes in human activities and the environment have had a significant impact on the prevalence of malaria.

In 1926, Europeans discovered copper ore near the Kafue River, and operations at the Roan Antelope Copper Mine began soon afterward. The town of Luanshya was established in 1930 to accommodate British engineers and overseers, as well as several thousand local people who worked the mines. Company records show that well over 50% of Luanshya's 5,000 plus citizens—white and black—contracted malaria in each of the next 2 years (**Figure 19.2**).

It was clear to mine owners that malaria was by far the greatest obstacle to the extraction of the region's copper, and they immediately began an aggressive program to stop transmission of this disease. Mining officials knew that the parasite that causes malaria is transmitted by mosquitoes of the genus *Anopheles* (**Figure 19.3**). The officials recruited hundreds of men to drain the swamps and marshy "dambos" where mosquito eggs are laid and their larvae mature. To prevent the entry of adult mosquitoes, the company put screens on the windows

▲ **Figure 19.1 Luanshya, Zambia**
Luanshya is located in the Copper Belt Province in north central Zambia. The swamps and seasonally wet savannas of this region provide ideal habitats for mosquito larvae.

and doors of houses. Workers and their families were given regular doses of chloroquinone, a drug that provided considerable resistance to the disease. With these measures in place, by 1939 only 13% of Luanshya's residents were contracting malaria each year; mortality from the disease had declined by 90%.

▼ **Figure 19.2 Malaria Incidence Rate**
The incidence of malaria among residents of Luanshya began to diminish in 1931 with the implementation of measures to control mosquito populations. Several factors, including the global ban on DDT use, have contributed to the resurgence of malaria since 1970.
Source: Utzinger, J., Tozan, Y. and Singer, BH. 2001. Efficacy and cost effectiveness of environmental management for malaria control. *Tropical Medicine and International Health* 6, 677–687.

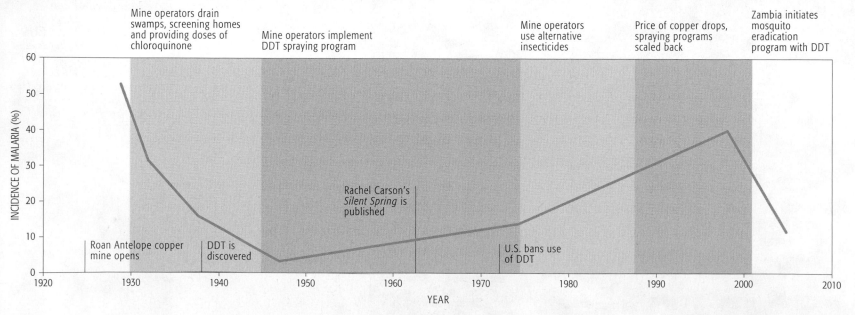

In 1946, Roan Antelope Mining began spraying the recently developed insecticide, DDT (dichlorodiphenyltrichloroethane). By 1948, the incidence of malaria had dropped to less than 2%. That year no mortalities from the disease were recorded in Luanshya.

In 1947, the World Health Organization's Expert Committee on Malaria reported that with DDT there was "at last a method of controlling malaria in many areas at costs within the economic means of the people." So it seemed. Throughout the 1950s and 1960s, malaria rates in central Africa remained low. Public health officials were optimistic that the disease might soon be eradicated.

Yet even then, evidence of the serious effects of DDT was accumulating. In l962, Rachel Carson's book *Silent Spring* called attention to the fact that this chemical accumulates in the tissues of many animals, causing serious environmental effects. She also noted that DDT causes cancer and developmental abnormalities in humans. In 1972, the United States banned all uses of DDT. Other countries soon followed suit. The World Health Organization withdrew its support for the use of DDT to control malaria in 1974. Alternative insecticides were available, but they were significantly more expensive. Initially, mining companies used their considerable wealth to maintain spraying programs using these alternatives. But in the late 1970s, the price of copper began to plummet, and spraying programs were scaled back. The incidence of malaria in the Copper Belt region began to rise once again.

By 1976, when Zambia began its first thorough national surveys of infection rates, the incidence of malaria in the Luanshya region was back up to 12% per year. In the following years, the incidence increased at an alarming rate. By 2000, the incidence of malaria in the Copperbelt region was 37.5%. More than 20% of children were dying before their fifth birthday, and 40% of those deaths were due to malaria.

In 2000, the Zambian government began to implement measures to control this epidemic. These included the distribution of insecticide-treated mosquito bed nets and the conservative use of DDT. Spraying of DDT was confined to the interior walls and ceilings of homes, where mosquitoes land at night (**Figure 19.4**). These measures reduced the incidence of malaria in Luanshya and nearby villages by more than 75%. Across Zambia, childhood mortality due to malaria declined by 33%. Recognizing these benefits, the World Health Organization reversed its policies and in 2006 began active support of the use of DDT to limit malaria.

Many public health officials worry that the benefits of using DDT do not offset the risks that this chemical poses to the environment and human health. For example, prenatal exposure to small amounts of DDT increases the likelihood of premature births; such births account for 17% of infant mortality in Zambia.

Scientists point out that the prohibition of DDT was probably not the sole cause of the recent malaria epidemic in places like Luanshya. Over the last 80 years, malaria parasites have become increasingly resistant to chloroquinone. Since 1970, the mean annual temperature in this region has increased by 2 °C; in many places, this increase in temperature appears to have improved breeding conditions for *Anopheles*. Since 1990, human populations in sub-Saharan Africa have doubled, and the higher population density facilitates the transmission of disease. In addition, 20%–30% of the population is infected with HIV/AIDS, which diminishes resistance to malaria infections.

Debate over the use of DDT to control malaria is likely to continue. It will be based in part on questions about the dispersion and uptake of this chemical and its effects on human health. It will also be informed by our changing understanding of environmental factors such as population growth, landscape alteration, and climate change that affect the spread of this disease. Most important, it will focus on the recognition that diminishing the risk of one health hazard can increase the risk of another. Questions and issues regarding changes in the environment and their impact on human health are among the important themes of this chapter.

◄ **Figure 19.3 *Anopheles* Mosquito**
This insect is the primary vector for the parasite that causes falciparum malaria. Larvae develop and mature in standing bodies of water. Females take multiple blood meals. In doing so, they transfer the parasite from infected to noninfected individuals.

▼ **Figure 19.4 Protecting Against Mosquitoes**
Programs to spray DDT on the interior walls and ceilings of homes have been implemented in many African countries. Female mosquitoes are active primarily at night and have a habit of lighting on walls before seeking out a blood meal.

19.1 Introduction to Public Health

The overall health of human populations is most often assessed with statistical measures such as life expectancy and infant mortality. Leading causes of death in populations depend on factors such as age structure and social and economic conditions. Environmental factors affect the **risk**, or likelihood, of particular hazards impairing health or causing death. However, human perceptions of the magnitude of risk associated with a particular hazard are often significantly different from the actual risk.

Measuring Public Health

Likelihood of death and causes of death are important measures of the health of human populations.

How does a doctor assess the status of your health? Perhaps by checking your body temperature, pulse rate, and blood pressure. Each of these measures provides information about your health because your doctor understands how they are connected to your basic bodily functions. Your body temperature is determined by your metabolic rate and the presence or absence of infections; your pulse and blood pressure are direct measures of your heart's function.

In a similar fashion, population-level statistics can be used to assess the health of human communities. Life expectancy at birth—the average number of years that a newborn child is likely to live—is probably the most widely used measure of the overall health of human populations (5.2). Since 1950, worldwide life expectancy at birth has increased from 46 years to more than 67 years.

Mortality rates also measure the health of a community. Infant mortality rate is the percentage of newborns that die during their first year of life. Although it focuses on the health of infants, public health officials consider this statistic a particularly sensitive indicator of factors that influence the overall health of a population, such as water quality, sanitation, nutrition, and health care. In 1950, the infant mortality rate of the world's human population was 15.3%; currently, it is 4.9%.

An elevated body temperature indicates illness, but tells a doctor relatively little about what is ailing a patient. In a similar way, mortality statistics alone tell little about what factors are causing rates to increase or decrease. Public health officials can infer much more about the health of a population from statistics on the leading causes of death. Worldwide, cardiovascular diseases, including heart attacks and strokes, account for nearly 30% of all human deaths (**Figure 19.5**). Infectious diseases are a close second, causing 25% of all deaths. Among the most important infectious diseases are diarrheal diseases, AIDS, tuberculosis, malaria, and respiratory infections, such as influenza and pneumonia.

Leading causes of death are greatly influenced by the age of the population. In the United States, for example, accidents are the leading cause of death for people between 1 and 44 years old. For people between 45 and 64, the leading cause of death is cancer. For people 65 and older, it is cardiovascular disease.

Changes in leading causes of death are directly related to changes in the environment. In 1900, for example, diarrheal diseases were responsible for more than 8% of all deaths in the United States. Today they account for less than 0.1%. Diarrheal diseases are caused by several different intestinal parasites, and their decreasing prevalence over the last century is a direct consequence of improved management of drinking water and waste.

The leading causes of death in poor countries differ significantly from those in wealthy countries (**Figure 19.6**). In less-developed countries, poor sanitation and nutrition result in a greater prevalence of infectious diseases. In wealthy countries, cardiovascular diseases are more prevalent, reflecting the older average age of their populations.

▼ **Figure 19.5 Worldwide Causes of Death**
Many cardiovascular, respiratory, and digestive diseases are caused by infectious organisms, as are some cancers. Thus the actual percentage of deaths caused by infectious disease is probably much higher than shown in this graph.
Source: World Health Organization. 2004. The World Health Report 2004: Changing History. Annex table 2. URL:http://www.who.int/whr/2004/en/09_annexes_en.pdf. Last accessed 10-6-2011.

- Cardiovascular diseases
- Infectious diseases
- Cancers
- Injuries
- Respiratory and digestive
- Other
- Maternal

▼ **Figure 19.6 Wealth and Causes of Death**
Differences in environmental and social conditions and the age structure of populations cause significant differences in the leading causes of death in wealthy and poor countries. *Source:* UC Atlas of Global Inequality. URL:http://ucatlas.ucsc.edu/cause.php. Last accessed 10-6-2011

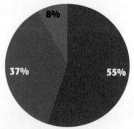

DEVELOPED REGIONS

9%
14%
77%

DEVELOPING REGIONS

8%
37%
55%

- Noncommunicable conditions
- Communicable diseases
- Injuries

- Communicable diseases
- Noncommunicable conditions
- Injuries

Hazards and Risk

Risks posed by environmental hazards may be immediate or delayed.

Most causes of death are influenced by hazards in the environment, although the specific relationships depend on the kind of hazard. For example, the prevalence of diarrheal diseases in developing countries is directly tied to inadequate sanitation for water and food. Respiratory diseases are exacerbated by poor indoor and outdoor air quality. The incidence of malaria is strongly influenced by climate.

The hazards that influence human health can be grouped into four categories—physical, chemical, biological, and cultural (**Figure 19.7**). Physical hazards include geological events such as earthquakes, weather events such as blizzards and floods, and the effects of radiation, such as that from ultraviolet light. Chemical hazards include pollutants in water, air, and food. Biological hazards are the result of a diverse array of organisms that cause or transmit infectious disease. Cultural hazards are a consequence of where people live, their occupations, and their lifestyle choices. Urban lifestyles, for example, increase the risk of being injured in automobile collisions. Lifestyle choices also affect an individual's exposure to hazards. Choosing to use cigarettes, for example, significantly increases exposure to the chemical toxins in tobacco smoke.

Risk is the mathematical probability of a harmful outcome, such as injury or death. Scientists who study environmental health discuss the risk posed by an environmental hazard in terms of both the likelihood of its occurrence and its consequences. Thus, the probability of a very large earthquake occurring near an urban center may be low, but its consequences in terms of injuries or lives lost may be high.

One widely used measure of environmental risk is the rate of mortality due to a particular hazard. In the most fundamental sense, death has the same consequence for everyone: It is the end of life. Yet in terms of years of life lost, the consequences of death are greater for young people than for old people. Therefore, many scientists and ethicists argue that mortality risk should be measured in terms of lost years of life. Using this approach, a hazard that affects children would be viewed as having a higher risk than an equally likely hazard that affects primarily older people.

Some effects of a hazard are immediate, while others are delayed; the delayed effects are called **latent consequences**. For example, more than 50 people died almost immediately as a consequence of the 1986 accident at the Chernobyl Nuclear Power Plant in the former Soviet Union. Since then, the World Health Organization estimates that there have been approximately 9,000 latent cancer deaths among the 7 million people who were most exposed to the radiation released by the accident.

▲ **Figure 19.7 Environmental Hazards**
Air pollution is a chemical hazard Ⓐ. Earthquakes such as the one that occurred in Sichuan, China, in 2008, are physical hazards Ⓑ. Measles and other infectious diseases are biological hazards Ⓒ. Cultural hazards refer to lifestyle choices such as cigarette smoking Ⓓ.

Risk Perception and Reality

Human perceptions of risk often differ from reality.

How much smoking do you have to do throughout your life to be considered "at-risk"?

Shayna Jefferson, University of Florida

Norm: *Smoking creates a variety of risks to your health. There is not a clear threshold for an amount of smoking or length of smoking below which there is no risk.*

1. How do the leading causes of death in developed countries differ from those in developing countries?

2. Using an example, explain what is meant by *latent risk*.

3. Describe five factors that can cause the perception of the risk associated with a hazard to differ from its actual risk.

Our world is filled with both natural and human-caused hazards. Risk measurement and assessment are powerful tools for making decisions about how to prioritize our responses to these hazards. Nevertheless, it is sometimes difficult to quantify risk. In addition, individual and societal responses to hazards are often determined by perceptions of risk that do not correspond to reality. Some of the key factors that influence perceptions of risk are described below.

The myth of zero risk. The U.S. EPA sets standards for acceptable levels of cancer-causing chemicals in the air and drinking water. Yet survey after survey shows that most people believe there is no safe level of exposure to a hazard such as a cancer-causing agent. Instead, they believe that the risk should be reduced to zero. In practice, it is virtually impossible to reduce the risk of most natural and human-caused hazards to zero. Furthermore, the incremental costs of reducing the risk of a hazard generally increase substantially as the level of risk from that hazard is decreased.

Public awareness. Publicity has a significant effect on perceptions of the degree of risk posed by hazards. For example, Alar is a synthetic chemical that was widely used by farmers between 1963 and 1989 to encourage ripening in several fruit crops, including apples (**Figure 19.8**). Studies between 1984 and 1989 found that Alar elevated the risk for cancer in laboratory animals. In 1989, a segment on the CBS television show *60 Minutes* portrayed Alar as a significant and imminent threat to public health.

Public reaction was swift, and sales of apples plummeted. The EPA banned the use of Alar very soon after the broadcast.

In retrospect, public response to this hazard was disproportionate to the actual risk. The laboratory tests that prompted the Alar scare used very large doses of the chemical. Even consuming large quantities of juice from Alar-treated apples would cause a lifetime cancer risk of only about 5 per million. This means that an individual who consumes large amounts of Alar-laden apples has 0.0005% greater chance of getting cancer than an individual who does not. This level of risk is far lower than that posed by many other commonly used chemicals.

Risk–risk trade-offs. The benefits of reducing the risk posed by one hazard are often considered in isolation from other, related risks. For example, the leaders of many African countries feel that the concerns of some environmental groups about the risks of DDT neglect the risk posed by malaria if DDT cannot be used. Certainly, the benefits of actions to reduce the risk of a hazard always incur costs or other risks. These are called **risk–risk trade-offs**. However, such costs are often poorly known. In addition, the people receiving the benefit of reducing one risk may not be the ones who must pay the costs associated with elevating another risk.

Control. People generally perceive a risk to be lower or more acceptable if they feel they have some level of control over that risk. For example, each year in the United States about 43,000 people die in auto collisions, a rate equal to about 1.4 deaths per 100 million miles traveled. The risk associated with commercial air travel based on miles traveled is considerably less than one-tenth of this amount. Even so, many people perceive the risk of automobile travel to be lower than that of air travel; this is partly because they believe that their driving skills and habits give them more control over that risk.

Risk and time. People tend to view latent risks as being less than immediate risks, even if they are equal or even greater. For example, it is now widely known that 90% of all lung cancers are the result of smoking. However, a large number of people in their teens and twenties ignore this risk because it is not likely to be realized until later in life.

In summary, our perception of the risk associated with a particular hazard is often quite different from the actual risk. This significantly influences how we as individuals and societies set priorities for and respond to environmental hazards.

◄ **Figure 19.8 Overreaction to Alar?**
Publicity about the possible health effects of apples sprayed with Alar led to its being banned by the EPA. Subsequent studies revealed that the risks to humans from Alar were minimal.

19.2 Physical Hazards in the Environment

Physical hazards are threats to human health that derive from physical processes in the environment. These include geologic hazards such as earthquakes and volcanic eruptions; weather hazards such as hurricanes, floods, and extreme temperatures; and wildfires. Although physical hazards are not among the 10 leading causes of death, individual events may claim hundreds or even thousands of lives in a very brief period of time. Such events are often difficult to predict or prevent. We often call these events "natural disasters," but their impacts on human life and property are greatly influenced by how we design and build our homes and cities and the ways in which we modify the ecosystems that surround us.

Geological Hazards

Earthquakes and tsunamis arise from natural causes, but their impacts are often increased by human actions.

Earthquakes are caused by movements in Earth's crust that are usually associated with geological faults (3.5). Around the world, crustal movements are occurring constantly, but the energy they release varies widely. This variation in energy is responsible for the great range in the magnitude, or intensity, of earthquakes.

The energy released by earthquakes is transmitted through Earth's crust and interior as waves that geologists detect and measure with seismometers. Earthquake energy is usually reported as a number on the **Richter magnitude scale**, in which each successive number indicates a 32-fold increase in energy. **Earthquake intensity** is a measure of the severity of shaking at a given location, and it is determined by its effects on people, structures and the natural environment. Earthquakes of similar magnitude may have different intensities depending on local geological conditions and the quality of building construction. Low-magnitude earthquakes are far more common than those of higher magnitudes. Thousands of earthquakes of magnitude 3 or less occur on Earth every day, but most go unnoticed.

Earth experiences two to three magnitude 5 earthquakes each day; these earthquakes release sufficient energy to damage poorly constructed buildings near the source, or epicenter, of the earthquake (**Figure 19.9**).

▼ Figure 19.9 **Earthquake Intensity**
The Richter magnitude scale expresses the energy released during an earthquake. The earthquake intensity scale indicates its effects on people and buildings. The 2011 earthquake in Japan had a magnitude of 9.0 and an intensity of X.

Richter magnitudes	Description of earthquake	Earthquake effects	Estimated frequency of occurrence
1.0–3.0	I	Generally not felt except under very favorable conditions	Thousands each day
3.0–3.9	II–III	Felt by most people indoors; vibrations similar to a passing truck	100,000 each year
4.0–4.9	IV–V	Felt by nearly everyone; dishes and windows disturbed or broken; Unstable objects overturned	13,000 each year
5.0–5.9	VI–VII	Frightening movement felt by all; heavy furniture moved; considerable damage to poorly built structures	1,300 each year
6.0–6.9	VII–IX	Considerable damage to many buildings; great damage to poorly constructed buildings; buildings may be shifted off foundations	100–130 each year
7.0 and higher	VIII–XII	Widespread damage, even to single story wood buildings; at extreme bridges destroyed and rails bent	10–15 each year

► Figure 19.10 **The Geography of Earthquake Hazards**
Severe earthquakes are most likely near the boundaries of Earth's major tectonic plates. This includes the west coast of the Americas and much of Asia and the Middle East.
Source: Global Seismic Hazard Assessment Program. URL: http://www.seismo.ethz.ch/static/gshap/

▼ Figure 19.11 **A Tale of Two Cities**
In 2010, a magnitude 7.0 earthquake in Haiti devastated its capitol city, Port au Prince Ⓐ. That year a magnitude 7.2 earthquake near Mexicali, Mexico, resulted in property damage and loss of human life that were modest in comparison Ⓑ. Building codes are far stricter in Mexico than in Haiti.

Over 90% of earthquakes of magnitude 6 or more occur in association with the boundaries of Earth's major tectonic plates (**Figure 19.10**). Earthquakes above magnitude 6 often cause considerable damage and loss of life. The effects of these earthquakes depend on their proximity to towns and cities and on the quality of human developments and building construction. Thus the risk of catastrophic disasters is increased by the rapid growth of very large cities in the vicinity of major faults, such as Mexico City, Los Angeles, and Tokyo.

The combined effects of location, poor construction, and population density are exemplified by comparing two recent earthquakes. In January 2010, a magnitude 7.0 quake struck Haiti near its densely populated capital, Port au Prince. This quake killed more than 225,000 people and left about 2 million homeless. Property damage was estimated to exceed $100 billion. Standards for the construction of buildings are notably absent in Haiti, and few structures were designed to withstand even moderate earthquakes. Three months later, Mexicali, Mexico, experienced a magnitude 7.2 earthquake. Although this quake was considerably stronger than the quake in Haiti, there were only 2 deaths, primarily because Mexico enforces strict building standards in earthquake-prone areas (**Figure 19.11**).

Efforts to reduce the risks associated with earthquakes have focused on the design of earthquake-resistant buildings and infrastructure. Modern materials, such as steel-reinforced concrete, are able to withstand the sideways forces of earthquakes. Building codes in many earthquake-prone areas now require that homes and apartments be designed to withstand earthquakes up to magnitude 7.8. Such standards help to protect both property and human lives.

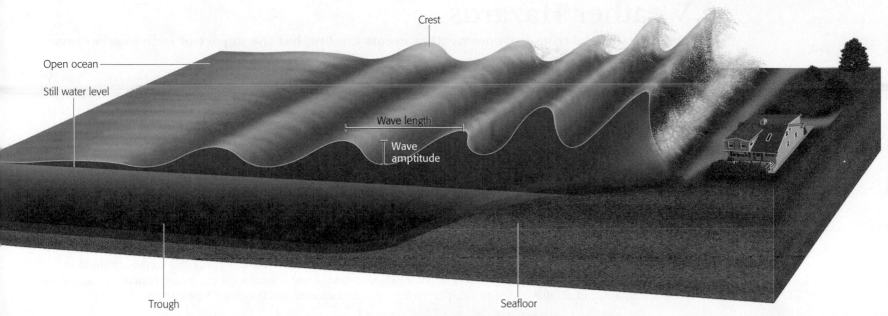

Crest

Open ocean

Still water level

Wave length

Wave amptitude

Trough

Seafloor

▲ **Figure 19.12 Forming a Tsunami**
Tsunami waves may be hardly noticed as they move across the open ocean, but their height is magnified as they move into the shallow water near coasts.

Earthquakes and the movement of sediment on the ocean floor can trigger tsunamis. Tsunamis are often called "tidal waves," although they are unrelated to tides. Tsunami waves may be only 1–3 m high as they move through deep ocean areas. As they move into shallow water, however, their height is magnified, and massive amounts of water may move onto shore (**Figure 19.12**).

On December 26, 2004, tectonic plates shifted beneath the Indian Ocean about 150 km (93 mi) west of the Indonesian island of Sumatra, causing an enormous earthquake. The shift in the seafloor associated with this event displaced the water above it, causing tsunami waves to spread out in all directions. Over a period of hours, these waves struck the coasts of many countries surrounding the Indian Ocean, destroying coastal cities and towns and killing more than 150,000 people (**Figure 19.13**).

Tsunamis are random events, but their consequences are significantly influenced by human modifications of coastal ecosystems. Developments along coastlines and the destruction of coastal mangrove swamps, which normally act as buffers, magnified the impact of the 2004 Indonesian tsunami in many locations.

Volcanic eruptions also create hazards to human health. Lava flows may envelop homes or villages, and explosive eruptions may directly threaten human life. For example, the violent 1980 eruption of Mount Saint Helens in the state of Washington killed 57 people. Plumes of ash from volcanoes can pose a threat to human communities. In 2010, an ash plume from a volcano in Iceland drifted over northern Europe. Because of the risk that the ash would cause jet engines to fail, airports across the region were forced to close for nearly a week.

▶ **Figure 19.13 The 2004 Tsunami**
These photos show the Indonesian coastal city of Banda Aceh before and after the major tsunami of 2004. Here and in many other locations, the impact of the tsunami was magnified by the destruction of coastal swamps and reefs and human developments built right up to the shoreline.

Weather Hazards

The overall risks from extreme weather events are low, but the impact of individual events can be very high.

Tropical storms, floods, blizzards, droughts, and heat waves are among the most important weather hazards to human well-being. Between 2001 and 2006, extreme weather events around the world claimed the lives of about 22,000 people per year, or about 0.03% of all deaths. This may seem low, but individual events such as major hurricanes may claim the lives of thousands of people at once.

Since 1990, tropical storms and their associated floods have accounted for more than 75% of weather-related deaths. Intense tropical storms in the Atlantic and eastern Pacific oceans are called hurricanes; similar storms in the western Pacific and Indian oceans are called typhoons or cyclones. These storms originate over warm ocean water north and south of the equator. Prevailing winds carry them westward, so they are most likely to make

landfall along the east coasts of continents (**Figure 19.14**). Tropical storms bring heavy rains and high water levels called **storm surges** that can cause devastating floods in low-lying regions and on islands.

Death tolls from tropical storms and floods are typically highest in poor countries that have undergone significant deforestation and lack the resources to protect their people during and after storms. For example, in May of 2008 a severe typhoon struck the country of Myanmar in Southeast Asia. At least 150,000 people died (six times the global annual average number of weather-related deaths) in the low-lying delta of the Irrawaddy River. This high death toll was partly due to deforestation of hillsides and coastal swamps, which resulted in landslides and accentuated flooding (**Figure 19.15**). It was also

▶ Figure 19.14 **Hurricane and Typhoon Tracks**
Powerful tropical storms are much more frequent in the Northern Hemisphere than in the Southern Hemisphere. They pose the greatest risk to eastern Asia, the Caribbean, and eastern North America. *Source:* National Oceanic and Atmospheric Administration. URL: http://csc.noaa.gov/hurricanes/

▶ Figure 19.15 **The 2008 Myanmar Typhoon**
These satellite photographs compare the reaches of the Irrawaddy River in Myanmar before Ⓐ and after Ⓑ a powerful storm in 2008. Extensive clearing of land in the river floodplain (beige in the photograph A) magnified the extent of flooding and the loss of human life.

due to the failure of the Myanmar government to provide aid to the people in hundreds of affected villages and towns. The lack of clean water facilitated the spread of disease, causing many deaths soon after the storm.

Hurricane Katrina, which struck the U.S. Gulf Coast in 2005, demonstrated that such events can cause high death tolls in developed countries, too. Katrina was an especially intense storm when it made landfall along the delta of the Mississippi River near New Orleans. The storm surge overwhelmed the levees and canals that normally prevent river water from flooding the parts of New Orleans that lie at or below sea level. Over a period of days, 80% of the city became flooded, and more than 1,800 people lost their lives.

Several factors contributed to the high death toll and widespread loss of property associated with Hurricane Katrina. Investigations revealed numerous flaws in the design and construction of the levees. Destruction of coastal wetlands and floodplain forests likely contributed to the remarkably high storm surge. Aid to the stranded population from federal, state, and local agencies was slow and disorganized. Indeed, this disaster resulted in the resignation of the director of the Federal Emergency Management Agency within days of the storm.

One of the important lessons from Hurricane Katrina and similar disasters is that the risks they pose to human health and well-being are not spread evenly across all segments of the population. For example, the risk of death or serious injury was much higher among New Orleans's poor and black people than among the wealthier, white population. The unevenness has continued in the years since this disaster; reconstruction and restoration have been much slower in the city's poorer areas than in wealthier areas (**Figure 19.16**).

Unlike tropical storms, extreme temperature events, such as blizzards and heat waves, generally account for a very small proportion of weather-related deaths. The 2003 heat wave in Europe was a major exception to this generalization. During August of that year, temperatures across southern Europe were routinely close to 40 °C (104 °F). Many people lacked air conditioning and were unaccustomed to such heat. It is estimated that 35,000 people, most of whom were elderly, died as a consequence (**Figure 19.17**).

Despite these recent events, both the worldwide number of deaths and mortality rates from weather hazards have decreased steadily over the past century. Between 1920 and 1950, about 430,000 people died each year as a consequence of extreme weather events.

During this period, the mortality rate from these events was about 200 deaths per million people each year. This compares with about 5 deaths per million people from such events each year since 1990.

Most of this decrease is due to diminished mortality associated with droughts, hurricanes, and floods. For example, during the period 1920–1929, drought was responsible for an average of 472,000 deaths worldwide each year. Since 1990, deaths directly attributable to drought have been fewer than 1,000 each year. (These statistics do not include deaths related to crop failures and starvation in drought-stricken areas.) These changes reflect improved infrastructure and water supply in many regions, improved forecasting of events such as hurricanes, as well as increased and improved relief efforts on the part of many organizations and governments.

▲ **Figure 19.16 Effects of Hurricane Katrina**
Wetland loss and poor levee construction magnified the impact of this event on the city of New Orleans. The effects of flooding from Hurricane Katrina were much more severe in poor parts of the city than in wealthier areas.

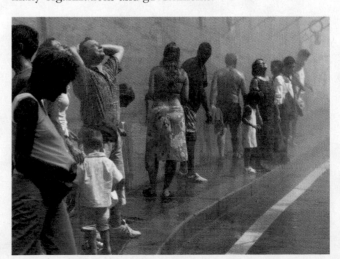

◄ **Figure 19.17 The 2003 Paris Heat Wave**
These Parisians are cooling down under jets of water at Paris Plage on the banks of the Seine River, protecting themselves against the scorching heat that spread across Europe during the summer of 2003.

Fire in the Environment

Patterns of human development and land management are increasing the risks posed by fire.

QUESTIONS

1. How is earthquake magnitude expressed?

2. How have human activities influenced the impact of tsunamis on many coastal communities?

3. Describe three kinds of weather events that can pose a risk to human well-being.

4. What factors have contributed to the increase in the loss of life and property associated with wildfires over the past two decades?

Although global data are lacking, it is well-known that fires of all kinds cause a great many human deaths each year. In the United States in 2005, for example, fires were responsible for 3,690 deaths. The statistics of deaths caused by fire include a disproportionate number of children and elderly people, who often have difficulty escaping burning structures.

Since 1990, deaths associated with wildfires in shrublands and forests have constituted an increasingly larger share of the total number of deaths caused by fire. This is partly due to the increased frequency of wildfires in many regions resulting from past management practices and climate change (13.6). It is also due to the expansion of human developments into wildland ecosystems that are naturally fire-prone.

Recent catastrophic wildfires in southern California exemplify this trend. In October 2007, a series of wildfires burned through the shrubby woodlands near San Diego. In total, they burned over 2,000 square kilometers (770 square miles), destroyed 1,500 homes, and caused the evacuation of more than 1 million people (**Figure 19.18**). Nine people died as a direct result of

these fires, and nearly 100 others were injured. These fires were ignited from several different sources and were fanned by hot, dry winds. However, their impact on life and property was greatly increased by suburban sprawl and housing developments built well into the highly flammable chaparral. Flammable landscaping and building materials, such as cedar roofing, carried fires from suburban areas into more urban areas that had historically been safe from fire. Events such as this are causing city and regional planners to pay more attention to standards for building and landscaping in what has come to be known as the wildland-urban interface.

Because geological, weather, and fire hazards are often a consequence of environmental processes that are outside human control, they are frequently called "natural disasters" or "acts of God." Yet in many places the destruction and mortality associated with these events have been reduced by improved stewardship of natural ecosystems and wise development practices. Enforcement of strict building codes and thoughtful development that anticipates risks can significantly diminish the human misery associated with these physical hazards.

▼ **Figure 19.18 Suburban Wildfire**
In southern California, the development of homes in fire-prone areas has increased the risk of wildfire to human life and property. Such development has also provided fuel to carry fires into more urban areas.

19.3 Chemical Hazards in the Environment

Toxicology is the scientific discipline that studies chemical poisons, or toxins, and their effects on human health. Toxins are classified according to their effects on human health; these effects include cancers, mutations, allergies, and neurological and developmental disorders. Risks to individual humans are determined by the dose and duration of exposure; they are also influenced by an individual's age, gender, genetic vulnerability, overall health, and socioeconomic status. Toxicological risks are influenced by the persistence, transport, and dispersal of chemicals in the environment. Testing is an important step in managing the risks from environmental toxins. However, testing is costly, and some forms of testing are controversial.

What Is a Toxin?

The toxicity of chemicals depends on dosage and duration of exposure.

The 16th-century Swiss physician and alchemist Paracelsus, who pioneered the use of chemicals and minerals in medicine, wrote the following: "Poison is in everything, and no thing is without poison. The dosage makes it either a poison or a remedy." Although this is a bit of an overstatement, Paracelsus was on to something important. When delivered in large amounts, the great majority of chemicals have adverse effects on human health. At low dosages, many chemicals have no effect on human health; some even have beneficial effects.

Consider some common examples. At doses below 1,000 mg, aspirin relieves pain and diminishes inflammation; larger doses of aspirin have adverse consequences, including internal bleeding and kidney failure. We depend on the oxygen in air to breathe. Normally 21% of air is composed of oxygen; air that is more than 50% oxygen is toxic.

Because dose is so important, scientists carry out dose-response studies to establish the toxicity of different chemicals. **Dose-response curves** show the relationship between increasing dosages of a chemical and some measure of health. One measure of health used to define toxicity is the percentage of mortality in an exposed population. More often, studies measure the proportion of individuals in a population who display specific symptoms or illnesses, such as respiratory difficulties or cancer. Because body size often affects dose responses, dosage is usually expressed in terms of the amount of the chemical per unit body weight, such as milligrams per kilogram (mg/kg).

Populations generally show little or no effect from low doses of chemicals. **Toxicity threshold** refers to the dose above which there is a measurable decline in the health measure. Scientists often characterize the toxicity of a chemical in terms of the dose at which 50% of an exposed population is killed or displays a specific symptom or illness. In the case of mortality, this is called the **median lethal dose** or **LD_{50}** (**Figure 19.19**).

In some cases, exposure to a toxin produces immediate effects. In other cases, the consequences are delayed. Toxins with delayed effects are said to pose latent risks. For example, latent cancer rates refer to the proportion of individuals in a population who contract cancer during some time period following exposure to a substance; latent death rates measure the number of individuals who die during that period.

Dose responses are also influenced by the duration of exposure to a chemical. Contact that occurs only once or for only a short time is called **acute exposure**. In humans, exposures of fewer than 14 days are considered to be acute. Contact with a chemical for periods of weeks, months, or years is called **chronic exposure**. Generally, the toxicity threshold for a chemical is lower for chronic than for acute exposures.

The effects of excessive alcohol consumption illustrate the difference between acute and chronic exposure. Drinking a large amount of alcohol in a very short period of time can elevate blood alcohol levels, causing poisoning and death within a few hours. The LD_{50} for blood alcohol content is about 0.45%. (In most states, individuals with a blood alcohol content at or above 0.08% are considered intoxicated.) In contrast, chronic consumption of large amounts of alcohol at levels below the LD_{50} can eventually cause liver damage and death.

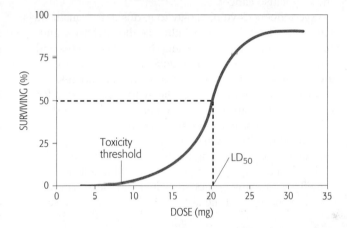

◄ Figure 19.19 Toxicity Threshold and LD_{50}
Toxicologists estimate the toxicity of a chemical by exposing organisms to increasing doses of it. The toxicity threshold is the dose that produces a measurable impact on survival. As the dose of a toxin is increased, the number of individuals affected by it increases. The median lethal dose, or LD_{50}, is the amount of a chemical required to kill half of the individuals in a population.

Human Vulnerability to Toxins

Genetics and environment cause differences in human vulnerability to toxins.

The U.S. Centers for Disease Control and Prevention estimates that more than 90% of all lung cancer is caused by cigarette smoking. Yet only 10–15% of individuals who smoke more than two packs of cigarettes each day develop lung cancer. Why are some people more vulnerable to certain toxins than others? The effects of toxic substances on different individuals vary widely, depending upon genetic makeup, age, health, and socioeconomic status (**Figure 19.20**).

Each person possesses a unique combination of genes that affects his or her vulnerability to toxins. As a consequence, the same exposure to potential toxins may affect individuals in different ways. For example, there is growing evidence that genetic differences among individuals play an important role in their susceptibility to chemicals that can alter the structure of DNA and thereby cause cancer.

Age influences vulnerability to toxins in four main ways. First, age-related behaviors influence the likelihood and route of toxin uptake. For example, crawling babies and toddlers spend much of their time close to floors; they are also likely to put foreign objects in their mouths. As a consequence, they are more likely to ingest lead-laden chips of paint than older children and adults. Second, developmental status influences the effects of toxins on individuals. The human nervous system undergoes enormous changes during the first 10 years of life, making children in this age group particularly vulnerable to the effects of many neurotoxins, including lead. Third, patterns and rates of metabolism vary with age. These

influence both the uptake and breakdown of toxins in the human body. Fourth, age influences the length of time that a person is exposed to a toxin and the amount that accumulates in the body. The longer a toxin is circulating in the body, the more time it has to cause a negative effect.

Gender also plays a role in vulnerability to toxins. In some cases, this is a consequence of gender-based differences in behavior or occupation. For example, cigarette smoking was historically more common among men than women, accounting for much higher incidence of lung cancer among men. These gender-based differences in smoking rates have diminished in recent years, as have gender-related differences in the incidence of smoking-related illness. Even with equivalent exposure, some toxins influence one sex more than the other. For example, methyl mercury and other forms of organic mercury are more toxic to men than women.

The overall health of individuals contributes significantly to their vulnerability to toxins. For example, people with lung diseases, such as emphysema, are at much greater risk of being harmed by exposure to moderate levels of photochemical smog and ozone than people with healthy lungs.

Socioeconomic status affects vulnerability to toxins in two ways. First, because poor people may have limited access to health care, they are more likely than their better-off peers to suffer from health problems that increase the impacts of toxins. Second, poor people are more likely to live in environments that increase their risk of exposure to toxins. For example, they are more likely to live in older homes with lead paint and plumbing. Poor people are also more likely to live close to industrial and waste facilities than is the population at large.

▼ Figure 19.20 **Population Response to Toxin Exposure**
The same amount of a chemical affects some individuals more severely than others. Genetic differences, age, and nutrition are among the factors that cause some individuals to be more sensitive to toxins and others to be more resistant.

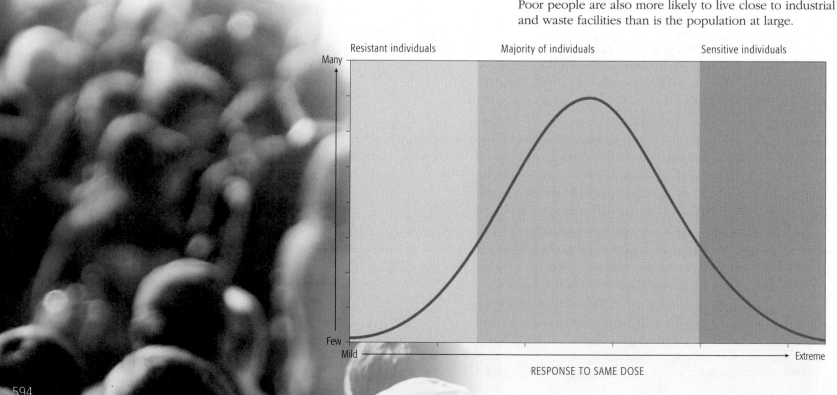

Resistant individuals | Majority of individuals | Sensitive individuals

Many

Few

Mild ——————————————————————————→ Extreme

RESPONSE TO SAME DOSE

Toxin Transport and Fate

Persistence, mobility, and uptake by organisms in the environment influence toxin dose and duration.

The impact of particular substances on human health depends to a large degree on factors that influence their movement through the environment. These factors include their persistence, volatility, and solubility in water, and their uptake and fate in other organisms.

The persistence of a toxin in the environment can be described in two different ways: residence time and half-life. Residence time describes the average length of time that a molecule of a substance spends in a particular part of the environment, such as the atmosphere or the ocean (7.1). Carbon monoxide, for example, has a residence time of about 60 days in the atmosphere, and mercury has a residence time of about 1 year.

Many substances undergo reactions that produce other substances with different levels of toxicity. This is particularly true of organic compounds that can be chemically altered by sunlight or metabolized by organisms. Toxicologists describe the persistence of a particular form of such a substance in terms of half-life. Half-life is the time that it takes for half of the molecules of a particular substance to be transformed into another form. For example, DDT breaks down to produce other organic compounds that vary in toxicity. Estimates of the global half-life of DDT released into the environment vary from 30 to 100 years. Because of high microbial activity, its half-life in soils is shorter, between 30 days and 2 years. Its half-life in aquatic ecosystems is much longer.

Transport in the environment is affected by a substance's volatility, or tendency to evaporate into the atmosphere, and its solubility, or tendency to dissolve in water. Volatile substances with long half-lives are likely to be dispersed widely by air circulating in the troposphere. This process has contributed to the global distribution of mercury. Chemicals that are soluble in water are more likely to pose problems in streams, rivers, and oceans.

The transport and persistence of toxins in the environment is significantly affected by their uptake and ultimate fate within individual organisms and food webs. **Bioaccumulation** refers to the process by which chemicals are stored in the tissues of living organisms. Fat-soluble chemicals are likely to bioaccumulate, as are chemicals that are not easily broken down or metabolized. Organisms that feed on other organisms whose tissues store a toxin are likely to store even higher concentrations of that toxin. Biological magnification is the process by which the concentration of a toxin increases at each higher level of a food web (13.8).

Toxicologists refer to the total amount of a chemical present in the tissues of an organism as **body burden**. The body burden for chemicals that bioaccumulate, such as DDT and mercury, tends to be higher than that of chemicals that are continuously excreted; the body burden of chemicals that bioaccumulate also tends to increase with age. However, chronic exposure to

chemicals that are easily excreted, such as arsenic, can produce a high body burden for these chemicals.

Each of us is exposed to thousands of chemicals, some of which become part of our total body burden. Nevertheless, little is known about the bioaccumulation or typical body burdens for most of these chemicals. Toxicologists suspect that high body burdens of some chemicals influence the toxicity of other chemicals, but there have been few thorough studies of such interactions.

Toxicologists designate particularly long-lived and widely dispersed organic chemicals as **persistent organic pollutants (POPs)**. Twelve substances called the "dirty dozen" are recognized as especially problematic because they are widely dispersed and their impacts on human health are well established (**Figure 19.21**).

▼ Figure 19.21 **The Dirty Dozen**
These 12 persistent organic pollutants (POPs) are among the most troublesome chemical hazards to human health. Ten of these chemicals are produced to kill either fungi or insects.

Persistant organic pollutants	Type of chemical
Aldrin	Insecticide
Chlordane	Insecticide
DDT	Insecticide
Dieldrin	Insecticide
Dioxins	Industrial by-product
Endrin	Insecticide
Furans	Industrial by-product
Heptachlor	Insecticide
Hexachlorobenzene	Fungicide, industrial by-product
Mirex	Insecticide, fire retardant
PCBs	Industrial chemical
Toxaphene	Insecticide

Kinds of Toxins

Toxins are classified according to their effects on specific organs and tissues.

Environmental toxins include a wide variety of inorganic and organic chemicals. Toxins may be classified by their typical route of entry into the human body. Air pollutants are typically absorbed through the respiratory tract. These pollutants may accumulate in the lungs or be carried to other parts of the body by the circulatory system. Toxins in both air and water may be absorbed through the skin. Toxins in food and water may be absorbed through the digestive tract.

Toxicologists generally classify toxic chemicals according to the way they affect the body, or their mode of toxicity. **Corrosive toxins** react with and directly destroy tissue. Their effect is usually immediate, so their toxicity is acute. Corrosive toxins include caustic chemicals, such as hydrochloric and sulfuric acids, household bleach, lye, and ammonia. They also include chemicals like chlorine, which reacts with tissues to form hydrochloric acid.

Asphyxiants are chemicals that deprive tissues of oxygen and cause suffocation. Carbon monoxide and cyanide are powerful asphyxiants that block the hemoglobin in blood from absorbing oxygen (**Figure 19.22**). Other kinds of asphyxiants interfere with the uptake of oxygen in the lungs or affect the transport and use of oxygen in other parts of the body.

Carcinogens are chemicals that cause or promote cancer. Cancer is a disease in which the growth of affected body cells is no longer controlled, leading to the formation of tumors. Carcinogens increase the risk of getting cancer by altering cellular DNA or metabolism in ways that encourage uncontrolled, malignant cell division. The environment contains thousands of potentially carcinogenic substances. Dozens of carcinogens have been identified from industrial and tobacco smoke alone. Dioxin and polychlorinated biphenyls (PCBs) are widespread and well-studied carcinogenic pollutants.

Substances that produce mutations and birth defects are called **teratogens**. Scientists estimate that about 10% of all birth defects are caused by prenatal infections or exposure to teratogenic substances. Teratogens, like carcinogens, often cause changes in cellular DNA. Thus, many carcinogens are also teratogens.

A great number of pollutants known as **allergens** stimulate a range of responses from the body's immune system. These responses can vary in intensity from a running nose and sneezing, to a skin rash and respiratory distress, to allergic shock and suffocation.

Because the immune system is especially sensitive to foreign proteins, other organisms and their products are often powerful allergens (**Figure 19.23**).

Neurotoxins affect the development or functioning of the nervous system. Important neurotoxins include heavy metals such as mercury, lead, and manganese, and pesticides such as DDT and various organophosphates.

▼ **Figure 19.22 Asphyxiants** Because their chemical structures are quite similar to oxygen, carbon monoxide and cyanide bind tightly to blood hemoglobin, thereby preventing the uptake of oxygen. In sufficient doses, carbon monoxide and cyanide cause asphyxiation.

O$_2$ (Oxygen)

CO (Carbon monoxide)

CN (Cyanide)

▲ **Figure 19.23 A Common Allergen** Microscopic dust mites, which are common in carpets and upholstery, cause allergic reactions in many people.

▶ Figure 19.24 **Declining Sperm Count**
In a synthesis of 101 studies since 1934, researchers have detected a significant decline in human sperm production. They speculate that this may be due to increasing amounts of endocrine-disrupting toxins in the environment. But they also note significant scatter in the data and suggest that many factors are likely to contribute to variation in sperm counts.
Source: Swan, S.H., E.P. Elkin and L. Fenster. 2000. The question of declining sperm density revisited: An analysis of 101 studies published 1934-1996. *Environmental Health Perspectives* 108: 961–966.

In recent years, toxicologists have discovered a number of substances called **endocrine disrupters** that interfere with the hormones that control growth and development, egg and sperm production, and reproductive function. One of the most important hormones influencing development and the production of egg and sperm is estrogen. Nearly every day, most humans are exposed to hundreds of chemicals that have been shown in laboratory studies to influence the production of estrogen. Whether these chemicals affect human health at the typical level of environmental exposure is hotly debated. However, there is growing evidence that chronic exposure to even low amounts of these chemicals is influencing the frequency of abnormal patterns of development in human populations (**Figure 19.24**).

The controversy surrounding the chemical bisphenol A demonstrates the challenges associated with proving that a chemical actually influences endocrine function in humans. Bisphenol A is found in plastic water bottles, baby bottles, and food containers, as well as in the linings of food and infant formula cans. Numerous animal studies have demonstrated that low concentrations of bisphenol A have the potential to affect the endocrine system. Such effects include increased reproductive problems, decreased sperm counts, accelerated sexual development, obesity, and elevated rates of some cancers. The U.S. Federal Drug Administration (FDA) and the chemical industry have argued that these studies are not relevant to human exposure and toxicity and that bisphenol A is safe at

the concentrations present in these plastic containers. In 2008, however, Canada announced plans to phase out the use of this chemical in products used by infants. Manufacturers, including Nalgene and Playtex, and retailers such as Wal-Mart have agreed to remove bisphenol A from their products. In late 2008, the FDA's Science Advisory Board concluded that its earlier assessment was flawed and recommended that the safety of this chemical be reevaluated, especially with regard to infants.

Tests reveal that drinking water in many cities contains a variety of pharmaceutical chemicals, including antibiotics, antidepressants, artificial hormones, and anti-inflammatory drugs. For example, a 2007 study of Philadelphia's drinking water identified 56 different medicinal chemicals. These drugs enter the water supply through the disposal of unused medicine and in human and animal wastes (**Figure 19.25**). To be sure, the concentrations of these drugs are very low, far below medicinal doses. Nevertheless, some of these drugs have the potential to bioaccumulate and biomagnify in food webs. Toxicologists are also concerned about their possible effects on fetuses and children.

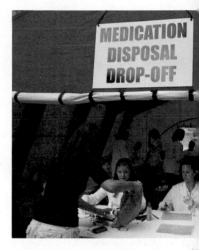

▲ Figure 19.25 **Avoiding Pollution Risk**
Pharmacists collect unwanted or expired prescription and over-the-counter medicines for safe disposal.

Toxin Testing and Regulation

Toxicity testing presents both ethical and practical challenges.

Ideally, whenever a new substance is proposed for commercial use, its toxicity and potential modes of transport in the environment would be fully evaluated. In practice, this is done for only a relative few chemicals. Historically, scientists have often determined that a chemical is toxic only after it is in wide use and its toxic effects have become apparent.

Scientists estimate that commercial substances contain between 150,000 and 200,000 different chemicals. Over 2,000 new chemicals are synthesized and introduced

each year. Some of these, such as nanomaterials, represent entirely new classes of compounds.

The U.S. EPA is charged with monitoring over 75,000 industrial chemicals, but at present it has evaluated and set standards for fewer than 10% of them. Chemicals are generally selected for testing based on their similarity to chemicals that are known to be toxic. Most chemicals are assumed to be nontoxic until proven otherwise.

Many government and environmental leaders have argued that the release of chemicals into the environment

QUESTIONS

1. Give an example of the adage "dosage makes a chemical either a remedy or a toxin."

2. Differentiate between acute and chronic exposure to a toxin.

3. What is the difference between bioaccumulation and biomagnification?

4. List and describe seven classes of toxins.

should be guided by this precautionary principle: If a chemical has the potential for severely harming the public or the environment, the responsibility for proving that it is safe should fall to those who advocate its use (2.5) (**Figure 19.26**). Those who manufacture and market chemicals and the products that contain them argue that most of these substances are benign; testing each chemical to establish this fact would considerably increase its price.

The thorough testing of chemicals for their toxicity to humans presents significant challenges. There is unanimous agreement that testing the effects of potential toxins on human subjects is unethical. Thus, nearly all chemical testing is done with animals or microorganisms. For some, this too raises serious ethical questions.

Many scientists worry that the effects of a chemical on a bacterium, fish, or rabbit may not mirror its effects on the human body. To counter this concern, toxicologists focus their testing on **model organisms** in which particular processes or metabolic functions have been demonstrated to be similar to those of humans. For example, the basic process of DNA replication in the bacterium *Escherichia coli,* a widely used model organism, has been shown to be similar to that of nearly all other organisms, including humans. Thus, a chemical that has a detrimental effect on the manufacture of DNA in *E. coli* might be suspected of being a carcinogen in humans. Similarly, the early development of fish embryos is similar to that of human embryos. Chemicals that affect development in zebra fish, another important model organism, might have similar effects in humans.

Toxicologists are quick to acknowledge that the absence of a toxic response in a model organism does

not guarantee that a chemical is nontoxic to humans. In addition, dose-response thresholds and LD_{50} of toxins in model organisms often differ from those in humans. Consequently, there is no simple formula for using the responses of a model organism to set standards for human exposure to a particular chemical.

The effects of chronic exposure to low levels of a chemical are particularly difficult to evaluate. Toxicologists typically administer relatively large doses of a chemical over a brief time and then use their response data to extrapolate to much lower, longer-term exposures. For example, if an acute dose of 100 mg/kg of a substance causes cancer in 50% of a population of zebra fish, a toxicologist might predict that 1 mg/kg would cause cancer in 0.5% of the human population. The toxicologist might further predict that chronic exposure to 1 mg/kg of the substance would increase the rate of cancer.

The standards established for levels of toxins in the environment should take into account how a chemical moves in the environment. A general understanding of properties such as volatility, water and fat solubility, and storage in animal tissues can provide the basis for predicting whether a chemical is likely to bioaccumulate and how it will be dispersed in ecosystems. These properties are, however, very difficult to measure.

A great many substances are potential toxins to animals and humans. Their actual toxicity depends on their dispersion, the dose received, and the vulnerability of individuals. Adequate toxicity testing for the great many substances in our environment is a daunting challenge.

▼ Figure 19.26 **How Much Testing Is Enough?**
The chart compares two approaches to managing the potential hazards of chemical products with uncertain risks.

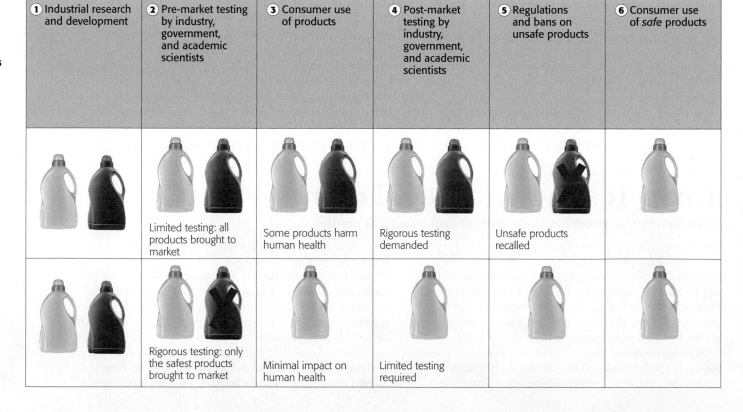

Sequence of events	① Industrial research and development	② Pre-market testing by industry, government, and academic scientists	③ Consumer use of products	④ Post-market testing by industry, government, and academic scientists	⑤ Regulations and bans on unsafe products	⑥ Consumer use of *safe* products
"Innocent until proven guilty" approach		Limited testing: all products brought to market	Some products harm human health	Rigorous testing demanded	Unsafe products recalled	
Precautionary principle approach		Rigorous testing: only the safest products brought to market	Minimal impact on human health	Limited testing required		

FOCUS ON SCIENCE: Nature's Chemicals vs. Synthetic Chemicals

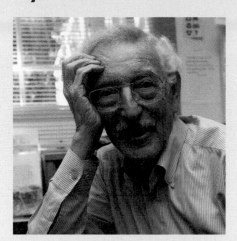

▲ Figure 19.27 **A Source of Challenging Ideas.**
Over his long career in toxicology, Bruce Ames's research and synthesis has catalyzed work by hundreds of students and colleagues.

In her book *Silent Spring,* Rachel Carson wrote, "for the first time in the history of the world, every human being is now subjected to contact with dangerous chemicals, from the moment of conception until death." She was expressing a widely shared view that the toxic chemicals that influence our health are mostly those that have been artificially synthesized by humans. Dr. Bruce N. Ames, professor of biochemistry and molecular biology at the University of California, Berkeley, thinks otherwise, and his thoughts carry great weight with toxicologists (**Figure 19.27**).

Over his long career, Ames has published more than 500 scientific papers and received numerous awards, including the Tyler Prize for Environmental Achievement and the National Medal of Science. He is particularly well-known for developing the Ames Test, an assay that uses bacteria to measure the cancer-causing potential of chemicals. Ames's challenge to conventional wisdom was published in a paper that demonstrated the importance of exhaustive review and synthesis of existing literature in science. Since its publication, this paper has had a significant impact on the field of toxicology.

Ames first pointed out that thousands of naturally occurring toxins have been present throughout the evolution of vertebrates. Indeed, we humans possess an array of defenses against this multitude of naturally occurring chemicals. For example, in our organs that are regularly exposed to toxins, such as the mouth, stomach, intestines, and lungs, layers of surface cells are continuously being shed. Our cells produce an array of chemicals, such as antioxidants, that can neutralize the effects of many toxins. Our cells can also repair strands of DNA that have been altered by chemicals. Perhaps most important, we possess keen senses of smell and taste that warn us not to consume natural toxins. Despite these adaptations, many natural chemicals remain quite toxic to us. For example, molds produce a variety of known carcinogens. Chemicals in comfrey tea and some herbal medicines are mutagens and can cause liver damage (**Figure 19.28**).

Natural chemicals often have the same mechanisms of transport and toxicity as synthetic toxins. For example, natural chemicals can bioaccumulate in the tissues of animals and biomagnify in food webs in much the same way as synthetic chemicals, such as DDT. Cabbage and broccoli contain chemicals that, at high concentrations, can stimulate rapid cell division in a fashion very similar to that of dioxin. Fortunately, the quantities of vegetables that we typically eat result in doses far below the toxic threshold of these chemicals. Ames points out that, at average levels of consumption in the United States, the mutagenic and carcinogenic effects of the alcohol in beer and wine far exceed the risks posed by exposure to dioxin at the dose limits set by the EPA.

Ames's arguments and assertions were not intended as a defense of synthetic chemicals or their haphazard release into the environment. Rather, he warned that we should not assume that naturally occurring chemicals are benign. Like many of their synthetic counterparts, most of these chemicals have not been tested for toxicity. Equally important, he pointed out that studies of naturally occurring chemicals have much to teach us about the transport, uptake, and toxic mechanisms of all chemicals.

Source: Ames, B.N., M. Profet and L.W. Gold. 1990. Nature's chemicals and synthetic chemicals: Comparative toxicology. *Proceedings of the National Academy of Sciences* 87:7782–7786.

Are the chemicals that naturally occur in plants and animals necessarily less toxic than chemicals synthesized by humans?

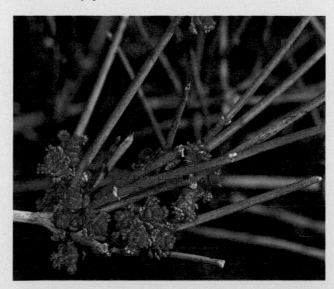

◄ Figure 19.28 **Natural Toxins**
Tea made from ephedra, commonly called Mormon tea, is used by many people to treat colds and hay fever. However, it can cause heart attacks and strokes in some people.

19.4 Biological Hazards in the Environment

Worldwide, infectious diseases cause more than 26% of human deaths. In the world's 50 poorest countries, such diseases cause more than 70% of deaths. Environmental factors exert a strong influence on the incidence, spread, and impact of infectious diseases in human populations. Climate, water availability, and water quality affect disease-causing organisms, as well as the animals that transmit diseases. The role of environmental factors varies greatly, depending on the manner by which a disease is transmitted. Infectious diseases can be managed by improving sanitation and water quality, by controlling organisms that cause and transmit disease, and by taking action to diminish vulnerability in human populations.

Infectious Disease and the Environment

Environmental factors influence populations of the organisms that cause disease, as well as those that spread disease.

Infectious diseases are caused by **pathogens**—viruses, bacteria, protozoa, fungi, or multicellular parasites that invade host organisms and impair their functioning. The extent to which a pathogen impairs the growth and reproduction of its host is called its **virulence**.

Virulence is heavily influenced by the nature of the interaction between a pathogen and its host. For example, a pathogen that destroys cells in its host's body is likely to be more virulent than a pathogen that is able to reproduce without destroying cells. A pathogen that has only recently come into contact with its host is likely to be more virulent than a pathogen that has a long history of coevolution with its host.

Virulence is also influenced by factors that influence the immune system and overall health of the host. Organisms that would typically cause no disease or only mild disease can be quite virulent in individuals whose immune systems have been impaired by malnutrition or another disease. The tuberculosis bacterium, for example, is much more virulent in individuals who are malnourished or have HIV/AIDS.

Pathogens are transmitted from person to person by a variety of mechanisms. Viruses and bacteria that cause respiratory diseases, such as the common cold and tuberculosis, are transmitted in the tiny droplets of saliva dispersed by sneezing and coughing. Diseases of the digestive tract, such as amoebic dysentery and cholera, are transmitted in contaminated food and water. Sexually transmitted diseases, such as syphilis and HIV/AIDS, are

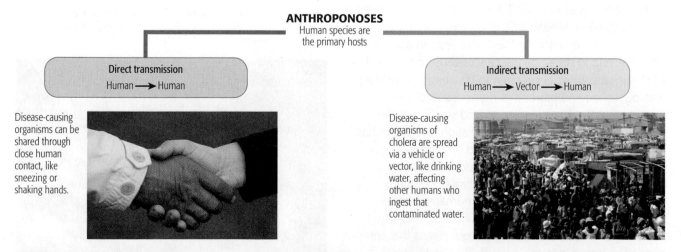

ANTHROPONOSES
Human species are the primary hosts

Direct transmission
Human ⟶ Human

Disease-causing organisms can be shared through close human contact, like sneezing or shaking hands.

Indirect transmission
Human ⟶ Vector ⟶ Human

Disease-causing organisms of cholera are spread via a vehicle or vector, like drinking water, affecting other humans who ingest that contaminated water.

▶ **Figure 19.29 Anthroponoses**
Diseases that are specific to humans are called *anthroponoses*. They may be transmitted directly or indirectly by an intermediate animal called a *vector*.

transmitted in bodily fluids, most often during sexual intercourse.

Many pathogens are transmitted by intermediary organisms called vectors (6.2). *Passive vectors* transmit diseases by simple contact. For example, a housefly alights on animal feces, picks up pathogenic bacteria on its appendages, and then deposits the bacteria on food that humans eat. The pathogen does not infect the passive vector or enter its body. *Active vectors* transmit pathogens that have entered their bodies; in some cases, the pathogens even complete part of their life cycle inside the vector. Active vectors often deliver pathogens to humans through a bite. Malaria, sleeping sickness, and viral encephalitis are blood-borne diseases that are transmitted by active vectors, such as mosquitoes, biting flies, fleas, and ticks.

The spread of an infectious disease is influenced by its virulence, its ease of transmission, and the mobility of infected individuals. For example, the highly virulent Ebola virus is relatively easily transmitted from person to person. However, once the disease is contracted, it incapacitates and kills its victims quickly. Since infected individuals are usually too ill to travel far from the source of their infection, outbreaks of Ebola tend to be confined to small geographic areas. In contrast, the influenza virus is widely dispersed by the sneezing and coughing of individuals who may be quite mobile in the early stages of the disease. As a consequence, strains of influenza can spread around the globe rapidly.

An **anthroponosis** is an infectious disease that occurs exclusively in humans. The pathogens that cause anthroponoses, such as falciparum malaria and yellow fever, have coevolved with humans. Humans are the sole host of these pathogens (**Figure 19.29**).

Many other infectious diseases are caused by pathogens that have hosts in addition to humans. For example, Lyme disease occurs in humans, deer, and small rodents. Various strains of flu are shared between humans and domestic animals, such as pigs and poultry. A disease that is shared by humans and other animals is called a **zoonosis** (**Figure 19.30**). The evolution and transmission of zoonoses are influenced by their multiple hosts.

Environmental factors such as climate, soil, and the availability of water affect the growth and reproduction of pathogens, as well as their vectors and host animals. As a consequence, the environment plays a significant role in the prevalence and spread of infectious disease. The following sections explore the role of the environment in specific kinds of respiratory, diarrheal, and blood-borne diseases.

ZOONOSES
Nonhuman species are the natural reservoir

Direct transmission
Animal ⟶ Animal ⟶ Human

Indirect transmission
Animal ⟶ Vector ⟶ Human

Strains of swine flu virus spread among domestic pigs and may directly infect humans.

Lyme disease-causing organisms spread first among animal populations (deer and rodents) before the infected tick spreads the disease to humans.

▲ **Figure 19.30 Zoonoses**
Diseases shared between humans and other animals are called *zoonoses*. These may be transmitted directly or indirectly through animal vectors.

Respiratory Disease

Influenza provides an example of a disease that is spread through the air.

Fever, chills, muscle pain, and general discomfort—we are all familiar with the symptoms of influenza, or the flu. Every year this common disease is responsible for several hundred thousand human deaths. In some years, the global death toll has been in the millions.

Flu is caused by several strains of the influenza virus and is most often transmitted from human to human in respiratory aerosols, the tiny droplets given off by coughing and sneezing. Human viral strains are often shared with birds, pigs, and some other mammals. Viral particles are typically transmitted from other animals to humans by direct contact with tissues or feces.

Influenza is most prevalent in human populations in the wintertime. This is probably because people tend to congregate indoors when it is cold, and their proximity encourages the transmission of the virus. Colder temperatures also allow the viral particles to survive on exposed surfaces, such as doorknobs. In addition, extensive holiday travel facilitates the spread of flu.

Each strain of influenza has a large range of genetic variability and therefore is able to undergo rapid evolution from year to year. It is this genetic variability that accounts for the great variability in the virulence of the flu. In 1918, a particularly virulent strain known as the Spanish flu suddenly appeared. Over the next 18 months, the Spanish flu killed more than 50 million people.

New genetic strains of influenza often appear first in other host animals, from which they may subsequently be transmitted to humans. In this regard, avian flu, formally known as strain H5N1, has been of particular concern (**Figure 19.31**). This virus was first identified in waterfowl in 1997; since then, it has been found in many bird species. This virus is easily transmitted among birds, and certain genetic forms of it cause high mortality. Scientists have noted a steady increase in the virulence of this virus, and many are concerned that it may accelerate the decline of populations of bird species that are already endangered.

A few hundred humans have contracted avian flu, apparently from handling infected waterfowl or poultry. The symptoms of this flu have been severe, and the mortality rate among infected individuals has been about 60%. Indeed, its effects on humans are considerably more severe than those of the virus responsible for Spanish flu. Fortunately, the virus that causes avian flu cannot be efficiently transmitted from human to human. Nevertheless, the World Health Organization and the U.S. Centers for Disease Control and Prevention have invested billions of dollars in developing vaccines and programs to limit the spread of this disease.

The situation with avian flu is quite different from the recent global outbreak of swine flu (virus H1N1). This viral strain is common in pigs but is rarely transmitted directly to humans. In 2009, however, a genetic strain of H1N1 that was easily transmitted from person to person appeared in the United States and Mexico. Early indications were that this virus might be quite virulent, and the World Health Organization declared a state of emergency. The swine flu did spread widely, but hospitalization and mortality rates were about the same as those of other seasonal flu viruses.

▼ **Figure 19.31 Avian Flu**
The micrograph shows the virus that causes the avian flu. The lab worker at right is testing for the presence of the virus.

Diarrheal Diseases

Poor sanitation and polluted water are responsible for most diseases that infect the stomach and intestines.

Worldwide, about 80% of all infectious diseases are caused by waterborne diarrheal pathogens. These pathogens include a variety of viruses, bacteria, and protozoans that infect and inflame the lining of the stomach and small intestine. In many developing countries, untreated sewage is simply discharged into surface waters. In India, for example, only about 10% of towns and cities have partial sewage-treatment facilities; only 1% of people live in towns with full wastewater-treatment facilities. Polluted water and poor sanitation contribute to more than 2 billion human infections of diarrhea each year. In developing countries, these infections are the leading cause of death among infants and children younger than five. The immediate cause of death among most victims of diarrheal diseases is dehydration.

Changes in temperature and rainfall cause the prevalence of some diarrheal diseases to vary from year to year. For example, the incidence of cholera in Bangladesh is closely correlated with changes in the temperature of its coastal waters. These temperatures, in turn, vary as a consequence of the distant El Niño Southern Oscillation (3.8). When temperatures in Bangladeshi estuaries are warm, populations of the cholera bacterium (*Vibrio cholerae*) in plankton, shellfish, and fish are also high, and the prevalence of cholera increases (**Figure 19.32**). A similar pattern of cholera prevalence has been observed in cities on the west coast of South America.

Outbreaks of diarrheal disease are not confined to developing countries. *Cryptosporidium* is a waterborne pathogen that occurs worldwide. This tiny protozoan escapes most filtration systems and is quite resistant to the chlorination that is widely used to kill microbes in drinking water. Serious outbreaks of this organism

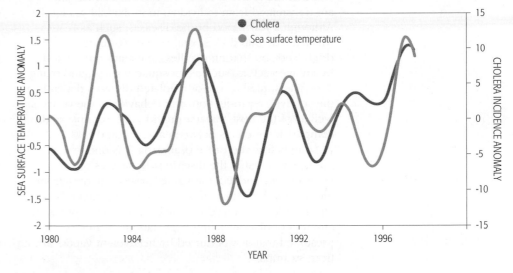

▲ **Figure 19.32 Cholera and Climate**
The incidence of cholera, corrected here for natural seasonal variation, in Bangladesh rises and falls in association with changes in the surface temperature of the sea. The increase in cholera associated with warmer water is likely due to flooding during El Niño events.
Source: Pascual, M. Rodó, X., S.P. Ellner, et al. 2000. Cholera dynamics and El Niño-Southern Oscillation. *Science* 289:1766-1769.

have occurred in a number of European and American cities over the past four decades. In 1993, for example, *Cryptosporidium* in the drinking water of Milwaukee, Wisconsin, resulted in 403,000 cases of diarrhea and 4,400 hospitalizations.

As recently as 1980, the annual global mortality due to diarrheal diseases exceeded 5 million. Since then, improved sanitation and the development of oral rehydration therapies have reduced this mortality to about 1 million deaths each year (**Figure 19.33**).

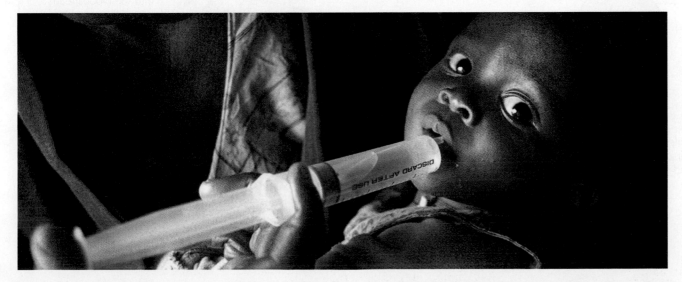

◄ **Figure 19.33 Rehydration Therapy**
Diarrheal diseases cause dehydration, which can be particularly fatal in infants. Oral treatment with relatively simple rehydryation fluids containing sugar and salts now saves thousands of lives.

Blood-Borne Diseases

Malaria is an example of a blood-borne disease that is spread by an insect vector.

Humans are afflicted with a number of diseases that are transmitted from individual to individual in blood. Although a few blood-borne diseases, such as HIV/AIDS, are acquired by sexual contact (or in the case of drug users, by sharing needles), most are transmitted by animal vectors, such as mosquitoes, ticks, and biting flies. The spread of vector-mediated diseases depends on the survival, reproduction, and behavior of the vector as well as of the host. Malaria provides an example of the complex interactions between a vector and host.

Long before the cause of malaria was understood, humans associated the disease with aspects of the environment. Indeed, the name "malaria" derives from the medieval Italian phrase *mala aria*, meaning "bad air." People recognized that malaria was more prevalent among populations living near standing water, so they presumed that it was caused by malevolent vapors arising from swamps.

In 1880, a French physician, Charles Laveran, discovered that malaria is caused by parasitic protozoa in the genus *Plasmodium*. In 1898, scientists proved that these parasites are transmitted by mosquitoes in the genus *Anopheles*.

The life cycle of *Plasmodium* is closely tied to the ecology of its vector, the female *Anopheles* mosquito. The cycle of malarial infection begins when a female mosquito bites a person infected with malaria and takes in blood containing *Plasmodium* (**Figure 19.34**). (Only the female mosquitoes feed on blood, which they need in order to reproduce.) Over about a week's time, the parasites in the mosquito undergo a complex process of differentiation and reproduction in which large numbers of parasites migrate to the mosquito's salivary glands. When the infected mosquito bites again, its saliva transmits the parasites to a new human victim. Each parasite invades a red blood cell, where it grows and divides.

▼ **Figure 19.34 Life Cycle of the Malaria Parasite**
Malaria is caused by the protozoan parasite *Plasmodium*, which has a complex life cycle involving the female *Anopheles* mosquito and humans.

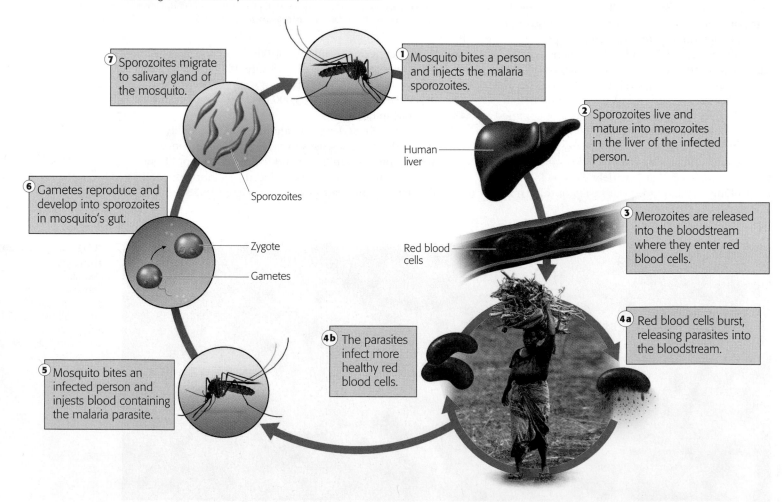

1 Mosquito bites a person and injects the malaria sporozoites.

2 Sporozoites live and mature into merozoites in the liver of the infected person.

3 Merozoites are released into the bloodstream where they enter red blood cells.

4a Red blood cells burst, releasing parasites into the bloodstream.

4b The parasites infect more healthy red blood cells.

5 Mosquito bites an infected person and injects blood containing the malaria parasite.

6 Gametes reproduce and develop into sporozoites in mosquito's gut.

7 Sporozoites migrate to salivary gland of the mosquito.

Sporozoites

Zygote

Gametes

Human liver

Red blood cells

Every 48 or 72 hours, the parasites burst out of their red blood cells and invade new blood cells. This synchronous activity produces the regular 48- or 72-hour fevers that are characteristic of the particular form of malaria.

The geographic distribution of malaria is closely tied to the habitat requirements of the *Anopheles*. Like all mosquitoes, *Anopheles* requires pools of standing water for reproduction and early larval development. It does not, however, tolerate temperatures close to or below freezing during any part of the year.

As recently as 150 years ago, malaria was common in warm temperate regions, such as those in the southeastern United States and Mediterranean Europe. At that time, human vulnerability to this disease could only be controlled by regular doses of quinine, a chemical extracted from the bark of cinchona trees native to South America (**Figure 19.35**).

When it was discovered that *Anopheles* mosquitoes were the sole vector of malaria, disease control began to focus on eliminating mosquito breeding habitat. Wetlands and swamps in many regions of the world were dredged and drained for this reason. In the 1940s, broadcast spraying of DDT was found to be quite effective in limiting *Anopheles* mosquitoes and malaria. Its widespread use was a major factor in the subsequent worldwide decline of this disease.

As discussed in the introductory essay on Zambia, the prevalence of malaria in some regions has increased dramatically over the past several decades. The global ban on DDT has probably contributed to this increase, but other changes in the environment have also been important. Through time, the *Plasmodium* parasite has become increasingly resistant to quinine and related chemicals, such as hydroquinone, while the mosquito

◀ Figure 19.35 **A Natural Defense Against Malaria**
The indigenous people of the lowland forests of the Andes discovered that the bark of the cinchona tree provided resistance to malaria. Scientists subsequently learned that it is the chemical quinine that gives cinchona bark this medicinal property.

vectors have become increasingly resistant to DDT and other insecticides. Changing climate and the growth and urbanization of human populations have likely facilitated the spread of this disease in some places.

Because the environment influences the spread and virulence of infectious diseases among humans, management of the environment is important in the control of such diseases. Limiting human contact and mobility are important ways to control respiratory disease. Improved sanitation and water quality have diminished the prevalence of diarrheal diseases. Alteration of vector habitats and direct control of vector populations are important strategies in the control of vector-borne diseases such as malaria.

Evolutionary Change

Coevolution between pathogens and their human hosts tends to diminish disease severity.

Pathogens exert very strong natural selection on their host populations. This leads to the evolution of genetically based traits that increase the host's resistance to the diseases caused by the pathogens. As the host becomes more resistant, natural selection exerts pressure on pathogen populations that favors the evolution of traits that increase its rate of transmission in host populations. This tit-for-tat selection is called coevolution (6.2). Such coevolution is evident in a number of human diseases.

Among the four different kinds of malaria that occur in humans, falciparum malaria is by far the most deadly. This kind of malaria is most prevalent in Africa, where it kills more than 1 million people each year. It is no coincidence that where falciparum malaria is prevalent, so is a human genetic disorder called sickle-cell disease. In this disease, red blood cells contain an abnormal form of hemoglobin that clumps together, causing the normally round cells to take on a sickle shape (**Figure 19.36**). The sickle-shaped cells tend to get

◀ Figure 19.36 **Sickle-Cell Disease**
In individuals with severe sickle-cell disease, some red blood cells are misshapen and prone to causing blockages in blood vessels. Blood cells in individuals with a mild form of the condition are normal, but resistant to invasion by the malaria parasite.

caught in blood vessels, causing blockages that may be fatal. Individuals with a mild form of sickle-cell disease have normal red blood cells and are not at risk for such blockages. However, when both parents have mild forms of the disease, their children have a 25% chance of inheriting the extreme form of sickle-cell disease.

You might expect that there would be very strong natural selection against a genetic trait such as sickle-cell disease. Yet where falciparum malaria is common, as much as 20% of the population have the mild form of this disease (**Figure 19.37**). Why is this trait so prevalent there? Malaria parasites cannot live in the red blood cells of people with mild forms of sickle-cell disease, making these people resistant to malaria. Thus natural selection maintains a high frequency of individuals who are resistant to malaria, even though their children have a significant risk of inheriting the extreme form of sickle-cell disease.

Recently, the evolution of the falciparum malaria parasite has been affected by the various medical treatments used to prevent the disease. For the past 50 years, doctors have prescribed a mixture of two drugs—chloroquine and sulfadoxine-pyrimethamine—to people living or traveling in areas where the falciparum parasite is prevalent. Over time, natural selection has favored the evolution of falciparum parasites that are resistant to these drugs. As a result, these drugs are no longer effective in much of Africa.

Over long spans of time, coevolution between a pathogen and its host tends to diminish the negative impacts of the disease caused by the pathogen. However, when hosts are exposed to a new pathogen, the resulting disease is often extreme or even fatal. Population growth and environmental changes are now exposing humans to a number of pathogens with whom they have not coevolved. HIV, for example, is thought to have arisen from a similar virus in chimpanzees, called SIV, which was transferred to humans sometime between 1880 and 1940. (That transfer was likely a consequence of hunting and eating chimps, although debate remains on this issue.) SIV is far less deadly in chimps than HIV is in humans.

Ebola hemorrhagic fever is another disease that has recently appeared in humans. More than 80% of the individuals infected with this disease die within a few weeks. This disease is also quite deadly in various species of monkeys. Its natural host appears to be several species of fruit-eating bats that live in tropical forests in central Africa. The sudden appearance of this disease in humans seems to be related to the movement of human populations into forested regions where these bats are common. Gorillas and some monkeys pick up the disease by eating bits of fruit dropped by the bats. Transmission to humans occurs when someone eats the meat of an infected gorilla or monkey.

In summary, human health and well-being are affected by infectious respiratory, diarrheal, and blood-borne diseases. The prevalence of these diseases is, in turn, influenced by variations in the environment. Diseases caused by parasites that have coevolved with humans tend to be less severe than those caused by parasites that have only recently appeared in human populations.

QUESTIONS

1. Why might zoonotic diseases be more difficult to eradicate than anthroponotic diseases?

2. What factors contribute to the higher prevalence of flu during the wintertime?

3. Why are mosquitoes becoming increasingly resistant to insecticides such as DDT in many regions?

Distribution of Malaria and Sickle-Cell Disease
⬤ Moderate to very high prevalence of malaria
◉ 4–25% of the population with mild or severe sickle-cell disease

▲ **Figure 19.37 Distribution of Malaria and Sickle-Cell Disease**
Falciparum malaria is most prevalent in equatorial east Africa, where more than 20% of the population have either the mild or severe form of sickle-cell disease.

19.5 Environmental Change and Human Health

The intimate and complex connections between our environment and our health are compelling reminders of the need to sustain key ecosystem processes and services. But Earth's environment is rapidly changing in ways that are affecting these ecosystem processes. As human populations increase, the risks associated with many physical, chemical, and biological hazards are increasing, particularly in urban centers. In many parts of the world, the negative impacts of air and water pollution are increasing. Changes in landscapes and waterways are altering the habitats of zoonoses and disease vectors, thereby increasing the prevalence of some infectious diseases. Climate change is also increasing the prevalence of some infectious diseases, as well as the risks of some weather hazards. Sustainable development and management will require us to understand how changes in the global environmental affect human health.

Human Population Size

Population growth and urbanization increases our vulnerability to many hazards.

Over the past century, the world's human population has grown from about 1.3 billion to about 7 billion. Much of that population is now concentrated in urban centers. In 1950, fewer than 25% of Earth's people lived in cities or towns. Today, more than 50% of the world's population is urban, and that proportion is likely to increase to more than 60% in the next 20 years.

Changes in human populations are altering the environment in ways that affect public health. The risks posed by nearly all physical hazards are magnified by population growth, especially the growth of urban centers. Much urban development, especially in poor countries, is susceptible to geological hazards such as earthquakes and landslides. City dwellers are also at greater risk of harm from extreme weather. Because cities tend to trap heat, temperatures in cities are much higher than those in rural areas (17.1). For example, more than 90% of the deaths caused by the European heat wave of 2003 occurred in cities of over 10,000 people.

Air and Water Pollution

Changes in air and water quality are increasing the incidence of respiratory and waterborne infectious diseases.

Population growth and increasing urbanization have reduced the air quality in many regions. The United Nations Environmental Program estimates that over 2 million people die each year from respiratory diseases directly related to urban air pollution, such as asthma, emphysema, and lower respiratory infections. More than 60% of these deaths occur in the urban centers of rapidly developing countries, such as China and India. Because developing countries often lack the resources to enforce air-quality standards or to mitigate air pollution, the proportion of total deaths caused by air pollution is likely to increase as urban growth accelerates.

Respiratory diseases caused by air pollution are also a problem in developed countries, as is clear from the increasing prevalence of asthma among children and adults in the United States during the past 20 years (**Figure 19.38**). During this period, air quality has

▼ **Figure 19.38 Asthma on the Rise**
The increasing prevalence of asthma in U.S. children and adults is a consequence of the increasing urbanization of the population and the diminishing air quality in many regions.
Source: U.S. EPA. URL = http://cfpub.epa.gov/eroe/index.cfm?fuseaction=detail.viewInd&lv=list.listbyalpha&r=235294&subtop=381. Last accessed 9/15/2011.

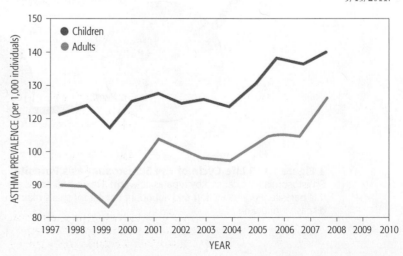

improved in some major urban centers, such as Los Angeles, but has become considerably worse in areas undergoing rapid urban development, such as the southeastern United States and California's Central Valley. The movement of people from rural to urban areas has probably contributed to the increasing prevalence of asthma.

Increasing populations also have an impact on water quality. Although technologies for the management of industrial and municipal wastewater have improved considerably, about a third of Earth's people still do not have access to clean water. This fraction is likely to increase because of the rapidly growing populations in poor countries that have a limited ability to adopt these technologies.

The development of water diversions and reservoirs alters ecosystems and often brings increased risks to human health. Next to malaria, schistosomiasis is the second-most important infectious disease in Africa. Schistosomiasis is caused by a parasitic fluke worm that also lives in freshwater snails (**Figure 19.39**). Over the past three decades, irrigation projects and the construction of dams have increased the habitat available to these snails, as well as human contact with water containing the parasitic worms. As a result, the incidence of schistosomiasis in Africa has increased significantly. For example, in the 25 years following the completion of the Aswan High Dam on the Nile River in Egypt, the rate of schistosomiasis infection in the surrounding population jumped from 5% to 77%.

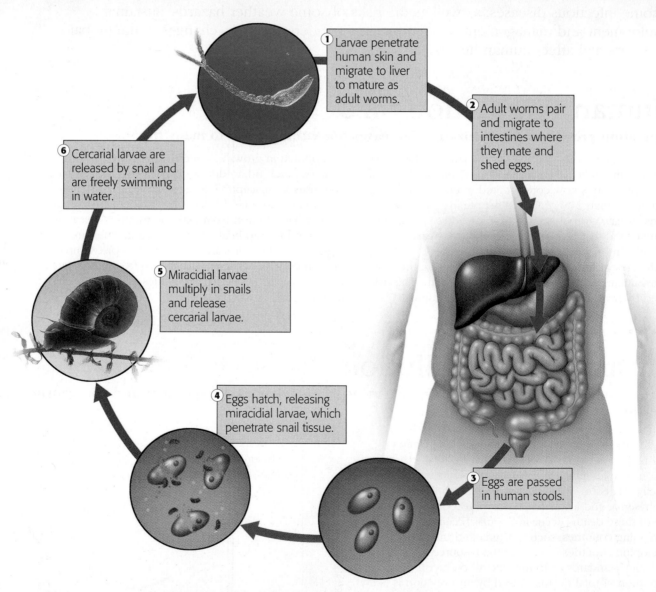

1. Larvae penetrate human skin and migrate to liver to mature as adult worms.

2. Adult worms pair and migrate to intestines where they mate and shed eggs.

3. Eggs are passed in human stools.

4. Eggs hatch, releasing miracidial larvae, which penetrate snail tissue.

5. Miracidial larvae multiply in snails and release cercarial larvae.

6. Cercarial larvae are released by snail and are freely swimming in water.

▲ Figure 19.39 **Life Cycle of the Schistosomiasis Parasite**
Schistosomiasis is caused by a parasitic worm. Freshwater snails are an intermediate host for this parasite. Poor sanitation and abundant habitat for snails maintain the prevalence of this disease worldwide.

SEEING SOLUTIONS: An Integrated Approach to Disease Management

North African countries, such as Egypt, are on track to eradicate schistosomiasis, but the same cannot be said of the poorer nations of sub-Saharan Africa. The World Health Organization estimates that the disease affects more than 100 million people in sub-Saharan Africa; in many communities, more than half the people are infected.

Programs to control schistosomiasis are often thwarted by poverty and poor governance. Given the complexity of the challenges presented by the disease, it is unlikely that a single organization will be able to alter the current situation. However, collaborations among nongovernmental organizations, pharmaceutical companies, and public agencies have begun to find solutions. The Bill and Melinda Gates Foundation has played a key role in catalyzing these collaborations.

In regions where the threat of schistosomiasis has been significantly reduced, collaborative groups have found it essential to use an integrated approach that involves three aspects of the problem: ecology, epidemiology, and education.

Ecology. In much of sub-Saharan Africa, access to water is a challenge. Unfortunately, many of the projects that have increased the availability of water have also created favorable habitat for snails, thereby increasing the prevalence of schistosomiasis. For example, reservoirs and irrigation canals often support large populations of the intermediate host snails.

This ecological challenge is a focus of research at the International Water Management Institute (IWMI) in Colombo, Sri Lanka. IWMI studies show that snail populations can be reduced by removing vegetation from canals and reservoirs and by applying molluscicides, chemicals that limit the growth and reproduction of snails. Where possible, irrigation canals should be lined with cement. Research by IWMI indicates that continued maintenance is essential for the success of ecological management (**Figure 19.40**).

Epidemiology. To improve the health outcomes for individuals and slow the spread of the parasite to other people, infected individuals should be identified and treated early. A single treatment with the drug praziquantel (PZQ) is often sufficient to eliminate the infection. However, the World Health Organization estimates that in 2008 fewer than 6% of infected Africans received treatment. To significantly reduce the spread of the disease, at least 30% of infected individuals must be treated. This level of treatment would require more than 200 million doses of PZQ per year.

The U.S. State Department's Agency for International Development and the United Kingdom's Department of International Development have committed to providing funding for more than 180 million doses of PZQ each year from 2010 to 2014. Merck-Serono, the manufacturer of PZQ, has agreed to donate an additional 20 million doses each year. The Bill and Melina Gates Foundation is providing significant funding to the Schistosomiasis Control Initiative (SCI) in London and the Carter Center in Atlanta, Georgia, for the distribution of PZQ.

Education. Public education is essential to the successful control of schistosomiasis (**Figure 19.41**). The Gates Foundation has provided funding to SCI and the Carter Center for educational programs aimed at improving public understanding of the importance of sanitation and treatment programs in limiting the spread of this disease.

Although schistosomiasis remains a serious threat to hundreds of millions of people in sub-Saharan Africa, an integrated approach holds the promise of change. How will we know if this integrated approach is successful? In 2008, the Gates Foundation provided funding to establish the Schistosomiasis Center for Operational Research and Evaluation (SCORE) at the University of Georgia. SCORE provides funding to a network of researchers to evaluate the effectiveness of ecological, epidemiological, and educational programs.

Can the prevalence of schistosomiasis in sub-Saharan Africa be reduced?

◄ **Figure 19.40 Long-Term Management Is Critical** This canal in Sudan has been lined to prevent the growth of vegetation that might provide habitat for the snails that carry schistosomiasis.

◄ **Figure 19.41 Education Is Essential** Students in a Nigerian primary school hold samples of their own urine. Blood in the urine is the first sign of Schistosomiasis. Left untreated, this disease will eventually damage intestines, liver, bladder and lungs.

Landscape Change

Altered landscapes have provided opportunities for some zoonotic diseases to expand.

Human activities have significantly altered many landscapes and waterways. These alterations have affected the habitats of many animals that are disease vectors or that share diseases with humans, thereby causing the frequency of some diseases to increase.

According to the Centers for Disease Control, one of the most important emerging diseases in the United States is Lyme disease. This disease was not discovered until the mid-1970s. In the 1980s, most cases were confined to New England. Since then, its prevalence has increased exponentially across much of the eastern United States.

Lyme disease often begins with a high fever, headache, and diagnostic skin rash. Left untreated, it can cause nervous disorders, heart problems, and arthritis. It is caused by bacteria in the genus *Borrelia*, which are transmitted to humans by the bite of infected deer ticks. It is a zoonosis that is shared among humans and forest mammals such as deer mice, chipmunks, and white-tailed deer.

Recent studies indicate that *Borrelia* and deer ticks were probably abundant in New England forests in precolonial times. By the middle of the 19th century, forest clearing had nearly removed the deer ticks and the disease they transmit. Since then, reforestation has allowed the populations of forest mammals to recover. Although lagging behind them, populations of deer ticks and the prevalence of *Borrelia* have gradually increased as well (**Figure 19.42**).

The recent, rapid increase in the incidence of Lyme disease is attributable to equally recent and rapid changes in the forested landscapes and distribution of people on those landscapes. Fragmentation of forests for suburban development has created habitats that are especially favorable for deer mice and white-tailed deer. Just as important, this development has brought large numbers of people into direct contact with these animals and their ticks (**Figure 19.43**).

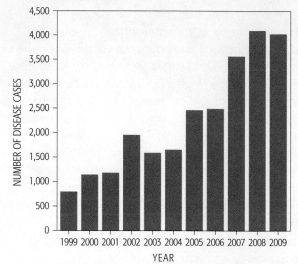

▲ **Figure 19.43 A Disease on the Increase**
This graph shows the number of confirmed cases of Lyme disease reported in the state of Massachusetts. The rapid increase corresponds to the rapid growth in the numbers of deer and to suburban development that has brought more people into forested habitats.
Source: http://www.malegislature.gov/committees/187/document/house/h46/lymediseasereport Last accessed 11/15/2011.

► **Figure 19.42 The Geography of Lyme Disease**
The deer tick is the vector for *Borrelia*, the pathogen that causes Lyme disease. The risk of Lyme disease is greatest in the suburban areas of southern New England, where fragmented forests and development provide ideal habitat for all of the animals that are part of this zoonosis. Those animals include deer mice, white-tailed deer, and humans.
Source: U.S. Centers for Disease Control. URL: http://www.cdc.gov/mmwr/preview/mmwrhtml/figures/R807A2F1.GIF. Last accessed 10/10/2011.

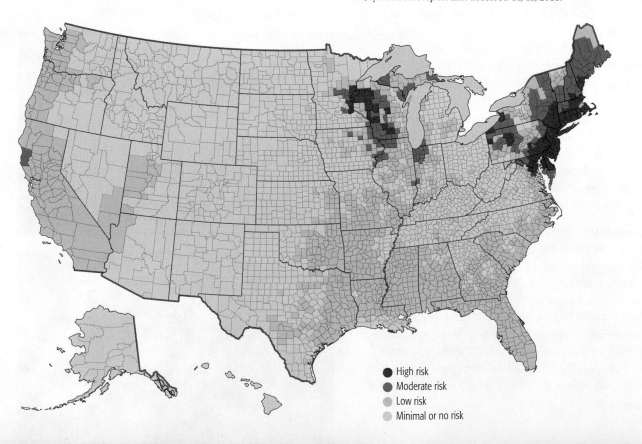

- ● High risk
- ● Moderate risk
- ● Low risk
- ● Minimal or no risk

Climate Change

Global warming is altering habitats for disease organisms and their vectors.

The effects of global warming differ considerably among the various regions of the world (8.5). These changes are already influencing factors that impact human health. In some areas, heat waves have become more frequent. There is strong evidence to support the hypothesis that warming will make severe hurricanes and typhoons more likely. Global warming is also affecting the habitat of some disease-causing organisms and their vectors, thereby increasing the risk of these diseases.

Public health scientists and officials are especially concerned with ongoing climate change at the transition between Earth's tropical and temperate zones. In general, this transition appears to be moving to higher latitudes. The frequency and severity of cold winter temperatures are decreasing, and in many places humidity and moisture levels are increasing. These changes have facilitated the expansion of populations of the mosquitoes that transmit diseases such as malaria and dengue fever.

The effect of global warming on the worldwide distribution of dengue fever is especially well documented. The virus that causes this debilitating disease is transmitted by *Aedes aegypti*, a mosquito that cannot tolerate freezing temperatures. The number of cases of dengue fever reported in North and South America has increased from 66,000 in 1980 to 552,000 in 2006. This increase is especially significant in countries in the transition region between the tropical and temperate zones. In Mexico, for example, cases of dengue fever have increased by 600% since 2000 (**Figure 19.44**).

In 2008, 233 cases of dengue fever in the United States were reported to the Centers for Disease Control and Prevention. Most of these cases were in individuals who had traveled to tropical countries, but an increasing number of cases in south Texas and Florida are appearing in individuals who have not traveled. If global warming continues at its current rate, the Intergovernmental Panel on Climate Change estimates that in 50 years an additional 3.4 billion people will be at risk for dengue fever.

In summary, human-caused environmental changes, including diminished air and water quality, modified landscapes, and global warming, are increasing the risks that we face from many physical, chemical, and biological hazards. People in the world's very poor and developing countries have been particularly hard-hit by these changes. Current trends threaten the well-being of future generations, regardless of their economic status. Reversing these trends will require an understanding of how our actions affect biotic interactions and ecosystem processes and a commitment to diminishing those impacts.

AVERAGE ANNUAL NUMBER OF DENGUE FEVER CASES REPORTED TO WHO & AVERAGE ANNUAL NUMBER OF COUNTRIES REPORTING DENGUE

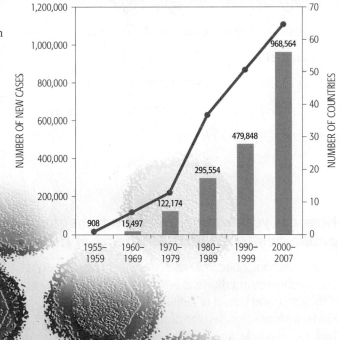

Are we prepared to deal with new epidemics as global climate change causes vector-borne diseases to become more widespread?

Chris Mills, North Carolina State University

Norm: *The U.S. Centers for Disease Control and Prevention is paying special attention to the potential impact of global warming on vector-borne diseases. Immunization and programs that manage the habitats of disease vectors, such as mosquitoes, will be important parts of climate-change adaptation.*

◄ **Figure 19.44 Climate Change and Prevalence of Dengue Fever**
Over the past 50 years, warmer and wetter conditions in many tropical and subtropical regions have provided favorable habitat for the mosquito that transmits dengue fever. This has increased the number of cases of dengue fever and contributed to a significant increase in its geographic range. The photomicrograph shows the virus that causes dengue fever.
Source: World Health Organization. URL = http://www.who.int/csr/disease/dengue/impact/en/. Last accessed 9/15/2011.

QUESTIONS

1. How is urbanization changing the risks associated with hazards such as earthquakes, air pollution, and infectious disease?

2. The connection between global warming and the increased prevalence of infectious diseases appears to be especially strong in zoonotic diseases and diseases spread by insect vectors. Why might this be the case?

AGENTS OF CHANGE: Taking Public Health to Rural Haiti

Will Perez is a medical student with a degree in public health from Brown University. After completing his undergraduate degree, Will deferred medical school for a year to live in rural Haiti. There he developed public health programs and trained youths to become community health workers who could bring health care to regions of Haiti where there are no doctors. Will's training programs are now in 12 schools in rural Haiti and reach 20,000 Haitians.

How did you first get the idea to begin public-health training programs in Haiti?

While in school, I became interested in changing medical policy—the way medicine is practiced, the way it is distributed, and how it is defined. After reading Tracy Kidder's *Mountains Beyond Mountains* and a *New York Times* article with a photo of a Haitian woman baking biscuits of clay and water to feed her children, I committed myself to becoming a doctor for the poor.

I wanted to improve the health of rural Haitians and relieve the death toll imposed on them by the constraints of poverty. Ninety percent of Haiti's children suffer from waterborne diseases and intestinal parasites. Half of the children in Haiti are unvaccinated, and just 40% of the population has access to basic health care.

What steps did you take to get started?

I began by researching my cause, contacting experts in the field to learn about Haiti from every possible angle. I found mentors who would support me and guide me in making difficult decisions. I raised funds and applied for every grant, fellowship, and award that I could. I also took every opportunity to educate people around me about what I was doing.

I learned Haitian Creole, the language spoken by rural Haitians. When I moved to Haiti, I lived in the village where I worked and spoke only Creole. By living in the village and speaking to the people in their own language, I learned about their true needs. No one knows the needs of the people better than the people themselves. Once I figured this out, a seemingly impossible mission became a somewhat more manageable task.

What are the biggest challenges you've faced with the project, and how did you solve them?

I learned my first lesson in foreign aid soon after I arrived in a small village in Haiti, prepared to execute a tuberculosis prevention program I had developed. The program was a complete flop because it didn't serve the most pressing needs of the village—daily battles with scabies, ringworm, fungal infections, and parasites.

Undeterred, I went door to door to figure out what issue I should tackle. The answer? Bedbugs. I set my sights on this less-than-glamorous foe and managed to eradicate bedbugs and ringworm from the village during my first year. This earned me the respect of the villagers, but led to new challenges.

Although the program was hugely successful, I had empty pockets. I could either pack up and leave or raise money to support my program. I decided to raise money. I started writing a blog, telling my story of living in Haiti. Through my blog, I raised over $200,000 to support my new goal of training community health workers.

I trained 16 orphaned youths and formed a public health team. The team went on to provide treatment to hundreds of children and adults who suffered from tuberculosis. Malaria rates in the village went down by 60%. Using simple preventative measures, the team also reduced the incidence of waterborne pathogens, bedbugs, and parasitic infections.

How has your work affected other people and the environment?

Through education, I am equipping the Haitian people with the tools they need to save their lives. These efforts are important because there are no doctors—the people are their own medical community.

Aside from engaging the community in public health issues and providing sustainable methods of treatment for hundreds of people, the youth I trained have become leaders in their communities and are now looked upon

with incredible respect. Their work is inspiring to their peers, as can be seen in the long list of young adults waiting for their chance to join the public health group.

My training program is the only one like it in the southwestern region of Haiti. Recognizing that prevention is by far the most cost-effective and surest way of eliminating the many diseases in the country, other health organizations throughout Haiti have begun to adopt my public-health training program.

What advice would you give to students who want to replicate your model on their campus?

Identifying what resources are available to you is the single most important thing you can do to make your ideas become a reality. Begin by making contacts. Talk to people who have been working on similar issues and are willing to mentor you or to collaborate with you. Professors, deans, and student groups are great places to start. Deans love to hear what students are doing, and they can steer you in the direction of fellowships, grants, and financial aid to support your work.

It takes courage to identify a need and then work to fulfill it. It won't happen overnight, and it may feel as though you keep running into walls. Don't be deterred. Take note of the obstacles and then target each one individually. You'll find that a seemingly impossible task can become achievable by simply taking it one step at a time.

Summary

Environmental changes, whether natural or human-caused, have significant consequences for human health. The health of human populations is often measured by statistics about the likelihood of death and leading causes of death; infant mortality is a particularly sensitive measure of the overall health of populations. Noncommunicable conditions, such as heart disease and cancer, account for more than 75% of the deaths in developed countries. Communicable diseases, such as malaria and diarrhea, are responsible for over half of the deaths in developing countries. Health risks are measured by the likelihood of a hazard occurring and by the consequences of that hazard. Public perceptions of the risks associated with a hazard are influenced by expectations that risks can be totally eliminated, by publicity, by neglect of risk-risk trade-offs, by perceived levels of control over the hazard, and by time delays in the consequences of the hazard.

Physical hazards in the environment include geological events, such as earthquakes and tsunamis, weather hazards, such as hurricanes and heat waves, and wildfires. Chemical hazards refer to toxins, chemicals that diminish human health. The toxicity of chemicals depends on dosage and duration of exposure. These factors are influenced by the manner in which toxins are transported in the environment and the way in which toxins are absorbed and stored in the body. Genetic and environmental factors also influence human vulnerability to toxins. Toxins are classified according to their effects on specific organs and tissues.

Biological hazards include infectious diseases caused by pathogenic organisms. Some pathogens infect only humans. Other pathogens are shared between humans and other animals, some of which play an active role in transmitting disease. Respiratory diseases are transmitted through the air. Diarrheal diseases are associated with poor water quality. Blood-borne diseases are generally transmitted by animal vectors, such as mosquitoes. Over time, coevolution between pathogens and their human hosts tends to diminish the severity of diseases. Environmental changes, such as increasing human populations and urbanization, diminished air and water quality, and global warming, are significantly affecting the risks posed by many environmental hazards.

19.1 Introduction to Public Health

- The health of human populations is generally assessed by statistics related to the likelihood of death and the leading causes of death.
- The risk of an environmental hazard is determined by the probability of its occurrence and its consequences.

KEY TERMS

risk, latent consequences, risk–risk trade-offs

QUESTIONS

1. Why do public health experts consider infant mortality to be an especially important measure of the health of a human population?

2. Over the past 60 years in the United States, the proportion of people dying of cardiovascular disease and cancer has increased compared to infectious disease. What trends are likely responsible for this change?

3. Two different environmental hazards may be equally likely but pose very different risks. How can this be?

4. What are risk–risk trade-offs?

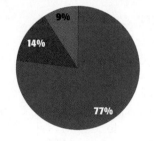

DEVELOPED REGIONS

9%
14%
77%

● Noncommunicable conditions
● Communicable diseases
● Injuries

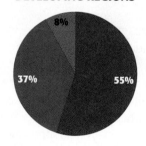

DEVELOPING REGIONS

8%
37%
55%

● Communicable diseases
● Noncommunicable conditions
● Injuries

19.2 Physical Hazards in the Environment

- Although physical hazards such as earthquakes, hurricanes, and wildfires are often called natural disasters, they are heavily influenced by the activities of humans.

KEY TERMS

Richter magnitude scale, earthquake intensity, storm surge

QUESTIONS

1. Describe three factors that influence the risks posed by earthquakes in different locations.

2. The consequences of natural hazards such as hurricanes are often different for poor people than for the wealthy. Do you consider this to be an environmental justice issue? Explain your answer.

3. What factors have contributed to the steady decrease in the proportion of deaths caused by weather hazards?

debated among energy e
If we define *availability*
by conventional technolo
there is no debate: We p
more than a decade ago.
reserves that we can tap
declined by over 85% sin

New and emerging te
from tar sands, oil shale,
course, more energy and
this oil. Oil shale and tar
large amounts of water a
of land surface. But as w
increased and supplies h
value of each barrel of o
has compensated for the
with new technologies.

Whatever technologie
of use will continue to p
increase greenhouse gas
the 2010 collapse of the
and devastating oil spill
technologies have signifi
costs. It may take decade
to recover from that disa

To reduce our individ
transportation, drive fue
simply walk and bicycle
we can use hydrogen fu
into renewable energy s
urban development can
use (**Figure 20.5**).

Given the high costs
our well-being, why do

Less Bioc

**Yes, we are likely t
provides, but we c**

The extinction rates of p
been increasing rapidly
those rates are likely to
extinction for the forese
and loss continue, espe
where species diversity
invasive species, air and
and global warming wil
to the loss of biodiversi

Most of us lament th
reasons, but their loss i
consequences for our o
cannot take the diverse
feed and medicate us fo
remains a reservoir of f

The loss of biodivers
and forests diminishes t
moderate the flow of w

19.3 Chemical Hazards in the Environment

- The toxicity of a chemical depends on the dose and duration of exposure; the vulnerability of individuals to particular toxins is influenced by genetic makeup and environmental factors.
- Toxins may influence one or a combination of organs and tissues.

KEY TERMS

toxicology, dose-response curve, toxicity threshold, median lethal dose (LD_{50}), latent risk, acute exposure, chronic exposure, bioaccumulation, biomagnification, body burden, persistent organic pollutants (POPs), corrosive toxins, asphyxiants, carcinogens, teratogens, allergens, neurotoxins, endocrine disrupters, model organisms

QUESTIONS

1. Some individuals can tolerate levels of a toxin that are above its LD_{50}. What factors might explain such high tolerance?

2. What factors influence the body burden for chemical toxins like DDT and arsenic?

3. Discuss the pros and cons for using model organisms to determine the toxicity of particular chemicals to humans.

19.4 Biological Hazards in the Environment

- Environmental factors influence populations of organisms that cause diseases, as well as the populations of organisms that transmit and contract diseases.
- The relative importance of an environmental factor to the prevalence of an infectious disease depends on the disease's mode of transmission and the organs or tissues that it affects.

KEY TERMS

pathogen, virulence, vector, anthroponosis, zoonosis

QUESTIONS

1. Why do some pathogens tend to be far more virulent among older individuals than younger individuals?

2. Why is it unlikely that we will ever be able to eradicate the common cold and the flu?

3. Explain why scientists view the prevalence of sickle-cell disease in some human populations and the occurrence of falciparum malaria to be an example of coevolution.

ANTHROPONOSES
Human species are the primary hosts

Direct transmission	Indirect transmission
Human → Human	Human → Vector → Human

Disease-causing organisms can be shared through close human contact, like sneezing or shaking hands.

Disease-causing organisms of cholera are spread via a vehicle or vector, like drinking water, affecting other humans who ingest that contaminated water.

19.5 Environmental Change and Human Health

- The risks posed by physical, chemical, and biological hazards are influenced by changes in the size and density of our populations, the quality of the air we breathe and the water we drink, and global climate.

QUESTIONS

1. Describe two factors that may be responsible for the increased prevalence of asthma among children and adults in the United States.

2. Explain the connection between suburban development and the increased prevalence of Lyme disease in the eastern United States.

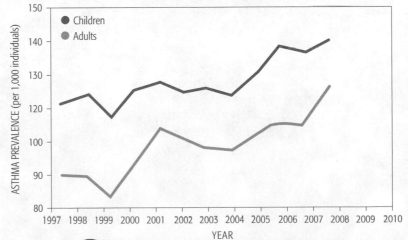

MasteringENVIRONMENTALSCIENCE

Go to **www.masteringenvironmentalscience.com** for homework assignments, practice quizzes, Pearson eText, and more.

We can reverse the trends in global biodiversity loss. Although we have much to learn, we already possess the knowledge and resources to halt biodiversity loss. We are protecting endangered species and ecosystems through an ever-increasing number of parks and protected areas. We are increasing the effectiveness of those reserves by making them as large as possible, by connecting them with corridors, and by buffering them with areas of low-impact use.

Importantly, we are learning how to accommodate the needs of people, especially poor people, while meeting conservation goals (**Figure 20.6**).

We can also provide economic incentives for conserving biodiversity, using tools such as the regulation of trade in endangered species, payments for ecosystem services, and ecotourism. But, in the end, we will protect what we value, and we will value what we understand. Therefore, education will be an important key to our success.

► Figure 20.6 **Stemming the Loss of Species**
A variety of actions can slow the rate or even halt the loss of forest habitats, which is a major cause of species loss worldwide.

What actions can we take to slow the loss of forest habitat

Reforestation to restore lost and degraded habitats

Buy wood products that are certified to have come from sustainably managed forests

Enforce international treaties to prevent illegal logging

Encourage sustainable harvest of non-timber products such as rubber

▼ Figure 20.4 **Sustaining a Renewable Resource**
The supply of water on land is limited by the rate at which rain falls. A variety of strategies are available to increase the efficiency with which we use th water and to ensure its quality.

Condensation

Precipitation

Wate

Water in ice and snow

Snowmelt and surface runoff

Groundwa

More People and Bigger Footprints

Yes, there will be more of us, but we can act to limit our growth in numbers and to diminish our ecological footprint.

Since 1900, the number of us has increased sixfold to about 7 billion. That number is certain to increase in the decades ahead. However, birth and death rates for the global human population have been declining for over 40 years, and the difference between these rates, the global population growth rate, is steadily getting smaller. Nevertheless, by 2050, more than 9 billion of us will populate this planet. How many more is less certain, as are the environmental consequences of our increased numbers.

Can Earth sustain so many of us? Some people are convinced that population growth is the "mother of all environmental challenges." But they are missing half of the story. It is at least as alarming that our ecological footprint is growing. Our ecological footprint is the area of Earth's surface needed to provide the resources that each of us uses and to absorb the waste and pollution that each of us creates. By some calculations, the sum

of the individual ecological footprints of all the people currently living in the world exceeds Earth's land area. Whether or not that is exactly true is a matter of debate, but nearly everyone agrees that the current growth of our population and ecological footprint is not sustainable.

Our situation is complicated significantly by the great differences in the rate of population growth and the size of ecological footprints between rich and poor regions. Population growth rates are high and ecological footprints are relatively small among the people of the world's poorest countries. In many ways, poverty encourages high population growth rates, and the demands of large populations tend to encourage poverty. Economic development, improved human well-being, and education—especially of women—are key factors in lowering population growth rates in these places. However, economic development is usually associated with rapid growth in ecological footprints.

By comparison, population growth rates are low and ecological footprints are large among the people of the world's wealthiest nations. In fact, ecological footprints of the average American or European are 5 to 10 times larger than those of people in poor countries. And it is especially worrisome that those footprints continue to grow.

Greenhouse gas emissions, resource extraction, biodiversity loss—each of these challenges increases with the size of our population and our ecological footprint. Because it diminishes poverty and environmental impacts and improves human well-being, reducing the rate of population growth is especially important in poor countries.

Limiting the growth of ecological footprints is important for both the poor and the rich. Some amount of footprint growth is probably needed to improve the well-being of people in poor countries. To diminish their disproportionate impact on Earth's ecosystems and to ensure opportunity for improved well-being of the poor, the consumption of resources by people in many of the world's wealthiest nations must diminish.

The growth in our numbers and in the size of our ecological footprints can be diminished while improving or maintaining our well-being. Policies aimed at controlling population growth are often controversial, especially in developing countries. However, noncontroversial strategies, such as increased educational opportunities for women and improved sanitation and health care, are proving to be very effective in lowering birth and population growth rates.

We can all monitor our own activities to reduce our individual demands and impacts on Earth's ecosystems. We can replace high-impact technologies with lower-impact technologies and use renewable resources in place of nonrenewable resources. Through increased conservation and efficiency— reduce, reuse, and recycle—we can shrink our demands for energy and materials. There are many ways to do these things without diminishing our well-being (**Figure 20.7**).

Climate change, diminishing resources, biodiversity loss, and increased human numbers and impacts are among the most important environmental challenges, but they are certainly not the only ones. It is possible to reverse current trends for all of these challenges. Such change will require that our individual actions embody our understanding of and commitment to sustainability. It will also require environmental leadership that encourages sustainable behavior in others.

QUESTIONS

1. What do you think will be the most challenging part of reducing the rate of global warming? Of adapting to global warming?

2. How is our use of oil and water resources similar?

3. What do you think are the most important steps that can be taken to conserve water and oil?

4. In many developing nations, the rate of population growth is declining. What factors are contributing to this trend?

The **IPAT equation** *works as a framework to reduce population growth and miminize our environmental impact.*

Provide educational opportunities for women

Provide counseling on contraception and reproductive health

Reduce our ecological footprint; monitor our energy, material use, transportation, and diet

Adopt technologies that reduce resource use and environmental impacts

Impact
loss of resources and degradation

Population
including size, growth, and distribution

Affluence
consumption per individual

Technology
things that demand resources

◄ **Figure 20.7 Sustainable Human Impact**
Humans can diminish the rate of growth of their populations, diminish the impacts of increasing influence, and develop environmentally sustainable technologies.

20.2 Sustainable Living

In 2011, human activities put 7.6 billion metric tons of greenhouse gases (CO_2e) into Earth's atmosphere, added nearly 20 million tons of nitrogen to Earth's rivers and streams, and were responsible for the extinction of as many as 20,000 species of microorganisms, plants, and animals. These numbers are so big that it is easy to fall into the trap of believing that individual actions can have little effect on them. But these numbers are, quite simply, the sum of each of our individual actions. Changes in these numbers—for better or worse—will ultimately be determined by the actions taken by each one of us. Each of us must take responsibility for his or her own actions. You can do this by understanding the challenges that confront you and how your activities affect those challenges. Mohandas Gandhi's words ring true: You must "be the change you wish to see in the world."

Continue to Learn and Improve Your Understanding

The knowledge you have acquired provides the basis for life-long learning.

Congratulations! You have taken a college-level course in environmental science. That alone means you already know more about the environment and its importance to human well-being than 99% of Earth's people. Undoubtedly, you have your own views about which environmental challenges are most compelling and which actions are most appropriate to meet those challenges. Nevertheless, you have the necessary knowledge

and vocabulary to read critically news accounts of environmental issues and to follow and weigh in on debates about environmental policy. So, do it. But also know that your knowledge brings with it additional responsibilities.

Learn more. More than anything else, this textbook and your course have prepared you to continue

▶ **Figure 20.8 Systems Thinking**
In recent years, West Nile virus has become increasingly common among humans in the United States. Managing this infectious disease requires an understanding of the complex environmental factors that interact to allow the disease to spread, as well as collaboration among scientists and decision-makers from many disciplines.

How can understanding **unlikely connections** *help manage diseases that spread across distant species and geography?*

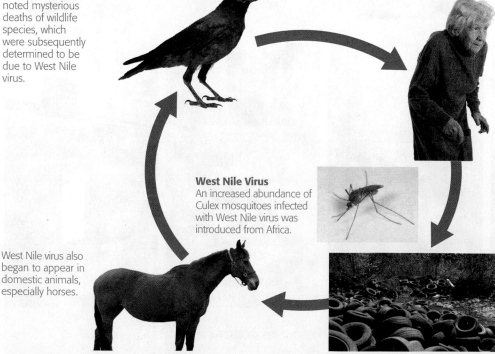

Wildlife biologists noted mysterious deaths of wildlife species, which were subsequently determined to be due to West Nile virus.

West Nile virus first appeared in the United States in elderly people in 1999.

West Nile Virus
An increased abundance of Culex mosquitoes infected with West Nile virus was introduced from Africa.

Warmer climates and increased mosquito breeding habitat associated with human impacts resulted in increased Culex mosquito abundance.

West Nile virus also began to appear in domestic animals, especially horses.

learning. You've only scratched the surface with regard to environmental topics such as human population growth, global warming, and energy. Although you can learn about current issues from news media and popular websites, you should read primary sources to understand environmental topics in depth. Major science journals, such as *Science* and *Nature*, publish reports of original research on environmental topics each week. In addition to primary research articles, these journals provide research news and policy perspectives on environmental issues. A variety of other journals are important sources of information on more specific topics.

Translate your knowledge into understanding.
Environmental science and public policy are rife with numbers, vocabulary, and acronyms. It's easy to get bogged down in these details and miss the important connections and broader perspectives. How do your eating habits influence global warming? How can reducing rural poverty slow the rate of species extinction? Which energy policies are likely to have the greatest immediate effect on greenhouse gas emissions? To find answers to these questions, you need knowledge of the basic processes in ecosystems and human social systems, but you also need to understand how these processes

are connected to one another. Knowledge often comes as separate pieces of information, but understanding requires that the pieces be placed in the context of integrated systems (**Figure 20.8**).

When uncertain, be a skeptic. But don't lose sight of what you know for certain.
All knowledge and understanding are open to scrutiny and revision—this is the most fundamental truism of science. It is also a peculiar property of science that as we learn more, we often expose new areas of uncertainty. Much is yet to be learned about all of the topics we have covered, and uncertainty abounds on many important issues.

It is easy to be overwhelmed by these uncertainties and lose sight of the things that are certain. For example, uncertainties remain regarding the exact extent of global warming in the next few decades, as well as the local and regional consequences of this warming. There is much left to learn about the contributions of different greenhouse gases to warming and about the factors that influence climate at different locations on Earth's surface. Even so, you can be quite certain that Earth's atmosphere is warming, and that human activities are directly responsible for this change (**Figure 20.9**).

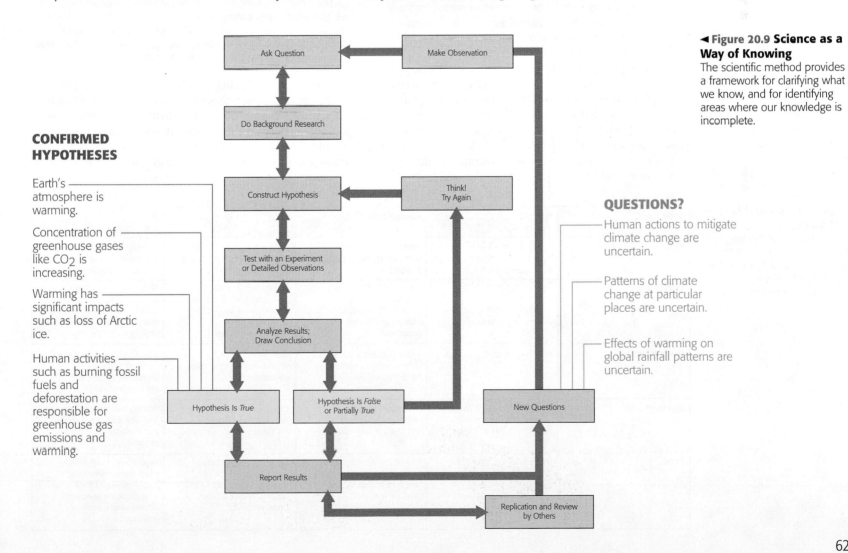

◄ Figure 20.9 **Science as a Way of Knowing**
The scientific method provides a framework for clarifying what we know, and for identifying areas where our knowledge is incomplete.

CONFIRMED HYPOTHESES

Earth's atmosphere is warming.

Concentration of greenhouse gases like CO_2 is increasing.

Warming has significant impacts such as loss of Arctic ice.

Human activities such as burning fossil fuels and deforestation are responsible for greenhouse gas emissions and warming.

QUESTIONS?

Human actions to mitigate climate change are uncertain.

Patterns of climate change at particular places are uncertain.

Effects of warming on global rainfall patterns are uncertain.

Ask Question — Make Observation — Do Background Research — Construct Hypothesis — Think! Try Again — Test with an Experiment or Detailed Observations — Analyze Results; Draw Conclusion — Hypothesis Is True — Hypothesis Is False or Partially True — New Questions — Report Results — Replication and Review by Others

Reduce Your Shoe Size

Monitor and reduce your impact on Earth's ecosystems.

Don't just stand there, do something! Given the number of challenges before us, this is apt advice. You can start by making some adjustments to your own activities. Reduce, reuse, and recycle are obvious steps, and you should make them happen. Actions that are less obvious might include paying closer attention to energy and water use in your dorm or choosing to eat meat less often. If you examine your activities systematically, you can prioritize changes to ensure the maximum benefit. You can also monitor the consequences of your actions to be sure you are meeting your goals.

Calculating your ecological footprint is an effective way to evaluate and adjust your actions. Several Internet calculators (e.g., www.ecologicalfootprint.com) allow you to estimate your personal ecological footprint, using information such as where you live, the size of your home, the kinds of food you eat, and how far you travel and by what means (**Figure 20.10**). The details of these calculators vary, but they all provide a means of ranking the consequences of changes in your various actions.

Much like getting on the scales now and then when you go on a diet, the very act of measuring and monitoring your environmental impacts is important. Without such measures, it is difficult to assess whether your actions are having any effect.

Ultimately, each of us must set criteria for our own well-being. However, we live in a consumer society that places high value on the possession of things. It is certainly true that among the world's poor, measures of human well-being improve as individual wealth increases. But in developed countries, overall measures of happiness do not increase as average wealth increases (**Figure 20.11**). Recent studies show that

▲ **Figure 20.11 Money Can't Buy Happiness**
The average rate of change in peoples' ratings of life satisfaction compared to the rate of change in their country's average wealth over 10 or more years. The survey includes 37 developed and undeveloped countries. None of the world's 50 poorest countries was included in this survey.
Source: Easterlin, R.A., L.A. McVey, M. Switek, O. Sawangfa and J.S. Zweig; 2010. The happiness-income paradox revisited. *Proceedings of the National Academy of Sciences* 107: 22463–22468.

our sense of well-being is influenced by less tangible things, such as the richness of relationships with other people and the quality of the environment. Does this mean you must live a Spartan lifestyle? No, but you should be aware that the lifestyle you choose has consequences for your environment and the well-being of others.

▶ **Figure 20.10 What Is Your Foot Size?**
Make a list of actions that influence your impact on the environment, and consider how you might alter them to diminish your ecological footprint.

Shoe size	Big foot 15 EEE	Intermediate 10 D	Small 7 B	Petite 5 A
Water use	Routinely leave water running; long showers; water used wastefully.	Some conservation concern; some water saving devices installed.	Water conservation a priority; effective use of water saving devices; gray water recycling.	Completely sustainable water use; rainwater capture and gray water recycling.
Energy use	No use of energy saving lighting or appliances; heating and cooling used excessively; poor insulation.	Some energy saving lighting and appliances; limited attention to energy use; heating and cooling used generously.	Energy saving lighting and appliances widely used and on only when required; heating and cooling carefully monitored.	All lighting and appliances are energy efficient; effective use of natural light and renewable energy sources; heating and cooling used sparingly.
Waste disposal	All waste simply thrown out.	Some products are recycled; large amount of waste ends up in the landfill.	Most products are recyclable; some materials are composted.	Attention is given to reuse; nearly all products are recyclable; organic wastes are composted.
Transportation	Drive a large vehicle nearly everywhere and pay little attention to fuel efficiency.	Drive most places in a vehicle with moderate size and average fuel efficiency.	You drive only 25% of the time in a fuel efficient vehicle.	You drive as little as possible and take advantage of public transportation whenever possible.
Food	Eat large amounts of meat, frozen and processed foods; pay no attention to where or how food is produced.	Several meatless meals each week with more fresh foods; try to buy local produce in season.	Limited consumption of meat and frozen or processed foods; try to buy sustainably and locally grown products.	Vegetarian or very limited meat consumption; eat locally produced fresh foods in appropriate seasons; choose only sustainably grown products.

Give What You Can

Invest your efforts in environmental programs that matter to you.

One of the best ways to effect change is by supporting one or more of the thousands of environmental programs and organizations around the world. The amount of time or money that you have to give is not an important issue. To do nothing because you can only do a little is a tragic mistake. Yet it is nearly as tragic to give without a sense of purpose or direction. Whether you are giving 10 hours of precious time or donating 10 dollars, you will find that your sense of accomplishment and your impact on the things you care most about will be magnified by informed priority setting (**Figure 20.12**). The Agents of Change features throughout this book provide wonderful examples and guidance on how to do this.

Environmental organizations vary widely with regard to their specific goals and their strategies for achieving those goals. Some organizations are truly global, and others are dedicated to particular places or regions. Certain organizations focus on a particular environmental issue, such as human population growth, water quality, or biodiversity, whereas others work on a wide range of issues. Organizations also vary in their approaches to meeting environmental challenges. Some are committed to education and advocacy; others, to legislative, regulatory, or judicial action. Organizations may act through grassroots activism or by facilitating market-based solutions. Your priorities should be based on the changes you would like to see in the world and on what you believe to be the most effective ways to make those changes happen. As you work on volunteer activities, your priorities may change as you discover interests and skills you never knew you had.

See your donations of time and resources as an investment in the future. Be interested in the return on your investment. Monitor the performance of organizations relative to your interests and priorities, and use this as a guide for your future commitments.

▼ **Figure 20.12 Give Wisely**
Your involvement in volunteer efforts can have a significant impact on the quality of our environment.

Think and Act for the Future

"Whatever you do may seem insignificant to you, but is most important that you do it."

(*Mohandas Gandhi*)

Think and act for the long term. Sustainable living is, most importantly, about your ability to meet your own needs without compromising the ability of others to meet their needs now or in the future.

Living sustainably is not an endpoint or destination; it is truly a journey. Your personal wants and needs will likely change, along with changes in your knowledge, values, and understanding. The social and technological world around you will likely change as well. You may feel that the changes you can make in your life today are limited, but you certainly can begin the journey now.

QUESTIONS

1. What kinds of organizations on your campus are carrying out activities to address environmental issues?

2. What environmental issue seems most important to you? Why?

3. What steps can you take now to help address that issue?

627

20.3 Environmental Leadership

You can be an agent of change; you can be an environmental leader. Leadership is nothing more or less than action by someone that changes the behavior of others. A high school student convinces local merchants to participate in a program to recycle small batteries. A young mother organizes a workshop aimed at reducing her community's greenhouse gas emissions. A college student develops a curriculum on water for third graders at a nearby elementary school. These are all examples of environmental leadership. Effective environmental leaders share five important qualities: (1) They are motivated by clearly articulated values that define their goals and determine the appropriate means of achieving them. (2) They value diversity. (3) They are willing to act and focused on outcomes. (4) They are humble and willing to learn. (5) They are committed to making a difference and confident that they can.

Articulate a Vision Based on Your Values

Effective leaders have a clear sense of the change they wish to see in the world.

▼ Figure 20.13 Multiple Strategies
These are some of the many actions your university could take to reduce its greenhouse gas emissions. The ones that make the biggest impact are often also the most costly or the ones that require the greatest change in behavior.

Imagine that you are a student trustee at your college or university. The board of trustees is considering a resolution to commit your school to reducing its net emissions of greenhouse gases to zero by the year 2025. This is consistent with your school's stated commitment to manage its campus in an environmentally sustainable fashion. To accomplish this, your school will need to spend several million dollars to upgrade its heating and cooling system and convert its coal-burning hot-water system to natural gas. Critics of this resolution point out that those funds could be used to hire new faculty or provide more student scholarships. Because you represent the student body and because you are knowledgeable about environmental issues, board members listen for your voice on this matter. Where do you stand?

By itself, a decision to act to meet environmental challenges can seem straightforward. If you are convinced that global warming is a significant threat and that universities should be models of sustainable management, then your answer must be yes. But you also live in a world of competing values and needs. In addition to its commitment to sustainability, your university is committed to building a strong and diverse faculty and student body.

Leadership requires decisions and action in the context of competing values and demands. It is tempting for leaders to yield to the loudest or most recent voice, or to set priorities based primarily on cost or ease of accomplishment. But decision making based only on these sorts of criteria is the antithesis of effective leadership. Posting reminders on light switches about using less energy is a visible signal of your university's interest in sustainability, but significant reductions in energy use and greenhouse gas emissions will require more costly investments in infrastructure (**Figure 20.13**). Effective environmental leaders have a clear vision of the goals they wish to achieve, and their decisions and actions are focused on that vision. Their decisions about appropriate goals and actions are guided by ethical principles or values.

So, how do you decide? If sustainability is an important value and part of your vision for your school's future, then action is essential. In most situations, however, the alternatives are rarely so stark as either sustainability or support for faculty and students.

If sustainability and greenhouse gas emissions were the only values of concern, you could argue that your college or university should simply close its doors and send everyone home. Of course, such a decision would be absurd, but it does make it clear that nearly all of our activities—education, home ownership, transportation, business, etc.—have environmental costs. Short of giving up these activities altogether, we must decide which environmental costs to accept and to what extent they are acceptable.

Yes, a commitment to reducing greenhouse gas emissions will have real costs—it would be a rather hollow commitment if it did not. But it also can have real benefits, such as long-term savings in the cost of energy, opportunities for research and teaching, and increasing the reputation of your school as a green institution. The exact time frame for reaching zero emissions and the allocation of costs in your school's budget might be matters for discussion and will depend upon your school's needs in other areas.

Cultivate Diversity

Human diversity is a reservoir of ideas.

Ecosystems and social systems are incredibly diverse and complex. Yet all too often, human management of these systems has centered on simplification. For example, diverse grasslands have been converted to agricultural ecosystems dominated by a single species, and complex, meandering rivers have been confined to simple, straight channels. Sustainable management seeks to maintain diversity and complexity and to maintain connections within and among ecosystems.

Effective environmental leadership also values diversity. While leaders may occasionally wish that everyone's vision and values were identical to theirs, they know that this will hardly ever be the case. They understand that civil argument and dispute often clarify choices and the values that underpin them. They find they can learn from critical and dissenting voices (**Figure 20.14**).

To craft an appropriate plan for reducing greenhouse emissions while also meeting other priorities, your school's board of trustees will need to listen to the voices of diverse stakeholders. Some people may feel that recycling programs are more important than energy-efficiency upgrades. Others may prefer to spend money on resources and additional faculty to enhance environmental course offerings.

Because environmental decisions include many trade-offs, they often cause strong emotions and even conflict among people who otherwise share similar goals. That is why conflict resolution and negotiation skills are important attributes of a leader. But good leaders also recognize that resolving conflict is not the most important goal in environmental decision making. Decisions that are focused on minimizing conflict or trying to make everyone happy run the serious risk of producing less-than-ideal solutions or even daunting problems.

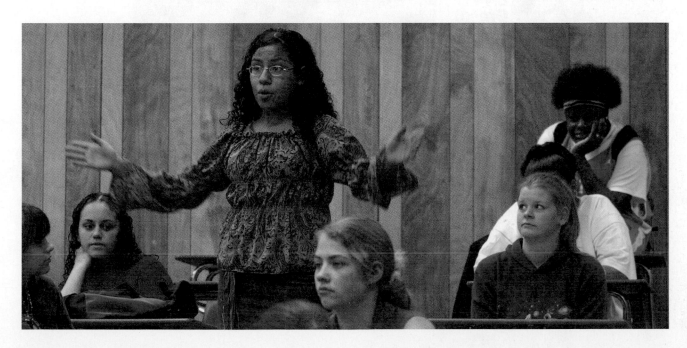

◄ **Figure 20.14 Many Voices**
Effective leaders encourage and listen to diverse points of view.

Focus on Outcomes

The environment is changing constantly, and action is inescapable.

A decision *not to act* to meet an environmental challenge is never neutral. For example, some members of your school's board of trustees may argue that action on greenhouse emissions should be deferred until uncertainties regarding their connections to global warming are resolved. But this decision not to act is itself a very powerful action. It means that the school's greenhouse emissions will continue to increase and assumes that the resources and support necessary for action will be available in the future.

Leaders emphasize their ultimate goals—the changes they wish to see in the world—not the means of reaching them. It is important that your school take actions such as converting the steam plant from coal to natural gas, but the real measure of success is the outcome—a reduction in greenhouse gas emissions.

Effective environmental leaders define their desired outcomes in objective and quantifiable terms that allow them to monitor the consequences of their actions. Just as important, they want others to understand and agree on the definitions of desired outcomes, as well as the approaches to monitoring them.

Be Humble and Adaptable

Uncertainty, new knowledge, and environmental change require adaptive leadership.

You may hear some people say, "You cannot manage what you don't understand." This statement seems logical, but it is not really true. Although our ability to manage a problem improves as our knowledge increases, some amount of uncertainty is inevitable. Indeed, what understanding we do have is constantly subject to change as we learn more.

Effective environmental leadership requires the humility to understand that your actions, or even your goals, may be incorrect. It is committed to making decisions and taking actions that can be modified as conditions change or new knowledge becomes available. To do this, leaders must be willing to learn from their successes and failures and to encourage such learning among those they lead **(Figure 20.15)**.

Adaptive leadership is likely to be important for actions taken by your school's board of trustees. You might, for example, encourage the board to identify the primary areas of uncertainty about how their actions

▶ **Figure 20.15 Measures of Success**
Leaders are willing to take action, but also constantly monitor the outcomes of those actions to ensure success. The author's university implemented it's Sustainable Duke program in 2007 with the goal of reducing its net greenhouse gas emissions to zero by 2025. Its CO_2-e emissions are closely monitored and published on the Sustainable Duke website.

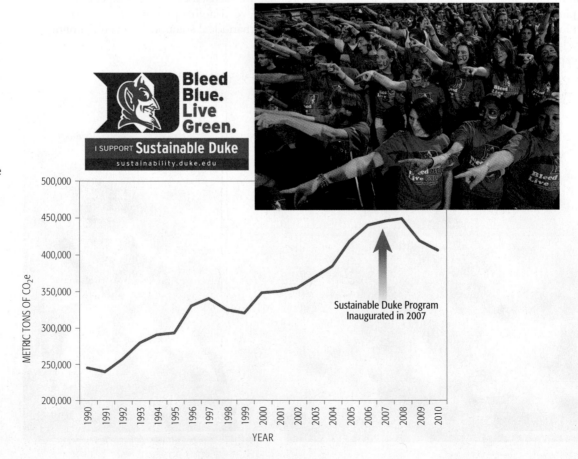

will affect greenhouse gas emissions and, ultimately, global warming. The board could then determine which of these uncertainties are most important or those areas where new knowledge might indicate the need for changes in their actions. Finally, you might recommend a regular review process to ensure that new knowledge is incorporated into future actions.

Adaptive leadership can be a precarious process. You must acknowledge your uncertainty and limited knowledge and, at the same time, maintain the confidence of others. Ultimately, your success as a leader depends on your ability to earn the trust of followers and stakeholders.

Be Confident, Committed, and Hopeful

"There is no greater mistake than to do nothing because you can only do a little."

(Edmund Burke)

Effective environmental leaders believe that their actions, large or small, matter. More than that, they are able to convey that same spirit to those they lead. It also requires a willingness to risk making someone unhappy—chances are good that you will (**Figure 20.16**).

Avoid both hopeless pessimism and naïve optimism. Extreme pessimism discourages action and may even encourage destructive actions: "If we're doomed, why not?" Naïve optimism understates the very formidable challenges that we face and is likely to underestimate the magnitude of actions needed to confront those challenges. A hopeful sense of the possible is the most appropriate approach. Through your individual actions and leadership, that hope and those possibilities become self-fulfilling prophecies.

QUESTIONS

1. What five qualities are essential for environmental leaders?
2. Why is it important to consider diversity when seeking solutions to environmental issues?

▼ Figure 20.16 **Agents of Change**
The author is grateful to each of you with the vision and commitment to make a difference in the world.

AGENTS OF CHANGE: One World Youth Project Connecting Classrooms Around the World

Jess Rimington is the founder and executive director of One World Youth Project (OWYP), a program that links students around the world in community service learning projects inspired by the UN Millennium Development Goals. OWYP trains college students from various countries to lead global education programs in local middle school and high school classrooms, providing educators with tools to prepare students for the challenges of the globalized 21st century.

How did you first get the idea for One World Youth Project?

In the sixth grade I had an amazing science teacher who taught me how human equity issues are related to the environment. I became really passionate about sustainable development and the intersection of humans and the environment. In high school, I got involved with Roots and Shoots, a program sponsored by the Jane Goodall Institute. Through Roots and Shoots, I received a scholarship to travel to Johannesburg, South Africa, for the World Summit on Sustainable Development. While there, it seemed as if world leaders were telling youth that they would have to solve all sorts of environmental problems and take action to create a more sustainable society. But environmental problems have no borders. I thought, "How are we supposed to come together to work around these issues if we don't know how to begin a global conversation?" I was really inspired by the idea of starting that conversation. That was the seed of inspiration for One World Youth Project.

What steps did you take to get started?

I set out with a vision of fostering relationships by connecting every middle school and high school classroom on Earth. I began by using the relationships I already had to connect classrooms in my region with classrooms abroad. As I did this, my community started to support me.

A girl from my high school helped me, and together we learned how to register as a 501(c)(3) nonprofit. We found a lawyer in our hometown who was willing to donate his time pro bono to help us file the paperwork, and suddenly we were a nonprofit. We got someone to donate a website, and it snowballed from there. Of course, OWYP looks very different today, but those were the first steps we took.

What are the biggest challenges you've faced with the project, and how did you solve them?

The number one challenge is money. It's always difficult to find funding. In some ways it's easier for young people to raise money, and in some ways it's more difficult. It is difficult to walk the line between getting money as a token and getting money that can make a real impact. In order to get more funding, I read articles about wealthy people in the United States, found their home addresses, sent them letters repeatedly, and even went to their doors in person to tell them about OWYP. At first I got rejected a lot, but after a while we were able to secure enough funds to hire our first full-time staff.

How has your work affected other people and the environment?

One of the classroom connections I am particularly proud of is that of the Hart Middle School in Washington, D.C., and the Yll Morina Middle School in Gjakova, Kosovo. In this connection, students shared elements of their cultural identity and learned about each others' lifestyles. Groups of students from each classroom talked to one another via Skype video chat, asking each other about their schools, fashion, music, and sports. Students in the two classrooms soon realized that even though they lived across the world from one another, they shared concerns about issues such as gender equality and air pollution. These kinds of understandings help youth view one another as individuals rather than the nations they represent, and that is the beginning of global cooperation and citizenship.

What advice would you give to students who want to replicate your model on their campus?

You can have a big vision, but start out small. Things will fall into place if you're passionate, committed, and have patience.

Summary

Greenhouse emissions are increasing and our climate is warming, renewable and nonrenewable resources are dwindling, and the loss of biodiversity is accelerating. These changes in Earth's environment are occurring not only because our population is growing but also because the average ecological footprint of humans is growing. Fortunately, we possess the knowledge and know-how to slow and even reverse these trends. To accomplish this, each of us will need to understand and make a personal commitment to sustainability. You can do this by continuing to learn about environmental issues, by translating your knowledge into understanding, by reducing your own ecological footprint, by committing your time and resources to the environmental issues that matter most to you, and by thinking and acting for the future. You can be an agent for change by clearly articulating your values and vision for the future, by cultivating diversity, by focusing on outcomes consistent with your values and vision, and by being humble, adaptable, confident, committed, and hopeful.

20.1 What Does the Future Hold?

- In the future, there are likely to be increasing numbers of humans, and Earth's environment is likely to be affected by human-caused changes, including global warming, scarce resources, diminished biodiversity, and pollution.

- We possess the knowledge and technologies to slow and reverse these trends.

QUESTIONS

1. Environmental scientists like to say, "Everything is connected to everything." Describe three ways in which climate change is connected to each of the following: the availability of oil, the availability of water, and biodiversity.

2. Even though the rate of human population growth is decreasing, human impacts on our planet are increasing. Why?

3. Describe three ways in which the loss of biodiversity might diminish your well-being.

4. Sustainability implies equitable treatment and justice among humans. How can we ensure that efforts to reduce ecological footprints in different countries are equitable and just?

The **IPAT equation** works as a framework to reduce population growth and miminize our environmental impact.

Provide educational opportunities for women

Provide counseling on contraception and reproductive health

Reduce our ecological footprint; monitor our energy, material use, transportation, and diet

Adopt technologies that reduce resource use and environmental impacts

Impact
loss of resources and degradation

Population
including size, growth, and distribution

Affluence
consumption per individual

Technology
things that demand resources

20.2 Sustainable Living

- Your own actions matter.
- You can be the change you wish to see in the world by continuing to learn, reducing your ecological footprint, giving your time and resources, and thinking and acting for the future.

QUESTIONS

1. What environmental challenge matters most to you? Explain how systems thinking is important to meeting that challenge.

2. Describe three changes in your own activities that you could make today that would diminish your ecological footprint.

3. What features or characteristics matter most to you in selecting an environmental project or organization in which to invest your time or resources?

4. How will you go about monitoring the success of that project or organization?

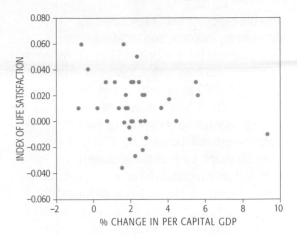

20.3 Environmental Leadership

- You can be an agent of change: Leadership is action by you that changes the behavior of others.
- Leadership is based on values and vision, commitment to diversity, focus on outcomes, humility, adaptation, confidence, commitment, and hope.

QUESTIONS

1. What core values are most important to your goals for environmental sustainability and your choice of actions to achieve those goals?

2. What things could you do to build a diverse organization or program?

3. How would you work in a project or organization to be sure that it was adaptable to changes in knowledge and circumstances?

4. Environmental philosopher David Orr describes hope as "a verb with its sleeves rolled up." How will you translate your hopes for the future into actions today?

Go to **www.masteringenvironmentalscience.com** for homework assignments, practice quizzes, Pearson eText, and more.

635

APPENDIX A

Graphs

The allocation of water resources for agriculture, industry, and residential use is very different for the United States compared to India. Countries with a low gross domestic product (GDP) tend to have high population growth rates. If past trends continue, the average annual temperature of Earth's atmosphere will increase by 3 °C over the next 50 years. Each of these statements is based on quantitative measurements of environmental features. Because they differ from time to time, place to place, or group to group, these features are called variables.

The statements about variables are verified by scientists using a variety of statistical methods.

It is hard, however, to visualize the relationships implied in these statements simply by looking at long lists of numbers. Instead, we rely on graphs, diagrams that compare the quantitative data for environmental features such as water use, economic activity, population growth rates, or average annual temperature. The kind of graph depends in part on the nature of the data and, most importantly, on the question we wish to answer with those data.

1. Pie Charts

How is a whole group or variable divided among categories?

Pie charts are used when we wish to represent the proportions of different categories within a whole variable. For example, how is water consumption in a country apportioned among different categories of use (**Figure A1**)? In this example, the relative amounts of water used by industry, agriculture, and households are quite different among these four countries. The absolute amount of water used by each country, listed in parentheses, is of secondary interest and is not part of the graph.

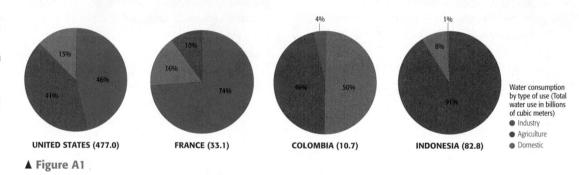

UNITED STATES (477.0)   **FRANCE (33.1)**   **COLOMBIA (10.7)**   **INDONESIA (82.8)**

Water consumption by type of use (Total water use in billions of cubic meters)
- Industry
- Agriculture
- Domestic

▲ **Figure A1**

2. Bar Charts

How does the amount of a variable differ among different groups, places, or times?

In bar charts, the amounts of a variable for each of several groups are indicated by the length of bars. Bars may be oriented horizontally or vertically. **Figure A2** shows that total reserves of coal and reserves of different types of coal differ among continents. **Figure A3** indicates the infant mortality rate for each of the world's 10 poorest countries. This graph shows clearly that infant mortality in Sierra Leone is about three times higher than in Eritrea. Bar charts can also be used to compare multiple data sets and group types. For example, **Figure A4** compares the number of women with the number of men in different 5-year age classes for Ethiopia and the United States. The proportion of the population that is represented by middle- and old-age individuals in Ethiopia is much less than in the United States. Furthermore, gender differences are small in Ethiopia, whereas in the United States women generally outnumber men.

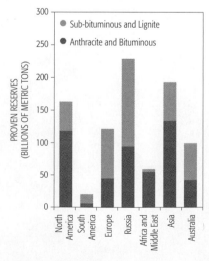

PROVEN RESERVES (BILLIONS OF METRIC TONS)

- Sub-bituminous and Lignite
- Anthracite and Bituminous

North America, South America, Europe, Russia, Africa and Middle East, Asia, Australia

▲ **Figure A2**

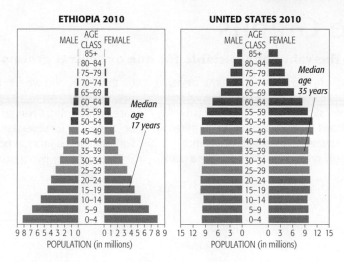

WORLD'S POOREST NATIONS

▲ Figure A3

▲ Figure A4

3. Scatter Plots

How do the relationships between variables vary among different groups or populations of things?

The relationship of one variable to another for different groups can be indicated as dots on a graph where values for one variable are indicated on the horizontal or *x* axis and values for the other variable are indicated on the vertical or *y* axis. In **Figure A5**, the percentage of girls among secondary school students is compared to birth rates for each of 78 countries, each represented by a dot. While there is considerable scatter among the dots, there is a clear

trend toward lower birth rates as the percentage of girls in secondary school increases. We can say, therefore, that birth rate is negatively correlated with girls' education for these 78 countries. **Figure A6** compares ratings of life satisfaction by people in different countries with recent change in average wealth. Here, there is also considerable scatter, but there is no clear trend.

▲ Figure A5

▲ Figure A6

4. Line Graphs

How does the value of a variable for one or several groups change through time?

In line graphs, changes in time or, in some cases, age are indicated along the *x* axis, and the values of a variable are indicated on the *y* axis. **Figure A7**, for example, shows the steady decrease in mean ozone concentrations in Los Angeles's air since 1975. Each yearly measurement is actually a point on this graph, and we connect the points with a line because they refer to a single group or place, in this case Los Angeles. Changes in a single variable can be compared for several groups on the same graph. For example, changes in population size are shown in **Figure A8** for two species of *Paramecium*.

▲ **Figure A7**

Line graphs can also be used to compare changes in two different variables through time. In such cases, the numerical values for one variable are indicated on the *y* axis on the left side of the graph and the values for the other are indicated on the *y* axis on the right side. Just such a comparison is shown in **Figure A9**. This graph shows that changes in CO_2 concentration in the atmosphere have closely tracked changes in global temperature over the past 450,000 years.

▲ **Figure A8**

▲ **Figure A9**

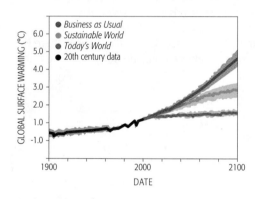

▲ **Figure A10**

Line graphs can also be used to indicate how values for a variable are likely to change in the future compared to past trends. In such cases, past trends are usually based on actual measurements. Forecasts may be based on a simple extension of such trend lines or on more complicated predictive models. In **Figure A10**, past changes in the temperature of Earth's atmosphere are indicated relative to that in 1900. The black line is based on real measurements up to the year 2000 from thousands of stations. The colored lines represent changes that are forecast from models based on different assumptions about future human actions. Business as Usual (red) implies no change in behavior, Sustainable World implies significant action to reduce greenhouse gas emissions, and Today's World implies that all greenhouse emissions cease immediately. The shaded area around each line indicates variation in measurements in the case of past trends and uncertainty about future trends in the case of the forecasts.

Metric System

Measurement	Unit and Abbreviation	Metric Equivalent	Metric to English Conversion Factor	English to Metric Conversion Factor
Length	1 kilometer (km)	=1,000 (10^3) meters	1 km = 0.62 mile	1 mile = 1.61 km
	1 meter (m)	=100 (10^2) centimeters	1 m = 1.09 yards	1 yard = 0.914 m
		=1,000 millimeters	1 m = 3.28 feet	1 foot = 0.305 m
			1 m = 39.37 inches	
	1 centimeter (cm)	= 0.01 (10^{-2}) meter	1 cm = 0.394 inch	1 foot = 30.5 cm
				1 inch = 2.54 cm
	1 millimeter (mm)	= 0.001 (10^{-3}) meter	1 mm = 0.039 inch	
Area	1 square meter (m^2)	= 10,000 square centimeters	1 m^2 = 1.1960 square yards	1 square yard = 0.8361 m^2
			1 m^2 = 10.764 square feet	1 square foot = 0.0929 m^2
	1 square centimeter (cm^2)	= 100 square millimeters	1 cm^2 = 0.155 square inch	1 square inch = 6.4516 cm^2
Mass	1 metric ton (t)	= 1,000 kilograms	1 t = 1.103 ton	1 ton = 0.907 t
	1 kilogram (kg)	= 1,000 grams	1 kg = 2.205 pounds	1 pound = 0.4536 kg
	1 gram (g)	= 1,000 milligrams	1 g = 0.0353 ounce	1 ounce = 28.35 g
	1 milligram (mg)	= 0.001 gram		
Volume (solids)	1 cubic meter (m^3)	= 1,000,000 cubic centimeters	1 m^3 = 1.3080 cubic yards	1 cubic yard = 0.7646 m^3
			1 m^3 = 35.315 cubic feet	1 cubic foot = 0.0283 m^3
	1 cubic centimeter (cm^3 or cc)	= 0.000001 cubic meter	1 cm^3 = 0.0610 cubic inch	1 cubic inch = 16.387 cm^3
		= 1 milliliter		
	1 cubic millimeter (mm^3)	= 0.000000001 cubic meter		
Volume (liquids and gases)	1 kiloliter (kl or kL)	= 1,000 liters	1 kL = 264.17 gallons	
	1 liter (l or L)	= 1,000 milliliters	1 L = 0.264 gallons	1 gallon = 3.785 L
			1 L = 1.057 quarts	1 quart = 0.946 L
	1 milliliter (ml or mL)	= 0.001 liter	1 mL = 0.034 fluid ounce	1 quart = 946 mL
		= 1 cubic centimeter	1 mL = approx. 1/4 teaspoon	1 pint = 473 mL
				1 fluid ounce = 29.57 mL
				1 teaspoon = approx. 5 mL
Temperature	Degrees Celsius (°C)		°C = 5/9(°F - 32)	
Energy and Power	1 kilowatt-hour	= 34,113 BTU = 860,421 calories		
	1 watt	= 3,413 BTU/hr = 14.34 calorie/min		
	1 calorie	= the amount of heat necessary to raise the temperature of 1 gram (1 cm^3) of water 1 degree Celsius		
	1 joule	= 9.481 x 10^{-4} BTU = 0.239 cal = 2.778 × 10^{-7} kilowatt-hour		
Pressure	1 pound per square inch (psi)	= 6894.757 pascal (Pa) = 0.068045961 atmosphere (atm) = 51.71493 millimeters of mercury (mm hg = Torr) = 68.94757 millibars (mbar) = 68.94757 hectopascal (hPa) = 6.894757 kilopascal (kPa) = 0.06894757 bar		
	1 atmosphere (atm)	= 101.325 kilopascal (kPa)		

BIBLIOGRAPHY

Sources and Further Reading

Chapter 1

Chapin III, F.S., P.A. Matson, and H.A. Mooney. 2002. *Principles of Terrestrial Ecosystem Ecology*. Springer-Verlag, New York, NY.

Chapin III, F.S., M.S. Torn, and M. Tateno. 1996. Principles of ecosystem sustainability. *American Naturalist* 148:1016–1037.

Daily, G.C. 1997. *Nature's Services: Societal Dependence on Natural Ecosystems*. Island Press, Washington, D.C.

Daily, G.C., S. Alexander, P.R. Ehrlich, et al. 1997. Ecosystem Services: Benefits supplied to human societies by natural ecosystems. *Issues in Ecology* 2:1–16.

Holdren, J.P. 2008. Science and technology for sustainable well-being. *Science* 319:424–434.

Kates, R.W., W.C. Clark, R. Corell, et al. 2003. Sustainability science. *Science* 292:641–642.

Kates, R.W. and T.M. Parris. 2003. Long-term trends and a sustainability transition. *Proceedings of the National Academy of Sciences* 100:8062–8067.

Kuhn, T.S. 1996. *The Structure of Scientific Revolutions* (3rd ed.). The University of Chicago Press, Chicago, IL.

Millennium Ecosystem Assessment. 2005. *Ecosystems and Human Well-Being*. Island Press, Washington, D.C. http://www.maweb.org/documents/document.356.aspx.pdf. Last accessed 10-1-2011.

National Research Council, Board on Sustainable Development. 1999. *Our Common Journey: A Transition Toward Sustainability*. National Academy Press, Washington, D.C.

Ostrom, E. 2009. A general framework for analyzing sustainability of social-ecological systems. *Science* 325:419–422.

Pimm, S.L. 2001. *The World According to Pimm*. McGraw Hill, New York, NY.

Speth, J.G. 2004. *Red Sky at Morning: America and the Crisis of the Global Environment*. Yale University Press, New Haven, CT.

United Nations. 1987. *Report of the World Commission on Environment and Development: Our Common Future*. http://www.un-documents.net/wced-ocf.htm. Last accessed 10-1-2011.

Vitousek, P.M., H.A. Mooney, J. Lubchenco, et al. 1997. Human domination of Earth's ecosystems. *Science* 277:494–499.

Chapter 2

Andrews, R.N.L. 2006. *Managing the Environment, Managing Ourselves: A History of American Environmental Policy*. Yale University Press, New Haven, CT.

Attfield, R. 2003. *Environmental Ethics: An Overview for the Twenty-First Century*. Polity Press, Cambridge, UK.

Bowler, P.J. 1993. *The Norton History of the Environmental Sciences*. W.W. Norton, New York, NY.

Callicott, J.B. 1994. *Earth's Insights: A Survey of Ecological Ethics from the Mediterranean Basin to the Australian Outback*. University of California Press, Berkeley, CA.

Carson, R. 1962. *Silent Spring*. Houghton Mifflin, Boston, MA.

Costanza, R., R. d'Arge, R. de Groot, et al. 1997. The value of the world's ecosystem services and natural capital. *Nature* 387:253–260.

Daly, H. 2005. Economics in a full world. *Scientific American* 293:100–107.

DesJardins, J.R. 1997. *Environmental Ethics: An Introduction to Environmental Philosophy* (2nd ed.). Wadsworth Publishing, Belmont, CA.

Ehrlich, P.R. 1968. *The Population Bomb*. Ballantine Books, New York, NY.

Freeman, A.M. 1993. *The Measurement of Environmental and Resources Values: Theory and Methods*. Resources for the Future, Washington, D.C.

Hardin, G. 1968. The tragedy of the commons. *Science* 162:1243–1248.

Hawkin, P. 1993. *The Ecology of Commerce: A Declaration of Sustainability*. HarperCollins, New York, NY.

Hawkin, P., A. Lovins, and H. Lovins. 1997. *Natural Capitalism: Creating the Next Industrial Revolution*. Little Brown, New York, NY.

IPCC. 2007. *Climate Change 2007: Synthesis Report. Contribution of Working Groups I, II, and III to the Fourth Assessment Report of the Intergovernmental Panel on Climate Change*. IPCC, Geneva, Switzerland.

Leopold, A. 1949. *A Sand County Almanac*. Oxford University Press, Oxford, UK.

Ostrom, E. 2009. A general framework for analyzing sustainability of social-ecological systems. *Science* 325:419–422.

Simpson, J.W. 2005. *Dam! Water, Power, and Preservation in Hetch Hetchy and Yosemite National Park*. Pantheon Books, New York, NY.

Wilson, E.O. 1999. *The Diversity of Life*. W.W. Norton, New York, NY.

World Watch Institute. 2008. *State of the World: Innovations for a Sustainable Economy*. W.W. Norton, New York, NY.

Chapter 3

Bonan, G. 2002. *Ecological Climatology: Concepts and Applications*. Cambridge University Press, Cambridge, UK.

Craig, J.R., D.J. Vaughn, and B.J. Skinner. 2010. *Earth Resources and the Environment* (4th ed.). Benjamin Cummings, San Francisco, CA.

Hidore, J.J., J.E. Oliver, M. Snow, et al. 2010. *Climatology: An Atmospheric Science* (3rd ed.). Prentice Hall, Upper Saddle River, NJ.

Keller, E.A. 2008. *Introduction to Environmental Geology* (4th ed.). Prentice Hall, Upper Saddle River, NJ.

Lewis, J.S. and R.G. Prinn. 1984. *Planets and Their Atmospheres: Origin and Evolution*. Academic Press, Orlando, FL.

Lovelock, J.E. 1979. *Gaia: A New Look at Life on Earth*. Oxford University Press, Oxford, UK.

Montgomery, C. 2010. *Environmental Geology* (9th ed.). McGraw-Hill, New York, NY.

Schlesinger, W.H. 1997. *Biogeochemistry: An Analysis of Global Change* (2nd ed.). Academic Press, Orlando, FL.

Wade, L.G. 2005. *Organic Chemistry* (5th ed.). Prentice Hall, Upper Saddle River, NJ.

Chapter 4

Campbell, N.A. and J.B. Reece. 2008. *Biology* (8th ed.). Benjamin Cummings, San Francisco, CA.

Darwin, C. 1859. *The Origin of Species by Means of Natural Selection*. John Murray, London, UK.

Deevey Jr., E.S. 1947. Life tables for natural populations of animals. *The Quarterly Review of Biology* 22:283–314.

Denholm, I., G.J. Devine, and M.S. Williamson. 2002. Insecticide resistance on the move. *Science* 297:2222–2223.

Endler, J.A. 1986. *Natural Selection in the Wild. Monographs in Population Biology 21*. Princeton University Press, Princeton, NJ.

Erickson, G.M., P.J. Currie, B.D. Inouye, et al. 2006. Tyrannosaur life tables: An example of nonavian dinosaur population biology. *Science* 313:213–217.

Grant, B.R. and P.R. Grant. 1989. Natural selection in a population of Darwin's finches. *The American Naturalist* 133:377–393.

Hua, X. and J.J. Wiens. 2010. Latitudinal variation in speciation mechanisms in frogs. *Evolution* 64:429–443.

Hutchinson, G.E. 1959. Homage to Santa Rosalia or why are there so many kinds of animals? *The American Naturalist* 93:145–159.

Hutchinson, G.E. 1978. *Introduction to Population Ecology*. Yale University Press, New Haven, CT.

Ridley, M. 1999. *Genome: The Autobiography of a Species in 23 Chapters*. HarperCollins, New York, NY.

Rundle, H.D. and P. Nosil. 2005. Ecological speciation. *Ecology Letters* 8:336–352.

Schilthuizen, M. 2001. *Frogs, Flies, and Dandelions, Speciation—The Evolution of New Species*. Oxford University Press, New York, NY.

Schluter, D. 2001. Ecology and the origin of species. *Trends in Ecology and Evolution* 16:372–380.

Smith, T.M. and R.L. Smith. 2009. *Elements of Ecology* (7th ed.). Benjamin Cummings, San Francisco, CA.

Van Dover, C.L. 2000. *The Ecology of Deep-Sea Thermal Vents*. Princeton University Press, Princeton, NJ.

Wilson, E.O. 1999. *The Diversity of Life*. W.W. Norton, New York, NY.

Chapter 5

Balter, M. 2006. The baby deficit. *Science* 312:1894–1897.

Bloom, D.E. 2011. 7 billion and counting. *Science* 333:562–569.

Bongaarts, J. and S. Sinding. 2011. Population policy in transition in the developing world. *Science* 333:574–576.

Central Intelligence Agency. 2009. *The World Fact Book*. https://www.cia.gov/library/publications/the-world-factbook/. Last accessed 10-1-2011.

Cohen, J.E. 1995. *How Many People Can the World Support?* W.W. Norton, New York, NY.

Cohen, J.E. 2003. Human population grows up. *Scientific American* 293:48–55.

Dasgupta, P. 1990. Well-being and the extent of its realization in poor countries. *The Economics Journal* 100:1–32.

Deevey Jr., E.S. 1961. The human population. *Scientific American* 203:195–204.

Ehrlich, P.R. 1968. *The Population Bomb*. Ballantine Books, New York, NY.

Ehrlich, P.R. and A.H. Ehrlich. 2009. The population bomb revisited. *The Electronic Journal of Sustainable Development* 1:63–71.

Holdren, J.P. and P.R. Ehrlich. 1974. Human population and the global environment. *American Scientist* 62:282–292.

James, K.S. 2011. India's demographic change: Opportunities and challenges. *Science* 333:576–580.

Lee, R. 2011. The outlook for population growth. *Science* 333:569–573.

Malthus, T.R. 1798. *An Essay on the Principle of Population*. J. Johnson, London, UK. Reprinted at: http://www.esp.org/books/malthus/population/malthus.pdf. Last accessed 10-1-2011.

Meadows, D.H. and D. Meadows. 1972. *The Limits of Growth: A Report for The Club of Rome's Project on the Predicament of Mankind*. Universe Books, New York, NY.

Meadows, D.H., J. Randers, and D. Meadows. 2004. *Limits to Growth: The 30-Year Update*. Chelsea Green Publishing Co., White River Junction, VT.

Monfreda, C., M. Wackernagel, and D. Deumling. 2004. Establishing national natural capital accounts based on detailed ecological footprint and biological capacity accounts. *Land Use Policy* 21:231–246.

Parker, J. 2011. The 9 billion-people question. *The Economist, Special Report*, February 26, 2011. http://www.economist.com/node/18200618. Last accessed 10-1-2011.

Pearl, R. and S. Gould. 1936. Human population growth. *Human Biology* 8:399–419.

Peng, X. 2011. China's demographic history and future challenges. *Science* 333:581–587.

Shah, A. 1998. *Ecology and the Crisis of Overpopulation: Future Prospects for Global Sustainability*. Edward Elgar Publishing, Cheltenham, UK.

Thompson, W. 1929. Population. *The American Journal of Sociology* 34:959–975.

United Nations. 2010. *World Population Prospects, the 2010 Revision*. http://esa.un.org/wpp/Other-Information/faq.htm. Last accessed 10-1-2011.

U.S. Census Bureau. http://www.census.gov/. Last accessed 10-1-2011.

Wackernagel, M. and W. Rees. 1996. *Our Ecological Footprint: Reducing Human Impact on the Earth*. New Society Publishers, Gabriola Island, BC, Canada.

Wackernagel, M., N.B. Schulz, D. Deumling, et al. 2002. Tracking the ecological overshoot of the human economy. *Proceedings of the National Academy of Sciences* 99:266–271.

Chapter 6

Carey, A.B., W. Colgan III, J.M. Trappe, et al. 2002. Effects of forest management on truffle abundance and squirrel diets. *Northwest Science* 76:148–157.

Carey, A.B., S.P. Horton, and B.L. Biswell. 1992. Northern spotted owls: Influence of prey base and landscape character. *Ecological Monographs* 62:233–250.

Cooper, W.S. 1923. The recent ecological history of Glacier Bay, Alaska: II. The present vegetation cycle. *Ecology* 4:223–246.

Cramer, V.A., R.J. Hobbs, and R.J. Standish. 2008. What's new about old fields? Land abandonment and ecosystem assembly. *Trends in Ecology and Evolution* 23:105–112.

Elton, C. and M. Nicholson. The ten-year cycle in numbers of lynx in Canada. *Journal of Animal Ecology* 11:215–244.

Estes, J.A., J. Terborgh, J.S. Brashares, et al. 2011. Trophic downgrading of planet Earth. *Science* 333:301–306.

Fortin, D., H.L. Beyer, M.S. Boyce, et al. 2005. Wolves influence elk movements: Behavior shapes a trophic cascade in Yellowstone National Park. *Ecology* 85:1320–1330.

Gause, G.F. 1932. Experimental studies on the struggle for existence. *Journal of Experimental Biology* 9:389–402.

Johnston, D.W. and E.P. Odum. 1956. Breeding bird populations in relation to plant succession on the piedmont of Georgia. *Ecology* 37:50–62.

Korstian, C.F. and T.S. Coile. 1938. *Plant Competition in Forest Stands*. Duke University School of Forestry, Bulletin 3, Durham, NC.

MacArthur, R.H. 1958. Ecology of some warblers of northeastern coniferous forests. *Ecology* 39:599–619.

Maser, C., A.W. Claridge, and J.M. Trappe. 2007. *Trees, Truffles and Beasts: How Forests Function*. Rutgers University Press, Brunswick, NJ.

Oosting, H.J. 1942. An ecological analysis of the plant communities of Piedmont, North Carolina. *American Midland Naturalist* 28:1–126.

Pickett, S.T.A. and P.S. White. 1985. *The Ecology of Natural Disturbance and Patch Dynamics*. Academic Press, Orlando, FL.

Pimm, S.L. 1979. The structure of food webs. *Theoretical Population Ecology* 16:144–158.

Ridley, M. 1999. *Genome: The Autobiography of a Species in 23 Chapters*. HarperCollins, New York, NY.

Ripple, W. J. and E.J. Larsen. 2000. Historic aspen recruitment, elk and wolves in northern Yellowstone National Park, USA. *Biological Conservation* 95:361–370.

Romme, W. H. and D. G. Despain. 1989. Historical perspective on the Yellowstone fires of 1988. *BioScience* 39:695–699.

Scheffer, M. and S.R. Carpenter. 2003. Catastrophic regime shifts in ecosystems: Linking theory to observation. *Trends in Ecology and Evolution* 18:648–656.

Smith, T.M. and R.L. Smith. 2009. *Elements of Ecology* (7th ed.). Benjamin Cummings, San Francisco, CA.

Stiling, P. 2002. *Ecology: Theories and Applications*. Prentice-Hall, Upper Saddle River, NJ.

Terborgh, J. and J. Estes. 2010. *Trophic Cascades: Predators, Prey and the Changing Dynamics of Nature*. Island Press, Washington, D.C.

Turner, M.G. 2005. Landscape ecology: What is the state of the science? *Annual Review of Ecology, Evolution and Systematics* 36:319–344.

Wilson, E.O. 1999. *The Diversity of Life*. W.W. Norton, New York, NY.

Chapter 7

Baskin, Y. 1997. *The Work of Nature: How Biodiversity Sustains Us*. Island Press, Washington, D.C.

Bormann, F.H. and G.E. Likens. 1979. *Pattern and Process in a Forested Ecosystem*. Springer-Verlag, New York, NY.

Chapin III, F.S., P.A. Matson, and H.A. Mooney. 2002. *Principles of Terrestrial Ecosystem Ecology*. Springer-Verlag, New York, NY.

Das, A. 1991. Human influence on river nitrogen. *Nature* 350:386–387.

Deacon, J.R., T.E. Smith, C.M. Johnston, et al. 2006. *Assessment of Total Nitrogen in the Upper Connecticut River Basin in New Hampshire, Vermont, and Massachusetts, December 2002–September 2005*. U.S. Geological Survey Scientific Investigations Report 2006–5144.

Galloway, J.N., R.W. Howarth, A.F. Michaels, et al. 1996. Nitrogen and phosphorus budgets of the North Atlantic Ocean and its watershed. *Biogeochemistry* 35:3–25.

Galloway, J.N., W.H. Schlesinger, H. Levy II, et al. 1995. Nitrogen fixation: Atmospheric enhancement-environmental response. *Global Biogeochemical Cycles* 9:235–252.

Golley, F.B. 1993. *A History of the Ecosystem Concept in Ecology: More Than the Sum of the Parts*. Yale University Press, New Haven, CT.

Howarth, R.W., G. Billen, D. Swaney, et al. 1996. Regional nitrogen budgets and riverine N & P fluxes for the drainages to the North Atlantic Ocean: Natural and human influences. *Biogeochemistry* 35:75–139.

Imhoff, M.L., L. Bounoua, T. Ricketts, et al. 2004. Global patterns in human consumption of net primary production. *Nature* 429:870–873.

Jackson, R.B., S.R. Carpenter, C.N. Dahm, et al. 2001. Water in a changing world. *Issues in Ecology* 9:1–16

Likens, G.E., F.H. Bormann, R.S. Pierce, et al. 1977. *Biogeochemistry of a Forested Ecosystem*. Springer-Verlag, New York, NY.

Long Island Sound Study. 2010. *Protection and Progress: 2009–2010*. Long Island Sound Study Biennial Report. http://longislandsoundstudy.net/wp-content/uploads/2011/08/GalleyFIN_lo_singles.pdf. Last accessed 10-1-2011.

Millennium Ecosystem Assessment. 2005. *Ecosystems and Human Well-Being*. Island Press, Washington, D.C. http://www.maweb.org/documents/document.356.aspx.pdf. Last accessed 10-1-2011.

Mullaney, J.R. 2004. Summary of water quality trends in the Connecticut River, 1968–1998. *American Fisheries Society Monograph* 9:273–286.

Pan, Y., R.A. Birdsey, and J. Fang. 2011. A large and persistent carbon sink in the world's forests. *Science* 333:988–993.

Postel, S.L., G.C. Daily, and P.R. Ehrlich. 1996. Human appropriation of renewable fresh water. *Science* 271:785–788.

Robinson, K.W., J.P. Campbell, and N.A. Jaworski. 2003. *Water-Quality Trends in New England Rivers During the 20th Century*. U.S. Geological Survey Water-Resources Investigations Report 03-4012.

Schlesinger, W.H. 1997. *Biogeochemistry: An Analysis of Global Change* (2nd ed.). Academic Press, Orlando, FL.

The Heinz Center. 2008. *The State of the Nation's Ecosystems 2008*. Island Press, Washington, D.C.

Vitousek, P.M., 2004. *Nutrient Cycling and Limitation: Hawai'i as a Model System*. Princeton University Press, Princeton, NJ.

Vitousek, P.M., J. Aber, R.W. Howarth, et al. 1997. Human alteration of the global nitrogen cycle: Causes and consequences. *Issues in Ecology*, 1:1–16.

Vitousek, P.M., H.A. Mooney, J. Lubchenco, et al. 1997. Human domination of Earth's ecosystems. *Science* 277:494–499.

Chapter 8

Behringer, W. 2010. *A Cultural History of Climate*. Polity Press, Malden, MA.

Boxall, A.B.A., A. Hardy, S. Beulke, et al. 2009. Impacts of climate change on indirect human exposure to pathogens and chemicals from agriculture. *Environmental Health Perspectives* 117:508–515.

Broecker, W.S. and G.M. Henderson. 1998. The sequence of events surrounding Termination II and their implications for the cause of glacial-interglacial CO_2 changes. *Paleoceanography* 13:352–364.

Brown, L.R. 2011. *World on the Edge: How to Prevent Environmental and Economic Collapse*. W.W. Norton, New York, NY.

Burke, M.B., D.B. Lobell, and L. Guarino. 2009. Shifts in African crop climates by 2050, and the implications for crop improvement and genetic resources conservation. *Global Environmental Change* 19:317–325.

Caldeira, K. and M.E. Wickett. 2003. Anthropogenic carbon and ocean pH. *Nature* 425:465.

Clausen, E. 2001. *Climate Change: Science, Strategies and Solutions*. The Pew Center for Global Climate Change, Arlington, VA.

Epstein, P.R. 2002. Climate change and infectious disease: Stormy weather ahead? *Epidemiology* 13:373–375.

Fagan, B. 2000. *The Little Ice Age: How Climate Made History (1300-1850)*. Basic Books, New York, NY.

Flannery, T. 2007. *The Weather Makers: How Man Is Changing the Climate and What It Means for Life on Earth*. Text Publishing, Melbourne, Australia.

Gore, A. 2006. *An Inconvenient Truth: The Planetary Emergency of Global Warming and What We Can Do About It*. Rodale, Emmaus, PA.

Hidore, J.J., J.E. Oliver, M. Snow, et al. 2010. *Climatology: An Atmospheric Science* (3rd ed.). Prentice Hall, Upper Saddle River, NJ.

Intergovernmental Panel on Climate Change (IPCC). 2007. *Climate Change 2007: Synthesis Report. Intergovernmental Panel on Climate Change*. http://www.ipcc.ch/publications_and_data/publications_ipcc_fourth_assessment_report_synthesis_report.htm. Last accessed 10-1-2010.

Karl, T.R., J.M. Melillo, and T.C. Peterson, editors. 2009. *Global Climate Change Impacts in the United States*. U.S. Global Change Research Program. Cambridge University Press, Cambridge, UK.

Kopp, R.J., and W.A. Pizer. 2007. *Assessing U.S. Climate Policy Options. Resources for the Future*, Washington, D.C. http://www.rff.org/cpfreport. Last accessed 10-1-2011.

Krupp, F. and M. Horn. 2009. *Earth: The Sequel*. W.W. Norton, New York, NY.

Loarie, S.R., P.B. Duffy, H. Hamilton, et al. 2009. The velocity of climate change. *Nature* 462:1052–1057.

Lomborg, B. 2001. *The Skeptical Environmentalist. Measuring the Real State of the World*. Cambridge University Press, Cambridge, UK.

Lomborg, B. 2010. *Smart Solutions to Climate Change: Comparing Costs and Benefits*. Cambridge University Press, Cambridge, UK.

Mudelsee, M. 2001. The phase relations among atmospheric CO_2 content, temperature and global ice volume over the past 420 ka. *Quaternary Science Reviews* 20:583–589.

National Geographic. 2008. Changing climate. *National Geographic Special Report*, 22 June 2008.

Orr, J.C., V.J. Fabry, O. Aumont, et al. 2005. Anthropogenic ocean acidification over the twenty-first century and its impact on calcifying organisms. *Nature* 437:681–686.

Pacala, S. and R. Socolow. 2004. Stabilization wedges: Solving the climate problem for the next 50 years with current technologies. *Science* 13:968–972.

Pan, Y., R.A. Birdsey, and J. Fang. 2011. A large and persistent carbon sink in the world's forests. *Science* 333:988–993.

Petit, J.R., J. Jouzel, D. Raynaud, et al. 1999. Climate and atmospheric history of the past 420,000 years from the Vostok ice core, Antarctica. *Nature* 399:429–436.

Pew Center on Global Climate Change. http://www.pewclimate.org/. Last accessed 10-1-2011.

Post, E., M.C. Forchhammer, M.S. Bret-Harte, et al. 2009. Ecological dynamics across the arctic associated with recent climate change. *Science* 325:1355–1358.

Rignot, E. and P. Kanagaratnam. 2006. Changes in the velocity structure of the Greenland Ice Sheet. *Science* 311:986–990.

Schlesinger, W.H. 1997. *Biogeochemistry: An Analysis of Global Change* (2nd ed.). Academic Press, Orlando, FL.

Schneider, S.H. and T.L. Root, editors. 2002. *Wildlife Responses to Climate Change: North American Case Studies*. Island Press, Washington, D.C.

Soverow, J.E., G.A. Wellenius, D.N. Fisman, et al. 2009. Infectious disease in a warming world: How weather influenced West Nile virus in the United States (2001–2005). *Environmental Health Perspectives* 117:1049–1052.

Speth, J.G. 2004. *Red Sky at Morning: America and the Crisis of the Global Environment*. Yale University Press, New Haven, CT.

United Nations. *Kyoto Protocol*. http://unfccc.int/kyoto_protocol/items/2830.php. Last accessed 10-1-2011.

United Nations. *U.N. Framework Convention on Climate Change*. http://unfccc.int/resource/docs/convkp/conveng.pdf. Last accessed 10-1-2011.

Vitousek, P.M., H.A. Mooney, J. Lubchenco, et al. 1997. Human domination of Earth's ecosystems. *Science* 277:494–499.

World Watch Institute. 2009. *State of the World: Into a Warming World*. W.W. Norton, New York, NY.

Chapter 9

Akimoto, H. 2003. Global air quality and pollution. *Science* 302:1716–1719.

Brooks, B.O. and W.F. Davis. 1992. *Understanding Indoor Air Quality*. CRC Press, Boca Raton, FL.

Godish, T. 2004. *Air Quality* (4th ed.). Lewis Publishers, Boca Raton, FL.

Jacobson, M.Z. 2002. *Atmospheric Pollution: History, Science, and Regulation*. Cambridge University Press, Cambridge, UK.

Metcalfe, S. and D. Derwent. 2005. *Atmospheric Pollution and Environmental Change*. Hodder Arnold, London, UK.

Molina, M.J. and F.S. Rowland. 1974. Stratospheric sink for chlorofluoromethanes: Chlorine atom catalyzed destruction of ozone. *Nature* 249:810–812.

National Atmospheric Deposition Program. *National Trends Network*. http://nadp.sws.uiuc.edu/. Last accessed 10-1-2011.

Parson, E.A. 2003. *Protecting the Ozone Layer: Science and Strategy*. Oxford University Press, Oxford, UK.

Schuster, P.F., D.P. Krabbenhoft, D.L. Naftz, et al. 2002. Atmospheric mercury deposition during the last 270 years: A glacial ice core record of natural and anthropogenic sources. *Environmental Science and Technology* 36:2303–2310.

Shotyk, W., D. Weiss, P.G. Appleby, et al. 1998. History of atmospheric lead deposition since 12,370 14C yr BP from a peat bog, Jura Mountains, Switzerland. *Science* 281:1635–1640.

Snyder, L.P. 1994. The death-dealing smog over Donora, Pennsylvania: Industrial air pollution, public health policy, and the politics of expertise, 1948–1949. *Environmental History Review* 18:117–139.

Sommerville, R.C.J. 1996. *The Forgiving Air: Understanding Environmental Change*. University of California Press, Berkeley, CA.

U.S. Environmental Protection Agency. *National Summary of Sulfur Dioxide Emissions*. http://www.epa.gov/air/emissions/so2.htm. Last accessed 10-1-2011.

U.S. Environmental Protection Agency. *NOx Budget Trading Program*. http://www.epa.gov/airmarkt/progsregs/nox/docs/NBPbasicinfo.pdf. Last accessed 10-1-2011.

World Health Organization. *Indoor Air and Household Energy*. http://www.who.int/heli/risks/indoorair/indoorair/en/index.html. Last accessed 10-1-2011.

Chapter 10

Archbold, O.W. 1995. *Ecology of World Vegetation*. Chapman-Hall, New York, NY.

Barbour, M.G. and W.D. Billings. 2000. *North American Terrestrial Vegetation*. Cambridge University Press, Cambridge, U.K.

Breckle, S-W. and G. Lawlor. 2002. *Walter's Vegetation of the Earth*. Springer-Verlag, New York, NY.

Chabot, B.F. and H.A. Mooney. 1985. *The Physiological Ecology of North American Plant Communities*. Chapman Hall, New York, NY.

Chapin III, F.S., M.W. Oswood, K. Van Cleve, et al. 2006. *Alaska's Changing Boreal Forest*. Oxford University Press, New York, NY.

Christensen, N.L. 2005. Fire in the parks: A case study for change management. *The George Wright Forum* 22:12–31.

Gibson, D.J. 2008. *Grasses and Grassland Ecology*. Oxford University Press, Oxford, UK.

Hari, P. and L. Kulmala. 2008. *Boreal Forest and Climate Change*. Springer-Verlag, Berlin, Germany.

Kellman, M.C. 1980. *Plant Geography*. Methuen & Co., London, UK.

Knight, D.H. 1994. *Mountains and Plains: The Ecology of Wyoming Landscapes*. Yale University Press, New Haven, CT.

Montagnini, F. and C.F. Jordan. 2005. *Tropical Forest Ecology: The Basis for Conservation and Management*. Springer-Verlag, Berlin, Germany.

Quinn, R.D., S.C. Keeley, and M.D. Wallace. 2006. *Introduction to California Chaparral*. University of California Press, Berkeley, CA.

Tscharntke, T., C. Leuschner, M. Zeller, et al. 2007. *Stability of Tropical Rainforest Margins: Linking Ecological, Economic and Social Constraints of Land Use and Conservation*. Springer-Verlag, Berlin, Germany.

Von Humboldt, A. and A. Bonpland. 1807. *Essay on the Geography of Plants*. Translation by S. Romanowski. Introduction by S.T. Jackson. Reprinted by University of Chicago Press, Chicago, IL.

Whitaker, R.H. 1975. *Communities and Ecosystems*. MacMillan, New York, NY.

Chapter 11

Balmford, A., A. Bruner, P. Cooper, et al. 2002. Economic reasons for conserving wild nature. *Science* 297:950–953.

Butman, C.A., J.T. Carlton, G.W. Boehlert, et al. 1995. *Understanding Marine Biodiversity*. National Academy Press, Washington, D.C.

Chapin III, S.F., E-D. Schulze, and H.A. Mooney. 1992. Biodiversity and ecosystem processes. *Trends in Ecology and Evolution* 7:107–108.

Daily, G.C. 1997. *Nature's Services: Societal Dependence on Natural Ecosystems*. Island Press, Washington, D.C.

Damschen, E.I., N.M. Haddad, J.L. Orrock, et al. 2006. Corridors increase plant species richness at large scales. *Science* 313:1284–1286.

Ehrlich, P.R. and A.H. Ehrlich. 1992. *Extinction: The Causes and Consequences of the Disappearance Species*. Random House, New York, NY.

Estes, J.A., J. Terborgh, J.S. Brashares, et al. 2011. Trophic downgrading of planet Earth. *Science* 333:301–306.

Groom, M.J., G.K. Meffe, and C.R. Carroll. 2006. *Principles of Conservation Biology* (3rd ed.). Sinauer Associates, Sunderland, MA.

Hoekstra, J.M., J.L. Molnar, M. Jennings, et al. 2010. *The Atlas of Global Conservation*. University of California Press, Berkeley, CA.

Hunter, Jr., M.L. and J. Gibbs. 2007. *Fundamentals of Conservation Biology*. Blackwell Publishing, Malden, MA.

Kinzig, A.P., S.W. Pacala, and D. Tilman. 2001. *The Functional Consequences of Biodiversity: Empirical Progress and Theoretical Extensions*. Monographs in Population Biology 33. Princeton University Press, Princeton, NJ.

Lovejoy, T.E. and L. Hannah. 2006. *Climate Change and Biodiversity*. Yale University Press, New Haven, CT.

MacArthur, R.H. and E.O. Wilson. 1967. *The Theory of Island Biogeography*. Monographs in Population Biology 1. Princeton University Press, Princeton, NJ.

Millennium Ecosystem Assessment. 2005. *Ecosystems and Human Well-Being*. Island Press, Washington, D.C. http://www.maweb.org/documents/document.356.aspx.pdf. Last accessed 10-1-2011.

Naeem, S., F.S. Chapin III, R. Costanza, et al. 1999. Biodiversity and ecosystem functioning: Maintaining natural life support processes. *Issues in Ecology* 4:1–11.

Naeem, S. and S. Li. 1997. Biodiversity enhances ecosystem reliability. *Nature* 390:507–509.

Norse, E.A., L.B. Crowder, and M.E. Soule. 2005. *Marine Conservation Biology: The Science of Maintaining the Sea's Biodiversity*. Island Press, Washington, D.C.

O'Riordan, T. and S. Stoll-Kleemann. 2002. *Biodiversity, Sustainability and Human Communities: Protecting Beyond the Protected*. Cambridge University Press, Cambridge, UK.

Pereira, H.M., P.W. Leadley, V. Proença, et al. 2010. Scenarios for global biodiversity in the 21st century. *Science* 330:1496–1501.

Pimm, S.L. and P. Raven. 2000. Extinction by numbers. *Nature* 403:843–845.

Pimm, S.L., G.J. Russel, J.L. Gittleman, et al. 1995. The future of biodiversity. *Science* 269:347–350.

Sellars, R.W. 1997. *Preserving Nature in the National Parks: A History*. Yale University Press, New Haven, CT.

Tilman, D. and J.A. Downing. 1994. Biodiversity and stability in grasslands. *Nature* 367:363–365.

Wilcove, D.S., D. Rothstein, J. Dubow, et al. 1998. Quantifying threats to imperiled species in the United States. *BioScience* 48:607–615.

Wilson, E.O. 1999. *The Diversity of Life*. W.W. Norton, New York, NY.

Wilson, E.O. 2002. *The Future of Life*. Alfred A. Knopf, New York, NY.

Chapter 12

Ballantine, K. and R. Schneider. 2009. Fifty-five years of soil development in restored freshwater depressional wetlands. *Ecological Applications* 19:1467–1480.

Bernhardt, E.S. and M.A. Palmer. 2007. Restoring streams in an urbanizing world. *Freshwater Biology* 52:738–751.

Brown, L.R. 2011. *World on the Edge: How to Prevent Environmental and Economic Collapse*. W.W. Norton, New York, NY.

Caldeira, K. and M.E. Wickett. 2003. Anthropogenic carbon and ocean pH. *Nature* 425:465.

Dahl, T.E. 2006. *Status and Trends of Wetlands in the Conterminous United States 1998 to 2004*. U.S. Department of the Interior; Fish and Wildlife Service, Washington, D.C. http://www.fws.gov/wetlands/_documents/gSandT/National Reports/StatusTrendsWetlandsConterminous US1998to2004.pdf. Last accessed 10-1-2011.

Ensign, S.H. and M.W. Doyle. 2006. Nutrient spiraling in streams and river networks. *Journal of Geophysical Research* 111, G04009, doi:10.1029/2005JG000114.

Food and Agriculture Organization (FAO). 2010. *State of World Fisheries and Aquaculture 2010*. Food and Agriculture Organization of the United Nations, Rome Italy. http://www.fao.org/docrep/013/i1820e/i1820e.pdf. Last accessed 10-1-2011.

Gleick, P.H. 2011. *The World's water, Volume 7: The Biennial Report on Freshwater Resources*. Island Press, Washington, D.C.

International Union of Geological Sciences. 2005. *Groundwater: Reservoir for a Thirsty Planet?* http://yearofplanetearth.org/content/downloads/Groundwater.pdf. Last accessed 10-1-2011.

Jackson, R.B., S.R. Carpenter, C.N. Dahm, et al. 2001. Water in a changing world. *Issues in Ecology* 9:

Maddigen, R.D., W.S. Chern, and C.G. Rizy. 1982. The irrigation demand for electricity. *American Journal of Agricultural Economics* 64:673–680.

McLeod, K. and H. Leslie. 2009. *Ecosystem-Based Management for the Oceans*. Island Press, Washington, D.C.

Millennium Ecosystem Assessment. 2005. *Ecosystems and Human Well-Being*. Island Press, Washington, D.C. http://www.maweb.org/documents/document.356.aspx.pdf. Last accessed 10-1-2011.

Mitsch, W.J., J.G. Gosselink, C.J. Anderson, et al. 2009. *Wetland Ecosystems*. John Wiley and Sons, Hoboken, NJ.

Nichols, F.H., J.E. Cloern, S.N. Luoma, et al. 1986. The modification of an estuary. *Science* 231:567–573.

Organization for Economic Cooperation and Development (OECD). 2008. *Environmental Outlook to 2030*. OECD, Paris, France. http://www.oecd.org/document/20/0,3746,en_2649_37465_39676628_1_1_1_37465,00.htm. Last accessed 10-1-2011.

Orr, J.C., V.J. Fabry, O. Aumont, et al. 2005. Anthropogenic ocean acidification over the twenty-first century and its impact on calcifying organisms. *Nature* 437:681–686.

Outwater, A. 1996. *Water: A Natural History*. Basic Books, New York, NY.

Postel, S. 1997. *Last Oasis: Facing Water Scarcity*. W.W. Norton, New York, NY.

Postel, S. 1999. *Pillar of Sand*. W.W. Norton, New York, NY.

Postel, S.L., G.C. Daily, and P.R. Ehrlich. 1996. Human appropriation of renewable fresh water. *Science* 271:785–788.

Reisner, M. 1993. *Cadillac Desert: The American West and Its Disappearing Water* (revised ed.). Penguin Books, New York, NY.

U.N. Environment Programme. 2008. *Africa: Atlas of Our Changing Environment*. United Nations Environment Programme, Nairobi, Kenya.

U.N. Food and Agriculture Organization. *AQUASTAT: FAO's Global Water Database*. http://www.fao.org/nr/water/aquastat/main/index.stm. Last accessed 10-1-2011.

U.S. Environmental Protection Agency. 2002. *Biological Assessments and Criteria: Crucial Components of Water Quality Programs*. http://water.epa.gov/scitech/swguidance/standards/criteria/aqlife/biocriteria/index.cfm. Last accessed 10-1-2011.

U.S. Environmental Protection Agency. 2004. *Primer for Municipal Wastewater Treatment Systems*. EPA 832-R-04-001. Office of Wastewater Management, Washington, D.C.

U.S. Environmental Protection Agency. 2009. *Water on Tap: What You Need to Know*. EPA 816-K-09-001. Office of Water, Washington, D.C.

U.S. Geological Service. 2004. Climatic fluctuations, drought, and flow in the Colorado River Basin. U.S. Geological Survey Fact Sheet 2004–3062. USGS, Washington, D.C.

U.S. Geological Service. 2005. Trends in the water budget of the Mississippi Basin. U.S. Geological Survey Fact Sheet 2005–3020. USGS, Washington, D.C.

Van Bers, C., D. Petry, and C. Pahl-Wostl, editors. 2007. *Global Assessments: Bridging Scales and Linking to Policy*. Report on the joint TIAS-GWSP workshop held at the University of Maryland University College, Adelphi, USA GWSP Issues in Global Water System Research, No. 2. GWSP IPO, Bonn, Germany.

Wu, J., J. Huang, X. Han, et al. 2004. The Three Gorges Dam: An Ecological Perspective. *Frontiers in Ecology and Environment* 2:241–248.

Chapter 13

Brown, L.R. 2011. *World on the Edge: How to Prevent Environmental and Economic Collapse*. W.W. Norton, New York, NY.

Collins, A. and R. Fairchild. 2007. Sustainable food consumption at a sub-national level: An ecological footprint, nutritional and economic analysis. *Journal of Environmental Policy and Planning* 9:5–30.

Cox, T.S., J.D. Glover, D.L. Van Tassel, et al. 2006. Prospects for developing perennial grain crops. *BioScience* 56:649–659.

Egan, T. 2006. *The Worst Hard Time: The Untold Story of Those Who Survived the Great American Dust Bowl*. Houghton Mifflin, New York, NY.

Ellis, E.C. and N. Ramankutty. 2008. Putting people in the map: Anthropogenic biomes of the world. *Frontiers in Ecology and Environment* 6:439–447.

Food and Agriculture Organization (FAO). 2006. *Livestock's Long Shadow: Environmental Issues and Options*. Food and Agriculture Organization of the United Nations, Rome, Italy. http://www.fao.org/docrep/010/a0701e/a0701e00.HTM. Last accessed 10-1-2011.

Food and Agriculture Organization (FAO). 2009. *The State of Food Insecurity in the World*. Food and Agriculture Organization of the United Nations, Rome, Italy. http://www.fao.org/publications/sofi/en/. Last accessed 10-1-2011.

Food and Agriculture Organization (FAO). 2010. *State of World Fisheries and Aquaculture 2010*. Food and Agriculture Organization of the United Nations, Rome, Italy. http://www.fao.org/docrep/013/i1820e/i1820e.pdf. Last accessed 10-1-2011.

Glover, J.D., C.M. Cox, and J.P. Reganold. 2007. Future farming: A return to roots? *Scientific American* 297:66–73.

Glover, J.D., J.P. Reganold, L.W. Bell, et al. 2010. Increased food and ecosystem security via perennial grains. *Science* 328:1638–1639.

Imhoff, M.L., L. Bounoua, T. Ricketts, et al. 2004. Global patterns in human consumption of net primary production. *Nature* 429:870–873.

Losey, J.E., L.S. Rayor, and M.E. Carter. 1999. Transgenic pollen harms monarch larvae. *Nature* 399:214.

Mann, C.C. 2002. Transgene data deemed unconvincing. *Science* 296:236–237.

Norris, R.F., E.P. Caswell-Chen, and M. Kogan. 2003. *Concepts in Integrated Pest Management*. Prentice Hall, Upper Saddle River, NJ.

Organic Trade Association. 2011. *The Organic Trade Association's 2011 Organic Industry Survey*. Organic Trade Association, Greenfield, MA.

Paoletti, M.G. and D. Pimentel. 1996. Genetic engineering in agriculture and the environment: Assessing risks and benefits. *BioScience* 46:665–673.

Pollan, M. 2006. *The Omnivore's Dilemma: A Natural History of Four Meals*. The Penguin Press, New York, NY.

Pollan, M. 2008. *In Defense of Food: An Eater's Manifesto*. The Penguin Press, New York, NY.

Shiva, V. 2000. *Stolen Harvest: The Hijacking of the Global Food Supply*. South End Press, Cambridge, MA.

Snow, A.A., D.A. Andow, P. Gepts, et al. 2005. Gentically engineered organisms and the environment: Current status and recommendations. *Ecological Applications* 15:377–404.

U.S. Census Bureau. 2011. *The 2011 Statistical Abstract: Agriculture*. U.S. Census Bureau, Washington, D.C. http://www.census.gov/compendia/statab/cats/agriculture.html. Last accessed 10-1-2011.

Wolfenbarger, L.L.R. 2000. The ecological risks and benefits of genetically engineered plants. *Science* 290:2088.

World Watch Institute. 2011. *State of the World: Innovations That Nourish the Planet*. W.W. Norton, New York, NY.

Chapter 14

Agrawal, A., D. Nepstad, and A. Chhatre. 2011. Reducing emissions from deforestation and forest degradation. *Annual Review of Environment and Resources* 36:373–396.

Bormann, F.H. and G.E. Likens. 1979. *Pattern and Process in a Forested Ecosystem*. Springer-Verlag, New York, NY.

Cashore, B., G. Auld, and D. Newsom. 2004. *Governing Through Markets: Forest Certification and the Emergence of Non-State Authority*. Yale University Press, New Haven, CT.

Chamberlain, J., R. Bush, and A.L. Hammett. 1998. Non-timber forest products: The *other* forest products. *Forest Products Journal* 48:10–19.

Christensen, N.L. 2009. Future forests, future fires. *Yellowstone Science*. 17:40–44.

Covington, W.W. and M.M. Moore. 1994. Southwestern ponderosa pine forest structure: Changes since Euro-American settlement. *Journal of Forestry* 92:39–47.

Diamond, J., 2006. *Collapse: How Societies Choose to Fail or Succeed*. Penguin Books, New York, NY.

Farmer, P. 1994. *The Uses of Haiti*. Common Courage Press, Monroe, ME.

Food and Agriculture Organization (FAO). 2010. *Global Forest Resources Assessment 2010*. Food and Agriculture Organization of the United Nations, Rome, Italy. http://www.fao.org/forestry/fra/fra2010/en/. Last accessed 10-1-2011.

Frelich, L.E. 2002. *Forest Dynamics and Disturbance Regimes: Studies from Temperate Evergreen-Deciduous Forests*. Cambridge University Press, Cambridge, UK.

Harris, L.D. 1984. *The Fragmented Forest: Island Biogeography Theory and the Preservation of Biotic Diversity*. University of Chicago Press, Chicago, IL.

Hunter, Jr., M.L. and F. Schmiegelow. 2010. *Wildlife, Forests and Forestry: Principles of Managing Forests for Biological Diversity* (2nd ed.). Prentice Hall, Upper Saddle River, NJ.

Keeley, J.E., Aplet, G., N.L. Christensen, Jr., et al. 2009. *Ecological Foundations of Fire Management*. U.S. Forest Service General Technical Report. PNW.GTR-779. USDA Forest Service, Washington, D.C.

Kohl, M., T. Baldauf, D. Plugge, et al. 2009. Reduced emissions from deforestation and forest degradation (REDD): A climate change mitigation strategy on a critical track. *Carbon Balance and Management* 4:10. http://www.cbmjournal.com/content/4/1/10. Last accessed 11-14-2011.

Kohm, K.A. and J.F. Franklin. 1997. *Creating a Forestry for the 21st Century: The Science of Ecosystem Management*. Island Press, Washington, D.C.

Laarman, J.G. and R.A. Sedjo. 1992. *Global Forests: Issues for Six Billion People*. McGraw Hill, New York, NY.

McCarthy, H.R., R. Oren, K.H. Johnsen, et al. 2010. Re-assessment of plant carbon dynamics at the Duke free-air CO_2 enrichment site: Interactions of atmospheric $[CO_2]$ with nitrogen and water availability over stand development. *New Phytologist* 185:514–528.

Medley, K.E., C.M. Pobocik, and B.W. Okey. 2003. Historical changes in forest cover and land ownership in a Midwestern U.S. landscape.

Annals of the Association of American Geographers 93:104–120.

Pan, Y., R.A. Birdsey, and J. Fang. 2011. A large and persistent carbon sink in the world's forests. *Science* 333:988–993.

Perry, D.A., R. Oren, and S.C. Hart. 2008. *Forest Ecosystems* (2nd ed.). Johns Hopkins Press, Baltimore, MD.

Romme, W.H. and D.G. Despain. 1989. Historical perspective on the Yellowstone fires of 1988. *BioScience* 39:695–699.

Sayer, J., J. McNeely, S. Maginnis, et al. 2007. *Local Rights and Tenure for Forests: Opportunity or Threat for Conservation*. Rights and Resources Initiative, Washington, D.C.

Smith, W.H. 1990. The health of North American forests: Stress and risk assessment. *Journal of Forestry* 88:32–35.

U.S. Department of Agriculture, Forest Service. 2002. *U.S. Forest Facts and Historical Trends*. USDA, Forest Service, Washington, D.C. http://fia.fs.fed.us/library/briefings-summaries-overviews/docs/2002_ForestStats_%20FS801.pdf. Last accessed 10-1-2011.

U.S. Department of Agriculture, Forest Service. 2008. *Forest Resources of the United States*. USDA, Forest Service, Washington, D.C. http://nrs.fs.fed.us/pubs/7334. Last accessed 10-1-2011.

Westerling, A.L., H.G. Hidalgo, D.R. Crayan, et al. 2006. Warming and earlier spring increase western wildfire activity. *Science* 313:940–943.

Westoby, J. 1989. *Introduction to World Forestry*. Basil Blackwell, Cambridge, UK.

Williams, M. 1989. *Americans and Their Forests: A Historical Geography*. Cambridge University Press, Cambridge, UK.

Chapter 15

British Petroleum. 2010. *BP Statistical Review of World Energy, June 2010*. British Petroleum, London, UK. http://www.bp.com/sectionbodycopy.do?categoryId=7500&contentId=7068481. Last accessed 10-1-2011.

Christie, B. 1999. Kern River field at 100: The city that oil built. *The Bakersfield Californian*, April 27, 1999.

Cleveland C.J., editor. 2009. Smart grid. In: *Encyclopedia of Earth*. Environmental Information Coalition, National Council for Science and the Environment, Washington, D.C. http://www.eoearth.org/article/Smartgrid. Last accessed 10-1-2011.

Deffeyes, K.S. 2001. *Hubbert's Peak: The Impending World Oil Shortage*. Princeton University Press, Princeton, NJ.

Energy Information Administration (EIA). 2011. *Annual Energy Review 2010*. U.S. Department of Energy, Energy Information Administration, Washington, D.C. http://205.254.135.24/totalenergy/data/annual/index.cfm. Last accessed 10-1-2011.

International Atomic Energy Agency (IAEA). 2009. *World Distribution of Uranium Deposits (UDEPO) with Uranium Deposit Classification*. International Atomic Energy Agency, Vienna, Austria. http://www-pub.iaea.org/MTCD/publications/PDF/TE_1629_web.pdf. Last accessed 10-1-2011.

International Energy Agency (IEA). 2008. *Energy Efficiency Indicators for Public Electricity Production from Fossil Fuels*. International Energy Agency, Paris, France. http://www.iea.org/papers/2008/En_Efficiency_Indicators.pdf. Last accessed 10-1-2011.

International Energy Agency (IEA). 2011. *World Energy Outlook 2011*. International Energy Agency, Paris, France. http://www.iea.org/ebc/feb2011/birol.pdf. Last accessed 10-2-2011.

Kerr, R.A. 2010. Natural gas from shale burst onto the scene. *Science* 328:1624–1626.

Krupp, F. and M. Horn. 2009. *Earth: The Sequel*. W.W. Norton, New York, NY.

Longwell, H.J. 2002. The future of the oil and gas industry: Past approaches, new challenges. *World Energy* 5:100–104

Lovins, A.B. 2004. Energy efficiency, taxonomic overview. *Encyclopedia of Energy* 2:382–401.

Nature. 2006. Special report: Chernobyl and the future. *Nature* 440:982–989.

Osborn, S.C., A. Vengosh, N.R. Warner, et al. 2011. Methane contamination of drinking water accompanying gas-well drilling and hydraulic fracturing. *Proceedings of the National Academy of Sciences* 108:8172–8176,

Pacala, S. and R. Socolow. 2004. Stabilization wedges: Solving the climate problem for the next 50 years with current technologies. *Science* 13:968–972.

Perez, C., A. Velando, I. Munilla, et al. 2008. Monitoring polycyclic aromatic hydrocarbon pollution in the marine environment after the Prestige Oil Spill by means of seabird blood analysis. *Environmental Science and Technology* 42:707–713.

Ristinen, R.A. and J.J. Kraushaar. 2005. *Energy and the Environment* (2nd ed.). John Wiley & Sons, Hoboken, NJ.

Scientific Committee on Emerging and Newly Identified Health Risks. 2007. *Possible Effects of Electromagnetic Fields (EMF) on Human Health*. European Commission, Health and Consumer Protection, Brussels, Belgium. http://ec.europa.eu/health/ph_risk/committees/04_scenihr/docs/scenihr_o_007.pdf. Last accessed 10-1-2011.

U.S. EPA. 2007. *Energy Trends in Selected Manufacturing Sectors: Opportunities and Challenges for Environmentally Preferable Energy Outcomes*. U.S. Environmental Protection Agency, Washington, D.C. http://www.epa.gov/sectors/pdf/energy/report.pdf. Last accessed 10-1-2011.

State of California. 2008. *2007 Preliminary Report of California Oil and Gas Production Statistics*. State of California Department of Conservation, Sacramento, CA.

Chapter 16

American Wind Energy Association (AWEA). 2011. *Annual Wind Industry Report*. American Wind Energy Association, Washington, D.C. http://www.awea.org/learnabout/publications/reports/AWEA-US-Wind-Industry-Market-Reports.cfm. Last accessed 10-1-2011.

BioFuels Journal. 2011. Annual and monthly U.S. ethanol production. http://www.biofuelsjournal.com/info/bf_articles.html?type=ec&ID=25474. Last accessed 10-1-2011.

Bradford, T. 2006. *Solar Revolution: The Economic Transformation of the Global Energy Industry*. MIT Press, Cambridge, MA.

British Petroleum. 2010. *BP Statistical Review of World Energy, June 2010*. BP, London, U.K. http://www.bp.com/sectionbodycopy.do?categoryId=7500&contentId=7068481. Last accessed 10-1-2011.

Chow, J. 2003. Energy resources and global development. *Science* 302:1528–1531.

Energy Information Administration (EIA). 2011. *Annual Energy Review 2010*. U.S. Department of Energy, Energy Information Administration, Washington, D.C. http://205.254.135.24/totalenergy/data/annual/index.cfm. Last accessed 10-1-2011.

Flavin, C. 2008. *Low-Carbon Energy: A Roadmap*. Worldwatch Report 178. Worldwatch Institute, Washington, D.C.

International Energy Agency (EIA). 2011. *World Energy Outlook 2011*. International Energy Agency, Paris, France. http://www.iea.org/ebc/feb2011/birol.pdf. Last accessed 10-2-2011.

Krupp, F. and M. Horn. 2009. *Earth: The Sequel*. W.W. Norton, New York, NY.

Pacala, S. and R. Socolow. 2004. Stabilization wedges: Solving the climate problem for the next 50 years with current technologies. *Science* 13:968–972.

Pearce, F. 2006. Fuels gold: Are biofuels really the greenhouse-busting answer to our energy woes? *New Scientist* 2570:36–41.

Perlack, R.D. 2005. *Biomass as Feedstock for a Bioenergy and Bioproducts Industry: The Technical Feasibility of a Billion Ton Annual Supply*. U.S. Department of Energy, Office of Scientific and Technical Information, Washington, D.C. http://www.osti.gov/energycitations/product.biblio.jsp?osti_id=885984. Last accessed 10-1-2011.

REN21. 2011. *Renewables Global Status Report*. Renewable Energy Policy Network for the 21st Century, Milan, Italy. http://www.ren21.net/REN21Activities/Publications/GlobalStatusReport/tabid/5434/Default.aspx. Last accessed 10-1-2011.

Ristinen, R.A. and J.J. Kraushaar. 2005. *Energy and the Environment* (2nd ed.). John Wiley and Sons, Hoboken, NJ.

Stoft, S. 1995. The economics of conserved-energy "supply" curves. University of California Energy Institute Working Paper PWP-028.

U.S. Department of Energy. 2010. *2009 Renewable Energy Data Book*. U.S. Department of Energy, Washington, D.C. http://www1.eere.energy.gov/maps_data/pdfs/eere_databook.pdf. Last accessed 10-1-2011.

Worldwatch Institute. 2007. *Biofuels for Transport: Global Potential and Implications for Sustainable Agriculture and Energy in the 21st Century*. Worldwatch Institute, Washington, D.C.

Chapter 17

Abbott, C. 2001. *Greater Portland: Urban Life and Landscape in the Pacific North West*. University of Pennsylvania Press, Philadelphia, PA.

Andrews, C.J. 2008. Greenhouse gas emissions along the rural-urban gradient. *Journal of Environmental Planning and Management* 51:1–20.

Beatley, T. and K. Manning. 1997. *The Ecology of Place: Planning for Environment, Economy and Community.* Island Press, Washington, D.C.

Benedict, M.A. and E.T. McMahon. 2006. *Green Infrastructure: Linking Landscapes and Communities.* Island Press, Washington, D.C.

Berkowitz, A.R., C.H. Nilon, and K.S. Hollweg. 2003. *Understanding Urban Ecosystems: A New Frontier for Science and Education.* Springer-Verlag, New York, NY.

Bradley, G. A. 1995. *Urban Forest Landscape: Integrating Multidisciplinary Perspectives.* University of Washington Press, Seattle.

Davis, B.N.K. 1979. The ground arthropods of London gardens. *London Naturalist* 58:15–24.

Diamond, H.L. and P.F. Noonan. 1996. *Land Use in America.* Island Press, Washington, D.C.

Duerksen, C. and C. Snyder. 2005. *Nature-Friendly Communities: Habitat Protection and Land Use Planning.* Island Press, Washington, D.C.

Faeth, S.H., P.S. Warren, E. Shochat, et al. 2005. Trophic dynamics in urban communities. *BioScience* 55:399–407.

Glaeser, E. 2011. Cities, productivity, and quality of life. *Science* 333:592–594.

Grey, G.W. 1996. *The Urban Forest: Comprehensive Management.* John Wiley & Sons, New York, NY.

Grimm, N.B., S.H. Faeth, N.E. Golubiewski, et al. 2008. Global change and the ecology of cities. *Science* 319:756–760.

Grimm, N.B., D. Foster, P. Groffman, et al. 2008. The changing landscape: Ecosystem responses to urbanization and pollution across climatic and societal gradients. *Frontiers in Ecology and the Environment* 6:264–272.

Grimm, N.B. and C.L. Redman. 2004. Approaches to the study of urban ecosystems: The case of central Arizona—Phoenix. *Urban Ecosystems* 7:199–213.

Hobbs, F. and N. Stoops. 2002. *Demographic Trends in the 20th Century.* U.S. Census Bureau Special Report CENSR-4.

Hopkins, R. 2008. *The Transition Handbook: From Oil Dependency to Local Resilience.* Chelsea Green Publishing, White River Junction, VT.

Kalnay, E. and M. Cai. 2003. Impact of urbanization and land-use change on climate. *Nature* 423:528–531.

Kareiva P., S. Watts, R. McDonald, et al. 2007. Domesticated nature: Shaping landscapes and ecosystems for human welfare. *Science* 316:1866–1869.

Kaye J.P., P.M. Groffman, N.B. Grimm, et al. 2006. A distinct urban biogeochemistry? *Trends in Ecology and Evolution* 21:192–199.

Mackun, P. and S. Wilson. 2011. *Population Distribution and Change 2000–2010.* U.S. Census Bureau 2010 Census Brief C2010BR-01.

Marzluff, J.M., E. Shulenberger, W. Endlicher, et al. 2008. *Urban Ecology: An International Perspective on the Interaction Between Humans and Nature.* Springer-Verlag, New York, NY.

McDonnell, M.J. and S.T.A. Pickett. 1990. Ecosystem structure and function along urban-rural gradients: An unexploited opportunity for ecology. *Ecology* 71:1232–1237.

McDonnell, M.J. and S.T.A. Pickett. 1993. *Humans as Components of Ecosystems: The Ecology of Subtle Human Effects and Populated Areas.* Springer-Verlag, New York, NY.

Newman, P. and I. Jennings. 2008. *Cities as Sustainable Ecosystems: Principles and Practices.* Island Press, Washington, D.C.

Pickett, S.T.A., M.L. Cadenasso, J.M. Grove, et al. 2001. Urban ecological systems: Linking terrestrial ecological, physical and socioeconomic components of metropolitan areas. *Annual Review of Ecology and Systematics* 32:127–157.

Price, J. 2008. Remaking American environmentalism: On the banks of the L.A. River. *Environmental History* 13:536–555.

Rosenzweig, M.L. 2003. *Win-Win Ecology: How Earth's Species Can Survive in the Midst of Human Enterprise.* Oxford University Press, Oxford, U.K.

Rosenzweig, R. and E. Blackmar. 1992. *The Park and the People: A History of Central Park.* Cornell University Press, Ithaca, NY.

United Nations. *United Nations World Urbanization Prospects: The 2009 Revision.* http://esa.un.org/unpd/wup/index.htm. Last accessed 10-1-2011.

Wear, D.N., M.G. Turner, and R.J. Naiman. 1998. Land cover along an urban-rural gradient: Implications for water quality. *Ecological Applications* 8:619–630.

World Watch Institute. 2007. *State of the World: Our Urban Future.* W.W. Norton, New York, NY.

Chapter 18

Arsova, L. 2008. The state of garbage in America. *BioCycle* 49:22.

Blackman Jr., W.C. 1993. *Basic Hazardous Waste Management.* Lewis Publishers, Boca Raton, FL.

Graedel, T.E. and B.R. Allenby. *Industrial Ecology* (2nd ed.). Prentice Hall, Upper Saddle River, NJ.

Japan International Cooperation Agency. 2005. *Supporting Capacity Development in Solid Waste Management in Developing Countries—Towards Improving Solid Waste Management Capacity of Entire Societies.* Japan International Cooperation Agency, Tokyo, Japan. http://www.jica.go.jp/english/publications/reports/study/topical/waste/. Last accessed 10-1-2011.

Lilienfeld, R. and W. Rathje. 1998. *Use Less Stuff: Environmental Solutions for Who We Really Are.* Ballantine, New York, NY.

McDonough, W. and M. Braungart. 2002. *Cradle to Cradle: Remaking the Way We Make Things.* North Point Press, New York, NY.

National Solid Wastes Management Association (NSWMA). 2004. *Modern Landfills: A Far Cry from the Past.* National Solid Wastes Management Association, Washington, D.C.

Rathje, W.L., W.W. Hughes, D.C. Wilson, et al. 1992. The archaeology of contemporary landfills. *American Antiquity* 57:437–447.

Rathje, W. and C. Murphy. 2001. *Rubbish! The Archeology of Garbage.* University of Arizona Press, Tucson, AZ.

U.S. Environmental Protection Agency. 2000. *EPA Guidelines for Management of Onsite/Decentralized Wastewater Systems.* U.S. Environmental Protection Fact Sheet EPA-832-F-00-012. U.S. EPA, Washington, D.C.

U.S. Environmental Protection Agency. 2004. *Primer for Municipal Wastewater Treatment Systems.* EPA 832-R-04-001. Office of Wastewater Management, Washington, D.C.

U.S. Environmental Protection Agency. 2008. *Management of Electronic Waste in the United States. U.S. Environmental Protection Fact Sheet.* http://www.epa.gov/osw/conserve/materials/ecycling/docs/fact7-08.pdf. Last accessed 10-1-2011.

U.S. Environmental Protection Agency. 2008. *Municipal Solid Waste Generation, Recycling, and Disposal in the United States: Facts and Figures for 2009.* http://www.epa.gov/osw/nonhaz/municipal/pubs/msw2009rpt.pdf. Last accessed 10-1-2011.

U.S. Environmental Protection Agency. 2008. *The National Biennial RCRA Hazardous Waste Report.* http://www.epa.gov/osw/inforesources/data/br07/list07.pdf. Last accessed 10-1-2011.

Saling, J.H. and A.W. Fentiman. 2002. *Radioactive Waste Management* (2nd ed.). Taylor and Francis, New York, NY.

Chapter 19

Akinbami, L.J., J.E. Moorman, P.L. Garbe, et al. 2009. Status of Childhood Asthma in the United States, 1980–2007. *Pediatrics* 123:S131–145.

Ames, B.N., M. Profet, and L.W. Gold. 1990. Nature's chemicals and synthetic chemicals: Comparative toxicology. *Proceedings of the National Academy of Sciences* 87:7782–7786.

Boelee, E. and H. Madsen. 2006. Irrigation and schistosomiasis in Africa: Ecological aspects. International Water Management Institute. *IWMI Research Report 99,* Colombo, Sri Lanka.

Boxall, A.B.A., A. Hardy, S. Beulke, et al. 2009. Impacts of climate change on indirect human exposure to pathogens and chemicals from agriculture. *Environmental Health Perspectives* 117:508–515.

Carlson, R. 1962. *Silent Spring.* Houghton Mifflin, Boston, MA.

Epstein, P.R. 2002. Climate change and infectious disease: Stormy weather ahead? *Epidemiology* 13:373–375.

Golanky, I.M. 2007. *Death and Death Rates due to Extreme Weather Events.* International Policy Network, London, UK.

Graham, J.D. and J.B. Wiener. 1995. *Risk vs. Risk: Tradeoffs in Protecting Health and the Environment.* Harvard University Press, Cambridge, MA.

Hamed, M.A. 2010. Strategic control of schistosome intermediate host. *Asian Journal of Epidemiology* 3:123–140.

Intergovernmental Panel on Climate Change (IPCC). 2007. *Climate Change 2007: Synthesis Report*. Intergovernmental Panel on Climate Change. http://www.ipcc.ch/publications_and_data/publications_ipcc_fourth_assessment_report_synthesis_report.htm. Last accessed 10-1-2011.

Millennium Ecosystem Assessment. 2005. *Ecosystems and Human Well-Being*. Island Press, Washington, D.C. http://www.maweb.org/documents/document.356.aspx.pdf. Last accessed 10-1-2011.

Moeller, D. 2004. *Environmental Health* (3rd ed.). Harvard University Press, Cambridge MA.

National Institutes of Health (NIH). 2008. *National Toxicology Program Monograph on the Potential Human Reproductive and Developmental Effects of Bisphenol A*. National Institutes of Health Publication 08-5994. Bethesda, MD.

Patz, J.A., H.K. Gibbs, J.A. Foley, et al. 2007. Climate change and global health: Quantifying a growing ethical crisis. *EcoHealth* 4:397–405.

Pimentel, D., M. Tort, L. D'Anna, et al. 1998. Ecology of increasing disease. *BioScience* 48:817–826.

Roberts, D.R., S. Manguin, and J. Mouchet. 2000. DDT house spraying and re-emerging malaria. *Lancet* 356:330–332.

Rosenzweig, M.L. 2003. *Win-Win Ecology: How Earth's Species Can Survive in the Midst of Human Enterprise*. Oxford University Press, Oxford, U.K.

Schistosomiasis Control Initiative. http://www3.imperial.ac.uk/schisto. Last accessed 10-1-2011.

Soverow, J.E., G.A. Wellenius, D.N. Fisman, et al. 2009. Infectious disease in a warming world: How weather influenced West Nile virus in the United States (2001–2005). *Environmental Health Perspectives* 117:1049–1052.

Stern, P.C. and H.V. Fineberg. 1996. *Understanding Risk: Informing Decisions in a Democratic Society*. National Academy Press, Washington, D.C.

Stockholm Convention on Persistent Organic Pollutants. http://chm.pops.int/default.aspx. Last accessed 10-1-2011.

Sunstein, C.R. 2002. *Risk and Reason: Safety, Law and the Environment*. Cambridge University Press, Cambridge, UK.

Swan, S.H., E.P. Elkin, and L. Fenster. 2000. The question of declining sperm density revisited: An analysis of 101 studies published 1934–1996. *Environmental Health Perspectives* 108:961–966.

United Nations. 2009. *World Urbanization Prospects, the 2009 Revision*. http://esa.un.org/unpd/wup/index.htm. Last accessed 10-1-2011.

United Nations. 2010. *World Population Prospects, the 2010 Revision*. http://esa.un.org/wpp/Other-Information/faq.htm. Last accessed 10-1-2011.

Utzinger, J., Y. Tozan, F. Doumani, et al. 2002. The economic payoffs of integrated malaria control in the Zambian copperbelt between 1930 and 1950. *Tropical Medicine and International Health* 7:657–677.

Utzinger, J., Y. Tozan, and B.H. Singer. 2001. Efficacy and cost effectiveness of environmental management for malaria control. *Tropical Medicine and International Health* 6:677–687.

World Bank. 2005. *Report of the Scientific Working Group Meeting on Schistosomiasis*. World Bank, Washington, D.C. http://apps.who.int/tdr/publications/tdr-research-publications/swg-report-schistosomiasis/pdf/swg_schisto.pdf. Last accessed 10-1-2011.

World Health Organization (WHO). 2009. *Dengue and Dengue Haemorrhagic Fever*. Fact Sheet No. 117. World Health Organization, Geneva, Switzerland. http://www.who.int/mediacentre/factsheets/fs117/en/. Last accessed 10-1-2011.

World Health Organization (WHO). 2009. *Global Health Risks: Mortality and Burden of Disease Attributable to Selected Major Risks*. World Health Organization, Geneva, Switzerland.

World Health Organization (WHO). 2011. *World Health Statistics 2011*. World Health Organization, Geneva, Switzerland. http://www.who.int/whosis/whostat/EN_WHS2011_Full.pdf. Last accessed 10-1-2011.

Chapter 20

Brown, L.R. 2011. *World on the Edge: How to Prevent Environmental and Economic Collapse*. W.W. Norton, New York, NY.

Campus Ecology. National Wildlife Federation. http://www.nwf.org/Global-Warming/Campus-Solutions.aspx. Last accessed 10-1-2011.

Carlson, S. 2006. In search of the sustainable campus: With eyes on the future, universities try to clean up their acts. *Chronicle of Higher Education* 53:A10.

Chapin III, F.S., Torn, M.S., and Tateno, M. 1996. Principles of ecosystem sustainability. *The American Midland Naturalist* 148: 1016–1037.

Dasgupta, P. 2007. The idea of sustainable development. *Sustainability Science* 2:5–11.

Easterlin, R.A., L.A. McVey, M. Switek, et al. 2010. The happiness-income paradox revisited. *Proceedings of the National Academy of Sciences* 107:22463–22468.

Gordon, J.C. and J.K. Berry. 2006. *Environmental Leadership Equals Essential Leadership: Redefining Who Leads and How*. Yale University Press, New Haven, CT.

Kahneman, D., A.B. Krueger, D. Schkade, et al. 2006. Would you be happier if you were richer? A focusing illusion. *Science* 312:1908–1910.

Kates, R.W. and T.M. Parris. Long-term trends and a sustainability transition. *Proceedings of the National Academy of Sciences* 100:8062–8067.

Millennium Ecosystem Assessment. 2005. *Ecosystems and Human Well-Being*. Island Press, Washington, D.C. http://www.maweb.org/documents/document.356.aspx.pdf. Last accessed 10-1-2011.

National Research Council, Board on Sustainable Development. 1999. *Our Common Journey: A Transition Toward Sustainability*. National Academy Press, Washington, D.C.

Ostrom, E. 2009. A general framework for analyzing sustainability of social-ecological systems. *Science* 325:419–422.

Speth, J.G. 2004. *Red Sky at Morning: America and the Crisis of the Global Environment*. Yale University Press, New Haven, CT.

United Nations. 1987. Report of the World Commission on Environment and Development: Our Common Future. http://www.un-documents.net/wced-ocf.htm. Last accessed 10-1-2011.

United Nations. 2010. *World Population Prospects, the 2010 Revision*. http://esa.un.org/wpp/Other-Information/faq.htm. Last accessed 10-1-2011.

University Leaders for a Sustainable Future. http://www.ulsf.org/. Last accessed 10-1-2011.

GLOSSARY

A

A horizon
A layer of soil containing a mixture of organic matter and mineral particles.

abiotic
Elements in the ecosystem that are nonliving.

acid deposition
The dispersion of acid-containing gases, aerosols, and rain onto soils, plants, buildings, and bodies of water.

active solar technology
The use of mechanical devices to heat water and buildings or to power electrical devices to generate electricity.

acute exposure
Contact with a toxin or pathogen that occurs only once or for only a short time.

adaptation
Inherited structures, functions, or behaviors that help organisms survive and reproduce.

adaptive radiation
The process by which several species evolve from a single ancestor to occupy new ecological niches.

aerosols
Very small solid and liquid particles suspended in the air.

age-specific birth rate
The number of children born in a year per 1,000 women within defined age groups.

age-specific fertility rate
The number of offspring produced by females during a particular age or range of ages.

age-structure pyramid
A graphical illustration of how populations are apportioned according to age and gender.

agriculture
The system of land management used to grow domesticated plants and animals for food, fiber, or energy.

agroecosystem
Crops and domestic animals, the physical environments in which they grow, and the communities of other organisms associated with them.

agronomy
The science that applies knowledge from other fields of study, such as genetics, physiology, chemistry, and ecology, to agriculture.

air pollution
Gases or particles that are present in high enough concentrations to harm humans, other organisms, or structures, such as buildings or pieces of art.

air quality
The amount of gases and small particles in the atmosphere that influences ecosystems or human well-being.

allergens
Substances that stimulate a range of responses from the body's immune system.

allocation
In forest management, decisions regarding the location and amounts of different forest uses.

amperes
A measure of electric current; the flow of electrically charged particles, usually electrons flowing through a wire.

anaerobic respiration
The process in which carbohydrates are partially broken down in the absence of oxygen to produce carbon dioxide and smaller carbohydrate molecules.

angiosperm
Plants whose seeds are borne in flowers.

animism
The belief that living and nonliving objects possess a spirit or a soul.

anthracite coal
With the highest energy content, this coal has a carbon content of 86–97%.

anthropocentric ethics
The assignation of intrinsic value only to humans; it defines right actions in terms of outcomes for human beings.

anthroponosis
An infectious disease that occurs exclusively in humans.

aphotic zone
A division of the pelagic zone lake and ocean waters that receive little or no sunlight.

aquifer
A layer of soil or rock that is saturated with groundwater.

arithmetic growth
Where the number of new individuals added at each generation is constant.

artesian well
A well drilled into a confined aquifer where the groundwater is under sufficient pressure to cause water to rise above the confining layer.

asexual reproduction
Simple cell division that produces offspring genetically identical to their single parent.

asphyxiants
Chemicals that deprive tissues of oxygen and cause suffocation.

atmosphere
The layer of gases above a planet's surface.

atmospheric pressure
The force caused by the pull of gravity on a column of air.

atom
The most basic subunit of an element; supplies an element's physical and chemical properties.

B

B horizon
The subsoil below A horizon that is especially rich in clay particles that form from minerals dissolved from the weathering of sand and silt.

baby boom
The period immediately following World War II in which the birth rate increased sharply.

base load
The minimum amount of electricity that a utility must continuously provide to meet consumer needs.

benthic zone
The water and sediment on the lake or ocean bottom.

bioaccumulation
Refers to the process by which chemicals are stored in the tissues of living organisms.

bioassessment
Surveys of aquatic biodiversity that are known to be sensitive to overall water quality.

biocapacity
A measure of the area and quality of land available to supply a population with resources.

biocentric ethics
A system of values based on the belief that the value of other living things is equal to the value of humans. It extends intrinsic value to *individual* organisms beyond human beings; organisms do not need to benefit humans in order to have value.

biodiesel
A biofuel derived from plant oils and animal fats that can be burned in diesel engines.

biodiversity
The variety of life in all its forms and combinations and at all levels of organization.

biodiversity hotspots
Places with high biodiversity and a large number of endemic species where human threats to habitats are high.

bioenergy
See biomass energy.

biofuels
Fuels derived directly from plant materials, such as corn or wood.

biogeochemical cycle
The flow of matter through an ecosystem.

biological magnification
The process by which the concentration of a chemical such as DDT or mercury increases through a food chain; also called biomagnification.

biological pest control
The use of predators and parasites to manage pests.

biomass energy
The energy contained in living and dead organic matter.

biome
Communities of similar organisms found in a particular climatic zone.

bioremediation
A process in which living organisms—microorganisms, plants, or fungi—are used to break down waste into less harmful components.

biosphere
All of the organisms on Earth and the nonliving environment with which they interact.

biostorage
The absorption and storage of CO_2 in plants and animals in Earth's ecosystems.

biota
Elements in the ecosystem that are living organisms.

birth rate
The number of births in the population per unit of time expressed as a percentage of the population size.

bituminous coal
Denser than either sub-bituminous coal or lignite, it has a carbon content of 45–86%.

black carbon
Solid carbon particles released during forest fires and the burning of fuel wood and charcoal that absorb light and warm the atmosphere; believed to be an important factor contributing to the worldwide loss of snow and ice.

body burden
The total amount of a chemical present in the tissues of an organism.

bog
Wetlands with peat deposits that support a variety of evergreen trees and shrubs; primary source of water is rainfall.

boreal forest
Coniferous forest in biomes where the growing season is less than 4 months long and winters are long, dry, and bitterly cold.

breeding barrier
An obstacle—geographic, temporal, behavioral, or structural—that prevents one species from breeding with another.

broad-spectrum pesticide
A pesticide that kills a wide variety of organisms.

C

C horizon
The lowest layer of soil or weathering zone.

calorie (cal)
The energy required to raise the temperature of 1 g of water by 1 °C (1 cal = 4.18 J).

cap and trade
A market-based policy in which a regulatory limit, or cap, is set on an action; the rights to exceed that limit are traded in markets.

capillary water
Water that is held in the soil by water-to-water hydrogen bonds.

capital
The total amount (mass) of an element or molecule in a pool. *See* pool.

carbohydrate
Organic molecules made of carbon, hydrogen, and oxygen.

carbon capture and storage (CCS)
Technologies that capture CO_2 produced by burning fossil fuels before it escapes to the atmosphere and then transfers it into a form that can be permanently stored underground.

carbon dioxide equivalents (CO_2e)
An expression of total greenhouse gas emissions, taking into account the global warming potential of each gas relative to CO_2.

carcinogens
Chemicals that cause or promote cancer.

carrying capacity
The population size at which a population uses resources equal to the rate at which those resources are supplied; the birth rate equals the death rate, and the population growth rate becomes zero.

case law
The various decisions made by the individual courts collectively that establish precedents and influence future court decisions.

catalyst
Substance that promotes chemical reactions without being consumed in the reaction.

catalytic converter
Part of an automobile exhaust system that causes conversion of harmful gases to less harmful products. Required on all cars manufactured after 1975 to reduce emissions of carbon monoxide, hydrocarbons, and nitrogen oxides.

cellular respiration
The process by which energy in carbohydrate molecules is retrieved and used to carry out cell functions and facilitate growth.

cellulose
Polysaccharides composed of hundreds of glucose molecules that are bonded in a way that prevents digestion by most animals.

cellulosic ethanol
Ethanol that is produced from agricultural residues, wood, grasses, or other cellulose-rich feedstocks.

channel
The waterway through which a stream normally flows.

chaparral
Biome of evergreen shrublands and low woodlands; occurs where winters are mild and moist and summers are hot and dry.

chemical energy
The potential energy associated with the formation or breakage of bonds between atoms.

chemosynthesis
A process by which energy is derived from inorganic chemical reactions.

chronic exposure
Contact with a toxin or pathogen over periods of months or even years.

clay
Smallest particles of minerals (diameters less than 0.002 mm) that are highly weathered, with unique physical and chemical properties.

clear-cutting
The complete removal of forest canopy trees; the most controversial harvest practice.

climate
The atmospheric conditions such as temperature, humidity, and rainfall that exist over large regions and relatively long periods of time.

climatograph
A graph that shows the pattern of seasonal changes in temperature and precipitation for a particular location.

climax community
Assemblages of plants and animals that are able to perpetuate themselves generation after generation.

closed basin
Lakes with no outlet stream.

closed watershed
Inland basins that do not drain to the sea.

closed-loop recycling
The material in a product is reused to produce a similar product; for example, aluminum cans are recycled to produce more aluminum cans.

coal seams
Layers of sediment that contain coal.

coevolution
Evolutionary change in which interactions between species select for adaptations in each species; for instance, plants evolve chemical defenses to thwart herbivores and herbivores become tolerant or even take advantage of plant chemical defenses.

combined-cycle power plant
A facility that creates power by burning natural gas to power turbines directly as well as steam boilers and steam turbines driven by the heat from the natural gas.

combustion by-products
Due to the incomplete combustion of organic carbon, many gases and particles are released into the air. These by-products include carbon monoxide, nitrogen and sulfur oxides, hundreds of volatile organic compounds (VOCs), and particulate matter.

commensalism
Refers to associations that benefit only one species, leaving the other species unaffected.

committed warming
Warming that would occur at a rate of about 0.1 °C per decade even if we were to stop all greenhouse gas emissions immediately.

compact fluorescent lights (CFLs)
Lights in which an electric current causes a gas to glow; CFLs emit three to five times more light per kilowatt-hour than do incandescent fixtures.

competitive exclusion principle
A postulate that two species that directly compete for essential resources cannot coexist; one species will eventually displace the other.

complementarity effect
Species within a community utilize resources at different times or from different locations; groups of species are thus able to use resources more efficiently than would any single species.

compost
Decayed organic matter that can be used to improve the fertility and water-holding capacity of soil, thereby improving plant growth.

compound
Molecules that are made of more than one element.

concentrating solar power (CSP) systems
A method of generating electricity from sunlight that relies on mirrors to concentrate the sun's rays on a tower or a series of pipes that hold water or another fluid. The concentrated sunlight heats the fluid to boiling, producing steam. The steam is then used to spin a turbine and generate electricity.

conduction
The direct transfer of heat by means of the collisions of molecules.

confined aquifer
Groundwater trapped between layers of comparatively impermeable rock or sediment.

conifer
A classification of trees and shrubs whose seeds are borne in cones.

consequence-based ethics
A system of beliefs that emphasizes the importance of *outcomes*. Right and wrong are defined in terms of pleasure or pain, benefit or harm, and satisfaction or dissatisfaction.

conservation supply curve
Graph used to compare the amount of energy saved by each action to the cost of that energy.

conservationist view
The opinion that public resources should be used and managed in a sustainable fashion to provide the greatest benefit to the greatest number of people.

consumers
Species that feed on other live organisms.

consumptive use
Water use by humans in which the water is not returned to streams or aquifers.

continental climate
The climate conditions in which daily and seasonal temperatures vary widely at locations within the interior of continents, away from large bodies of water.

contingent valuation
The estimates of the value of an ecosystem service gained by surveying people's willingness to pay for that service.

convection
Warm regions in a gas or liquid become less dense and rise, causing the gas or liquid to circulate.

conventional international laws
Laws established by formal, legally binding conventions or treaties among countries.

convergent boundary
The point at which tectonic plates collide.

convergent evolution
The process by which natural selection favors similar features among otherwise unrelated species.

core
The center of Earth; it is composed of a mixture of nickel and iron, with smaller amounts of other heavy elements.

Coriolis effect
The apparent change in wind direction due to Earth's rotation. Air moving over the ground in a straight path toward the equator appears to be deflected from east to west. Air moving toward the poles appears to be deflected west to east.

corrosive toxins
Chemicals that react with and directly destroy tissue; the effect is usually immediate.

cost of saved energy (CSE)
A measure that compares the costs and benefits of energy alternatives by dividing the cost of an energy investment by the energy savings resulting from that investment.

covalent bond
A strong bond formed when atoms share electrons.

crust
A thin veneer of solid and relatively light rocks above Earth's mantle.

cultural services
The spiritual and recreational benefits that ecosystems provide.

customary international laws
Accepted norms of behavior or rules that countries follow as a matter of a long-standing precedent.

cycling time
The average time that it takes an element or molecule to make its way through an entire biogeochemical cycle.

cytoplasm
A plasma membrane that regulates materials flowing in and out of the cell.

D

dead zone
Regions of ocean water near the mouths of rivers with very low oxygen in which few marine animals can survive; generally a consequence of eutrophication.

death rate
The number of individuals dying per unit of time expressed as a percentage of the population size.

debt-for-nature swaps
International arrangements in which poor nations are forgiven a portion of their debts in exchange for the establishment of permanent parks and preserves.

decomposers
Organisms that feed on nonliving organic matter.

deep ecology movement
The view that all elements of the environment have equal intrinsic value and deserve moral respect in their own right.

deforestation
The process by which humans cut a forest and replace it with a non-forest use, such as agriculture, pasture, or pavement.

delta
The landform at the mouth of a river where it flows into a lake or into the sea.

demographers
Those who study factors influencing the size, density, and distribution of human populations.

demographic transition model
Describes the four stages of change in birth rates, death rates and growth rates of human populations associated with economic development.

denitrification
Nitrogen in soil and water is returned to the atmosphere;

specialized bacteria convert NO_3^- to N_2 gas. The total nitrogen in Earth's organisms, soils, and waters is determined by the balance between nitrogen fixation and denitrification.

deoxyribonucleic acid (DNA)
The hereditary material that is passed from generation to generation in all organisms.

desalination
The process of removing salts and other chemicals from seawater.

desert biome
Land with a sustained and significant moisture deficit.

design for disassembly (DfD)
The development of materials, components, construction techniques, and management systems that permit the easy recovery of parts and materials from discarded products.

detritus
Organic matter such as dead leaves and nutrient-rich soil that wash in from adjacent terrestrial ecosystems.

dew point
The temperature at which the relative humidity is 100%.

dipole bond
Weaker bonds that form between atoms and molecules as a consequence of shifts in electrical charge.

directional selection
Natural selection that favors the survival of individual organisms at one extreme end of the range of variation within a population.

discharge zone
Groundwater flowing to the surface from places where a confined aquifer meets the soil surface.

discount rate
The rate at which economic value declines with time.

disruptive selection
Situations in which individuals with genetic traits that fall at the extreme ends of the range are more likely to survive than individuals with average traits.

distillation
A method of desalination by boiling water and condensing the steam.

divergent boundary
Separate tectonic plates that are moving apart; boundaries coincide with ocean ridges.

dose-response curve
The relationship between increasing dosages of a chemical and some measure of health.

doubling time
The length of time required for a population to double in size.

drip irrigation
A system that provides water to plants through pipes with small openings at the base of each plant that feed water to the root zone.

dry deposition
Nonliquid particles are removed from the atmosphere by gravity.

duty-based ethics
A set of beliefs that the rightness or wrongness of actions should be determined by a set of rules or laws.

dynamic homeostasis
The process by which systems adjust to changes in ways that minimize how much features or processes vary from their normal values.

E

Earthquake Intensity
A measure of the severity of shaking at a given location; it is determined by its effects on people, structures and the natural environment.

ecocentric ethics
A system of values based on the belief that *collections* of organisms or critical features in the environment have intrinsic value.

ecofeminism
The view that exploitation and abuse of both the environment and women derive from male domination.

ecological community
Species that interact within a specific area.

ecological disturbance
Events such as fires, hurricanes, or the logging of forests that may result in the loss of many or all of the species in an ecological community.

ecological footprint
The land area required to produce the ecosystem goods and services for entire human populations or for the average individual in such populations.

ecological legacy
Environmental features, such as soil and woody debris, that are left behind following a disturbance.

ecological niche
The various activities that define that organism's role in an ecosystem.

ecological valuation
Determination of the ecological value of the potential cost of the loss or degradation of an ecosystem service.

ecology

The branch of environmental science that focuses on the abundance and distribution of organisms in relation to their environment.

economic energy intensity

A measure of overall energy use efficiency by a country or group of countries that considers the amount of energy used per dollar of economic output.

economic system

The institutions and interactions in a society that influence the production, distribution, and consumption of goods and services.

economic value

A commodity's value determined by the price that consumers are willing to pay for it.

economies of scale

The cost per unit of a good or service declines as the level of production increases.

ecosystem

All of the organisms and their physical and chemical environment within a specific area.

ecosystem functions

The flow of matter and energy and the processes influencing the distribution and abundance of organisms.

ecosystem integrity

The web of interactions that regulate ecosystem functions and ecosystem services.

ecosystem services

The multitude of resources and processes that ecosystems supply to humans.

ecotourism

A form of tourism in which visitors come to see the natural beauty and biodiversity of an area in a sustainable fashion that supports local economies.

El Niño/Southern Oscillation (ENSO)

Weather conditions that vary as a consequence of changes in temperature of surface waters in the tropical regions of the Pacific Ocean.

electric generator

A system of spinning magnets that generates an electrical field.

electric power grid

A network of transmission lines and transformers that distributes electricity from generating facilities to end users.

electromagnetic field

An area of electromagnetic force created by spinning magnets in a generator; used to provide the

voltage that causes an electric current to flow.

electromagnetic radiation

The energy of light.

electromagnetic spectrum

The full range of electromagnetic radiation wavelengths; the energy of photons decreases from short wavelength gamma rays to long wavelength radio waves.

electrons

Negatively charged particles surrounding the nucleus of an atom.

element

A chemical that cannot be broken down or separated into other chemicals.

emigration rate

The number of organisms moving out of an area calculated as a percentage of population size.

end use

The final application or use of a primary energy source.

endemic

Species that are found only in specific environments or in restricted localities.

endemism

Species that thrive within very restricted geographic ranges and are a consequence of evolution that has produced very specialized ecological niches.

endocrine disrupters

Substances that interfere with the hormones that control growth and development, egg and sperm production, and reproductive function.

energy

The capacity to do work.

energy budget

A system of accounting that measures all the energy entering and leaving an ecosystem or the entire Earth.

energy conversion efficiency

The percentage of energy in one form that is captured when it is converted to another form.

energy conversion

Process of changing primary energy into other forms.

energy end-use efficiency

The product of the efficiencies of all the energy conversions from the primary source to the end use.

energy flow

The transfer and transformation of high-energy organic molecules within ecological communities.

energy return on investment (EROI)

Measure of the useful energy produced from an oil well, natural gas field, or coal mine divided by the amount of energy it took to obtain it.

entropy

Describes the disorder or disorganization in a system.

environment

All of the physical, chemical, and biological factors and processes that influence the growth and survival of organisms.

environmental ethics

A system of beliefs based on the moral relationship of humans to the environment and its nonhuman contents. Systems of environmental ethics vary regarding who or what has value.

environmental justice

The principle that, in the management of natural resources and the environment, people should be treated fairly regardless of race, gender, or economic status.

environmental policy

The decisions and actions that influence environmental conditions or processes.

environmental science

The study of all aspects of the environment, including physical, chemical, and biological factors.

enzymes

Proteins that serve as catalysts.

equilibrium

The condition when competing influences in a system are balanced. For example, when inputs of an element equal outputs from a pool, the capital of the element in that pool remains constant, and the pool is said to be in equilibrium.

equilibrium theory of island biogeography

States that the diversity of species on an island is largely determined by the relationship between two factors: the rate at which new species migrate to the island and the rate at which species disappear from the island or become extinct.

establishment stage

The stage of forest development in which tree seedlings germinate on grounds recently cleared.

estuary

Region of transition from freshwater to marine ecosystems resulting in a mix of fresh water and saltwater.

eukaryotes

Organisms whose cells contain an organized nucleus and membrane-

bound organelles. They range from single-celled algae and protozoa to plants, fungi, and animals that are made of millions of cells.

eutrophication

High concentrations of rapidly growing algae in streams and lakes caused by elevated amounts of nutrients like nitrogen. Abundant organic matter from the algae encourages high rates of respiration by decomposing organisms.

evapotranspiration

The process in which water evaporates to the atmosphere from soils, plant leaves, and water bodies.

even-aged management

Use of clear-cutting to produce trees of uniform age.

e-waste

Electronic devices disposed of by businesses and households.

exploitation competition

In the competition for resources, the ability to take up or use resources more efficiently.

exponential growth

The number of new individuals added to a population in each generation as a multiple of the number present in the previous generation.

extended product responsibility (EPR)

The belief that those who make a product should assume responsibility for that product throughout its life cycle.

externality

The costs and benefits associated with the production of a commodity that affect people other than buyers and sellers.

exurb

Areas where patches of commercial and residential development are interspersed in a matrix of rural land.

F

facilitation

Species alter the environment around them in ways that make it more habitable for other species. It is especially important in the early stages of succession.

feedback

The ability of a system to adjust based on changes in the system itself.

fen

Wetlands that are fed primarily by groundwater.

Ferrel cells
Atmospheric convection cells between latitudes of 30 and 60 degrees.

fertility rate
The number of offspring produced by an average female in a population.

fertility transition
The third stage of the demographic transition in which continued economic development produces social and cultural changes that lead to lower birth rates.

field capacity
The amount of water that a given volume or weight of soil can hold against the force of gravity.

first law of thermodynamics
Energy can be transformed from one form to another, but it can be neither created nor destroyed; also called the law of energy conservation.

fitness
A measure of an organism's reproductive success.

floodplain
Land along rivers that experiences periodic flooding.

flushing time
The time required for an estuary to completely cycle its contents from its river source to the ocean.

flux
The rate at which matter moves from one pool to another. *See* pool.

fly ash
Solid material left when coal is burned.

food chain
Feeding relationships among organisms are often depicted as simplified food webs. In a food chain, one species or trophic level is eaten by another, which is eaten by yet another, and so on.

food web
The feeding interactions among species in a community.

forest degradation
Changes in forest structure and composition diminish biodiversity or key ecosystem services.

free market economy
An economy in which the government does not influence the marketplace with subsidies, taxation, or regulation.

fuel recycling
Process in which nuclear fuel is reprocessed and the elements that can undergo fission chain reactions are reconcentrated.

fundamental niche
The complete range of environmental conditions, such as requirements for temperature, food, and water, over which the species might possibly exist.

G

gametes
Cells that fuse with other cells to form zygotes in reproduction.

gas giant
Large planets composed mainly of hydrogen and helium.

gasoline-electric hybrid automobile
Vehicle equipped with both an internal combustion engine and an electric motor powered by batteries.

gas-turbine power plant
A facility that creates energy using natural gas to turn turbines that are directly connected to electrical generators.

gene
Segment of DNA that carries the code used to synthesize, or build, specific proteins.

generation time
The average difference in age between mothers and their offspring.

genetic diversity
The genetic variation among individuals within a population.

genetic drift
A change in the frequency of an inherited trait in a population that is brought about by chance events.

genetically modified organism (GMO)
An organism whose DNA has been altered by the insertion of a gene from a different species into the organism's cells.

genome
An organism's complete set of DNA.

genuine progress indicator (GPI)
An alternative measure of national wealth using GDP plus or minus the economic value of enhancements or degradations to the environment and to human communities.

genus
The taxonomic term used for a group of closely related organisms. Each genus comprises a number of species. *See* species.

geologic nuclear repository
A layer of rock in which nuclear waste and all of its toxic and radioactive elements are permanently sealed from the biosphere.

geothermal energy
Heat energy that is generated and stored in Earth itself.

global circulation model (GCM)
Computer programs that forecast climate change using mathematical equations to simulate the physical processes that determine Earth's energy budget.

global warming
Refers specifically to the increase in atmospheric temperature since the late 1800s.

global warming potential (GWP)
A measure of an individual molecule's long-term impact on atmospheric temperature relative to a molecule of CO_2.

gradient
Slope of the land that determines the speed of a stream as it flows downhill.

gravity water
Water that flows through the soil as it percolates into the soil and is drawn downward by the force of gravity.

graywater
The water from washers, sinks, and baths.

Green Revolution
The period since about 1950 during which global agricultural productivity increased many times over.

greenhouse effect
The absorption of radiation by greenhouse gases and trapping of that heat in the troposphere.

greenhouse gases
Gases that efficiently capture heat in the troposphere: water vapor and carbon dioxide, methane, nitrous oxide and ozone

gross domestic product (GDP)
Determination of the total value of goods and services produced by the citizens of a country.

gross primary production (GPP)
The total amount of CO_2 that photosynthetic organisms convert to organic carbon each year.

ground-source heat pumps (GSHPs)
Method of using Earth's heat for energy by taking advantage of the fact that the upper 10 ft of Earth's crust maintains a nearly constant temperature of 10–15 °C (50–60 °F), regardless of the time of year.

groundwater
Liquid water that has percolated through the soil and into underlying rocks.

gyre
A pattern of circulation in surface ocean waters that is clockwise in the Northern Hemisphere and counterclockwise in the Southern Hemisphere.

H

Haber-Bosch process
A nonbiological method of nitrogen fixation adapted to the production of chemical fertilizers.

habitat
The complex environment in which an organism is found and upon which it depends for survival.

Hadley cells
The two atmospheric convection cells on either side of the equator.

half-life
The length of time that it takes for half of a collection of atoms to undergo a chemical transformation. For example, the length of time required for half of the molecules of a radioisotope to decay.

harvest
The extraction of goods or services from a forest or cropland.

harvest index
The fraction of total production that can be used by people.

hazardous waste
Waste that is flammable, corrosive, explosive, or toxic.

heat
Least usable form of energy, which is created by the random motion of molecules.

hedonic valuation
Economic value that is determined by the difference in the market price of real estate that is affected by different environmental alternatives.

herbivore
Species that eat plants.

high-grading
In forest harvest, the removal of a particular species of trees for fuel and construction.

high-level radioactive waste
Waste generated by nuclear power plants and weapons labs that requires long-term storage and management.

Holocene epoch
A name for the last 10,000 years of the Pleistocene epoch when Earth was in a warm interglacial period.

human well-being
A multifaceted concept that includes life's basic necessities,

such as food and shelter, as well as good health, social stability, and personal freedom.

humus
Leaves or crop residue and the organic products of their decomposition.

hydraulic fracturing
Process to extract oil using high-pressure liquids to fracture the rocks; also called hydraulic fracking.

hydrocarbon
Organic molecules composed entirely of carbon and hydrogen atoms.

hydroelectric power
Electricity generated by using the kinetic energy of moving water to turn a turbine.

hydrologic cycle
The distribution and flux of water through Earth's biogeochemical system.

hydrologist
Scientist who studies the movement of Earth's water.

hydrophytes
Species of plants that have adaptations that allow them to grow in anaerobic wetland soils.

hydropower
The energy of water moving under the force of gravity.

hygroscopic water
Water that is bound to soil particles and cannot be easily used by plants.

hyporheic zone
The region of saturated sediment beneath a stream and its adjacent riparian zone.

hypothesis
A proposition that can be tested by careful observation or experimentation.

I

igneous rocks
Rocks that form as magma solidifies.

immigration rate
The number of individuals entering an area per unit of time as a percentage of population size.

impervious surface cover (ISC)
The percentage of land covered with sealed surfaces, such as pavement and rooftops.

impoundment hydroelectricity plant
Electricity generated by impounded water, such as water behind a dam, flowing downhill through a turbine.

inbreeding
Mating between closely related individuals.

incandescent lights
Lights in which an electric current causes a thin metal filament to glow.

incentive
Policies that are implemented by encouraging action through the offering of something appealing to take that action.

industrial ecology
Industrial waste systems are redesigned to mimic the cycling of materials in natural ecosystems.

industrial smog
Formed primarily of pollutants released in coal burning. It includes carbon monoxide (CO), sulfur dioxide (SO_2), and particles of carbon soot.

industrial solid waste (ISW)
Waste produced in industrial processes before a finished product reaches the consumer.

infant mortality rate
The percentage of infants within a population who die before age 1.

inorganic compounds
Compounds that are not made up of carbon and hydrogen.

instrumental value
The view that a thing is valued as a means to some other end.

integrated conservation and development projects
Programs that promote both the protection of wildlife and the improvement of living standards for people who live within and around areas that are managed for biodiversity.

integrated pest management (IPM)
The effective combination of all three kinds of pest control—chemical, biological, and cultural—carefully designed to minimize environmental damage.

interference competition
In the competition for resources, the fending off of would-be competitors with aggressive or territorial behavior.

intermittent stream
Streams in which water flows only during certain seasons or following major rain events.

internal combustion engine
An engine in which fuel is explosively ignited in a closed cylinder, driving a piston that turns a crankshaft.

interplanting
The practice of planting multiple crop species in the same field.

interspecific competition
Interactions among different species as they compete for shared resources.

intertidal zone
The area along seacoasts that is submerged at high tides and exposed during low tides.

intertropical convergence zone
The band of rising air at the equator.

intraspecific competition
Where members of the same species pursue limited resources.

intrinsic value
The view that a person, organism, or object is valued as an end unto itself.

investment
In forest management, the array of strategies used to restore, maintain, or protect a forest's capacity to provide goods and ecosystem services.

ions
Electrically charged atoms or molecules for which the atom donating the electron has a net positive charge and the atom receiving the electron has a net negative charge.

ionic bonds
Chemical bonds that form between positively charged ions and negatively charged ions.

IPAT equation
An equation that summarizes the relationship between human impact and population size, affluence, and technology: $I = P \times A \times T$, where I represents environmental impact, P represents population, A represents affluence, and T represents technology.

irrigation efficiency
The percentage of the water applied to fields that is actually used by crop plants.

isotope
Atoms of an element having different numbers of neutrons.

J

joule (J)
The energy required to support a 1-kg mass against the force of gravity. Named in honor of a 19th-century scientist, James Prescott Joule.

judicial international laws
Standards for the actions of countries based on the decisions of international courts and tribunals.

K

kerogen
A waxy organic substance that is a precursor to oil and natural gas.

keystone species
Species that play a particularly important role regarding the abundance of other species and that are considered essential to the health of the ecosystem.

kilowatt hours
Electric power measurement that is calculated by multiplying electrical potential (volts) by current (amperes): power (kWh) = voltage (volts) $\times$ current (amperes).

kinetic energy
The energy of motion.

L

land trusts
Nonprofit organizations that are dedicated to conserving land for the protection of biodiversity and ecosystem services.

landscape diversity
The differences in the variety and abundance of species from place to place.

latent consequences
Effects of a hazard that are delayed.

latent heat transfer
Molecules with the highest kinetic energy evaporate, leaving behind molecules with lower kinetic energy and temperature.

leach field
An area of ground where wastewater released from perforated pipes percolates through the soil, allowing microorganisms to break down the waste materials. *See* septic system.

lentic
Ecosystems associated with lakes and ponds.

levee
Embankment along stream channels in wide floodplains.

life expectancy
The average age to which a baby born at a given time will live; provides a summary of differences in survivorship.

life-cycle assessment (LCA)
Methodology used to evaluate the environmental impact of a product throughout all the stages of its life from extraction as a natural resource to disposal.

light-emitting diodes (LEDs)
Lights in which an electric current passes through silicone-rich materials; more efficient and long-lasting than CFLs and incandescent lights.

lignite
A type of coal that is generally associated with younger deposits that have not been subjected to as much heat and pressure as other types of coal; contains 25–35% carbon.

lipids
Organic molecules made of long chains of carbon and hydrogen atoms and a shorter region with one to several oxygen molecules; usually fats and oils.

liquefied natural gas (LNG)
The liquid form of natural gas created when natural gas is chilled to –160 °C (–260 °F).

lithosphere
An area of interaction between the crust and the upper reaches of the mantle; the zone of geologic activity that has shaped Earth's surface and continues to do so.

Little Ice Age
The time between A.D. 1400 and about 1750 when a cold and wet period set in over the Northern Hemisphere.

littoral zone
The shallow, sloped area of a lake or the ocean that is relatively close to shore.

logistic growth curve
Changes in population size that follow a predictable and repeatable S-shaped trajectory.

lotic
Ecosystems dominated by flowing fresh water.

low-level radioactive waste
Waste that contains comparatively low amounts of radioisotopes.

M

macromolecule
A large complex molecule composed of small organic molecules linked to one another.

macronutrients
Elements that organisms require in comparatively large amounts, such as carbon, oxygen, hydrogen, nitrogen, phosphorus, and sulfur.

magma
A liquid-like state of rocks in the upper mantle.

mantle
The layer of less dense rock above Earth's core. The mantle, which is 2,900 km (1800 mi) thick, occupies about 70% of Earth's total volume.

marginal value
The difference in people's willingness to pay for one action compared to an alternative.

maritime climate
Regions near oceans or large lakes that experience little variation in temperature.

market-based policy
The use of economic markets to guide environmental actions.

market economy
The production and consumption of goods and services in markets guided by prices based on a system of currency.

marsh
Areas, either tidal or nontidal, that are periodically or continuously flooded.

mass
The property responsible for the gravitational attraction of all matter to all other matter.

mass-balance accounting
A method that allows scientists to account for changes in the abundance of an element within the ecosystem.

meander
The wide loops in the courses of streams and rivers.

median lethal dose (LD$_{50}$)
Toxicity of a chemical in terms of the dose at which 50% of an exposed population is killed or displays a specific symptom or illness.

Medieval Warm Period
The years from A.D. 1000 to 1300 when the average global temperatures may have increased by 0.5 °C, with greater warming in northern latitudes.

megacities
Urban areas with more than 10 million inhabitants.

mesosphere
The layer of the atmosphere where the air temperature can reach lows of –173 °C (–140 °F) at about 90 km (55 mi).

metamorphic rock
Rocks that form when great heat and pressure transform the physical and chemical properties of sedimentary or igneous rocks.

metropolitan area
A large central city and the adjacent communities that have a high degree of social and economic integration.

micronutrients
Elements that organisms require in very small amounts, such as molybdenum, manganese, and boron.

migration corridors
Connections among habitat fragments that facilitate migration of species; e.g., strips of forest connecting larger forest patches.

mine tailings
Rock and debris from mining operations; often contain high concentrations of pollutants.

minimum dynamic area
The area necessary to maintain all different individual forest stand types and populations of their associated species, given the typical patterns of disturbance.

model organism
Organism in which particular processes or metabolic functions have been demonstrated to be similar to those of humans.

molecule
Two or more atoms held together by chemical bonds.

monocultures
Plantings of a single crop species.

mortality rate
Another term used for the death rate.

mortality transition
The second stage of the demographic transition in which improved economic conditions relieve food shortages, produce better living conditions and health care, and expand access to education. The death rate decreases while the birth rate stays high or even increases.

multinational development banks
Institutions that provide financial and technical assistance to countries for economic, social, and environmental development.

municipal sewage treatment plant
Facilities run by a city or town that use a stepwise process to remove wastes and chemicals from the water.

municipal solid waste (MSW)
Materials that are discarded from homes, small businesses, and institutions such as hospitals and universities; what most of us call trash or garbage.

mutation
Chance changes in the structure of an organism's DNA that produces new forms of inherited features.

mutualism
A symbiotic relationship between two species in which both benefit.

N

natural capital
All of Earth's resources that are necessary to produce the ecosystem services on which we depend.

natural selection
The process by which individuals in a population that are most fit survive and leave more offspring, causing their adaptations to become more common. Individuals that are less fit leave fewer offspring, so their features become less common or cease to exist.

negative feedback
Directional change in a process that alters the system and reverses the direction of that change.

neoclassical economic theory
Determination of the price and production of goods and services through the interaction of supply and demand.

neo-Malthusian movement
Activities related to the belief that as the human population increased, it would soon outpace the supply of food.

net ecosystem production (NEP)
The amount of organic carbon left each year after subtracting the respiration of nonphotosynthetic organisms from the net primary production.

net migration rate
A calculation of the difference between immigration and emigration per 1,000 individuals in the population.

net primary production (NPP)
Gross primary production minus the organic carbon respired by green plants; the amount of organic carbon available to all the nonphotosynthetic organisms, or consumers, in an ecosystem.

neurotoxins
Substances that affect the development or functioning of the nervous system.

neutrons
The electrically neutral particles in the nucleus of atoms.

new urbanism
A social movement that accepts that growth can only be economically, socially, and environmentally sustainable if a combination of smart growth strategies is employed.

niche differentiation
The dividing up of the resources that make up a fundamental niche by potential competitors so that they are able to coexist.

nitrification
A process in which specialized bacteria in the soil convert NH^+ to nitrite (NO_2^-) and nitrate (NO_3^-).

nitrogen fixation
A process in which atmospheric N_2 is converted to NH_3; most nitrogen fixation is carried out by nitrogen-fixing bacteria, which convert N_2 to NH_3 and then into nitrogen-containing organic molecules, such as amino acids and nucleic acids.

nonconsumptive use
Activities that use water and then return it to streams or aquifers.

non-point source
Pollution that comes from a variety of activities occurring at different places, such as urban runoff or fertilizer-enriched runoff.

nonrenewable
The amount a resource in an ecosystem declines with virtually any level of use.

nonrenewable energy
The use of an energy source at rates faster than the rate of replenishment.

nuclear energy
The potential energy contained in the structure of matter itself.

nuclear fission
Occurs when the nucleus of an atom is split, producing two or more smaller nuclei and a great deal of electromagnetic and kinetic energy. The sum of the masses of the resulting products is slightly less than the mass of the original atom. The difference in the mass of the system before and after fission is equal to the electromagnetic and kinetic energy that is released.

nuclear fusion
Occurs when two atoms collide with so much energy that their nuclei fuse, forming an atom of a new element. The fusion of relatively small atoms, such as hydrogen, carbon, and oxygen, releases energy as very small amounts of mass are converted to energy.

nucleic acid
Polymers of chemical subunits called nucleotides.

O

O horizon
The top layer of natural soil consisting of humus.

ocean thermal energy conversion
A technique that exploits the sharp differences in temperature between surface water and deeper water to generate electricity.

old-field succession
The reforestation of abandoned farmland.

old-growth stage
The stage of forest development characterized by very large, old trees and abundant standing and fallen woody debris.

oil reservoir
Oil trapped in a rocks between layers of impermeable rock.

open basin
Lakes drained by a well-defined stream or river.

open-loop recycling
A material is reused in a different product that is discarded after it is used; for example, plastic soda bottles may be recycled into other plastic products.

open watershed
The area of land drained by rivers that eventually make their way to the sea.

opportunity cost
The cost of buying a thing is equal to the economic value of the best alternative use of that money.

organelle
A membrane-enclosed part of a cell in a eukaryote that performs a specialized function.

organic foods
Foods produced to meet strict standards that limit the use of fertilizers and pesticides.

organic molecule
Molecules composed of carbon atoms that are covalently bonded to hydrogen and other atoms; the primary structural and functional molecules of organisms.

outbreeding
Mating between individuals that are not closely related.

ozone layer
The zone in the stratosphere that contains comparatively large amounts of oxygen in the form of ozone (O_3).

P

paramo
High mountain ecosystems in the tropics that are dominated by a tundra-like collection of grasses; temperature at night is frosty, with warm daytime temperatures.

parasite
Organism that lives in or on other plants or animals without killing them directly.

passive solar technology
The use of the energy of sunlight without relying on electrical or mechanical devices, such as by orienting buildings so they receive maximum sunlight and using building materials that absorb sunlight to keep interior spaces warm.

pathogens
Viruses, bacteria, protozoa, fungi, or multicellular parasites that invade host organisms and impair their functioning.

peak load
The maximum amount of electricity demanded by consumers during a day.

peak oil
A term used to describe the point of maximum production in an oil field.

peat
Organic material broken down anaerobically into a wet, partially decomposed mixture.

pelagic zone
Waters between the surface of the ocean or lakes and bottom or benthic zone waters.

per capita land consumption
The average amount of land used by each urban citizen; the size of a metropolitan area divided by its population.

perennial stream
A stream in which water flows year-round.

permafrost
Permanently frozen soil at depths below 30–100 cm (11–39 in.) in boreal forest and tundra ecosystems.

permanent retrieval storage site
Specially designed landfills in which wastes are monitored to ensure they do not leak into the surrounding environment.

persistent organic pollutants (POPs)
Particularly long-lived and widely dispersed organic chemicals.

pH scale
A quantitative representation of the relative amounts of hydrogen and hydroxyl ions in a substance.

pheromones
Volatile chemical signals used to send a message from one individual organism to another.

phloem
A thin layer of specialized tissue that transports carbohydrates down the stem to the roots.

photic zone
The upper portion of the pelagic zone that receives sufficient sunlight to support photosynthesis.

photochemical smog
Formed when certain primary pollutants interact with sunlight. Nitrogen oxide (NO) and volatile organic compounds (VOCs) are among the most important of these primary pollutants.

photochemicals
Secondary pollutants whose formation is facilitated by sunlight.

photosynthesis
Formation of carbohydrates from carbon dioxide and a source of hydrogen that occurs in the chlorophyll-containing tissues of plants exposed to sunlight.

photovoltaic (PV) cell
Devices that directly convert sunlight to electrical current.

photovoltaic (PV) technology
A method of generating solar power that converts light energy into electricity through the photovoltaic effect, in which light energy causes certain materials to emit electrons, thereby generating an electric current.

phylogenetic tree
A diagram that depicts scientific understanding of the evolutionary ancestry of organisms.

phytoplankton
The various algae suspended in the water in the pelagic zone; uses the light energy to carry out photosynthesis.

pioneer species
The earliest colonists of an ecosystem. They are widely dispersed, they are able to grow under very harsh and resource-poor conditions, and as they grow, they alter their environment in ways that allow other plants to become established.

planetoids
Planet-like objects composed of ice and rock.

planned economy
An economy in which the government regulates the prices of goods and services and the level of production.

Pleistocene epoch
A time of alternating cold and warm periods starting about 2 million years ago; each cycle of cold and warm lasted about 100,000 years.

plug-in hybrids
Vehicles operating much like gasoline electric hybrids, except their batteries can be recharged by plugging the car into an electric outlet.

plutonium
An element that can undergo fission and generate heat; it is a by-product in uranium-fueled nuclear reactors.

point source
Pollution coming from a specific location, such as an industrial facility or municipal sewage treatment plant.

polar cell
Convection cells between latitudes of 60° and 90°.

polar zone
The climate zone that falls above 60° latitude north and south; its climate is determined by the polar wind cells.

policy cycle
The stepwise process by which governments and communities develop and revise policies and laws.

polyculture
See interplanting.

polymer
Simple molecules linked together to form much larger chains or networks.

polysaccharide
Polymers of simple sugar molecules.

pool
In reference to nutrient cycling, a part of an ecosystem in which matter may reside, such as the atmosphere or soil.

population growth rate
The multiple by which an exponentially growing population increases.

population viability
The probability that a species will not go extinct in a given number of years.

positive feedback
Directional change in a process that alters the ecosystem to reinforce the change.

potential energy
Energy stored in a system and available to do work in the future.

potential evapotranspiration (PET)
An estimate of the average amount of water that would evaporate from a hypothetical agricultural field over the course of a year.

precautionary principle
When there is reasonable evidence that an action or policy may place human health or the environment at risk, precautionary measures should be taken, even in the context of uncertainty.

predators
Species that hunt and kill their prey.

preservationist view
The opinion that parks and public lands should *preserve* wild nature in its pristine state for its own value.

preserves
Areas that are protected by various means to maintain biodiversity.

prey switching
As predation depletes a species' numbers, predators switch their attention to other, more abundant species, which provides predators with a steady supply of food and ensures that none of the prey species will be totally eliminated.

primary air pollutants
Chemicals or particles that are directly emitted from identifiable sources.

primary energy
Contained in natural resources, such as coal, oil, sunlight, wind, and uranium.

primary forest
Forest that has been largely unaffected by humans; worldwide only about 19 million square kilometers (7.3 million square miles) remain.

primary growth
The upward and downward growth due to rapid cell division and growth at the very tips of plant shoots and roots.

primary oil recovery
The extraction of crude oil pushed to the surface by built-up pressure in the reservoir.

primary producers
Those organisms that transform energy from sunlight or certain inorganic chemicals into high-energy carbohydrates.

primary succession
After a disturbance has removed virtually all ecological legacies, plants, animals, and microscopic organisms slowly colonize the disturbed area. Because it occurs slowly, often over centuries or even millennia, primary succession is difficult to study directly.

primary treatment
In reference to wastewater treatment, a process that produces a relatively homogeneous liquid that can be treated biologically and a sludge that can be processed separately.

prior appropriation water rights
Water law in which the person who first uses a quantity of water from a water source for a beneficial use has the right to continue to use that quantity of water for that purpose indefinitely. Rights to the remaining water are similarly allocated to subsequent users. When water runs short, users with the earliest appropriation get their full allocation; later users may get no water.

profit
The difference between the cost to produce a commodity and its price in the marketplace.

prokaryotes
Organisms, including bacteria and their relatives, whose cells lack an organized nucleus and membrane-bound organelles.

protein
Polymers made of nitrogen containing organic molecules called amino acids.

proton
Positively charged particles in the central nucleus of an atom.

proven reserves
The quantities of a primary energy source that could be recovered from known deposits using current technology at current prices.

provisioning service
An ecosystem service that supplies humans with resources, such as food, water, and air.

pull factors
Conditions that encourage people to immigrate into a country.

push factors
Conditions that force people to emigrate from a country.

R

radiation
The release of electromagnetic energy.

radioactive
Isotopes that have an unstable nuclei; they spontaneously "decay," emitting various combinations of high-energy protons, neutrons, electrons, and radiation.

radon
A gas produced from the radioactive decay of the element radium-226 (226 Ra).

rain shadow
The dry region on the leeward side of mountains; descending air on leeward slopes produces less rain and snow than on the windward side.

range of tolerance
The span between the minimum and maximum amounts of an environmental feature such as temperature or the amount of a chemical that an organism can tolerate; the range of conditions over which an organism can maintain population growth rates.

rationing
In forest management, a process of determining the rate at which forest goods or services can be used by humans.

reactor core
Where fuel assemblies for a nuclear power plant are placed and where the fission reaction takes place.

realized niche
The range of conditions over which a species is able to compete effectively against other species.

recharge zone
An area where water flows through the soil surface and percolates to the water table.

regulating services
The ways that ecosystems control important conditions and processes, such as climate, the flow of water, and the absorption of pollutants.

regulations
A set of specific rules that establish standards for performance, programs to ensure compliance, and protocols for enforcement.

regulatory mandate
Policies or laws that set legal standards for actions.

relative humidity
Measure of the extent to which air is saturated, expressed as a percentage; vapor pressure/ saturation vapor pressure × 100.

renewable energy
Energy derived from sources that are not depleted when they are used, such as sunlight and wind, or that can be replenished in a short period of time, such as fuel wood.

renewable portfolio standard (RPS)
A government mandate that a certain percentage of energy use

come from renewable energy resources, the objective of which is to help build market demand for renewable technologies, increasing their scale of production, and driving down costs.

reproductive isolation
Members of one species do not interbreed with members of other species, even those that are closely related.

reserves-to-production ratio (R/P)
Proven reserves figure for a given primary energy source divided by a particular year's level of production or use.

residence time
The average time that an atom of an element or molecule of a compound spends in a pool; easily calculated as the size of the equilibrium pool divided by the flux through the pool.

return time
The average time between disturbances at any given place.

reverse osmosis
A method of desalination by filtering water through a selective membrane.

ribonucleic acid (RNA)
Various nucleic acids that contain ribose and uracil as structural elements and that play a central role in the synthesis of proteins using information encoded in DNA.

Richter magnitude scale
The numerical scale used to report the energy released in an earthquake in which each successive number indicates a 32-fold increase in energy.

riparian water rights
A system for assigning water possession derived from English common law whereby all landowners with property adjacent to a body of water have a right to make reasonable use of it.

riparian zone
The region of transition between a stream and the terrestrial ecosystems that surround it.

risk
Likelihood of particular hazards impairing health or causing death.

risk–risk trade offs
The benefits of actions to reduce the risk of a hazard incur other costs or risks.

rock cycle
The process in which elements within Earth's crust and mantle are slowly converted from one type of rock to another.

rotation time
The average interval between successive cuts in a forest.

runoff
The action of rainwater on land flowing across the surface until it eventually enters lakes and streams.

run-of-river hydroelectric plant
A hydropower facility that uses the force of river water diverted into a series of pipes to spin a turbine to generate electricity.

S

sampling effect
The increased likelihood as species diversity increases that a species has a big effect on a process in an ecosystem.

sand
Mineral particles 0.05–2 mm in diameter; the chemical structure of these particles still resembles their parent bedrock.

sanitary landfill
A system of municipal solid waste disposal designed to prevent the contamination of groundwater and to reduce the odors and aesthetic problems associated with open dumps.

saturation vapor pressure
A measure of the maximum amount of water that air can hold at a given temperature.

science
A process that poses and answers questions objectively in order to increase knowledge and lessen uncertainty by using careful observation and experimentation to produce results that can be duplicated by others.

sclerophyllous
Vegetation with thick, hard leaves resistant to water loss and wilting.

scrubbers
Devices containing chemicals that react with pollutants such as SO_2 and filter them from industrial exhausts.

second law of thermodynamics
Every energy transformation increases disorder, or entropy. In any energy transformation, energy always tends to go from more usable or higher quality forms, to less usable or lower quality forms, such as heat.

secondary air pollutants
Chemicals or particles that are produced in the atmosphere as a result of reactions among other chemicals or aerosols.

secondary energy
Forms of energy such as electricity or the kinetic energy of an automobile that are derived from energy contained in natural resources such as flowing water or oil.

secondary forest
Forests created by natural succession or reforestation following human disturbance.

secondary growth
Cell growth and division that increases the diameter of plant stems and roots.

secondary oil recovery
Injection of water to extract another 10–20% of the oil reservoir after the primary recovery stage.

secondary succession
The process of change following disturbances that leave behind legacies such as soil, woody debris, or plant seeds.

secondary treatment
In reference to wastewater treatment, a process using bacteria and other microorganisms to break down the organic material dissolved in the wastewater.

sedimentary rock
Rock that forms when sediments, such as sand, silt, and the remains of dead organisms, become "glued together" under pressure.

seed
A plant embryo embedded in nutritive tissue and surrounded by a protective coat.

septic system
System to treat household sewage and wastewater; solids settle to the bottom of the tank and wastewater flows to a series of perforated underground pipes.

sexual reproduction
Mating between two parents that produces offspring that are genetically unique.

shifting agriculture
The process in which forests are cut and the slash is burned to release nutrients to the newly cleared soil to create soil for agriculture. When crop production declines, the fields are abandoned and the process is repeated in other forested areas.

silt
Mineral particles that have experienced more weathering than sand and that have diameters ranging from 0.002 to 0.5 mm.

soil
The uppermost layer of Earth's crust that supports plant growth; a mixture of organic matter and mineral particles.

soil fertility
The variety of soil characteristics that support plant growth.

soil profile
The distinctive vertical structure of soil layers.

soil salinization
The process that occurs when salts contained in irrigation water concentrate and crystallize on the soil surface due to evaporation.

soil wilting point
The point at which there is no longer enough water in the soil for plants to replace the water that is being lost to transpiration.

solar photovoltaic farms
In areas with high solar potential, thousands of PV panels are arranged to generate electricity on a large scale.

solar power
The use of sunlight to generate electricity.

solubility
The ability of a chemical to dissolve in a particular liquid.

sovereignty
The concept that a country may behave as it pleases within its borders as long as it does not violate international laws to which it has agreed.

species
A group of similar organisms that can potentially mate and produce fertile and are reproductively isolated from other groups.

species evenness
A measure of the relative abundance of the different species in a community.

species richness
The total number of species in each community; the simplest measure of biodiversity.

spent nuclear fuel
Nuclear fuel rods that are depleted in ^{235}U; but still contain large amounts of hazardous and radioactive chemicals.

stability transition
The fourth stage of the demographic transition that is characterized by low birth, death, and population growth rates.

stabilizing selection
Selection that favors traits that fall somewhere in the middle of the range of variation within a population.

starch
Polysaccharides composed of hundreds of glucose molecules. The bonds between these glucose molecules are easily broken in animal digestion.

storm surge
Storm-caused increase in water levels that can produce devastating floods in low-lying regions and on islands.

stratosphere
The layer of the atmosphere that extends from 15 to about 48 km (9–30 mi) above Earth's surface.

stream order
Classification based on the number of streams that feed into another. A first-order stream flows from a single headwater, and two first-order streams converge to form a second-order stream, etc.

structural complexity
The three-dimensional distribution of species and biological features within an ecosystem.

sub-bituminous coal
Coal containing 35–45% carbon; its energy content is somewhat higher than lignite.

subsistence economy
A basic economic system in which a society meets its needs from its environment without accumulating wealth.

suburbs
The parts of the metropolitan area located outside of the central city.

succession
The process of post-disturbance change in an ecological community.

succulent
Plant with thick, fleshy stems or leaves that store large quantities of water.

sugar
Carbohydrates with the general chemical formula $(CH_2O)_n$, where n is between 3 and 7. A basic source of energy in most organisms and an ingredient in the synthesis of numerous other organic molecules.

supporting services
The basic ecosystem processes, such as nutrient cycles and soil formation, that are needed to maintain other services.

survivorship
The probability of surviving to a particular age.

sustainability
Meeting the needs of the present in an equitable and fair fashion without compromising the ability of future generations to meet their own needs.

sustainability science
The study of the interactions between ecosystems and human social systems, with a special focus on long-term changes in environment and human populations and society.

swamp
A wetland dominated by shrubs or trees and fed by flowing water.

symbiosis
Intimate interdependencies between species.

T

taiga
Russian word meaning "little sticks" and used synonymously with boreal forest.

take-back regulation
Requires that electronic manufacturers take back their products at the end of their useful life, at no charge to the consumer.

taxonomy
Classification of organisms through description, identification, and naming.

tectonic plate
A piece or plate of Earth's crust that floats on top of the mantle. The positions of the plates are not fixed; rather, they slowly move in relation to each other.

temperate deciduous forest
A biome in regions with moderate summers and cold winters that is dominated by broad-leaved trees that lose their leaves in the autumn and grow a new set in the spring. Precipitation is relatively even throughout the year.

temperate evergreen forest
In this biome with less precipitation and warmer temperatures than the temperate deciduous forest, trees keep their leaves throughout the year.

temperate grassland
Grasses interspersed with a diverse array of herbs in a biome too dry to support forest, woodland, or shrubland, but wet enough to prevent the land from becoming desert.

temperate rain forest
Lush and diverse evergreen forest in a region of heavy winter rains, summer fogs, and ample year-round moisture and mild temperatures.

temperate zone
The climate zone that falls between 25° and 60° latitude north and south, corresponding to the area of prevailing westerly winds in the Ferrel wind cells.

temperature
A measure of the average kinetic energy of a collection of molecules.

temperature anomaly
The difference between each year's average temperature and the average annual temperature over 20–30 benchmark years. A positive anomaly indicates a year that is warmer than the benchmark average, and a negative anomaly represents a year that is cooler than the benchmark average.

temperature inversion
A pattern of warmer air formed above colder air. Temperature inversions prevent air close to the ground from mixing with the air above it.

teratogens
Substances that produce mutations and birth defects.

terrestrial planets
The four inner planets—Mercury, Venus, Earth, and Mars. Each has a dense metallic core surrounded by thick layers of less dense rock.

tertiary oil recovery
Methods used to stimulate the flow of additional oil by injecting CO_2, steam, or hot water into reservoirs.

tertiary treatment
Referring to wastewater treatment, a process to remove inorganic nutrients from wastewater by passing them through sand and charcoal filters to remove residual solids and toxins.

texture
The relative amount of sand, silt, and clay in soil.

thermohaline circulation
The three-dimensional circulation of ocean water around the world due to the differences in temperature and salinity, causing an exchange of surface water and water from the depths.

thermosphere
The layer of the atmosphere that extends about 480 km (300 mi) into space.

thinning stage
A stage of forest development during which intense competition for light and water results in the death of smaller trees. The largest trees continue to grow and the total biomass in the forest continues to increase.

tidal barrage
A dam across the narrow outlet of a tidal basin that uses the ebb and flow of tidal waters to power turbines and dynamos to generate electricity.

tidal fence
An array of underwater turnstiles or revolving doors set across a narrow channel. As water moves through the channel, its energy is used to generate electricity.

tidal power
The kinetic energy of ocean tides used to generate electricity.

tidal turbines
Turbines operating underwater that are spun by the movement of tides and currents.

tilth
The physical arrangement of soil particles that facilitates aeration, seedling emergence, and the growth of roots.

total fertility rate
The sum of age-specific fertility rates across all ages; the *potential* number of offspring that an average female in a population can produce if she survives to old age.

toxicity threshold
The dose above which there is a measurable decline in a health measure.

toxicology
The scientific discipline that studies chemical poisons or toxins and their effects on human health.

trace gases
Gases contained in the atmosphere, most of which are present in concentrations expressed in parts per million (1 ppm = 0.0001%) or parts per billion (1 ppb = 0.0000001%).

tragedy of the commons
The overexploitation of common resources that occurs when individual interests prevail over the common good.

transcription
DNA code for a particular gene rewritten as a segment of RNA.

transform fault boundary
The zone along which tectonic plates slide past one another. This sliding may generate earthquakes.

transformer
A device used to adjust amperage and voltage so that electricity can move efficiently over long distances through transmission lines and be delivered to end users.

transition stage
The stage in which tree seedlings become established in the gaps created by the death of large trees; also called the gap phase.

translation
The process of forming a protein molecule using the information contained in RNA, which serves as the template for the assembly of a specific protein.

transpiration
The action of water evaporating from leaves; a major component of evaporation in terrestrial ecosystems.

travel-cost valuation
A valuation method based on the amount of money that people are willing to pay for transportation and lodging to visit a particular place.

tree line
Region of transition from boreal forest to tundra-like vegetation on a mountain or in polar regions.

trophic cascade
The influence of carnivores across multiple trophic levels; for example, loss of carnivores may allow populations of herbivores to increase, resulting in the overconsumption of primary producers.

trophic level
Classification of organisms based on their source of food; for example, primary producers, herbivores, and carnivores.

trophic level efficiency
The fraction of energy that the organisms in one trophic level store as biomass and make available to the next trophic level.

tropical rain forest
An area in which plentiful rainfall (greater than 2,000 mm per year) and warm climate support an enormous diversity of plants and animals.

tropical savanna
An ecosystem of open grasslands and a few scattered trees, which occurs where rainfall is highly seasonal and drought conditions generally persist for more than half of the year.

tropical seasonal forest
Tropical forests and woodlands dominated by trees that lose their leaves during seasonal dry periods.

tropical zone
The climate zone that straddles the equator and extends to about 25° latitude north and south, corresponding to the Hadley wind cells in the troposphere.

troposphere
The lowest layer of the atmosphere based on differences in temperature and chemical properties.

tundra
Treeless landscapes dominated by grasses, herbs, and low shrubs in polar regions with short growing seasons.

U

umbrella species
Species whose ecological requirements are highly correlated with the needs of many other species.

unconfined aquifer
A zone of water-saturated rock or sediment that is open to water recharge from the rock and sediment layers above it.

uneven-aged management
Maintaining biodiversity and ecosystem services through less intensive methods of harvest in which trees, dead snags, and woody debris are retained.

uranium
An element used to produce fuel rods for nuclear reactors. The decay of uranium isotope ^{235}U releases large amounts of energy.

urban
An area in which the average population density is at least 500 people per square mile.

urban decay
Decline in economic and community activity and deterioration of infrastructure that has occurred in many cities.

urban ecosystem
A region of Earth that supports the commerce, industry, and community interactions of a large number and high density of humans.

urban growth boundary (UGB)
A form of regional zoning that delineates areas suitable for high-density urban development from those suitable for lower density development.

urban heat island
The tendency for cities to be warmer than surrounding rural areas due to urban features that affect the absorption of solar energy, the rate of heat loss, and patterns of air movement.

urban planning
The profession that unites disciplines such as engineering, architecture, ecology, and sociology to design and build environments and transportation systems for urban communities.

urban sprawl
A widespread growth in the amount of land included in metropolitan areas.

urbanization
The increasing concentration of human populations in densely populated cities.

utilitarianism
An example of consequence-based ethics that defines right actions as those that deliver the greatest good to the greatest number.

V

vapor pressure
The fraction of atmospheric pressure that is a consequence of water molecules.

vectors
Intermediary organisms that transmit a parasite from host to host.

virtue ethics
A system of values based on the belief that an action is right if it is motivated by virtues, such as kindness, honesty, loyalty, and justice.

virulence
The extent to which a pathogen impairs the growth and reproduction of its host.

volatile organic compounds (VOCs)
Diverse array of organic chemicals that can vaporize into the air; many VOCs can impair the health of organisms, including humans.

volts
An electrical potential energy measurement; the greater the difference in electrical potential between the terminals, the greater the voltage of the current.

volunteerism
Work performed freely on behalf of a community.

W

waste-to-energy facility
Incinerators that use the heat produced by burning waste to generate electricity.

water balance coefficient
A general index of the availability of water in a particular region calculated as rainfall minus evapotranspiration.

water table
The underground depth where rock and sediment are completely saturated with water.

watershed
The area of land from which rainfall drains into a river or lake; also called drainage basins.

watt-hour (Wh)
The energy unit most commonly used to measure the everyday use of electricity.

wave power
The kinetic energy of ocean waves is used to generate electricity.

weather
Short-term variations in local atmospheric conditions.

wet deposition
Trace gases and particles are captured and removed from the atmosphere in raindrops, snowflakes, or droplets of fog.

wind farms
Clusters of turbines built close together in areas with the strongest and most regular winds.

wind power
Using wind to generate electricity.

wind turbine
A mechanical assembly that is designed to transform air movement into electricity.

work
The force applied to an object over a distance.

X

xylem
Specialized tissue that transports water and mineral nutrients up the plant stem.

Z

zoning
The designation of specific areas for different categories of land use.

zoonosis
A disease that is shared by humans and other animals.

zooplankton
Diverse array of tiny protists and animals suspended in pelagic waters.

zygote
A new, genetically unique cell formed by the union of an egg and a sperm.

PHOTO CREDITS

Chapter 1
Chapter Opener: Aspen Stock/.agefotostock; **Opening Essay:** 1.1 Michael Doolittle / Alamy; 1.2 Fancy / Alamy; 1.3 Nigel Hicks, Dorling Kindersley; 1.4 Chris Stowers, Dorling Kindersley; **Module 1:** 1.6 Albert Bierstadt/SuperStock; 1.7 Andre Jenny / Alamy; 1.8 Henny Ray Abrams/Reuters; 1.9 NASA; **Module 2:** 1.10 Richard Walters/iStockphoto; 1.11a Exactostock/SuperStock; 1.11b NASA; 1.11c thumb/Shutterstock; 1.12 Richard Walters/iStockphoto; **Module 3:** 1.13a Donald R. Swartz/Shutterstock; 1.13b Aimin Tang/iStockphoto; 1.14 NASA; 1.15 Eryk Jaegermann/agefotostock; 1.16 UN Photo/John Olsson; 1.17 rabh images / Alamy; **Module 4:** 1.18 fotog/agefotostock; 1.19 Dave Rudkin, Dorling Kindersley; 1.20 Stacy Gold/ National Geographic Stock; **Module 5:** 1.21 Image Source/agefotostock; 1.23 North Wind Picture Archives/AP Images; **Module 6:** 1.24 Max Alexander, Dorling Kindersley; 1.25 Petty Officer 2nd Class Kyle Niemi/U.S. Coast Guard; 1.26 Bruno Morandi/agefotostock; 1.27 Stuart Pearce/agefotostock; **Agents of Change:** 1.1 Rachel Barge; 1.2 Stock Connection Blue / Alamy; 1.3 AP Photo/Rich Pedroncelli

Chapter 2
Chapter Opener: Asia Images Group Pte Ltd / Alamy; **Opening Essay:** 2.1 Wadsworth Atheneum Museum of Art / Art Resource, NY; 2.2 Library of Congress; 2.3 Anthony Dunn / Alamy; **Module 1:** 2.4 Holton Collection/SuperStock; 2.5 McPHOTO/agefotostock; 2.6 The Print Collector/agefotostock; 2.7 Mick Roessler/agefotostock; 2.8 Robert Adrian Hillman/Shutterstock; **Module 2:** 2.9 Science Photo Library/agefotostock; 2.10 Ariadne Van Zandbergen / africanpictures.net / The Image Works; Frank Greenaway, Dorling Kindersley; David Peart, Dorling Kindersley; 2.11 David Reed/agefotostock; 2.12 AP Photo/Mary Ann Chastain; **Module 3:** 2.13 Jacques Jangoux / Alamy; 2.14 TAO Images Limited / Alamy; 2.16 Alan Hills, Dorling Kindersley; GM Productions/ agefotostock; 2.17 IndexStock/ SuperStock; **Module 4:** 2.18 TOPIC PHOTO AGENCY IN/agefotostock; 2.22 Edward Parker / Alamy; **Focus on Science:** 2.20 Jaime Florez/World WildlifeFund; 2.21 Fax Sinclair; **Module 5:** 2.24 Bill Clark/Roll Call Photos/Newscom; 2.25 First Light / Alamy; 2.26 Photoshot/ agefotostock; 2.27 Wade H. Massie/ Shutterstock; **Module 6:** 2.29 Sam Fried / Photo Researchers, Inc.; **Module 7:** 2.30 PCL / Alamy; 2.31: UN Photo/Paulo Filgueiras; **Seeing Solutions:** 2.34 GUIZIOU Franck/Photolibrary

Chapter 3
Chapter Opener: Pekka Parviainen/ Photo Researchers, Inc.; **Opening Essay:** 3.1–3 NASA; **Module 1:** 3.7 David Murray, Dorling Kindersley; 3.8 Karl Shone, Dorling Kindersley; 3.9 Volodymyr Goinyk/Shutterstock; 3.10 Andy Crawford, Dorling Kindersley; Dorling Kindersley; design56 / Shutterstock; **Module 3:** 3.17 fStop / Alamy; 3.18 Thomas & Pat Leeson / Photo Researchers, Inc.; 3.20 Dave King, Dorling Kindersley; 3.21 Roman Sigaev/ Shutterstock; 3.22 Jeffrey Greenberg / Photo Researchers, Inc.; 3.23 BestThinking.com; 3.24 NASA; **Module 4:** 3.26 Science Source / Photo Researchers, Inc.; 3.27 Mark Garlick / Photo Researchers, Inc.; 3.28 The International Astronomical Union / Martin Kornmesser / Photo Researchers, Inc.; **Module 5:** 3.32 Kip Evans / Alamy; 3.34 M-Sat Ltd / Photo Researchers, Inc.; **Module 6:** 3.35a NASA; 3.35b NASA / Science Source / Photo Researchers, Inc.; 3.37 Jim Henderson / Alamy; **Module 7:** 3.41 Lloyd Park /Dorling Kindersley; 3.42 Ultimate Group, LLC / Alamy

Chapter 4
Chapter Opener: Jean-Paul Chatagnon/ Biosphoto/Photo Researchers Inc.; **Opening Essay:** 4.1 the Everett Collection; 4.2 Embassy of Australia; 4.3 blickwinkel / Alamy; **Module 1:** 4.5 Dorling Kindersley; 4.6 National Oceanic and Atmospheric Administration/ Department of Commerce; 4.7 Dorling Kindersley; 4.8 Monty Rakusen/ agefotostock; **Module 2:** 4.9 Dr. Alexey Khodjakov / Photo Researchers, Inc.; 4.10a David Scharf / Photo Researchers, Inc.; 4.10b Oxford Scientific/Photolibrary; 4.10c Kenneth W. Fink / Photo Researchers, Inc.; 4.11 Aleksandar Zoric/Shutterstock; **Module 3:** 4.12 Dorling Kindersley; Jerry Young, Dorling Kindersley; Frank Greenaway, Dorling Kindersley; M. I. Walker, Dorling Kindersley; 4.14 imagebroker / Alamy; **Focus on Science:** 4.17 Gregory Erickson; 4.18 De Agostini / SuperStock; 4.19 Gregory Erickson; **Module 4:** 4.23 WaterFrame / Alamy; blickwinkel / Alamy; Tom McHugh / Photo Researchers, Inc.; 4.24 Ian West/Photolibrary; All Canada Photos / SuperStock; 4.25 Johner Images / Alamy; Exactostock / SuperStock; **Module 5:** 4.26 Ralph Lee Hopkins/National Geographic Stock; 4.27 Stubblefield Photography / Shutterstock; Judy O'Connor; Brumm H, Farrington H, Petren K, Fessl B (2010) Evolutionary Dead End in the Galápagos: Divergence of Sexual Signals in the Rarest of Darwin's Finches. PLoS ONE 5(6): e11191.; David Hosking / Alamy; Andrzejas/Dreamstime; Frank Greenaway, Dorling Kindersley; Dave Rudkin, Dorling Kindersley; Neil Fletcher, Dorling Kindersley; Peter Gardner, Dorling Kindersley; 4.28 James L. Stanfield/National Geographic Stock; 4.30 Flirt/SuperStock; **Module 6:** 4.35 Danita Delimont / Alamy; Bruce Corbett / Alamy; Design Pics/SuperStock; 4.36 Jean-Paul Chatagnon/Biosphoto/Photo Researchers Inc.; 4.37 Bob Gibbons / Alamy; Accent Alaska.com / Alamy; 4.38 Christian Charisius/Reuters; **Module 7:** 4.39 Shattil & Rozinski/Nature Picture Library; 4.40 Kim Taylor, Dorling Kindersley; Jerry Young, Dorling Kindersley; Laura Wickenden, Dorling Kindersley; Karl Shone, Dorling Kindersley; Dave King, Dorling Kindersley; Geoff Brightling, Dorling Kindersley; Sergey Goruppa/Shutterstock; Geoff Dann, Dorling Kindersley; Steve Gorton, Dorling Kindersley; D. Kucharski & K. Kucharska/Shutterstock; Frank Greenaway, Dorling Kindersley; David Peart, Dorling Kindersley; Robin Hunter, Dorling Kindersley; Matthew Ward, Dorling Kindersley; M. I. Walker, Dorling Kindersley; Eye of Science / Photo Researchers, Inc.; Dr Julian Thorpe / Dorling Kindersley

Chapter 5
Chapter Opener: Kazuyoshi Nomachi/ CORBIS; **Opening Essary:** 5.1 Ladi Kirn / Alamy; 5.2a Buyenlarge/Getty Images; 5.2b Emin Kuliyev/Shutterstock; 5.3 Photononstop/SuperStock; **Module 1:** 5.9 Radius/CORBIS inset Michael Maslan Historic Photographs/CORBIS; **Module 2:** 5.11b Flirt/SuperStock; 5.16 Stock Connection Blue / Alamy; 5.20 TARIQ MAHMOOD/AFP/Getty Images/ Newscom; **Module 3:** 5.21 Corbis inset Photos.com; **Module 4:** 5.25 Dorling Kindersley; 5.26 Aurora Photos / Alamy inset Kazuyoshi Nomachi/CORBIS; 5.28 Michael J. Balick/Photolibrary; David R. Frazier Photolibrary, Inc. / Alamy; Ashley Cooper / Alamy; Chris Cooper-Smith / Alamy; 5.29 Les Stone/Sygma/CORBIS; 5.30 NASA; **Focus on Science:** 5.31 Sir Partha Dasgupta; 5.32 DESHAKALYAN CHOWDHURY/AFP/Getty Images/ Newscom; **Module 5:** 5.33 Getty Images Inc.– Hulton Archive Photos; 5.34 David Turnley/CORBIS; 5.36 Tom Nebbia/ CORBIS; **Seeing Solutions:** 5.37 Jorgen Schytte/Photolibrary

Chapter 6
Chapter Opener: Eric Baccega/Nature Picture Library; **Opening Essay:** 6.1 Radius / SuperStock; 6.2a Christoph Hähnel/ Dreamstime; Biophoto Associates / Photo Researchers, Inc.; 6.2b Stephen Dalton / Photo Researchers, Inc.; 6.2c All Canada Photos / SuperStock; **Module 1:** 6.3 Anyka/Shutterstock; 6.4 Eric V. Grave / Photo Researchers, Inc.; Michael Abbey / Photo Researchers, Inc.; 6.7 Chris Curtis/Shutterstock; 6.8 Anup Shah/Nature Picture Library; **Module 2:** 6.9 Peter Anderson, Dorling Kindersley; 6.10 NHPA / SuperStock; 6.11 Paul Souders/ Getty Images; 6.12 Poxford/Photolibrary; 6.15a Irochka/Dreamstime; 6.15b All Canada Photos / SuperStock; 6.15c Visions of America, LLC / Alamy; **Module 3:** 6.16 Gregory G. Dimijian, M.D. / Photo Researchers, Inc.; 6.17 Martin Shields / Photo Researchers, Inc.; 6.18 Frank Greenaway, Dorling Kindersley; **Module 4:** 6.19 Cynthia Kidwell /Shutterstock; GraÃ§a Victoria/Shutterstock; artcasta / Shutterstock; Radaway/Dreamstime; Martin Pateman /Shutterstock;6.20 Geoff Dann, Dorling Kindersley; Steve Shott, Dorling Kindersley; 6.22 airphoto.gr / Shutterstock; Roger Phillips, Dorling Kindersley; Frank Greenaway, Dorling Kindersley; sobur/ Shutterstock; Cyril Laubscher, Dorling Kindersley; Philip Dowell, Dorling Kindersley; Geoff Brightling, Dorling Kindersley; Clive Streeter and Patrick McLeavy, Dorling Kindersley; Dave King, Dorling Kindersley; Biology Pics / Photo Researchers, Inc.; Science Source / Photo Researchers, Inc.; 6.23 George Burba/Shutterstock; hvoya / Shutterstock; Steve Byland /Shutterstock; Shutterstock; Ornitolog82/Fotolia; fivespots /Shutterstock; Jane Burton, Dorling Kindersley; Kim Taylor, Dorling Kindersley; Martin Oeggerli / Photo Researchers, Inc.; Chris leachman / Shutterstock; Peter Anderson, Dorling Kindersley; Dorling Kindersley; 6.24 Larry Malvin/Photolibrary; 6.25 Paul Gill/Alamy; Ronnie Howard / Shutterstock; **Focus on Science:** 6.26 Jon Chase; 6.27 sojourns in nature/ Photolibrary; 6.28 LianeM / Shutterstock; **Module 5:** 6.29 Kathie Thomas/Dreamstime; 6.30a Alexander Ryabchun/Dreamstime; 6.30b WILDLIFE GmbH / Alamy; 6.30c Joe Klune/Shutterstock; 6.30d Design Pics / SuperStock; 6.31 Kenn Stilger/ Dreamstime; 6.32 Norm Christensen;

TEXT CREDITS

2.19: "Figure 2: Global map of the value of ecosystem services" from "The Value of the World's Ecosystem Services and Natural Capital" in NATURE, Volume 387, page 257, May 15, 1997. Copyright © 1997 by Macmillan Publishers Ltd. Reprinted by permission of Macmillan Publishers Ltd.

4.29: From "Figure 1" from "Intense Natural Selection in a Population of Darwin's Finches (Geospizlnae) in the Galapagos" by Peter T. Boag and Peter R. Grant from SCIENCE 214: 82–84, October 2, 1981. Reprinted by permission of AAAS, P. T. Boag and P. R. Grant.

5.4: "Figure 3: The migration of modern Homo sapiens" from "The application of molecular genetic approaches to the study of human evolution" by L. Luca Cavalli-Sforza and Marcus W. Feldman from NATURE GENETICS 33:266–275, March 2003. Copyright 2003 © by Macmillan Publishers Ltd. Reprinted by permission of Macmillan Publishers Ltd.

5.27: "Map 6: Living on more 2001" from ASIA-PACIFIC 2005 - THE ECOLOGICAL FOOTPRINT AND NATURAL WEALTH, http://assets.panda.org/downloads/asialpr2005.pdf. Reprinted by permission of the World Wildlife Fund.

6.33: From "Figure 1" and "Figure 2" from "Breeding bird populations in relation to plant succession on the Georgia Piedmont" by D. W. Johnston and E. P. Odum from ECOLOGY 37:50–62, January 1956. Copyright © 1956. Reprinted by permission of Ecological Society of America via Copyright Clearance Center.

7.15: "Net Primary Productivity" from ATLAS OF BIOSPHERE from http://www.sage.wisc.edu/atlas/maps/npp/atl_npp.jpg. Reprinted by permission of The Nelson Institute Center for Sustainability and the Global Environment at the University of Wisconsin-Madison.

7.21: From "Figure 14 - Global Trends in the Creation of Reactive Nitrogen on Earth by Human Activity, with Projection to 2050 (R9 Fig 9.1)" from UN MILLENNIUM ASSESSMENT SYNTHESIS REPORT, ECOSYSTEMS AND HUMAN WELL-BEING; GENERAL SYNTHESIS, Page 17, May 23, 2005. Reprinted by permission of World Resources Institute.

7.29: From Figure 28 from BIOGEOCHEMISTRY OF A FORESTED ECOSYSTEM by G. E. Likens, et al. Copyright © 1977 by Springer Science+Business Media. Reprinted by permission from Springer Science+Business Media and Gene E. Likens.

7.31: Figure 5-7 from PATTERN AND PROCESS IN A FORESTED ECOSYSTEM by G. E. Likens and F. H. Bormann. Copyright © 1979 by Springer Science+Business Media. Reprinted by permission of Springer Science+Business Media and G. E. Likens.

8.30: Adapted from "Figure 3.2 (left panel)" from CLIMATE CHANGE 2007: SYNTHESIS REPORT. CONTRIBUTION OF WORKING GROUPS I, II AND III TO THE FOURTH ASSESSMENT REPORT OF THE INTERGOVERNMENTAL PANEL ON CLIMATE CHANGE. Reprinted by permission of IPCC, Geneva, Switzerland.

8.31: Adapted from "Figure 3.2 (right panel)" from CLIMATE CHANGE 2007: SYNTHESIS REPORT. CONTRIBUTION OF WORKING GROUPS I, II AND III TO THE FOURTH ASSESSMENT REPORT OF THE INTERGOVERNMENTAL PANEL ON CLIMATE CHANGE. Reprinted by permission of IPCC, Geneva, Switzerland.

8.32: "Figure 3.3" from CLIMATE CHANGE 2007: SYNTHESIS REPORT. CONTRIBUTION OF WORKING GROUPS I, II AND III TO THE FOURTH ASSESSMENT REPORT OF THE INTERGOVERNMENTAL PANEL ON CLIMATE CHANGE. Reprinted by permission of IPCC, Geneva, Switzerland.

8.33: "Figure 3.6" from CLIMATE CHANGE 2007: SYNTHESIS REPORT. CONTRIBUTION OF WORKING GROUPS I, II AND III TO THE FOURTH ASSESSMENT REPORT OF THE INTERGOVERNMENTAL PANEL ON CLIMATE CHANGE. Reprinted by permission of IPCC, Geneva, Switzerland.

8.34: From "Figure 1b" from "Stabilization wedges: Solving the climate problem for the next 50 years with current technologies" by S. Pacala and R. Socolow from SCIENCE 305: 968–972, August 13, 2004. Reprinted by permission of AAAS.

9.20: "Deaths From Indoor Smoke from Solid Fuels" from http://www.who.int/heli/risks/indoorair/indoorair/en/index.html. Reprinted by permission of the World Health Organization Press, Geneva, Switzerland.

9.22: Adapted from http://blog.geoiq.com/wp-content/uploads/2008/12/bans.jpg. Powered by GeoIQ. Reprinted by permission of GeoIQ.

10.6: From "Figure 4.10" from COMMUNITIES AND ECOSYSTEMS, 1st Edition, by R. H. Whittaker. Copyright © 1900. Adapted by permission of Pearson Education, Inc., Upper Saddle River, NJ.

10.9: "Figure 4.3 (a)" from CLIMATE CHANGE 2007: IMPACTS, ADAPTATION AND VULNERABILITY. WORKING GROUP II CONTRIBUTION TO THE FOURTH ASSESSMENT REPORT OF THE INTERGOVERNMENTAL PANEL ON CLIMATE CHANGE. Cambridge University Press. Reprinted by permission of IPCC, Geneva, Switzerland.

10.17: From "The Relation of Geography to Timber Supply" by W. B. Greeley from ECONOMIC GEOGRAPHY, Volume 1, March 1925. Copyright © 1925 by Clark University. Reprinted by permission of John Wiley & Sons, Inc.

11.5: From "Figure 21" from "The Vegetation of the Great Smoky Mountains" from ECOLOGICAL MONOGRAPHS 26:1–80, January 1956. Copyright © 1956. Reprinted by permission of Ecological Society of America via Copyright Clearance Center.

12.7: "Large Scale" from www.watersheds.org. Reprinted by permission.

12.60: "MWRA Water System Demand 1986–2009." Reprinted by permission of Massachusetts Water Resources Authority.

12.37: From "Figure 7" from "Fifty-Five years of soil development in restored freshwater depressional wetlands" by K. Ballantine and R. Schneider from ECOLOGICAL APPLICATIONS 19:1467–1480, September 2009. Copyright © 2009. Reprinted by permission of Ecological Society of America via Copyright Clearance Center.

13.9: "City of Ur" from http://dome.mit.edu/bitstream/handle/1721.3/2063/100359_cp.jpg?sequence=1, 10/03/2006. Copyright © 2006 by MIT Libraries' Dome. Reprinted by permission of Massachusetts Institute of Technology. All rights reserved.

13.15: From "Figure 4.10" from COMMUNITIES AND ECOSYSTEMS, 1st Edition, by R. H. Whittaker. Copyright © 1900. Adapted by permission of Pearson Education, Inc., Upper Saddle River, NJ.

15.5: From "Individual Energy Consumption" from http://www.world-nuclear.org/education/whyu.htm. Reprinted by permission of World Nuclear Association.

15.6: "World Primary Energy Consumption, 2005" from "Can't we put the carbon dioxide in a big hole and just forget it?" by Jean-Marc Jancovici from http://www.manicore.com/anglais/documentation_a/greenhouse/sequestration_a.html. Reprinted by permission of Manicore.

15.21: "Typical Oil and Gas Reservoir" from GLOBAL CLIMATE CHANGE AND ENERGY: CARBON DIOXIDE CAPTURE AND STORAGE, http://www.planetseed.com/node/15250. Copyright © 2011 by Schlumberger Excellence in Education Development (SEED) Inc. All rights reserved. Reprinted by permission of SEED. For more information, visit our website at www.planetseed.com.

17.13: "Figure 2: Urban heat island profile" from CLIMATE CHANGE IMPACTS AND ADAPTATION: A CANADIAN PERSPECTIVE, http://adaptation.nrcan.gc.ca/perspective/pdf/report_e.pdf. Copyright © Department of Natural Resources Canada. All rights reserved. Reprinted by the GeoGratis License Agreement.

17.36: From "The ground arthropods of London gardens" by B. N. K. Davis from THE LONDON NATURALIST 58:15–24. Copyright © 1979. Reprinted by permission of the London Natural History Society.

19.6: "Deaths by Broad Cause Group" from INFECTIOUS DISEASES KILL 1/3 WORLDWIDE; AIDS IS TOP CAUSE OF DEATH IN DEVELOPING REGION, http://ucatlas.ucsc.edu/cause.php. Reprinted by permission of UC Atlas of Global Inequality.

19.12: "Figure 1: Change in tsunami wave shape as it approaches land" from http://www.waikatoregion.govt.nz/Services/Regional-services/Regional-hazards-and-emergency-management/Coastal-hazards/Tsunami/. Reprinted by permission of Waikato Regional Council.

19.32: From "Cholera Dynamics and El Nino-Southern Oscillation" by Mercedes Pascual, Xavier Rodo, Stephen P. Ellner, et al. from SCIENCE 289:1766–1769, September 8, 2000. Reprinted by permission of AAAS.

19.43: From "Figure 1: Number of Confirmed Lyme Disease Cases Reported in MA, 1999–2008" from LYME DISEASE SURVEILLANCE IN MASSACHUSETTS, 2008, http://www.mass.gov/Eeohhs2/docs/dph/cdc/lyme/lyme_disease_surveillance_2008.pdf. Massachusetts Department of Public health, Bureau of Infectious Disease; Data are current as of May 29, 2009. Reprinted by permission of Massachusetts Department of Public Health.

19.44: "Figure - Average Annual Number of DF/DHF Cases Reported to WHO & Average Annual Number of Countries Reporting Dengue" from GLOBAL ALERT AND RESPONSE (GAR): IMPACT OF DENGUE, http://

www.who.int/csr/disease/dengue/impact/en/, last accessed 9/15/2011. Reprinted by permission of the World Health Organization Press, Geneva, Switzerland.

EOC 19.1: "Deaths by Broad Cause Group" from INFECTIOUS DISEASES KILL 1/3 WORLDWIDE; AIDS IS TOP CAUSE OF DEATH IN DEVELOPING REGION, http://ucatlas.ucsc.edu/cause.php. Reprinted by permission of UC Atlas of Global Inequality.

20.2: From "Figure 1b" from "Stabilization wedges: Solving the climate problem for the next 50 years with current technologies" by S. Pacala and R. Socolow from SCIENCE 305: 968–972, August 13, 2004. Reprinted by permission of AAAS.

20.3: "Figure 3.6" from CLIMATE CHANGE 2007: SYNTHESIS REPORT. CONTRIBUTION OF WORKING GROUPS I, II AND III TO THE FOURTH ASSESSMENT REPORT OF THE INTERGOVERNMENTAL PANEL ON CLIMATE CHANGE. Reprinted by permission of IPCC, Geneva, Switzerland.

INDEX